Gunstream's Anatomy & Physiology

With Integrated Study Guide

Jason LaPres
Beth Kersten
Yong Tang

SIXTH EDITION

Mc Graw Hill Education

GUNSTREAM'S ANATOMY & PHYSIOLOGY: WITH INTEGRATED STUDY GUIDE, SIXTH EDITION

This book is printed on acid-free paper.

2 3 4 5 6 7 8 9 0 RMN/RMN 1 0 9 8 7 6 5

ISBN 978-0-07-809729-4
MHID 0-07-809729-0

Senior Vice President, Products & Markets: *Kurt L. Strand*
Vice President, General Manager, Products & Markets: *Marty Lange*
Vice President, Content Design & Delivery: *Kimberly Meriwether David*
Managing Director: *Michael S. Hackett*
Brand Manager: *Amy Reed*
Director, Product Development: *Rose Koos*
Product Developer: *Mandy C. Clark*
Marketing Manager: *Jessica Cannavo*
Director of Digital Content Development: *Michael Koot*
Digital Product Developer: *John J. Theobald*
Director, Content Design & Delivery: *Linda Avenarius*
Program Manager: *Angela R. FitzPatrick*
Content Project Managers: *Vicki Krug/Christina Nelson*
Buyer: *Sandy Ludovissy*
Design: *Matt Diamond*
Content Licensing Specialists: *John Leland/Leonard J. Behnke*
Cover Image: *© Getty Images/Brigitte Sporrer*
Compositor: *Laserwords Private Limited*
Printer: *R. R. Donnelley*

Library of Congress Cataloging-in-Publication Data

Kersten, Beth.
Gunstream's anatomy & physiology : with integrated study guide / Beth Kersten, State College of Florida, Jason LaPres, Lone Star Community College-North Harris, Yong Tang, Front Range Community College.–Sixth edition.
pages cm
title: Anatomy & physiology : with integrated study guide
title: Anatomy and physiology : with integrated study guide
ISBN 978-0-07-809729-4 (alk. paper)
1. Human physiology–Textbooks. 2. Human physiology–Study guides. 3. Human anatomy–Textbooks. 4. Human anatomy–Study guides. I. LaPres, Jason. II. Tang, Yong (Teacher of human anatomy & physiology) III. Gunstream, Stanley E. Anatomy & physiology. IV. Title. V. Title: Anatomy & physiology : with integrated study guide. VI. Title: Anatomy and physiology : with integrated study guide.

QP34.5.G85 2016
612–dc23

2014026221

www.mhhe.com

ABOUT THE AUTHORS

Jason LaPres

Lone Star College-North Harris

Jason LaPres received his Master's of Health Science degree with an emphasis in Anatomy and Physiology from Grand Valley State University in Allendale, Michigan

Over the past 12 years, Jason has had the good fortune to be associated with a number of colleagues who have mentored him, helped increase his skills, and trusted him with the responsibility of teaching students who will be caring for others. Jason began his career in Michigan, where from 2001-2003 he taught as an adjunct at Henry Ford Community College, Schoolcraft College, and Wayne County Community College, all in the Detroit area. Additionally, at that time he taught high school chemistry and physics at Detroit Charter High School. Jason is currently Director of The Honors College and Professor of Biology at Lone Star College-University Park in Houston, Texas. He has been with LSC since 2003. In his capacity with LSC he has served as Faculty Senate President for two of the six LSC campuses. His academic background is diverse and, although his primary teaching load is in the Human Anatomy and Physiology program, he has also taught classes in Pathophysiology and mentored several Honor Projects.

Prior to authoring this textbook, Jason produced dozens of textbook supplements and online resources for many other Anatomy and Physiology textbooks.

Beth Kersten

State College of Florida

Beth Ann Kersten is a tenured professor at the State College of Florida (SCF). Though her primary teaching responsibilities are currently focused on Anatomy and Physiology I and II, she has experience teaching comparative anatomy, histology, developmental biology, and non-major human biology. She authors a custom A&P I laboratory manual for SCF and sponsors a book scholarship for students enrolled in health science programs. She coordinates a peer tutoring program for A&P and is working to extend SCF's STEM initiative to local elementary schools. Beth employs a learning style specific approach to guide students in the development of study skills focused on their learning strengths, in addition to improving other student skills such as time management and note taking. She graduated with a PhD from Temple University where her research focused on neurodevelopment in zebrafish. Her post-doctoral research at the Wadsworth Research Center focused on the response of rat nervous tissue to the implantation of neural prosthetic devices. At Saint Vincent College, she supervised senior research projects on subjects such as the effects of retinoic acid on heart development in zebrafish and the ability of vitamin B12 supplements to regulate PMS symptoms in ovariectomized mice. Beth also maintains memberships in the Society for Neuroscience and the Human Anatomy & Physiology Society.

Beth currently lives in North Port FL with her husband John and daughter Melanie. As former Northerners, they greatly enjoy the ability to swim almost year round both in their pool and in the Gulf of Mexico.

Yong Tang

Front Range Community College

Dr. Tang is an Izaak Walton Killam scholar. He received his M.Sc. in Anatomy and Ph.D. in Physiology from Dalhousie University. He has also received post-doctoral training at the University of British Columbia and physical therapy training at Dalhousie University. Dr. Tang had taught a wide variety of biology courses at Dalhousie University, Saint Mary's University, Northeastern Illinois University, and University of Colorado at Boulder. He is currently a biology professor at Front Range Community College, where he teaches Human Anatomy and Physiology, Human Biology, and Pathophysiology. His research interest focuses on comparative physiology, particularly, the exercise physiology of animals. He has authored many research articles in scientific journals including Journal of Experimental Biology and American Journal of Physiology. He is also very active in developing teaching and learning materials and has written numerous ancillaries of Anatomy and Physiology textbooks.

CONTENTS

PREFACE

GUNSTREAM'S ANATOMY & PHYSIOLOGY WITH INTEGRATED STUDY GUIDE, Sixth Edition, is designed for students who are enrolled in a one-semester course in human anatomy and physiology. The scope, organization, writing style, depth of presentation, and pedagogical aspects of the text have been tailored to meet the needs of students preparing for a career in one of the allied health professions.

These students usually have diverse backgrounds, including limited exposure to biology and chemistry, and this presents a formidable challenge to the instructor. To help meet this challenge, this text is written in clear, concise English and simplifies the complexities of anatomy and physiology in ways that enhance understanding without diluting the essential subject matter.

Themes

There are two unifying themes in this presentation of normal human anatomy and physiology: (1) the relationships between structure and function of body parts, and (2) the mechanisms of homeostasis. In addition, interrelationships of the organ systems are noted where appropriate and useful.

Organization

The sequence of chapters progresses from simple to complex. The simple-to-complex progression is also used within each chapter. Chapters covering an organ system begin with anatomy to ensure that students are well prepared to understand the physiology that follows. Each organ system chapter concludes with a brief consideration of common disorders that the student may encounter in the clinical setting. An integrated study guide, unique among anatomy and physiology texts, is located between the text proper and the appendices.

Study Guide

The *Study Guide* is a proven mechanism for enhancing learning by students and features full-color line art. There is a study guide of four to nine pages for each chapter. Students demonstrate their understanding of the chapter by labeling diagrams and answering completion, matching, and true/false questions. The completion questions "compel" students to write and spell correctly the technical terms that they must know. Each chapter study guide concludes with a few critical-thinking, short-answer essay questions where students apply their knowledge to clinical situations.

Answers to the *Study Guide* are included in the *Instructor's Manual* to allow the instructor flexibility: (1) answers may be posted so students can check their own responses, or (2) they may be graded to assess student progress. Either way, prompt feedback to students is most effective in maximizing learning.

Chapter Opener and Learning Objectives

Each chapter begins with a list of major topics discussed in the chapter along with an opening vignette and image, which introduces and relates the content theme of the chapter. Under each section header within every chapter, the learning objectives are noted. This informs students of the major topics to be covered and their minimal learning responsibilities.

Key Terms

Several features have been incorporated to assist students in learning the necessary technical terms that often are troublesome for beginning students.

1. A list of *Selected Key Terms* with definitions, and including derivations where helpful, is provided at the beginning of the chapter to inform students of some of the key terms to watch for in the chapter.
2. Throughout the text, key terms are in bold or italic type for easy recognition, and they are defined at the time of first usage. A *phonetic pronunciation* follows for students who need help in pronouncing the term. Experience has shown that students learn only terms that they can pronounce.
3. *Keys to Medical Terminology* in appendix A explains how technical terms are structured and provides a list of prefixes, suffixes, and root words to further aid an understanding of medical terminology.

Figures and Tables

Over 350 high quality, full-color illustrations are coordinated with the text to help students visualize anatomical features and physiological concepts. Tables are used throughout to summarize information in a way that is more easily learned by students.

Clinical Insight

Numerous boxes containing related clinical information are strategically placed throughout the text. They serve to provide interesting and useful information related to the topic at hand. The Clinical Insight boxes are identified by a *medical cross* for easy recognition.

Check My Understanding

Review questions at the end of major sections challenge students to assess their understanding before proceeding.

Chapter Summary

The summary is conveniently linked by section while it briefly states the important facts and concepts covered in each chapter.

Self-Review

A brief quiz, composed of completion questions, allows students to evaluate their understanding of chapter topics. Answers are provided in appendix B for immediate feedback.

Critical Thinking

Each chapter concludes with several critical thinking questions, which further challenge students to apply their understanding of key chapter topics.

Changes in the Sixth Edition

The sixth edition has been substantially improved to help beginning students understand the basics of human anatomy and physiology. Many of the changes are based on reviewer feedback.

Global Changes

- Added chapter opening vignette and chapter outline.
- Updated terminology based on *Terminologia Anatomica (TA), Terminologia Histologica (TH) and Terminologia Embryologica (TE).*
- Revised selected key terms lists to include most relevant terms.
- Revised learning objectives that have been moved from the chapter outline to the beginning of major sections.
- Revised self-review and critical thinking questions.
- Revised study guides to match chapter content changes.
- Updated the art throughout for a more vibrant and consistent style.

CHAPTER 1

- Revised planes and sections for clarity, terminology, and inclusion of "longitudinal section" and "cross-section."
- Updated and revised homeostasis discussion.
- Added figures 1.7 and 1.8 (serous membranes), 1.14 (positive-feedback mechanism), and four figures illustrating negative-feedback mechanisms.

CHAPTER 2

- Added Figure 2.1 containing the periodic table with the 12 most abundant elements in humans.

ONLINE TEACHING AND LEARNING RESOURCES

Help Your Students Prepare for class

Mc Graw Hill Education | LEARNSMART® ADVANTAGE | LearnSmartAdvantage.com

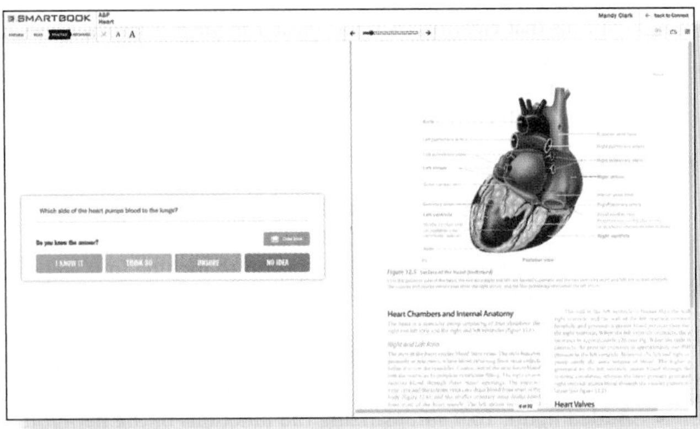

Mc Graw Hill Education | SMARTBOOK®

SmartBook is the first and only adaptive reading experience available for the higher education market. Powered by an intelligent diagnostic and adaptive engine, SmartBook facilitates the reading process by identifying what content a student knows and doesn't know through adaptive assessments. As the student reads, the reading material constantly adapts to ensure the student is focused on the content he or she needs the most to close any knowledge gaps.

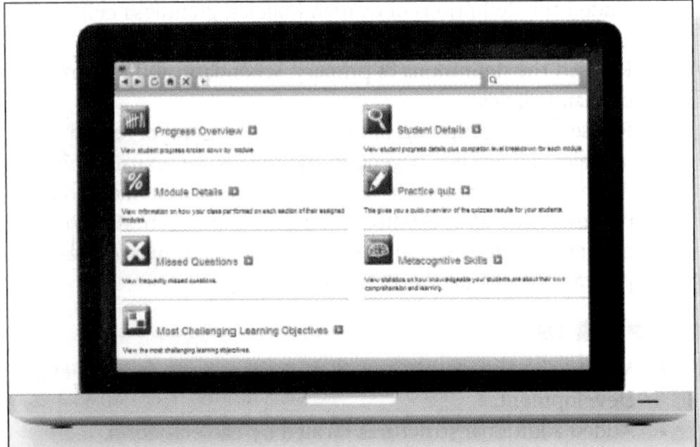

Mc Graw Hill Education | LEARNSMART®

LearnSmart is the only adaptive learning program proven to effectively assess a student's knowledge of basic course content and help them master it. By considering both confidence level and responses to actual content questions, LearnSmart identifies what an individual student knows and doesn't know and builds an optimal learning path, so that they spend less time on concepts they already know and more time on those they don't. LearnSmart also predicts when a student will forget concepts and introduces remedial content to prevent this. The result is that LearnSmart's adaptive learning path helps students learn faster, study more efficiently, and retain more knowledge, allowing instructors to focus valuable class time on higher-level concepts.

LEARNSMART Mc Graw Hill Education | LABS®

LearnSmart Labs is a super adaptive simulated lab experience that brings meaningful scientific exploration to students. Through a series of adaptive questions, LearnSmart Labs identifies a student's knowledge gaps and provides resources to quickly and efficiently close those gaps. Once the student has mastered the necessary basic skills and concepts, they engage in a highly realistic simulated lab experience that allows for mistakes and the execution of the scientific method.

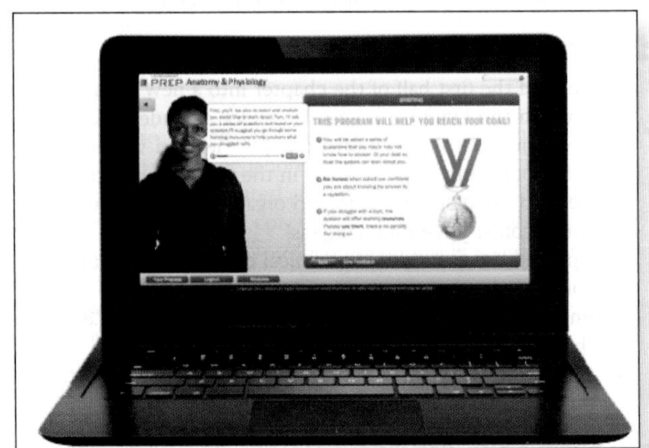

The primary goal of LearnSmart Prep is to help students who are unprepared to take college level courses. Using super adaptive technology, the program identifies what a student doesn't know, and then provides "teachable moments" designed to mimic the office hour experience. When combined with a personalized learning plan, an unprepared or struggling student has all the tools they need to quickly and effectively learn the foundational knowledge and skills necessary to be successful in a college level course.

MCGRAW-HILL CONNECT® ANATOMY & PHYSIOLOGY

McGraw-Hill Connect® Anatomy & Physiology integrated learning platform provides auto-graded assessments, a customizable, assignable eBook, an adaptive diagnostic tool, and powerful reporting against learning outcomes and level of difficulty–all in an easy-to-use interface. Connect Anatomy & Physiology is specific to your book and can be completely customized to your course and specific learning outcomes, so you help your students connect to just the material they need to know.

Save time with auto-graded assessments and tutorials

Fully editable, customizable, auto-graded interactive assignments using high quality art from the textbook, animations, and videos from a variety of sources take you way beyond multiple choice. Assignable content is available for every Learning Objective in the book.

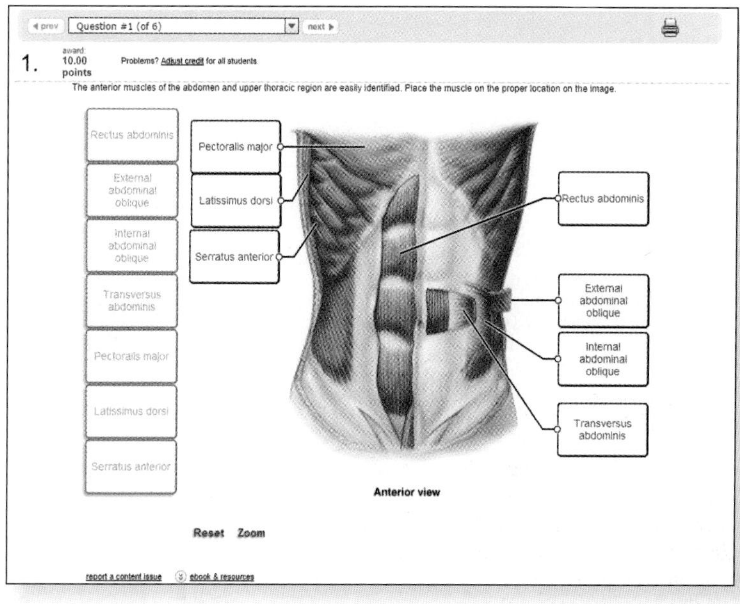

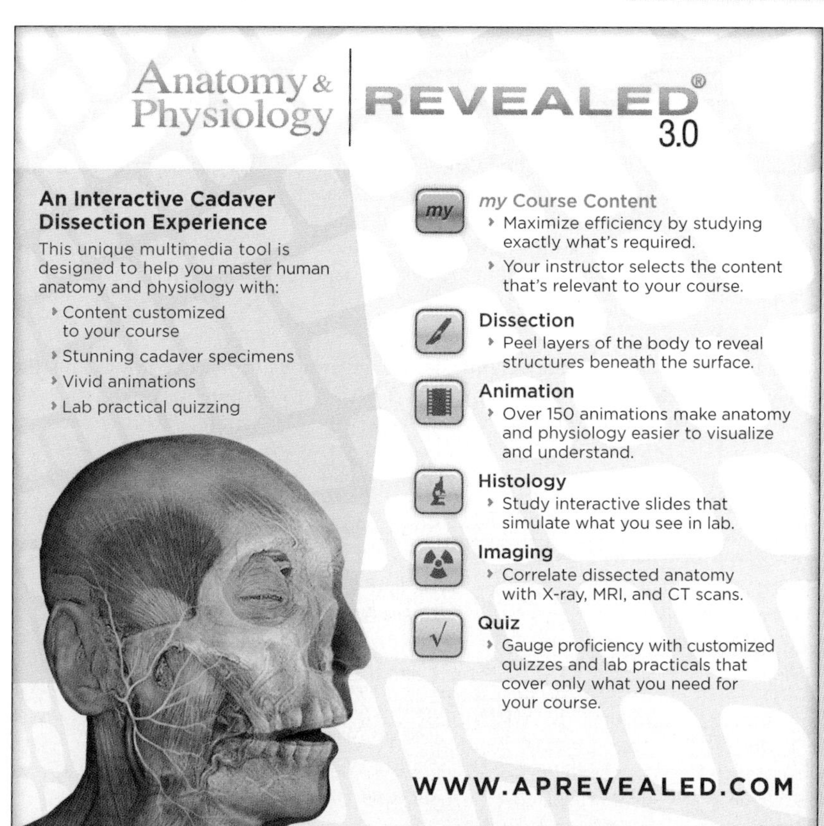

Anatomy & Physiology Revealed® is now available in cat and fetal pig versions!

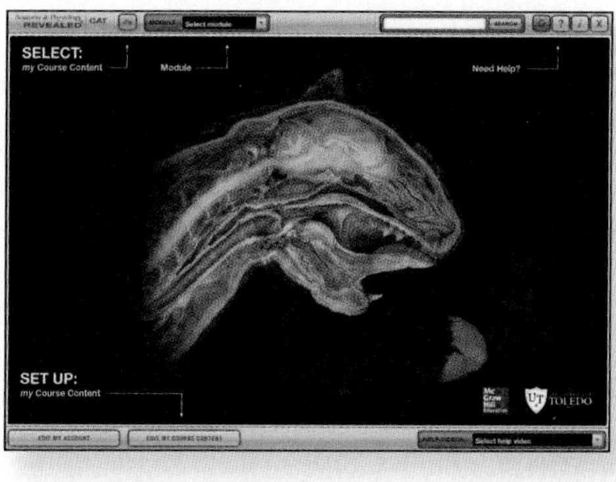

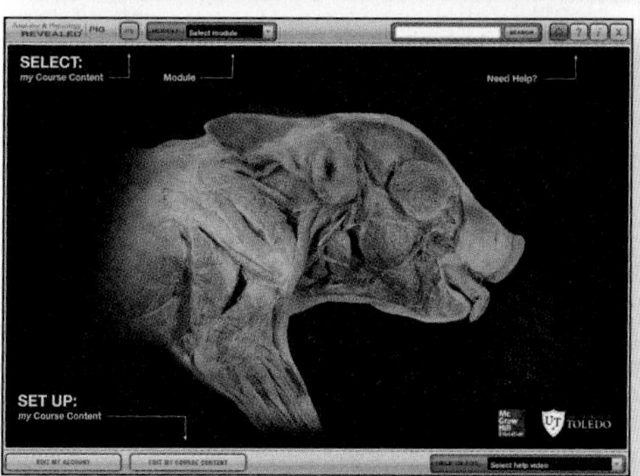

PRESENTATION TOOLS ALLOW INSTRUCTORS TO CUSTOMIZE LECTURE

Everything you need, in one location

Enhanced Lecture Presentations contain lecture outlines, FlexArt-adjustable leader lines and labels, art, photos, tables, and embedded animations where appropriate. Fully customizable, but complete and ready to use, these presentations will enable you to spend less time preparing for lecture!

Animations–over 100 animations bringing key concepts to life, available for instructors and students.

Animation PPTs–animations are truly embedded in PowerPoint for ultimate ease of use! Just copy and paste into your custom slideshow and you're done!

Take your course online—*easily*—with one-click digital Lecture capture

McGraw-Hill Tegrity campus™ records and distributes your lecture with just a click of a button. Students can view them anytime/anywhere via computer, iPod, or mobile device. Tegrity Campus indexes as it records your slideshow presentations and anything shown on your computer so **students can see keywords to find exactly what they want to study.**

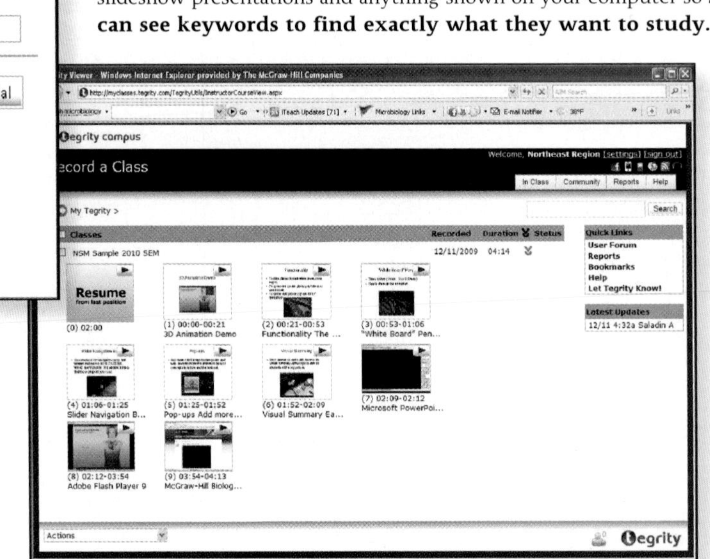

Laboratory Manual

Anatomy & Physiology Laboratory Textbook, Essentials Version, by Stanley E. Gunstream, Harold J. Benson, Arthur Talaro, and Kathleen Talaro, all of Pasadena City College. Kyla Ross of Georgia State University made significiant contributions to the sixth edition of the laboratory manual. This excellent lab text presents the fundamentals of human anatomy and physiology in an easy-to-read manner that is appropriate for students in allied health programs. It is designed especially for the one-semester course; it features a simple, concise writing style, self directing exercises, full-color photomicrographs in the Histology Atlas, and numerous illustrations in each exercise.

Acknowledgments

The development and production of this sixth edition has been the result of a team effort. Our dedicated and creative teammates at McGraw-Hill have contributed greatly to the finished product. We gratefully acknowledge and applaud their efforts. It has been a pleasure to work with these gifted professionals at each step of the process. We are especially appreciative of the support of Amy Reed, Brand Manager, Mandy Clark, Product Developer, and Vicki Krug, Content Project Manager.

The following instructors have served as critical reviewers:

Dr. Cecilia Bianchi-Hall
Lenoir Community College

David Evans
Penn College

Eleanor K. Flores, R.N., B.S.N., M.Ed
Lincoln College of New England

Karen Sue Frederick
Terra State Community College

Caroline Garrison
Carroll Community College

Daniel G. Graetzer, PhD
Northwest University

Dawn Hilliard
Northeast Mississippi Community College

Dale R. Horeth
Tidewater Community College

Patricia Jean Hubel
Thomas College

Scott Jones
Victor Valley College

Allart Kok
Community College of Baltimore County

Ryan D. Morris
Pierce College Military Program

Jean L. Mosley
Surry Community College

Dr. Raul E. Rivero
Online Adjunct Professor

Tisha Vestal
Centura College

Martin Zahn
Thomas Nelson Community College

Donald W. Zakutansky
Gateway Technical College

1 CHAPTER

Introduction to the Human Body

Michael, a freshman in college, overslept and is late for his first anatomy and physiology class. He has been dreading this class but it is necessary for his graduation requirements. Because he does not want to get off to a bad start, he sprints across campus. The combination of the warm day and physical exertion raises his body temperature and, as he throws himself into the nearest seat, sweat is pouring out across his body. Michael begins to feel cooler as he relaxes and he stops sweating within a few minutes. As his first lecture begins, he is introduced to the concept of homeostasis, which describes the condition of balance within the body, and the feedback cycles responsible for maintaining his internal "normal." He thinks about his morning, the sweat that cooled his body, and realizes just how amazing the human body really is. What a great semester this is going to be!

CHAPTER OUTLINE

Module 1
Body Orientation

SELECTED KEY TERMS

Anatomy (ana = apart; tom = to cut) The study of the structure of living organisms.
Appendicular (append = to hang) Pertaining to the upper and lower limbs.
Axial (ax = axis) Pertaining to the longitudinal axis of the body.
Body region (regio = boundary) A portion of the body with a special identifying name.
Directional term (directio = act of guiding) A term that references how the position of a body part relates to the position of another body part.
Effector (efet = result) A structure that functions by performing an action that is directed by an integrating center.
Homeostasis (homeo = same; sta = make stand or stop) Maintenance of a relatively stable internal environment.
Integrating center (integratus = make whole) A structure that functions to interpret information and coordinate a response.
Metabolism (metabole = change) The sum of the chemical reactions in the body.
Parietal (paries = wall) Pertaining to the wall of a body cavity.
Pericardium (peri = around; cardi = heart) The membrane surrounding the heart.
Peritoneum (peri = around; ton = to stretch) The membrane lining the abdominal cavity and covering the abdominal organs.
Physiology (physio = nature; logy = study of) The study of the functioning of living organisms.
Plane (planum = flat surface) Imaginary two-dimensional flat surface that marks the direction of a cut through a structure.
Pleura (pleura = rib) The membrane lining the thoracic cavity and covering the lungs.
Receptor (recipere = receive) A structure that functions to collect information.
Section (sectio = cutting) A flat surface of the body produced by a cut through a plane of the body.
Serous membrane (serum = watery fluid; membrana = thin layer of tissue) A two-layered membrane that lines body cavities and covers the internal organs.
Visceral (viscus = internal organ) Pertaining to organs in a body cavity.

YOU ARE BEGINNING a fascinating and challenging study–the study of the human body. As you progress through this text, you will begin to understand the complex structures and functions of the human organism.

This first chapter provides an overview of the human body to build a foundation of knowledge that is necessary for your continued study. Like the chapters that follow, this chapter introduces a number of new terms for you to learn. It is important that you start to build a vocabulary of technical terms and continue to develop it throughout your study. This vocabulary will help you reach your goal of understanding human anatomy and physiology.

1.1 Anatomy and Physiology

Learning Objective

1. Define anatomy and physiology.

Knowledge of the human organism is obtained primarily from two scientific disciplines–anatomy and physiology–and each consists of a number of subdisciplines.

Human **anatomy** (ah-nat′-ō-mē) is the study of the structure and organization of the body and the study of the relationships of body parts to one another. There are two subdivisions of anatomy. *Gross anatomy* involves the dissection and examination of various parts of the body without magnifying lenses. *Microanatomy*, also known as *histology*, consists of the examination of tissues and cells with various magnification techniques.

Human **physiology** (fiz-ē-ol′-ō-jē) is the study of the function of the body and its parts. Physiology involves observation and experimentation, and it usually requires the use of specialized equipment and materials.

In your study of the human body, you will see that there is always a definite relationship between the anatomy and physiology of the body and body parts. Just as the structure of a knife is well suited for cutting, the structure (anatomy) of a body part enables it to perform specific functions (physiology). For example, the arrangement of bones, muscles, and nerves in your hands enables the grasping of large objects with considerable force and also the delicate manipulation of small objects. Correlating the relationship between structure and function will make your study of the human body much easier.

1.2 Levels of Organization

Learning Objectives

2. Describe the levels of organization in the human body.
3. List the major organs and functions for each organ system.

The human body is complex, so it is not surprising that there are several levels of structural organization, as shown in figure 1.1. The levels of organization from simplest to most complex are chemical, cellular, tissue, organ, organ system, and organismal (the body as a whole).

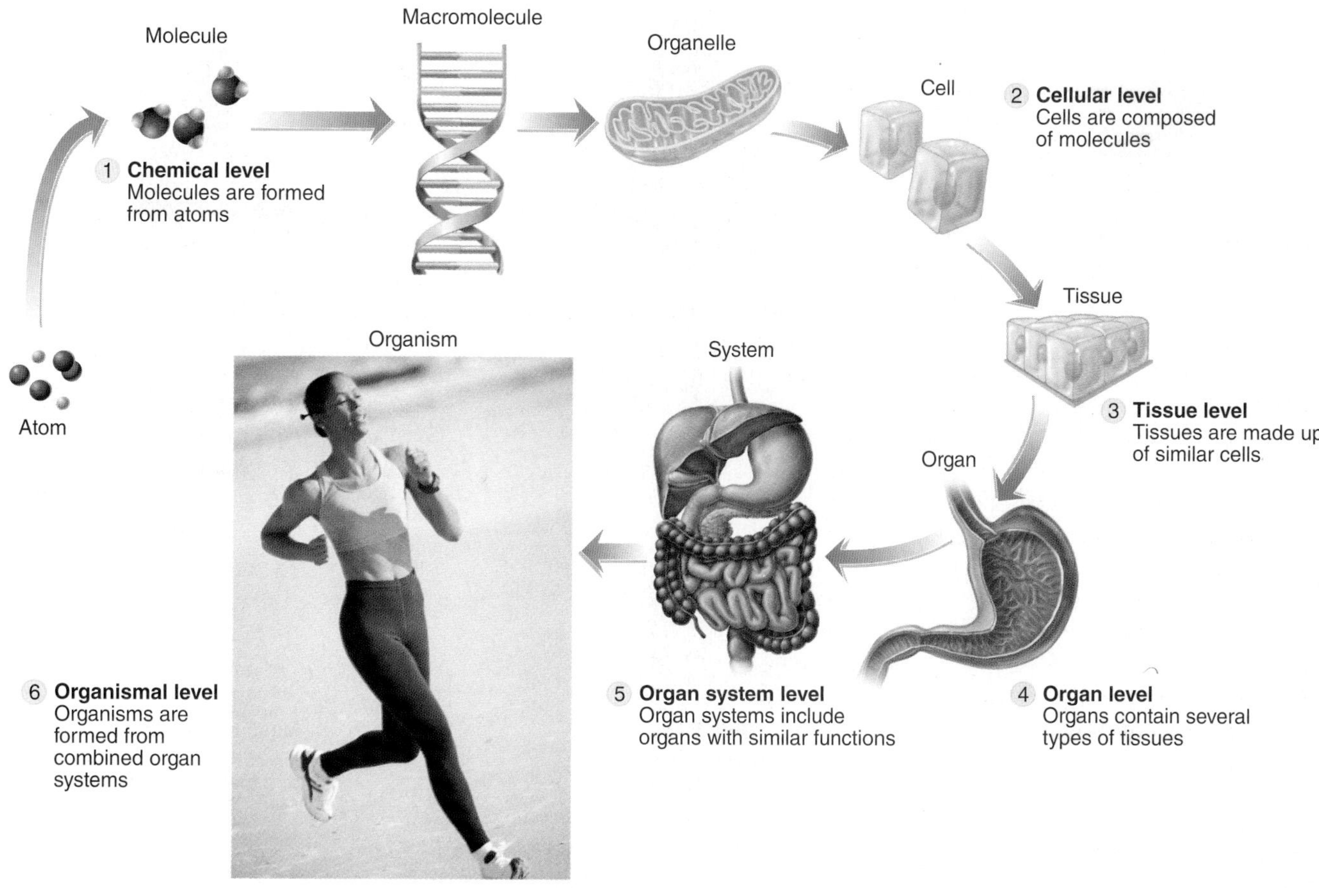

Figure 1.1 Six levels of organization in the human body range from chemical (simplest) to organismal (most complex).

Chemical Level

The *chemical level* consists of *atoms, molecules,* and *macromolecules.* At the simplest level, the body is composed of chemical substances that are formed of atoms and molecules. *Atoms* are the fundamental building blocks of chemicals, and atoms combine in specific ways to form *molecules.* Some molecules are very small, such as water molecules, but others may be very large, such as the macromolecules of proteins. Various small and large molecules are grouped together to form organelles. An **organelle** (or″-ga-nel′) is a microscopic subunit of a cell, somewhat like a tiny organ, that carries out specific functions within a cell. Nuclei, mitochondria, and ribosomes are examples.

Cellular Level

Cells are the basic structural and functional units of the body because all of the processes of life occur within cells. A cell is the lowest level of organization that is alive. The human body is composed of trillions of cells and many different types of cells, such as muscle cells, blood cells, and nerve cells. Each type of cell has a unique structure that enables it to perform specific functions.

Tissue Level

Similar types of cells are usually grouped together in the body to form a tissue. Each body **tissue** consists of an aggregation of similar cells that perform similar functions. There are four major classes of tissues in the body: epithelial, connective, muscle, and nervous tissues.

Organ Level

Each **organ** of the body is composed of two or more tissues that work together, enabling the organ to perform its specific functions. The body contains numerous organs, and each has a definite form and function. The stomach, heart, brain, and even bones are examples of organs.

Organ System Level

The organs of the body are arranged in functional groups so that their independent functions are coordinated to perform specific system functions. These coordinated, functional groups are called **organ systems.** The digestive and nervous systems are examples of organ systems. Most organs belong to a single organ system, but a few organs are assigned to more than one organ system. For

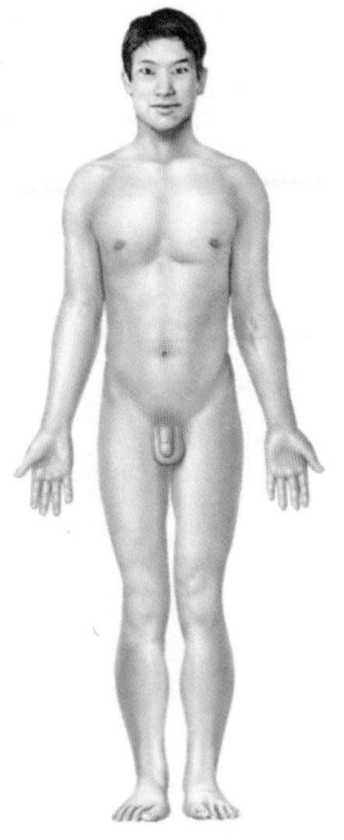

Integumentary system

Components: skin, hair, nails, and associated glands
Functions: protects underlying tissues and helps regulate body temperature

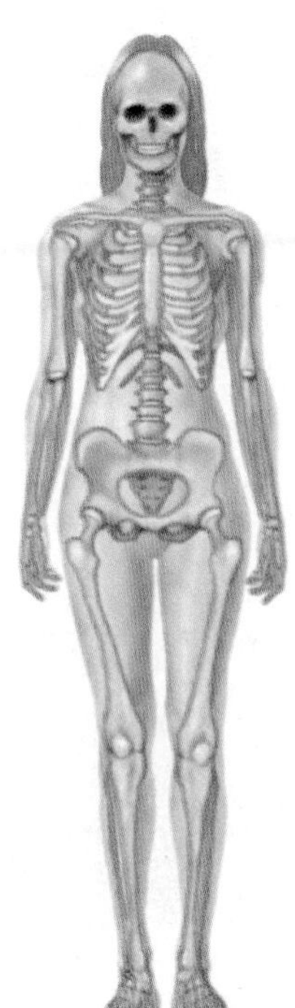

Skeletal system

Components: bones, ligaments, and associated cartilages
Functions: supports the body, protects vital organs, stores minerals, and produces formed elements

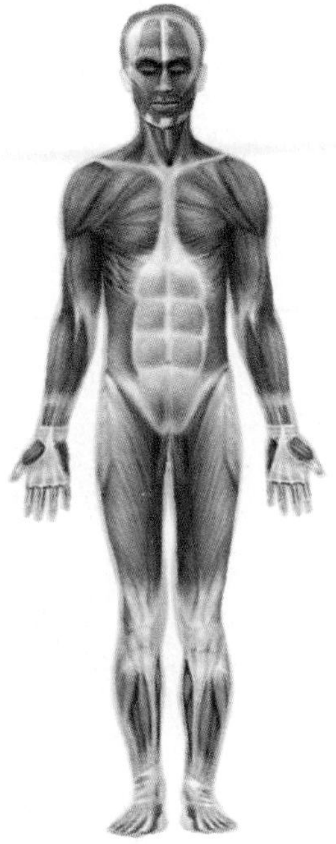

Muscular system

Components: skeletal muscles
Functions: moves the body and body parts and produces heat

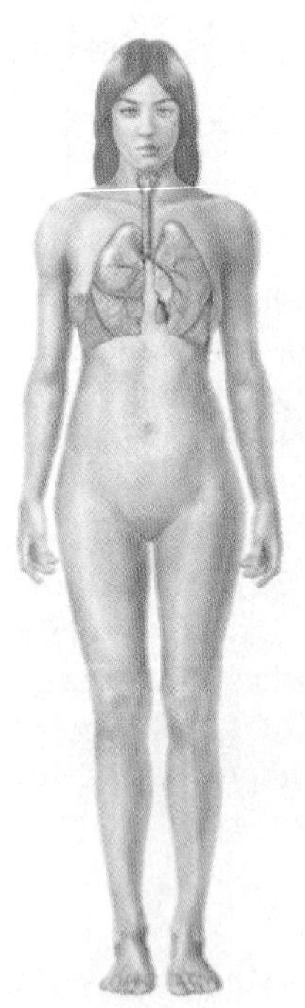

Respiratory system

Components: nose, pharynx, larynx, trachea, bronchi, and lungs
Functions: exchanges O_2 and CO_2 between air and blood in the lungs, pH regulation, and sound production

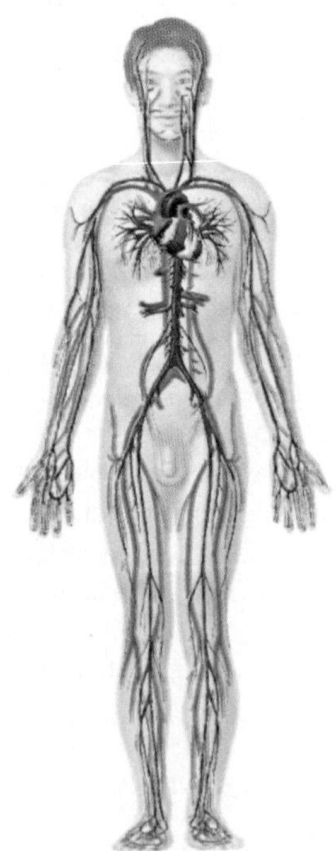

Cardiovascular system

Components: blood, heart, arteries, veins, and capillaries
Functions: transports heat and materials to and from the body cells

Figure 1.2 The 11 Organ Systems of the Body. APR

example, the pancreas belongs to both the digestive and endocrine systems.

Figure 1.2 illustrates the 11 organ systems of the human body and lists the major components and functions for each system. Although each organ system has its own unique functions, all organ systems are interdependent on one another. For example, all organ systems rely on the cardiovascular system to transport materials to and from their cells. Organ systems work together to enable the functioning of the human body.

Organismal Level

The highest organizational level dealing with an individual is the *organismal level,* the human organism as a whole. It is composed of all of the interacting organ systems. All of the organizational levels from chemicals to organ systems contribute to the functioning of the entire body.

Check My Understanding

1. What are the organizational levels of the human body?
2. What are the major organs and general functions of each organ system?

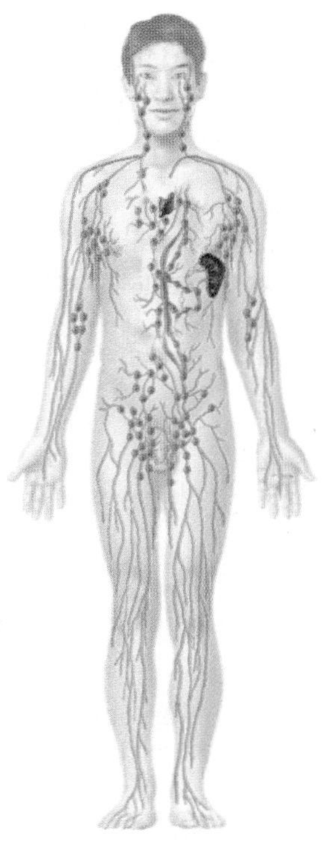

Lymphoid system

Components: lymph, lymphatic vessels, and lymphoid organs and tissues
Functions: collects and cleanses interstitial fluid, and returns it to the blood; provides immunity

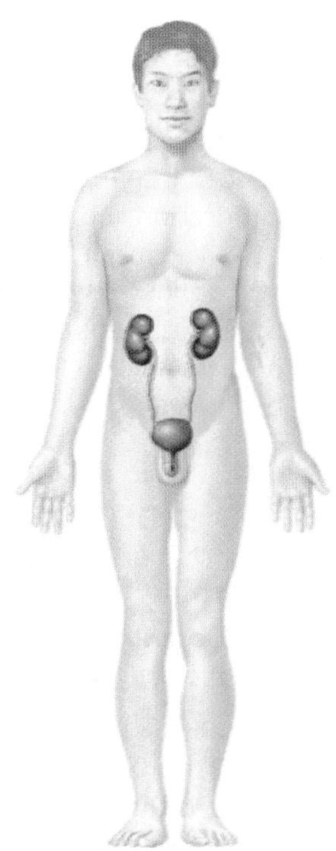

Urinary system

Components: kidneys, ureters, urinary bladder, and urethra
Functions: regulates volume and composition of blood by forming and excreting urine

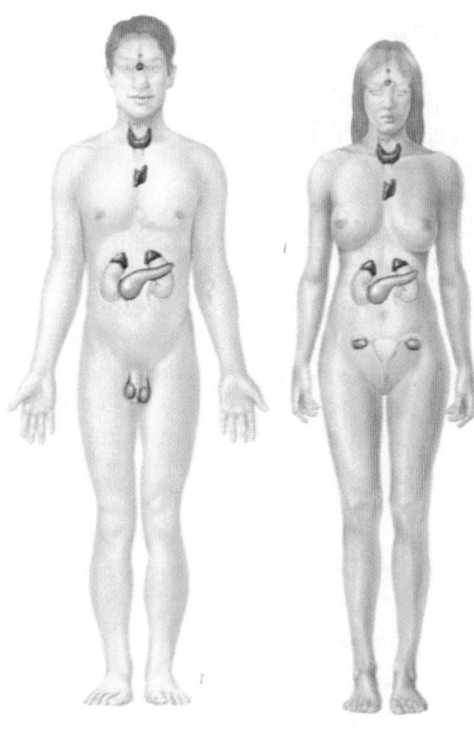

Endocrine system

Components: hormone-producing glands, such as the pituitary and thyroid glands
Functions: secretes hormones that regulate body functions

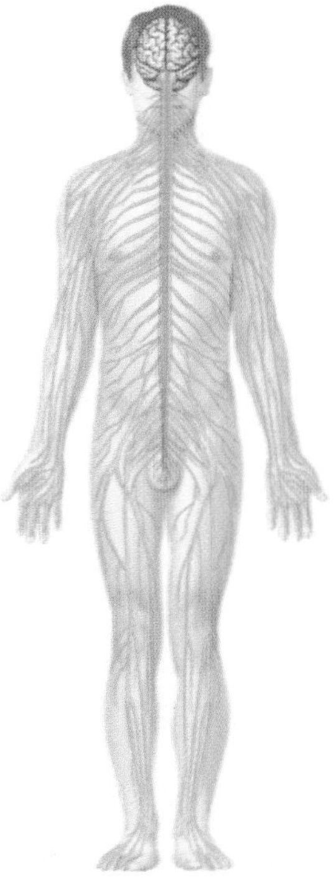

Nervous system

Components: brain, spinal cord, nerves, and sensory receptors
Functions: rapidly coordinates body functions and enables learning and memory

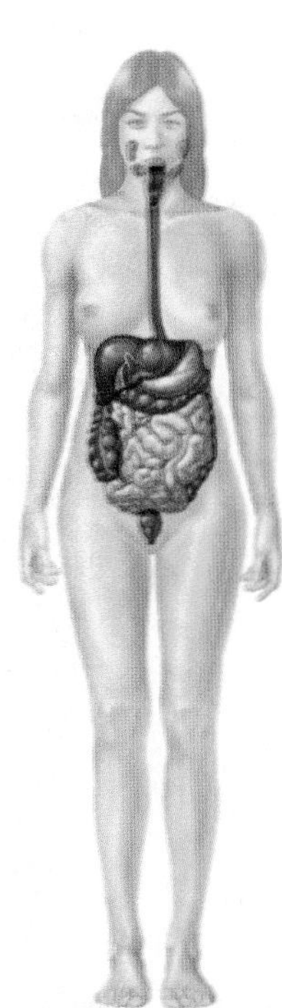

Digestive system

Components: mouth, pharynx, esophagus, stomach, intestines, liver, pancreas, gallbladder, and associated structures
Functions: digests food and absorbs nutrients

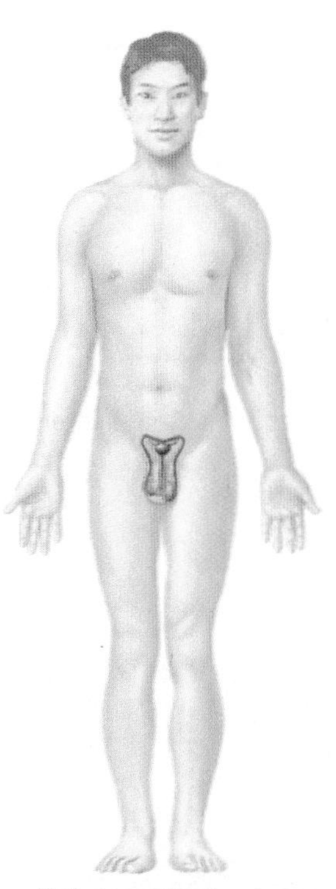

Male reproductive system

Components: testes, epididymides, vasa deferentia, prostate gland, bulbo-urethral glands, seminal vesicles, and penis
Functions: produces sperm and transmits them into the female vagina during sexual intercourse

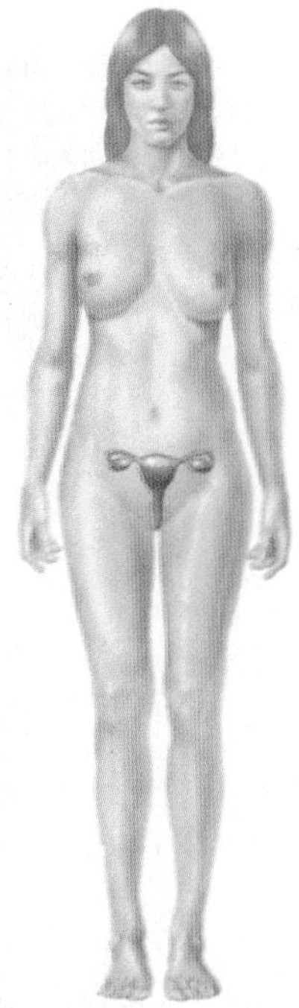

Female reproductive system

Components: ovaries, uterine tubes, uterus, vagina, and vulva
Functions: produces oocytes, receives sperm, provides intrauterine development of offspring, and enables birth of an infant

1.3 Directional Terms

Learning Objective

4. Use directional terms to describe the locations of body parts.

Directional terms are used to describe the relative position of a body part in relationship to another body part. The use of these terms conveys a precise meaning enabling the listener or reader to locate the body part of interest. It is always assumed that the body is in a standard position, the *anatomical position,* in which the body is standing upright with upper limbs at the sides and palms of the hands facing forward, as in figure 1.3. Directional terms occur in pairs, and the members of each pair have opposite meanings, as noted in table 1.1.

1.4 Body Regions

Learning Objective

5. Locate the major body regions on a chart or anatomical model.

The human body consists of an **axial** (ak′-sē-al) **portion,** the head, neck, and trunk, and an **appendicular** (ap-pen-dik′-ū-lar) **portion,** the upper and lower limbs and their girdles. Each of these major portions of the body is divided into regions with special names to facilitate communication and to aid in locating body components.

The major **body regions** are listed in tables 1.2 and 1.3 to allow easy correlation with figure 1.4, which shows the locations of the major regions of the body. Take time to learn the names, pronunciations, and locations of the body regions.

1.5 Body Planes and Sections

Learning Objective

6. Describe the four planes used in making sections of the body or body parts.

In studying the body or organs, you often will be observing the flat surface of a **section** that has been produced by a cut through the body or a body part. Such sections are made along specific **planes.** These well-defined planes–transverse,

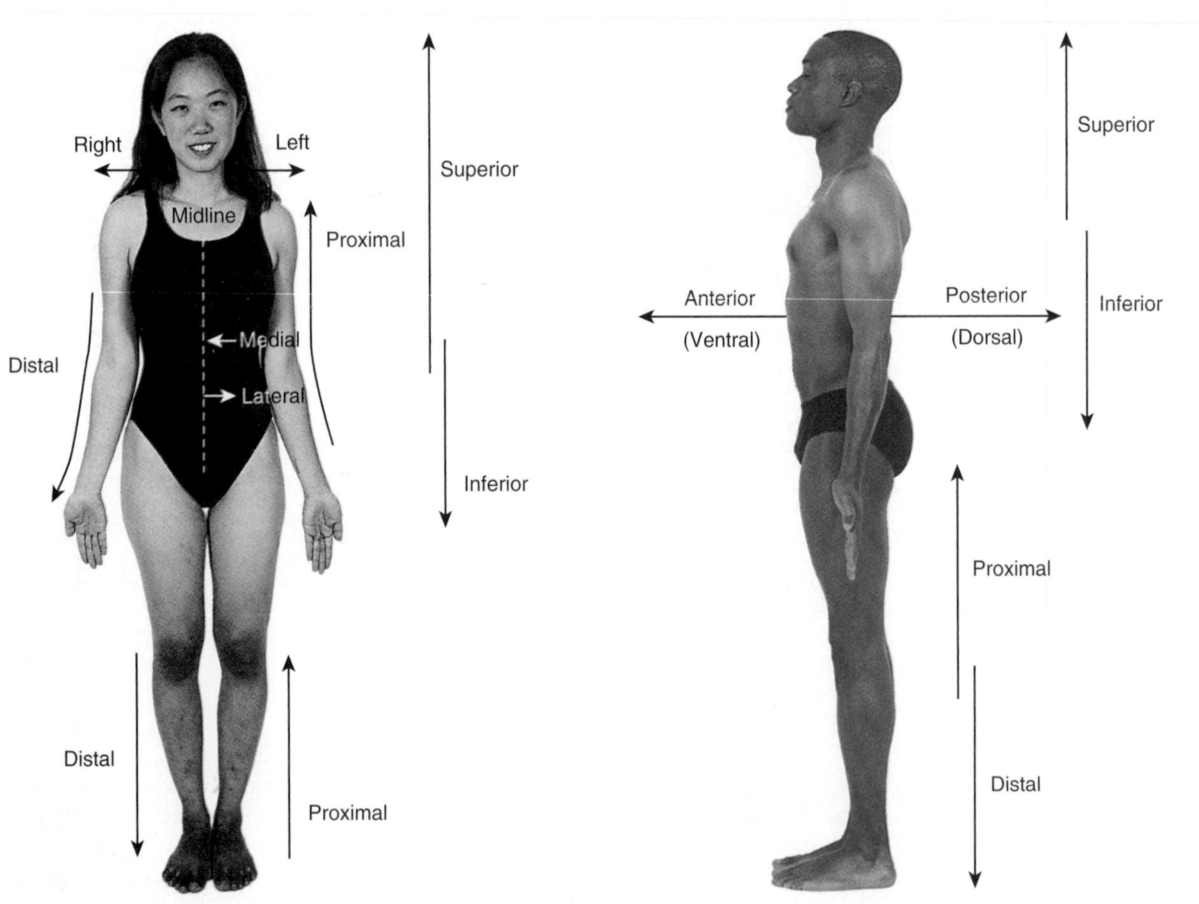

Figure 1.3 Anatomical Position and Directional Terms. AP|R

Table 1.1 Directional Terms

Term	Meaning	Example
Anterior (ventral)	Toward the front or abdominal surface of the body	The abdomen is anterior to the back.
Posterior (dorsal)	Toward the back of the body	The spine is posterior to the face.
Superior (cephalic)	Toward the top/head	The nose is superior to the mouth.
Inferior (caudal)	Away from the top/head	The navel is inferior to the nipples.
Medial	Toward the midline of the body	The breastbone is medial to the nipples.
Lateral	Away from the midline of the body	The ears are lateral to the cheeks.
Parietal	Pertaining to the outer boundary of body cavities	The parietal pleura lines the pleural cavity.
Visceral	Pertaining to the internal organs	The visceral pleura covers the lung.
Superficial (external)	Toward or on the body surface	The skin is superficial to the muscles.
Deep (internal)	Away from the body surface	The intestines are deep to the abdominal muscles.
Proximal	Closer to the beginning	The elbow is proximal to the wrist.
Distal	Farther from the beginning	The hand is distal to the wrist.
Central	At or near the center of the body or organ	The central nervous system is in the middle of the body.
Peripheral	External to or away from the center of the body or organ	The peripheral nervous system extends away from the central nervous system.

Table 1.2 Major Regions of the Head, Neck, and Trunk

Region			
Head and Neck	***Anterior Trunk***	***Posterior Trunk***	***Lateral Trunk***
Buccal (bu-kal)	Abdominal (ab-dom′-i-nal)	Dorsum (dor′-sum)	Axillary (ak′-sil-lary)
Cephalic (se-fal′-ik)	Abdominopelvic (ab-dom-i-nō-pel′-vik)	Gluteal (glu′-tē-al)	Coxal (kok′-sal)
Cervical (ser′-vi-kal)	Inguinal (ing′-gwi-nal)	Lumbar (lum′-bar)	***Inferior Trunk***
Cranial (krā′-nē-al)	Pectoral (pek′-tōr-al)	Sacral (sāk′-ral)	Genital (jen′-i-tal)
Facial (fā′-shal)	Pelvic (pel′-vik)	Vertebral (ver-tē′-bral)	Perineal (per-i-nē′-al)
Nasal (nā-zel)	Sternal (ster′-nal)		
Oral (or-al)	Umbilical (um-bil′-i-kal)		
Orbital (or-bit-al)			
Otic (o-tic)			

Table 1.3 Major Regions of the Limbs

Region		
Upper Limb	Digital (di′-ji-tal)	Femoral (fem′-ōr-al)
Antebrachial (an-tē-brā′-kē-al)	Olecranal (ō-lēk′-ran-al)	Patellar (pa-tel′-lar)
Antecubital (an-tē-kū-bi-tal)	Palmar (pal′-mar)	Pedal (pe′-dal)
Brachial (brā′-kē-al)	***Lower Limb***	Plantar (plan′-tar)
Carpal (kar′-pal)	Crural (krū′-ral)	Popliteal (pop-li-tē′-al)
Deltoid (del-tȯid)	Digital (di′-ji-tal)	Sural (sū′-ral)

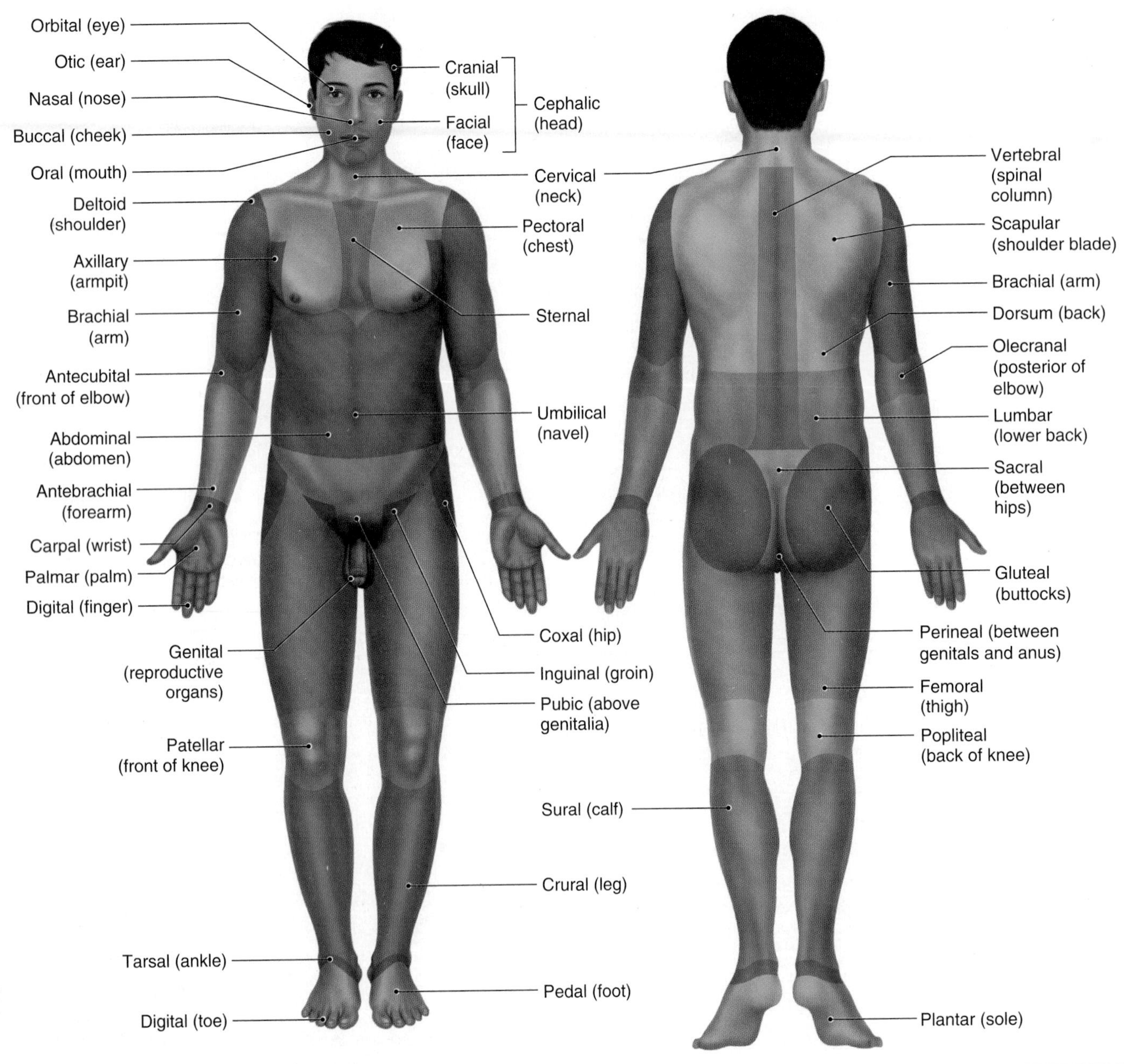

Figure 1.4 Major Regions of the Body. AP|R

sagittal, and frontal planes–lie at right angles to each other as shown in figure 1.5. It is important to understand the nature of the plane along which a section was made in order to understand the three-dimensional structure of an object being observed.

Transverse, or horizontal, **planes** divide the body into superior and inferior portions and are perpendicular to the longitudinal axis of the body.

Sagittal planes divide the body into right and left portions and are parallel to the longitudinal axis of the body. A **median** (midsagittal) **plane** passes through the midline of the body and divides the body into equal left and right halves. A **parasagittal plane** does not pass through the midline of the body.

Frontal (coronal) **planes** divide the body into anterior and posterior portions. These planes are perpendicular to sagittal planes and parallel to the longitudinal axis of the body.

Cuts made through sagittal and frontal planes, which are parallel to the longitudinal axis of the body, produce *longitudinal sections*. However, the term longitudinal section also refers to a section made through the longitudinal axis of an individual organ, tissue, or other structure. Similarly, cuts made through the transverse plane produce *cross*

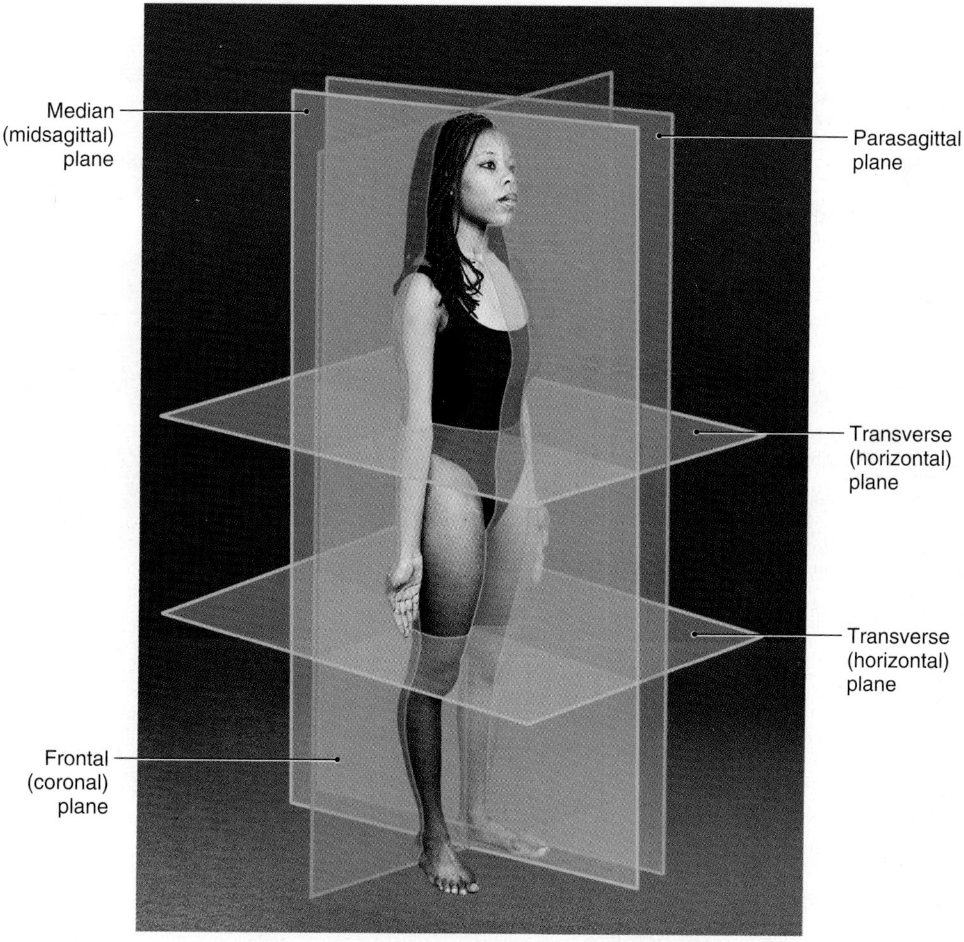

Figure 1.5 Anatomical Planes of Reference. AP|R

sections of the body and can also be produced in organs and tissues when cutting at a 90° angle to the longitudinal axis. *Oblique sections* are created when cuts are made in between the longitudinal and cross-sectional axes.

Check My Understanding

3. How do sagittal, transverse, and frontal planes differ from one another?

1.6 Body Cavities

Learning Objectives

7. Name the two major body cavities, their subdivisions and membranes.
8. Locate the body cavities, their subdivisions and membranes on a diagram.
9. Name the organs located in each body cavity.

There are two major cavities of the body that contain internal organs: the dorsal (posterior) and ventral (anterior) cavities. The body cavities protect and cushion the contained organs and permit changes in their size and shape without impacting surrounding tissues. Note the locations and subdivisions of these cavities in figure 1.6.

The **dorsal cavity** is subdivided into the **cranial cavity,** which houses the brain, and the **vertebral canal,** which contains the spinal cord. Note in figure 1.6 how the cranial bones and the vertebral column form the walls of the dorsal cavity and provide protection for these delicate organs.

The **ventral cavity** is divided by the *diaphragm,* a thin dome-shaped sheet of muscle, into a superior **thoracic cavity** and an inferior **abdominopelvic cavity.** The thoracic cavity is protected by the *rib cage* and contains the heart and lungs. The abdominopelvic cavity is subdivided into a superior **abdominal cavity** and an inferior **pelvic cavity,** but there is no structural separation between them. To visualize the separation, imagine a transverse plane passing through the body just superior to the pelvis. The abdominal cavity contains the stomach, intestines, liver, gallbladder, pancreas, spleen, and kidneys. The pelvic cavity contains the urinary bladder, sigmoid colon, rectum, and internal reproductive organs.

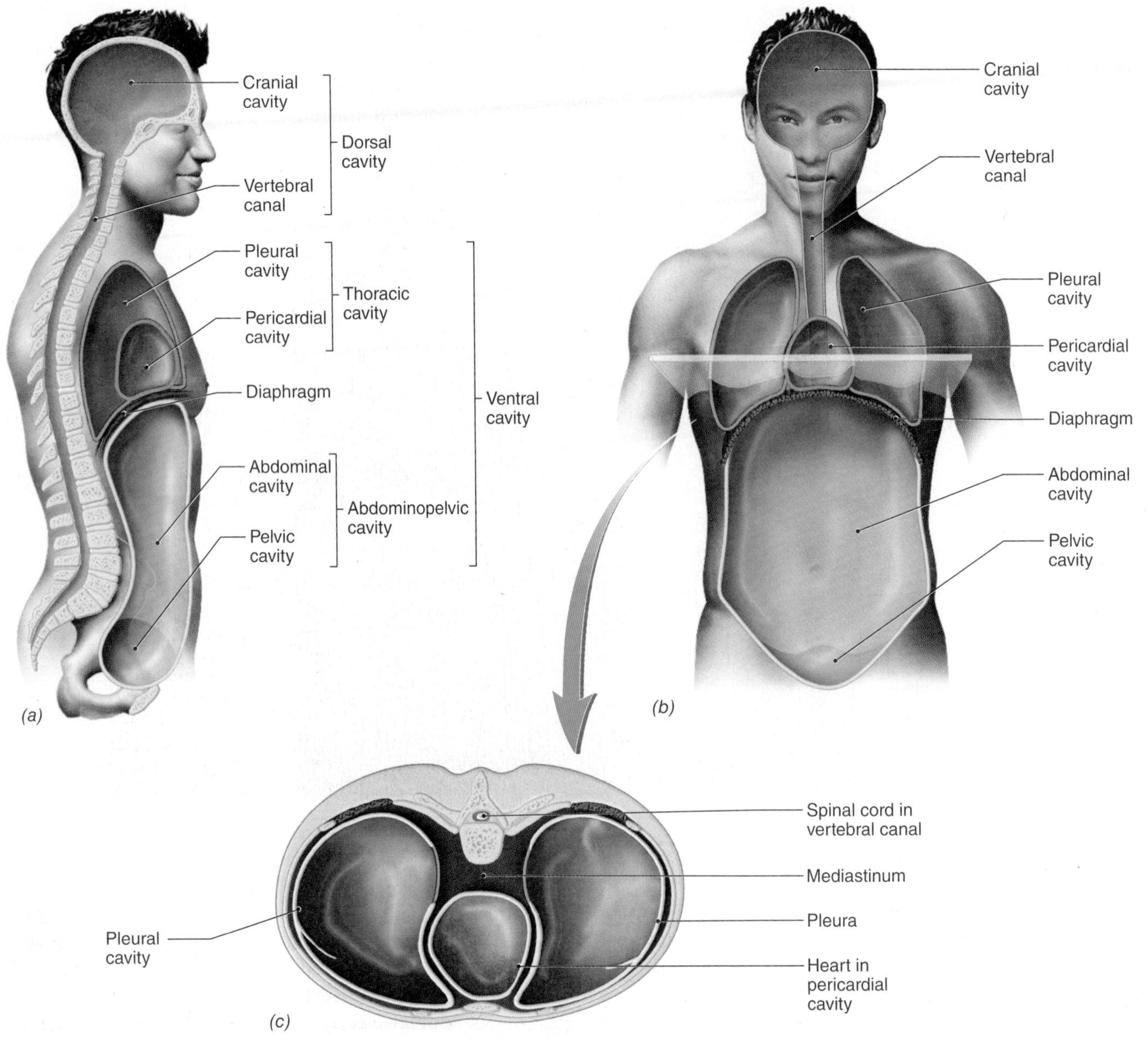

Figure 1.6 Body Cavities and Their Subdivisions.
(a) Sagittal section. *(b)* Frontal section. *(c)* Transverse section through the thoracic cavity. AP|R

Check My Understanding

4. What organs are located in each subdivision of the dorsal cavity?
5. What organs are located in each subdivision of the ventral cavity?

Membranes of Body Cavities

The membranes lining body cavities support and protect the internal organs in the cavities.

Dorsal Cavity Membranes

The dorsal cavity is lined by three layers of protective membranes that are collectively called the **meninges** (me-nin′-jēz; singular, *meninx*). The most superficial membrane is attached to the wall of the dorsal cavity, and the deepest membrane tightly envelops the brain and spinal cord. The meninges will be covered in chapter 8.

Ventral Cavity Membranes

The ventral body cavity organs are supported and protected by **serosae** (singular, *serosa*), or **serous membranes.**

Clinical Insight

Physicians use certain types of diagnostic imaging systems, for example, *computerized tomography (CT), magnetic resonance imaging (MRI),* and *positron emission tomography (PET),* to produce images of sections of the body to help them diagnose disorders. In computerized tomography, an X-ray emitter and an X-ray detector rotate around the patient so that the X-ray beam passes through the body from hundreds of different angles. X-rays collected by the detector are then processed by a computer to produce sectional images on a screen for viewing by a radiologist. A good understanding of sectional anatomy is required to interpret CT scans. Transverse sections, such as the image on the left, are always shown in the same way. Convention is to use supine (face up), inferior views as if looking up at the section from the foot of the patient's bed. What structures can you identify in the CT image shown on the right? APR

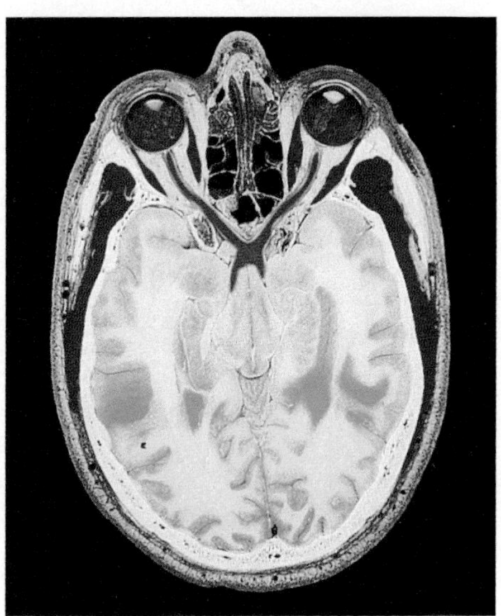

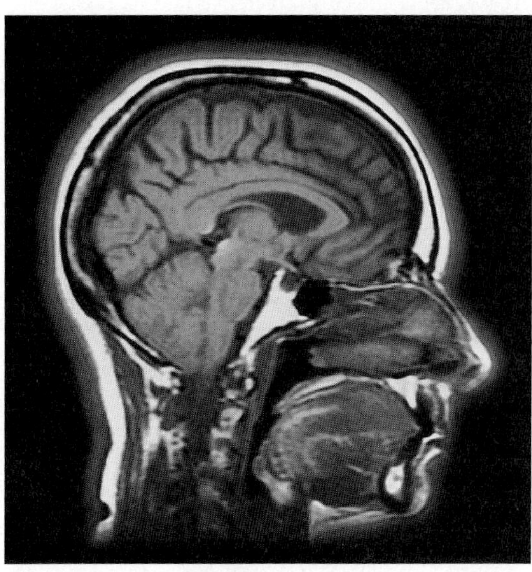

The serous membranes are thin layers of tissue that line the body cavity and cover the internal organs. Serous membranes have a superficial *parietal* (pah-rī′-e-tal) *layer* that lines the cavity and a deep *visceral* (vis′-er-al) *layer* that covers the organ. The parietal and visceral layers secrete a watery lubricating fluid that is generically called *serous fluid* into the cavity formed between the layers. This arrangement is similar to that of a fist pushed into a balloon (figure 1.7). The serous membranes of the body are the pleura, pericardium, and peritoneum.

The serous membranes lining the thoracic cavity are called **pleurae** (singular, *pleura*), or **pleural membranes.** The walls of the left and right portions of the thoracic cavity are lined by the *parietal pleurae.* The surfaces of the lungs are covered by the *visceral pleurae.* The parietal and visceral pleurae are separated by a thin film of serous fluid called pleural fluid, which reduces friction as the pleurae rub against each other as the lungs expand and contract during breathing. The potential space (not an actual space) between the parietal and visceral pleurae is known as the **pleural cavity.**

The left and right portions of the thoracic cavity are divided by a membranous partition, the *mediastinum* (mē-dē-a-stī′-num). Organs located within the mediastinum include the heart, thymus, esophagus, and trachea.

The heart is enveloped by the **pericardium** (per-i-kar′-dē-um), which is formed by membranes of the mediastinum. The thin *visceral pericardium* is tightly adhered to the surface of the heart. The *parietal pericardium* lines the deep surface of a loosely fitting sac around the heart. The potential space between the visceral and parietal pericardia is the **pericardial cavity,** and it contains serous fluid, called pericardial fluid, that reduces friction as the heart contracts and relaxes.

The walls of the abdominal cavity and the surfaces of abdominal organs are lined with the **peritoneum** (per-i-tō-nē′-um), or **peritoneal membrane.** The *parietal peritoneum* lines the walls of the abdominal cavity but not the pelvic cavity. It descends only to cover the superior portion of the urinary bladder. The kidneys, pancreas, and parts of the intestines are located posterior to the parietal peritoneum in a space known as the *retroperitoneal space.* The *visceral peritoneum,* an extension of the parietal peritoneum, covers the surface of the abdominal organs. Double-layered folds of the visceral peritoneum, the *mesenteries* (mes′-en-ter″-ēs), extend between the abdominal organs

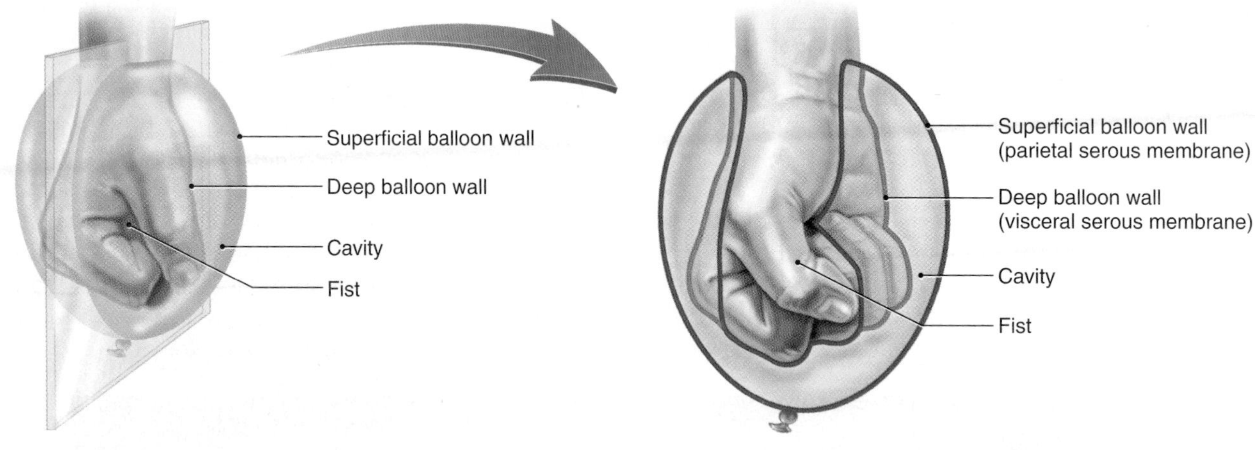

Figure 1.7 Illustration of a fist pushed into a balloon as an analogy to serous membranes. APR

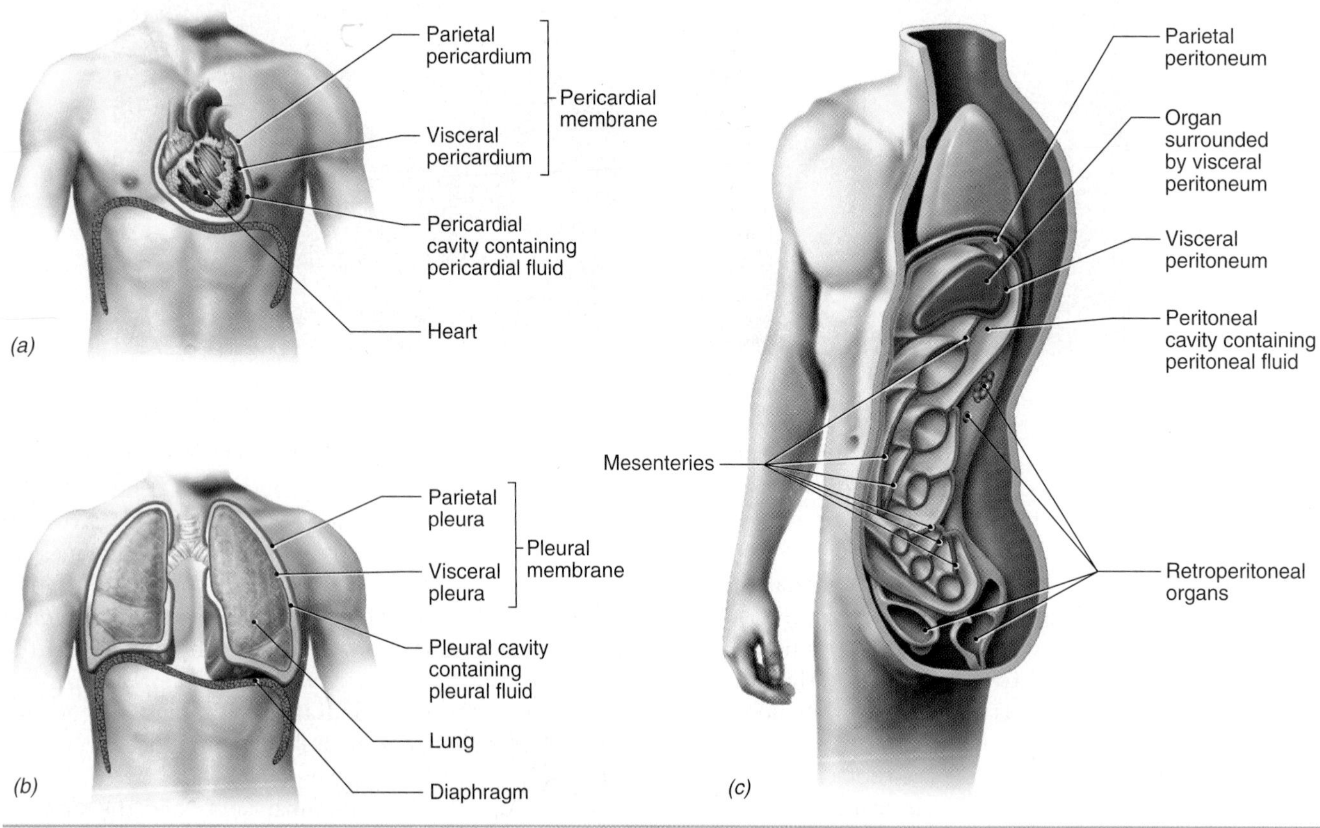

Figure 1.8 Serous Membranes of The Ventral Cavity.
(a) Anterior view of pericardium. *(b)* Anterior view of pleura. *(c)* Sagittal view of peritoneum. APR

and provide support for them (see figure 1.8*c*). The potential space between the parietal and visceral peritoneal membranes is called the **peritoneal cavity** and contains a small amount of serous fluid called peritoneal fluid (figure 1.8).

Check My Understanding

6. What membranes line the dorsal and ventral cavities?
7. What is the function of serous fluid?

1.7 Abdominopelvic Subdivisions

Learning Objectives

10. Name the abdominopelvic quadrants and regions.
11. Locate the abdominopelvic quadrants and regions on a diagram.

The abdominopelvic cavity is subdivided into either four quadrants or nine regions to aid health care providers in locating underlying organs in the abdominopelvic cavity. Physicians may feel (palpate) or listen to (auscultate) the abdominopelvic region to examine it. Changes in firmness or sounds may indicate abnormalities in the structures of a quadrant or region.

The four quadrants are formed by two planes that intersect just superior to the umbilicus (navel), as shown in figure 1.9*a*. Note the organs within each quadrant.

The nine regions are formed by the intersection of two sagittal and two transverse planes as shown in figure 1.9*c*. The sagittal planes extend inferiorly from the midpoints of the collarbones. The superior transverse plane lies just inferior to the borders of the 10th costal cartilages, and the inferior transverse plane lies just inferior to the superior border of the hip bones.

Study figures 1.8 and 1.9 to increase your understanding of the locations of the internal organs and associated membranes.

Now examine the colorplates that follow this chapter. They show an anterior view of the body in progressive stages of dissection that reveals major muscles, blood vessels, and internal organs. Study these plates to learn the normal locations of the organs of the ventral cavity. Also, check your understanding of the organs within each abdominopelvic quadrant and region.

Check My Understanding

8. What are the four quadrants and nine regions of the abdominopelvic region?

1.8 Maintenance of Life

Learning Objectives

12. Define metabolism, anabolism, and catabolism.
13. List the five basic needs essential for human life.
14. Define homeostasis.
15. Explain how homeostasis relates to both healthy body functions and disorders.
16. Describe the general mechanisms of negative feedback and positive feedback.

Humans, like all living organisms, exhibit the fundamental processes of life. **Metabolism** (me-tab′ō-lizm) is the term that collectively refers to the sum of all of the chemical reactions that occur in the body.

There are two phases of metabolism: anabolism and catabolism. **Anabolism** (ah-nab′-ō-lizm) refers to processes that use energy and nutrients to build the complex organic molecules that compose the body. **Catabolism** (kah-tab′-ō-lizm) refers to processes that release energy and break down complex molecules into simpler molecules.

Life is fragile. It depends upon the normal functioning of trillions of body cells, which, in turn, depends upon factors needed for survival and the ability of the body to maintain relatively stable internal conditions.

Survival Needs

There are five basic needs that are essential to human life:

1. **Food** provides chemicals that serve as a source of energy and raw materials to grow and to maintain cells of the body.
2. **Water** provides the environment in which the chemical reactions of life occur.
3. **Oxygen** is required to release the energy in organic nutrients, which powers life processes.
4. **Body temperature** must be maintained close to 36.8°C (98.2°F) to allow the chemical reactions of human metabolism to occur.
5. **Atmospheric pressure** is required for breathing to occur.

Homeostasis

Homeostasis is the maintenance of a relatively stable internal environment by self-regulating physiological processes. Homeostasis keeps body temperature and the composition of blood and interstitial fluids within their normal range. This relatively stable internal environment is maintained in spite of the fact that internal and external factors tend to alter body temperature, and materials are continuously entering and exiting the blood and interstitial fluid.

All of the organ systems work in an interdependent manner to maintain homeostasis. For example, changes in one system tend to affect one or more other body systems. Therefore, any disruption in one body system tends to be corrected but may disrupt another body system. The internal environment is maintained via a *dynamic equilibrium* where there is constant fluctuation taking place in order to maintain homeostasis. Malfunctioning or overcompensation in a homeostatic mechanism can lead to disorders and diseases.

The dynamic equilibrium of homeostasis is primarily maintained by physiologic processes called **negative-feedback mechanisms.** Body fluid composition and other physiological variables fluctuate near a normal value, called a *set point*, and negative-feedback mechanisms are

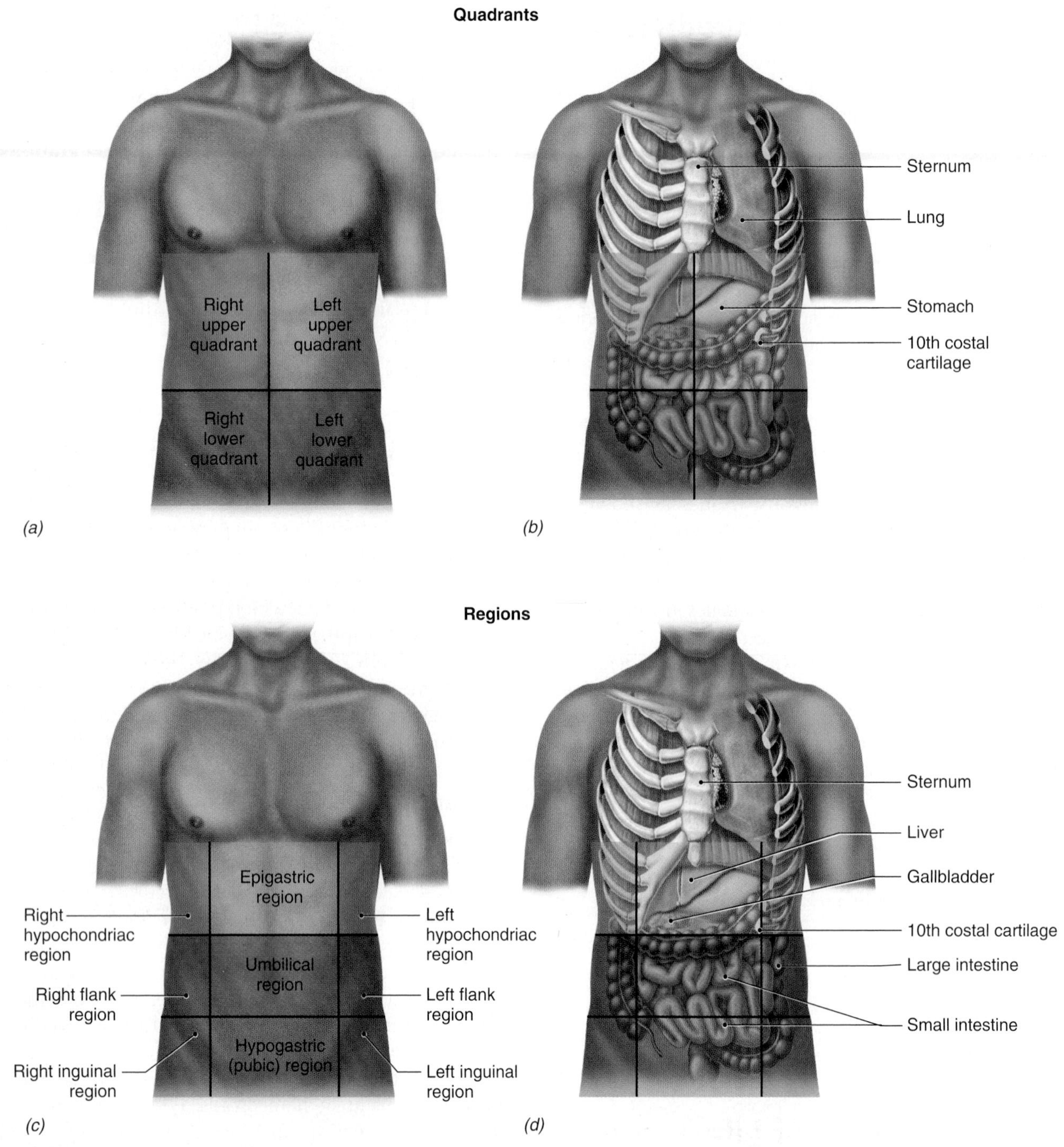

Figure 1.9 The four quadrants and nine regions of the abdominopelvic cavity. APR

used to keep these variables within their normal range (figure 1.10). For a negative-feedback mechanism to work, it needs to be able to monitor and respond to any changes in homeostasis. The structure of the negative-feedback mechanism allows it to function in exactly this manner and is a great example of how anatomical structure complements function. To monitor a physiological variable, a negative-feedback mechanism utilizes a **receptor** to detect deviation from the set point and send a signal notifying the integrating center about the deviation. The **integrating center,** which is the body region that knows the set point for the variable, processes the information from a receptor and determines the course of action that is needed. It then sends a signal that activates an **effector.**

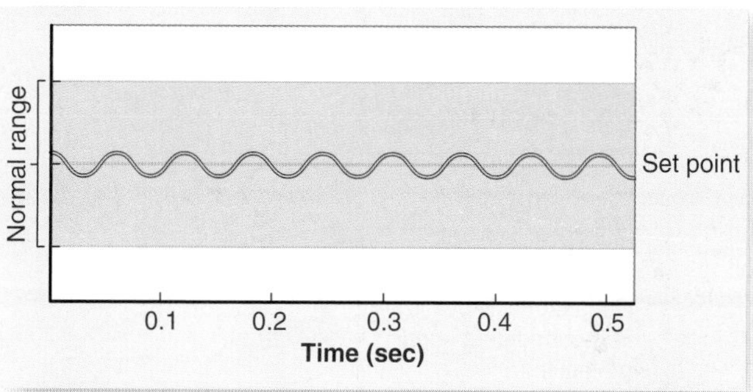

(a) High sensitivity

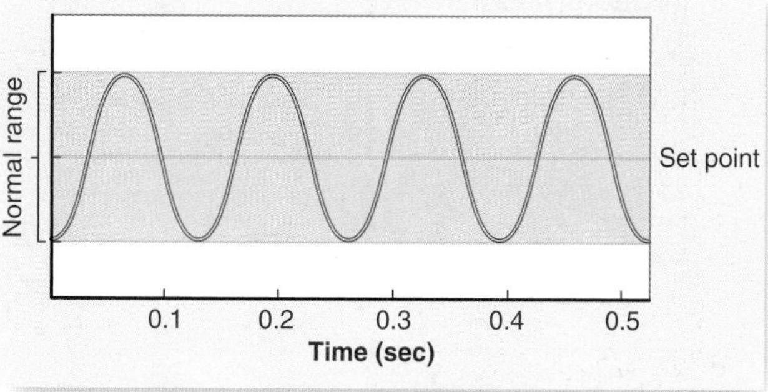

(b) Low sensitivity

Figure 1.10 *(a)* A negative-feedback mechanism with high sensitivity. *(b)* A negative-feedback mechanism with low sensitivity.

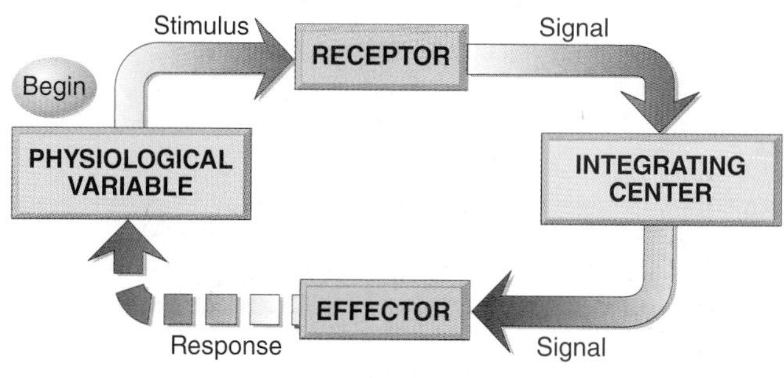

Figure 1.11 A negative-feedback mechanism controlling homeostasis.

The effector will carry out the necessary response according to the directions of the integrating center and return the variable back toward the set point. In a negative-feedback mechanism the response of the effector will always be the opposite of the change detected by the receptor (figure 1.11). Once the set point is reached, the negative-feedback mechanism will automatically turn off.

Our body's ability to maintain relatively constant blood glucose levels relies on negative-feedback mechanisms. When blood glucose levels begin to rise, as they do after a meal, there are receptors in the pancreas that can detect this *stimulus* (change). The beta cells of the pancreas act as an integrating center and release the hormone insulin in response to this change. Insulin travels through the blood to several effectors, one of which is the liver. Insulin causes the liver cells to take excess glucose out of the bloodstream and thus decrease the blood glucose level back toward normal. The pancreas possesses other receptors that can detect decreases in blood glucose, such as occurs between meals. The alpha cells of the pancreas, acting as the integrating center, release the hormone glucagon. Glucagon causes the liver to release glucose into the bloodstream, which will increase blood glucose back toward normal (figure 1.12).

It is important to note that the response of the integrating center will be stronger if the original stimulus is farther from normal. For example, if the blood glucose level rises sharply out of the normal range, causing *hyperglycemia* (blood glucose level above normal), the amount of insulin the beta cells release will be more than the amount released if the blood glucose level is elevated but is still within the normal range. This type of response is called a *graded response* because it can respond on different levels (figure 1.13).

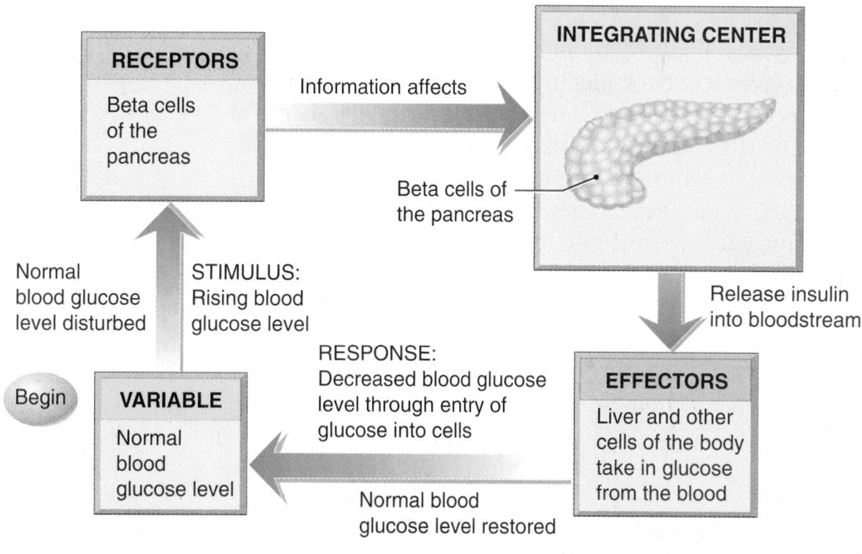

Figure 1.12 The negative-feedback mechanism that regulates blood glucose levels.

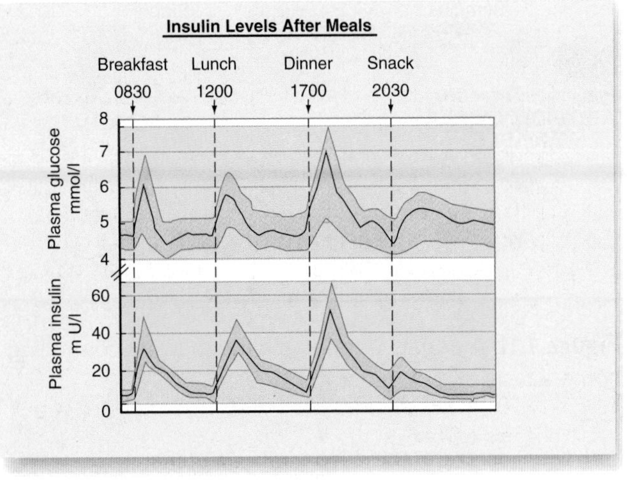

Figure 1.13 The graded response of insulin release is based on amount of blood glucose elevation.

Positive-feedback mechanisms utilize the same basic components as negative-feedback mechanisms. However, the outcome of a positive-feedback mechanism is very different from that of a negative-feedback mechanism. A positive-feedback mechanism is used when the originating stimulus needs to be amplified and continued in order for the desired result to occur. A few examples of positive-feedback mechanisms include fever, activation of the immune response, formation of blood clots, certain aspects of digestion, and uterine contractions of labor. If you think about blood clot formation, blood clots do not form "normally"; when they begin to form, this occurs quickly and completely in order to stop blood loss. This is a necessary mechanism for overall homeostasis. Figure 1.14 illustrates the specific steps of the positive-feedback mechanism of saliva production.

Positive-feedback mechanisms can be harmful because they lack the ability to stop on their own. They will continue to amplify the effect of the original stimulus, which can push the body dangerously out of homeostasis, until the cycle is interrupted by an outside factor. For example, an uncontrolled fever can increase body temperature to a point that is fatal. For this reason, positive-feedback mechanisms are used for rare events within the body, rather than for the daily maintenance of homeostasis.

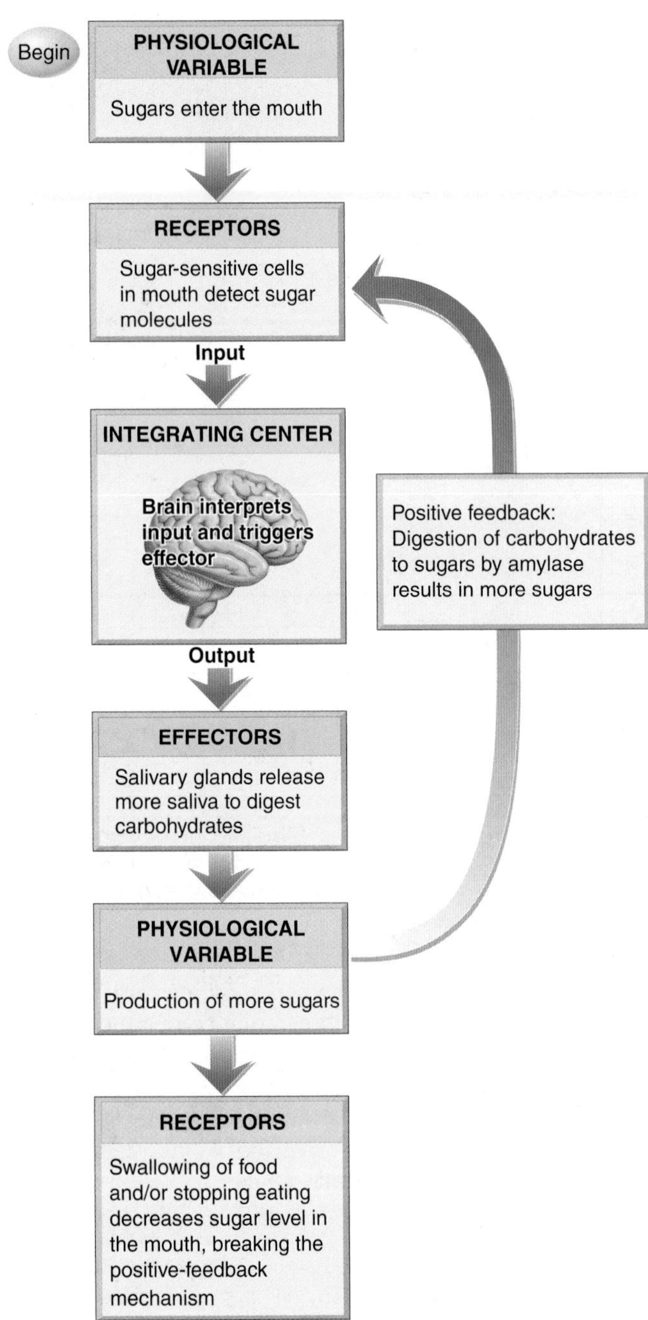

Figure 1.14 A positive-feedback mechanism illustrating the production of saliva.

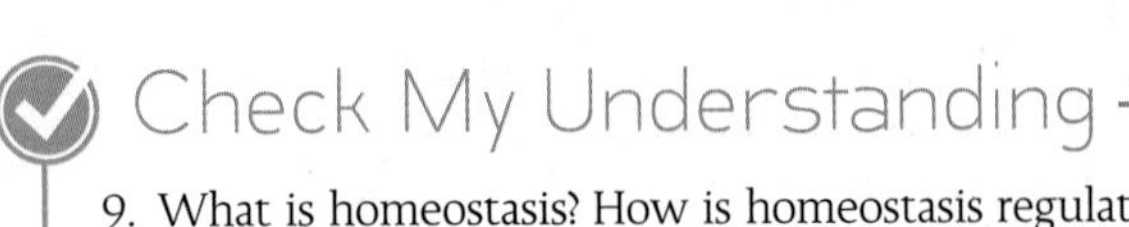

Check My Understanding

9. What is homeostasis? How is homeostasis regulated?

Chapter Summary

1.1 Anatomy and Physiology

- Human anatomy is the study of body structure and organization.
- Human physiology is the study of body functions.

1.2 Levels of Organization

- The body consists of several levels of organization of increasing complexity.
- From simple to complex, the organizational levels are chemical, cellular, tissue, organ, organ system, and organismal.
- The organs of the body are arranged in coordinated groups called organ systems.
- The 11 organ systems of the body are

integumentary	cardiovascular
skeletal	lymphoid
muscular	respiratory
nervous	urinary
endocrine	reproductive
digestive	

1.3 Directional Terms

- Directional terms are used to describe the relative positions of body parts.
- Directional terms occur in pairs, with the members of a pair having opposite meanings.

anterior–posterior	proximal–distal
superior–inferior	external–internal
medial–lateral	parietal–visceral
central–peripheral	

1.4 Body Regions

- The body is divided into two major portions: the axial portion and the appendicular portion.
- The axial portion is subdivided into the head, neck, and trunk.
- The head and neck contain cervical, cranial, and facial regions. The cranial and facial regions combine to form the cephalic region.
- The facial region consists of orbital, nasal, oral, and buccal regions.
- The trunk consists of anterior, posterior, lateral, and inferior regions.
- Anterior trunk regions include the abdominal, inguinal, pectoral, pelvic, and sternal regions. The abdominal and pelvic regions combine to form the abdominopelvic region.
- Posterior trunk regions include the dorsal, gluteal, lumbar, sacral, and vertebral regions.
- Lateral trunk regions are the axillary and coxal regions.
- Inferior trunk regions are the genital and perineal regions.
- The appendicular portion of the body consists of the upper and lower limbs.
- The upper limb is attached to the trunk at the shoulder. Regions of the upper limb are the antebrachial, antecubital, brachial, carpus, digital, olecranal, and palmar regions.
- The lower limb is attached to the trunk at the hip. Regions of the lower limb are the crural, digital, femoral, patellar, pedal, plantar, popliteal, sural, and tarsal regions.

1.5 Body Planes and Sections

- Well-defined planes are used to guide sectioning of the body or organs.
- The common planes are transverse, sagittal, and frontal.
- The common planes produce longitudinal sections and cross sections of the body.

1.6 Body Cavities

- There are two major body cavities: dorsal and ventral.
- The dorsal cavity consists of the cranial cavity and vertebral canal.
- The ventral cavity consists of the thoracic and abdominopelvic cavities.
- The thoracic cavity lies superior to the diaphragm. It consists of two lateral pleural cavities and the mediastinum, which contains the pericardial cavity.
- The abdominopelvic cavity lies inferior to the diaphragm. It consists of a superior abdominal cavity and an inferior pelvic cavity.
- The body cavities are lined with protective and supportive membranes.
- The meninges consist of three membranes that line the dorsal cavity and enclose the brain and spinal cord.
- The parietal pleurae line the internal walls of the rib cage, while the visceral pleurae cover the external surfaces of the lungs.
- The pleural cavity is the potential space between the parietal and visceral pleurae.
- The parietal pericardium is a saclike membrane in the mediastinum that surrounds the heart. The visceral pericardium is attached to the surface of the heart.
- The pericardial cavity is the potential space between the parietal and visceral pericardia.
- The parietal peritoneum lines the walls of the abdominal cavity but does not extend into the pelvic cavity. The visceral peritoneum covers the surface of abdominal organs.
- The peritoneal cavity is the potential space between the parietal and visceral peritoneal membranes.
- The mesenteries are double-layered folds of the visceral peritoneum that support internal organs.
- Kidneys, pancreas, and parts of the intestines are located posterior to the parietal peritoneum in the retroperitoneal space.

1.7 Abdominopelvic Subdivisions

- The abdominopelvic cavity is subdivided into either four quadrants or nine regions as an aid in locating organs.
- The four quadrants are

right upper	left upper
right lower	left lower

- The nine regions are

epigastric	right flank
left hypochondriac	hypogastric (pubic)
right hypochondriac	left inguinal
umbilical	right inguinal
left flank	

1.8 Maintenance of Life

- Metabolism is the sum of all of the body's chemical reactions. It consists of anabolism, the synthesis of body chemicals, and catabolism, the breakdown of body chemicals.
- The basic needs of the body are food, water, oxygen, body temperature, and atmospheric pressure.
- Homeostasis is the maintenance of a relatively stable internal environment.
- Homeostasis is regulated by negative-feedback mechanisms.
- Negative-feedback mechanisms consist of three components: receptors, integrating center, and effectors.
- Positive-feedback mechanisms promote an ever-increasing change from the norm.

Self-Review

Answers are located in Appendix B.

1. A study of body functions is called ______.
2. Blood, the heart, and blood vessels compose the ______ system.
3. Rapid coordination of body functions is the function of the ______ system.
4. The fingers are located ______ to the wrist.
5. The upper and lower limbs compose the ______ portion of the body.
6. The posterior surface of the knee is known as the ______ region.
7. The thigh is known as the ______ region.
8. The ______ body cavity is divided into the cranial cavity and ______ canal.
9. The gallbladder is located in the ______ quadrant and the ______ region.
10. The ______ separates the left and right portions of the thoracic cavity.
11. The abdominal cavity is lined by the ______.
12. The maintenance of a relatively stable internal environment is called ______.

Critical Thinking

1. A hypoglycemic (low blood glucose level) patient is given orange juice to drink. Explain how this increases blood glucose level and the organ systems involved.
2. Describe the location of the kneecap in as many ways as you can using directional terms.
3. Describe where serous membranes are located in the body, name the three types of serous fluid, and explain the function of serous fluid.
4. Explain how negative-feedback mechanisms regulate homeostasis.

ADDITIONAL RESOURCES

COLORPLATES OF THE HUMAN BODY

The five colorplates that follow show the basic structure of the human body. The first plate shows the anterior body surface and the superficial anterior muscles of a female. Succeeding plates show the internal structure as revealed by progressively deeper dissections.

Refer to these plates often as you study this text in order to become familiar with the relative locations of the body organs. ■

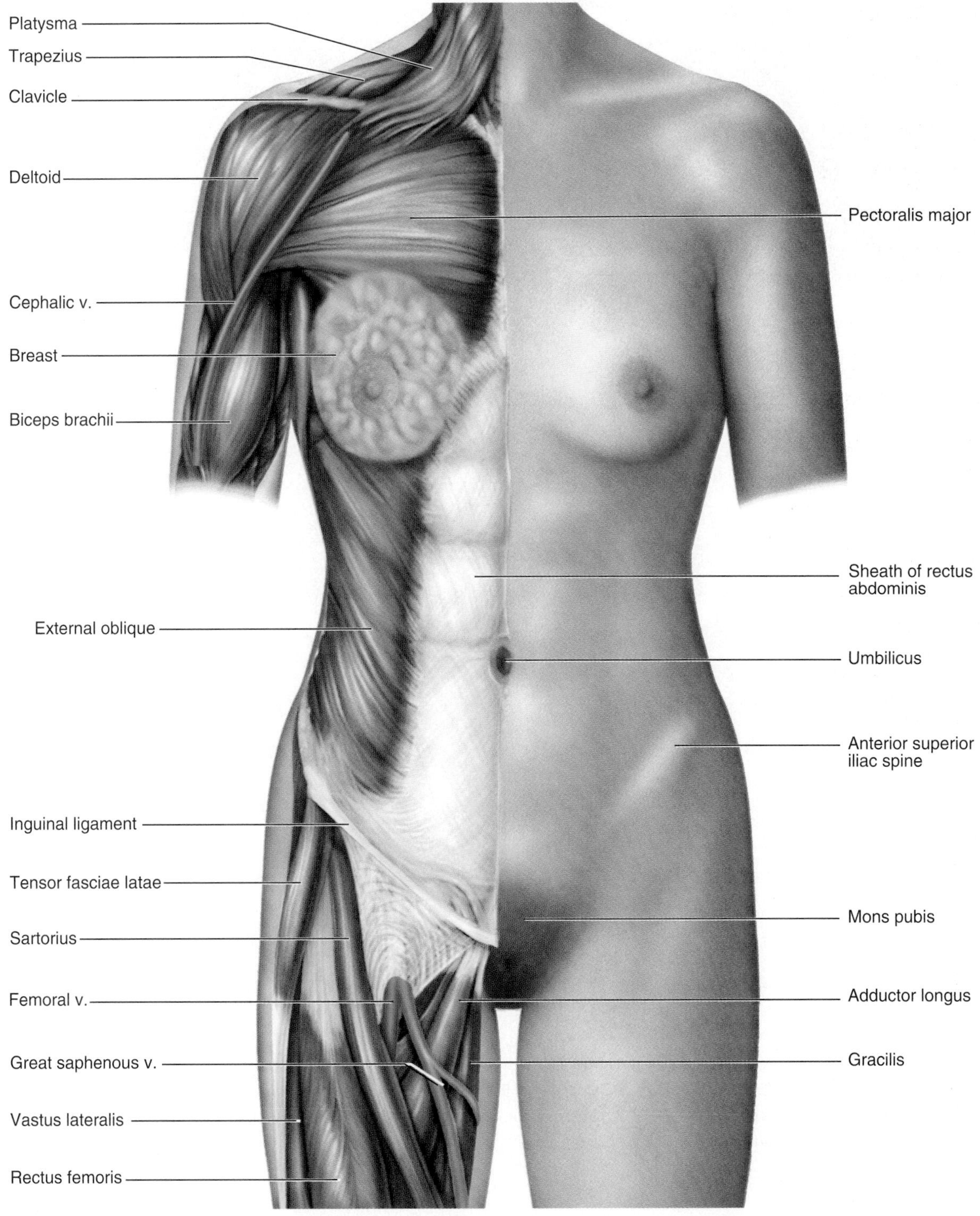

Plate 1 Superficial Anatomy of the Trunk (Female).
Surface anatomy is shown on the anatomical left, and structures immediately deep to the skin on the right (v. = vein).

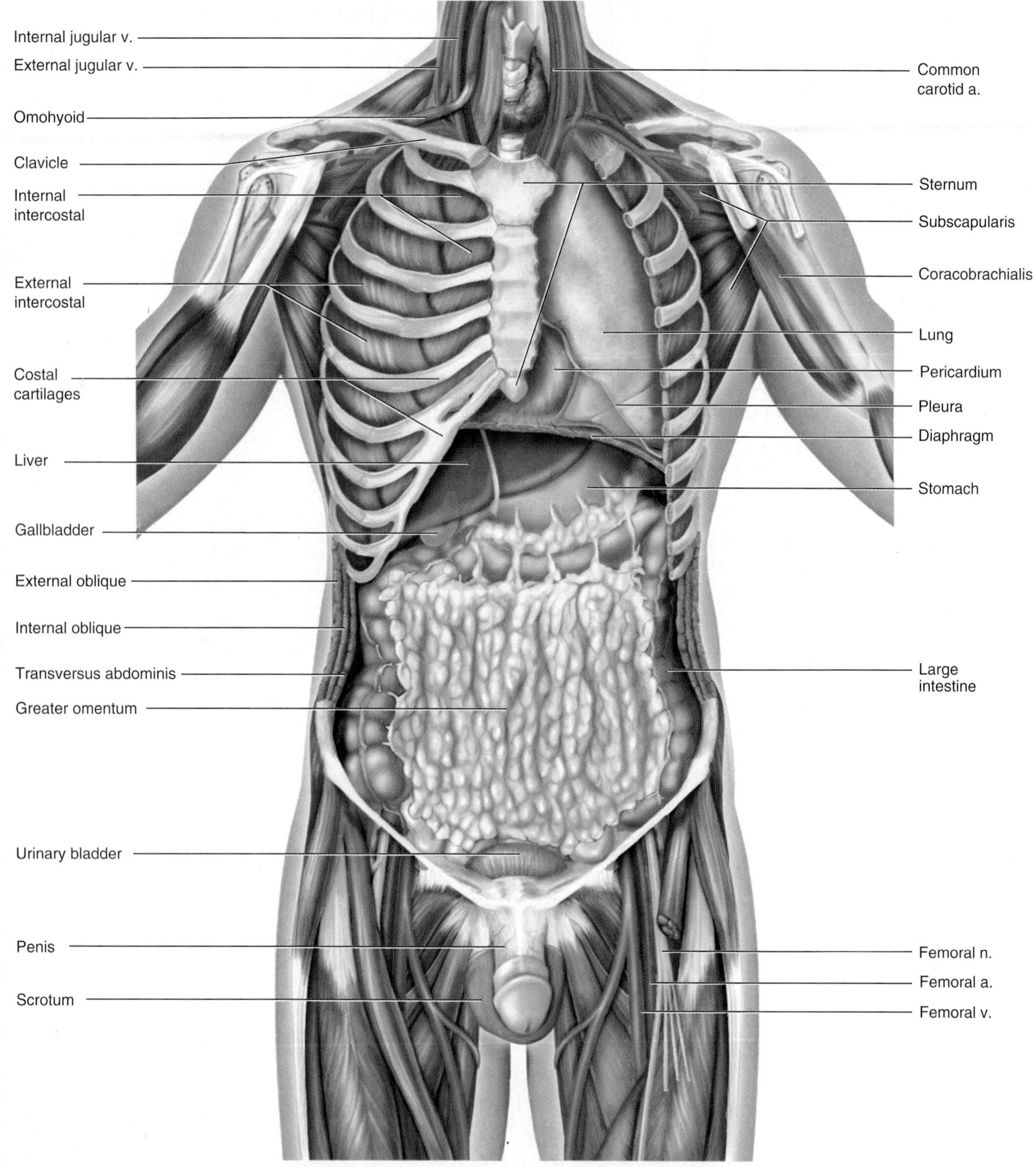

Plate 2 Anatomy at the Level of the Rib Cage and Greater Omentum (Male).
The anterior body wall is removed, and the ribs, intercostal muscles, and pleurae are removed from the anatomical left (a. = artery; v. = vein; n. = nerve).

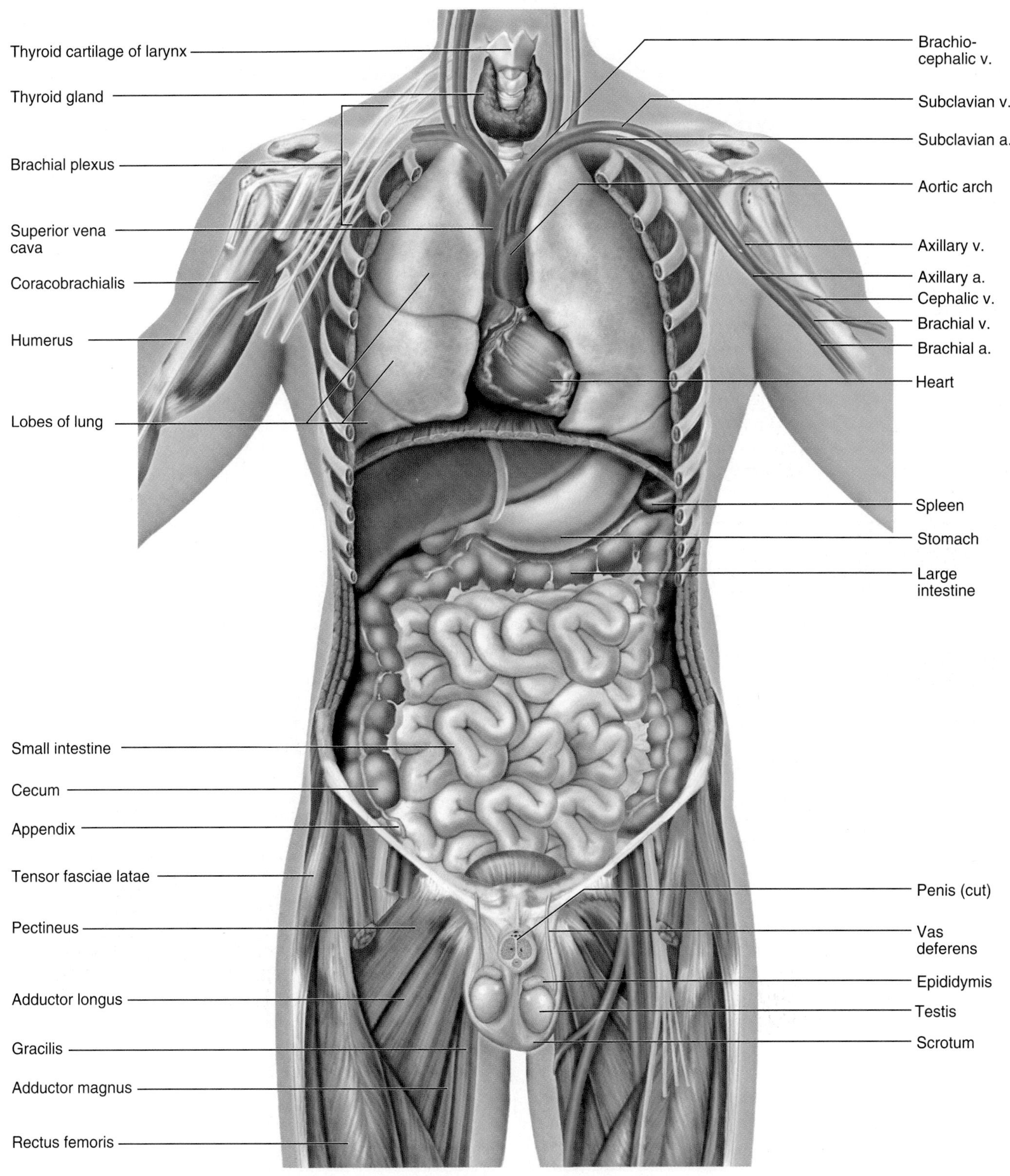

Plate 3 Anatomy at the Level of the Lungs and Intestines (Male).
The sternum, ribs, and greater omentum are removed (a. = artery; v. = vein).

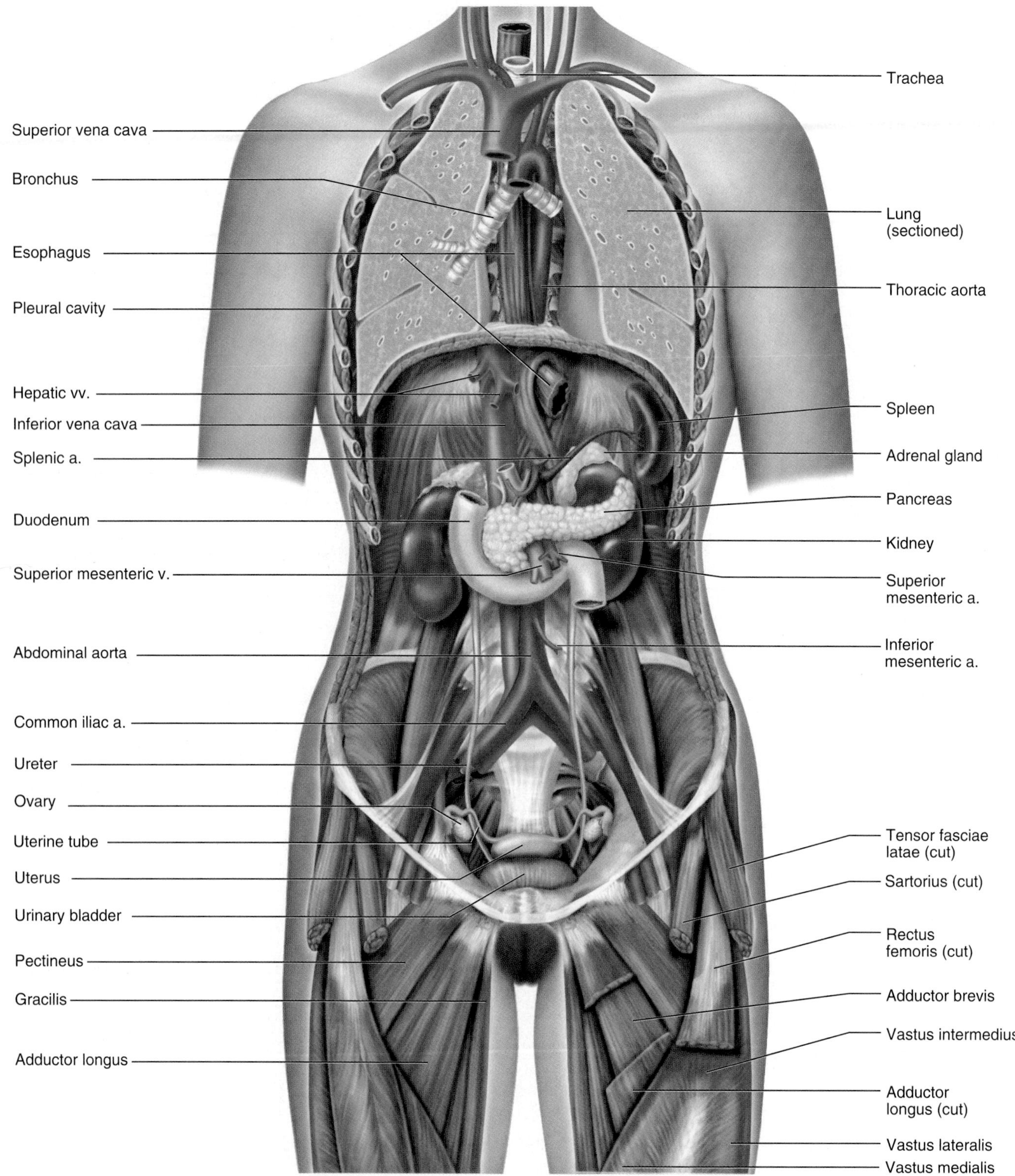

Plate 4 Anatomy at the Level of the Retroperitoneal Viscera (Female).
The heart is removed, the lungs are frontally sectioned, and the viscera of the peritoneal cavity and the peritoneum itself are removed (a. = artery; v. = vein; vv. = veins).

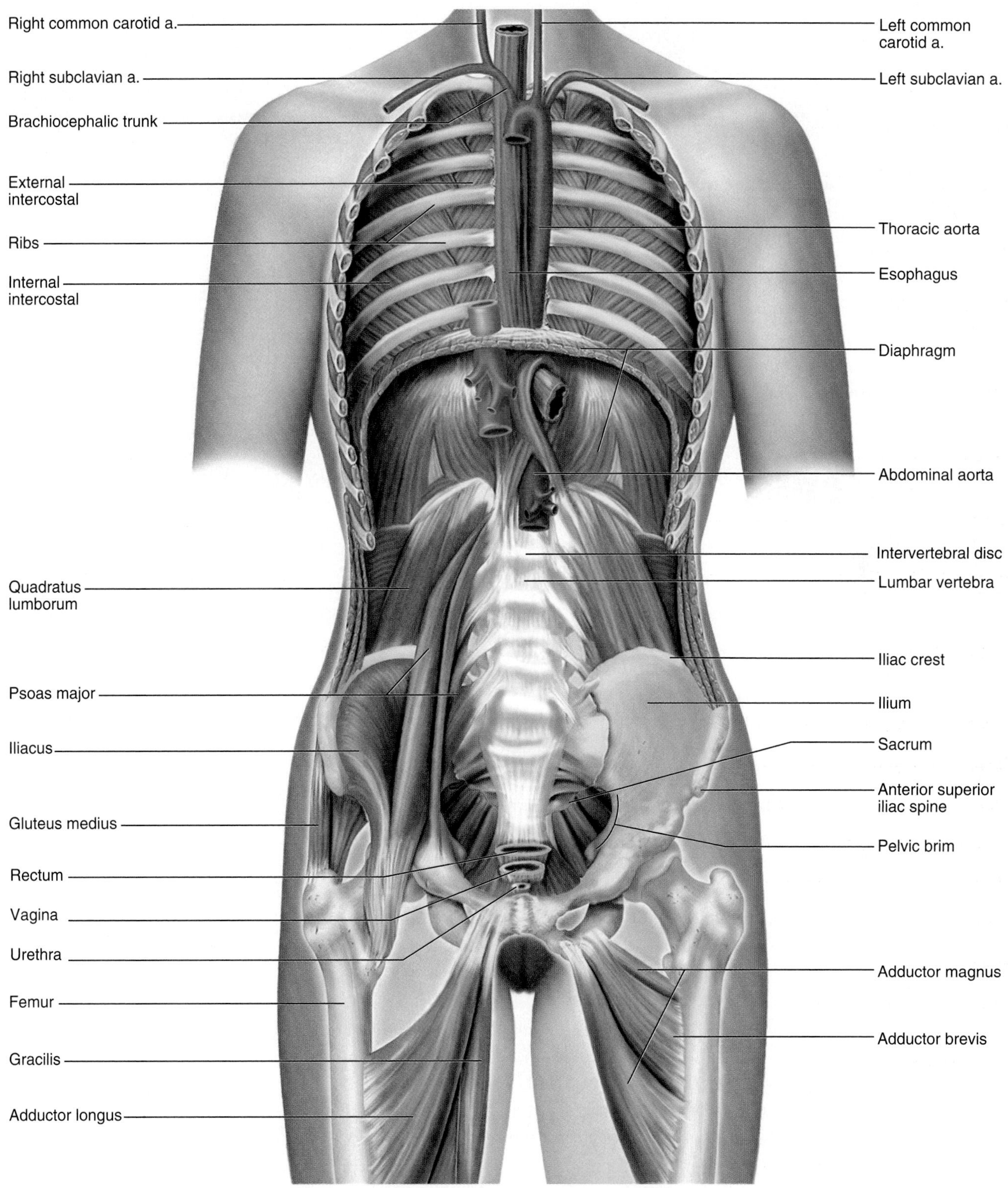

Plate 5 Anatomy at the Level of the Posterior Body Wall (Female).
The lungs and retroperitoneal viscera are removed (a. = artery).

2 CHAPTER

Chemicals of Life

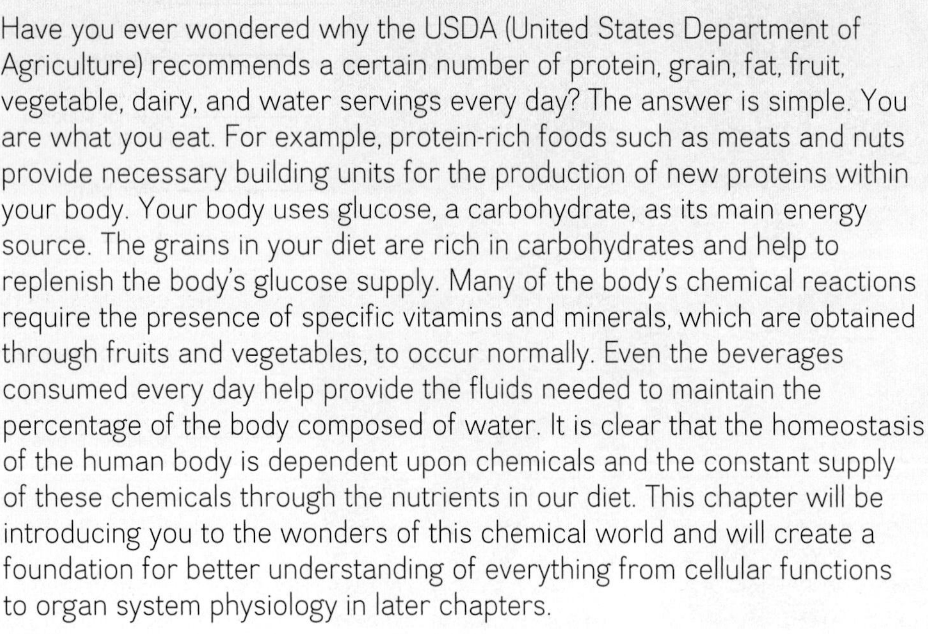

Have you ever wondered why the USDA (United States Department of Agriculture) recommends a certain number of protein, grain, fat, fruit, vegetable, dairy, and water servings every day? The answer is simple. You are what you eat. For example, protein-rich foods such as meats and nuts provide necessary building units for the production of new proteins within your body. Your body uses glucose, a carbohydrate, as its main energy source. The grains in your diet are rich in carbohydrates and help to replenish the body's glucose supply. Many of the body's chemical reactions require the presence of specific vitamins and minerals, which are obtained through fruits and vegetables, to occur normally. Even the beverages consumed every day help provide the fluids needed to maintain the percentage of the body composed of water. It is clear that the homeostasis of the human body is dependent upon chemicals and the constant supply of these chemicals through the nutrients in our diet. This chapter will be introducing you to the wonders of this chemical world and will create a foundation for better understanding of everything from cellular functions to organ system physiology in later chapters.

CHAPTER OUTLINE

Module 2
Cells & Chemistry

SELECTED KEY TERMS

Atom (atomos = indivisible) The smallest unit of an element.
Carbohydrate (carbo = carbon; hydr = water) An organic compound composed of carbon, hydrogen, and oxygen, with hydrogen and oxygen at a 2:1 ratio.
Chemical bond (bond from band = fasten) Joining of chemical substances using attractions between electrons.
Chemical formula (formula = draft or small form) Shorthand notation showing the type and number of atoms in a molecule or compound.
Chemical reaction (re = again; actionem = put into motion) Process involving the formation and/or breakage of chemical bonds resulting in new combinations of atoms.
Compound (componere = to place together) A substance formed by atoms from two or more elements.
Element A substance that cannot be broken down into simpler substances by ordinary chemical means.
Enzyme (en = in; zym = ferment) A protein that catalyzes chemical reactions.
Inorganic compound (in = not) Small, simple substance that usually does not have carbon and hydrogen in the same substance.
Lipid (lip = fat) An organic compound containing mostly carbon and hydrogen, with small amounts of oxygen. These compounds do not mix with water.
Molecule (molecula = little mass) A substance formed by two or more atoms bonded together by covalent bonds.
Nucleic acid (nucle = kernel) A complex organic molecule composed of nucleotides.
Organic compound (organon = from living things) Large, complex substances that contain both carbon and hydrogen in the same molecule, usually with oxygen too.
Protein A group of nitrogen-containing organic compounds formed of amino acids.

ANYTHING THAT OCCUPIES SPACE IS **MATTER**. Chemistry is the scientific study of matter and the interactions of matter. A basic knowledge of chemistry is necessary for health-care professionals because the human body is composed of chemicals and the processes of life are chemical interactions.

2.1 Atoms and Elements

Learning Objectives

1. Describe the basic structure of an atom.
2. Distinguish between atoms, isotopes, and radioisotopes.

The entire physical universe, both living and nonliving, is composed of matter. All matter is composed of **elements,** substances that cannot be broken down into simpler substances by ordinary chemical means. Carbon, hydrogen, and nitrogen are examples of chemical elements.

New elements are being discovered relatively frequently as technology continues to advance. As of the writing of this textbook, there were 118 elements in the periodic table. Most scientists consider 92 of these elements to be "naturally occurring," which generally means they can be found in samples of soil, air, and water. The remaining elements in the periodic table are man-made. The average person has detectable traces of approximately 60 elements in his or her body, but by most current definitions only 24 are recognized as being involved in maintaining life.

Figure 2.1 highlights the 12 elements of the human body that occur in significant amounts (totaling 99.9%). The four elements isolated in figure 2.1 (oxygen, carbon, hydrogen, and nitrogen) make up approximately 96% of the human body and are found making up the body's major organic molecules, discussed later in the chapter. Other remaining elements occur in very small amounts and are referred to as *trace elements*.

Atomic Structure APIR

An **atom** (a'-tom) is the smallest single unit of an element. Atoms of a given element are similar to each other, and they are different from atoms of all other elements. Atoms of different elements differ in size, mass, and how they interact with other atoms.

Atoms are composed of three types of subatomic particles: protons, neutrons, and electrons. Each **proton** has a positive electrical charge. Each **electron** has a negative electrical charge. Each **neutron** has no electrical charge. Protons and neutrons are located in the **nucleus** at the center of an atom.

Electrons orbit, or revolve around, the nucleus at high speeds in *electron shells* that are located at various distances from the nucleus. The first shell of electrons, the shell closest to the nucleus, can hold a maximum of two electrons even if it is the only electron shell. An atom with two or more electron shells reacts with other atoms to fill its **valence** (outermost) shell with eight electrons. Atoms always fill the lowest electron shells first. See the diagram of the atomic structures of hydrogen and carbon in figure 2.2.

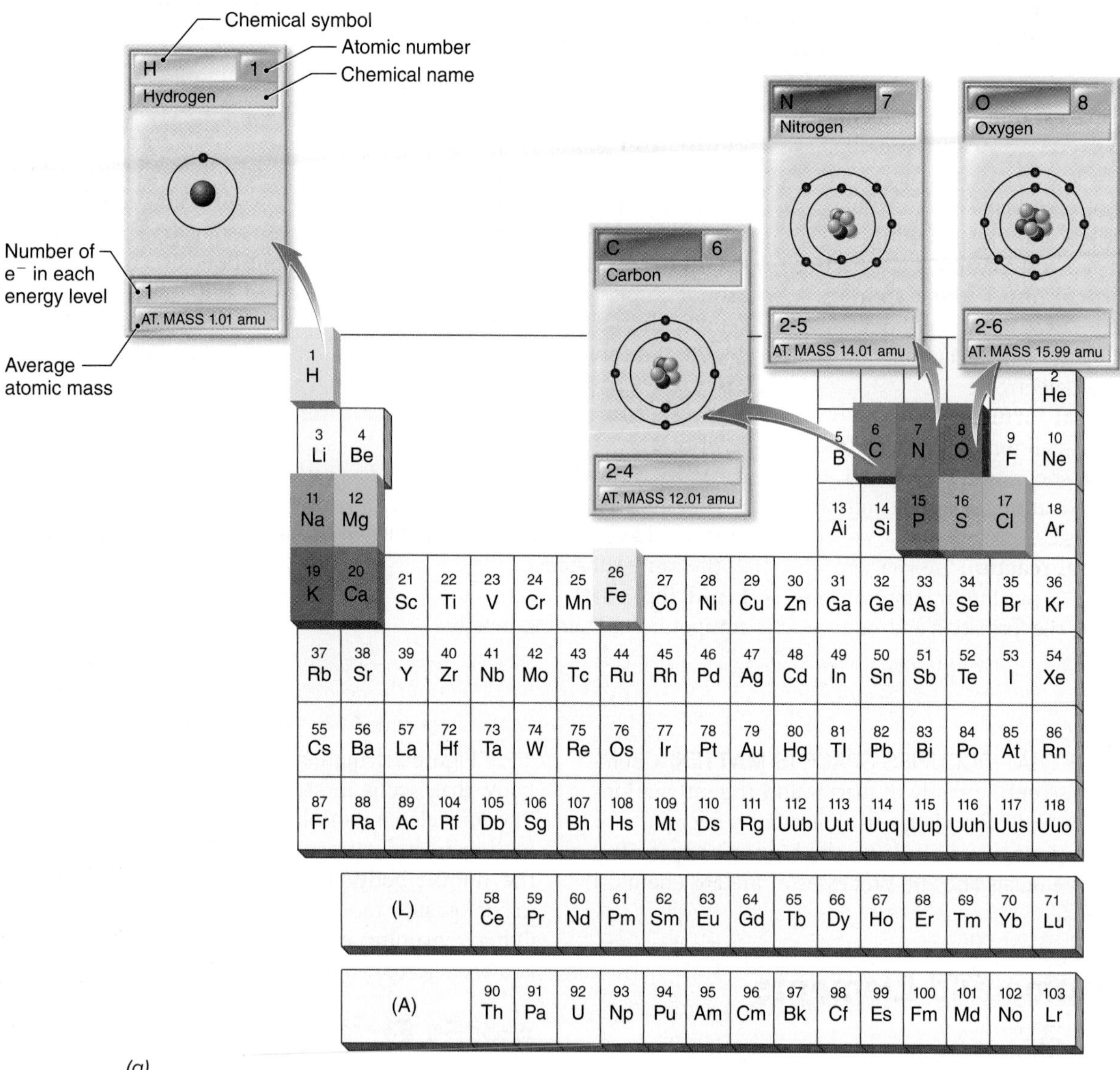

Most Common Elements of the Human Body

Major elements (collectively compose more than 98% of body weight)			Lesser elements (collectively compose less than 1% of body weight)		
Chemical symbol		% Body weight	Chemical symbol		% Body weight
O	Oxygen	65.0	S	Sulfur	0.25
C	Carbon	18.0	K	Potassium	0.25
H	Hydrogen	10.0	Na	Sodium	0.15
N	Nitrogen	3.0	Cl	Chlorine	0.15
Ca	Calcium	1.5	Mg	Magnesium	0.05
P	Phosphorus	1.0	Fe	Iron	0.006

(b)

Figure 2.1 *(a)* The periodic table of elements. *(b)* The 12 most abundant elements in the human body.

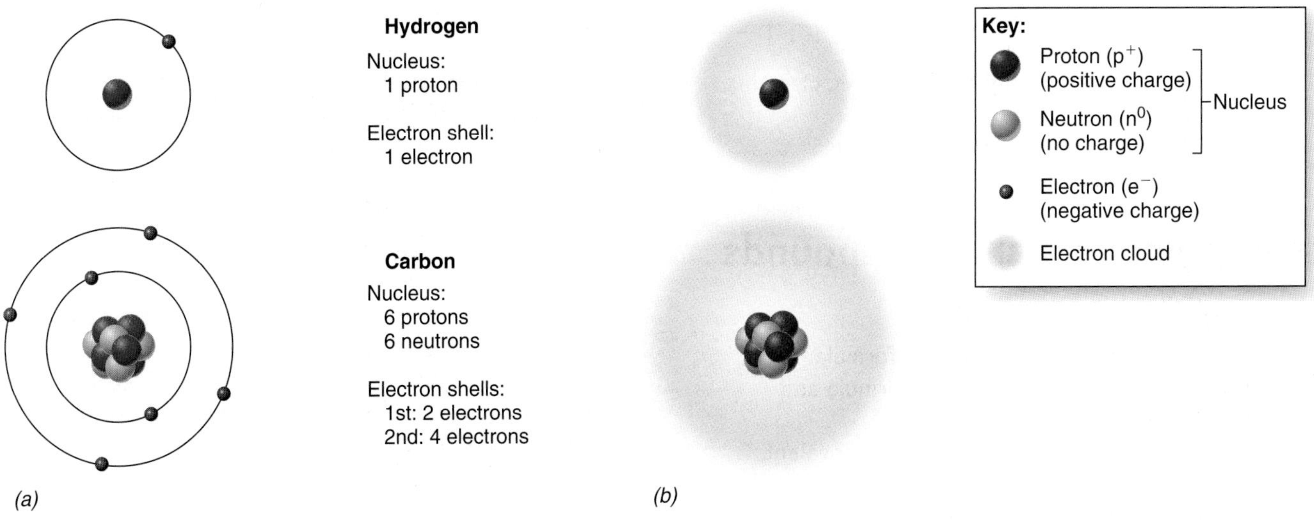

Figure 2.2 Atomic structures of hydrogen and carbon in *(a)* electron shell model, and *(b)* electron cloud model. These models show the most likely locations of the electrons.

An atom is electrically neutral because it has the same number of protons as electrons, although the number of neutrons may vary. Most atoms are not stable in this state and have characteristic ways of losing, gaining, or sharing electrons to achieve stability, which is key to forming chemical bonds.

The atoms of each element are characterized by a specific atomic number, chemical symbol, and atomic mass. These characteristics are used to identify the element. The **atomic number** indicates the number of protons and also the number of electrons in each atom. The *chemical symbol* is a shorthand way of referring to an element or to an atom of the element. The mass of either a proton or a neutron is defined as one *atomic mass unit* (amu). Because of this, the **atomic mass** of an atom is simply the sum of the number of protons plus the number of neutrons in each atom. For example, an atom of carbon has an atomic number of 6, a chemical symbol of C, and an atomic mass of 12. From this information, you know that an atom of carbon has six protons, six electrons, and six neutrons.

Isotopes

As mentioned in the preceding section, all atoms of an element have the same number of protons and electrons. However, some atoms may have a different number of neutrons. An atom of an element with a different number of neutrons is called an **isotope** (ī′-so-tōp). For example, hydrogen has three isotopes: ^{1}H, ^{2}H, and ^{3}H (figure 2.3). All isotopes of an element have the same chemical properties because they have the same number of protons and electrons.

Certain isotopes of some elements have an unstable nucleus that emits high-energy radiation as it breaks down to form a more stable nucleus. Such isotopes are called *radioisotopes*. Certain radioisotopes are used in the diagnosis of disorders and in the treatment of cancer. See the clinical insight box later in this chapter.

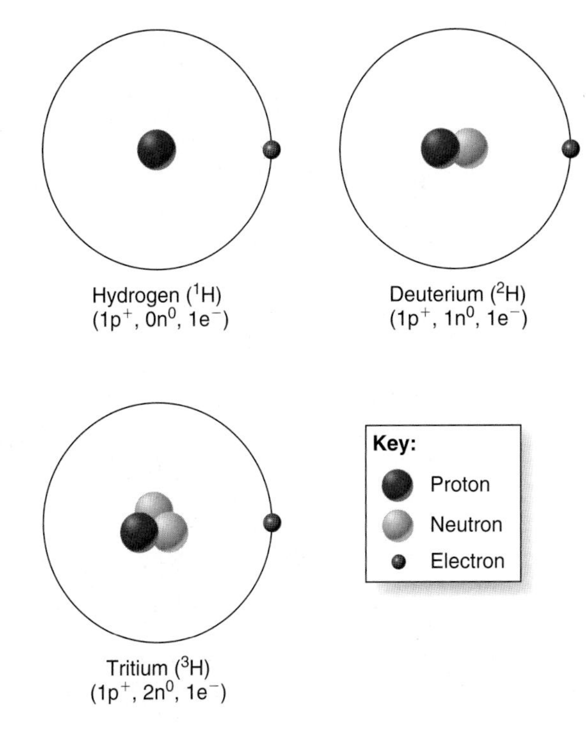

Figure 2.3 The three isotopes of hydrogen. Notice that only the number of neutrons changes.

Check My Understanding

1. What is the relationship among matter, elements, and atoms?
2. What is the basic structure of an atom?

2.2 Molecules and Compounds

Learning Objectives

3. Explain the meaning of a chemical formula.
4. Compare and contrast molecular formula and structural formula.
5. Compare and contrast ionic, nonpolar covalent, polar covalent, and hydrogen bonds.
6. Compare synthesis, decomposition, exchange, and reversible reactions.

A few elements exist separately in the body, but most are chemically bound to others to form **molecules.** Some molecules are composed of like elements–an oxygen molecule (O_2), for example. Others, such as water (H_2O), are composed of different kinds of elements. **Compounds** are substances composed of atoms from two or more different elements. Thus, the chemical structure of water may be referred to as both a molecule and a compound. Whether a substance is called a molecule or a compound also depends upon the type of chemical bond used to build the substance. Molecules are built by covalent bonds only, while compounds are built by either ionic or covalent bonds. Chemical bonds are discussed in more detail shortly.

Chemical Formulae

A **chemical formula** expresses the chemical composition of a molecule or compound. Two major types of chemical formulae exist, the molecular formula and the structural formula. A *molecular formula* expresses both the composition of a single molecule and the composition of a compound. In a molecular formula, chemical symbols indicate the elements of the atoms involved, while subscripts identify the number of atoms of each element in the molecule. For example, the molecular formula for water is H_2O, which indicates that two atoms of hydrogen combine with one atom of oxygen to form a water molecule. The molecular formula does not describe how the two hydrogen atoms and one oxygen atom in a water molecule are attached to each other. There are many possibilities, H–H–O, H–O–H, or O–H–H, for example. Even if the order in which the atoms are attached is known, the atoms may not be arranged in a straight line as indicated above. A *structural formula* is a diagram that both indicates the composition and number of atoms and illustrates how the atoms are

Clinical Insight

Nuclear medicine is the medical specialty that uses radioisotopes in the diagnosis and treatment of disease. Very small amounts of weak radioisotopes may be used to tag biological molecules in order to trace the movement or metabolism of these molecules in the body. Special instruments can detect the radiation emitted by the radioisotopes and identify the location of the tagged molecules.

In *nuclear imaging,* the emitted radiation creates an image on a special photographic plate or computer screen. In this way, it is possible to obtain an image of various organs or parts of organs where the radioisotopes accumulate. Positron emission tomography (PET) uses certain radioisotopes that emit positrons (positively charged electrons), and it enables precise imaging similar to computerized tomography (CT) scans. PET can be used to measure processes, such as blood flow, rate of metabolism of selected substances, and effects of drugs on body functions. It is a promising technique for both the diagnosis of disease and the study of normal physiological processes.

Another form of nuclear medicine involves the use of radioisotopes to kill cancerous cells. Certain radioisotopes may be attached to specific biological molecules and injected into the blood. When these molecules accumulate in cancerous tissue, the emitted radiation kills the cancerous cells. A similar effect is obtained by implanting pellets of radioactive isotopes directly in cancerous tissue.

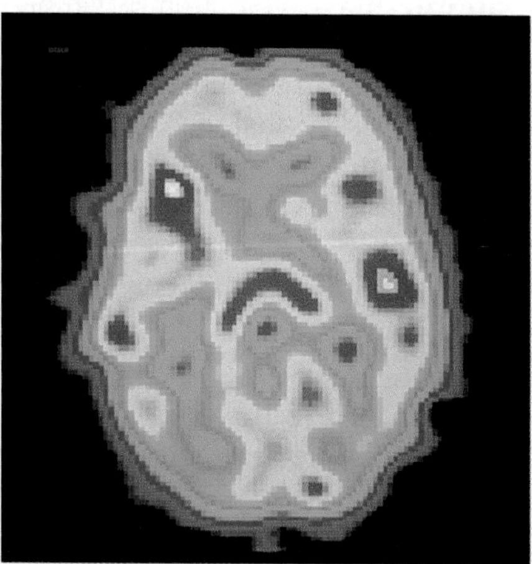

Positron Emission Tomography (PET).
Transverse section through the head. The highest level of brain activity is indicated in red, with successively lower levels represented by yellow, green, and blue.

linked to one another. Many figures in the text will use structural formulae. Figures 2.5 and 2.7 are good examples.

Up to this point it has been mentioned that molecules are composed of atoms that are "chemically combined." However, no mention has been made as to how this occurs. We will explore this next.

Chemical Bonds APR

Chemicals are combined when electrons interact to form **chemical bonds,** which join atoms together to form a molecule. A chemical bond is a force of attraction between two atoms. An atom combines with another atom in order to fill its valence shell. A full valence shell makes an atom more stable. To do this, atoms either (1) receive or lose electrons, which results in the formation of an ionic bond; or (2) share electrons, which leads to the formation of a covalent bond.

Ionic Bonds

Consider the interaction of sodium and chlorine in the formation of sodium chloride (table salt), as shown in figure 2.4. Sodium has a single electron in its valence shell, while chlorine has seven electrons in its valence shell. Note in step 2 of figure 2.4 that, after transferring an electron from sodium to chlorine, sodium now has 11 protons (+) and 10 electrons (−), while chlorine has 17 protons (+) and 18 electrons (−). Thus, the transfer of an electron from sodium to chlorine causes the sodium atom to have a net electrical charge of +1 and the chlorine atom to have a net electrical charge of −1.

Atoms with a net electrical charge, either positive or negative, are called **ions.** Thus, the transfer of an electron from sodium to chlorine has (1) resulted in the valence shell of each atom being filled with electrons and (2) produced a sodium ion (Na^+) and a chloride ion (Cl^-). Positively charged ions, such as Na^+, are called *cations.* Negatively charged ions, such as Cl^-, are called *anions.* The force of attraction that holds cations and anions together is an **ionic bond.**

Covalent Bonds

Atoms that form molecules by sharing electrons are joined by **covalent bonds.** The shared electrons orbit around each atom for part of the time so that they may be counted in the outer shell of each atom. Thus, the valence shell of each atom is filled.

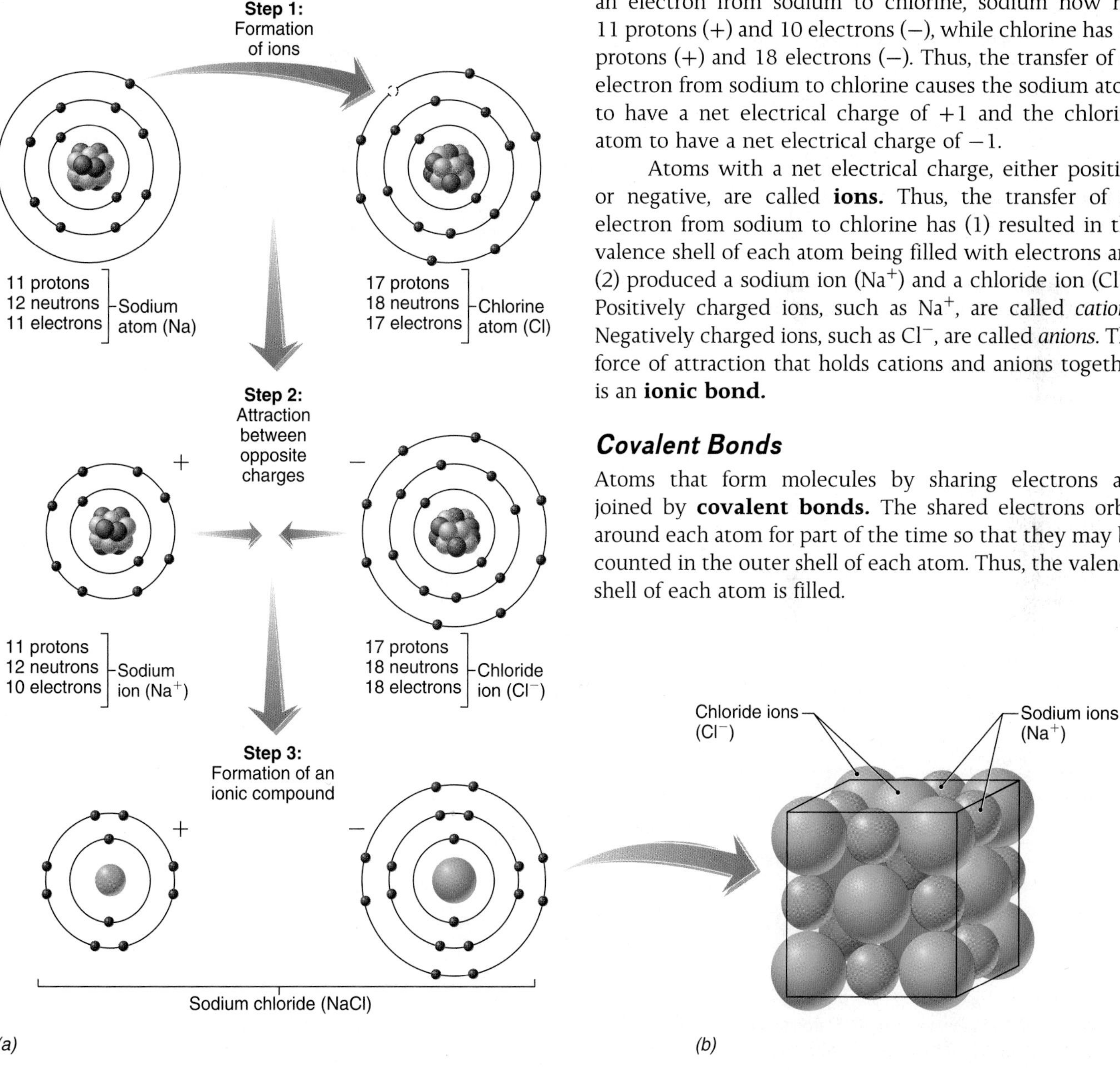

Figure 2.4 The synthesis of sodium chloride by the formation of an ionic bond.
(a) The transfer of an electron from sodium to chlorine converts sodium to a cation and chlorine to an anion.
(b) The attraction between these oppositely charged ions is an ionic bond.

A simple example of covalent bonding is found in a molecule of hydrogen gas. A hydrogen atom with a single electron requires one more electron to fill its valence shell. Two hydrogen atoms can form a molecule of hydrogen gas (H_2) by sharing their electrons. In this way, the valence shell of both atoms is complete and a single covalent bond is formed. The single covalent bond is shown in a structural formula as a single straight line between chemical symbols for hydrogen (H–H) as is illustrated in figure 2.5*a*.

Double and triple covalent bonds can also form. A molecule of gaseous oxygen (O_2) is formed when two oxygen atoms share two pairs of electrons. Each oxygen atom requires two electrons to complete its valence shell, so by sharing two pairs of electrons the valence shell of both atoms is complete and a double covalent bond (O=O) is formed (figure 2.5*b*). Similarly, some compounds contain carbon atoms that are triple bonded. A triple bond is formed when two atoms

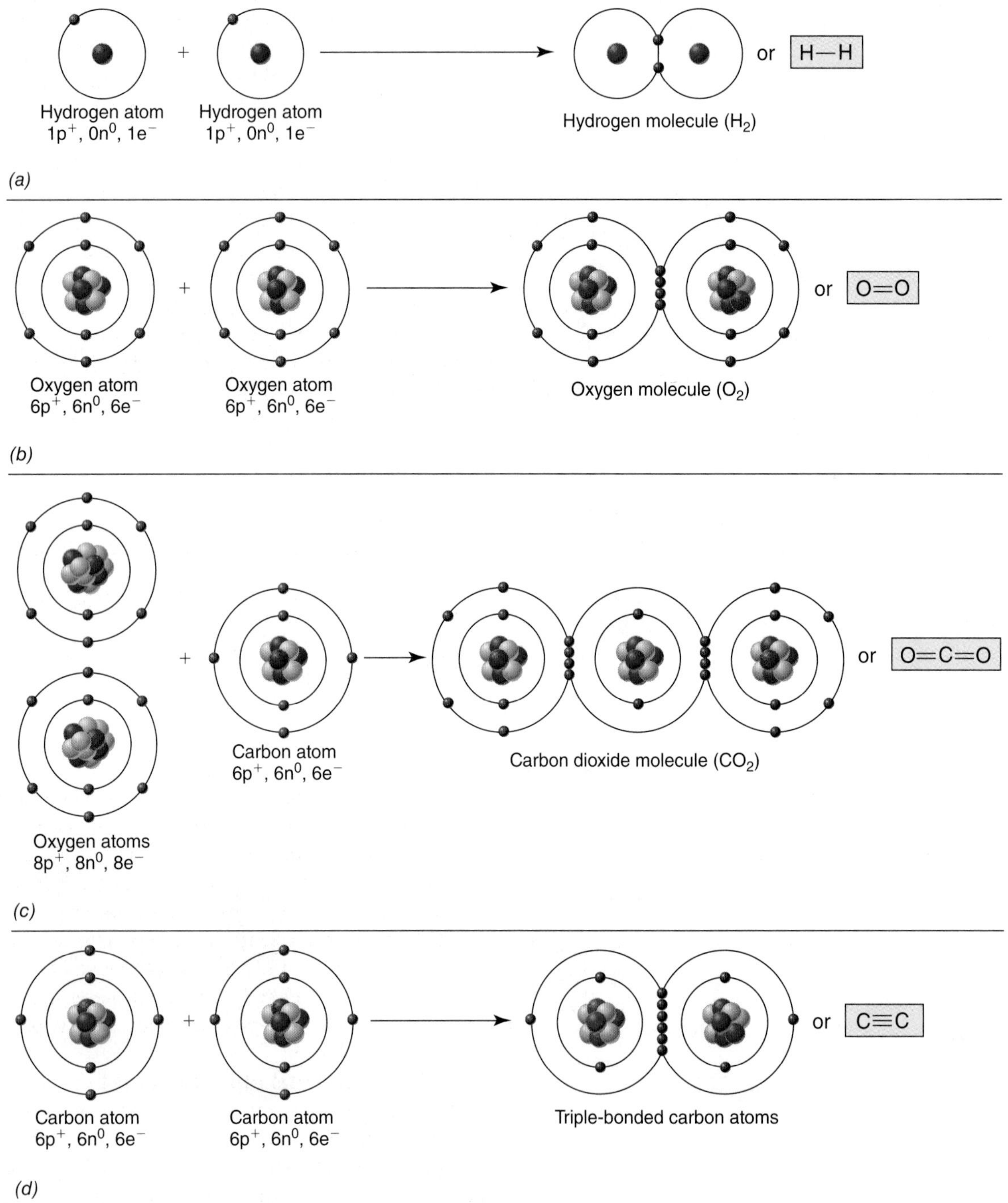

Figure 2.5 Formation of nonpolar covalent bonds.

share three pairs of electrons. Nitrogen gas (N_2) is also formed of triple bonds (N≡N). Figure 2.5*d* shows a triple bond.

There are two types of covalent bonds: nonpolar covalent and polar covalent. *Nonpolar covalent bonds* are commonly found between atoms of the same type and between C and H. In a nonpolar covalent bond, the shared electrons spend equal time revolving between the two atoms. The equal sharing forms a molecule that is electrically neutral. These **nonpolar molecules** do not mix well with water and are referred to as being **hydrophobic** (hydro = water; phobos = fear). *Polar covalent bonds* involve an unequal sharing of electrons between two atoms. For example, when a hydrogen atom is covalently bonded to an oxygen atom, the shared electrons spend less time near the hydrogen atom and more time near the oxygen atom. This occurs because the oxygen atom has a stronger pull on the electrons, which is referred to as *electronegativity*. In this situation the hydrogen atom becomes slightly positively charged, notated as δ^+, and the oxygen becomes slightly negatively charged, notated as δ^-, (figure 2.6). Most molecules formed by polar covalent bonds are called **polar molecules** because different areas of the molecule have a different electrical charge. Polar molecules and ions tend to mix well with water and are thus referred to as **hydrophilic** (philos = loving). A good rule of thumb in determining what whether a substance is hydrophobic or hydrophilic is "like mixes with like." For a substance to mix with water, it must be like water, meaning it must also be electrically charged.

It is important to note that a molecule may contain polar covalent bonds and still be a nonpolar molecule. The carbon dioxide molecule in figure 2.5*c* is a great example; while the C=O bonds are polar covalent bonds, the molecule is linear and symmetrical so that the opposing bonds cancel one another, making the entire structure similar (nonpolar) throughout.

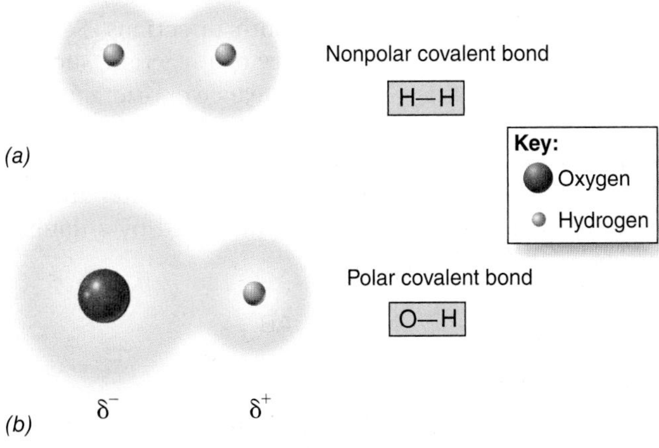

Figure 2.6 Comparison of electron locations in *(a)* nonpolar versus *(b)* polar covalent bonds.

Hydrogen Bonds

A **hydrogen bond** is a weak attractive force between a slightly positive area and a slightly negative area of a polar molecule. These attractions may occur between different sites within the same molecule or between different molecules. They may also occur between polar molecules and ions. Figure 2.7*c* illustrates how the slightly negative oxygen atom of one water molecule attracts the slightly positive hydrogen atom of a different water molecule to form a hydrogen bond. It is important to note that the bonds between oxygen and hydrogen within one molecule are polar covalent bonds. The hydrogen bonds between water molecules are responsible for many of water's unique characteristics that help support life, which is why space exploration focuses so heavily on identifying other planets with water. For example, based on atomic mass, water should be a gas at room temperature. However, the hydrogen bonds keep water liquid at room temperature. Hydrogen bonds also play an important role in protein and nucleic acid structure, as you will see later in this chapter.

Check My Understanding

3. How do ionic and covalent bonds join atoms to form compounds?
4. How are nonpolar covalent bonds and polar covalent bonds different?
5. What are hydrogen bonds?

Chemical Reactions

In **chemical reactions,** bonds between atoms are formed or broken, and the result is a new combination of atoms. There are four basic types of chemical reactions: synthesis, exchange, decomposition, and reversible reactions. Such reactions occur continuously within the body. In general, a chemical reaction begins with one or more substance(s) referred to as the *reactant(s)*. The reaction occurs and the new combination of atoms is (are) called the *product(s)*. Figure 2.8 shows a few examples of chemical reactions.

Synthesis (anabolic) reactions form new chemical bonds and energy is required for the reactions to occur. Atoms or simple molecules combine to form a more complex product. The reaction of hydrogen and oxygen (reactants) to form water (product) is an example. Figure 2.8*b* shows how energy may be used to combine amino acids to form a protein. Synthesis reactions produce

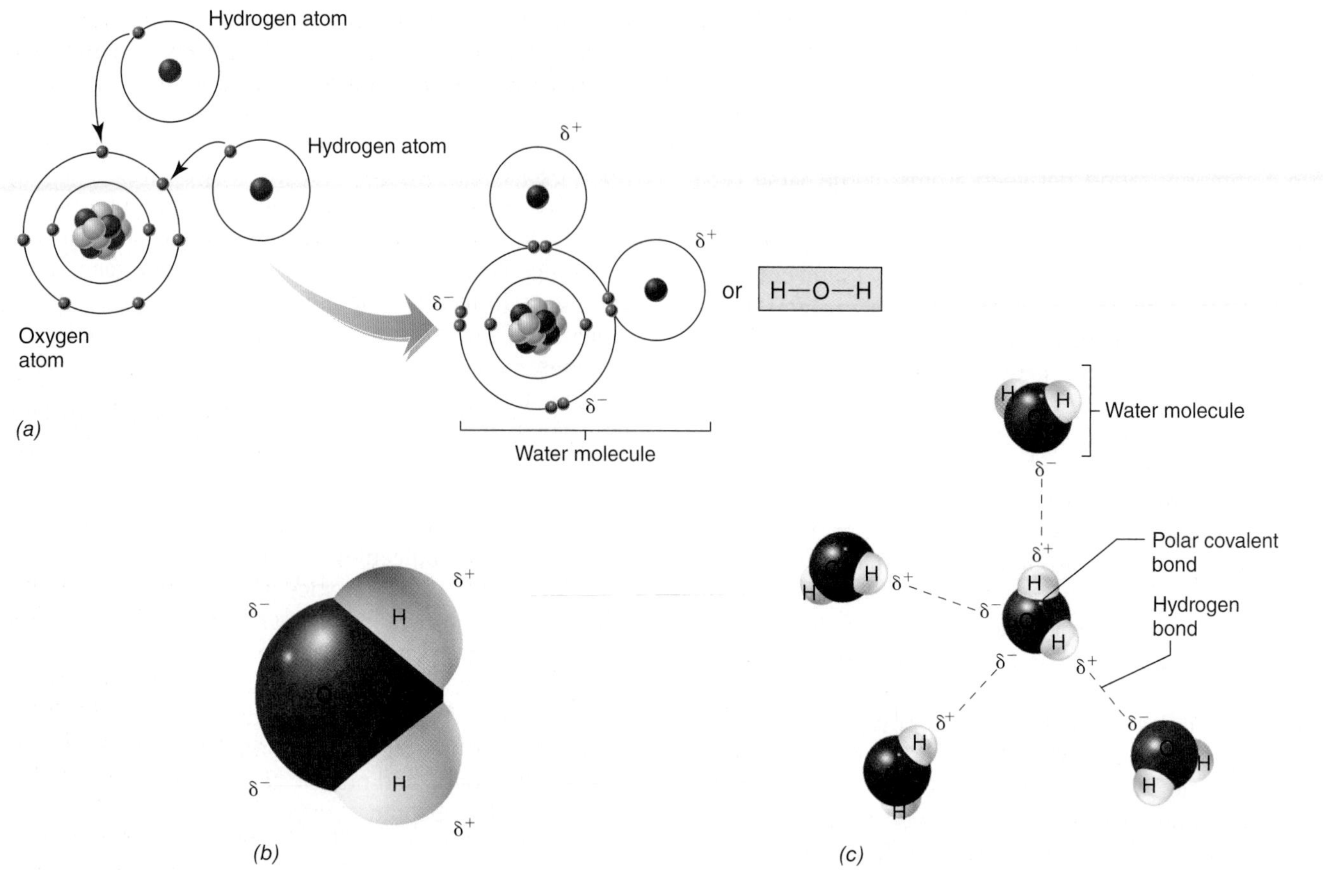

Figure 2.7 *(a)* The synthesis of a water molecule by the formation of covalent bonds. *(b)* Space-filling model of water molecule. *(c)* Hydrogen bonds forming between adjacent water molecules.

complex molecules used in the growth and repair of body parts. Synthesis reactions may be generalized as

$$A + B \rightarrow AB$$

A **decomposition (catabolic) reaction** is the opposite of a synthesis reaction. Chemical bonds of a complex molecule are broken to form two or more simpler molecules, releasing energy in the process. For example, water can decompose to form hydrogen and oxygen. Decomposition reactions are used to break down food molecules to form nutrients usable by body cells. Figure 2.8*c* shows how glycogen may be decomposed to form glucose, thus releasing energy. Decomposition reactions may be generalized as

$$AB \rightarrow A + B$$

Exchange (rearrangement) reactions occur when two different reactants exchange components, resulting in the breakdown of the reactants and the formation of two new products. Thus, exchange reactions involve both decomposition of the reactants and synthesis of the new products. The dehydration synthesis and hydrolysis reactions dicussed later in this chapter are examples of exchange reactions. Exchange reactions may be generalized as

$$AB + CD \rightarrow AD + CB$$

Reversible reactions exist in which the reactants and products may convert in both directions. Several factors determine the reversibility of a reaction, such as energy available and relative abundance of reactants and products. The reactions of chemical buffers are great examples of exchange reactions that are also reversible reactions. Chemical buffers are discussed later in this chapter. Reversible reactions are indicated by a double arrow:

$$A + B \rightleftharpoons AB$$

6. How do synthesis, decomposition, exchange, and reversible reactions differ?

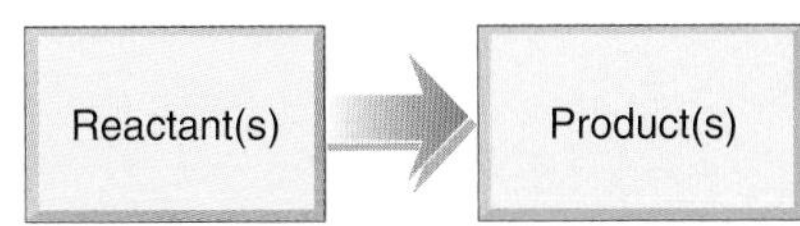

(a) Chemical reaction

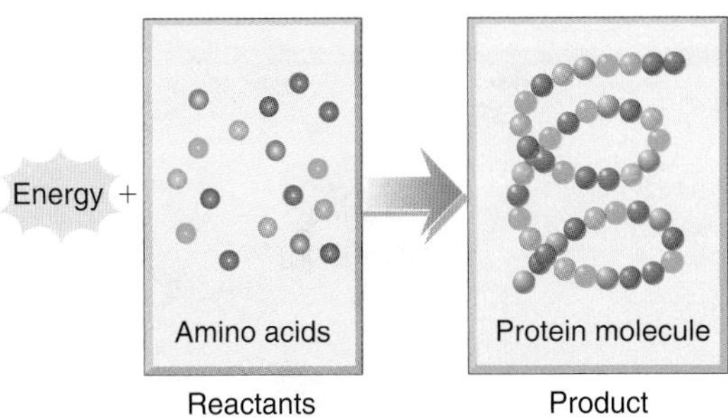

(b) Anabolic (or synthesis) reaction

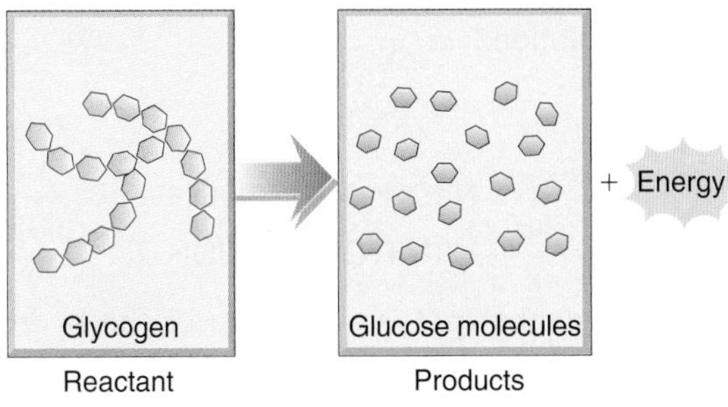

(c) Catabolic (or decomposition) reaction

Figure 2.8 *(a)* General example of a chemical reaction. *(b)* A synthesis reaction uses energy to form bonds resulting in one or more larger products. *(c)* A decomposition reaction breaks bonds, releasing energy and resulting in numerous smaller products.

2.3 Compounds Composing the Human Body

Learning Objectives

7. Distinguish between inorganic and organic substances.
8. Explain the importance of water and its locations in the body.
9. Compare and contrast electrolytes and nonelectrolytes, and acids and bases.
10. Explain the use of the pH scale.
11. Explain the importance of buffers.
12. Distinguish between carbohydrates, lipids, proteins, and nucleic acids and their roles in the body.
13. Explain the role of enzymes.
14. Describe the mechanism of enzymatic action.
15. Describe the structure and function of adenosine triphosphate (ATP).

Compounds composing the human body include both inorganic and organic substances. Molecules of **inorganic compounds** may contain either carbon or hydrogen in the same molecule, *but not both*. Bicarbonates, such as sodium bicarbonate ($NaHCO_3$), are an exception to this rule. Molecules of **organic compounds** always contain *both* carbon and hydrogen, and they usually also contain oxygen. The carbon atoms form the "backbone" of organic molecules.

Major Inorganic Compounds

The major inorganic compounds in the body are water, most acids and bases, and inorganic salts. Acids, bases, and mineral salts **ionize** to release ions when dissolved in water, as you will see shortly.

Water

Water is by far the most abundant compound found within cells and in the extracellular fluid. Water composes about two-thirds of the body weight. Water generally occurs within the body as part of a mixture called an **aqueous solution.** A *solution* is a mixture composed of a solvent and one or more solutes. A *solvent* is a liquid used to dissolve or suspend substances. *Solutes* are the substances dissolved or suspended in the solvent. In this aqueous solution, water is acting as a solvent. Recall from earlier in the chapter that water is a solvent for electrically charged substances, such as polar molecules and ions. Chemical reactions that occur in the body take place in this aqueous solution. Water is used to transport many solutes throughout the body, through the plasma membrane of a cell, or from one part of the cell to another. Another function of water is that it is a great lubricant in the body. Water is also important in maintaining a constant cellular temperature, and thus a constant body temperature, because it absorbs and releases heat slowly. Evaporative cooling (sweating) through the skin also involves water. Another function of water is as a reactant in the breakdown (hydrolysis) of organic compounds.

The specific locations where water is found in the body are called *water compartments*. They are

- *Intracellular fluid (ICF):* fluid within cells; about 65% of the total body water.
- *Extracellular fluid (ECF):* all fluid not in cells; about 35% of the total body water.
  - *Interstitial fluid (tissue fluid):* fluid in spaces between cells.
  - *Plasma:* fluid portion of blood.
  - *Lymph:* fluid in lymphatic vessels.
  - *Transcellular fluids:* fluid in more limited locations, such as serous fluid, cerebrospinal fluid (CSF–located within and around the brain and spinal cord), and synovial fluid (in certain joints).

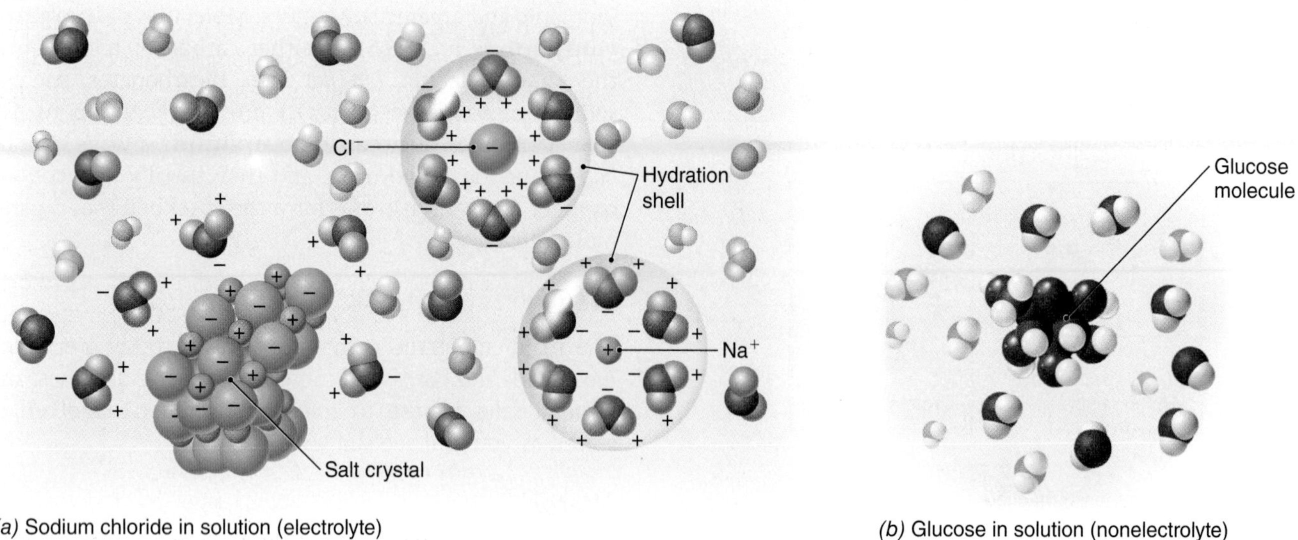

Figure 2.9 Comparison of electrolyte and nonelectrolyte in aqueous solution.

Electrolytes APR

When ionic compounds are dissolved in water, their molecules tend to ionize (dissociate), releasing ions. The process of ionization involves the formation of *hydration spheres* by water using its polar nature to separate ions from each other (figure 2.9*a*). Such compounds are called **electrolytes** (ē-lek′-trō-lītz) because when dissolved they can conduct an electrical current. Acids, bases, and salts are types of electrolytes (figure 2.10). Substances that dissolve in water without ionizing and therefore do not conduct electrical current are termed **nonelectrolytes,** such as the glucose molecule in figure 2.9*b*. Nonelectrolytes are usually organic compounds, which are discussed later in the chapter. The composition and concentration of electrolytes and nonelectrolytes in the body must be kept within narrow limits to maintain homeostasis.

Acids and Bases

An **acid** is a chemical that releases **hydrogen ions (H^+)** into solution.

The stronger the acid, the greater is the degree of dissociation, which results in a greater concentration of H^+. Hydrochloric acid (HCl) is a strong acid. It ionizes into hydrogen and chloride ions (H^+ and Cl^-).

$$HCl \rightarrow H^+ + Cl^-$$

A **base** decreases the concentration of H^+ in a solution by combining with the H^+. The stronger the base, the greater is its ability to combine with H^+. Some bases combine directly with H^+. Other bases, such as sodium hydroxide (NaOH), ionize to release hydroxide ions (OH^-), which combine with H^+ to form water.

$$NaOH \rightarrow Na^+ + OH^-$$

$$H^+ + OH^- \rightarrow HOH$$

The symbol **pH** is a measure of the hydrogen ion concentration in a solution.

Measurement of pH Chemists have developed a pH scale that is used to indicate the measure of acidity or alkalinity (basicity) of a solution, meaning the relative concentrations of hydrogen ions (H^+) and hydroxide ions (OH^-) in a solution. The **pH scale** ranges from 0 to 14. The concentration of H^+ decreases and the concentration of OH^- increases as the pH values increase. These ions are equal in concentration at pH 7, so a solution with a pH of 7 is neither an acid nor a base and is referred to as neutral. For example, pure water has a pH of 7.

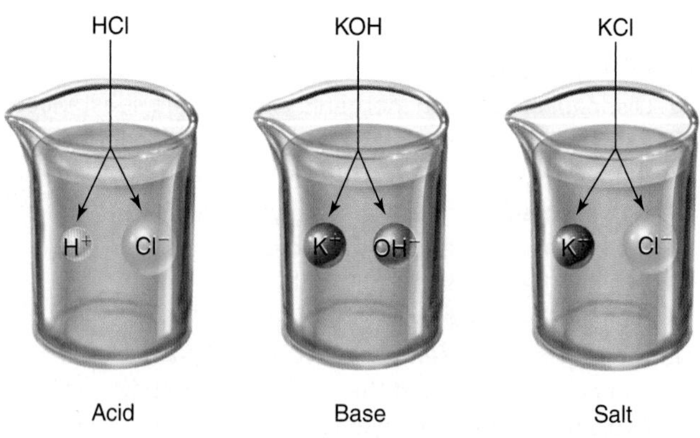

Figure 2.10 Ionization of an acid, a base, and a salt in water.

Solutions with a pH less than 7 are acids, and those with a pH greater than 7 are bases. The lower the pH value below 7, the more acidic is the solution, and the higher the pH value above 7, the more alkaline is the solution. There is a tenfold difference in the concentrations of H^+ and OH^- when the pH changes by one unit. For example, an acid with a pH of 4 has a concentration of H^+ that is 10 times greater than that of an acid with a pH of 5. Some examples of pH values for body fluids include blood (7.4), stomach acid (1-2), urine (6), and intestinal fluid (8).

Buffers Cells of the body are especially sensitive to pH changes. Even slight changes can be harmful. The hydrogen ion concentration (pH) of blood and other body fluids is maintained within narrow limits by the lungs, kidneys, and buffers in body fluids. A **buffer** is a chemical or a combination of chemicals that either picks up excess H^+ or releases H^+ to keep the pH of a solution rather constant. The carbonic acid-bicarbonate buffer system illustrated in figure 2.11 is the most important buffer system in the body. Notice that carbonic acid and bicarbonate ions react in a reversible reaction whose direction is determined by pH.

Buffers are extremely important in maintaining the normal pH of body fluids but they can be overwhelmed by a disruption of homeostasis. The normal pH of the arterial blood is 7.35 to 7.45. In *acidosis*, the pH is less than 7.35, and a patient could go into a coma. In *alkalosis*, the pH range is greater than 7.45, and a patient may have uncontrolled muscle contractions. Extreme variations, outside of 6.8-8.0, may be fatal.

Salts

Like acids and bases, *salts* are ionic compounds that ionize in an aqueous solution, but they do not produce hydrogen or hydroxide ions. The most important salts in the body are sodium, potassium, and calcium salts. Calcium phosphate is the most abundant salt because it is a main component of bones and teeth. Sodium chloride (NaCl), a common salt in body fluids, ionizes into sodium and chloride ions (Na^+ and Cl^-).

$$NaCl \rightarrow Na^+ + Cl^-$$

Salts provide ions that are essential for normal body functioning. Physiological processes in which ions play an essential role include blood clotting, muscle and nervous functions, and pH and water balance (table 2.1).

Check My Understanding

7. Where is water located in the body and why is it so important?
8. What is the relationship between acids, bases, and pH?
9. What are salts and why are they important?

Major Organic Compounds

The major organic compounds of the body are carbohydrates, lipids, proteins, and nucleic acids (table 2.2). Another compound, adenosine triphosphate (ATP), is also considered here because it plays such a vital role in the transfer of energy within cells.

Before beginning a study of the various organic compounds, it is important to understand one reversible reaction that will help to clarify a great deal about how biochemistry (chemistry in living things) in general works. The reaction involves two processes, dehydration

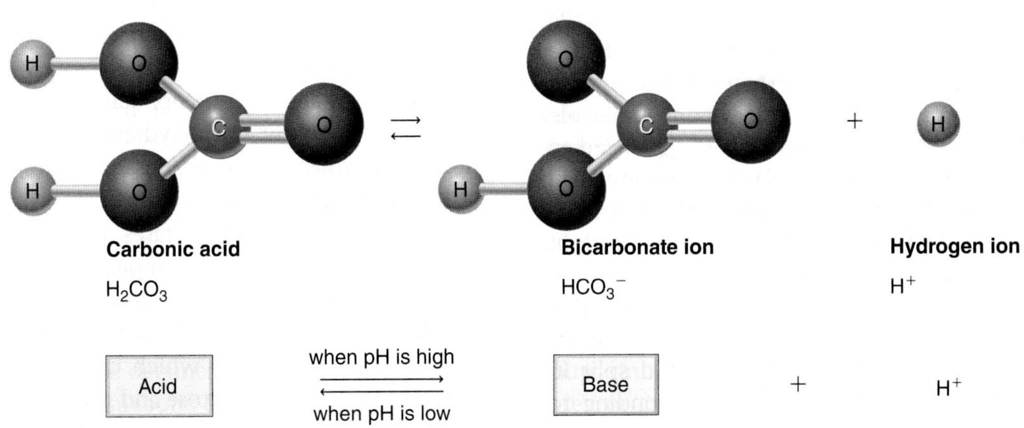

Figure 2.11 Carbonic acid ionizes to release H^+ when pH is high. Bicarbonate ion combines with H^+ when pH is low.

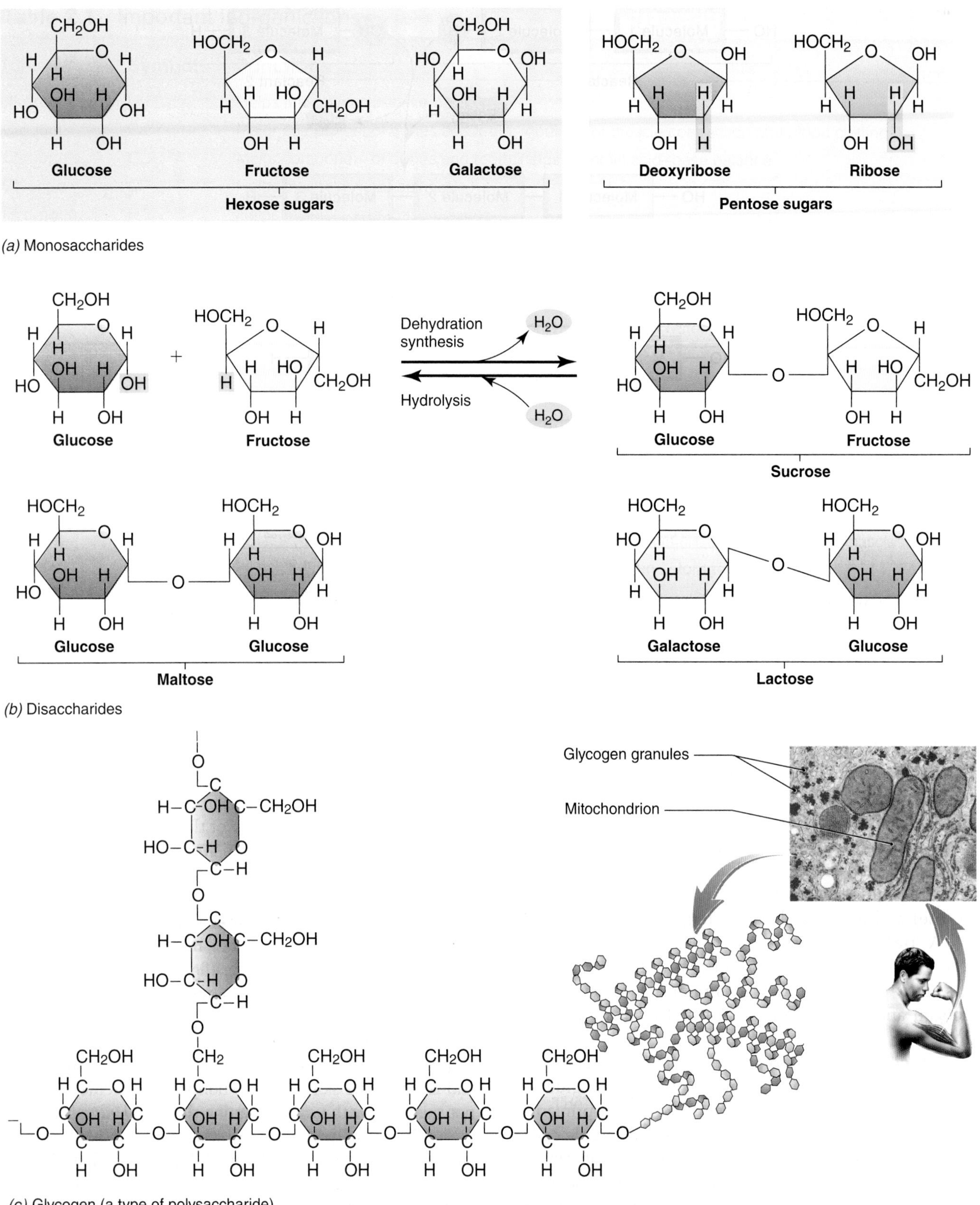

Figure 2.13 *(a)* The building units of carbohydrates are monosaccharides. *(b)* Two monosaccharides combine to form a disaccharide. *(c)* The combination of many monosaccharide units forms a polysaccharide.

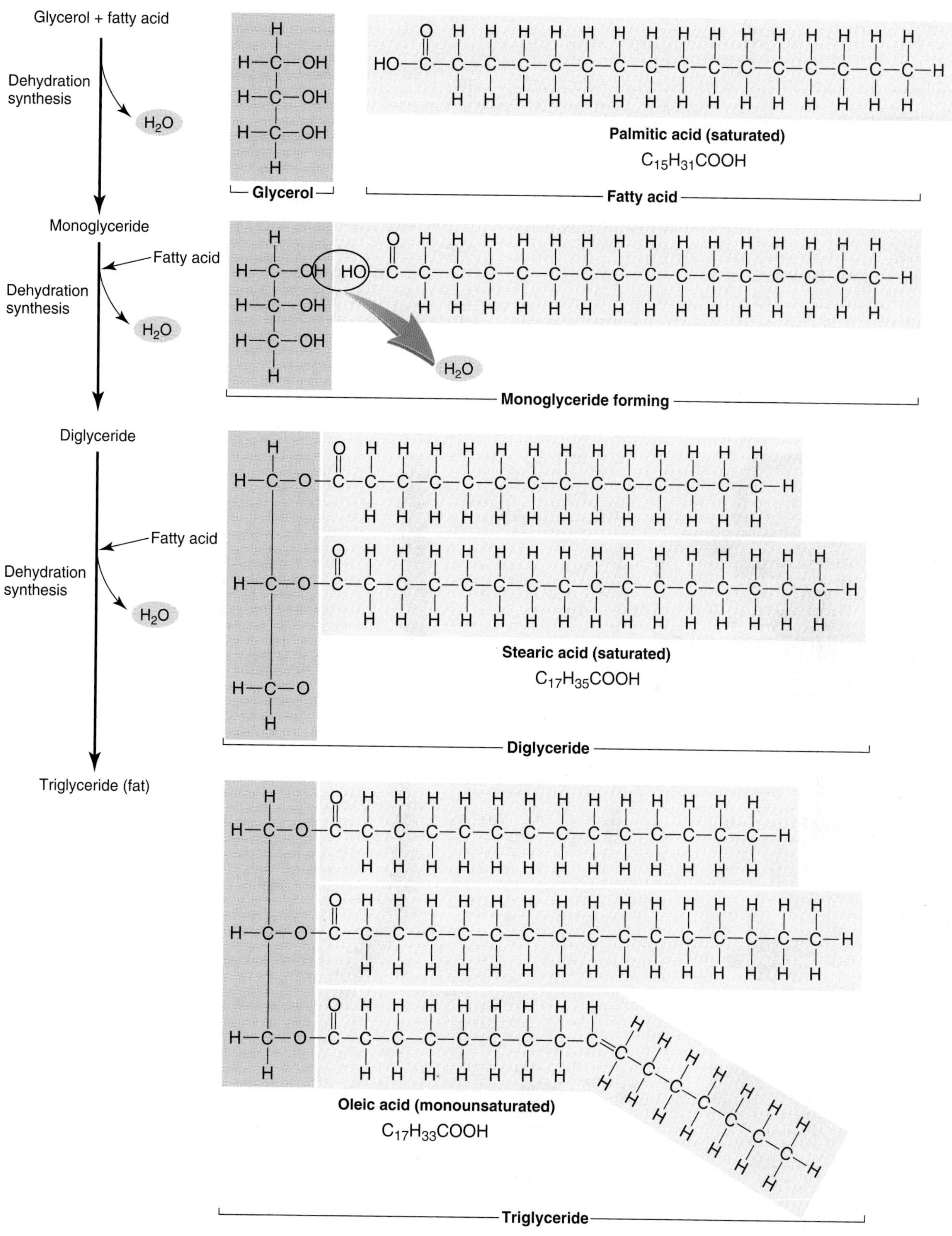

Figure 2.14 A triglyceride molecule consists of three fatty acids joined to a glycerol molecule by three successive dehydration synthesis reactions.

Clinical Insight

Triglycerides may be classified as either saturated or unsaturated fats. In *saturated fats* (e.g., animal fats), the bonds of the carbon atoms in the fatty acids are saturated (filled) by hydrogen atoms so that the carbon–carbon bonds are all single bonds. Saturated fats, such as butter and lard, are solid at room temperature. Excessive saturated fats in the diet are associated with an increased risk of heart disease (coronary artery disease).

In *unsaturated fats* (plant oils), not all carbon bonds in the fatty acids are filled with hydrogen atoms, and one or more double carbon–carbon bonds are present. Fatty acids of monounsaturated fats (e.g., olive, peanut, and canola oils) have one carbon–carbon double bond; those of polyunsaturated fats (e.g., corn, safflower, and soy oils) have two or more carbon–carbon double bonds. Unsaturated fats occur as oils at room temperature.

Hydrogenation, the process of adding hydrogen atoms to unsaturated fats, converts most carbon–carbon double bonds to carbon–carbon single bonds and changes vegetable oil to a solid (e.g., margarine) at room temperature. This process also changes the bonding pattern of some fatty acids to form *trans fats,* which increase the risk of coronary artery disease even more than do saturated fats. It is better to cook with oils than with lard or margarine because saturated and trans fats are more easily converted into the "bad" cholesterol associated with heart disease.

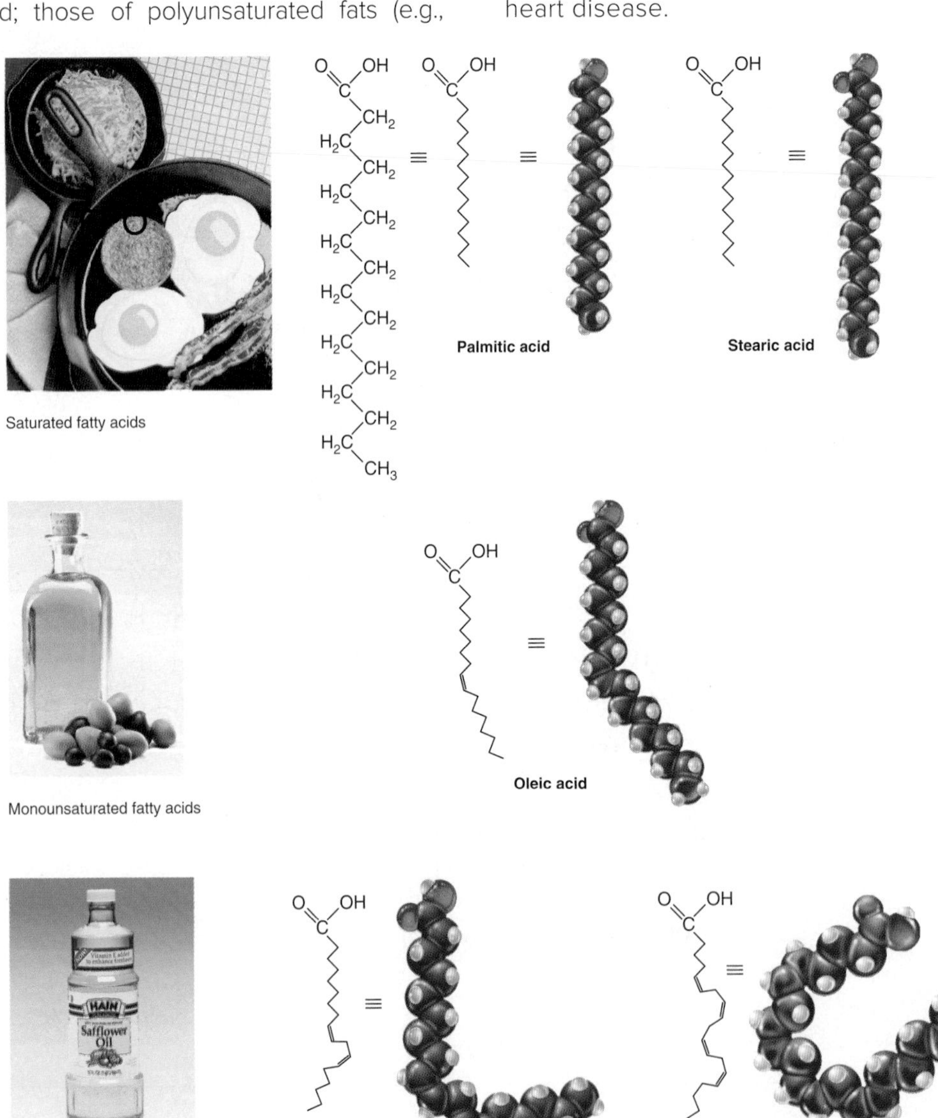

Saturated fatty acids

Palmitic acid

Stearic acid

Monounsaturated fatty acids

Oleic acid

Polyunsaturated fatty acids

Linoleic acid

Arachidonic acid

but insoluble in water. Thus, phospholipids can join–or serve as an interface between–a water environment on one side and a lipid environment on the other. They are major components of plasma membranes, that surround cells and certain organelles within the cell (see chapter 3). Figure 2.15 shows the basic structure of a phospholipid molecule.

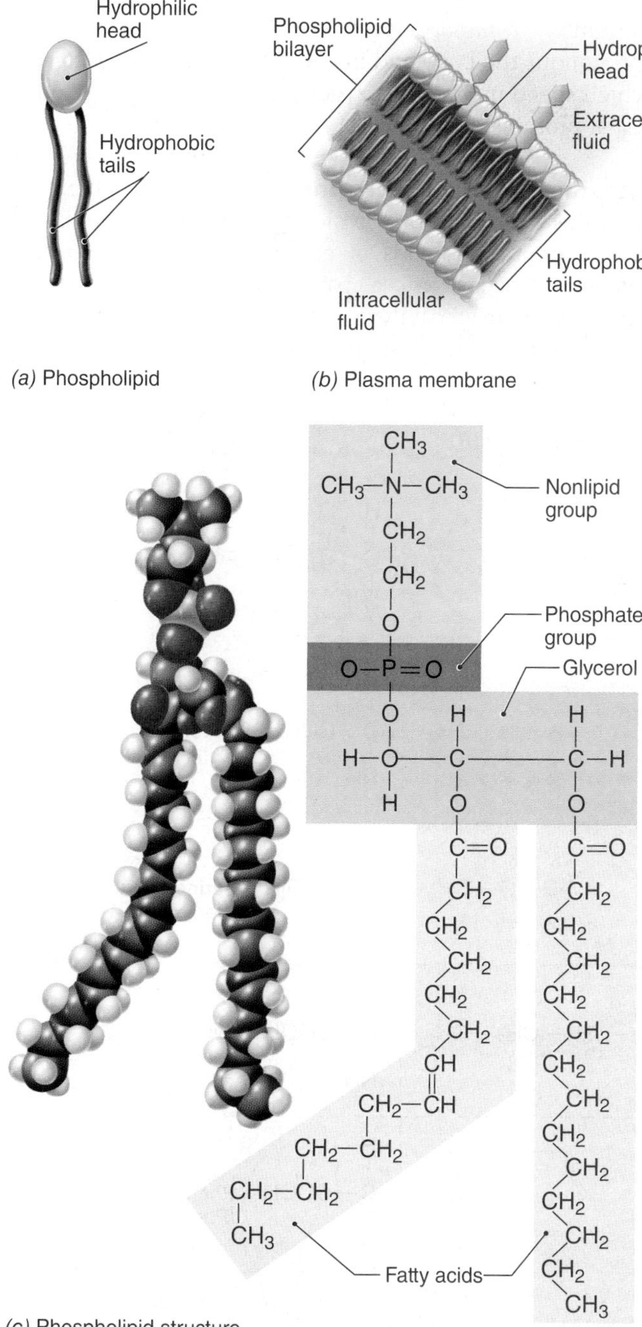

Figure 2.15 In a phospholipid, one fatty acid is replaced by a polar phosphate-containing group.

Steroids constitute another group of lipids, and their molecules characteristically contain four carbon rings. Cholesterol (kō-les′-ter-ol), vitamin D, certain adrenal hormones, and sex hormones are examples of steroids. Figure 2.16 shows several examples of steroids. Cholesterol is an essential component of body cells and serves as the raw material for the synthesis of other steroid molecules.

Proteins

Proteins (pro′-tēns) are large, complex molecules composed of smaller molecules (building units) called **amino acids.** There are 20 different kinds of amino acids used in building proteins, and each is composed of carbon, hydrogen, oxygen, and nitrogen. Each amino acid consists of a central C atom which is attached to 4 separate components. Three components are the same in all amino acids. The first is simply a hydrogen atom, and the other two are the components for which an amino acid is named, an *amine group* (–NH_2) and an *acid group* (-COOH).

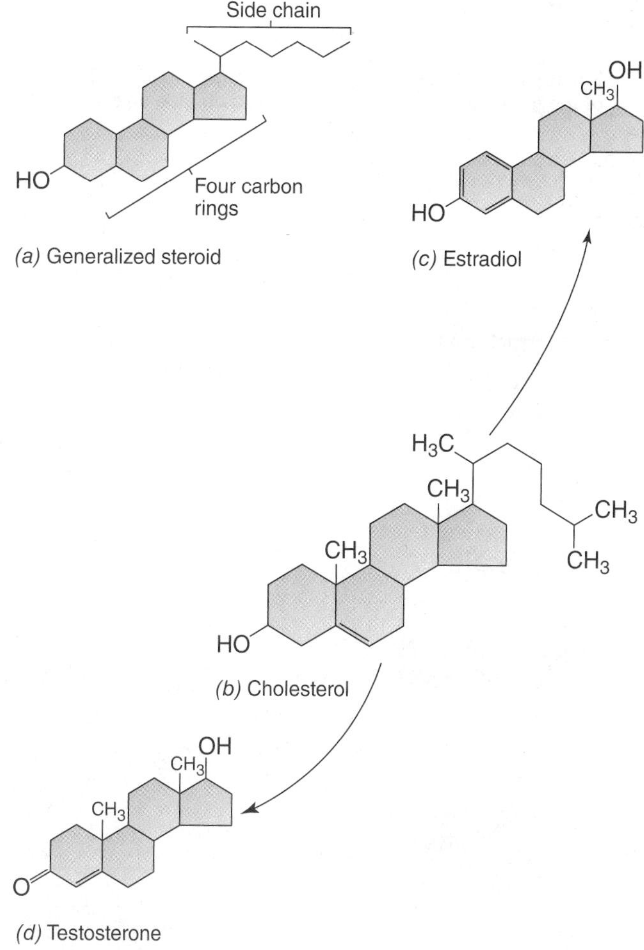

Figure 2.16 Note the four carbon rings, which are characteristic of steroids. Carbon and hydrogen atoms are not all shown in these shorthand structures.

However, it is the side chain, called the *R group,* that distinguishes the different kinds of amino acids and their chemical properties. Figure 2.17 depicts the basic structure of an amino acid as well as examples of the four types of amino acids.

Amino acids are joined by **peptide bonds.** Two amino acids bonded together form a **dipeptide.** A chain of many amino acids forms a **polypeptide.** A chain of more than 50 amino acids forms a protein. As figure 2.18 shows, the sequence of amino acids is the *primary structure* of the protein. The long chain of amino acids forms hydrogen bonds between polar parts of the polypeptide, causing various areas of the molecule to twist (helix) and/or fold (pleated sheet). Helices (plural of helix) and pleated sheets are examples of *secondary structures.* The polypeptide then folds in response to the watery environment of the human body. The folding results in a *tertiary structure* with mostly hydrophobic amino acids at its interior and hydrophilic amino acids exposed to the watery exterior. The result is that each specific polypeptide or protein has a unique three-dimensional shape.

In some instances multiple polypeptides combine to form a protein. In these cases the protein is said to have *quaternary structure.*

Proteins may be classified as either structural or functional proteins. *Structural proteins* compose parts of body cells and tissues, where they provide support and strength in binding parts together. Ligaments, tendons, and contractile fibers in muscles are composed of structural proteins. *Functional proteins* perform a variety of different functions in the body. *Antibodies,* which provide immunity, *transport proteins,* which carry substances throughout the body and in and out of cells, and *enzymes,* which speed up chemical reactions, are examples of functional proteins.

Enzymes Without **enzymes,** the body's chemical reactions would occur too slowly to maintain life. Body cells contain thousands of enzymes, and each enzyme *catalyzes* (speeds up) a particular chemical reaction. An enzyme may catalyze synthesis, decomposition, or exchange reactions. A single enzyme may also catalyze a

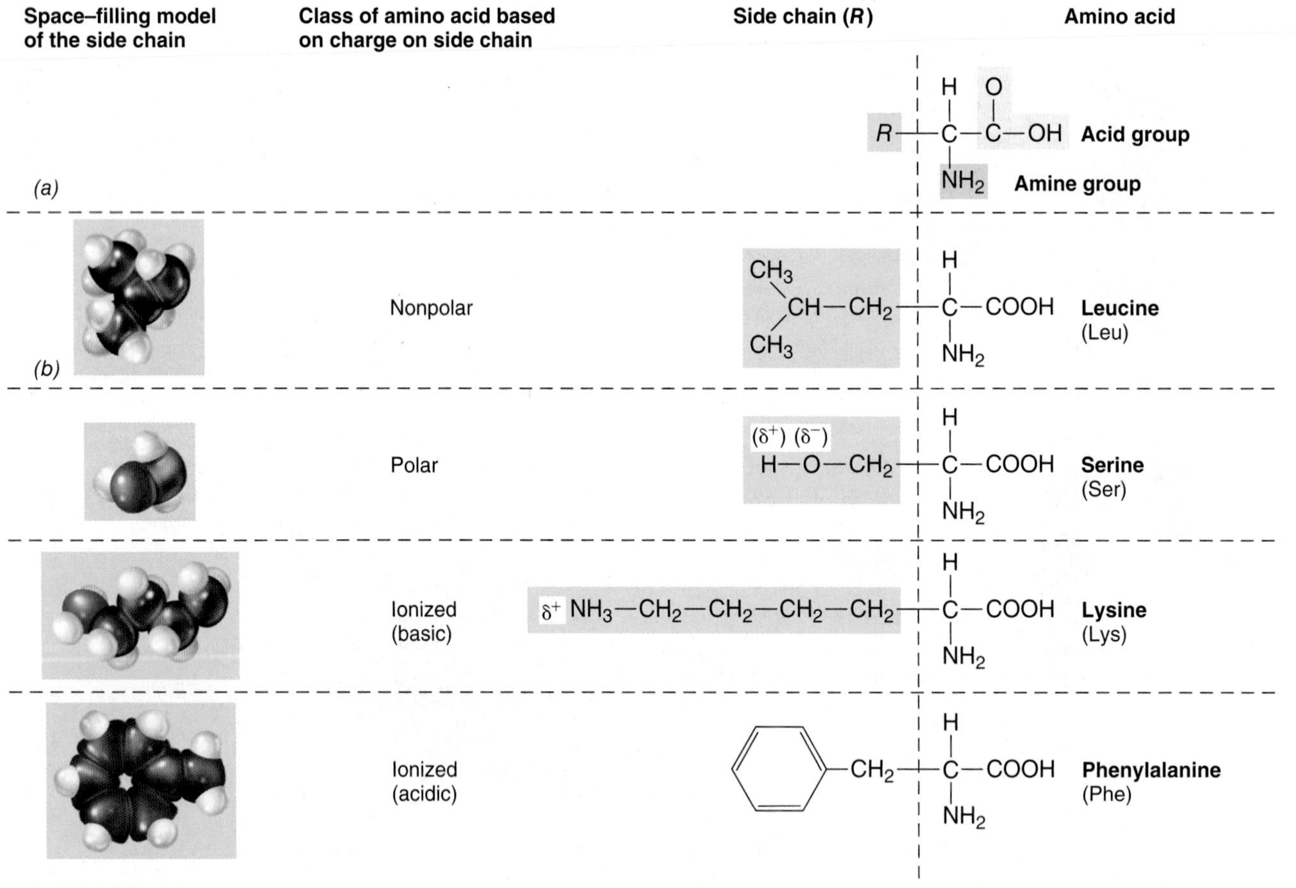

Figure 2.17 Amino Acids.
(a) The basic structure of an amino acid. *(b)* Representative amino acids. Note that the hydrogen atom, the amine group, and acid group are the same in all amino acids. It is only the R group that is different among different amino acids.

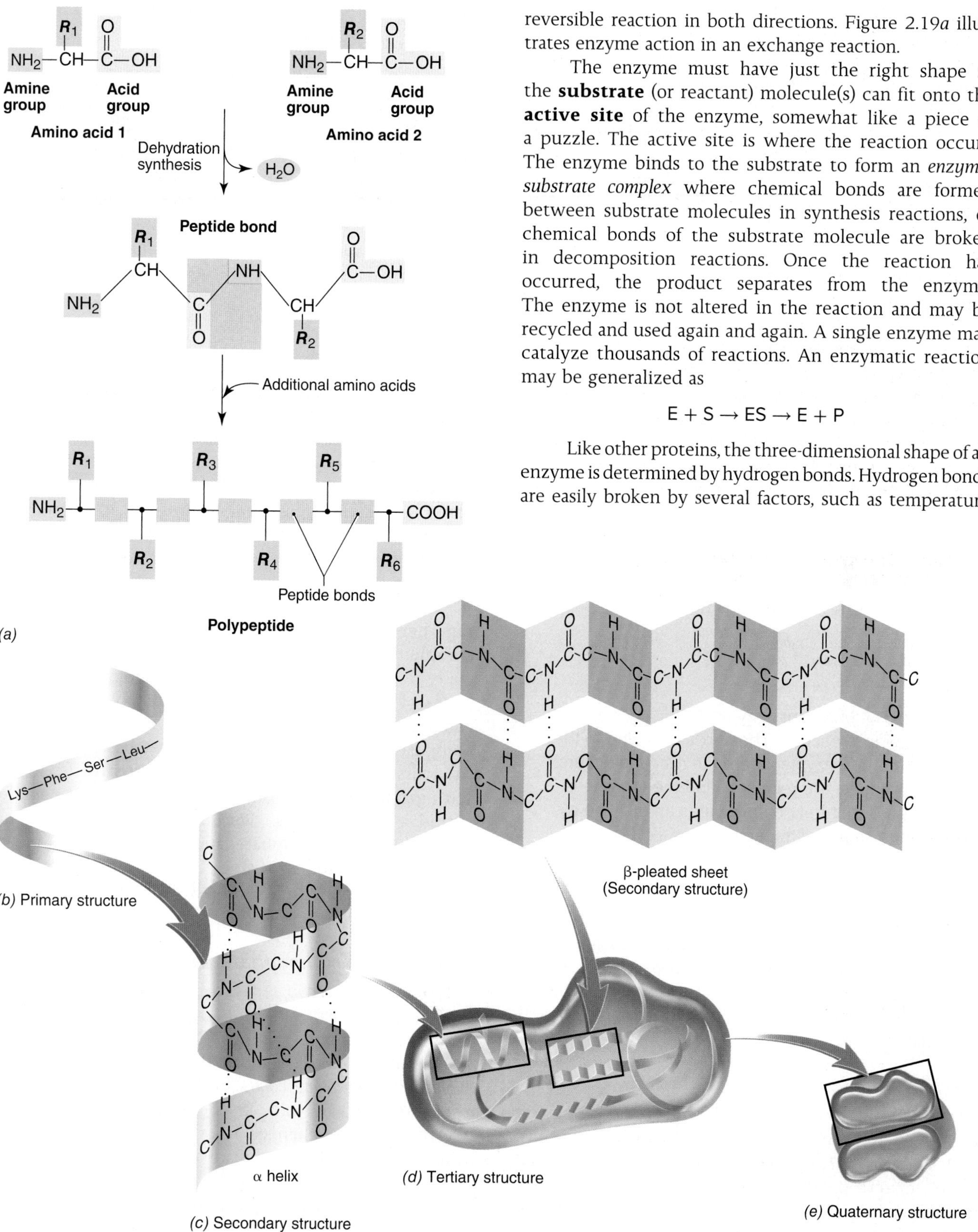

Figure 2.18 *(a)* The formation of a peptide bond. *(b-e)* The primary structure of a polypeptide ultimately determines how it folds into a more complex structure.

reversible reaction in both directions. Figure 2.19*a* illustrates enzyme action in an exchange reaction.

The enzyme must have just the right shape so the **substrate** (or reactant) molecule(s) can fit onto the **active site** of the enzyme, somewhat like a piece of a puzzle. The active site is where the reaction occurs. The enzyme binds to the substrate to form an *enzyme-substrate complex* where chemical bonds are formed between substrate molecules in synthesis reactions, or chemical bonds of the substrate molecule are broken in decomposition reactions. Once the reaction has occurred, the product separates from the enzyme. The enzyme is not altered in the reaction and may be recycled and used again and again. A single enzyme may catalyze thousands of reactions. An enzymatic reaction may be generalized as

$$E + S \rightarrow ES \rightarrow E + P$$

Like other proteins, the three-dimensional shape of an enzyme is determined by hydrogen bonds. Hydrogen bonds are easily broken by several factors, such as temperature

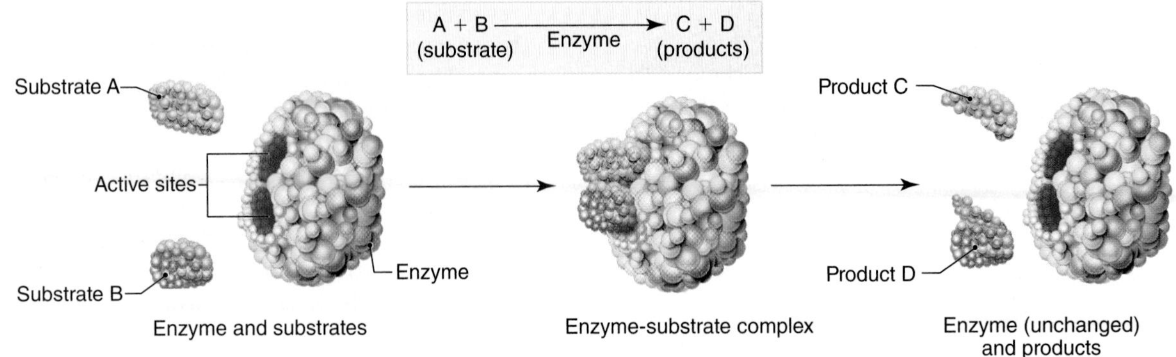

(a) Enzymatic reaction

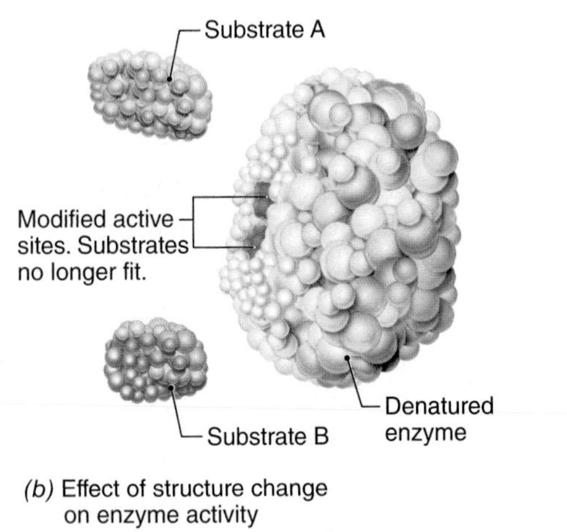

(b) Effect of structure change on enzyme activity

Figure 2.19 A Model of Enzyme Action.
(*a*) An exchange reaction. The substrates bind to the active site of the enzyme where the reaction occurs. Then the products are released, and the enzyme is recycled. (*b*) Denaturation inactivates an enzyme by changing the shape of the active sites. APR

and pH changes, poisons, and radiation. If an enzyme's hydrogen bonds are altered, its shape is changed, and the enzyme is denatured (inactivated) (see figure 2.19*b*). Because it cannot bind to the substrate, the reaction it catalyzes will not occur. If the reaction is vital, the result can be fatal.

Nucleic Acids

Nucleic (nū-klā′-ic) **acids** are the human body's largest molecules. Two types of nucleic acids occur in cells: DNA and RNA. **Deoxyribonucleic** (dē-ok″-se-rī″-bō-nū-klā′-ic) **acid (DNA)** composes the hereditary portion of chromosomes in the cell nucleus. DNA contains the *genetic code,* encoded information that determines hereditary traits and cellular functions. One way the genetic code does this is by determining the structure of proteins. The portions of DNA that encode for specific proteins are called **genes** (jēns). **Ribonucleic acid (RNA)** carries the coded instructions from DNA to the cellular machinery involved in protein synthesis.

Both DNA and RNA consist of repeating building units called **nucleotides** (nū′-klē-ō-tīds). Each nucleotide consists of three parts: a five-carbon sugar, a phosphate group, and a nitrogenous base. Figure 2.20 shows the typical structure of a deoxyribonucleotide and a ribonucleotide as well the five possible nitrogenous bases used in nucleic acids.

DNA consists of two strands of nucleotides joined together by hydrogen bonds that form between the complementary pairing nitrogenous bases. It superficially resembles a "twisted ladder," see figure 2.21. The "sides of the ladder" are formed of deoxyribose sugars and phosphate groups. The "rungs of the ladder" are composed of the paired nitrogenous bases: **adenine** (A) pairs with **thymine** (T), and **cytosine** (C) pairs with **guanine** (G). In contrast, RNA consists of a single strand of nucleotides. The backbone is formed of ribose sugar and phosphate, and the nitrogenous bases are adenine, uracil, cytosine, and guanine. Note that **uracil** is present in RNA instead of thymine, which occurs in DNA.

Adenosine Triphosphate (ATP)

Adenosine (ah-den′-ō-sēn) **triphosphate** (trī-fos′-fāt), or **ATP,** is a modified nucleotide that consists of adenosine and three phosphate groups. The last two phosphate groups are joined to the molecule by special bonds called *high-energy phosphate bonds.* In figure 2.22*b,* these bonds are represented by wavy lines. Energy in these bonds is released to power chemical reactions within a cell. In this way, ATP provides immediate energy to keep cellular processes operating.

Energy extracted from nutrient molecules by cells is temporarily held in ATP and then released to power chemical reactions. When the terminal high-energy phosphate bond of ATP is broken and energy is transferred, ATP is broken down into **adenosine diphosphate (ADP)** and a low-energy phosphate group (Pi). The addition of

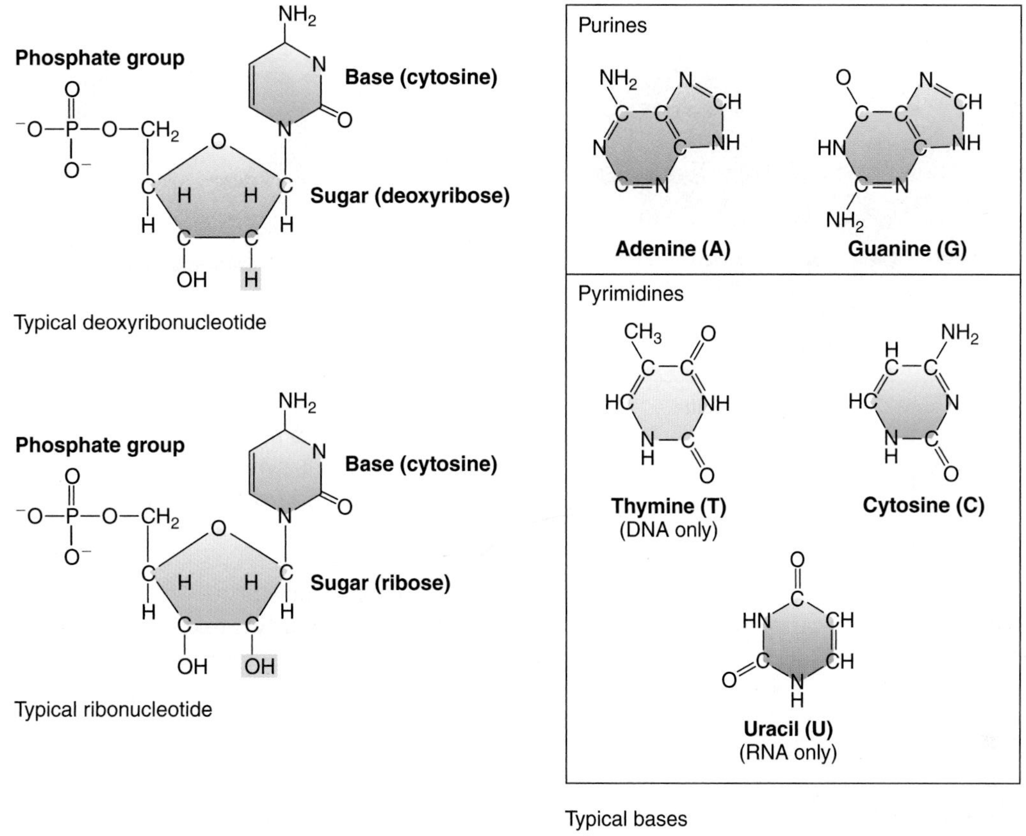

Phosphate group
Base (cytosine)
Sugar (deoxyribose)
Typical deoxyribonucleotide
Phosphate group
Base (cytosine)
Sugar (ribose)
Typical ribonucleotide
Purines
Adenine (A)
Guanine (G)
Pyrimidines
Thymine (T)
(DNA only)
Cytosine (C)
Uracil (U)
(RNA only)
Typical bases

Figure 2.20 Nucleotide structure and the five nitrogenous bases.

Sugar-phosphate backbone
Sugar-phosphate backbone
Complementary base pairing
Hydrogen bond
(a)
(b)
(c)

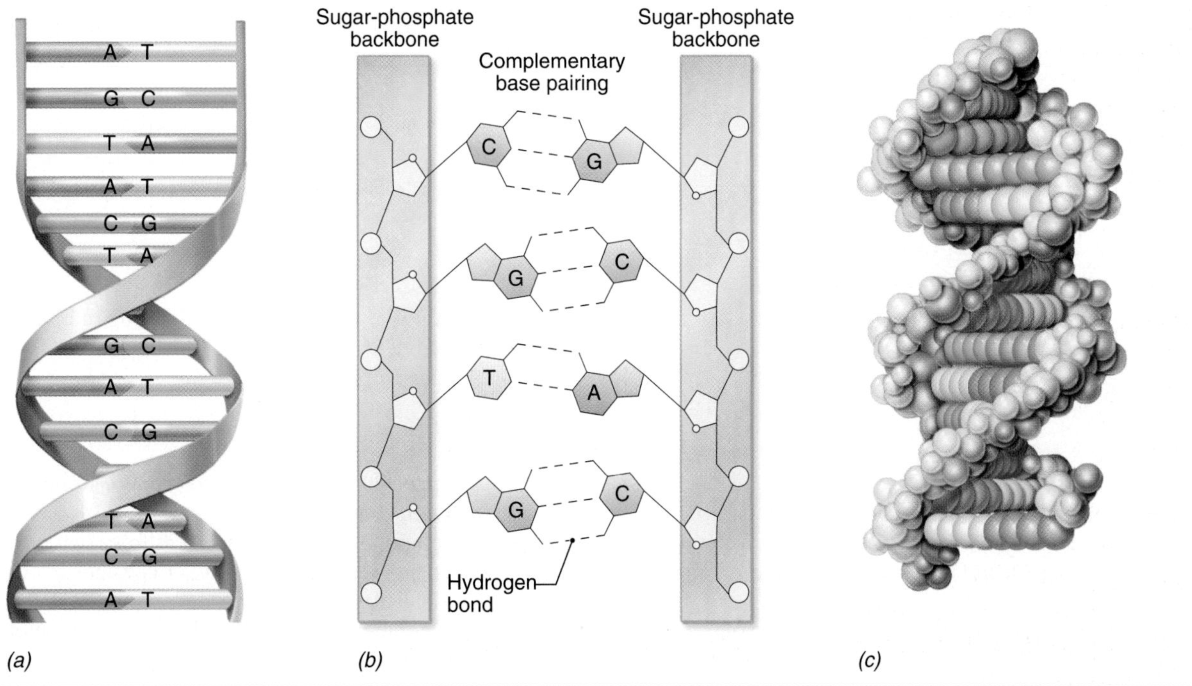

Figure 2.21 The structure of DNA.
(*a*) A model showing the double helix or "twisted ladder" structure of DNA. (*b*) A small segment of DNA showing sugar–phosphate molecules forming the backbone of each strand, and the strands joined together by hydrogen bonds formed between the paired nitrogenous bases. (*c*) A more complex molecular model. AP|R

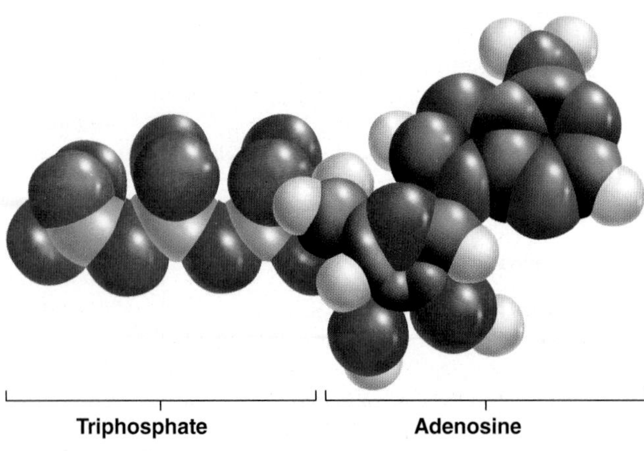

(a) A molecular model of ATP

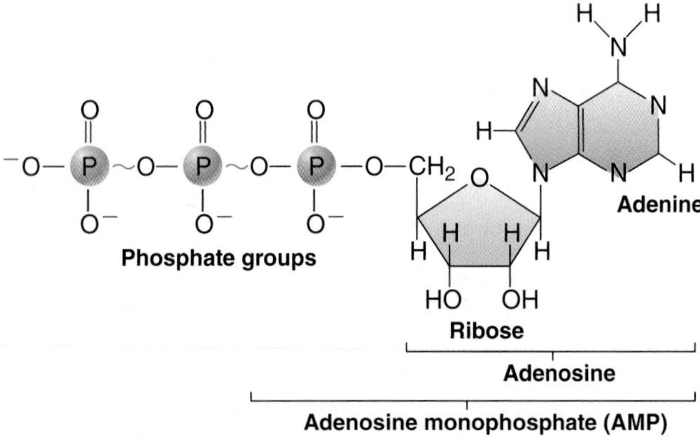

(b) ATP structure

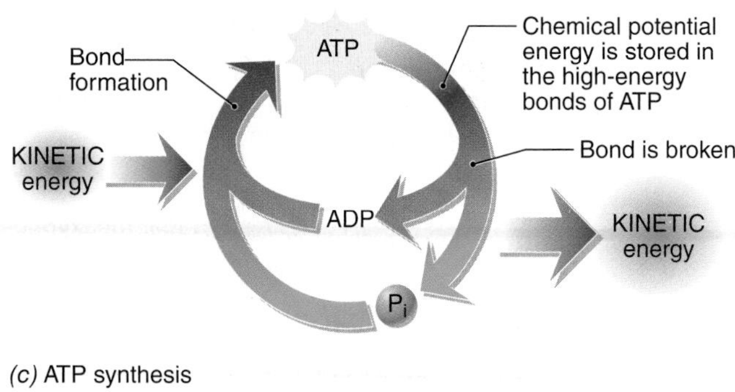

(c) ATP synthesis

Figure 2.22 The breakdown of ATP forms ADP and a phosphate group and releases energy to power cellular reactions. ATP is synthesized from ADP, phosphate, and energy extracted from nutrients.

a high-energy phosphate group to ADP re-forms ATP. ATP is continuously broken down into ADP to release energy, and it is re-formed as energy is made available from nutrients (figure 2.22*c*).

Check My Understanding

10. What distinguishes the chemical structure and functions of carbohydrates, lipids, proteins, and nucleic acids?
11. What is the role of enzymes in body cells?
12. What is ATP, and what is its role in body cells?

Chapter Summary

2.1 Atoms and Elements

- Matter is composed of elements, substances that cannot be broken down into simpler substances by chemical means.
- Oxygen, carbon, hydrogen, and nitrogen form 96% of the human body by weight.
- An atom is the smallest unit of an element.
- An atom consists of a nucleus formed of protons (+1) and neutrons (0) and electrons (−1) that orbit around the nucleus.
- Electrons fill electron shells from inside to outside.
- The outermost shell containing electrons is the valence shell.
- Elements are characterized by their atomic numbers, chemical symbols, and atomic mass.
- Isotopes of an element have differing numbers of neutrons.
- Radioisotopes emit radiation.

2.2 Molecules and Compounds

- A molecule is formed of two or more atoms joined by covalent bonds.
- A compound is formed of atoms from two or more elements combined by ionic or covalent bonds.
- A molecular formula indicates the types of elements and number of atoms of each element in a molecule or compound.
- A structural formula adds to a molecular formula by also showing how the atoms fit together.
- Chemical bonds join atoms to form molecules.

- An ionic bond is the force of attraction between two ions with opposite electrical charges. It results from one atom donating one or more electrons to another atom.
- A covalent bond is formed between two atoms by the sharing of electrons in the valence shell.
- Nonpolar covalent bonds share electrons equally; polar covalent bonds share electrons unequally.
- Nonpolar substances are hydrophobic (water fearing); polar substances and ions are hydrophilic (water loving).
- A hydrogen bond is a weak force of attraction between a slightly positive H atom and a slightly negative atom either in the same molecule or in different molecules, or between ions and polar molecules.
- Synthesis reactions combine simpler substances to produce more complex substances.
- Decomposition reactions break down complex substances into simpler substances.
- Exchange reactions involve both decomposition of the reactants and synthesis of new products.
- Reversible reactions may occur in either direction depending on the environment.

2.3 Compounds Composing the Human Body

- Inorganic compounds do *not* contain both carbon and hydrogen. Organic compounds contain *both* carbon and hydrogen.
- Water (H_2O) is the most abundant inorganic molecule in the body, and it is the solvent of living systems.
- There are two major water compartments: intracellular fluid (65% of body water) and extracellular fluid (35% of body water).
- Electrolytes ionize (dissociate) in water, producing ions. The resulting solution can conduct electricity.
- Nonelectrolytes are substances that do not produce ions in water, and they do not conduct electricity.
- An acid releases H^+ in an aqueous solution, and a base releases OH^- or absorbs H^+ in an aqueous solution.
- pH is a measure of the relative concentrations of H^+ and OH^- in a solution.
- A buffer keeps the pH of a solution relatively constant by picking up or releasing H^+.
- A salt releases positively and negatively charged ions in an aqueous solution, but they are neither H^+ nor OH^-.
- Organic molecules are synthesized by dehydration synthesis and broken down by hydrolysis.
- Carbohydrates are composed of C, H, and O with a 2:1 ratio between H and O.
- Monosaccharides are the building units of carbohydrates. Disaccharides consist of two monosaccharides. Polysaccharides are formed of many monosaccharides.
- Lipids are a diverse group of organic compounds that include triglycerides, phospholipids, and steroids.
- Triglycerides (fats) consist of three fatty acids bonded to glycerol. Unsaturated fats differ from saturated fats by having one or more double carbon-carbon bonds in their fatty acids. Excess nutrients are stored as fats.
- Phospholipids consist of two fatty acids and a phosphate-containing group bonded to glycerol.
- Steroids are an important group of lipids that includes sex hormones and cholesterol.
- Proteins are large molecules formed of many amino acids. The 20 different kinds of amino acids are distinguished by their R groups.
- Amino acids are joined by peptide bonds.
- Structural proteins form parts of cells and tissues. Functional proteins include enzymes, transporters, and antibodies.
- Enzymes catalyze chemical reactions.
- Nucleic acids are very large molecules formed of many nucleotides.
- A nucleotide consists of a five-carbon sugar (ribose in RNA and deoxyribose in DNA), a phosphate group, and a nitrogenous base.
- There are two types of nucleic acids: deoxyribonucleic acid (DNA) and ribonucleic acid (RNA). DNA determines hereditary traits and controls cellular functions. RNA works with DNA in the synthesis of proteins.
- Adenosine triphosphate (ATP) is a modified nucleotide that temporarily holds energy in high-energy phosphate bonds and releases that energy to power chemical reactions in a cell.

Self-Review

Answers are located in Appendix B.

1. An atom contains the same number of electrons as ______.
2. An ______ is an atom with a net electrical charge.
3. A ______ is composed of two or more atoms from different elements chemically bonded to each other.
4. When atoms share electrons, a ______ bond is formed.
5. The most abundant compound in the body is ______.
6. An acid has a pH ______ than 7.
7. In a ______ reaction, smaller molecules are combined to form a larger molecule.
8. The building units of a carbohydrate are ______.
9. The building units of proteins are ______.
10. The building units of nucleic acids are ______.
11. Glycerol and fatty acids combine to form ______.
12. ______ is the energy carrier that releases energy to power cellular processes.

Critical Thinking

1. Draw at least four isomers of the molecule $C_6H_{12}O_6$.
2. Potassium (K) has an atomic number of 19. How many electrons are in its valence shell? What type of chemical bond is it likely to form?
3. The pH of an aqueous solution is 6.0. Explain how the addition of the following substances would alter (or not alter) the pH of the solution.
   a. HCl b. NaOH c. HOH
4. Explain how changes in pH might affect protein structure.

ADDITIONAL RESOURCES

3 CHAPTER

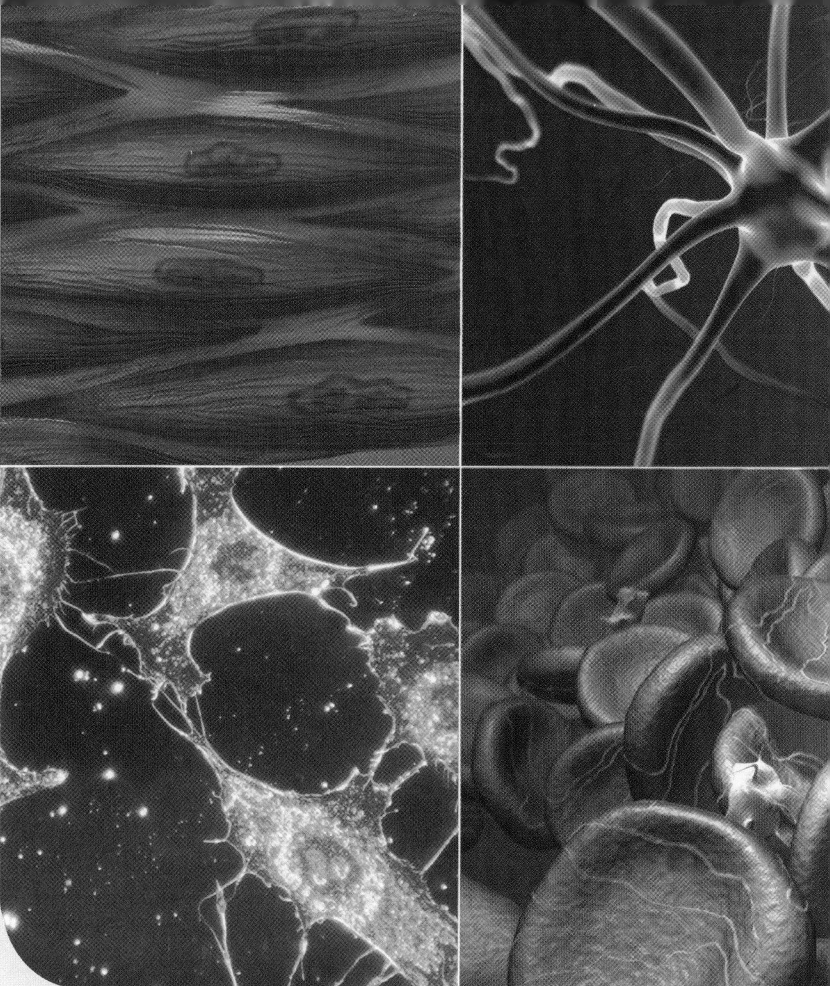

Cell

As a living structure, a cell possesses numerous organelles that carry out the chemical processes necessary to maintain homeostasis. Learning the relationships between organelle structure and function can be quite challenging for students owing to their microscopic nature and the complexity of the chemical processes involved. So for a moment, imagine that you are a cell, a small living unit within the body. Every day, numerous challenges must be overcome to survive but, within you, there are structures that do just that. For example, if you need to obtain nutrients or remove wastes, you have an outer boundary that controls the movement of chemicals. If energy is needed to carry out a chemical process, you have structures that make the energy in your nutrients accessible. In order to respond to changes in your surroundings, you have the ability to communicate with other cells by detecting or producing chemical and electrical signals. By making cellular structures and functions relate to you and your real-world experiences, you will create a way to better "visualize" and understand the microscopic world within your body.

CHAPTER OUTLINE

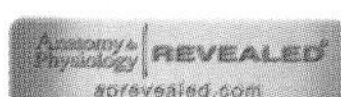

Module 2
Cells & Chemistry

SELECTED KEY TERMS

Active transport Movement of substances across a plasma membrane, requiring the expenditure of energy by the cell.
Cell (cella = room, cell) The simplest structural and functional living unit of organisms.
Cellular respiration Breakdown of organic nutrients in cells, to release energy and form ATP.
Centrioles (centr = center) Paired cylindrical organelles that form the spindle during cell division.
Chromosome (chrom = color; soma = body) A threadlike or rodlike structure in the nucleus that is composed of DNA and protein.
Cytoplasm (cyt = cell; plasma = molded) The semifluid substance located between the nucleus and the plasma membrane.
Cytosol (sol = soluble) The gel-like fluid of the cytoplasm.
Diffusion Passive movement of substances from an area of higher concentration to an area of lower concentration.
Endocytosis (end = inside; cyt = cell; sis = condition) The process by which a cell engulfs substances by invagination of the plasma membrane.
Exocytosis (exo = outside) The process by which a cell releases substances by fusion of a vesicle with the plasma membrane.
Mitosis (mit = thread; sis = condition) Separation and distribution of chromosomes to daughter cells during mitotic cell division.
Nucleus (nucle = kernel) Spherical organelle containing chromosomes and controlling cellular functions.
Organelle (elle = little) A specific structure within a cell that performs a specific function.
Osmosis The passive movement of water across a selectively permeable membrane.
Passive transport Movement of substances across a plasma membrane without expenditure of energy by the cell.
Plasma membrane Outer boundary of a cell.
Selectively permeable membrane A membrane that allows some, but not all, substances to pass across it.

THE HUMAN BODY is composed of about 75 trillion **cells,** the smallest living units that exist. Body cells can be classified into about 300 types, such as neurons, epithelial cells, muscle cells, and red blood cells. Each type of cell has a unique structure for performing specific functions. Although these cells vary in size, shape, and function, they exhibit many structural and functional similarities.

Human cells are very small and are visible only with a microscope. Knowledge of cell structure is based largely on the examination of cells with an electron microscope, a type of microscope that provides magnifications up to 200,000× or more.

3.1 Cell Structure

Learning Objective

1. Describe the structure and function(s) of each part of a generalized cell.

Although human cells are small, they are amazingly complex with many specialized parts. The composite cell in figure 3.1 illustrates the major structures known to compose human cells. These structures are shown as they appear in electron microscope images. Most, but not all, of these structures are found in each human cell. The three common parts found in all the cells are the plasma membrane, cytoplasm, and nucleus. The other structures may or may not be present, depending on cell type. As each part of a cell is discussed, note its structure and relationship to other structures in figure 3.1.

The Plasma Membrane

The **plasma membrane** forms the outer boundary of a cell. It maintains the integrity of the cell and separates the intracellular fluid from the extracellular fluid surrounding the cell. The plasma membrane consists of two layers of phospholipid molecules, aligned back-to-back, with their fatty acid tails forming the internal layer of the membrane and their polar heads facing the extracellular and intracellular fluids (figure 3.2). Cholesterol molecules are scattered among the phospholipids, where they serve to increase the stability of the plasma membrane. The fatty acid tails of the plasma membrane allow lipid-soluble substances to pass across the membrane but prevent the passage of water-soluble substances. Thus, the plasma membrane serves as a barrier between water-soluble substances in the intracellular and extracellular fluids.

Many different types of protein molecules are embedded in the plasma membrane, and each type has specific functions. Some proteins form channels or pores through which water and water-soluble substances move across the membrane. Some of these proteins allow a variety of substances to pass across; others permit only specific molecules or ions to enter or exit a cell. Some proteins serve as receptors for substances, such as hormones,

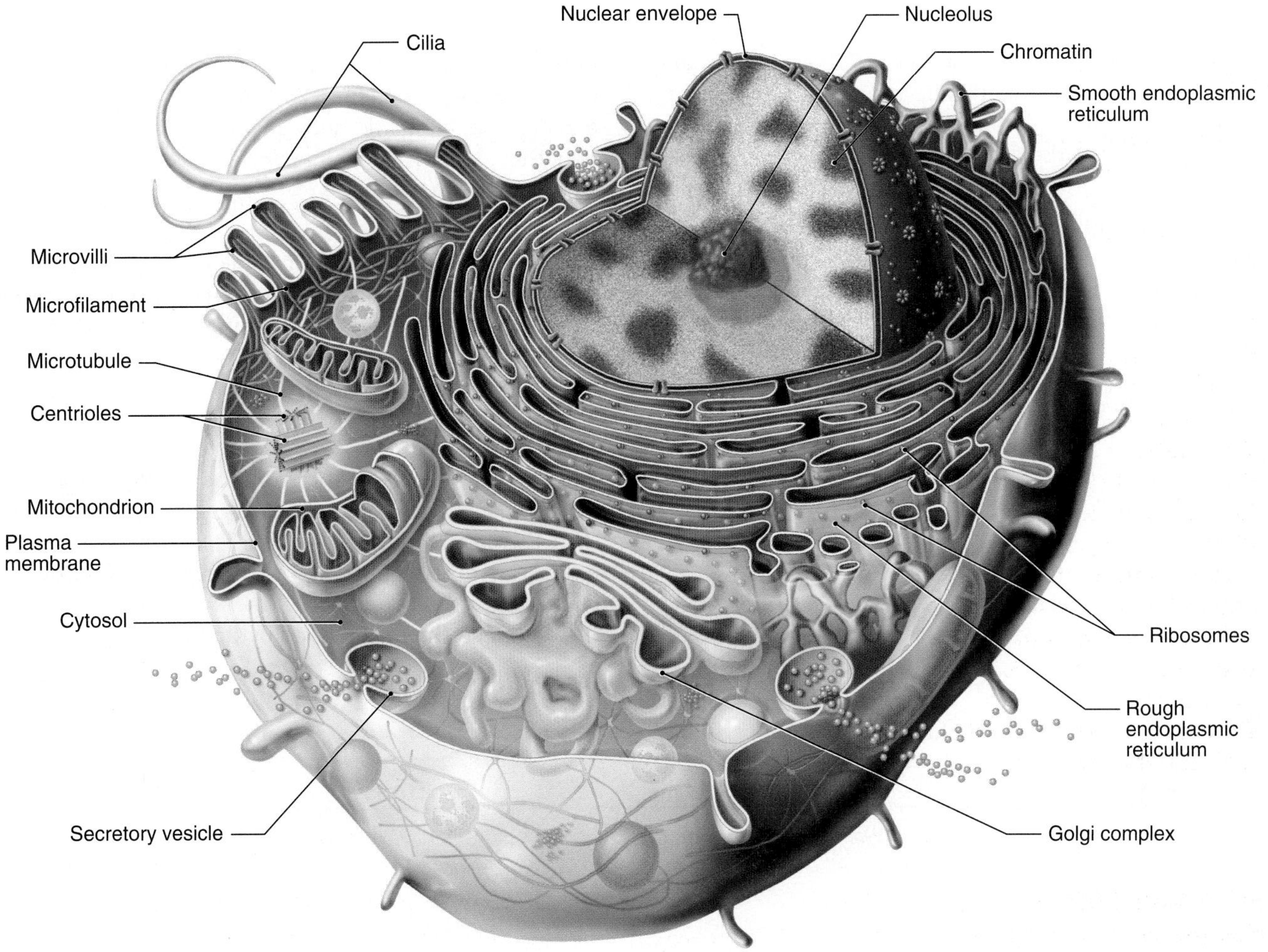

Nuclear envelope
Nucleolus
Chromatin
Cilia
Smooth endoplasmic reticulum
Microvilli
Microfilament
Microtubule
Centrioles
Mitochondrion
Plasma membrane
Cytosol
Ribosomes
Rough endoplasmic reticulum
Secretory vesicle
Golgi complex

Figure 3.1 A composite human cell showing the major organelles. No cell contains all of the organelles shown. APIR

Carbohydrate chains
Outer membrane surface
Receptor protein
Polar regions of phospholipid bilayer
Extracellular fluid
Phospholipid bilayer
Nonpolar region of phospholipid bilayer
Intracellular fluid
Cytoskeleton
Membrane channel protein
Cholesterol
Protein

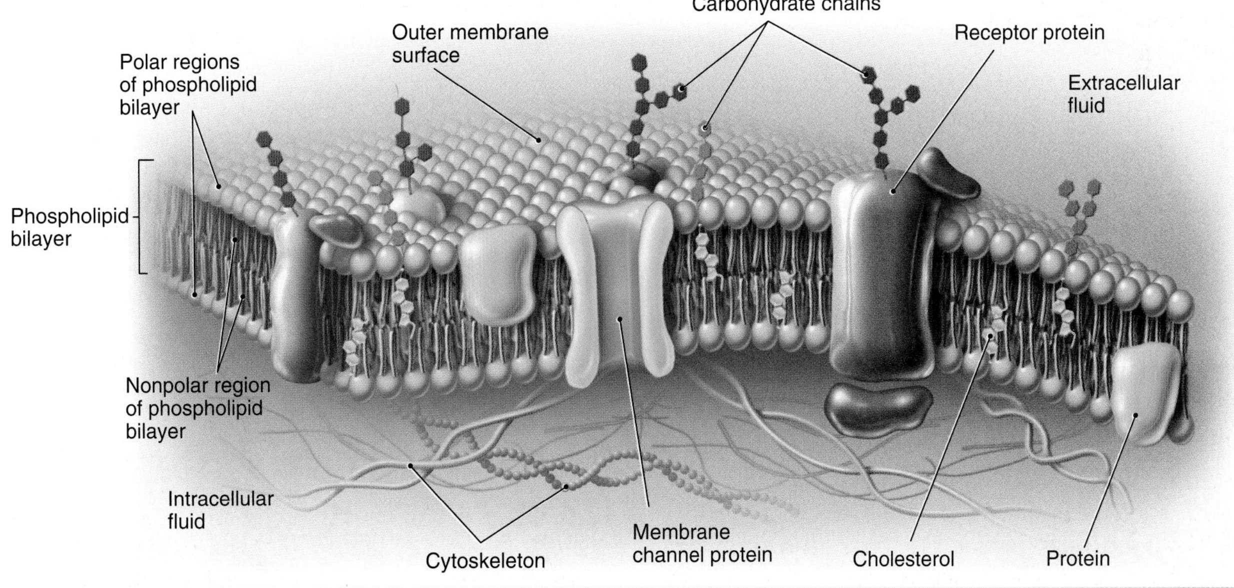

Figure 3.2 The plasma membrane is composed of two layers of phospholipid molecules with scattered embedded protein and cholesterol molecules. The hydrophilic heads of the phospholipids face the extracellular and intracellular fluids, and the hydrophobic tails form the internal layer of the membrane. APIR

that influence the function of a cell. Other proteins are enzymes that catalyze metabolic reactions. Certain proteins, in combination with carbohydrate molecules, serve as identification markers allowing cells to recognize each other. These identification markers allow the lymphoid system to recognize "self" cells from "nonself" (foreign) cells, a distinction essential in fighting pathogens.

All materials that enter or exit a cell must pass across the plasma membrane. The plasma membrane is a **selectively permeable membrane** because it allows only certain substances to enter or exit the cell. Whether or not a substance can pass across the membrane is determined by a number of factors that include the substance's size, solubility, electrical charges, and attachment to carrier proteins (discussed later in the chapter).

Cytoplasm

The interior of a cell between the plasma membrane and the nucleus is filled with a semifluid material called **cytoplasm** (sī′-tō-plasm). It is composed of a gel-like fluid called **cytosol,** which is 75–90% of water and contains organic and inorganic substances, and small subcellular structures known as organelles.

Organelles

A variety of **organelles** (or-gah-nel′z), or little organs, are surrounded by cytosol. Organelles are distinguished by size, shape, structure, and specific function. Table 3.1 summarizes the structure and functions of the major parts of a cell.

Nucleus

The largest organelle is the **nucleus** (nū′klē-us), a spherical or egg-shaped structure that is slightly more dense than the surrounding cytoplasm. It is separated from the cytoplasm by a double-layered **nuclear envelope** containing numerous pores that allow the movement of materials between the nucleus and cytoplasm.

Table 3.1 Summary of Cell Parts

Component	Structure	Function
Plasma membrane	Phospholipid bilayer with proteins and cholesterol molecules embedded in it	Selectively controls movement of materials into and out of the cell; maintains integrity of the cell; has receptors for hormones
Cytosol	Gel-like fluid surrounding organelles	Site of numerous chemical reactions
Organelles		
Nucleus	Largest organelle; contains chromosomes and nucleoli	Controls cellular functions
Endoplasmic reticulum (ER)	System of membranes extending through the cytoplasm; RER has ribosomes on the membrane; SER does not	Serves as sites of chemical reactions; channels for material transport within cell
Ribosomes	Tiny granules of rRNA and protein either associated with RER or free in cytoplasm	Sites of protein synthesis
Golgi complex	Series of stacked membranes near nucleus; associated with ER	Sorts and packages substances in vesicles for export from cell or use within cell; forms lysosomes
Mitochondria	Contain a folded internal membrane within a smaller external membrane	Sites of aerobic respiration that form ATP from breakdown of nutrients
Lysosomes	Small vesicles containing strong digestive enzymes	Digest foreign substances or worn-out parts of cells
Microfilaments	Thin rods of protein dispersed in cytoplasm	Provide support for cell; contraction causes cell movement
Microtubules	Thin tubules dispersed in cytoplasm	Provide support for cell, cilia, and flagella; form spindle during cell division
Microvilli	Numerous, tiny extensions of the plasma membrane on certain cells	Increase the surface area, which aids absorption
Centrioles	Two short cylinders formed of microtubules; located near nucleus	Form spindle fibers during cell division
Cilia	Numerous short, hairlike projections from certain cells	Move materials along the free surface of cells
Flagella	Long, whiplike projections from sperm	Enable movement of sperm
Vesicles	Tiny membranous sacs containing substances	Transport or store substances

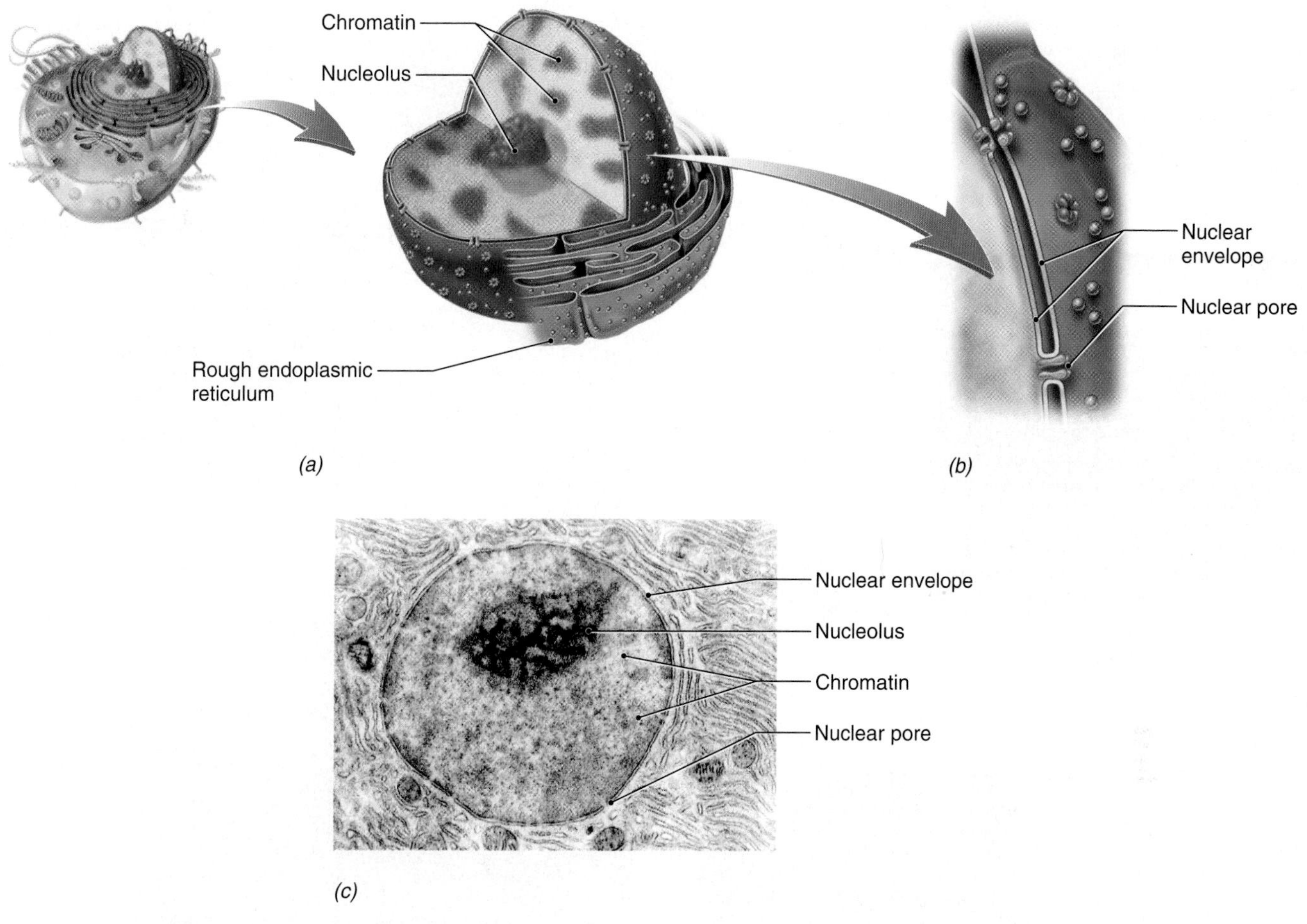

Figure 3.3 *(a)* The nuclear envelope is selectively permeable and allows certain substances to pass. *(b)* Details of the nuclear envelope. *(c)* Transmission electron photomicrograph of a cell nucleus (8,000×). APR

Chromosomes (krō′-mō-sōms), the most important structures within the nucleus, consist of DNA and proteins. The DNA of chromosomes contains coded instructions, called genes, that determine the functions of the cell (see chapter 18 for the details). When a cell is not dividing, chromosomes are extended to form thin threads that appear as *chromatin* (krō′-mah-tin) *granules* when viewed microscopically, as in figure 3.3. During cell division, the chromosomes coil, shorten, and become rod-shaped (see figure 3.19). Each human body cell contains 23 pairs of chromosomes, with a total of 46 in all.

One or more dense spherical bodies, called the **nucleolus** (nū-klē-ō-lus) or **nucleoli** (nū-klē-ō-lē), are also present in the nucleus. A nucleolus consists of RNA and protein and is the site of ribosome production.

Ribosomes

Ribosomes are tiny organelles that appear as granules within the cytoplasm even in electron photomicrographs. They are composed of ribosomal RNA (rRNA) and proteins, which are preformed in a nucleolus before migrating from the nucleus into the cytoplasm. Ribosomes are the sites of protein synthesis in cells. They may occur singly or in small clusters and are located either on the endoplasmic reticulum (figure 3.4) or as free ribosomes in the cytoplasm.

Endoplasmic Reticulum

The numerous membranes that extend from the nucleus throughout the cytoplasm are collectively called the **endoplasmic reticulum** (en″-dō-plas′-mik rē-tik′-ū-lum), or **ER** for short. These membranes provide some support for the cytoplasm and form a network of channels that facilitate the movement of materials within the cell.

There are two types of ER: rough ER and smooth ER. *Rough endoplasmic reticulum (RER)* is characterized by the presence of numerous ribosomes located on the outer surface of the membranes. *Smooth endoplasmic reticulum (SER)* lacks ribosomes and serves as a site for the synthesis of lipids (see figure 3.4).

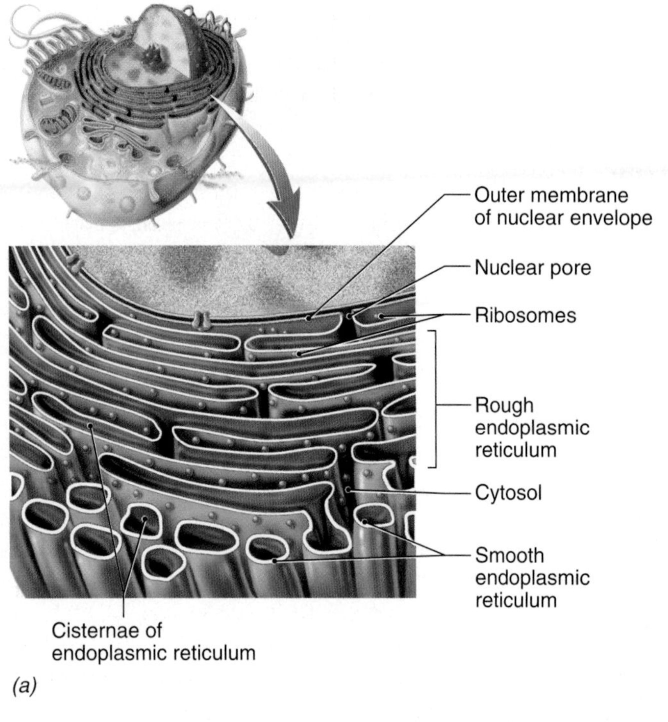

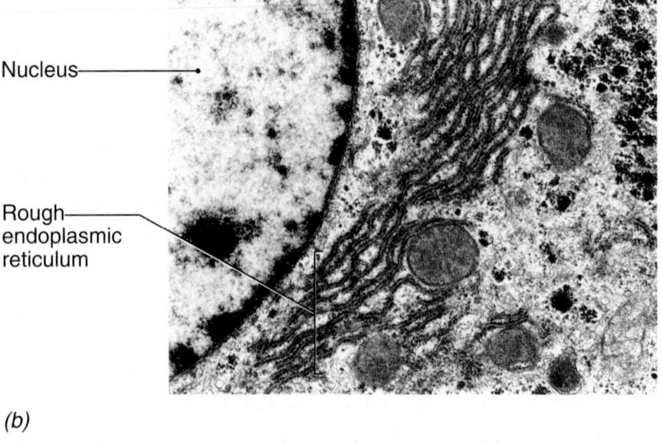

Figure 3.4 *(a)* Rough ER is dotted with ribosomes, and smooth ER lacks ribosomes. *(b)* Transmission electron photomicrograph of ER (100,000×). APIR

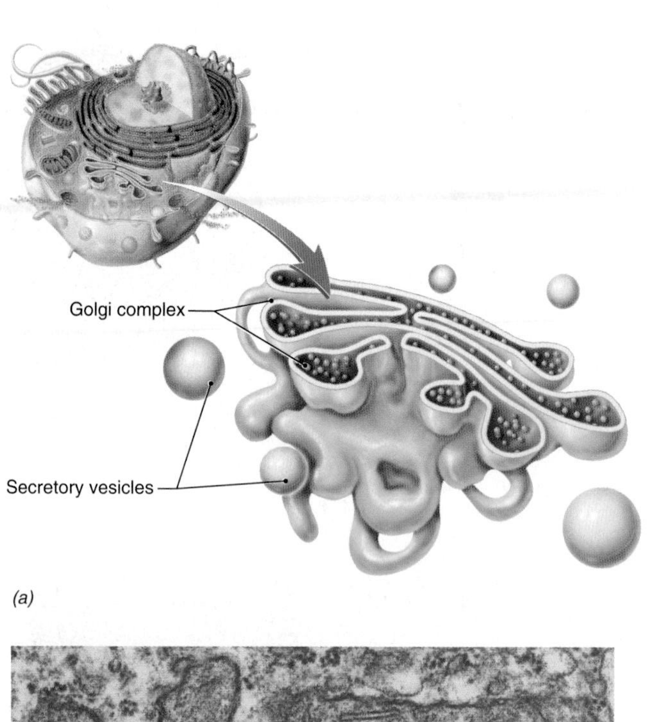

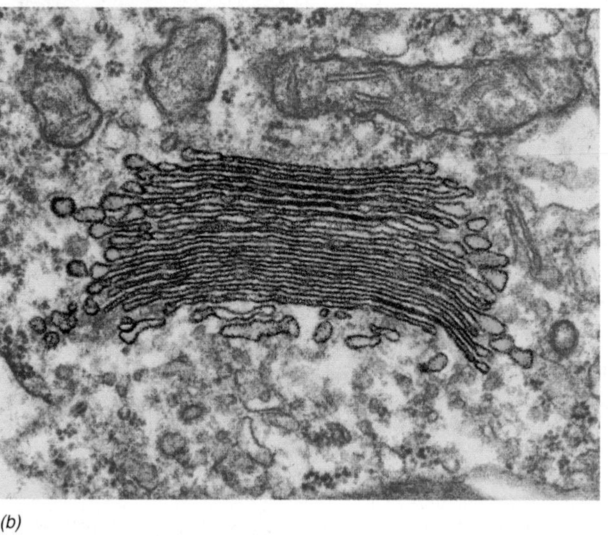

Figure 3.5 *(a)* The Golgi complex packages substances in vesicles that move within the cell or to the plasma membrane to release the substances outside the cell. *(b)* Transmission electron photomicrograph of Golgi complex (100,000×).

Golgi Complex

This organelle appears as a stack of flattened membranous sacs that are usually located near the nucleus and in close association with the nucleus and ER. The **Golgi** (Gol′-jē) **complex** processes and sorts synthesized substances, such as proteins, into vesicles. **Vesicles,** or "little bladders," are tiny membranous sacs that carry substances from place to place within a cell. *Secretory vesicles* transport substances to the plasma membrane and release them outside the cell (figure 3.5).

Mitochondria

The **mitochondria** (mī″-to-kon′-drē-ah, singular, *mitochondrion*) are relatively large organelles that are characterized by having a folded *internal membrane* surrounded by a smooth *external membrane.* The internal membrane folds, called *cristae* (singular, crista), possess the enzymes involved in aerobic respiration.

The release of energy from nutrients and the formation of ATP by aerobic respiration occur within mitochondria. For this reason, mitochondria are sometimes

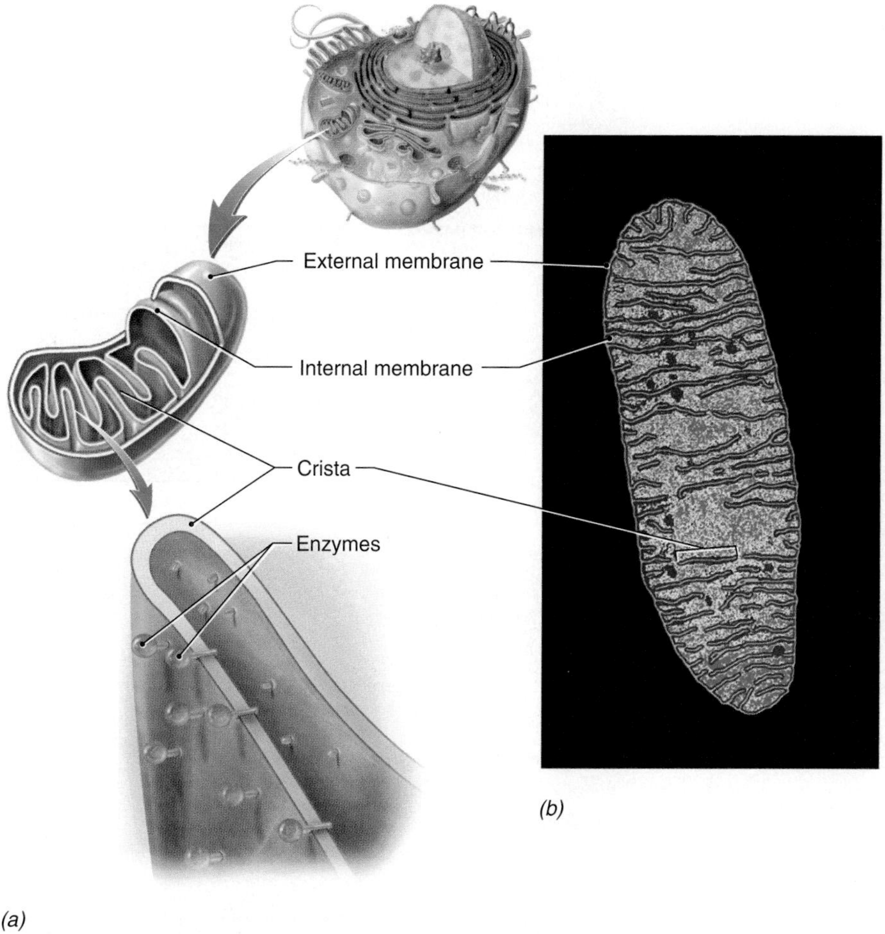

Figure 3.6 *(a)* A mitochondrion and its internal membrane. *(b)* Transmission electron photomicrograph of a mitochondrion (40,000×). APR

called the "powerhouses" of the cell. Mitochondria can replicate themselves if the need for additional ATP production increases (figure 3.6).

In addition to the nucleus, mitochondria also contain a small amount of DNA, known as mitochondrial DNA. The genes carried by this DNA account for less than 0.2% of the total genes in the human body, and are responsible only for the functions of the mitochondria. Mitochondrial DNA cannot be used to establish paternity as with nuclear DNA, because only maternal mitochondrial DNA is passed on to offspring.

Lysosomes APR

Lysosomes (lī′-sō-sōms) are formed by the Golgi complex. They are small vesicles that contain powerful digestive enzymes (see figure 3.1). These enzymes are used to digest (1) bacteria that may have entered the cell, (2) cell parts that need replacement, and (3) entire cells that have become damaged or worn out. Thus, they play an important role in cleaning up the cellular environment. (figure 3.7*a*).

The Cytoskeleton

Microtubules and microfilaments compose the cytoskeleton. **Microtubules** are long, thin protein tubules that provide support for the cell and are involved in the movement of organelles. The thinner **microfilaments** are tiny rods of contractile protein that not only support the cell but also play a role in cell movement and cell division. (figure 3.7).

Centrioles

The **centrioles** (sen′-trē-olz) are two short cylinders that are located near the nucleus and are oriented at right angles to each other. Nine triplets of microtubules are arranged in a circular pattern to form the wall of each cylinder (see figure 3.1). Centrioles form and organize the *spindle fibers* during cell division (see figure 3.19), and they are involved in the formation of microtubules found in cilia and flagella.

Cilia, Flagella, and Microvilli

Cilia and flagella (singular, flagellum) are small, hairlike projections from cells that are capable of wavelike movement. **Cilia** (sil′-ē-ah) are numerous, short, hairlike projections from cells that, in humans, are used to move substances along the free cell surfaces in areas such as the respiratory and reproductive tracts (figure 3.8). **Flagella** (flah-jel′-ah) are long, whiplike projections from cells. In humans, only sperm possess flagella, and each sperm has a single flagellum that enables movement. Both cilia and flagella contain microtubules that originate from centrioles positioned at the base of these flexible structures.

Microvilli are extensions of the plasma membrane that are smaller and more numerous than cilia. They do not move like cilia or flagella, but they increase the surface area of the plasma membrane and, therefore, aid absorption of substances. Microvilli are abundant on the free surface of the cells lining the intestines (see figure 3.7*a*).

Check My Understanding

1. What are the distinguishing features and functions of a mitochondrion, a nucleus, the Golgi complex, and rough endoplasmic reticulum?
2. What organelles enable cell movement or movement of substances along the free surface of the cells?

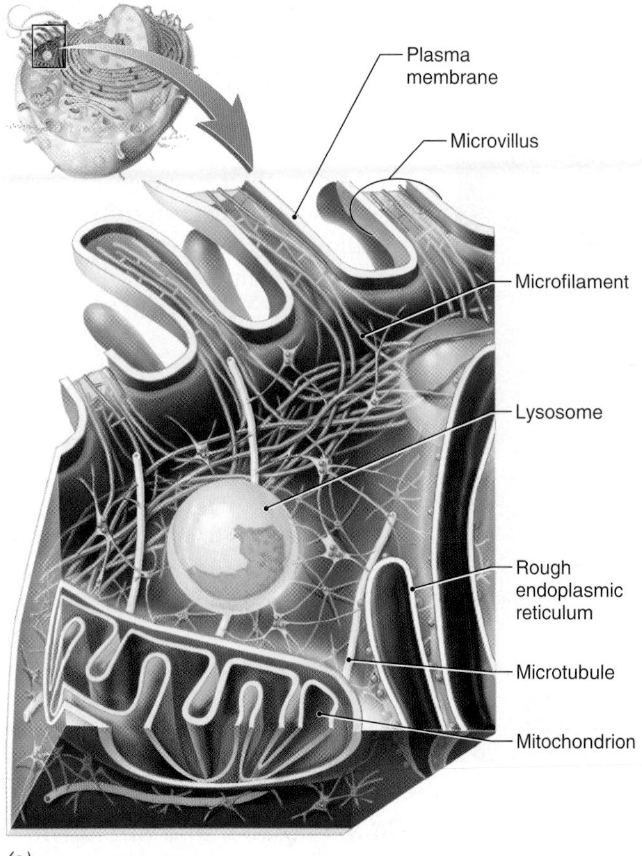

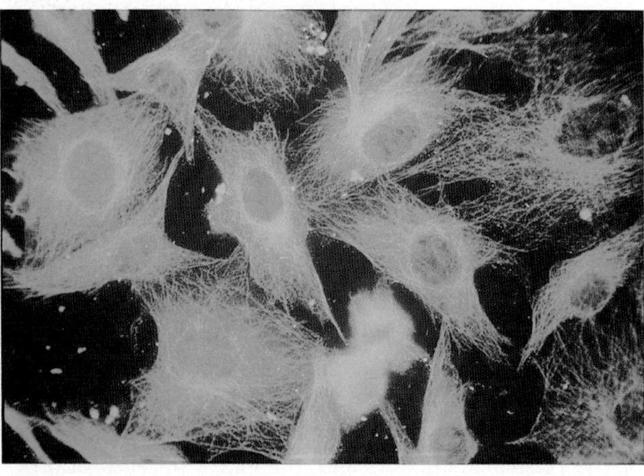

Figure 3.7 *(a)* Microtubules and microfilaments. *(b)* A false-color electron photomicrograph (750×) shows the microtubules and microfilaments of the cytoskeleton in green. APR

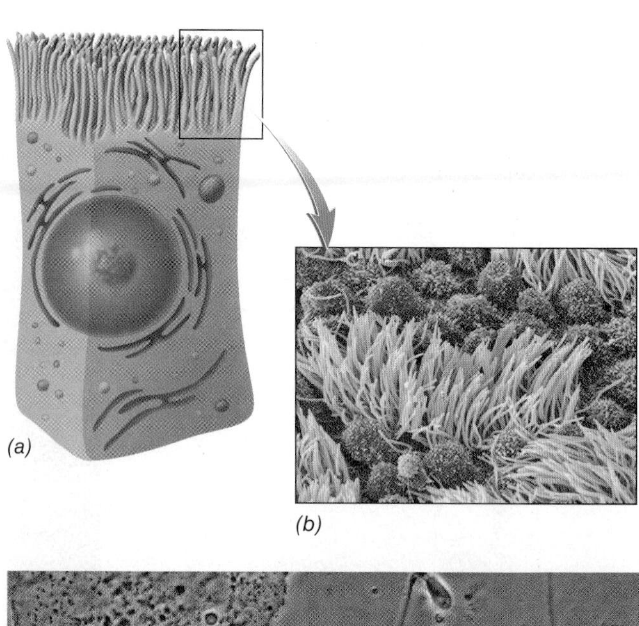

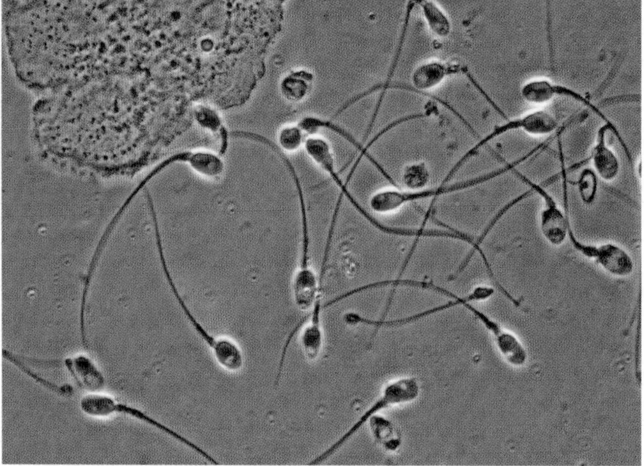

Figure 3.8 *(a)* Cilia are located on the free surface of certain cells. Because these cells are stationary, beating cilia move substances along the free surface of the cells. *(b)* An electron photomicrograph of cilia (10,000×). *(c)* A light photomicrograph of human sperm (1,000×).

3.2 Transport Across Plasma Membranes

Learning Objectives

2. Compare the mechanisms of passive and active transport of substances across the plasma membrane.
3. Describe osmosis and tonicity, and the effect of tonicity on the cells.

A cell maintains its homeostasis primarily by controlling the movement of substances across the selectively permeable plasma membrane. Some substances pass across the plasma membrane by **passive transport,** which requires no expenditure of ATP by the cell. Other substances move across the plasma membrane by **active transport,** which requires the cell to expend ATP.

Passive Transport

There are three major types of passive transport: diffusion, osmosis, and filtration. Filtration is described in chapters 12 and 16.

Diffusion

Diffusion (di-fū′-zhun) is the net movement of substances from an area of higher concentration to an area of lower concentration. Thus, the movement of substances is along a **concentration gradient,** the difference between the concentration of the specific substances in the two areas.

Diffusion occurs in both gases and liquids and results from the constant, random motion of substances. Diffusion is not a living process; it occurs in both living and nonliving systems. For example, if a pellet of a water-soluble dye is placed in a beaker of water, the dye molecules will slowly diffuse from the pellet (the area of higher concentration) throughout the water (the area of lower concentration) until the dye molecules are equally distributed, that is, at equilibrium (figure 3.9). In a similar way, the molecules of cologne, on the skin of a student sitting in the corner of a classroom, will spread throughout the room.

Lipid-soluble molecules, such as lipids, oxygen, carbon dioxide, and lipid-soluble vitamins, are able to diffuse across a plasma membrane along concentration gradients because they can dissolve in the phospholipid molecules of the plasma membrane. This type of diffusion is called **simple diffusion** (figure 3.10*a*) because it does not require the help of the membrane proteins. For example, the exchange of respiratory gases occurs by simple diffusion. Air in the lungs has a greater concentration of oxygen and a lower concentration of carbon dioxide than the blood does (figure 3.10*a*). Therefore, oxygen diffuses from air in the lungs into the blood, and carbon dioxide diffuses from the blood into the air in the lungs.

Figure 3.9 An example of diffusion. As a drop of ink gradually dissolves in a beaker of water, the ink molecules diffuse from the region of their higher concentration to a region of their lower concentration. APR

Water-soluble molecules, such as glucose, amino acids, water-soluble vitamins, and ions, cannot be transported by simple diffusion because they cannot dissolve in the phospholipids. Some water-soluble substances are transported through channel proteins. **Channel proteins** are tunnel-shaped membrane proteins that create pores or openings, which allow for specific substances to pass across the plasma membrane along their concentration

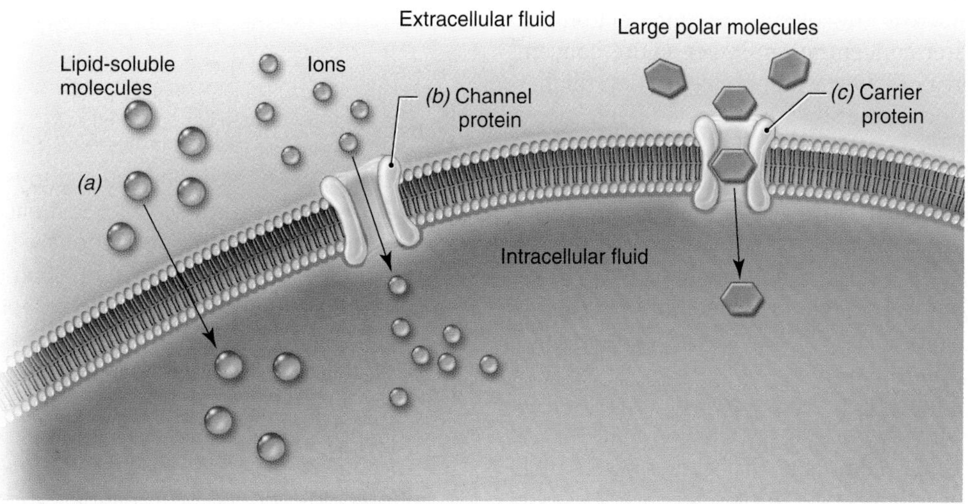

Figure 3.10 Diffusion.
(a) Simple diffusion. *(b)* Channel-mediated diffusion. *(c)* Carrier-mediated diffusion. APR

Clinical Insight

Dialysis involves the application of diffusion to remove small solute molecules across a selectively permeable membrane from a solution containing both small and large molecules. Dialysis is the process that is used in artificial kidney machines. As blood is passed through a chamber with a selectively permeable membrane, small waste molecules diffuse from the blood across the membrane into an aqueous solution that has a low concentration of these waste molecules. In this way, waste products in the blood are reduced to normal levels.

gradient. Channel proteins are generally selective; this means they tend to allow limited substances to pass across based mostly on size and charge. This type of transport is called *channel-mediated diffusion* (figure 3.10*b*). Other water-soluble substances use carrier proteins. **Carrier proteins** are membrane proteins that physically bind to and transport specific substances across the plasma membrane; this means that one type of carrier protein binds only one type of substance. This type of transport is called *carrier-mediated diffusion* (figure 3.10*c*). Carrier-mediated diffusion is a type of **facilitated transport,** which uses carrier proteins to facilitate the movement of substances across the plasma membrane. Carrier-mediated active transport, another type of facilitated transport, will be discussed later.

Osmosis

The passive movement of water across a selectively permeable membrane is called **osmosis** (os-mo′-sis). Water molecules move across the plasma membrane from an area of higher water concentration (lower solute concentration) into an area of lower water concentration (higher solute concentration), either by crossing the plasma membrane directly or by moving through a channel protein. Osmosis plays a very important role in the functions of the cells and the whole body. Water molecules are the dominant components of cells and serve as the solvent of the other chemicals. Also, the movement of water molecules into and out of the cells has the ability to significantly affect the volume of cells and the concentration of the chemicals within them.

Figure 3.11 illustrates the process of osmosis. The beaker is divided into two compartments (A and B) by a selectively permeable membrane that allows water molecules but not sugar molecules to pass across it. Because the higher concentration of water is in compartment A, water moves from compartment A into compartment B. Sugar molecules cannot pass across the membrane, so

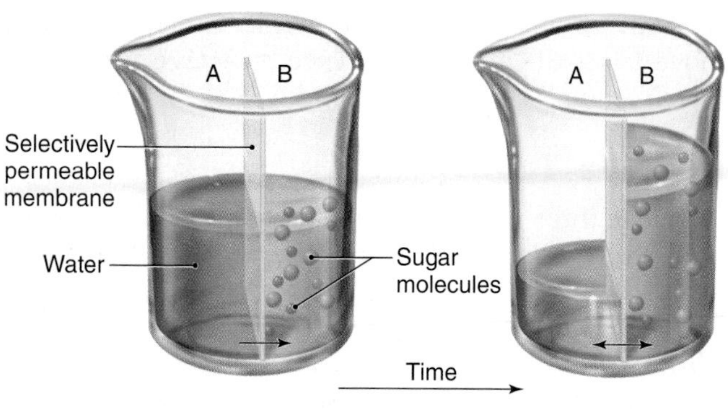

Figure 3.11 Osmosis. APR

water molecules from compartment A continue to move into compartment B, causing the volume of the solution in compartment B to increase as the volume of water in compartment A decreases.

Like compartment B in figure 3.11, living cells also contain many substances to which the plasma membrane is impermeable. Therefore, any change in the concentration of water across the plasma membrane will result in net gain or loss of water by the cell and a change in cell volume and shape.

The ability of a solution to affect the tone or shape of living cells by altering the cells' water content is called *tonicity*. A solution with a lower concentration of solutes (higher concentration of water) than the cell is called a **hypotonic solution.** A cell placed in this solution will gain water and increase in size, which may eventually lead to rupture of the cell (figure 3.12*a*). A solution with a higher concentration of solutes (lower concentration of water) than the cell is known as a **hypertonic solution.** A cell placed in this solution will lose water and shrink, which may lead to cell death (figure 3.12*c*).

Clinical Insight

Solutions that are administered to patients intravenously usually are isotonic. Sometimes hypertonic solutions are given intravenously to patients with severe edema, or an accumulation of excess fluid in body tissues. The hypertonic solution will help to draw the excess fluid out of the body tissue and into the blood, where it can be removed by the kidneys and excreted in urine. Severely dehydrated patients may be given a hypotonic solution orally or intravenously to increase the water concentration of blood and tissue fluid, by increasing water movement from the digestive tract into the blood and from the blood into body tissues.

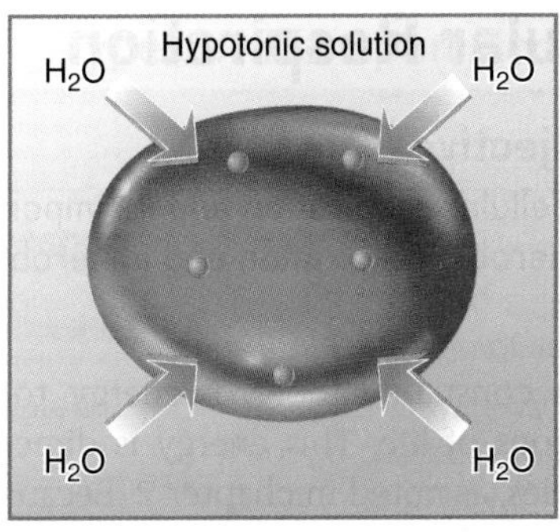

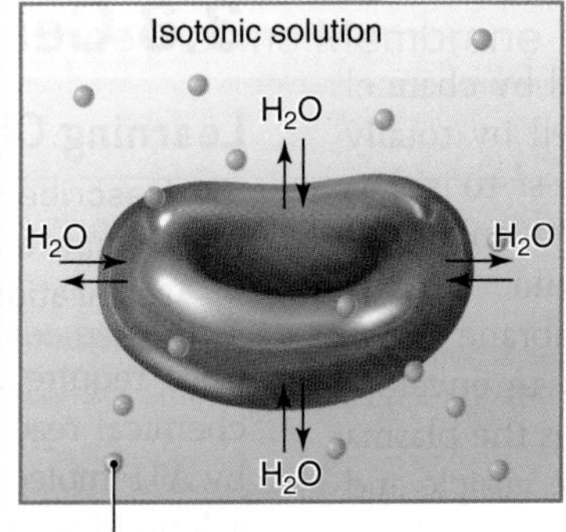

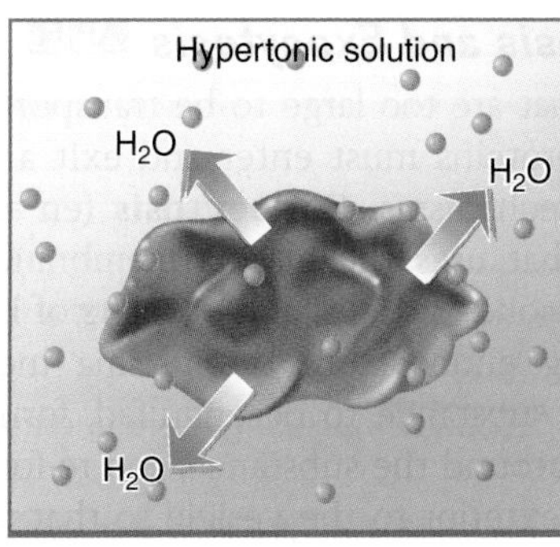

(a) In a hypotonic solution, there is a net gain of water by the cell, causing it to swell. Ultimately, the cell may burst.

(b) In an isotonic solution, there is no net gain or loss of water by the cell; the shape of the cell remains unchanged.

(c) In a hypertonic solution, there is a net loss of water from the cell, causing it to shrink.

Figure 3.12 The effect of tonicity on human red blood cells. APR

A solution that has the same concentration of solutes (same concentration of water) as the cell is an **isotonic solution.** When surrounded by this solution, a cell exhibits no net gain or loss of water and no change in volume (figure 3.12*b*).

Active Transport

Unlike passive transport, active transport requires the cell to expend energy (ATP) to move substances across a plasma membrane. There are three basic active transport mechanisms: carrier-mediated active transport, endocytosis, and exocytosis.

Carrier-Mediated Active Transport

Carrier-mediated active transport uses carrier proteins to move substances across the plasma membrane, usually opposite to (against) their concentration gradient, using energy provided by ATP. Figure 3.13 shows how a carrier protein, called the *sodium-potassium pump* (Na^+/K^+ pump), moves three sodium ions and two potassium ions against their concentration gradients. The action of this pump causes a sodium gradient from outside to inside the cell and a potassium gradient from the inside of the cell to the outside. The gradients established are highly important in the overall functioning of the entire human body.

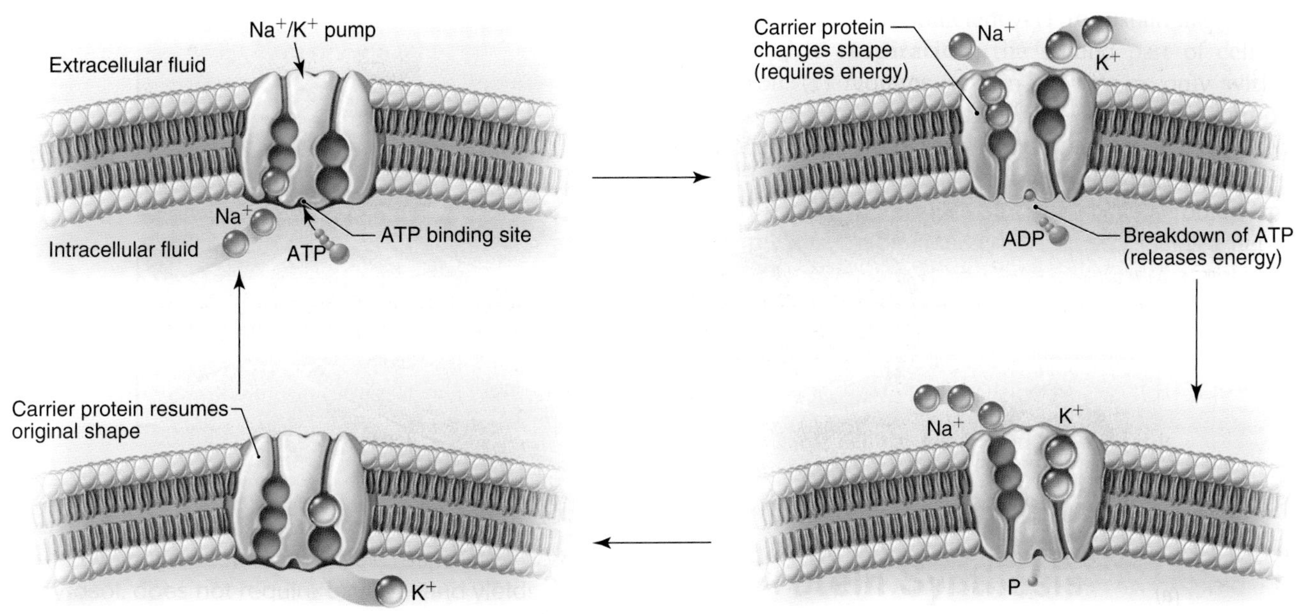

Figure 3.13 Carrier-mediated active transport. Sodium and potassium ions are moved across the plasma membrane against the concentration gradient by carrier-mediated active transport. APR

Proteins play a vital role in the body. Structural proteins compose significant portions of all cells, and functional proteins, such as enzymes and hormones, directly regulate cellular activities. Remember that a protein is formed of a long chain of amino acids joined together by peptide bonds. Protein synthesis involves placing a specific amino acid in the correct position in the amino acid chain.

DNA and RNA are intimately involved in the synthesis of proteins.

The Role of DNA

Recall the structure of DNA described in chapter 2. The two coiled strands of nucleotides are joined by hydrogen bonds between the nucleotide bases in each strand by *complementary base pairing.* Adenine (A) pairs with thymine (T), and cytosine (C) pairs with guanine (G).

The sequence of bases in a DNA molecule encodes information that determines the sequence of amino acids in a protein. More specifically, a sequence of three nucleotide bases (a triplet) in DNA encodes for a specific amino acid. For example, a sequence of ACA encodes for the amino acid cysteine, while AGG encodes for serine. In this way, inherited information that determines the structure of proteins is encoded in DNA.

The Role of RNA

In contrast to DNA, RNA consists of a single strand of nucleotides. Each nucleotide contains one of four nitrogenous bases: adenine, cytosine, guanine, or *uracil* (U). Note that uracil is present in RNA instead of thymine, which occurs in DNA. RNA is synthesized in a cell's nucleus by using a strand of DNA as a template. Complementary pairing of RNA bases with DNA bases produces a strand of RNA nucleotides whose bases are complementary to those in the DNA molecule. Uracil (U) in RNA pairs with adenine (A) in DNA; adenine (A) in RNA pairs with thymine (T) in DNA.

There are three types of RNA, and each plays a vital role in protein synthesis.

Messenger RNA (mRNA) carries the genetic information from DNA into the cytoplasm to the ribosomes, the sites of protein synthesis. This information is carried by the sequence of bases in mRNA, which is complementary to the sequence of bases in the DNA template.

Ribosomal RNA (rRNA) and protein compose ribosomes, the sites of protein synthesis. Ribosomes contain the enzymes required for protein synthesis.

Transfer RNA (tRNA) carries amino acids to the ribosomes, where the amino acids are joined like a string of beads to form a protein. There is a different tRNA for transporting each of the 20 kinds of amino acids used to build proteins.

Table 3.3 summarizes the characteristics of DNA and RNA.

Table 3.3 Distinguishing Characteristics of DNA and RNA

	DNA	RNA
Strands	Two strands joined by the complementary pairing of their nitrogenous bases	One strand
Sugar	Deoxyribose	Ribose
Bases	Adenine	Adenine
	Thymine	Uracil
	Cytosine	Cytosine
	Guanine	Guanine
Shape	Helix	Straight

Transcription and Translation

The process of protein synthesis involves two successive events: **transcription,** which occurs in the nucleus, and **translation,** which takes place in the cytoplasm.

In *transcription,* the sequence of bases in DNA determines the sequence of bases in mRNA due to complementary base pairing. Thus, transcription transfers the encoded information of DNA into the sequence of bases in mRNA. For example, if a triplet of DNA bases is AGG, which encodes for the amino acid serine, the complementary paired triplet of bases in mRNA is UCC. A triplet of bases in mRNA is known as a **codon,** and there is a codon for each of the 20 amino acids composing proteins. Messenger RNA consists of a chain of codons. Once it is synthesized, mRNA moves out of the nucleus into the cytoplasm where it combines with a ribosome, the site of protein synthesis. AP|R

In *translation,* the encoded information in mRNA is used to produce a specific sequence of amino acids to form the protein. As the ribosome moves along the mRNA strand, tRNA molecules bring amino acids to the ribosome and place them in the correct sequence in the forming polypeptide chain (protein) as specified by the mRNA codons. AP|R

Each tRNA molecule has a triplet of RNA bases called an **anticodon** at one end of the molecule. Because there are 20 different kinds of amino acids composing proteins, there are at least 20 kinds of tRNA whose anticodons can bind with codons of mRNA. A tRNA molecule can only transport the specific amino acid that is encoded by the codon to which its anticodon can bond. For example, a tRNA transporting the amino acid serine has the anticodon AGG that can bond with the mRNA codon UCC to place serine in the correct position in the forming amino acid chain. See figure 3.16.

By transcription and translation, DNA determines the structure of proteins, which, in turn, determines the functions of proteins. Transcription and translation may be summarized as follows:

$$\text{DNA} \xrightarrow[\textit{Transcription}]{} \text{mRNA} \xrightarrow[\textit{Translation}]{} \text{Protein}$$

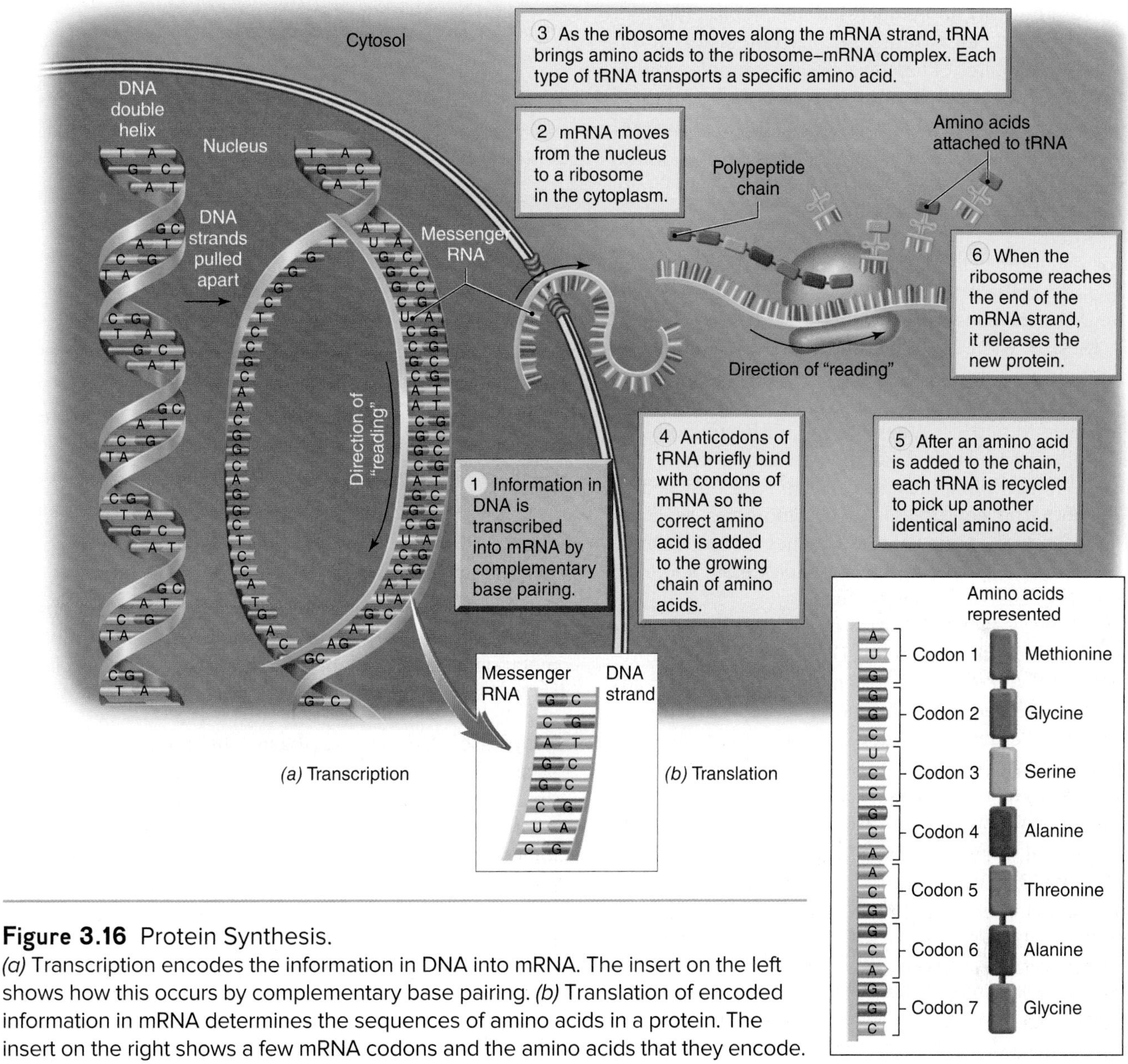

Cytosol

DNA double helix

Nucleus

DNA strands pulled apart

Messenger RNA

Direction of "reading"

1 Information in DNA is transcribed into mRNA by complementary base pairing.

2 mRNA moves from the nucleus to a ribosome in the cytoplasm.

3 As the ribosome moves along the mRNA strand, tRNA brings amino acids to the ribosome–mRNA complex. Each type of tRNA transports a specific amino acid.

4 Anticodons of tRNA briefly bind with condons of mRNA so the correct amino acid is added to the growing chain of amino acids.

5 After an amino acid is added to the chain, each tRNA is recycled to pick up another identical amino acid.

6 When the ribosome reaches the end of the mRNA strand, it releases the new protein.

Polypeptide chain

Amino acids attached to tRNA

Direction of "reading"

Messenger RNA | DNA strand

(a) Transcription

(b) Translation

Amino acids represented

AUG	Codon 1	Methionine
GGC	Codon 2	Glycine
UCC	Codon 3	Serine
GCA	Codon 4	Alanine
ACG	Codon 5	Threonine
GCA	Codon 6	Alanine
GGC	Codon 7	Glycine

Figure 3.16 Protein Synthesis.
(a) Transcription encodes the information in DNA into mRNA. The insert on the left shows how this occurs by complementary base pairing. *(b)* Translation of encoded information in mRNA determines the sequences of amino acids in a protein. The insert on the right shows a few mRNA codons and the amino acids that they encode.

Check My Understanding

5. How does chromosomal DNA determine the structure of proteins?

3.5 Cell Division

Learning Objectives

8. Describe the two types of cell division and their roles.
9. Describe each phase of mitosis.

Cells replicate themselves through a process called **cell division.** Two types of cell division occur in the body: mitotic cell division and meiotic cell division. Somatic cells (cells other than sex cells) divide by **mitotic** (mī-tot′-ik) **cell division,** during which a parent cell divides to form two new daughter cells that have the same number (46) and composition of chromosomes as the parent cell. It enables growth and the repair of tissues. **Meiotic** (mī-ot′-ik) **cell division** occurs only in the production of ova and sperm. In meiosis, a single parent cell divides to form four daughter cells that contain only half the number of chromosomes (23) found in the parent cell. In this chapter, we consider mitotic cell division only. Meiotic cell division is studied in chapter 17.

Mitotic Cell Division

Starting with the first division of the fertilized egg, mitotic cell division is the process that produces new cells for growth of the new individual and the replacement of worn or damaged cells. Mitotic cell division occurs at different rates in different kinds of cells. For example, epithelial cells undergo almost continuous division but muscle cells lose the ability to divide as they mature. Three processes are involved in mitotic cell division: (1) replication (production of exact copies) of chromosomes, (2) mitosis, and (3) division of the cytoplasm.

In dividing cells, the time period from the separation of daughter cells of one division to the separation of daughter cells of the next division is called the **cell cycle. Mitosis** constitutes only 5% to 10% of the cell cycle. Most of the time, a cell merely is carrying out its normal functions (figure 3.17).

Interphase is defined as the phase when the cell is not involved in mitosis. When viewed with a microscope, a cell in interphase is identified by its intact nucleus containing chromatin granules. In cells that are destined to divide, both the centrioles and chromosomes replicate during interphase, while other organelles are synthesized and assembled. There is a growth period before and after replication of the 46 chromosomes.

A chromosome consists of a very long DNA molecule coated with proteins. During interphase, chromosomes are uncoiled and resemble very thin threads within the cell nucleus. Chromosomes replicate during interphase in order to provide one copy of each chromosome for each of the two daughter cells that will be formed by mitotic cell division. Chromosome replication is dependent upon the replication of the DNA molecule in each chromosome. Figure 3.18 illustrates the process of DNA replication.

The two original DNA strands "unzip," and new nucleotides are joined in a complementary manner by their bases to the bases of the separated DNA strands. When the new nucleotides are in place and joined together, each new DNA molecule consists of one "new" strand of nucleotides joined to one "old" strand of nucleotides. In this way, a DNA molecule is precisely replicated so that both new DNA molecules are identical.

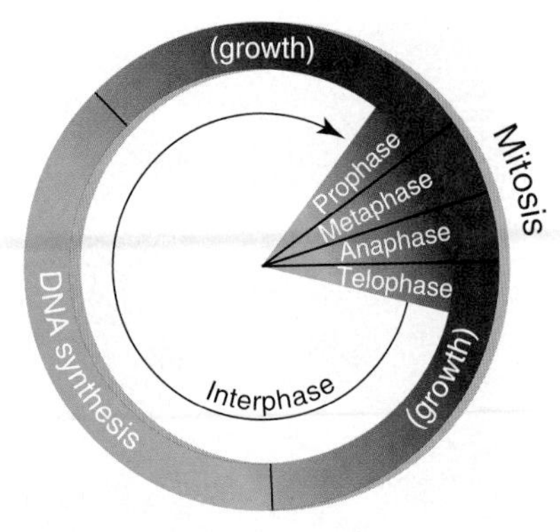

Figure 3.17 The Cell Cycle.
Interphase occupies most of the cell cycle. Only 5% to 10% of the time is used in mitosis. DNA and chromosome replication occur during interphase. AP|R

Mitotic Phases

Once it begins, mitosis is a continuous process that is arbitrarily divided into four sequential phases: prophase, metaphase, anaphase, and telophase. Each phase is characterized by specific events that occur.

Prophase During **prophase,** the replicated chromosomes coil, appearing first as threadlike structures and finally shortening sufficiently to become rod-shaped. Each replicated chromosome consists of two **chromatids** joined at their **centromeres.** Simultaneously, the nuclear envelope gradually disappears, and each pair of centrioles migrate toward opposite ends of the cell. A **spindle** is formed between the migrating centrioles. The spindle consists of spindle fibers that are formed of microtubules (figure 3.19*a*).

Metaphase During the brief **metaphase,** the replicated chromosomes line up at the equator of the spindle. The centromeres are attached to spindle fibers (figure 3.19*b*).

Clinical Insight

Mitotic cell division is normally a controlled process that ceases when it is not necessary to produce additional cells. Occasionally, control is lost and cells undergo continuous division, which leads to the formation of tumors. Tumors may be benign or malignant. *Benign tumors* do not spread to other parts of the body and may be surgically removed if they cause health or cosmetic problems. *Malignant tumors,* or *cancers,* may spread to other parts of the body by a process called *metastasis* (me-tas′-ta-sis). Cells break away from the primary tumor and are often carried by blood or lymph to other areas, where continued cell divisions form secondary tumors.

Treatment of malignant tumors involves surgical removal of the tumor, if possible, and subsequent chemotherapy and/or radiation therapy. Both chemotherapy and radiation therapy tend to kill malignant cells because dividing cells are more sensitive to treatment, and malignant cells are constantly dividing.

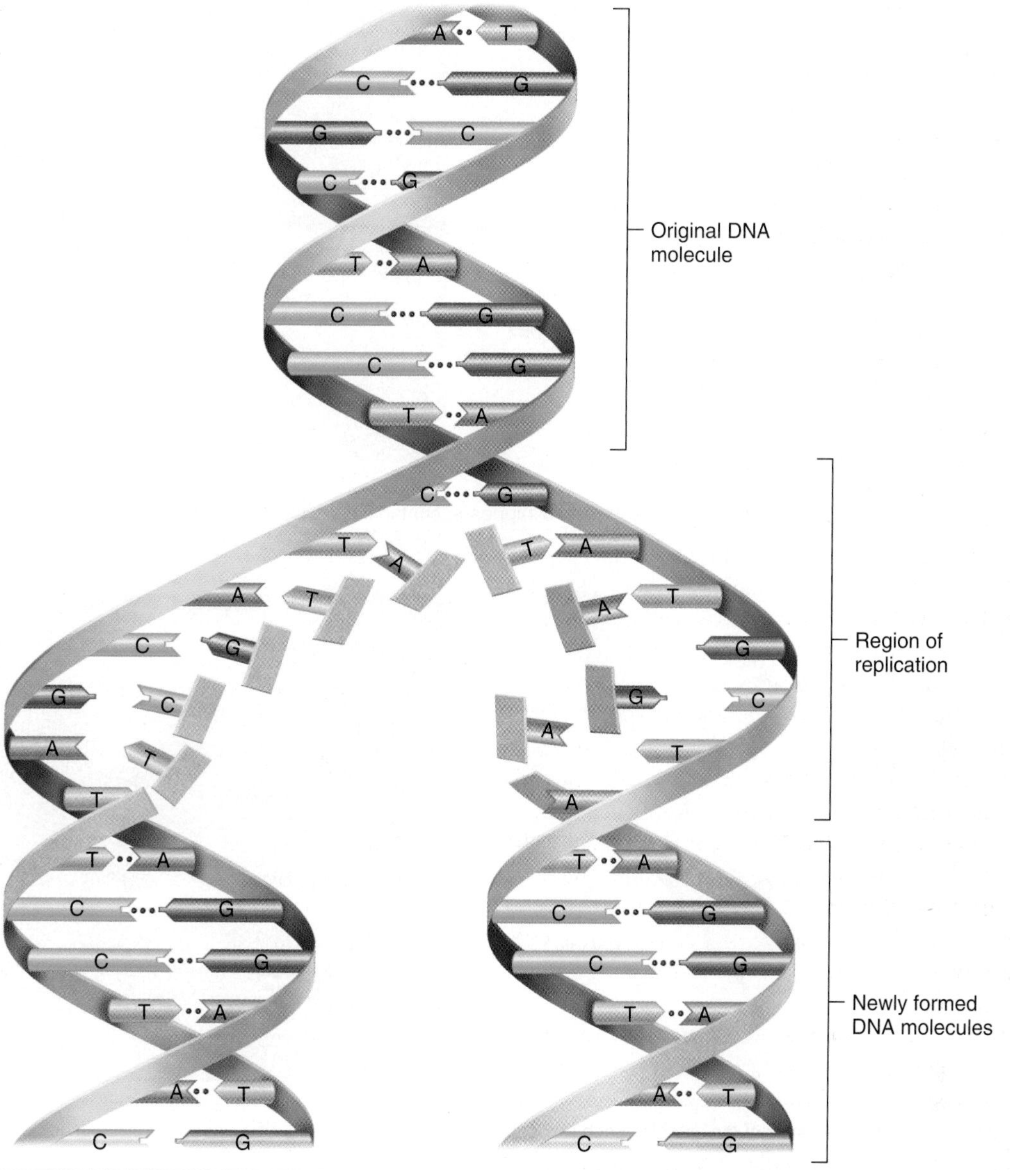

Original DNA molecule

Region of replication

Newly formed DNA molecules

Figure 3.18 When a DNA molecule replicates, the two strands unzip. Then a new complementary strand of nucleotides forms along each "old" strand to produce two new DNA molecules. APR

Anaphase During **anaphase,** separation of the centromeres results in the separation of the paired chromatids. The members of each pair are pulled by spindle fibers towards opposite sides of the cell. The separated chromatids are now called chromosomes, and each new set of chromosomes is identical (figure 3.19*c*).

Telophase During **telophase,** the spindle fibers disassemble and a new nuclear envelope starts forming around each set of chromosomes as the new nuclei begin to take shape. The chromosomes start to uncoil, and they will ultimately become visible only as chromatin granules. The new daughter nuclei are completely formed by the end of telophase.

Usually during late anaphase and telophase, the most obvious change is the division of the cytoplasm, which is called **cytokinesis** (si″-to-ki-nē′-sis). It is characterized by a furrow that forms in the plasma membrane across the equator of the spindle and deepens until the parent cell is separated into two daughter cells. The formation of two daughter cells, each having identical chromosomes in the nuclei, marks the end of mitotic cell division (figure 3.19*d*).

6. What are the phases of mitosis and how is each phase distinguished?

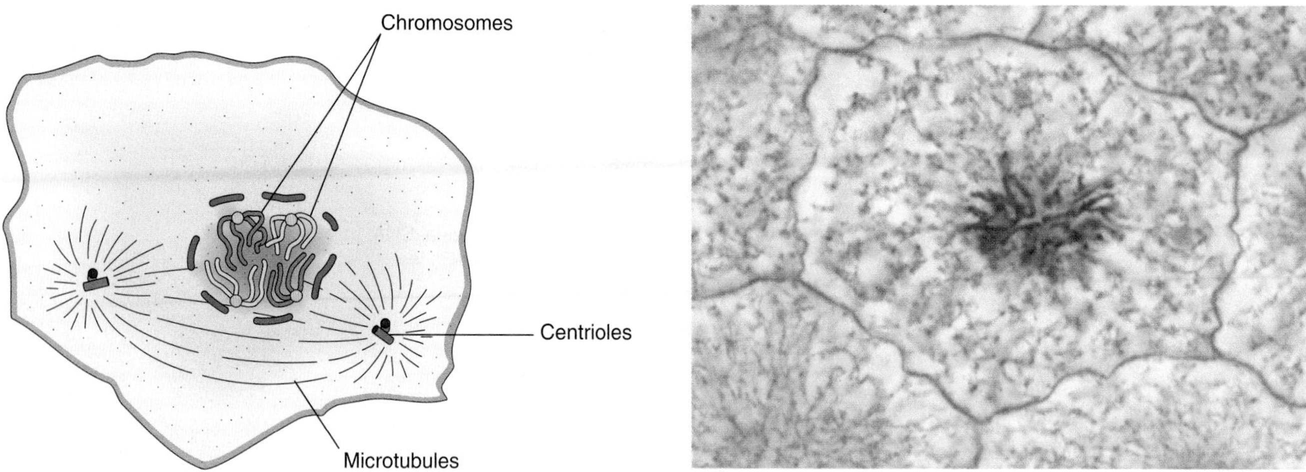

(a) Prophase
Replicated chromosomes coil and shorten; nuclear envelope disappears; centriole pairs move toward opposite sides of the cell, forming a spindle between them.

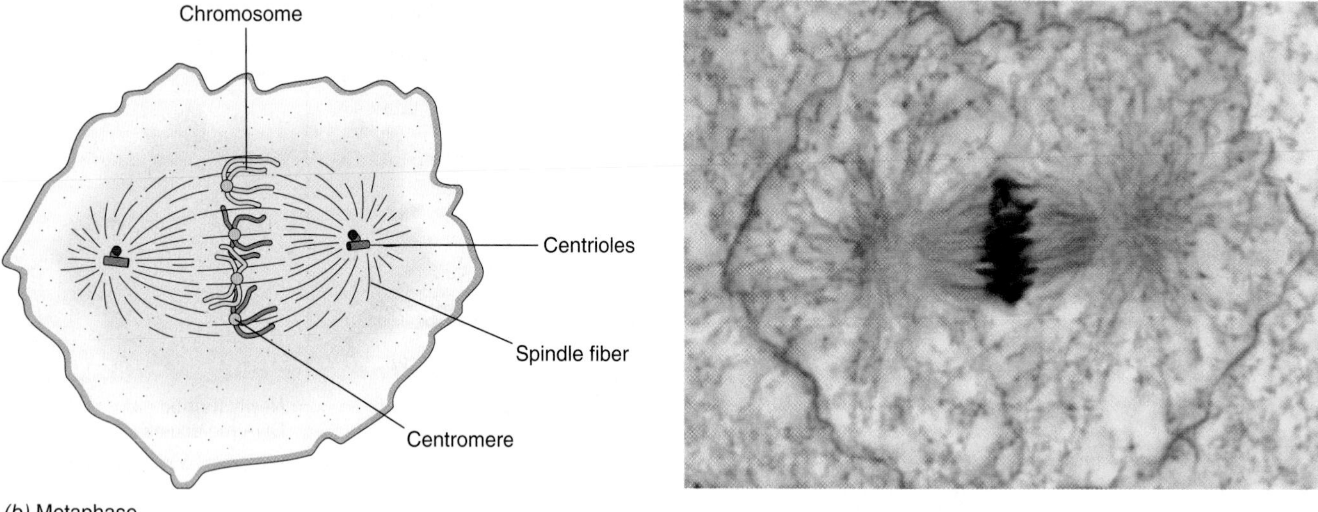

(b) Metaphase
Chromosomes line up at the equator of the spindle. Each replicated chromosome consists of a pair of chromatids joined by centromere.

Figure 3.19 Drawings and photomicrographs (1,000×) of mitosis.

Chapter Summary

3.1 Cell Structure

- The plasma membrane is composed of a double layer of phospholipid molecules along with associated cholesterol and protein molecules. It is selectively permeable and controls the movement of the materials into and out of cells.
- The cytoplasm, which is composed of cytosol and organelles, lies external to the nucleus and is enveloped by the plasma membrane.
- The nucleus is a large, spherical organelle surrounded by the nuclear envelope.
- Chromosomes, composed of DNA and protein, are found in the nucleus. The uncoiled chromosomes appear as chromatin granules in nondividing cells.
- The nucleolus is the site of ribosome synthesis.
- Ribosomes are tiny organelles formed of rRNA and protein. They are sites of protein synthesis.
- The endoplasmic reticulum (ER) consists of membranes that form channels for transport of materials within the cell. RER is studded with ribosomes that synthesize proteins for export from the cell. SER lacks ribosomes and is involved in lipid synthesis.

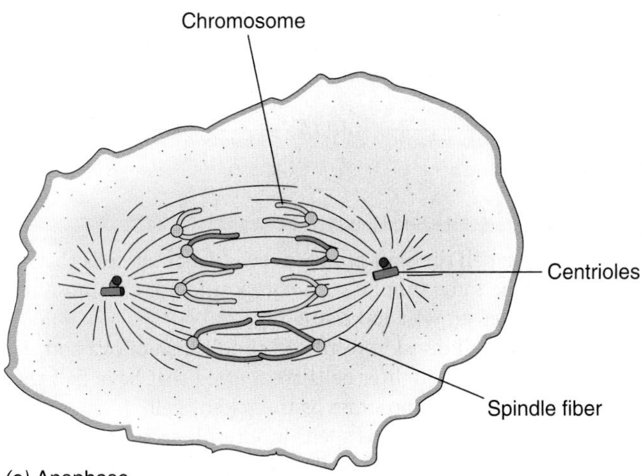

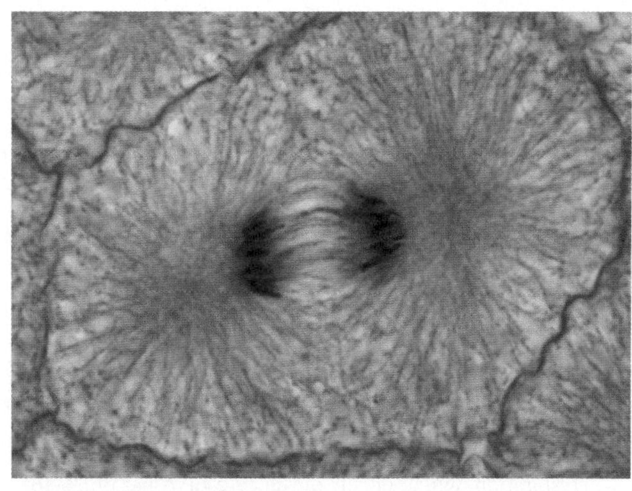

(c) Anaphase
Chromatids separate, members of each chromatid pair move toward opposite ends of the spindle.

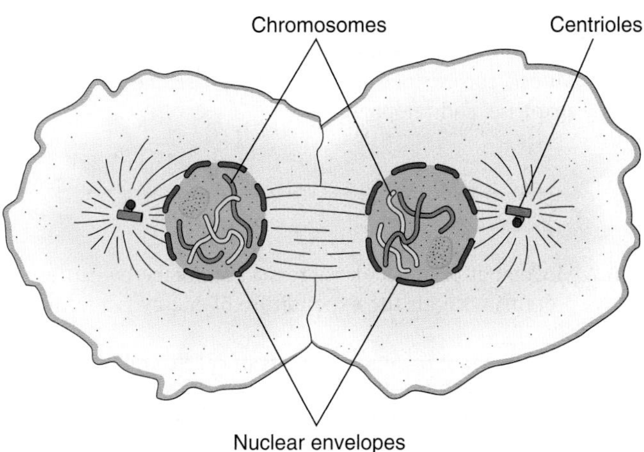

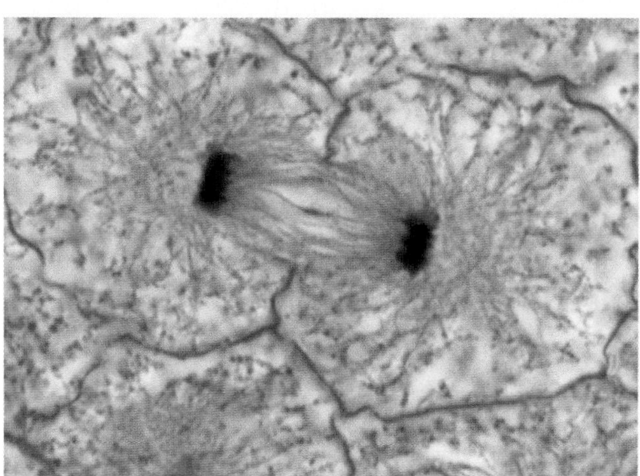

(d) Telophase
Nuclear envelopes form around each set of chromosomes; spindle fibers disappear; chromosomes uncoil and extend; cytokinesis produces two daughter cells.

- The Golgi complex packages materials into vesicles for secretion from the cell or transport within the cell.
- Mitochondria are large, double-membraned organelles within which aerobic respiration occurs.
- Lysosomes are small vesicles that contain digestive enzymes used to digest foreign particles, worn-out parts of a cell, or an entire damaged cell.
- The cytoskeleton is formed by microtubules and microfilaments, and is used in maintaining cell structure and cell movement.
- A pair of centrioles, used in cell division, is present near the cell's nucleus. The wall of each centriole is composed of microtubules arranged in groups of three.
- Cilia are short, hairlike projections on the free surface of certain cells. The beating of cilia moves materials along the cell surface.
- Each sperm swims by the beating of a flagellum, a long, whiplike organelle.

3.2 Transport Across Plasma Membranes

- Passive transport does not require the expenditure of energy by the cell.
- Diffusion is the movement of substances from an area of higher concentration to an area of lower concentration. It is caused by the constant motion of substances in gases and liquids.
- Substances diffuse across plasma membrane by simple diffusion, channel-mediated diffusion, and carrier-mediated diffusion.
- Osmosis is the passive movement of water across a selectively permeable membrane.
- Hypotonic solutions have a higher water concentration than the cells. Hypertonic solutions have a lower water concentration than the cells. Isotonic solutions have the same water concentration as the cells.

- Cells in hypotonic solutions have a net gain of water. Cells in hypertonic solutions have a net loss of water. Cells in isotonic solutions have no net change of water content.
- Active transport requires the cell to expend energy.
- Active transport mechanisms include carrier-mediated active transport, endocytosis, and exocytosis.

3.3 Cellular Respiration

- Cellular respiration is the breakdown of nutrients in cells to release energy and form ATP molecules, which power cellular processes.
- Cellular respiration of glucose involves anaerobic respiration and aerobic respiration.
- Cellular respiration of a glucose molecule yields a net of 36–38 ATP. A net of 2 ATP is produced during anaerobic respiration, which occurs in the cytosol. A net of 34–36 ATP is produced during aerobic respiration, which occurs in mitochondria.

3.4 Protein Synthesis

- Protein synthesis involves the interaction of DNA, mRNA, rRNA, and tRNA.
- The sequence of bases in DNA determines the sequence of codons in mRNA, which, in turn, determines the sequence of amino acids in a protein.

$$\text{DNA} \xrightarrow[\text{Transcription}]{} \text{mRNA} \xrightarrow[\text{Translation}]{} \text{Protein}$$

3.5 Cell Division

- Mitotic cell division produces two daughter cells that have the same number and composition of chromosomes. It enables growth and tissue repair.
- Meiotic cell division results in production of ova and sperm. Four daughter cells are formed that have half the number of chromosomes as the parent cell.
- Most of a cell cycle is spent in interphase, where cells carry out normal metabolic functions. In cells destined to divide, chromosomes and centrioles are replicated in interphase.
- After chromosome replication, mitosis is the orderly process of separating and distributing chromosomes equally to the daughter cells.
- Mitosis consists of four phases: prophase, metaphase, anaphase, and telophase.

Self-Review

Answers are located in Appendix B.

1. Movement of materials in and out of cells is controlled by the ______.
2. Molecules of ______ located in chromosomes control the activities of cells.
3. Aerobic respiration occurs within ______.
4. The sites of protein synthesis are ______.
5. The ______ assembles protein and RNA to form ribosomes.
6. The ______ consists of intracellular membranous channels for material transport.
7. Movement of molecules from an area of their higher concentration to an area of their lower concentration is known as ______.
8. Movement of molecules across a membrane by carrier proteins without the expenditure of energy is a form of ______.
9. Breakdown of organic nutrients in cells to release energy and form ATP is called ______.
10. Instructions for synthesizing a protein are carried from DNA to ribosomes by ______.
11. The equal distribution of chromosomes to daughter nuclei occurs by ______.

Critical Thinking

1. How do the characteristics of a substance determine the transport mechanism that will be used to move it across the plasma membrane?
2. How is the correct sequence of amino acids in proteins determined?
3. How are glucose, pyruvic acid, mitochondria, oxygen, ADP, and ATP involved in cellular respiration?
4. How does mitotic cell division yield daughter cells with the same DNA content?

ADDITIONAL RESOURCES

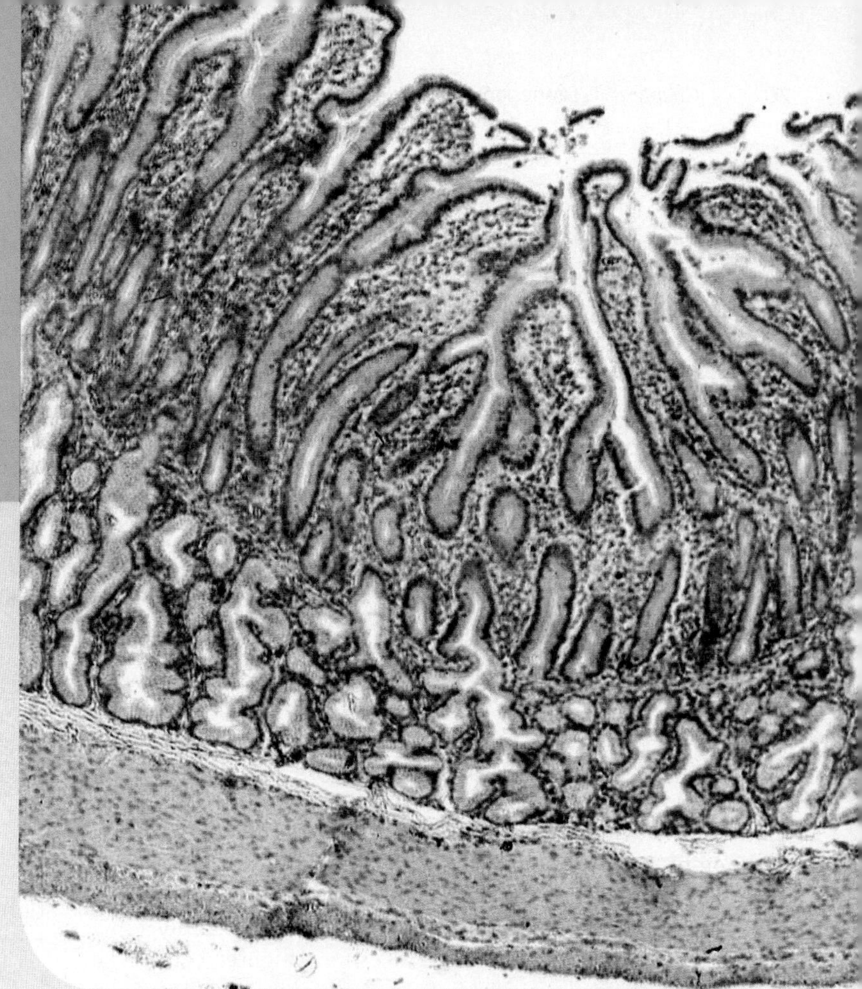

4 CHAPTER

Tissues and Membranes

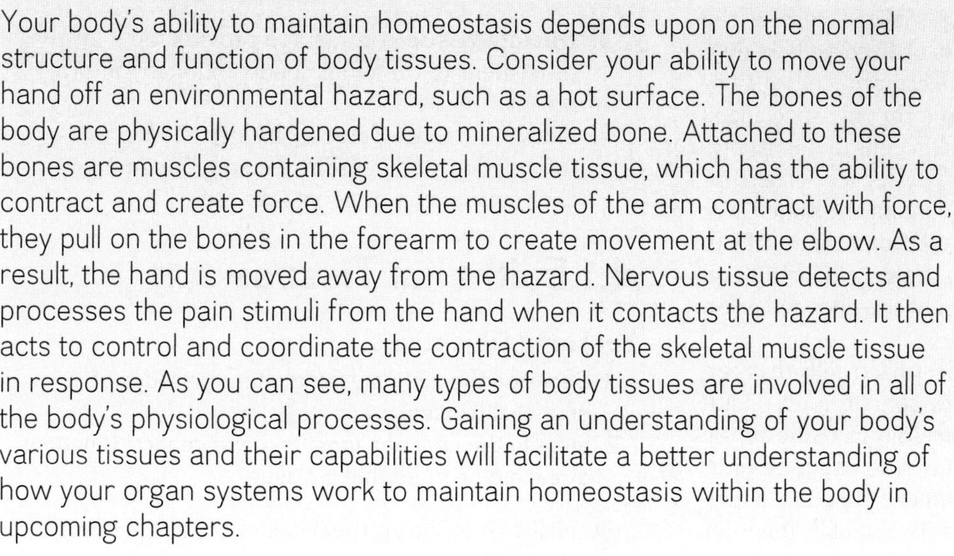

Your body's ability to maintain homeostasis depends upon on the normal structure and function of body tissues. Consider your ability to move your hand off an environmental hazard, such as a hot surface. The bones of the body are physically hardened due to mineralized bone. Attached to these bones are muscles containing skeletal muscle tissue, which has the ability to contract and create force. When the muscles of the arm contract with force, they pull on the bones in the forearm to create movement at the elbow. As a result, the hand is moved away from the hazard. Nervous tissue detects and processes the pain stimuli from the hand when it contacts the hazard. It then acts to control and coordinate the contraction of the skeletal muscle tissue in response. As you can see, many types of body tissues are involved in all of the body's physiological processes. Gaining an understanding of your body's various tissues and their capabilities will facilitate a better understanding of how your organ systems work to maintain homeostasis within the body in upcoming chapters.

CHAPTER OUTLINE

Module 3
Tissues

SELECTED KEY TERMS

Adipose tissue (adip = fat) A connective tissue that stores fat.
Bone A hard connective tissue with a rigid matrix of calcium salts and fibers.
Cartilage A connective tissue with a relatively rigid, semisolid matrix.
Connective tissue (connect = to join) A tissue that binds other tissues together.
Epithelial tissue (epi = upon, over; thel = delicate) A thin tissue that covers body and organ surfaces and lines body cavities, and forms secretory portions of glands; epithelium.
Fibroblast (fibro = fiber; blast = germ) A cell that produces fibers and ground substance in connective tissue.
Matrix The extracellular substance in connective tissue.
Mucous membrane Epithelial membrane that lines tubes and cavities that have openings to the external environment.
Muscle tissue (mus = mouse) A tissue whose cells are specialized for contraction.
Nervous tissue A tissue that forms the brain, spinal cord, and nerves.
Serous membrane Epithelial membrane that lines the external surfaces of organs and the body wall in the ventral cavity.
Tissue (tissu = woven) A group of similar cells performing similar functions.

THE DIFFERENT KINDS OF CELLS COMPOSING the human body result from the specialization of cells during embryonic development. **Embryonic stem cells** of an early embryo are unspecialized cells containing encoded information in their DNA that enables them to form all types of specialized cells. As these cells divide repeatedly producing many generations of cells, the daughter cells become partially specialized. Such cells can produce daughter cells for only certain related types of specialized cells. This trend of decreasing potential (increasing specialization) continues through many generations of cells, ultimately producing the highly specialized cells of the human body plus a few partially specialized cells known as **adult stem cells.** Once fully specialized, cells may or may not divide. If they do, they can form only specialized cells like themselves; for example, skin cells divide to produce only skin cells. Because all of a person's cells (except red blood cells, which lack nuclei) contain the same DNA, the transition from unspecialized embryonic stem cells to fully specialized body cells results from cellular mechanisms that turn off specific portions of the encoded information in DNA.

A **tissue** is a group of fully specialized cells that perform similar functions. Most tissues contain a few adult stem cells, which play an important role in tissue repair. Each type of tissue is distinguished by the structure of its cells, its extracellular substance, and the function it performs. The structure of a tissue is a reflection of its function.

The different tissues of the body are classified into four basic types: epithelial, connective, muscle, and nervous tissues.

1. **Epithelial** (ep″-i-thē′-lē-al) **tissue** covers the surfaces of the body, lines body cavities and covers organs, and forms the secretory portions of glands.
2. **Connective tissue** binds organs together and provides protection and support for organs and the entire body.

Clinical Insight

Adult stem cells from a variety of tissues are used in medical therapies. For example, stem cells in red bone marrow are used to treat leukemia. Medical scientists think that, with more research, stem cells may be used to treat cancer, brain and spinal cord injuries, multiple sclerosis, Parkinson disease, and other injuries and disorders.

3. **Muscle tissue** contracts to provide force for the movement of the whole body and many internal organs.
4. **Nervous tissue** detects changes, processes information, and coordinates body functions via the transmission of nerve impulses.

4.1 Epithelial Tissues AP|R

Learning Objectives

1. Describe the distinguishing characteristics of epithelial tissues.
2. Identify the common locations and general functions of each type of epithelial tissue.

Epithelial tissues, or epithelia (singular, epithelium), may be composed of one or more layers of cells. The number of cell layers and the shape of the cells provide the basis for classifying epithelial tissues (figures 4.1 and 4.2). Epithelial tissues are distinguished by the following five characteristics:

1. Epithelial cells are packed closely together with very little extracellular material between them.
2. The sheetlike tissue is firmly attached to the deeper connective tissue by a thin layer of

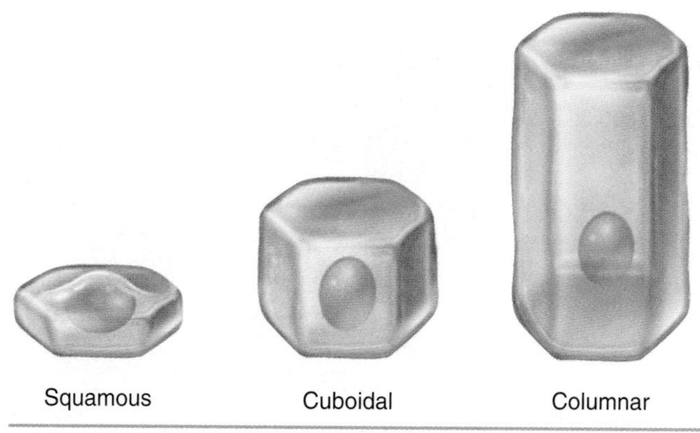

Figure 4.1 Classes of epithelium based on cell shape.

proteins and carbohydrates called the **basement membrane.**

3. The surface of the tissue (free surface) opposite the basement membrane is not attached to any other type of tissue and is located on a surface or next to an opening.
4. Blood vessels are absent, so epithelial cells must rely on diffusion to receive nourishment from blood vessels in the deeper connective tissue. Because these tissues are on surfaces, they are prone to damage. The lack of blood vessels prevents unnecessary bleeding.
5. Epithelial tissues regenerate rapidly by mitotic cell division of the cells. Large numbers of epithelial cells are destroyed and replaced each day.

The functions of epithelial tissues vary with the specific location and type of tissue, but generally they include *protection, diffusion, osmosis, absorption, filtration,* and *secretion.* Certain epithelial cells form *glandular epithelium,* the cells in glands that produce secretions. Two basic types of glands are contained in the body: exocrine and endocrine glands. **Exocrine glands** (exo = outside of; crin = to secrete) have ducts (small tubes) that carry their secretions to specific areas; sweat glands and salivary glands are examples. **Endocrine glands** (endo = within) lack ducts. Their secretions, called hormones, are carried by the blood supply to organs within the body to regulate their function. The thyroid gland and adrenal glands are examples of endocrine glands. Endocrine glands and their hormones will be discussed in more detail in chapter 10.

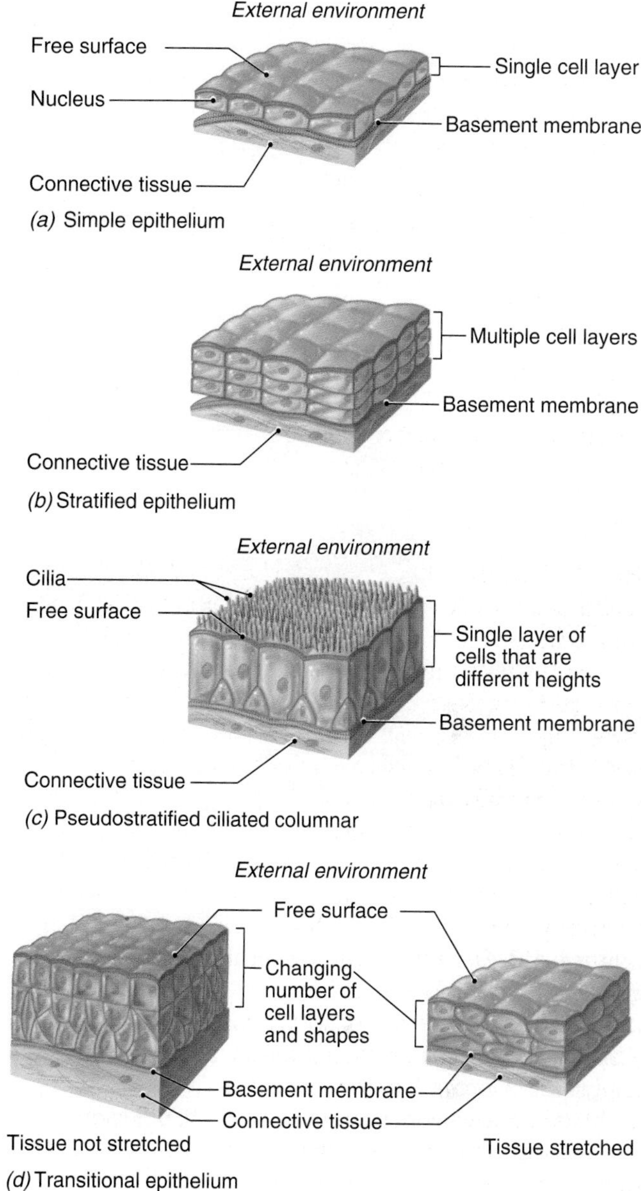

Figure 4.2 Classes of epithelium based on the number of cell layers.

Simple Epithelium

Simple epithelium consists of a single layer of cells that may be flat (squamous), cube-like (cuboidal), and column-like (columnar) in shape (figures 4.1 and 4.2). These tissues are located where rapid diffusion, secretion, or filtration occur in the body.

Simple squamous (skwā′-mus) *epithelium* consists of thin, flat cells that have an irregular outline and a flat, centrally located nucleus. In a surface view, the cells somewhat resemble tiles arranged in a mosaic pattern. Simple squamous epithelium performs a diverse set of functions that include diffusion, osmosis, filtration, secretion, absorption, and friction reduction. Its locations in the body include (1) the air sacs in the lungs, where O_2 and CO_2 diffuse into and out of the blood, respectively; (2) special structures in the kidney called *glomeruli* (glo-mer′u-li), where blood is filtered during urine production (see chapter 16); (3) the **mesothelium,** which is part of the serous membranes lining the ventral cavity; (4) and the **endothelium,** which lines the internal surfaces of the heart, blood vessels, and lymphatic vessels (figure 4.3).

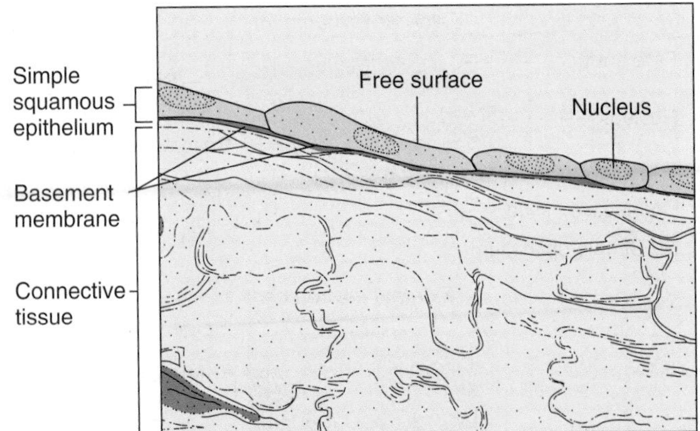

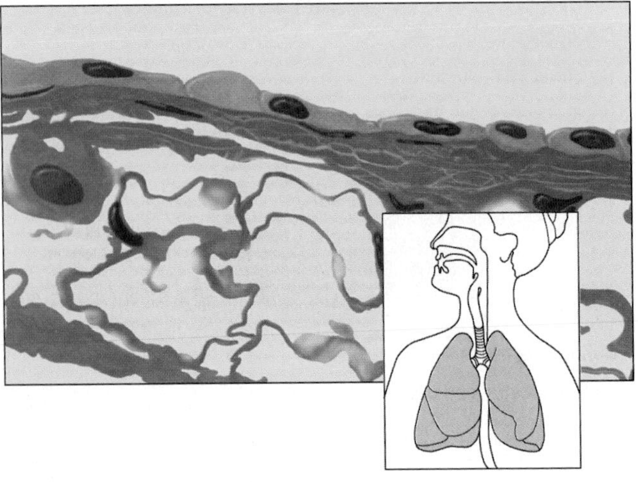

Figure 4.3 Simple Squamous Epithelium (250×). APR
Structure: A single layer of squamous cells.
Location: Endothelium, mesothelium, air sacs of the lungs, and glomeruli of the kidneys.
Function: Absorption, secretion, filtration, diffusion, osmosis, and friction reduction.

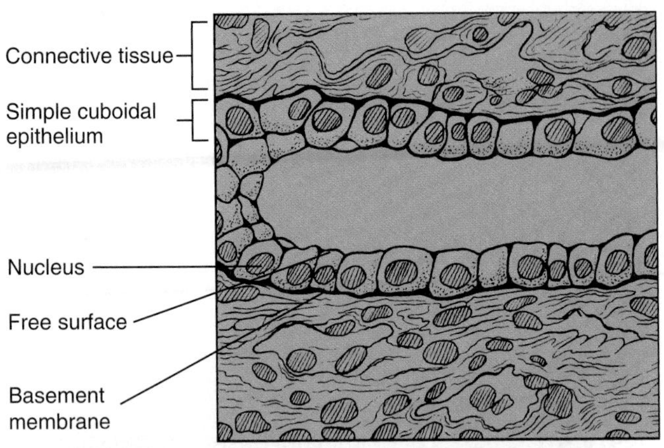

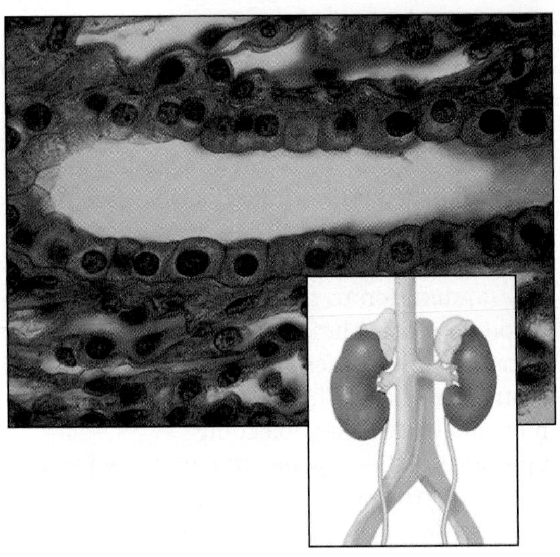

Figure 4.4 Simple Cuboidal Epithelium (250×). APR
Structure: A single layer of cuboidal cells.
Location: Forms kidney tubules, secretory portion of some glands, and the superficial layer of the ovaries.
Function: Absorption and secretion.

Simple cuboidal epithelium consists of a single layer of cube-shaped cells. The cells have a single, round, centrally located nucleus. Its basic functions are absorption and secretion. Locations for simple cuboidal epithelia include (1) the secretory portion of glands, such as the thyroid and salivary glands; (2) the kidney tubules where secretion and reabsorption of materials occur; and (3) the superficial layer of the ovaries (figure 4.4).

Simple columnar epithelium consists of a single layer of elongated, columnar cells with oval nuclei usually located near the basement membrane. Scattered among the columnar cells are **goblet cells,** specialized mucus-secreting cells with a goblet or wine glass shape. Their purpose is to secrete a protective layer of mucus on the free surface of the epithelium. Secretion and absorption are the major functions of this tissue in areas such as the stomach and intestines. The cells lining the intestine possess numerous microvilli on their free surface, often called a "brush border" because of its bristle-like appearance, which greatly increases their absorptive surface area. In areas such as uterine tubes, paranasal sinuses, and ventricles of the brain, this tissue possesses cilia that allow for movement of materials across the tissue surface (figure 4.5).

Pseudostratified ciliated columnar epithelium consists of a single layer of cells. It is said to be **pseudostratified** (pseudo = false) because its structure creates a visual illusion of being multilayered but it really is a simple epithelium. The layered effect results in part from the nuclei being located at various levels within the cells. Also, even though all of the cells are attached to the basement membrane, not all of them reach the free surface (see figure 4.2). Just as in simple columnar epithelium, goblet cells are scattered throughout the tissue. This epithelium lines the internal surfaces of many of the respiratory passageways, where it collects and removes airborne

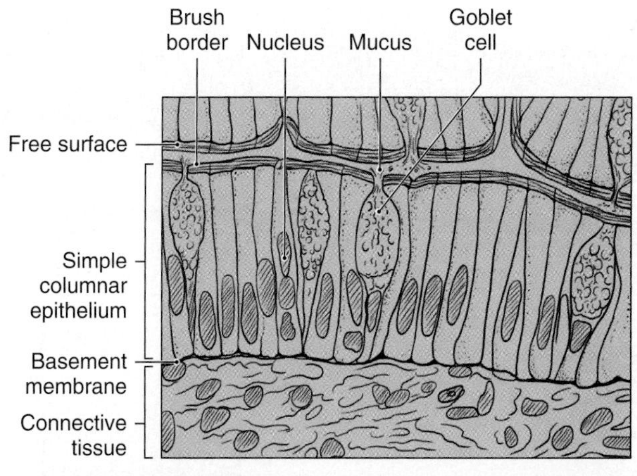

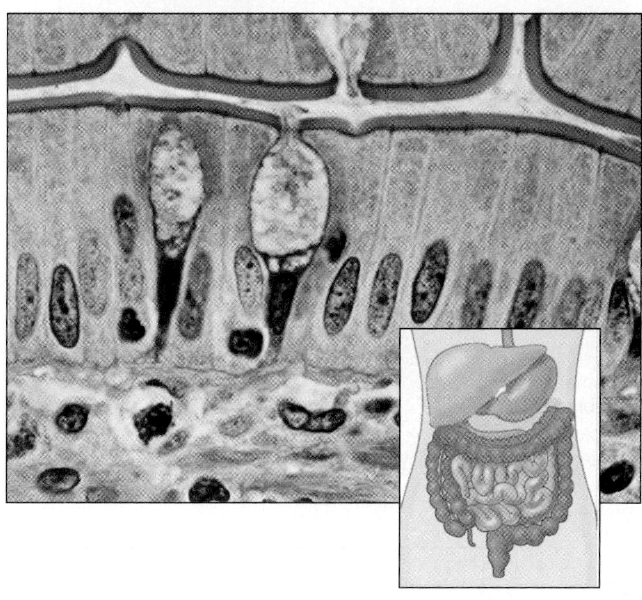

Figure 4.5 Simple Columnar Epithelium (400×). APR
Structure: A single layer of columnar cells; contains scattered goblet cells.
Location: Lines the internal surfaces of the stomach and intestines, the ducts of many glands, uterine tubes, paranasal sinuses, and ventricles in the brain.
Function: Absorption, secretion, and protection.

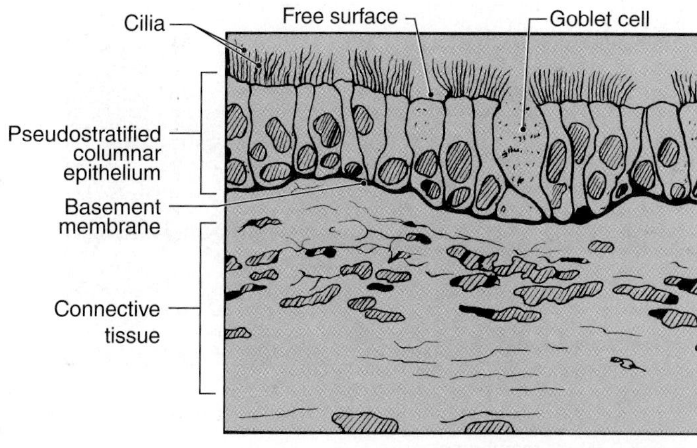

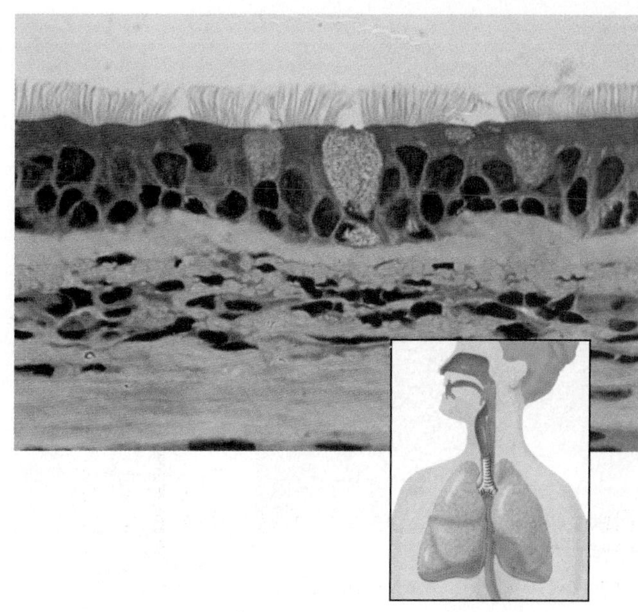

Figure 4.6 Pseudostratified Ciliated Columnar Epithelium (500×). APR
Structure: A single layer of ciliated columnar cells that appears to be more than one layer of cells; contains scattered goblet cells.
Location: Lines many respiratory passageways.
Function: Secretion of mucus; beating cilia remove secreted mucus and entrapped particles.

particles. The particles are trapped in the secreted mucus, which is moved by the beating cilia to the throat, where it is either swallowed or expectorated (figure 4.6).

Stratified Epithelium

Stratified epithelium consists of more than one layer of cells, which makes them more durable to abrasion (see figure 4.2). Only the deepest layer of cells produces new cells by mitotic cell division. The cells are pushed toward the free surface of the tissue as more new cells are formed deep to them. Cells in the superficial layer are continuously lost as they die and are rubbed off by abrasion. Protection of underlying tissues is an important function of stratified epithelia. These tissues are named according to the shape of cells on their free surfaces.

Stratified squamous epithelium occurs in two distinct forms: keratinized and nonkeratinized. The keratinized type forms the superficial layer (epidermis) of the skin. Its cells become impregnated with a waterproofing protein, **keratin** (ker′-ah-tin), as they migrate to the free surface of the tissue. This specific type of epithelium is discussed

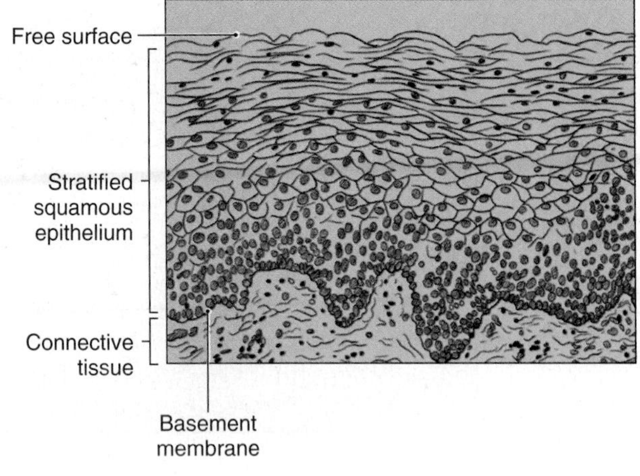

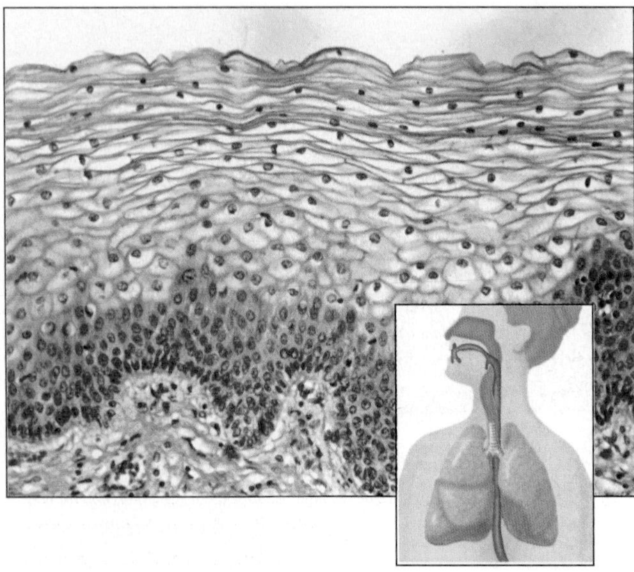

Figure 4.7 Stratified Squamous Epithelium (70×). APIR
Structure: Several cell layers; cells in the deepest layer are cuboidal in shape but gradually become flattened as they migrate to the surface of the tissue.
Location: The keratinized type forms the epidermis of the skin; nonkeratinized type lines the mouth, esophagus, vagina, and rectum.
Function: Protection.

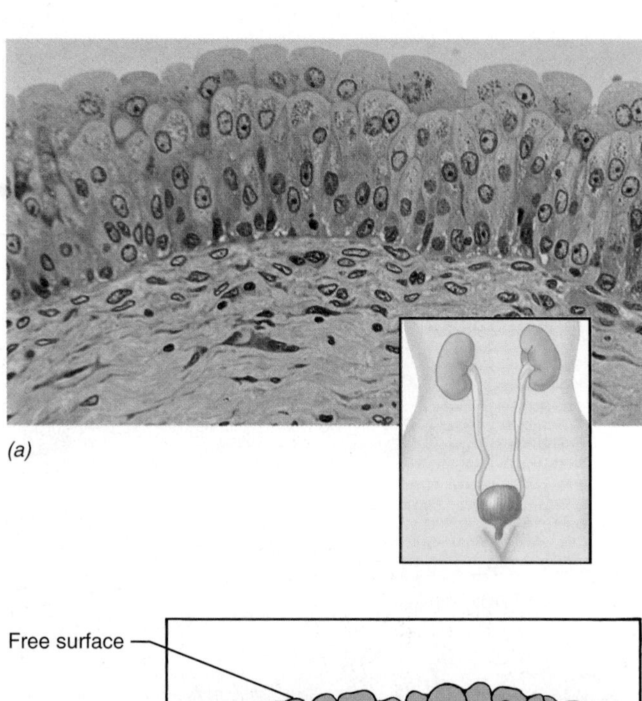

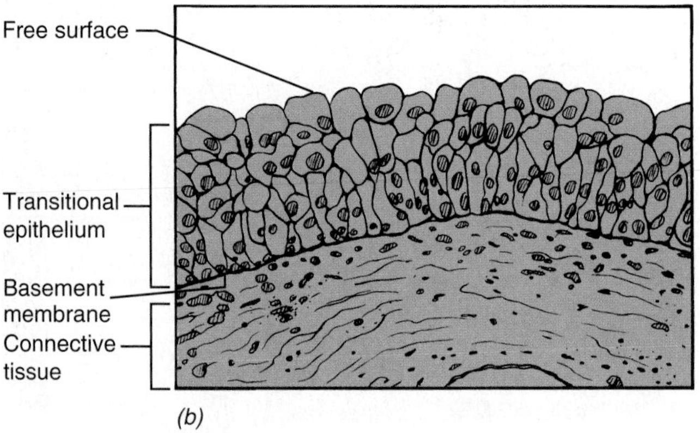

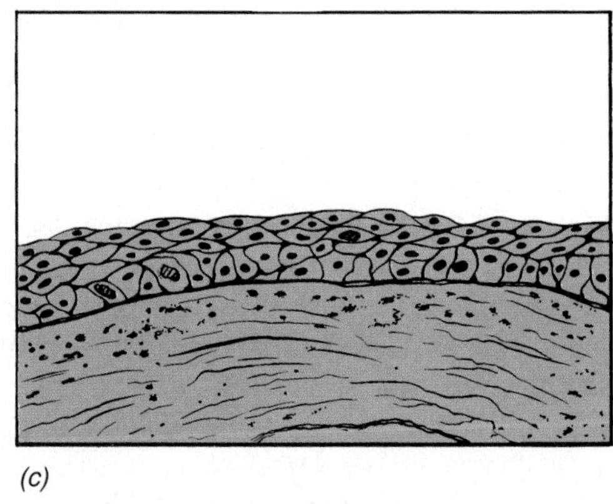

Figure 4.8 Transitional Epithelium. APIR
(a) A photomicrograph (250×) and *(b)* drawing showing several layers of rounded cells when the urinary bladder wall is contracted. *(c)* When the bladder wall is stretched, the tissue and cells become flattened.
Structure: Several layers of large, rounded cells that become flattened when stretched.
Location: Lines the internal surface of the urinary tract.
Function: Protection; permits stretching of the wall of the urinary tract.

further in chapter 5. The nonkeratinized type lines the mouth, esophagus, vagina, and rectum. Both types provide resistance to abrasion (figure 4.7).

Transitional epithelium lines most of the urinary tract and stretches as these structures fill with urine. It consists of multiple layers of cells, with the free surface cells of the unstretched tissue possessing a large and rounded shape. When stretched, the free surface cells become thin, flat cells resembling squamous epithelial cells (figure 4.8, see figure 4.2).

Two relatively rare types of stratified epithelial tissues are not shown. *Stratified cuboidal epithelium* lines larger ducts of certain glands (e.g., mammary and salivary glands). *Stratified columnar epithelium* lines parts of the pharynx and male urethra.

Check My Understanding

1. What are the general characteristics and functions of epithelial tissues?
2. How are the various epithelial tissues different in terms of structure, location, and function?

4.2 Connective Tissues AP|R

Learning Objectives

3. Describe the distinguishing characteristics of each type of connective tissue.
4. Identify the common locations and general functions of each type of connective tissue.

Connective tissues are the most widely distributed and abundant tissues in the body. As the name implies, connective tissues support and bind together other tissues so they are never found on exposed surfaces. Like epithelial cells, most connective tissue cells have retained the ability to reproduce by mitotic cell division.

Connective tissues consist of a diverse group of tissues that can be divided into three broad categories: (1) loose connective tissues, (2) dense connective tissues, and (3) connective tissues with specialized functions–cartilage, bone, blood, and lymph. Loose and dense connective tissues are sometimes referred to as "connective tissue proper" because they are common tissues that function to bind other tissues and organs together.

All connective tissues consist of relatively few, loosely arranged cells and a large amount of extracellular substance called **matrix** (ma'-triks). Matrix, which is produced by the cells, is used to classify connective tissues. It is composed of ground substance and protein fibers. **Ground substance,** which is composed of water and both inorganic and organic compounds, can be fluid, semifluid, gelatinous, or calcified.

Three types of protein fibers are found in the matrix of connective tissues. **Collagen fibers,** composed of collagen protein, are relatively large fibers resembling cords of a rope. They provide strength and flexibility but not elasticity. **Reticular fibers,** also made of collagen, are very thin and form highly branched, delicate, supporting frameworks for tissues. **Elastic fibers** are made of elastin protein and possess great elasticity, which means they can stretch up to 150% their resting length without damage and then recoil back to their resting length.

Clinical Insight

Because epithelial and connective tissue cells are active in cell division, they are prone to the formation of tumors when normal control of cell division is lost. The most common types of cancer arise from epithelial cells, possibly because these cells have the most direct contact with *carcinogens,* cancer-causing agents in the environment. A cancer derived from epithelial cells is called a *carcinoma.* Malignant tumors that originate in connective tissue are also common types of cancer. A cancer of connective tissue is called a *sarcoma.*

Loose Connective Tissue

Loose connective tissues help to bind together other tissues and form the basic supporting framework for organs. Their matrix consists of a semifluid or jelly-like *ground substance* in which fibers and cells are embedded. The word "loose" describes how the fibers are widely spaced and intertwined between the cells. **Fibroblasts** are the most common cells and they are responsible for producing the ground substance and protein fibers. There are three types of loose connective tissue: areolar connective tissue, adipose tissue, and reticular tissue.

Areolar Connective Tissue

Areolar (ah-rē′-ō-lar) **connective tissue** is the most abundant connective tissue in the body. Fibroblasts are the most numerous cells, but macrophages are present to help protect against invading pathogens (see chapters 11 and 13). A semifluid ground substance fills the spaces between the cells and fibers. Areolar connective tissue (1) attaches the skin to underlying muscles and bones as part of the subcutaneous tissue (see chapter 5); (2) provides a supporting framework for internal organs, nerves, and blood vessels; (3) is a site for many immune reactions; and (4) forms the superficial region of the dermis, which is the deep layer of the skin (figure 4.9).

Adipose Tissue

Large accumulations of fat cells, or **adipocytes**, form **adipose** (ad′-i-pōs) **tissue,** a special type of loose connective tissue. It occurs throughout the body but is more common deep to the skin, within the subcutaneous tissue, and around internal organs. Adipocytes are filled with fat droplets that push the nucleus and cytoplasm to the edge of the cells. In addition to fat storage, adipose tissue serves as a protective cushion for internal organs, especially around the kidneys and posterior to the eyeballs. It also helps to insulate the body from abrupt temperature changes and, as part of the subcutaneous tissue, to attach skin to underlying bone and muscle (figure 4.10).

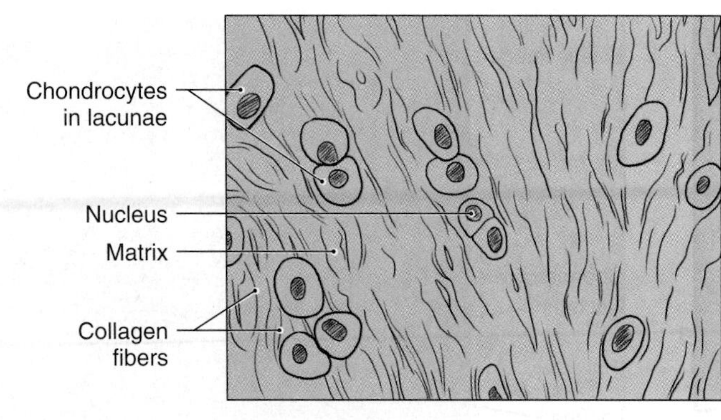

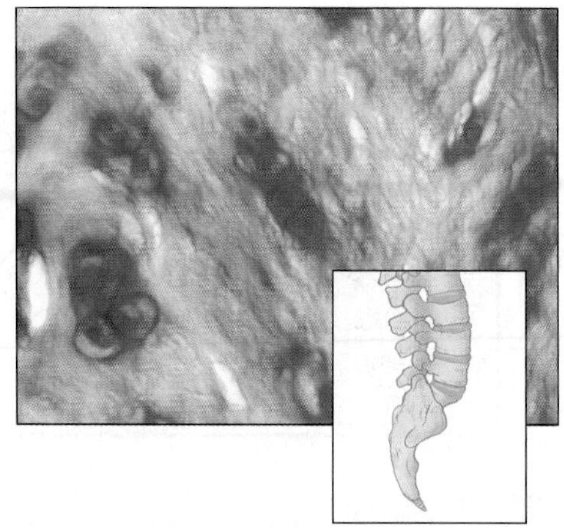

Figure 4.17 Fibrocartilage (250×). APR
Structure: Consists of rows or clusters of chondrocytes occupying lacunae in a matrix containing tightly packed collagen fibers.
Location & Function: Composes the intervertebral discs between vertebrae, the pubic symphysis, and cartilaginous pads in the knee joint where it serves as a protective shock absorber.

(2) the cartilaginous pads in the knee joints, and (3) the protective cushion of the pubic symphysis (anterior union of the pelvic bones). Fibrocartilage is especially tough, and the dense collagen fibers enable it to absorb greater shocks and pressure without permanent damage (figure 4.17).

Bone

Of all the supportive connective tissues, **bone,** also called *bone tissue* or *osseous tissue,* is the hardest and most rigid. This results from the minerals, mostly calcium salts, that compose the matrix along with some collagen fibers. Bone provides the rigidity and strength necessary for the skeletal system to support and protect the body. There are two types of bone: **compact bone** and **spongy bone** (figure 4.18).

In compact bone, bone matrix is deposited in concentric rings, called **lamellae** (lah-mel'-e), around microscopic tubes called **central** (or **osteonic**) **canals.** These canals contain blood vessels and nerves. A central canal and the lamellae surrounding it form an **osteon,** the structural unit of compact bone. Spongy bone does not possess osteons; rather, the lamellae are organized into thin, interconnected bony plates called **trabeculae.** The spaces between trabeculae are filled with highly vascular red or yellow bone marrow. Bone cells, or **osteocytes,** are located in lacunae that are located between lamellae in both types of bone. The tiny, fluid-filled canals that extend outward from the lacunae are called **canaliculi** (kan"-ah-lik'-u-li; singular, *canaliculus*) and they contain cell processes from osteocytes. Canaliculi serve as passageways for the movement of materials between osteocytes and the blood supply within the central canals and bone marrow. Bone will be discussed further in chapter 6.

Blood

Blood is a specialized type of connective tissue, called a fluid connective tissue. It consists of numerous **formed elements** that are suspended in the plasma, the liquid matrix of the blood. There are three basic types of formed elements: red blood cells, white blood cells, and platelets (figure 4.19).

Blood plays a vital role in carrying materials and gases throughout the body. For example, blood is used to carry nutrients absorbed by the digestive tract to cells throughout the body and wastes produced by body cells to the kidneys for elimination. The formed elements also perform crucial body functions. **Red blood cells** are used primarily to transport O_2 molecules and, to a lesser degree, CO_2 molecules. **White blood cells** carry out various defensive and immune functions throughout the body. **Platelets** play a crucial role in the process of blood clotting. Blood is discussed in more depth in chapter 11.

Check My Understanding

3. What are the general characteristics and functions of connective tissue?
4. What are the characteristics, locations, and functions of the different connective tissues?

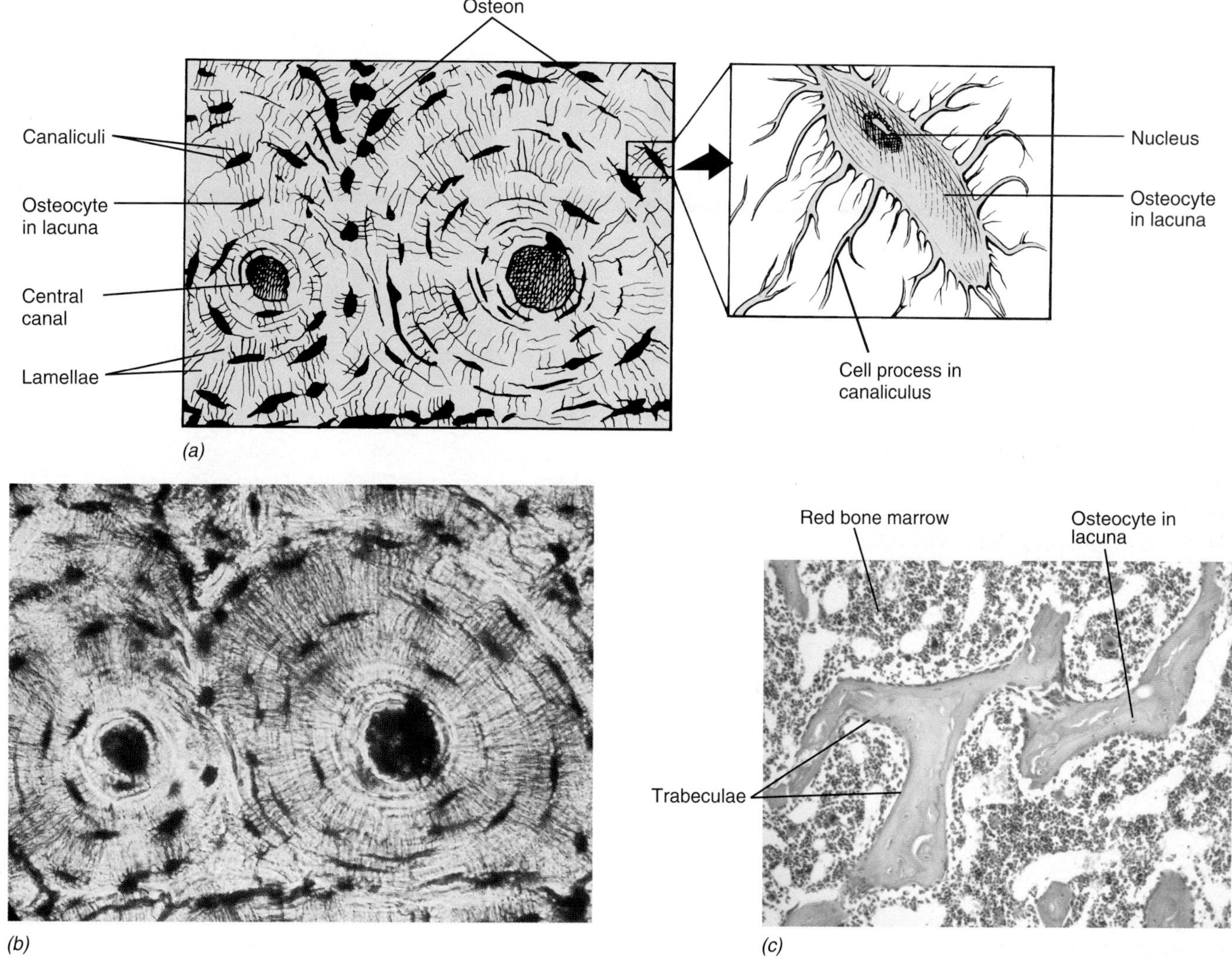

Figure 4.18 *(a) (b)* Compact bone (160×) *(c)* Spongy Bone (100×). AP|R
Structure: In compact bone, matrix is arranged in concentric layers around central canals. In spongy bone, the bone layers form thin, bony plates called trabeculae. Osteocytes are found within lacunae located between layers of matrix. Canaliculi, minute channels between lacunae, enable movement of materials between osteocytes in both compact and spongy bone.
Location & Function: Forms bones of the skeleton that provide support for the body and protection for vital organs.

4.3 Muscle Tissues AP|R

Learning Objectives

5. Describe the distinguishing characteristics and locations of each type of muscle tissue.
6. Identify the general functions of each type of muscle tissue.

Muscle tissue consists of muscle cells. Muscle cells have lost the ability to divide, so destroyed muscle cells cannot be replaced. In skeletal muscle tissue, muscle cells are called **muscle fibers** owing to their long, cylindrical appearance. The cells in smooth and cardiac muscle tissue are not long and cylindrical, so they are referred to as muscle cells but not muscle fibers. The cells within all three types of muscle tissue are specialized for contraction (shortening). Contraction is enabled by the interaction of specialized protein fibers. The contraction of these tissues enables the movement of the whole body and many internal organs, in addition to producing heat energy.

Three types of muscle tissue–skeletal, cardiac, and smooth muscle tissue–are classified according to their (1) location in the body, (2) structural features, and (3) functional characteristics.

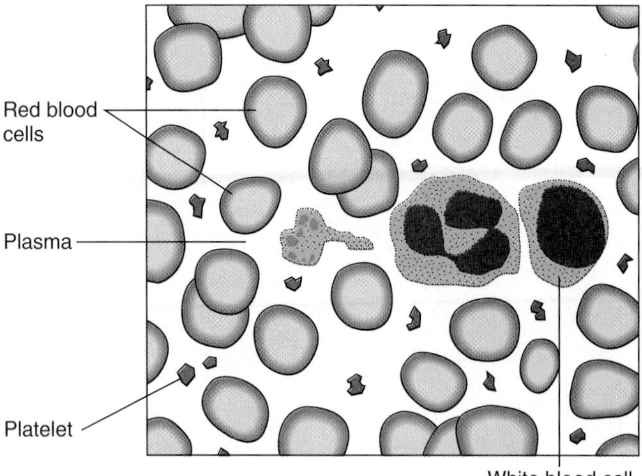

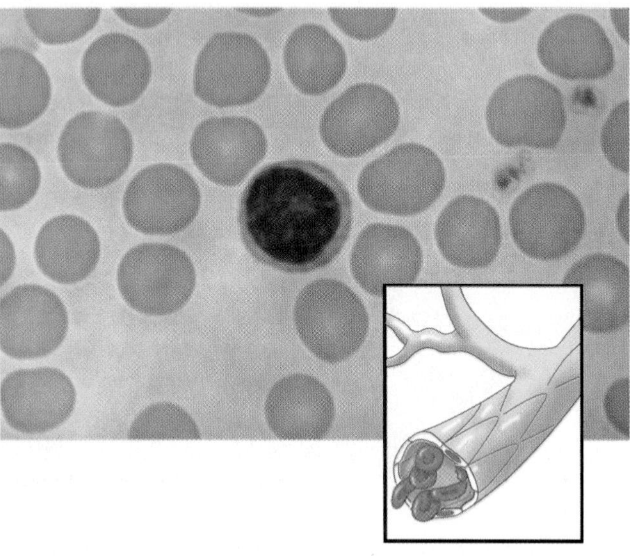

Figure 4.19 Blood (1,000×). APIR
Structure: Consists of red blood cells, white blood cells, and platelets that are carried in a liquid matrix called plasma.
Location & Function: Located within blood vessels and the heart; transports materials and gases throughout the body, participates in blood clotting process, provides defense against disease.

Skeletal Muscle Tissue

Named for its location, **skeletal muscle tissue** is usually attached to bones and skin. Its contractions enable movement of the head, trunk, and limbs. The muscle fibers are wide, elongated, and cylindrical. Each skeletal muscle fiber contains multiple nuclei, which are located along the periphery of the fiber. *Striations*, alternating light and dark bands, extend across the width of the fibers.

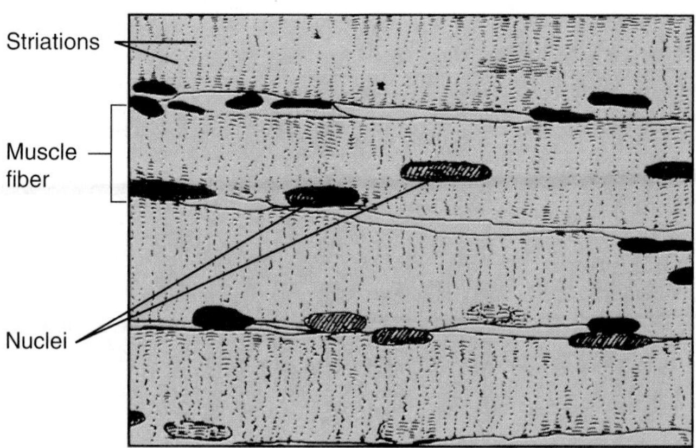

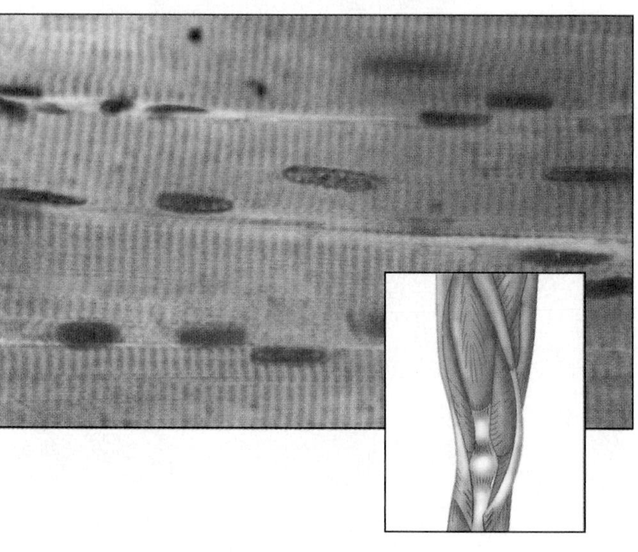

Figure 4.20 Skeletal Muscle Tissue (250×). APIR
Structure: Consists of cylindrical muscle fibers that have striations and multiple, peripherally located nuclei.
Location & Function: Composes skeletal muscles that attach to bones and skin; voluntary, rapid contractions.

Functionally, skeletal muscle tissue is considered to be **voluntary** muscle because its rapid contractions can be consciously controlled (figure 4.20). Skeletal muscle tissue is discussed in greater detail in chapter 7.

Cardiac Muscle Tissue

The muscle tissue located in the walls of the heart is **cardiac** (kar′-dē-ak) **muscle tissue.** It consists of branching cells that interconnect in a netlike arrangement. **Intercalated** (in-ter-kah′-lā-ted) **discs** are present where the cells join together. Cardiac muscle cells are striated like skeletal muscle fibers but possess only one centrally located nucleus per cell. The rhythmic

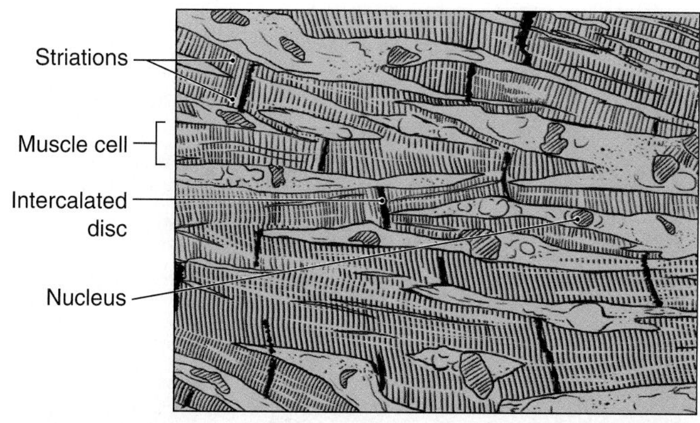

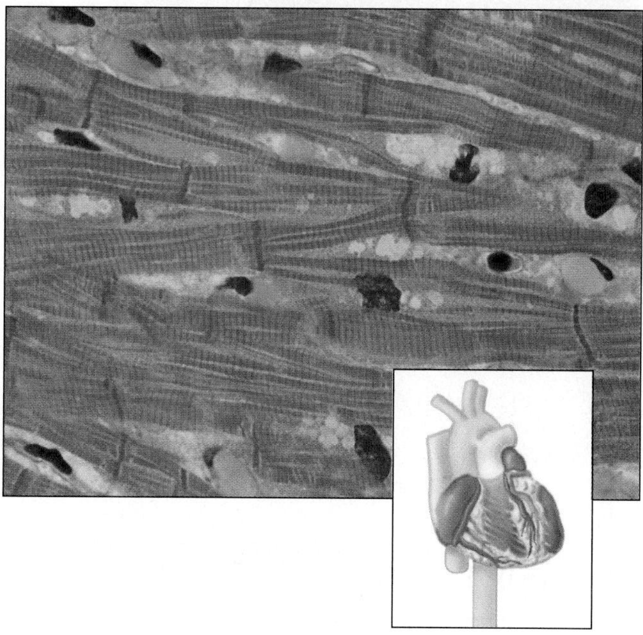

Figure 4.21 Cardiac Muscle Tissue (400×). AP|R
Structure: Consists of striated cells that are arranged in an interwoven network. Intercalated discs are present at the junctions between cells. A single, centrally located nucleus is present in each cell.
Location & Function: Forms the muscular walls of the heart; involuntary, rhythmic contractions.

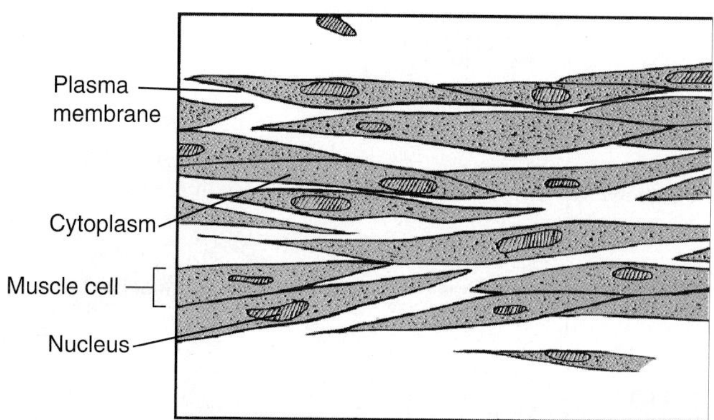

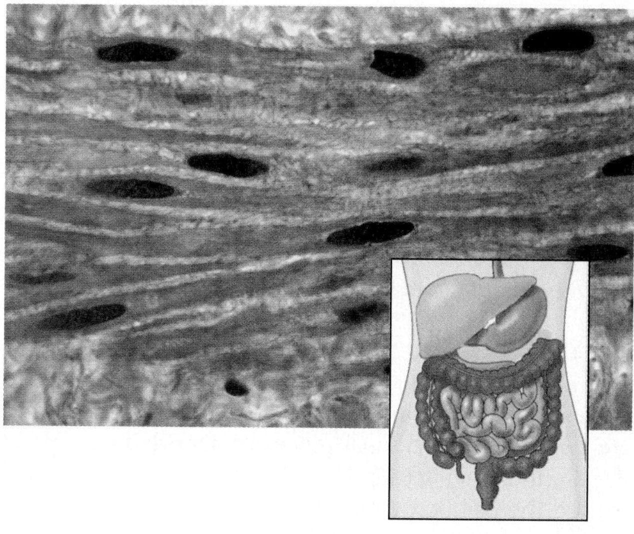

Figure 4.22 Smooth Muscle Tissue (250×). AP|R
Structure: Consists of elongate, tapered cells that lack striations and have a single, centrally located nucleus.
Location & Function: Forms muscle layers in the walls of hollow internal organs; involuntary, slow contractions.

contractions of cardiac muscle are **involuntary** because they cannot be consciously controlled (figure 4.21).

Smooth Muscle Tissue

Smooth muscle tissue derives its name from the absence of striations in its cells. It occurs in the walls of hollow internal organs, such as the stomach, intestines, urinary bladder, and blood vessels. The cells are long and spindle-shaped with a single, centrally located nucleus. The slow contractions of smooth muscle tissue are *involuntary* (figure 4.22).

4.4 Nervous Tissue

Learning Objectives

7. Describe the distinguishing characteristics and general functions of nervous tissue.
8. Identify the common locations of nervous tissue.

The brain, spinal cord, and nerves are composed of **nervous tissue,** which consists of **neurons** (nū′-ronz), or nerve cells, and numerous supporting cells that are collectively called **neuroglia** (nū-rog′-lē-ah). Neurons are the functional units of nervous tissue. They are specialized to detect and respond to environmental changes by generating and transmitting nerve impulses. Neuroglia nourish, insulate, and protect the neurons.

A neuron consists of a **cell body,** the portion of the cell containing the nucleus, and one or more *neuronal*

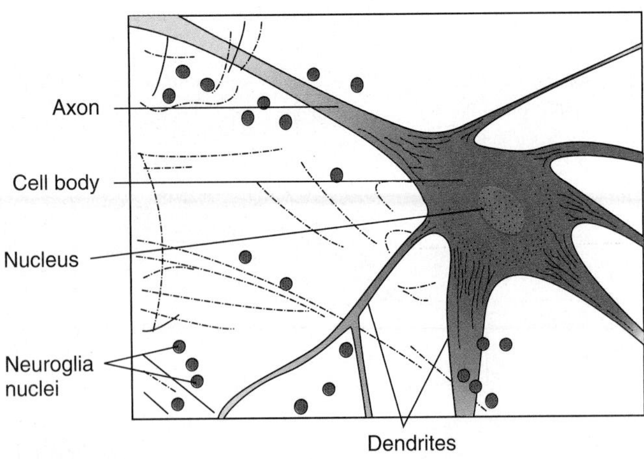

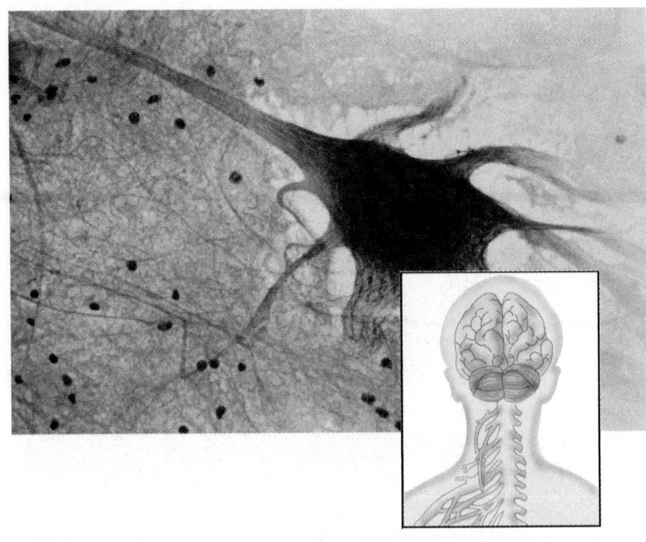

Figure 4.23 Nervous Tissue (50×). 
Structure: Consists of neurons and neuroglia. Each neuron consists of a cell body, which houses the nucleus, and one or more neuronal processes extending from the cell body.
Location & Function: Forms the brain, spinal cord, and nerves; nerve impulse formation and transmission.

processes extending from the cell body (figure 4.23). There are two types of neuronal processes. **Dendrites** respond to stimuli by generating impulses and transmitting them toward the cell body. An **axon** transmits nerve impulses away from the cell body and dendrites. A neuron may have many dendrites but only one axon. The complex interconnecting network of neurons enables the nervous system to coordinate body functions. Nervous tissue is discussed in more detail in chapter 8.

Following minor injuries, tissues repair themselves by *regeneration*—the division of the remaining intact cells. The capacity to regenerate varies among different tissues. For example, epithelial tissues, loose connective tissues, and bone readily regenerate, but cartilage and skeletal muscle have little capacity for regeneration. Cardiac muscle never regenerates, and neurons in the brain and spinal cord usually do not regenerate.

After severe injuries, repair involves **fibrosis,** the formation of scar tissue. Scar tissue is formed by an excess production of collagen fibers by fibroblasts. Scar tissues that join together tissues or organs abnormally are called *adhesions,* which sometimes form following abdominal surgery.

Check My Understanding

5. What are the distinguishing characteristics, locations, and functions of the three types of muscle tissue?
6. What types of cells form nervous tissue and what are their functions?

4.5 Body Membranes

Learning Objectives

9. Compare epithelial and connective tissue membranes.
10. Describe the locations and functions of each type of epithelial membrane.
11. Identify examples of connective tissue membranes.

Membranes of the body are thin sheets of tissue that line cavities, cover surfaces, or separate tissues or organs. Some are composed of both epithelial and connective tissues; others consist of connective tissue only.

Epithelial Membranes

Sheets of epithelial tissue overlying a thin supporting framework of areolar connective tissue form the epithelial membranes in the body. Blood vessels in the connective tissue serve both connective and epithelial tissues. There are three types of epithelial membranes: serous, mucous, and cutaneous membranes (figure 4.24).

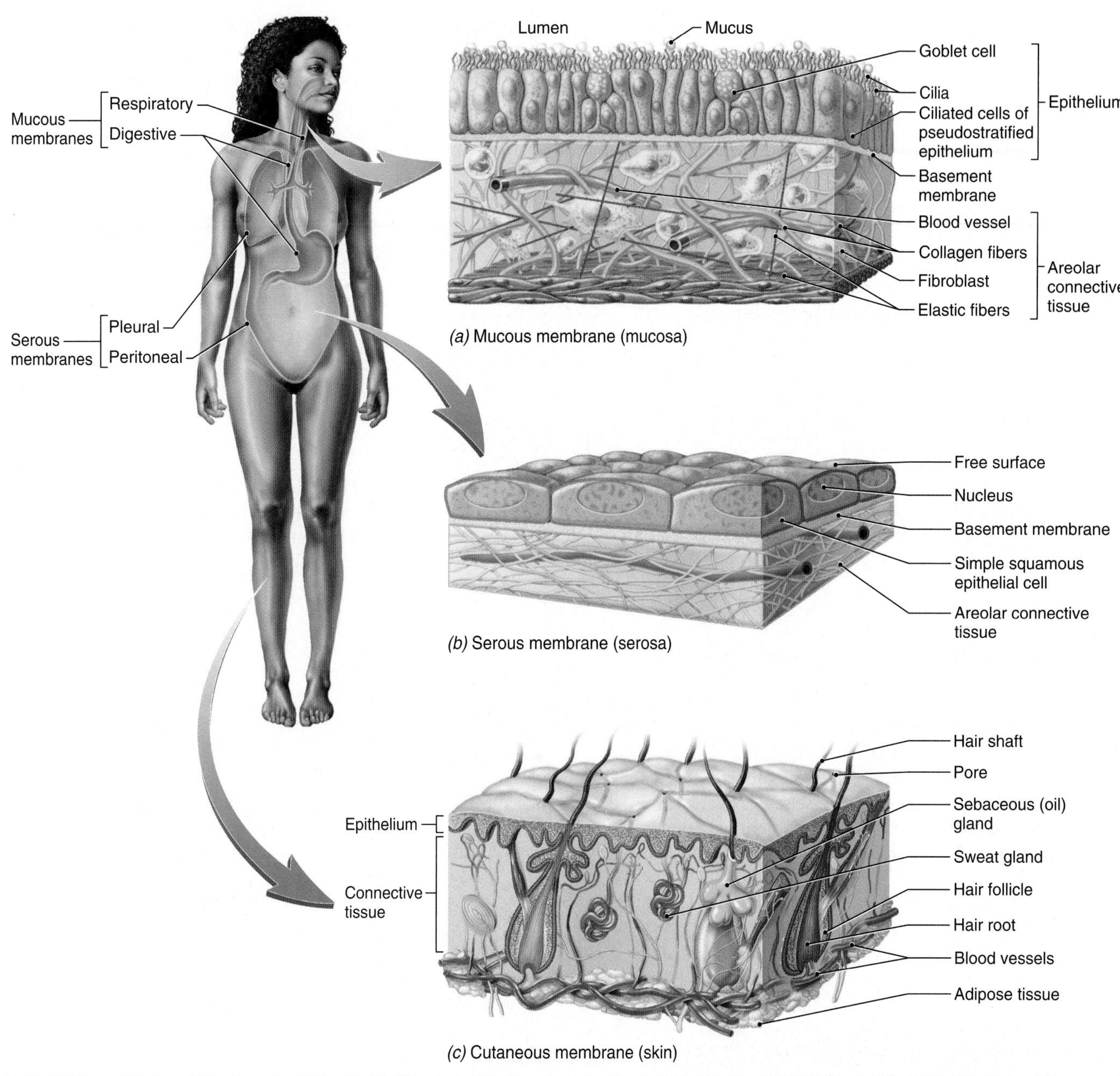

(a) Mucous membrane (mucosa)

(b) Serous membrane (serosa)

(c) Cutaneous membrane (skin)

Figure 4.24 Epithelial Membranes.

Serous membranes, or *serosae,* line the ventral body cavity and cover most of the internal organs. They secrete **serous fluid,** a watery fluid, which reduces friction between the membranes. The pleurae, pericardium, and peritoneum are serous membranes. Recall that the epithelium of a serous membrane is a special tissue called mesothelium.

Mucous membranes, or *mucosae,* line tubes or cavities of organ systems, which have openings to the external environment. Their goblet cells secrete **mucus,** which coats the surface of the membranes to keep the cells moist and to lubricate their surfaces. The mucus also helps to trap foreign particles and pathogens, which limits their ability to enter the body. The digestive, respiratory, reproductive, and urinary tracts are lined with mucous membranes.

The **cutaneous membrane** is the skin that covers the body. Unlike other membranes, its free surface is dry, composed of nonliving cells, and exposed to the external environment. The skin is discussed in detail in chapter 5.

Connective Tissue Membranes

Some specialized membranes are formed only of connective tissue, usually dense irregular connective tissue. These are considered in future chapters with their respective organ systems, but here are four examples:

1. The **meninges** are three connective tissue membranes that envelop the brain and spinal cord.
2. The **perichondrium** is a connective tissue membrane covering the surfaces of cartilage. It contains blood vessels, which supply cartilage through diffusion.
3. The **periosteum** is a connective tissue membrane that covers the surfaces of bones. It contains blood vessels that enter and supply the bone.
4. **Synovial membranes** line the cavities of freely movable joints, such as the knee joint. They secrete watery synovial fluid, which reduces friction in the joint.

Check My Understanding

7. What are the two kinds of body membranes?
8. How do the structures, locations, and functions of the epithelial membranes differ?

Chapter Summary

- As cells specialize during embryonic development, they form groups of similar cells called tissues.
- The body is formed of four basic types of tissues: epithelial, connective, muscle, and nervous tissues.

4.1 Epithelial Tissues

- Epithelial tissue covers surfaces of organs and of the body and lines the body cavities.
- Epithelial tissue is composed of closely packed cells with little extracellular material.
- Epithelial tissues are attached to underlying connective tissue by a noncellular basement membrane.
- Epithelial tissue lacks blood vessels.
- Epithelial tissues function in absorption, secretion, filtration, diffusion, osmosis, protection, and friction reduction.
- Epithelial tissues are classified according to the number of cell layers and the shape of the free surface cells. The epithelial tissues are

  Simple Epithelium
  - squamous
  - cuboidal
  - columnar
  - pseudostratified ciliated columnar

  Stratified Epithelium
  - stratified squamous
  - transitional

4.2 Connective Tissues

- Connective tissue is composed of relatively few cells located within a large amount of matrix.
- All but cartilage are supplied with blood vessels.
- Connective tissue binds other tissues together and provides support and protection for organs and the body.
- Connective tissue is classified according to the nature of the matrix. The connective tissues are

  Loose Connective Tissue
  - Areolar connective tissue
  - Adipose tissue
  - Reticular tissue

  Dense Connective Tissue
  - Dense regular connective tissue
  - Dense irregular connective tissue
  - Elastic connective tissue

  Cartilage
  - Hyaline
  - Elastic
  - Fibrocartilage

  Bone

  Blood

4.3 Muscle Tissues

- Muscle tissue is composed of muscle cells that are specialized for contraction.
- Contraction of muscle tissue enables movement of the body and internal organs.
- Muscle tissue is classified according to its location in the body, the characteristics of the muscle cells, and the type of contractions (voluntary or involuntary).
- Three types of muscle tissue are skeletal, cardiac, and smooth muscle tissue.

4.4 Nervous Tissue

- Nervous tissue consists of neurons and neuroglia.
- Neurons consist of a cell body and long, thin neuronal processes, and are adapted to form and conduct nerve impulses.
- Nervous tissue forms the brain, spinal cord, and nerves.

4.5 Body Membranes

- Membranes in the body are either epithelial membranes or connective tissue membranes.
- Epithelial membranes are composed of both epithelial and connective tissues, while connective tissue membranes are composed of connective tissue only.
- There are three types of epithelial membranes: serous, mucous, and cutaneous.
- Examples of connective tissue membranes are meninges, perichondrium, periosteum, and synovial membranes.

Self-Review

Answers are located in Appendix B.

1. Simple epithelial tissues consist of ______ layer(s) of cells.
2. Epithelial tissue contains ______ (little/much) extracellular material.
3. Many respiratory passageways are lined with ______ epithelium.
4. The stomach and intestines are lined with ______ epithelium.
5. Protection from abrasion is the function of ______ epithelium.
6. The extracellular substance in connective tissues is called ______.
7. ______ fibers provide tissues with great strength and flexibility.
8. Triglycerides are stored within the cells of ______ tissue.
9. Cushioning pads in the knee joint are composed of ______.
10. Muscle tissue in the wall of the heart is ______ muscle tissue.
11. Muscle tissue lacking striations and found in walls of the digestive tract is ______ muscle tissue.
12. Nervous tissue consists of neurons and supporting ______.
13. Membranes lining digestive, respiratory, and urinary tracts are classified as ______ membranes.

Critical Thinking

1. Explain why healing a torn ligament can be problematic.
2. Why is stratified squamous epithelium not found within the lungs?
3. Why is it important for homeostasis that cardiac and smooth muscle tissue are involuntary?
4. How are skeletal muscle tissue, dense regular connective tissue, and bone involved in movement of limbs?

ADDITIONAL RESOURCES

5 CHAPTER

Integumentary System

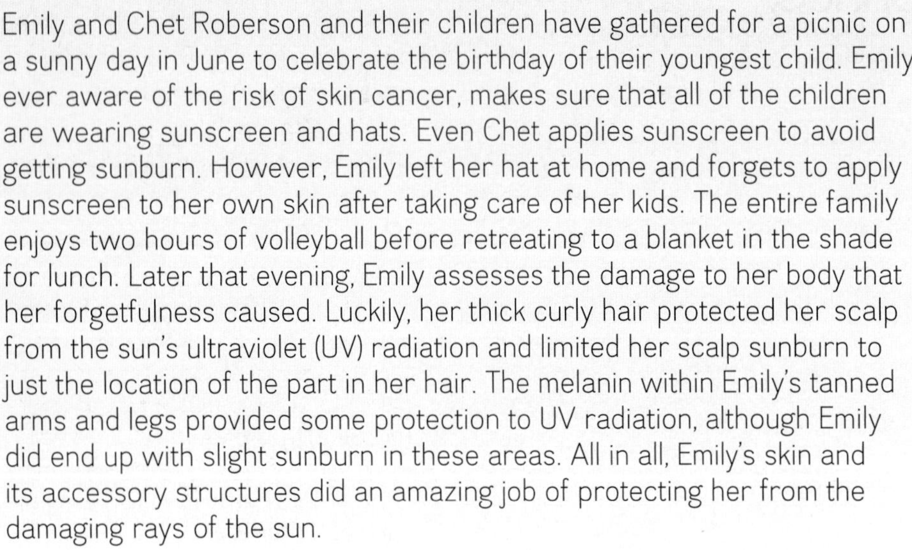

Emily and Chet Roberson and their children have gathered for a picnic on a sunny day in June to celebrate the birthday of their youngest child. Emily, ever aware of the risk of skin cancer, makes sure that all of the children are wearing sunscreen and hats. Even Chet applies sunscreen to avoid getting sunburn. However, Emily left her hat at home and forgets to apply sunscreen to her own skin after taking care of her kids. The entire family enjoys two hours of volleyball before retreating to a blanket in the shade for lunch. Later that evening, Emily assesses the damage to her body that her forgetfulness caused. Luckily, her thick curly hair protected her scalp from the sun's ultraviolet (UV) radiation and limited her scalp sunburn to just the location of the part in her hair. The melanin within Emily's tanned arms and legs provided some protection to UV radiation, although Emily did end up with slight sunburn in these areas. All in all, Emily's skin and its accessory structures did an amazing job of protecting her from the damaging rays of the sun.

CHAPTER OUTLINE

Module 4
Integumentary System

SELECTED KEY TERMS

Apocrine sweat gland (apo = detached; crin = separate off) A sweat gland opening into a hair follicle.
Cutaneous (cutane = skin) Pertaining to the skin.
Dermal papillae (papilla = nipple) Nipple-like projections of the dermis at the dermis-epidermis border.
Dermis (derm = skin) The deep layer of the skin.
Eccrine sweat gland (ec = out from) A sweat gland opening on the skin surface.
Epidermis (epi = upon) The superficial layer of the skin.
Hair follicle (folli = bag) A saclike epidermal ingrowth in which a hair develops.
Integument (integere = to cover) The skin.
Keratin (kerat = horny, hard) Waterproofing, abrasion-resistant protein produced by keratinocytes.
Melanin (melan = black) The brown-black pigment formed by melanocytes.
Sebaceous gland (seb = grease, oil) A sebum-producing gland associated with a hair follicle.
Subcutaneous tissue (sub = below) The loose connective tissue deep to the skin.
Sweat gland A sweat-producing gland.

THE SKIN AND THE STRUCTURES that develop from it—hair, glands, and nails—form the *integumentary* (in-teg-ū-men′-tar-ē) *system*. The skin, or **integument,** is also known as the **cutaneous membrane,** one of the three types of epithelial membranes as noted in chapter 4. The skin is a pliable, tough, waterproof, self-repairing barrier that separates deeper tissues and organs from the external environment. Although it often gets little respect, the skin is vital for maintaining homeostasis.

5.1 Functions of the Skin

Learning Objective

1. Explain the functions of the skin.

The skin performs six important functions:

1. **Protection.** The skin provides a physical barrier between internal tissues and the external environment. It provides protection from abrasion, dehydration, ultraviolet (UV) radiation, chemical exposure, and pathogens.
2. **Excretion.** Perspiration, produced by sweat glands, removes small amounts of organic wastes, salts, and water.
3. **Temperature regulation.** During periods of excessive heat production by the body, blood vessels near the body surface dilate to increase heat loss and cool the body. Sweat production and evaporation also aid in heat loss. During periods of excessive heat loss, blood vessels near the body surface constrict to conserve body heat.
4. **Sensory perception.** The skin contains nerve endings and sensory receptors that detect stimuli associated with touch, pressure, temperature, and pain.
5. **Synthesis of vitamin D.** Exposure to UV radiation stimulates the production of precursor molecules that are needed for the body to form active vitamin D.
6. **Absorption.** The skin is capable of absorbing lipid-soluble vitamins (A, D, E, and K), in addition to lipid-soluble drugs (e.g., topical steroids, nicotine patches) and toxins (e.g., acetone, lead, mercury).

5.2 Structure of the Skin and Subcutaneous Tissue

Learning Objectives

2. Describe the structures and functions of the two tissue layers forming the skin.
3. Describe the structure and functions of the subcutaneous tissue.

The skin is thickest in areas subjected to wear and tear (abrasion), such as the soles of the feet, where it may be 6 mm in thickness. It is thinnest on the eyelids, eardrums, and external genitalia, where it averages about 0.5 mm in thickness.

The skin consists of two major layers: the epidermis and the dermis. The *epidermis,* the thinner superficial layer, is composed of an epithelium. The *dermis,* the thicker deep layer, is composed of connective tissue. The *subcutaneous tissue,* located deep to the dermis, is not part of the skin but is considered here because of its close association with the skin. Figure 5.1 shows the arrangement of the epidermis, dermis, and subcutaneous tissue as well as accessory organs of the skin. Table 5.1 summarizes these three tissue layers.

Epidermis

The **epidermis** is a keratinized stratified squamous epithelium. Recall from chapter 4 that an epithelium is avascular, meaning it lacks blood vessels. Since the epidermis is prone to injury, the lack of blood vessels prevents unnecessary

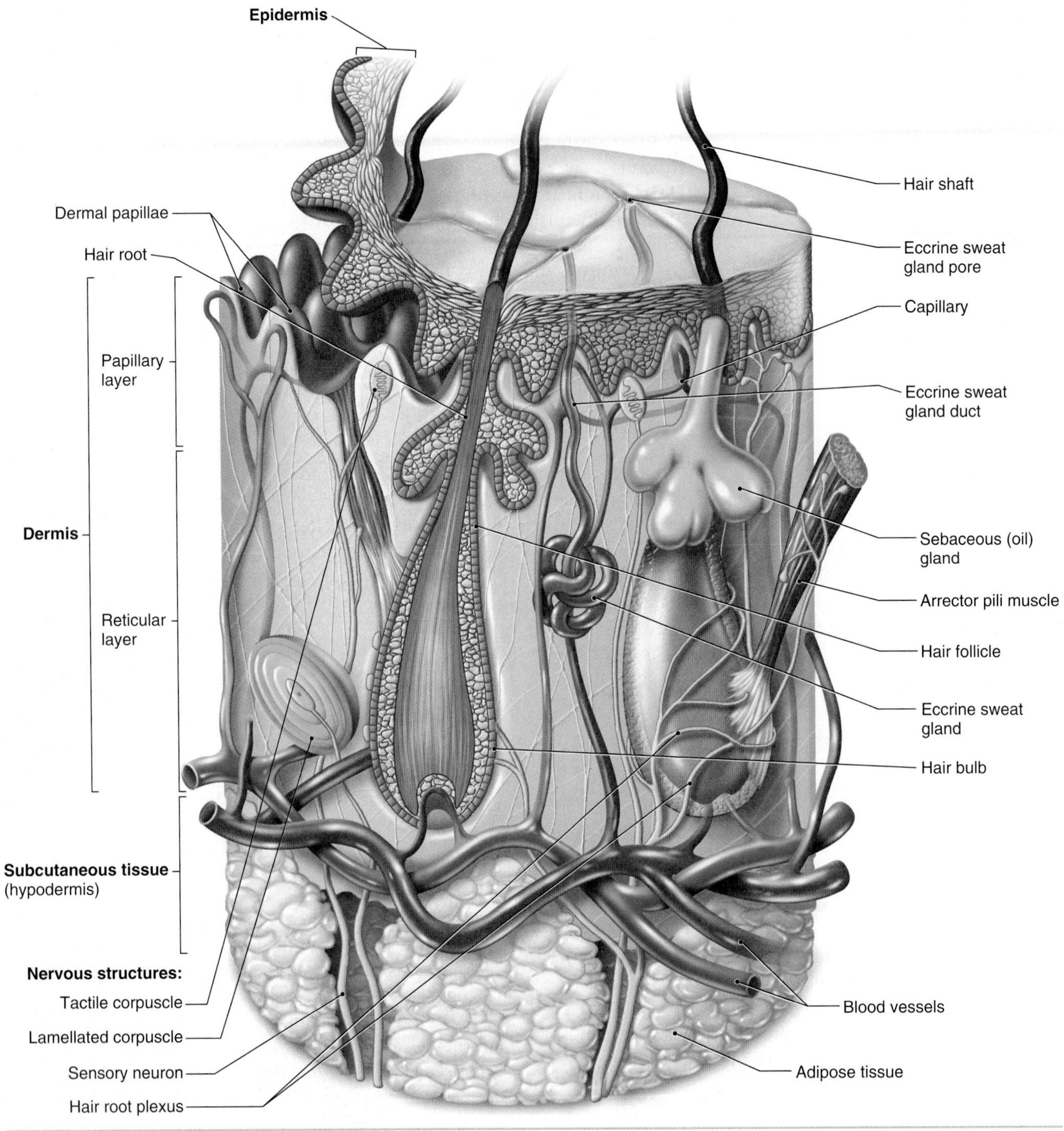

Figure 5.1 A Section of Skin. APR

bleeding. It also means that epidermal tissue must obtain nutrients and remove wastes through diffusion with blood vessels in the dermis. However, the epidermis does contain numerous nerve endings, which allows it to easily detect changes on the body's surface. The epidermis is the boundary between the body and the external environment. It protects the body against (1) the entrance of pathogens, (2) UV radiation, (3) excessive water loss, (4) exposure to environmental chemicals, and (5) abrasion. The epidermis is also involved in the production of the precursor molecules necessary for active vitamin D production.

The epidermis is composed primarily of cells called **keratinocytes** and these cells are organized into distinct layers (figure 5.2). In the thin skin covering the majority of the body's surface, the epidermis is organized into four layers. Thick skin, which is organized into five layers, is

Table 5.1 The Skin and Subcutaneous Tissue

Layer	Structure	Function
Epidermis	Keratinized stratified squamous epithelium; forms hair and hair follicles, sebaceous glands, and sweat glands that penetrate into the dermis or subcutaneous tissue	Protects against abrasion, evaporative water loss, pathogen invasion, chemical exposure, and UV radiation
Dermis	Areolar and dense irregular connective tissues containing blood vessels, nerves, and sensory receptors	Provides strength and elasticity; sensory receptors enable detection of touch, pressure, pain, cold, and heat; blood vessels supply nutrients to the epidermis
Subcutaneous tissue	Areolar connective tissue and adipose tissue; contains abundant blood vessels and nerves	Provides insulation, protection from impact, and fat storage; attaches skin to deeper organs

found only in areas subjected to high levels of abrasion, such as the palms and soles (figure 5.3). The keratinocytes within both thin and thick skin are so named because they undergo a process called *keratinization,* which leads to the eventual hardening and flattening of the cells through the production of keratin. **Keratin** (ker′-ah-tin) is a tough, fibrous protein that provides waterproofing and abrasion protection for the skin.

The deepest layer of cells, the **stratum basale** (ba-sah′-le), continuously produces new keratinocytes by mitotic cell division. As new cells are produced, the older cells superficial to them are gradually pushed toward the surface. The constant division of cells in the stratum basale enables the epidermis to repair itself when damaged. Upon leaving the stratum basale, keratinocytes enter the **stratum spinosum.** In this layer, keratinocytes begin to produce keratin within their cytoplasm. Neighboring

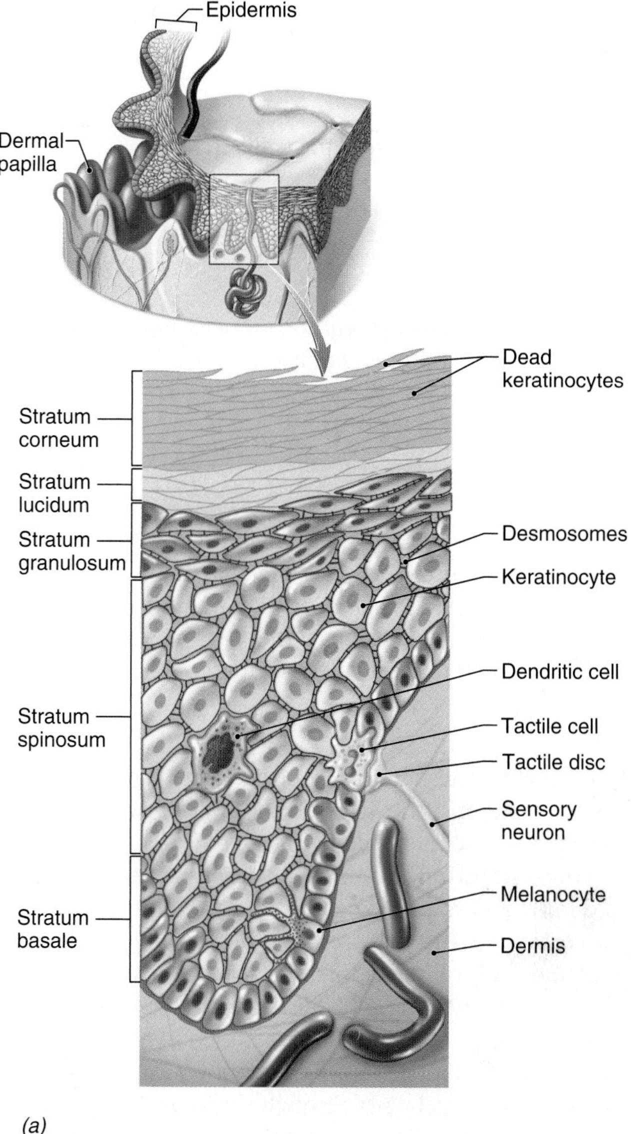

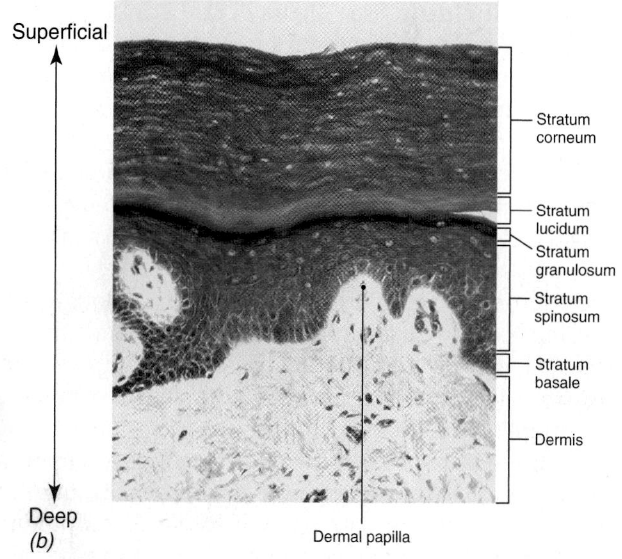

Figure 5.2 Anatomy of the Epidermis.
(a) The five layers of the epidermis in thick skin and the four types of cells that compose the epidermis. *(b)* Photomicrograph of the epidermis (200×) in thick skin. APIR

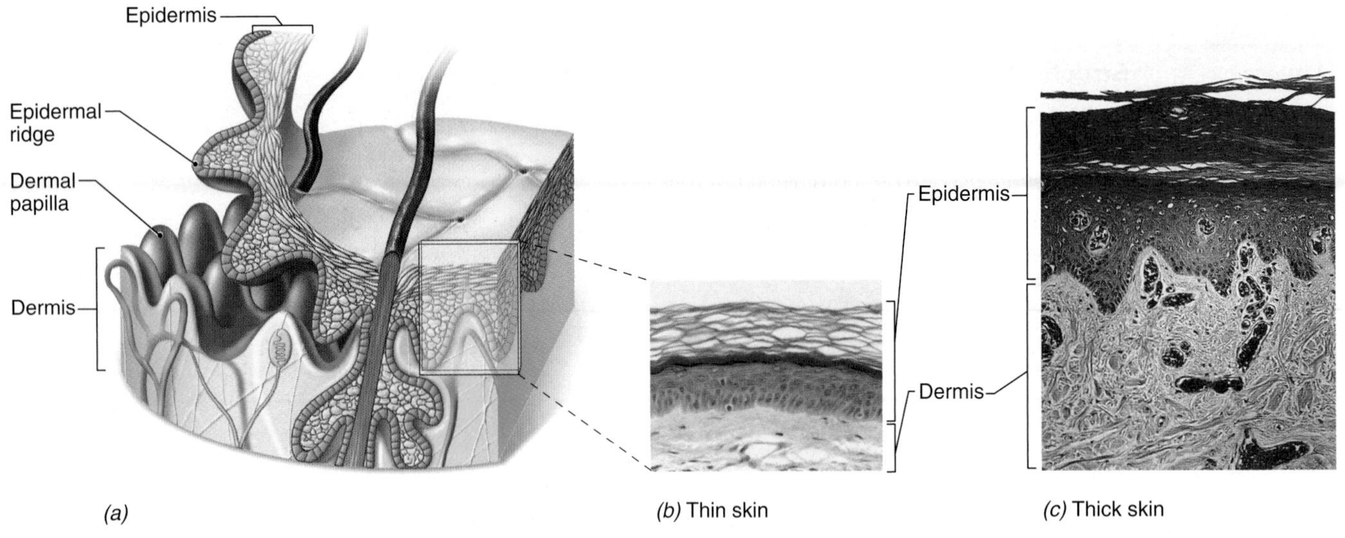

Figure 5.3 Comparison of the Epidermis in Thin and Thick Skin.
(a) A section of the skin. *(b)* Photomicrograph of thin skin (400×). *(c)* Photomicrograph of thick skin (100×).

cells also become physically connected, which creates a "spiny" appearance microscopically in fixed tissue.

As the keratinocytes continue to be pushed towards the epidermal surface, they leave the stratum spinosum and enter the **stratum granulosum.** The cells within this layer possess numerous granules that stain darkly in tissue section and are essential for the formation of keratin inside the cells. The cells within the stratum granulosum also undergo *apoptosis*, which is programmed cell death and involves the destruction of the nucleus and other organelles.

In thick skin, cells move superficially into the **stratum lucidum** (see figure 5.2). Because the cells in this layer do not have nuclei or organelles, this layer often appears "lucid" or transparent in section. The stratum lucidum is absent in thin skin, meaning cells move directly from the stratum granulosum into the most superficial layer of the epidermis, the **stratum corneum** (koŕ-nē-um). The stratum corneum is so named because it consists of approximately 20–40 layers of dead, squamous, and keratinized (cornified) cells. The superficial cells of the stratum corneum are continually being sloughed off and replaced by underlying cells moving towards the surface. The journey from stratum basale to stratum corneum usually takes seven to ten days, with cells remaining in the stratum corneum for another two weeks on average.

Three types of specialized cells are also present in the epidermis (see figure 5.2). **Melanocytes** (mel-an′-ō-sītz) are located in the stratum basale. They produce the brown-black pigment that is primarily responsible for skin color. **Dendritic** (*Langerhans*) **cells** are located in the strata spinosum and granulosum of the epidermis and are derived from monocytes, a type of white blood cell (see chapter 11). These cells migrate throughout the epidermis where they use phagocytosis to remove pathogens trying to enter the body and alert the lymphoid system to launch an attack. **Tactile** (*Merkel*) **cells** in the stratum basale work with tactile discs in the dermis in touch sensation detection (see chapter 9).

Clinical Insight

Whenever possible, surgeons try to make incisions parallel to the dominant direction of collagen fibers in the dermis because the healing of such incisions forms little scar tissue.

Dermis

The **dermis,** the deep layer of the skin, can be divided into two regions: the superficial papillary layer and the deeper reticular layer (see figure 5.1).

The **papillary layer** of the dermis is adjacent to the epidermis and is composed of areolar connective tissue. The most notable features of this region are **dermal papillae** (pah-pil′-ē), nipple-like projections of the dermis that extend superficially into the epidermis. The dermal papillae contain numerous blood vessels that are used to supply nutrients to and remove wastes from the adjacent epidermal cells through diffusion. They contain touch receptors called the **tactile** *(Meissner)* **corpuscles** (see figure 5.1 and chapter 9). The *epidermal ridges* and grooves that produce the fingerprints and toe prints unique to each person are formed by the dermal papillae (figure 5.4, see figure 5.3). Epidermal ridges provide a

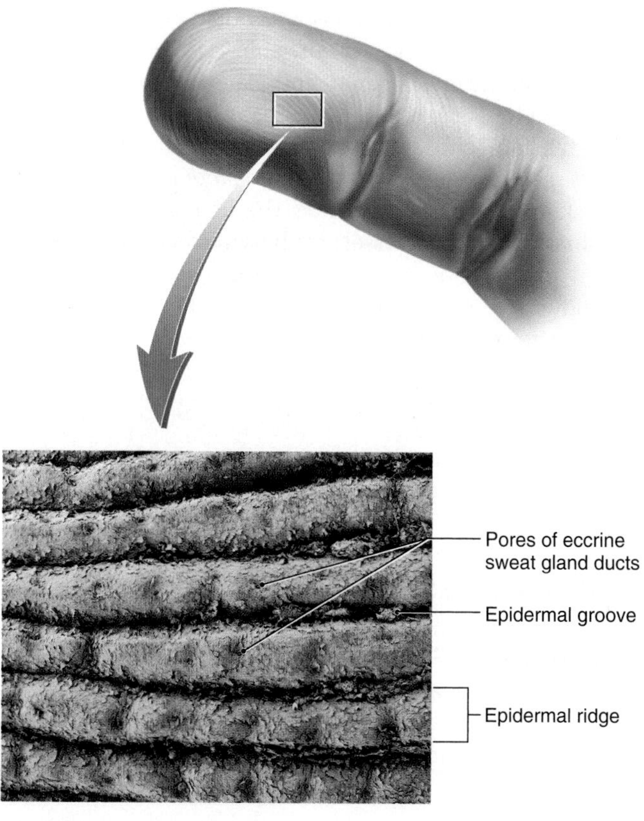

Figure 5.4 Epidermal ridges and grooves. The pattern of epidermal ridges and grooves is established by the dermal papillae of the papillary layer of the dermis.

textured surface that increases traction on these gripping surfaces, in addition to the man-made application of personal identification. The dermal papillae and epidermal ridges also help to interlock the epidermis and dermis, so that they move as a unit.

The **reticular layer** of the dermis is deeper and thicker than the papillary layer, making up 70–80% of the total thickness of the dermis. The dense irregular connective tissue within this region possesses an abundance of collagen and elastic fibers. The collagen provides the dermis with strength and toughness, while the elastic fibers provide extensibility (ability to stretch) and elasticity (ability to return to its original shape). Numerous pressure, pain, and temperature receptors are located here. For example, the **lamellated** (*Pacinian*) **corpsucles** that are used to detect pressure are found within the deeper areas of the reticular layer (see figure 5.1 and chapter 9). Free nerve endings responsible for touch, pain, and temperature are located throughout both the dermis and the epidermis (see chapter 9). The blood vessels found within this region play an important role in temperature regulation, which is considered later in this chapter.

Subcutaneous Tissue

The **subcutaneous tissue,** also called the *hypodermis,* attaches the skin to deeper tissues and organs. It consists primarily of areolar connective tissue and adipose tissue. It is the site used for subcutaneous injections and where white blood cells attack pathogens that have penetrated the skin. Subcutaneous adipose tissue absorbs the forces created by impact to the skin, which protects deeper structures, and serves as a storage site for fat. It insulates the body by conserving body heat and limits the penetration of external heat into the body. Blood vessels and nerves within the subcutaneous tissue give off branches that supply the dermis.

Check My Understanding

1. What are the general functions of the skin?
2. What changes occur in epidermal cells after they are formed?
3. What are the functions of the dermis and subcutaneous tissue?

5.3 Skin Color

Learning Objectives

4. Describe how skin color is determined.
5. Explain how the skin provides protection from ultraviolet radiation.

Skin color results from the interaction of three different pigments: hemoglobin, carotene, and melanin. **Hemoglobin** (hémō-glō″-bin) is the red pigmented protein in red blood cells that is used to carry oxygen and carbon dioxide in the blood. **Carotenes** (kair̍-ō-tēns) are a group of lipid-soluble plant pigments that range in color from violet, to red-yellow, to orange-yellow. Beta-carotene, which is the most abundant carotene, is found in yellow-orange and green leafy fruits and vegetables. The human body uses carotenes for vitamin A production, which is needed for the maintenance of epithelial tissues. Excess carotenes are stored in and add color to the body's fatty areas, such as the subcutaneous tissue, and the stratum corneum. **Melanin** (mel̍-ah-nin) is a brown-black pigment that is formed by melanocytes (figure 5.5). Melanocytes insert melanin into adjacent keratinocytes where it forms a protective UV radiation shield over their nuclei.

Most people have the same number and distribution of melanocytes. Generally, melanocytes are equally distributed throughout the epidermis. However, there are areas that have higher numbers of melanocytes and, as a result, greater amounts of melanin and darker coloring. Areas of the skin subjected to more sun exposure, such as the face, neck, and limbs, have greater numbers of melanocytes to

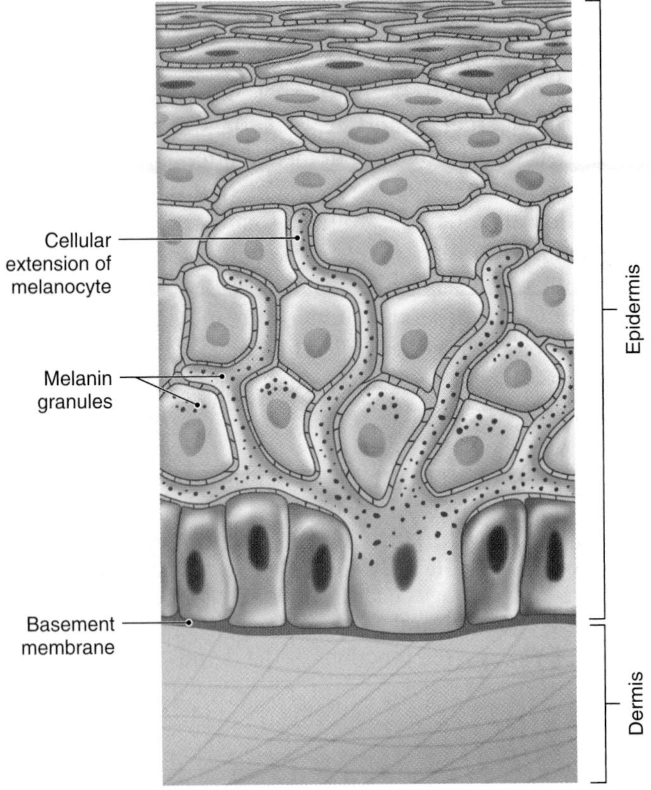

Figure 5.5 A melanocyte with cellular extensions that transfer melanin granules to adjacent epidermal cells. 

provide more UV protection to these areas. Greater melanocyte numbers are also found in structures such as the areola surrounding the nipple and the external genitalia, creating darker coloration in body areas that indicate reproductive readiness. Also, melanocytes can occur randomly in small clumps and create *freckles* due to the concentration of melanin in small patches.

The amount of melanin that a person can produce is determined genetically. Some people are genetically predisposed to make melanin at a faster rate, which results in a darker skin color. Other people are predisposed to make melanin at a slower rate, which results in a lighter skin color. A person's rate of melanin production can be influenced by exposure to UV radiation. Exposure to UV radiation will increase melanin production and produce a tanned appearance in lighter-skinned individuals. The increased melanin production is a protective, homeostatic mechanism. When exposure to UV radiation decreases, melanin production also decreases and the tan is lost in a few weeks as the "tanned" cells migrate to and are sloughed off the stratum corneum.

The skin colors characteristic of the various human races result primarily from varying amounts of carotene and melanin in the skin and are inherited. The effect of hemoglobin is relatively constant, but it is often masked by high concentrations of melanin. Dark-skinned races produce abundant melanin and have a greater protection from UV radiation. Caucasians produce relatively little melanin and are more susceptible to the harmful effects of UV radiation. The reduced amount of melanin in light-skinned Caucasians allows the hemoglobin of blood within dermal blood vessels to show through and give the skin a pinkish hue. Some people of Asian descent have a yellowish-tinged skin color due to the presence of a different form of melanin.

Check My Understanding

4. What relationship exists between skin color and protection against UV radiation?

5.4 Accessory Structures

Learning Objectives

6. Describe the anatomy of each accessory structure formed by the epidermis.
7. Explain the function of each epidermal accessory structure.

The accessory structures of the skin–hair, glands, and nails–develop from the epidermis. These structures originate in either the dermis or the subcutaneous tissue because they develop from inward growths of the epidermis. Table 5.2 summarizes these accessory structures.

Hair

A *hair* is formed of keratinized cells and consists of two parts: a shaft and a root. The *hair shaft* is the portion that projects above the skin surface. The *hair root* lies below the skin surface in a **hair follicle,** an inward, tubular extension of the epidermis. The follicle penetrates into the dermis and usually into the subcutaneous tissue (figure 5.6; see figure 5.1). The region of cell division, where the stratum basale forms new hair cells, is located in the *hair bulb* (base) of the follicle. The hair bulb is enlarged where it fits over a dermal papilla, which contains blood vessels that nourish the dividing epidermal cells. The hair cells become keratinized, die, and become part of the hair root. The continuing production of new cells causes growth of the hair.

Each hair follicle has an associated *arrector pili muscle* that is attached at one end to the deeper portion of a hair follicle and to the papillary layer of the dermis at the other end (see figure 5.6). Each muscle is a small group of smooth muscle cells. When a person is frightened or very cold, the arrector pili muscles contract and raise the hairs on end, producing goose bumps or chicken skin. Though of little value in humans, the erect hairs express

Table 5.2 Accessory Structures

Structure	Origin	Function
Hair	Fused keratinized epidermal cells formed at base of hair follicle	Scalp hair: protects scalp from excessive heat loss, mechanical injury, and UV radiation
		Eyelashes and eyebrows: protect eyes from sunlight and dust
		Ear and nasal hairs: keep dust and insects out of the external acoustic meatus and nasal cavity
Sebaceous glands	Formed as an epidermal outpocketing from a hair follicle	Produce sebum to prevent excessive dryness of hair and skin, inhibit microbial growth on skin surface, and reduce evaporative water loss
Sweat glands	Apocrine sweat glands formed as an epidermal outpocketing from hair follicles in axillary, areolar, beard, and genital regions	Produce a sweat that contains human pheromones
	Eccrine sweat glands formed as epidermal ingrowth	Produce sweat that cools the body through evaporation, protects from pathogens through lysozyme, has an acidic pH, provides a flushing action, and makes minor contribution to waste removal
Ceruminous glands	Modified sweat glands	Produce waxy cerumen to keep eardrum moist and capture particles and insects entering external acoustic meatus
Nails	Fused keratinized epidermal cells formed within an ingrowth of the epidermis	Protect the distal surfaces of fingers and toes; manipulating small objects

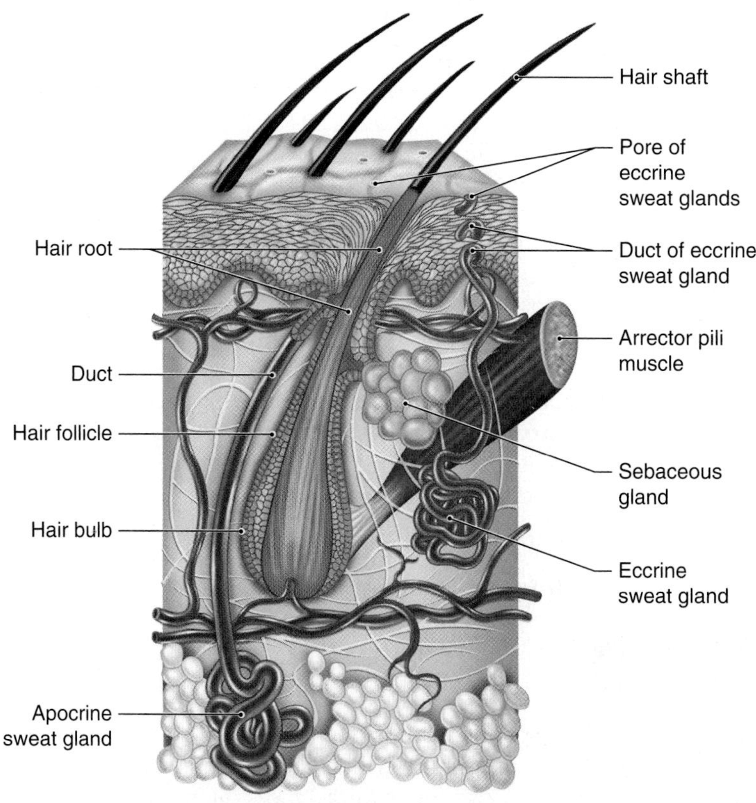

Figure 5.6 A section of skin showing the association of a hair follicle, arrector pili muscle, sebaceous gland, apocrine sweat gland, and eccrine sweat gland. APIR

rage to enemies or increase the thickness of the insulating layer of hair in cold weather in many mammals.

Hair is present over most of the body, but it is absent on the soles, palms, nipples, lips, and portions of the external genitalia. Its primary function is protection, although the tiny hairs over much of the body have little function. Eyelashes and eyebrows shield the eyes from sunlight and foreign particles. Hair in the nostrils and external acoustic meatuses (ear canals) act as filters to protect against the entrance of foreign particles and insects. Hair protects the scalp from sunlight, mechanical injury, and heat loss. The *hair root plexus* wrapping around the hair follicle detects movement of the hair shaft when skin contact is made (see figure 5.1).

Glands

Exocrine glands associated with the skin are of three types: sebaceous (oil), sweat, and ceruminous (wax) glands. Each type of gland is formed by an inward growth of the epidermis during embryonic development.

Sebaceous Glands APIR

Sebaceous (se-bā-shus) **glands** are oil-producing glands that usually empty their oily secretions into hair follicles (see figures 5.1 and 5.6). The oily

Clinical Insight

Normal hair loss from the scalp is about 75 to 100 hairs per day. Hair loss is increased and regeneration is decreased by poor diet, major illnesses, emotional stress, high fever, certain drugs, chemical therapy, radiation therapy, and aging. Baldness, an inherited trait, is much more common in males than in females.

secretion, called **sebum,** moves superficially along the hair root until it reaches the skin surface. Sebum helps to protect the body by inhibiting the growth of some pathogens on the skin surface and by keeping the hair and skin pliable and soft. Moisturizing the skin keeps it from becoming dry and cracked, which creates passageways for pathogens to enter the body. It also aids in preventing dehydration by reducing evaporative water loss.

Sweat Glands APIR

Sweat glands play an important role in maintaining homeostasis. The two types of sweat glands are *apocrine sweat glands* and *eccrine* (or *merocrine*) *sweat glands.* Both types of glands are tubular in shape, similar to a garden hose. The secretory portion of both types of sweat glands is coiled and located within the dermis and subcutaneous tissue, with the apocrine type located deeper than the eccrine type. In addition to being richly supplied with blood vessels, each sweat gland possesses a relatively straight, narrow duct that carries the glandular secretion to either the skin surface or a hair follicle (see figures 5.1 and 5.6).

Eccrine (ek′-rin) **sweat glands** are the most abundant skin glands, with over two million scattered across the body surface. The clear, watery sweat produced is delivered through a narrow duct directly to the skin surface. Eccrine sweat glands are active from birth and are stimulated to produce sweat when body temperature starts to rise during activities such as exercise. The watery nature of eccrine sweat assists in cooling the body through evaporation from the skin surface. Eccrine sweat is capable of protecting the body from environmental hazards. It contains a chemical called *lysozyme,* which has the ability to kill certain types of bacteria on the skin surface. The slightly acidic nature of eccrine sweat limits pathogen growth on the skin surface. Also the large volume of sweat that can be produced creates a flushing action that can wash chemicals, pathogens, dirt, etc. from the skin surface. Although the kidneys and lungs are the primary organs of excretion, eccrine sweat does contain small amounts of salts and wastes, in addition to other substances that happen to be in excess within the blood. For example, glucose (blood sugar) can be detected in the sweat of individuals with diabetes mellitus.

Apocrine (ap′-ō-krin) **sweat glands** have ducts that empty secretions into hair follicles (see figure 5.6). They occur primarily in the axillary, areolar, beard, and genital regions and become active at puberty under the influence of the sex hormones. These glands become activated during times of emotional stress and sexual excitement. Apocrine sweat is viscous and milky in color, due to the addition of lipid and proteins. Although the secretions are essentially odorless, decomposition by bacteria produces waste products that cause body odor. Apocrine sweat also contains *pheromones,* chemicals with the ability to alter physiological processes in nearby organisms. Studies have demonstrated the ability of human pheromones to affect reproductive function in males and females.

Ceruminous Glands

Ceruminous (se-rū′mi-nus) **glands** are modified apocrine sweat glands that are located in the external acoustic meatus and produce a waxy secretion called *cerumen.* The sticky, waxy nature of cerumen helps to keep foreign particles and insects out of the external acoustic meatus. Occasionally, excessive production of cerumen causes a buildup of wax that becomes impacted in the external acoustic meatus. This condition may cause a slight hearing loss as well as pain. Once the impacted wax is removed by irrigation and/or mechanical means, hearing returns to normal.

Nails

Hard, hooflike *nails,* composed of dead keratinized epidermal cells, cover the distal surfaces of the fingers and toes (figure 5.7). A nail consists of a *nail body,* the portion that is visible, and a *nail root,* the proximal portion that is inserted into the dermis. Nails are colorless but they normally appear pinkish due to the blood vessels in the *nail bed,* which is the skin deep to the nail body. Nails appear bluish in persons suffering from severe anemia or oxygen deficiency. Near the nail root is a whitish, crescent-shaped area that is called the *lunula.* The *cuticle* is a band of epidermis attached to the proximal border of the nail body. The major function of nails is protection, but fingernails are also useful in manipulating small objects.

Clinical Insight

Sebaceous glands increase their production of sebum at puberty. Accumulated sebum in enlarged hair follicles may form blackheads, whose color comes from oxidized sebum and melanin and not from dirt as is commonly believed. Invasion of certain bacteria may result in pimples or boils.

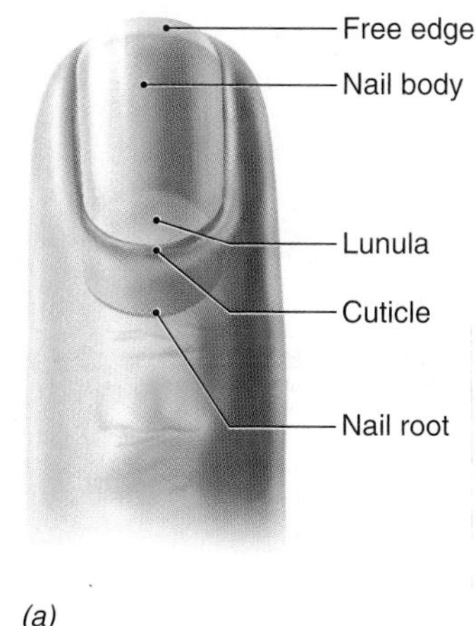

(a)

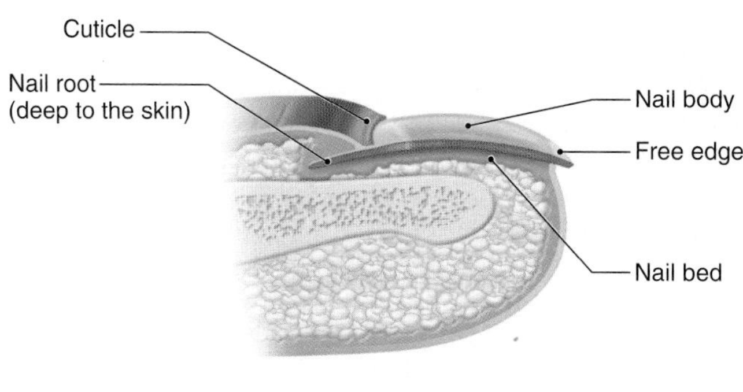

(b)

Figure 5.7 The structure of a fingernail. A fingernail is composed of dead, heavily keratinized epidermal cells that are formed and undergo keratinization in specialized tissue deep to the nail root. AP|R

Check My Understanding

5. Explain the functions of hair.
6. What are the functions of sebaceous and sweat glands?

5.5 Temperature Regulation

Learning Objectives

8. Describe how the skin aids in the regulation of body temperature.
9. Contrast hypothermia and hyperthermia, including the causes and bodily effects of each.

According to a 1992 study published in the *Journal of the American Medical Association,* humans are able to maintain an average healthy body temperature near 36.8°C (98.2°F), although the surrounding environmental temperature may vary widely. Variations in body temperature of 0.5° to 1.5°F during a 24-hour activity cycle are normal, but variations of more than a few degrees can be life threatening. The brain controls the regulation of body temperature, while the skin plays a key role in conserving or dissipating heat. Decomposition reactions (see chapter 2), especially in metabolically active tissues such as the liver and skeletal muscles, are the source of body heat. Figure 5.8 illustrates the major aspects of temperature regulation.

When body temperature begins to rise above normal, the brain triggers dilation (widening) of the blood vessels within the skin. The resulting increase in blood flow to the skin increases heat loss from the skin surface. When body temperature becomes excessively high, the brain also activates eccrine sweat glands. These glands release sweat onto the skin surface and its evaporation aids in the removal of excess heat. Once body temperature returns to normal, these changes in blood flow and sweat production cease.

When body temperature begins to fall below normal, the brain triggers constriction (narrowing) of the blood vessels within the skin. The resulting decrease in blood flow to the skin decreases heat loss from the skin surface. Eccrine sweat glands are not activated, so heat is not lost through sweat evaporation. If heat loss becomes excessive, the brain stimulates small groups of skeletal muscles to produce involuntary, rapid, small contractions (shivering). The increase in skeletal muscle activity increases cellular respiration and ATP hydrolysis, which in turn generates additional heat to raise body temperature. Once body temperature returns to normal, the changes in blood flow and muscle activity return to normal.

Sometimes the temperature-regulating mechanism is insufficient to counter environmental extremes of temperature. *Hypothermia,* a body temperature below 35.0°C (95.0°F), can result from prolonged exposure to a cold environment, which overwhelms the body's temperature-regulating mechanism. Without treatment, an initial feeling of coldness and shivering can progress to mental confusion, lethargy, loss of consciousness, and death. Persons with little subcutaneous fat (e.g., elderly or thin persons) are more susceptible to hypothermia. In treating a person with hypothermia, the body temperature must be raised gradually to stabilize the cardiovascular and respiratory systems.

In *hyperthermia,* the temperature-regulating mechanism cannot prevent a dangerous increase in body temperature. A person with hyperthermia has a body temperature over 40°C (104°F). It can be caused by environmental factors, trauma, or drug exposure. Consider a person in an environment with both a high air temperature and high humidity level. The high humidity prevents perspiration from evaporating and cooling the skin surface. The high air temperature also decreases heat loss and, in situations where the environmental temperature is higher than body temperature, heat is actually gained from surrounding air. In such an environment, excessive physical exertion is a

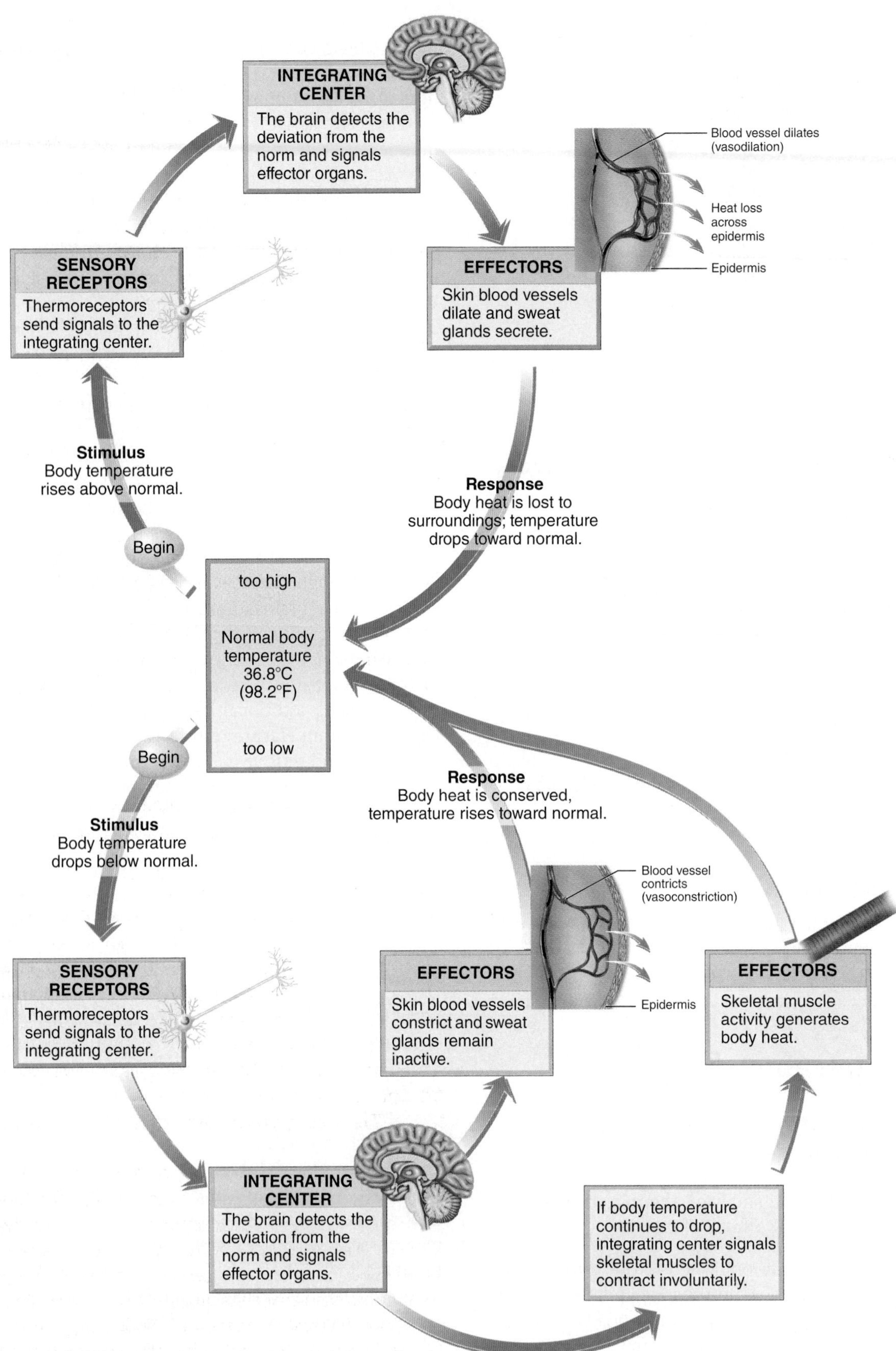

Figure 5.8 The Negative-Feedback Mechanism that Regulates Body Temperature.

Clinical Insight

Physicians have known for a long time that the UV radiation in sunlight produces damaging changes in the skin. The damage is cumulative, although a single day's exposure can produce noticeable changes. Short-term changes include sunburn and a tan as the body tries to protect itself from UV radiation by producing more melanin. Long-term UV damage ranges from increased wrinkling and loss of elasticity to liver spots and skin cancer.

Because the long-term effects of UV radiation do not appear for a number of years, the summer tans so important to some people will accelerate aging of the skin and may produce skin cancers in later life. The advent of tanning salons only increases potential problems for unwary users.

Skin cancer is by far the most common type of cancer. Fortunately, most skin cancers are *carcinomas* involving basal or squamous cells and are usually curable by surgical removal. However, *melanoma,* cancer of melanocytes, tends to spread rapidly to other organs and can be lethal if it is not detected and removed in an early stage. Melanoma is fatal in about 45% of the cases.

The undesirable effects of overexposure to UV radiation can easily be prevented by reducing the exposure of the skin to sunlight and by the liberal use of sunblock. UV radiation is at its peak between 11:00 A.M. and 3:00 P.M., so avoiding exposure during these hours is especially helpful.

Sunblock is available with different levels of protection, and they are labeled according to the sun protection factor (SPF) provided. This allows for the selection of a sunblock that is appropriate for a particular type of skin and duration of exposure. For example, sunscreens with an SPF of 30 or higher are available for fair-skinned persons who burn easily. Sunblock with an SPF of 15 may give adequate protection for olive-skinned persons who rarely burn.

Because of its deadly nature, the American Cancer Society has devised the "ABCD rule" for distinguishing malignant melanoma from a mole: *A* for asymmetry (one side has a different shape than the other); *B* for border irregularity (the border is not uniform); *C* for color (color is not uniform, often a mixture of black, brown, tan, or red); *D* for diameter (more than ¼ inch).

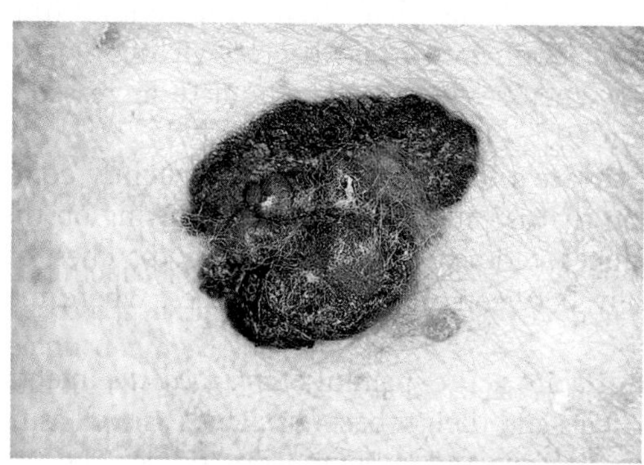

A Cutaneous Malignant Melanoma.

common trigger for hyperthermia. Persons who are dehydrated or overweight are also easily susceptible. Without treatment, progressive symptoms may include nausea, headache, dizziness, confusion, loss of consciousness, and death. Lying in a tub of cool (not cold) water is an effective treatment except in severe cases.

Check My Understanding

7. How is the skin involved in the regulation of body temperature?

5.6 Aging of the Skin

Learning Objective

10. Describe how aging affects the skin.

A newborn infant's skin is very thin, and there is not a lot of subcutaneous fat present. During infancy, an infant's skin thickens and more subcutaneous fat is deposited, producing the soft, smooth skin typical of infants.

As a child grows into adulthood, the skin continues to become thicker because it is subjected to numerous harmful conditions: sunlight, wind, abrasions, chemical irritants, and invasions of bacteria. The continued exposure of the skin through the adult years produces damaging effects. However, noticeable changes usually are not apparent until a person approaches 50 years of age. Thereafter, continued aging of the skin is more noticeable.

Typical changes in aging skin are as follows: (1) a breakdown of collagen and elastic fibers (hastened by exposure to sunlight) causes wrinkles and sagging skin; (2) a decrease in subcutaneous fat makes a person more sensitive to temperature changes; (3) a decrease in sebum production by sebaceous glands may cause dry, itchy skin; (4) a decrease in melanin production produces gray hair and sometimes a splotchy pattern of pigmentation; and (5) a decrease in hair replacement results in thinning hair or baldness, especially in males.

5.6 Aging of the Skin

- After 50 years of age, wrinkles and sagging skin become noticeable.
- The effects of aging are caused by a breakdown of collagen and elastic fibers, a decrease in sebum production, a decrease in melanin production, and a decrease in subcutaneous fat.

5.7 Disorders of the Skin

- Infectious disorders of the skin include acne, athlete's foot, boils, fever blisters, and impetigo.
- Noninfectious disorders of the skin include alopecia, bed bugs, bedsores, blisters, burns, calluses and corns, common moles, dandruff, eczema, hives, and psoriasis.

Self-Review

Answers are located in Appendix B.

1. New epidermal cells are formed by the stratum ______.
2. Resistance to abrasion and waterproofing of the epidermis are due to the presence of ______.
3. The strength and elasticity of the skin are due to protein fibers within the ______.
4. The skin is attached to deeper tissues and organs by the ______.
5. Epidermal cells are nourished by blood vessels located in dermal ______.
6. The ______ glands produce an oily secretion that keeps the hair and skin moist, soft, and pliable.
7. Watery perspiration is produced by ______ sweat glands.
8. Goose bumps (also called chicken skin) are produced when ______ muscles contract.
9. Constriction of dermal blood vessels ______ heat loss.
10. As the skin ages, a breakdown of collagen and elastic fibers leads to the formation of ______.
11. ______ is a common skin disorder caused by a fungus.
12. ______ and ______ are thickened areas of epidermis resulting from chronic pressure.

Critical Thinking

1. Consider the functions of hair and explain how it contributes to maintaining homeostasis.
2. Why does keratinized stratified squamous epithelium form the epidermis and not one of the other epithelial tissues or a connective tissue?
3. Skin cancers in dark-skinned people are relatively rare. Explain why.
4. Young people of Caucasian ancestry seem to enjoy sunbathing to get a "tan." Why is a "tan" a temporary state? Why does a "tan" develop more quickly on the upper and lower limbs compared to the torso?

ADDITIONAL RESOURCES

6 CHAPTER

Skeletal System

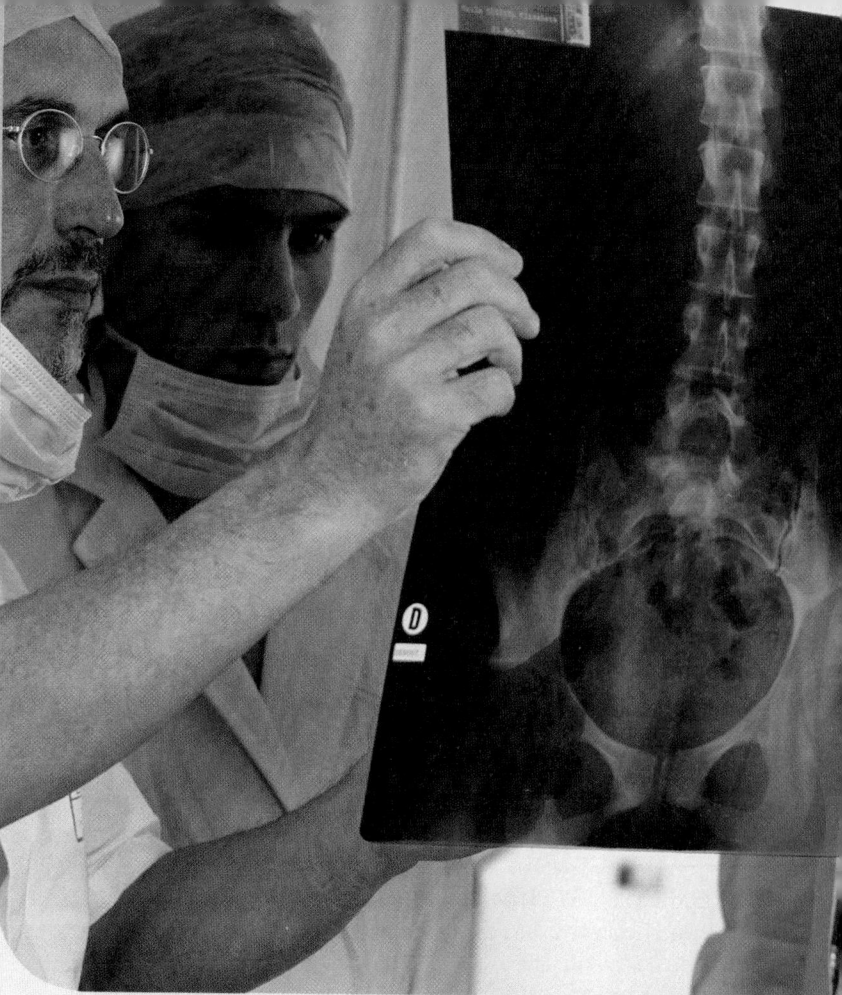

Steven and his friends are amateur skateboarders hanging out at their city skate park. At the urging of his friends, Steven decides to try a 360° spin for the very first time even though he has forgotten his helmet and other safety gear. As he reaches the top of the ramp and begins his spin, his right foot slips off the skateboard, disturbing his balance. Steven throws his arms out to brace his fall but he is unable to keep his head from impacting the ramp as he rolls to the bottom. He sits up and rubs his head, stunned from the fall. His friends race down and begin to check Steven's limbs for fractures. Thankfully, the dense minerals in the bones of his arms and legs were able to resist fracturing. His skull was also hard enough to protect his brain from damage when it hit the ground. Covered in minor cuts and abrasions, Steven stands up and grabs his skateboard. Without even a thought, he fearlessly walks back to the top of the ramp, convinced that he will conquer the 360° spin today.

CHAPTER OUTLINE

Module 05
Skeletal System

SELECTED KEY TERMS

Amphiarthrosis (amphi = two-sided; arthrosis = joint) A slightly movable joint.
Articulation (articul = joint) A joint formed between two bones or between a bone and a tooth.
Compact bone Dense bone formed of numerous tightly packed osteons.
Diaphysis (dia = through, apart; physis = to grow) The shaft of a long bone.
Diarthrosis (dia = apart, through) A freely movable joint.
Endochondral ossification (endo = inside; chondr = cartilage; oss = bone) The formation of bone within cartilage.
Epiphysial (growth) plate The hyaline cartilage between the epiphysis and diaphysis of a growing long bone.
Epiphysis (epi = upon) The enlarged ends of a long bone.
Intramembranous ossification (intra = inside) The formation of bone within embryonic connective tissue.
Ligament A band or cord of dense regular connective tissue that joins bones together at joints.
Medullary cavity (medulla = marrow) The cavity within the shaft of a long bone that is filled with yellow bone marrow.
Spongy (trabecular) bone Bone composed of interconnected bony plates surrounded by red or yellow bone marrow.
Synarthrosis (syn = together) An immovable joint.
Synovial joint (syn = with; ov = egg) A freely movable joint containing a joint cavity filled with synovial fluid.

THE SKELETAL SYSTEM SERVES as the supporting framework of the body, and performs several other important functions as well. The body shape, mechanisms of movement, and erect posture observed in humans would be impossible without the skeletal system. Two very strong tissues, bone and cartilage, compose the skeletal system.

6.1 Functions of the Skeletal System

Learning Objective

1. Describe the basic functions of the skeletal system.

The skeletal system performs five major functions:

1. **Support.** The skeleton serves as a rigid supporting framework for the soft tissues of the body.
2. **Protection.** The arrangement of bones in the skeleton provides protection for many internal organs. The thoracic cage provides protection for the internal thoracic organs including the heart and lungs; the cranial bones form a protective case around the brain, ears, and all but the anterior portion of the eyes; vertebrae protect the spinal cord; and the pelvic girdle protects some reproductive, urinary, and digestive organs.
3. **Attachment sites for skeletal muscles.** Skeletal muscles are attached to bones and extend across articulations. Bones function as levers, enabling movement at joints when skeletal muscles contract.
4. **Blood cell production.** The red bone marrow in spongy bone produces formed elements.
5. **Mineral storage.** The matrix of bones serves as a storage area for large amounts of calcium salts, which may be removed for use in other parts of the body when needed.

6.2 Bone Structure

Learning Objectives

2. List the types of bones based on their shapes.
3. Describe the gross structure and microstructure of a long bone and a flat bone.

There are approximately 206 bones in an adult and each bone is an organ composed of a number of tissues. Bone tissue forms the bulk of each bone and consists of both living cells and a nonliving matrix formed primarily of calcium salts. Other tissues include cartilage, blood, nervous tissue, adipose tissue, and dense irregular connective tissue.

There are six basic types of bones based on their shapes (figure 6.1). *Short bones* are the bones within the wrist and ankle and they possess a small, boxy appearance. *Long bones* possess a long, skinny shape and are found in the upper and lower limbs, with the exception of the wrists, ankles, and patella (kneecap). *Sutural bones* are small bones that form within the sutures of the skull and they vary in number and location from person to person. Certain skull bones, the ribs, and the sternum (breastbone) are classified as *flat bones,* meaning they are thin, flat, and slightly curved. *Irregular bones,* such as the vertebrae (spinal column), coxal bones (hip bones), and some skull bones, possess irregular shapes with numerous projections. *Sesamoid bones* are unique because

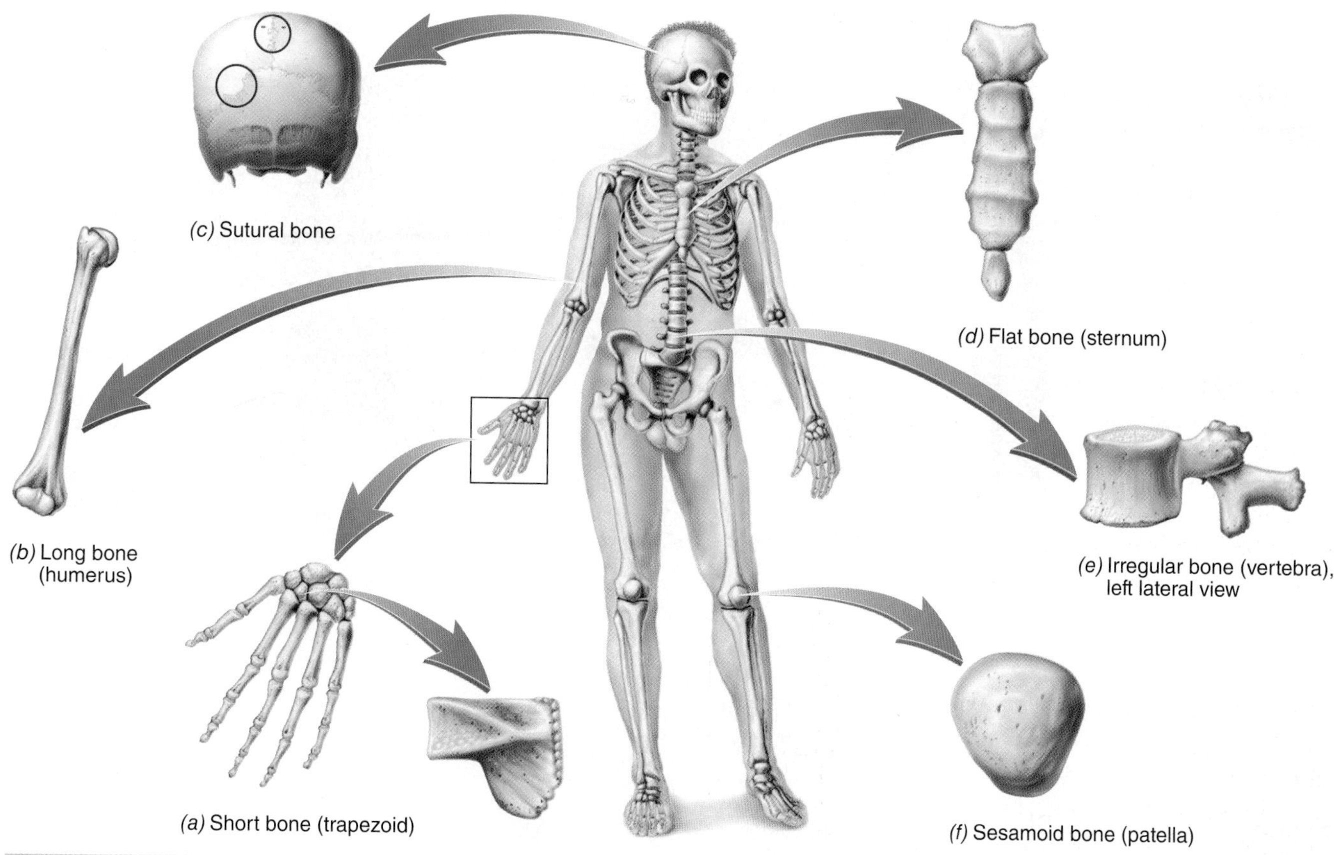

Figure 6.1 Basic Types of Bones.

of their sesame seed shape and the fact that they form inside muscle tendons. The patella is a sesamoid bone.

Gross Structure of a Long Bone

The femur, the bone of the thigh, will be used as an example in considering the structure of a long bone. Refer to figure 6.2 as you study the following section.

At each end of the bone, there is an enlarged portion called an **epiphysis** (ē-pif′-e-sis). The *epiphyses* (plural) articulate with adjacent bones to form joints. The **articular cartilage,** which is composed of hyaline cartilage, covers the articular surface of each epiphysis. Its purpose is to protect and cushion the end of the bone, in addition to providing a smooth surface for movement of joints. The long shaft of bone that extends between the two epiphyses is the **diaphysis** (dī-af′-e-sis). Each epiphysis is joined to the diaphysis by an **epiphysial (growth) plate** of hyaline cartilage in immature bones or by an **epiphysial line,** a line of fusion, in mature bones.

Except for the region covered by articular cartilages, the entire bone is covered by the **periosteum** (per-ē-os′-tē-um), a dense irregular connective tissue membrane that is firmly attached to the underlying bone. The periosteum provides protection and also is involved in the formation and repair of bone. Tiny blood vessels from the periosteum help to nourish the bone.

The internal structure of a long bone is revealed by a longitudinal section. **Spongy (trabecular) bone** forms the internal structure of the epiphyses and the internal surface of the diaphysis wall. It consists of thin rods or plates called *trabeculae* (trah-bek′-u-lē) that form a meshlike framework containing numerous spaces. The trabeculae are covered by a thin connective tissue membrane called **endosteum** (en-dos′-te-um) that is involved in forming and repairing bone. Spongy bone reduces the weight of a bone without reducing its supportive strength. In an adult's long bones, **red bone marrow** fills the spaces between trabeculae within the proximal epiphyses of the humerus and femur. In other epiphyses of the limbs, the spaces between trabeculae are filled with **yellow bone marrow,** which is composed of adipose tissue. The blood-forming properties of red bone marrow will be discussed in chapter 11.

Compact bone forms the wall of the diaphysis and a thin superficial layer over the epiphyses. As the name implies, compact bone is formed of tightly packed bone that lacks the spaces found in spongy bone. Compact bone is very strong, and it provides the supportive strength of long bones. The cavity that extends the length of the

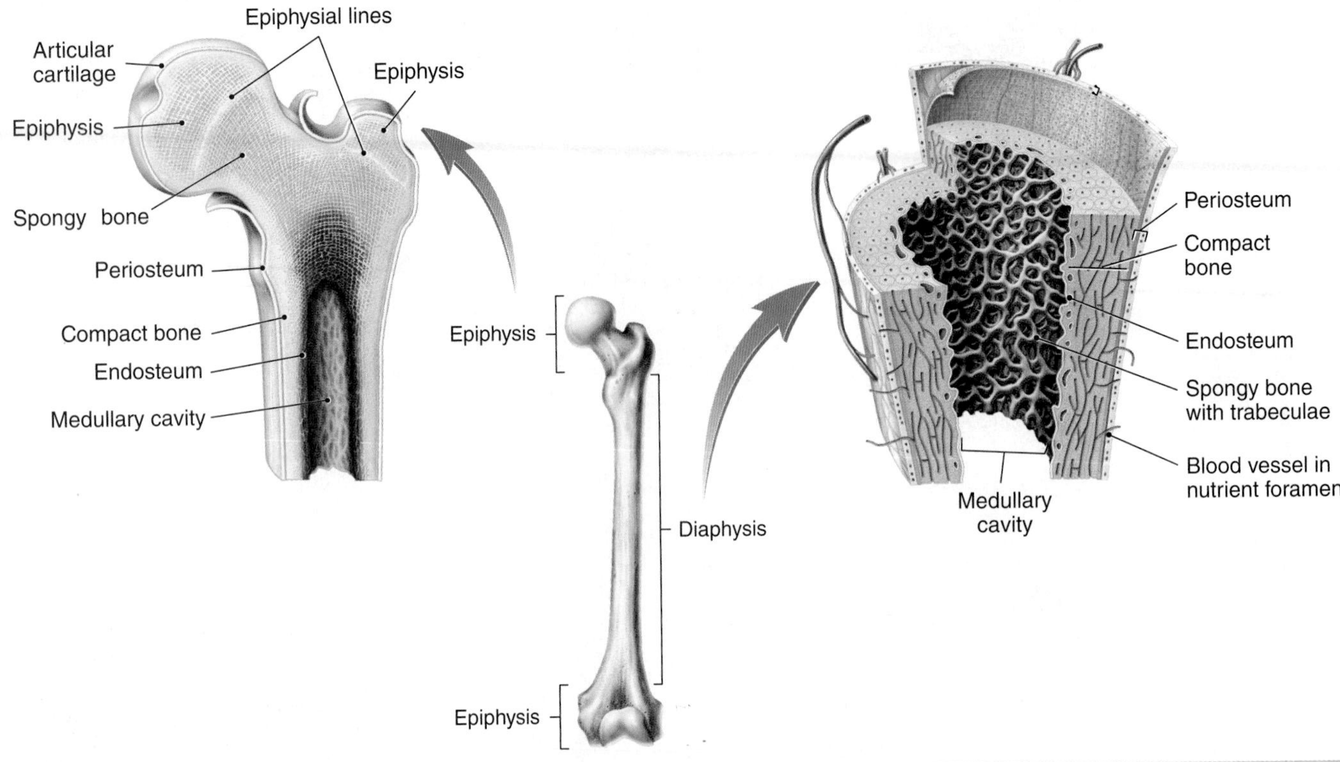

Figure 6.2 The Gross Structure of a Long Bone. APR

diaphysis is the **medullary cavity**. It is lined by the endosteum and is filled with yellow bone marrow.

Microscopic Structure of a Long Bone

As noted earlier, there are two types of bone: compact bone and spongy bone. When viewed microscopically, compact bone is formed of a number of subunits called osteons (figure 6.3). An **osteon** (os′-tē-on) is composed of a **central canal** containing blood vessels and nerves, surrounded by the **lamellae** (singular, *lamella*), concentric layers of bone matrix. Bone cells, the **osteocytes** (os′-tē-ō-sītz), are arranged in concentric rings between the lamellae and occupy tiny spaces in the bone matrix called **lacunae.**

Blood vessels and nerves enter a bone through a **nutrient foramen** (fō-rā′-men; plural, *foramina*), a channel entering or passing through a bone. The blood vessels form branches that pass through *perforating canals* and enter the central canals to supply nutrients to the osteocytes. **Canaliculi** the tiny tunnels radiating from the lacunae, interconnect osteocytes with each other and the blood supply.

The trabeculae of spongy bone lack osteons, so osteocytes receive nutrients by diffusion of materials through canaliculi from blood vessels in the bone marrow surrounding the trabeculae (figure 6.3).

The structure of the other bone types is like that of the epiphyses of long bones. Their external structure is a thin layer of compact bone covered with periosteum; the internal structure is spongy bone covered with endosteum. In most of these bones, red bone marrow fills in the spaces between the trabeculae.

Check My Understanding

1. What are the general functions of the skeletal system?
2. What are the major gross anatomical structures of a long bone?
3. How are compact and spongy bone histologically different?

6.3 Bone Formation

Learning Objectives

4. Compare intramembranous and endochondral ossification.
5. Compare the functions of osteoblasts and osteoclasts.

The process of bone formation is called **ossification** (os-i-fi-kā′-shun). It begins during the sixth or seventh week

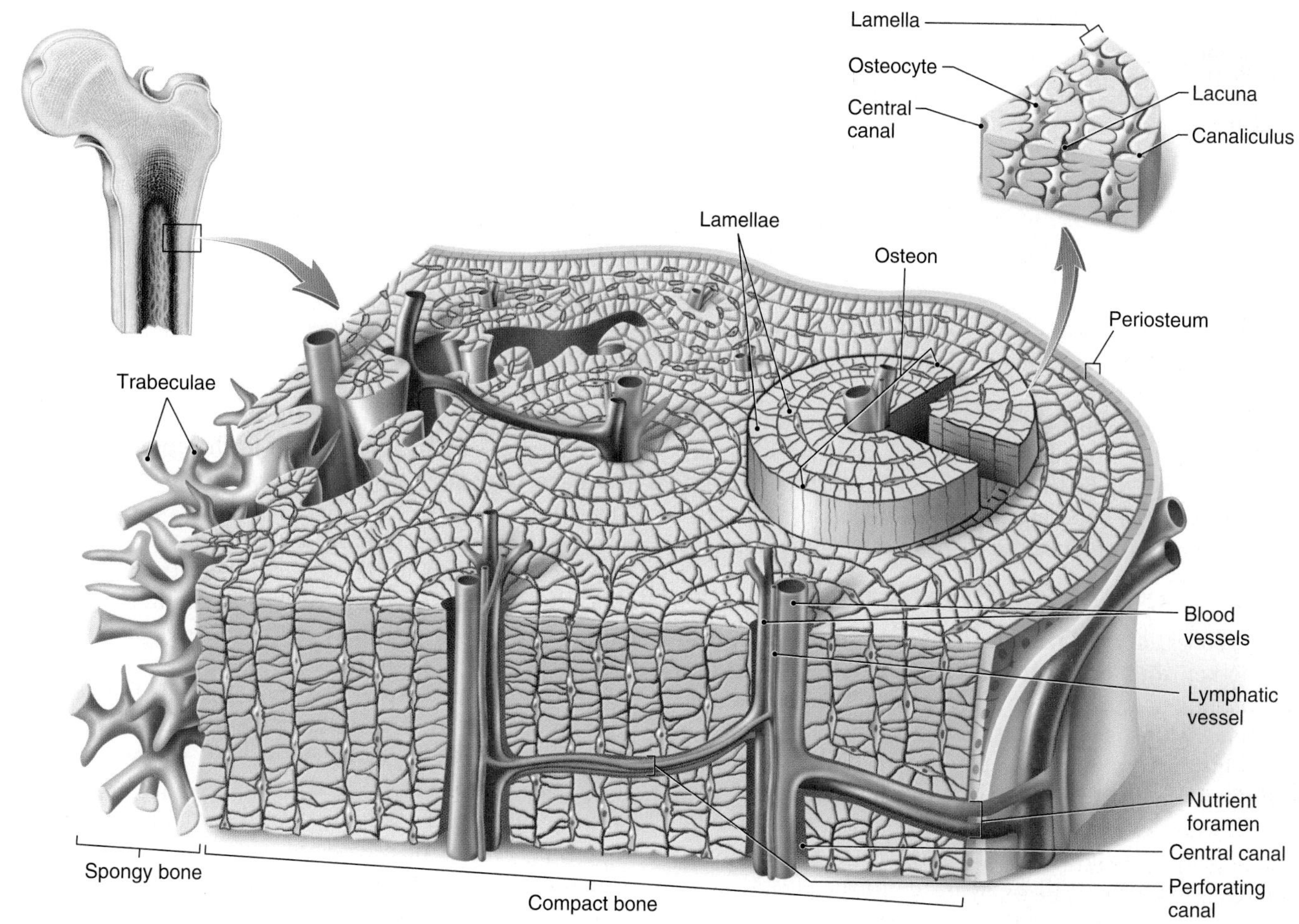

Figure 6.3 Microstructure of a Long Bone.

of embryonic development. Bones are formed by the replacement of existing connective tissues with bone (figure 6.4). There are two types of bone formation: intramembranous ossification and endochondral ossification. Table 6.1 summarizes these.

In both types of ossification, some primitive connective tissue cells are changed into bone-forming cells called **osteoblasts** (os′-tē-ō-blasts). Osteoblasts deposit bone matrix around themselves and soon become imprisoned in lacunae. Once this occurs, they are called osteocytes.

Intramembranous Ossification

Most skull bones are formed by **intramembranous ossification**. Connective tissue membranes form early in embryonic development at sites of future intramembranous bones. Later, some connective tissue cells become osteoblasts and deposit spongy bone within the membranes starting in the center of the future bone. Osteoblasts from the periosteum deposit a layer of compact bone over the spongy bone.

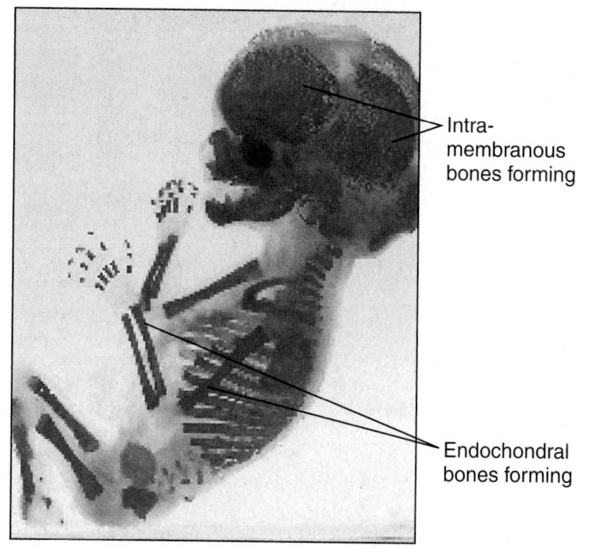

Figure 6.4 The stained developing bones of a 14-week fetus.

Table 6.1 Comparison of Intramembranous and Endochondral Ossification

Intramembranous	Endochondral
1. Membranes of embryonic connective tissue form at sites of future bones.	1. Bone is preformed in hyaline cartilage.
2. Some connective tissue cells become osteoblasts, which deposit spongy bone within the membrane.	2. Osteoblasts of periosteum form a collar of compact bone that thickens and grows toward each end of the bone.
3. Osteoblasts from the enclosing membrane, now called the periosteum, deposit a layer of compact bone over the spongy bone.	3. Cartilage is calcified, and osteoblasts derived from the periosteum form spongy bone, which replaces cartilage in ossification centers. The spongy bone is later removed in the diaphysis to form the medullary cavity.

Some bone must be removed and re-formed in order to produce the correct shape of the bone as it develops and grows. Cells that remove bone matrix are called **osteoclasts.** The opposing actions of osteoblasts and osteoclasts ultimately produce the shape of the mature bone.

Endochondral Ossification

Most bones of the body are formed by **endochondral** (en-dō-kon′-drul) **ossification**. Future endochondral bones are preformed in hyaline cartilage early in embryonic development. Figure 6.5 illustrates the ossification of a long bone.

In long bones, a new periosteum develops around the diaphysis of the hyaline cartilage template. Osteoblasts from the periosteum form a collar of compact bone around the diaphysis. A *primary ossification center* also forms in the middle of the cartilage shaft due to the enlargement of chondrocytes and a loss of cartilage matrix between lacunae. Calcification, which involves the depositing of calcium salts, occurs within the primary ossification center and leads to the death of chondrocytes. Blood vessels and nerves penetrate into the primary ossification center carrying along osteoblasts from the periosteum. As *secondary ossification centers* form in the epiphyses of the cartilage template, osteoclasts begin to remove spongy bone from the diaphysis to form the medullary cavity. The bone continues to grow as ossification progresses. As cartilage continues to be replaced, the cartilage between the primary and secondary ossification centers decreases until only a thin plate of cartilage, the *epiphysial plate,* separates the epiphyses from the diaphysis.

Subsequent growth in diameter results from continued formation of compact bone by osteoblasts from the periosteum. Growth in length occurs as bone replaces cartilage on the diaphysis side of each epiphysial plate while new cartilage is formed on the epiphysis side. The opposing actions of osteoblasts and osteoclasts continually reshape the bone as it grows.

Growth usually continues until about age 25, when the epiphysial plates are completely replaced by bone. After this, growth in the length of a bone is not possible. The visible lines of fusion between the epiphyses and the diaphysis are called *epiphysial lines.*

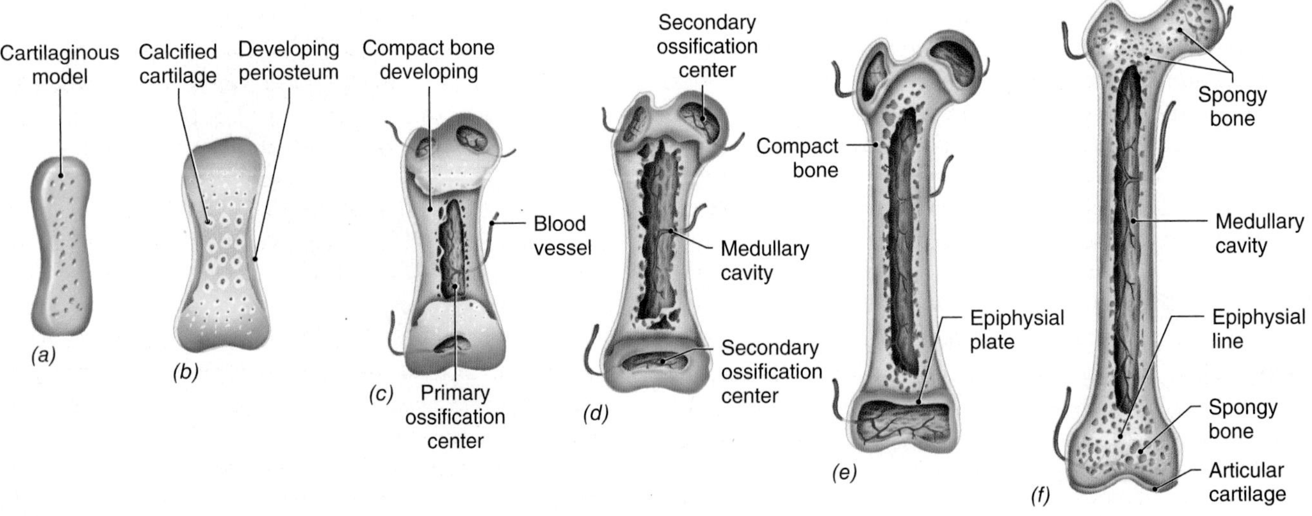

Figure 6.5 Major stages (*a–f*) in the development of an endochondral bone. (Bones are not shown to scale.)

Homeostasis of Bone

Bones are dynamic, living organs, and they are continually restructured throughout life. This occurs by the removal of bone matrix by osteoclasts and by the deposition of new bone matrix by osteoblasts. Physical activity causes the density and volume of bones to be maintained or increased, though inactivity results in a reduction in bone density and volume.

Calcium salts may be removed from bones to meet body needs anytime blood calcium levels are low, such as when dietary intake is inadequate. When dietary calcium intake increases blood calcium to a sufficient level, some calcium is used to form new bone matrix.

Children have a relatively large number of protein fibers in their bone matrix, which makes their bones somewhat flexible. But as people age, the amount of protein gradually decreases. This trend causes older people to have brittle bones that are prone to fractures. Older persons may also experience a gradual loss of bone matrix, which reduces the strength of the bones. A severe reduction in bone density, and therefore increased risk of fracture, is called *osteoporosis.*

 Check My Understanding

4. How do intramembranous ossification and endochondral ossification differ?
5. How does physical activity affect the homeostasis of bones?

6.4 Divisions of the Skeleton

Learning Objectives

6. Name the two divisions of the skeleton.
7. Describe the major surface features of the bones and their importance.

The human adult skeleton is composed of two distinct divisions: the axial skeleton and the appendicular skeleton. The **axial** (ak′-sē-al) **skeleton** consists of the bones along the longitudinal axis of the body that support the head, neck, and trunk. The **appendicular** (ap-en-dik′-ū-lar) **skeleton** consists of the bones of the upper limbs and pectoral girdle and of the lower limbs and pelvic girdle (figure 6.6).

A study of the skeleton includes the various surface features of bones, such as projections, depressions, ridges, grooves, and holes. Specific names are given to each type of surface feature. Knowledge of surface bony features is essential for understanding the origins and insertions of skeletal muscles discussed in the muscular system, and is important in locating internal structures in clinical practice. The names of the major surface features are listed in table 6.2 and shown in figure 6.7 for easy reference as you study the bones of the skeleton.

6.5 Axial Skeleton

Learning Objectives

8. Identify the bones of the axial skeleton.
9. Compare the skulls of an infant and an adult.
10. Compare cervical, thoracic, lumbar, and sacral vertebrae.
11. Compare true, false, and floating ribs.

The major components of the axial skeleton are the skull, vertebral column, and thoracic cage. Bones of the axial skeleton are shown in figure 6.6.

Skull APR

The **skull** is subdivided into the *cranium,* which is formed of eight bones encasing the brain, and 14 *facial bones.* With the exception of the mandible, all the skull bones are joined by immovable joints, called **sutures** (sū′-churs) because they resemble stitches. The cranial cavity formed by the cranium protects the brain. The facial bones surround and support the openings of the digestive and respiratory systems. Several bones in the skull contain air-filled spaces called **paranasal sinuses** (figure 6.9) that are connected to the nasal cavity. These sinuses reduce the weight of the skull, add resonance to a person's voice, and produce mucus, which helps to moisten and purify the air within the sinuses and within the nasal cavity. The bones of the skull are shown in figures 6.8 to 6.12. Locate the bones on these figures as you study this section.

Cranium

The **cranium** is formed of one frontal bone, two parietal bones, one sphenoid, two temporal bones, one occipital bone, and one ethmoid.

The **frontal bone** forms the anterior part of the cranium, including the superior portion of the orbits (eye sockets), the forehead, and the roof of the nasal cavity. There are two large *frontal sinuses* in the frontal bone, one located superior to each eye.

The two **parietal** (pah-rī′-e-tal) **bones** form the sides and roof of the cranium. They are joined at the midline by the *sagittal suture* and to the frontal bone by the *coronal suture.*

The **occipital** (ok-sip′-i-tal) **bone** forms the posterior portion and floor of the cranium. It contains a large opening, the *foramen magnum,* through which the brainstem extends to join with the spinal cord. On each side of the foramen magnum are the *occipital condyles* (kon′-dīls), large knucklelike surfaces that articulate with the first vertebra of the vertebral column. The occipital bone is joined to the parietal bones by the *lambdoid* (lam′doyd) *suture.*

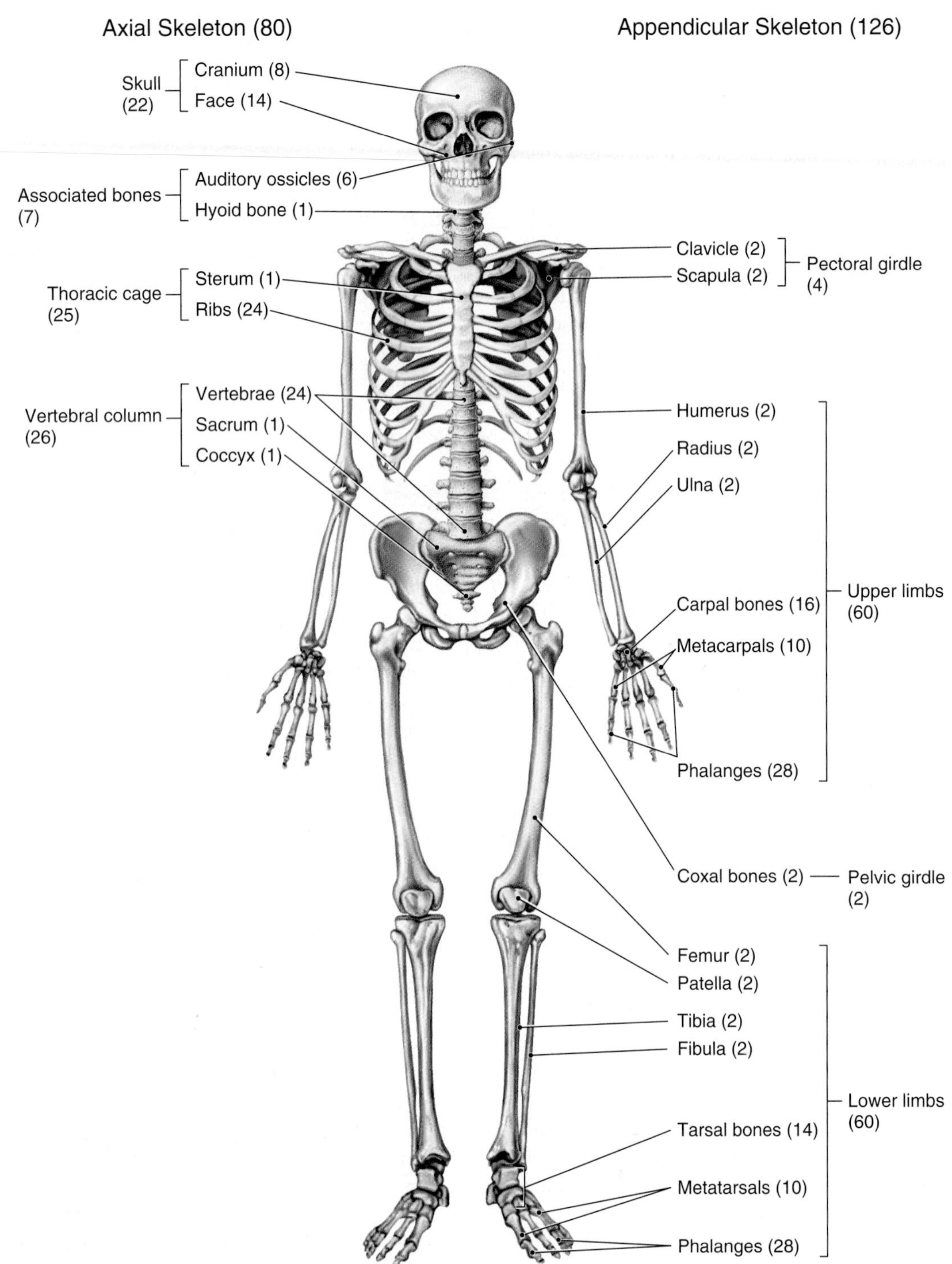

Figure 6.6 Bones of axial skeleton (colored gold) and appendicular skeleton (colored blue). APR

The **temporal bones** are located inferior to the parietal bones on each side of the cranium. They are joined to the parietal bones by *squamous* (skwā'mus) *sutures* and to the occipital bone by the lambdoid suture. In each temporal bone, an **external acoustic meatus** leads inward to the eardrum. Just anterior to the external acoustic meatus is the *mandibular fossa,* a depression that receives the mandibular condyle to form the *temporomandibular joint.*

Three processes are located on each temporal bone. The *zygomatic* (zī-gō-mat'-ic) *process* projects anteriorly to join with the zygomatic bone. The *mastoid* (mas'-toyd) *process* is a large, rounded projection that is located inferior

Table 6.2 Surface Features of Bones

Feature	Description
Processes Forming Joints	
Condyle	A rounded or knucklelike process
Head	An enlarged rounded end of a bone supported by a constricted neck
Facet	A smooth, nearly flat surface
Processes for Attachment of Ligaments and Tendons	
Crest	A prominent ridge or border
Epicondyle	A prominence superior to a condyle
Spine	A sharp or slender process
Trochanter	A very large process found on the femur
Tubercle	A small, rounded process
Tuberosity	A large, roughed process
Depressions and Openings	
Alveolus	A deep pit or socket
Canal, Meatus	A tubelike passageway into or through a bone
Foramen	An opening or passageway through a bone
Fossa	A small depression
Groove	A furrowlike depression
Sinus	An air-filled cavity within a bone

to the external acoustic meatus. It serves as an attachment site for some neck muscles. The *styloid process* lies just medial to the mastoid process. It is a long, spikelike process to which muscles and ligaments of the tongue and neck are attached.

The **sphenoid** (sfē′-noid) forms part of the floor of the cranium, the posterior portions of the orbits, and the lateral portions of the cranium just anterior to the temporal bones. Because it articulates with all other cranial bones, the sphenoid is referred to as the "keystone" of the cranium. On its superior surface at the midline is a saddle-shaped structure called the *sella turcica* (ter′-si-ka), or turkish saddle. It has a depression that contains the pituitary gland. Two *sphenoidal sinuses* are located just inferior to the sella turcica.

The **ethmoid** (eth′-moid) forms the anterior portion of the cranium, including part of the medial surface of each orbit and part of the roof of the nasal cavity. The lateral portions contain several air-filled sinuses called *ethmoidal cells*. The *perpendicular plate* extends inferiorly to form most of the nasal septum, which separates the right and left portions of the nasal cavity. It joins the sphenoid and vomer posteriorly and the nasal and frontal bones anteriorly.

The *superior* and *middle nasal conchae* (kong′-kē, singular, concha) extend from the lateral portions of the ethmoid toward the perpendicular plate. These delicate, scroll-like bones support the mucous membrane and increase the surface area of the nasal wall. The roof of the nasal cavity is formed by the *cribriform plate* of the ethmoid; the olfactory nerves enter the cranial cavity through foramina in the cribriform plate. On the superior surface where these plates join at the midline is a prominent projection called the *crista galli*, or cock's comb. The meninges that envelop the brain are attached to the crista galli.

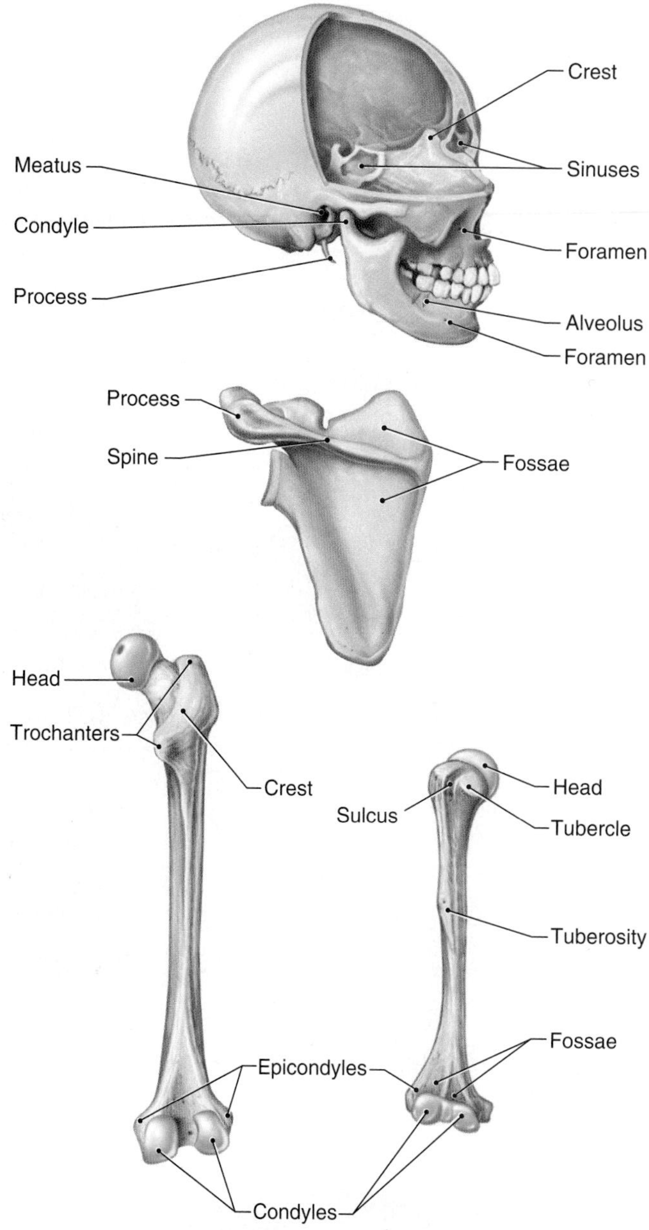

Figure 6.7 Surface Features of Bones.

Facial Bones

The paired bones of the face are the maxillae, palatine bones, zygomatic bones, lacrimal bones, nasal bones, and inferior nasal conchae. The single bones are the vomer and mandible.

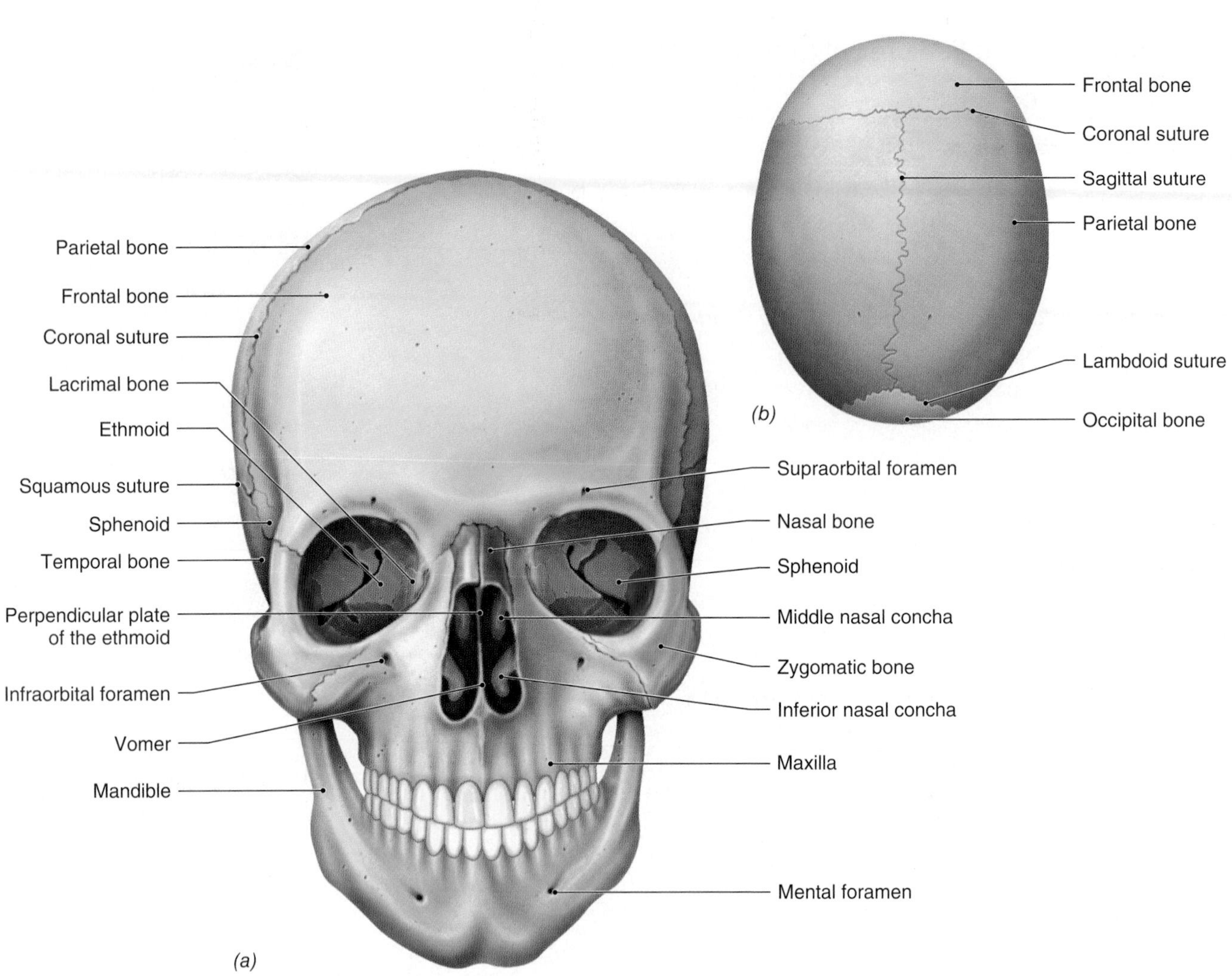

Frontal bone
Coronal suture
Sagittal suture
Parietal bone
Lambdoid suture
Occipital bone
(b)
Parietal bone
Frontal bone
Coronal suture
Lacrimal bone
Ethmoid
Squamous suture
Sphenoid
Temporal bone
Perpendicular plate of the ethmoid
Infraorbital foramen
Vomer
Mandible
Supraorbital foramen
Nasal bone
Sphenoid
Middle nasal concha
Zygomatic bone
Inferior nasal concha
Maxilla
Mental foramen
(a)

Figure 6.8 *(a)* Anterior view of the skull. *(b)* Superior view of the skull. APR

Frontal sinus
Ethmoidal cells
Sphenoidal sinus
Maxillary sinus

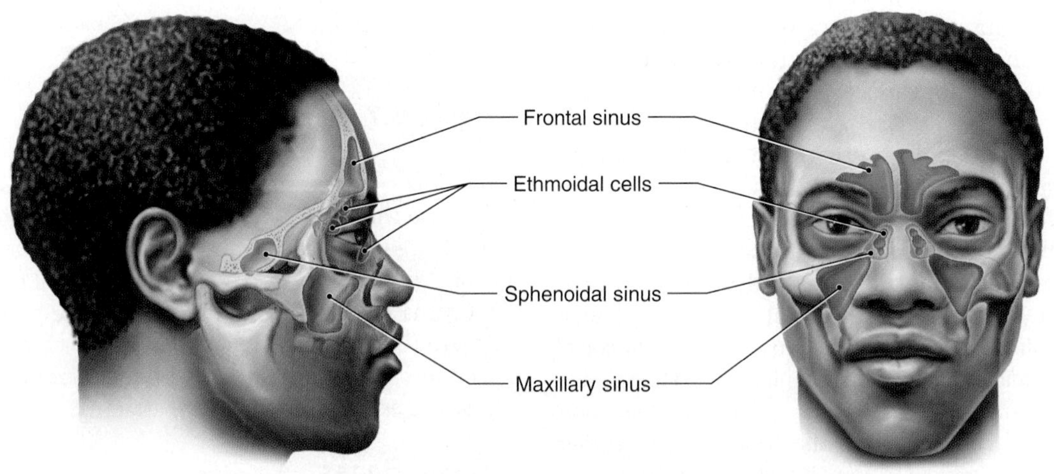

Figure 6.9 Paranasal sinuses are located in the frontal bone, the ethmoid, the sphenoid, and the maxillae. They are connected with the nasal cavity and increase the surface area of the nasal cavity.

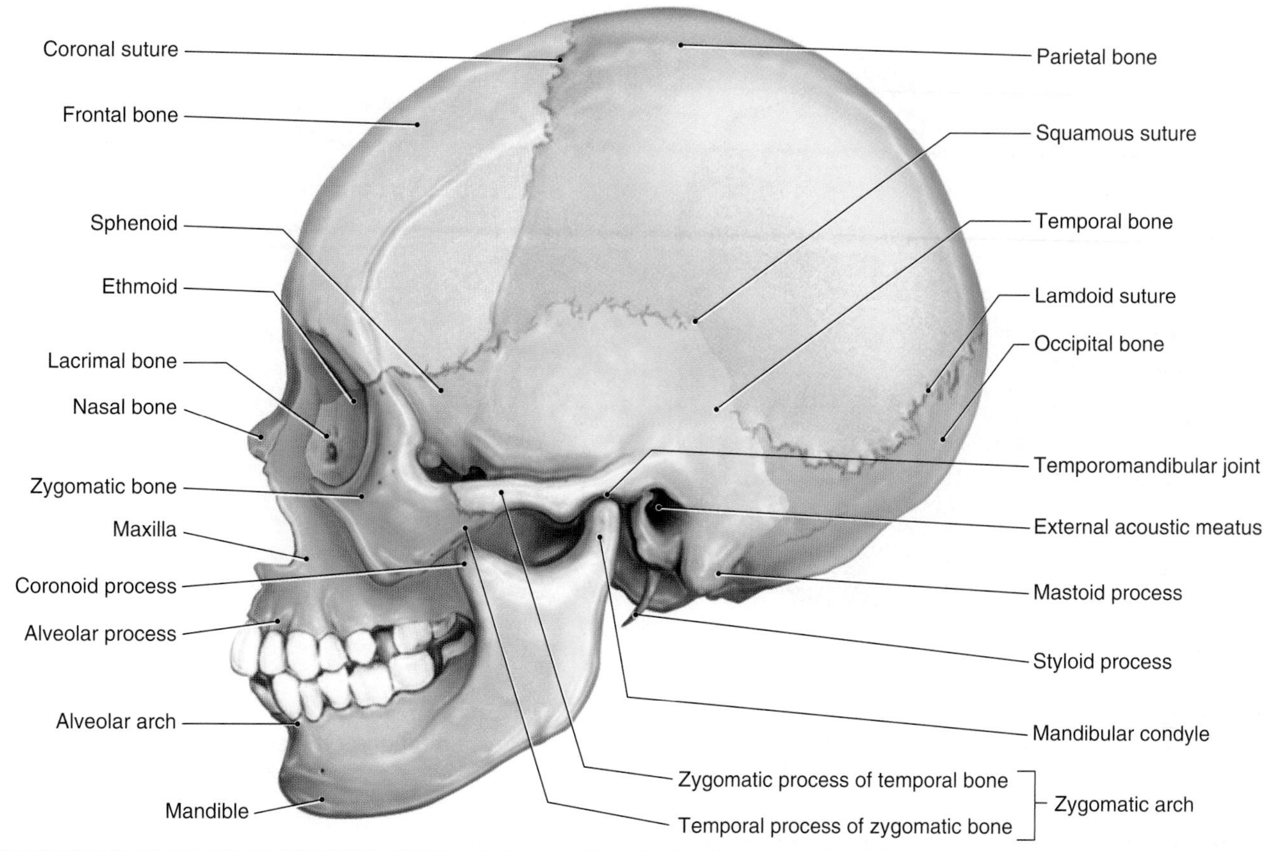

Figure 6.10 Lateral View of the Skull. APR

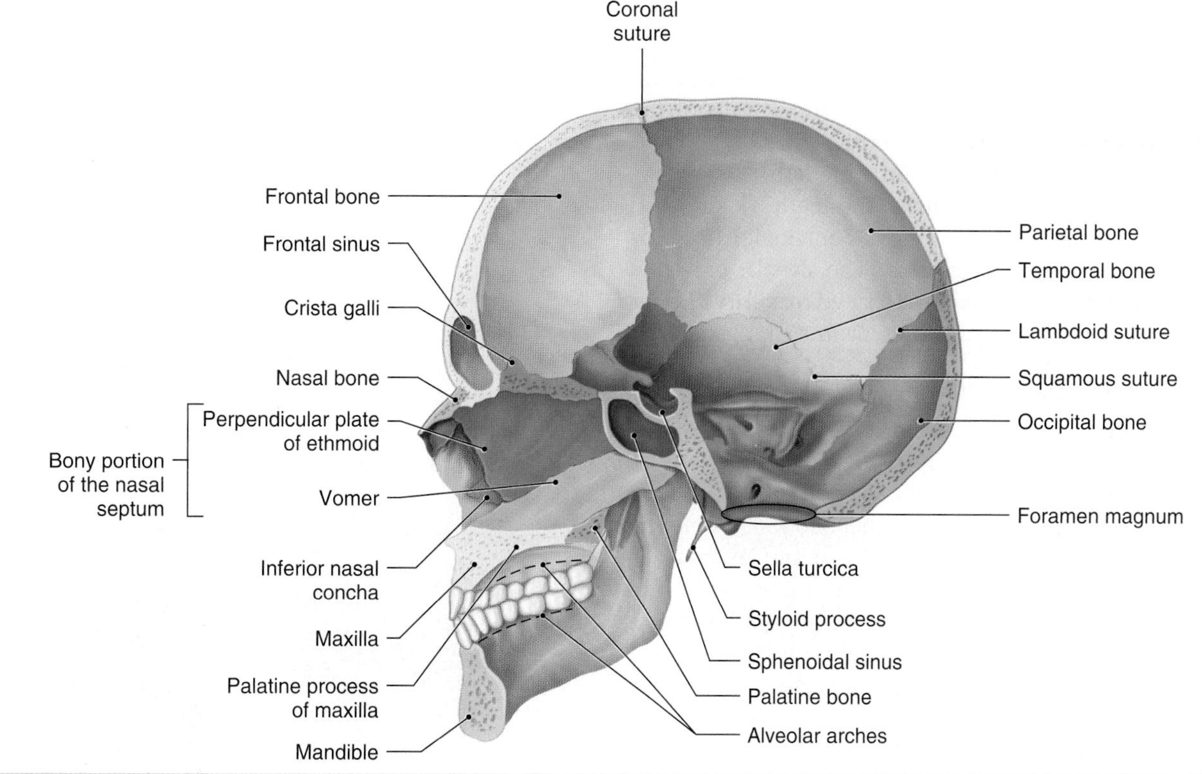

Figure 6.11 Median View of the Skull. APR

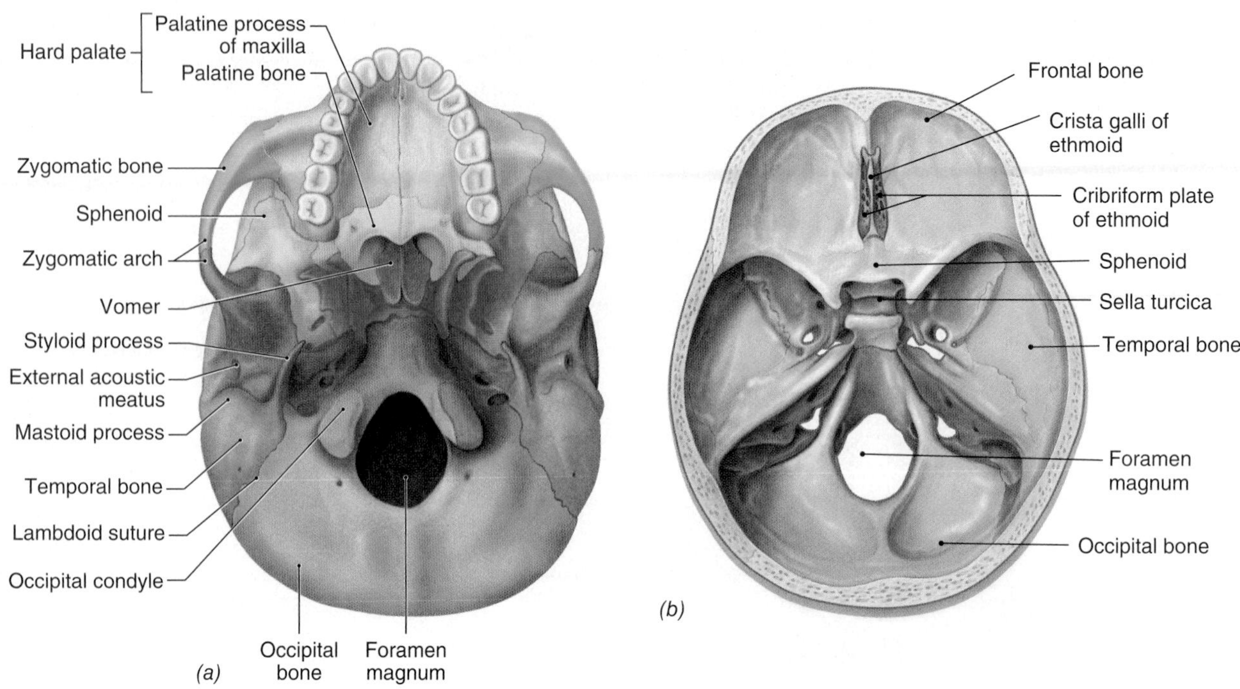

Figure 6.12 *(a)* Inferior view of the skull. *(b)* Superior view of the transverse section of the skull. AP|R

The **maxillae** (mak-sil′-ē) form the upper jaw. Each maxilla is formed separately, but they are joined at the midline during embryonic development. The maxillae articulate with all of the other facial bones except the mandible. The *palatine processes* of the maxillae form the anterior portion of the *hard palate* (roof of the mouth and floor of the nasal cavity), part of the lateral walls of the nasal cavity, and the floors of the orbits.

Each maxilla possesses an inferiorly projecting, curved ridge of bone that contains the teeth. This ridge is the *alveolar process,* and the sockets containing the teeth are called *alveoli* (singular, alveolus). The alveolar processes unite at the midline to form the U-shaped maxillary *alveolar arch.* A large *maxillary sinus* is present in each maxilla just inferior to the orbits.

The **palatine** (pal′-ah-tīn) **bones** are fused at the midline to form the posterior portion of the hard palate. Each bone has a lateral portion that projects superiorly to form part of a lateral wall of the nasal cavity.

The **zygomatic bones** (cheekbones) form the prominences of the cheeks and the floors and lateral walls of the orbits. Each zygomatic bone has a posteriorly projecting process, the *temporal process,* that extends to unite with the zygomatic process of the adjacent temporal bone. Together, they form the *zygomatic arch.*

The **lacrimal** (lak′-ri-mal) **bones** are small, thin bones that form part of the medial surfaces of the orbits. Each lacrimal bone is located between the ethmoid and maxilla.

The **nasal** (nā′-zal) **bones** are thin bones fused at the midline to form the bridge of the nose.

The **vomer** is a thin, flat bone located on the midline of the nasal cavity. It joins posteriorly with the perpendicular plate of the ethmoid, and these two bones form the bony part of the nasal septum.

The **inferior nasal conchae** are scroll-like bones attached to the lateral walls of the nasal cavity inferior to

Clinical Insight

The hard palate separates the nasal cavity from the oral cavity, which allows for chewing and breathing to occur at the same time. A *cleft palate* results when the palatine processes of the maxillae and the palatine bones fail to join before birth to form the hard palate. A *cleft lip,* a split upper lip, is often associated with a cleft palate. These congenital deformities can be corrected surgically after birth.

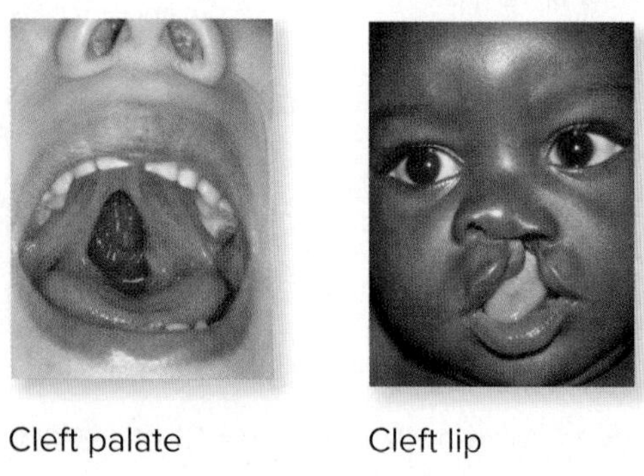

Cleft palate — Cleft lip

the middle nasal conchae of the ethmoid. They project medially into the nasal cavity and serve the same function as the superior and middle nasal conchae of ethmoid.

The **mandible** (lower jaw) is the only movable bone of the skull. It consists of a U-shaped *body* with a superiorly projecting portion, a *ramus,* extending from each end of the body. The superior portion of the body forms the mandibular *alveolar arch,* which contains the alveoli for the teeth. The superior part of each ramus is Y-shaped and forms two projections: an anterior *coronoid process* and a posterior *mandibular condyle.* The coronoid process is a site of attachment for muscles used in chewing. The mandibular condyle articulates with the mandibular fossa of the temporal bone to form a *temporomandibular joint.* These joints are sometimes involved in a variety of dental problems associated with an improper bite.

Associated Bones to the Skull

The **hyoid** (hī′-oyd) **bone** and **auditory ossicles** are known as associated bones to the skull because they are located in or near the skull but are not directly connected with any skull bones. The hyoid bone is a small, U-shaped bone located in the anterior portion of the neck, inferior to the mandible. It does not articulate with any bone. Instead, it is suspended from the styloid processes of the temporal bones by muscles (figure 6.13). Muscles of the tongue are attached to the hyoid bone. The auditory ossicles (malleus, incus, stapes) are the smallest bones in the human body. They articulate with each other in the middle ears and assist in sound conduction and amplification (see chapter 9).

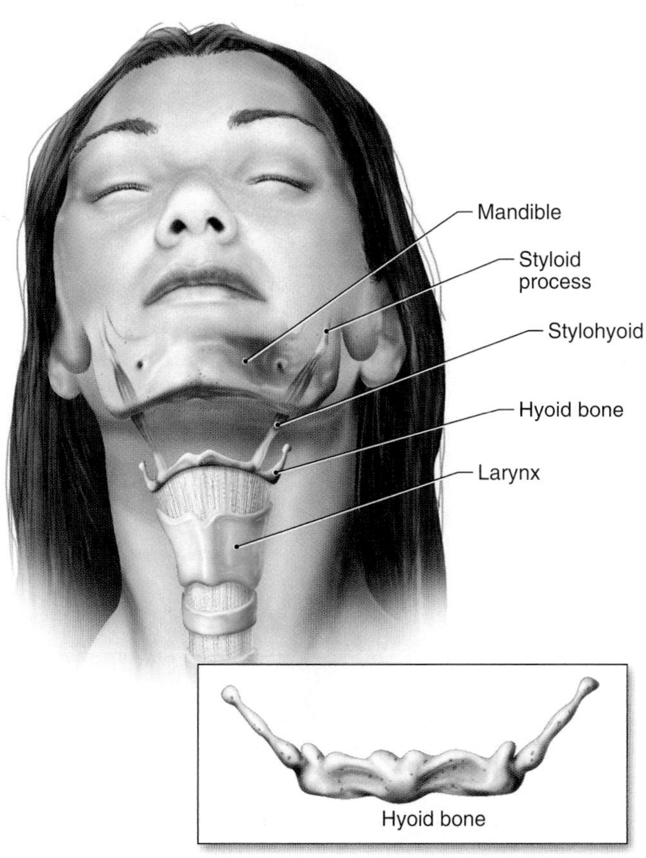

Figure 6.13 The Anterior View of the Hyoid Bone.

The Infant Skull

The skull of a newborn infant is incompletely developed. The face is relatively small with large orbits, and the bones are thin and incompletely ossified. The bones of the cranium are separated by dense connective tissue, with six rather large, nonossified areas called **fontanelles** (fon′-tah-nels) or soft spots (figure 6.14). The frontal bone is formed of two separate parts that fuse later in development. Incomplete ossification of the skull bones and the abundance of dense irregular connective tissue make the skull somewhat flexible and allow for partial compression of the skull to facilitate easier vaginal delivery.

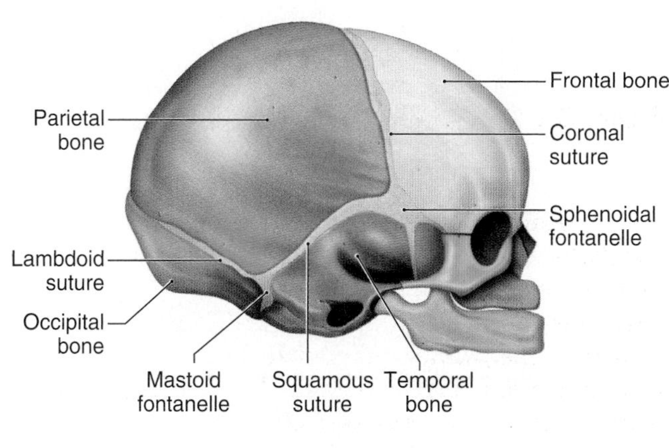

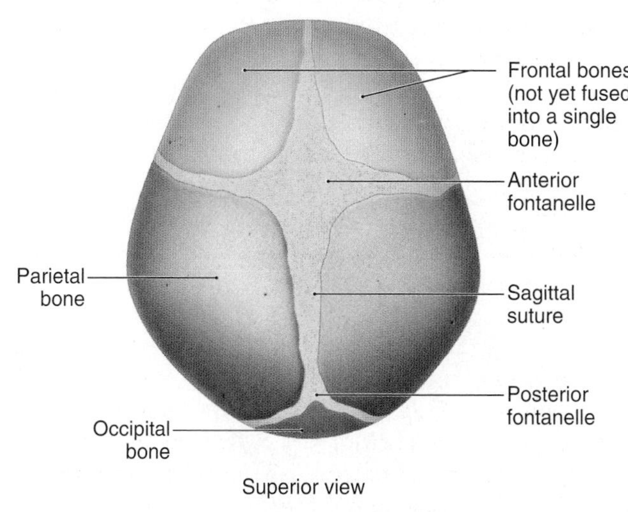

Figure 6.14 There are one anterior, two mastoid, one posterior, and two sphenoidal fontanelles between the cranial bones in the infant's skull. Note the fontanelles and the membranes between the cranial bones.

After birth, they allow for the skull to expand easily and accommodate the rapidly growing brain. Compare the infant skull in figure 6.14 with the adult skull in figure 6.8 and figure 6.10.

Check My Understanding

6. What bones form the cranium?
7. What bones form the face?

Vertebral Column

The vertebral column (spine or backbone) extends from the skull to the pelvis and forms a somewhat flexible but sturdy longitudinal support for the trunk. It is formed of 24 slightly movable vertebrae, the sacrum, and the coccyx. The vertebrae are separated from each other by **intervertebral discs** that serve as shock absorbers and allow bending of the spinal column. Four distinct curvatures can be seen on the lateral view of the vertebral column (figure 6.15). From superior to inferior they are the *cervical, thoracic, lumbar,* and *sacral curvatures.* These curvatures provide flexibility and cushion, and allow the vertebral column to bear body weight more efficiently.

Structure of a Vertebra

Vertebrae are divided into three groups: cervical, thoracic, and lumbar vertebrae. Although each type has a distinctive anatomy, they have many features in common (figure 6.16).

The anterior, drum-shaped mass is the *body,* which serves as the major load-bearing portion of a vertebra. A bony *vertebral arch* surrounds the large *vertebral foramen* through which the spinal cord and nerve roots pass. A

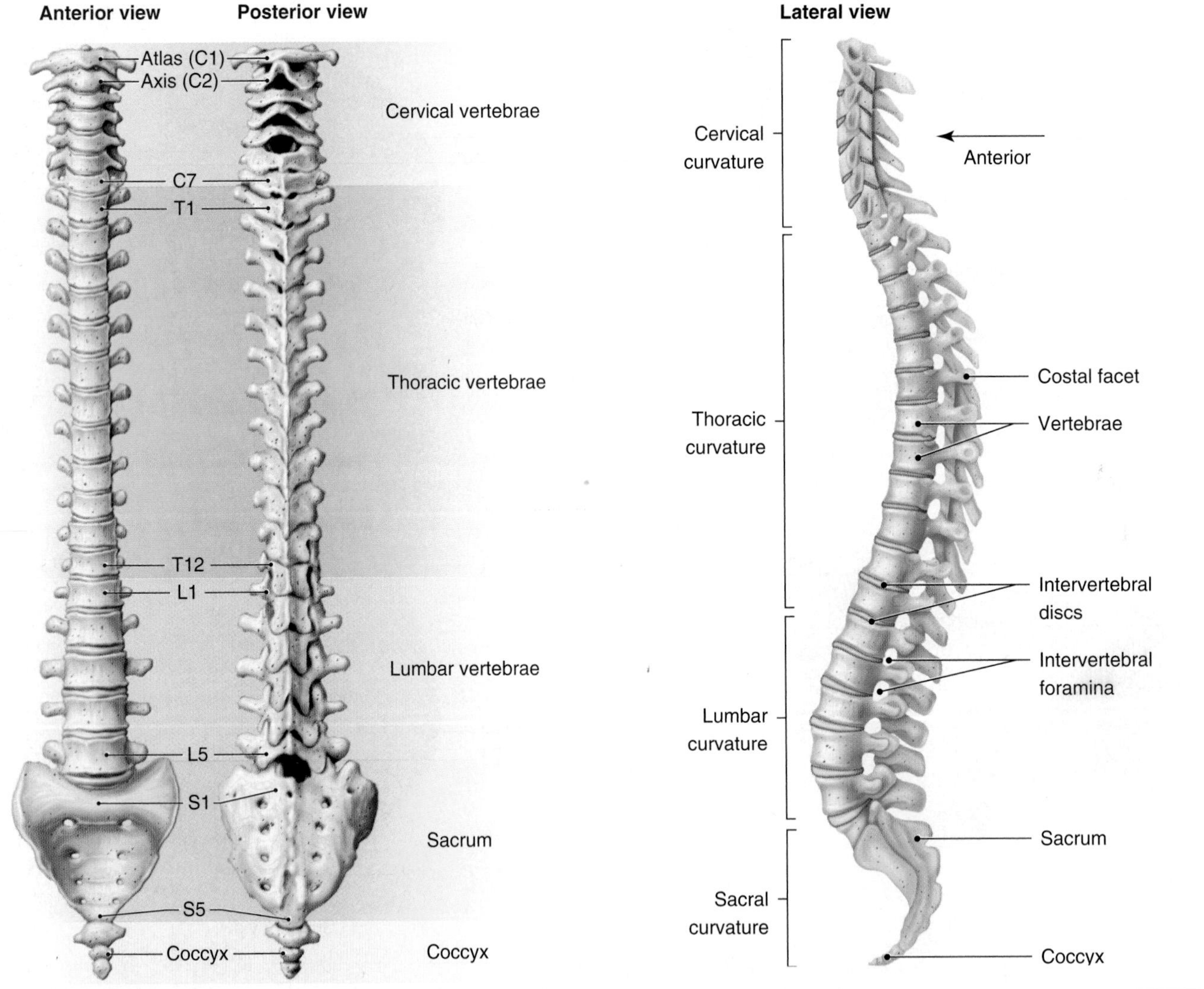

Figure 6.15 The vertebral column consists of 24 movable vertebrae, separated by intervertebral discs, sacrum, and coccyx. APIR

spinous process projects posteriorly and *transverse processes* project laterally from each vertebral arch.

A pair of *superior articular processes* projects superiorly and a pair of *inferior articular processes* projects inferiorly from the vertebral arch. The *articular facet* (fa′-set) of each superior articular process articulates with the articular fact of the inferior articular process of the adjacent vertebra superior to it. When joined by ligaments, the vertebrae form the *vertebral canal* that protects the spinal cord.

Small *intervertebral foramina* occur between adjacent vertebrae. They serve as lateral passageways for spinal nerves that exit the spinal cord (see figure 6.15 and figure 6.16).

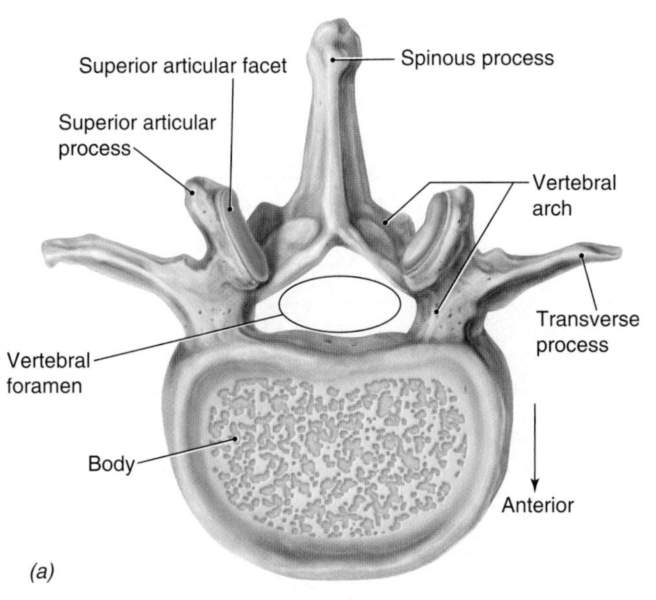

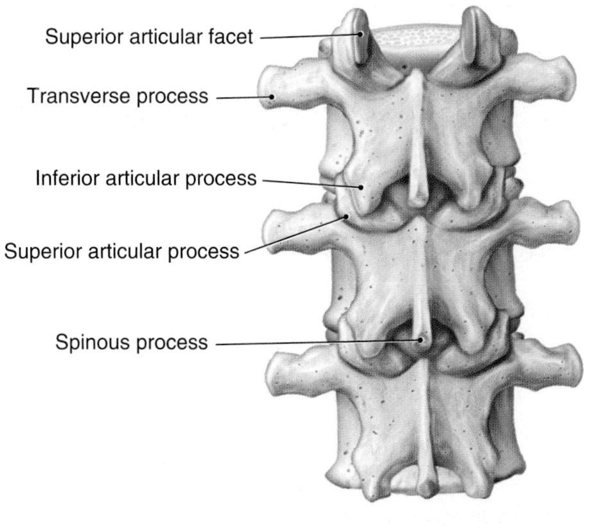

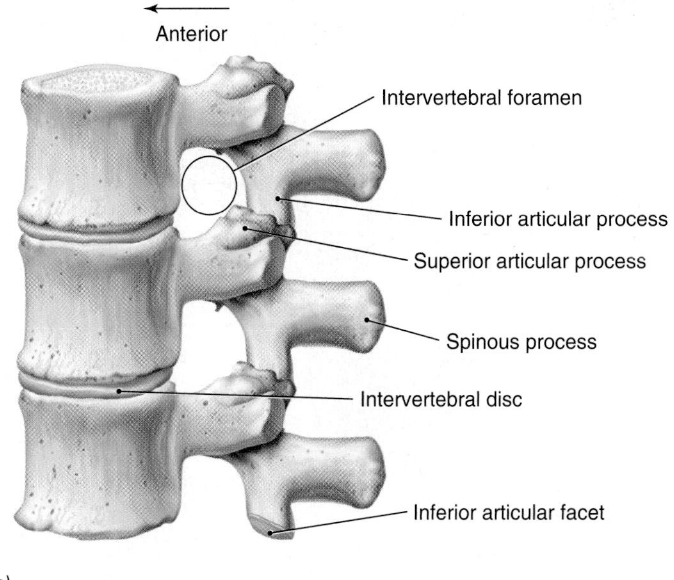

Figure 6.16 *(a)* Superior view of a lumbar vertebra. *(b)* Posterior view of lumbar vertebrae. *(c)* Lateral view of lumbar vertebrae.

Cervical Vertebrae

The first seven vertebrae are the **cervical** (ser′-vi-kul) **vertebrae** (C1–C7) that support the neck. They are unique in having a *transverse foramen* in each transverse process. It serves as a passageway for the vertebral arteries and veins, blood vessels involved in blood flow to and from the brain (figures 6.17*a–d*).

The first two cervical vertebrae are distinctly different from the rest. The first vertebra (C1), or **atlas,** whose superior articular facets articulate with the occipital condyles, supports the head. The second vertebra (C2), which is called the **axis,** has a prominent *dens* that projects superiorly from the vertebral body, providing a pivot point for the atlas. When the head is turned, the atlas rotates on the axis (see figure 6.17*a–c*).

Thoracic Vertebrae

The 12 **thoracic vertebrae** (T1–T12) are larger than the cervical vertebrae, and their spinous processes are longer and slope inferiorly. The ribs articulate with *costal facets* on the transverse processes and bodies of thoracic vertebrae (figures 6.17*e* and 6.19*b*).

Lumbar Vertebrae

The five **lumbar vertebrae** (L1–L5) have heavy, thick bodies to support the greater stress and weight that is placed on this region of the vertebral column. The spinous processes are blunt and provide a large surface area for the attachment of heavy back muscles (see figures 6.16, 6.17*f*).

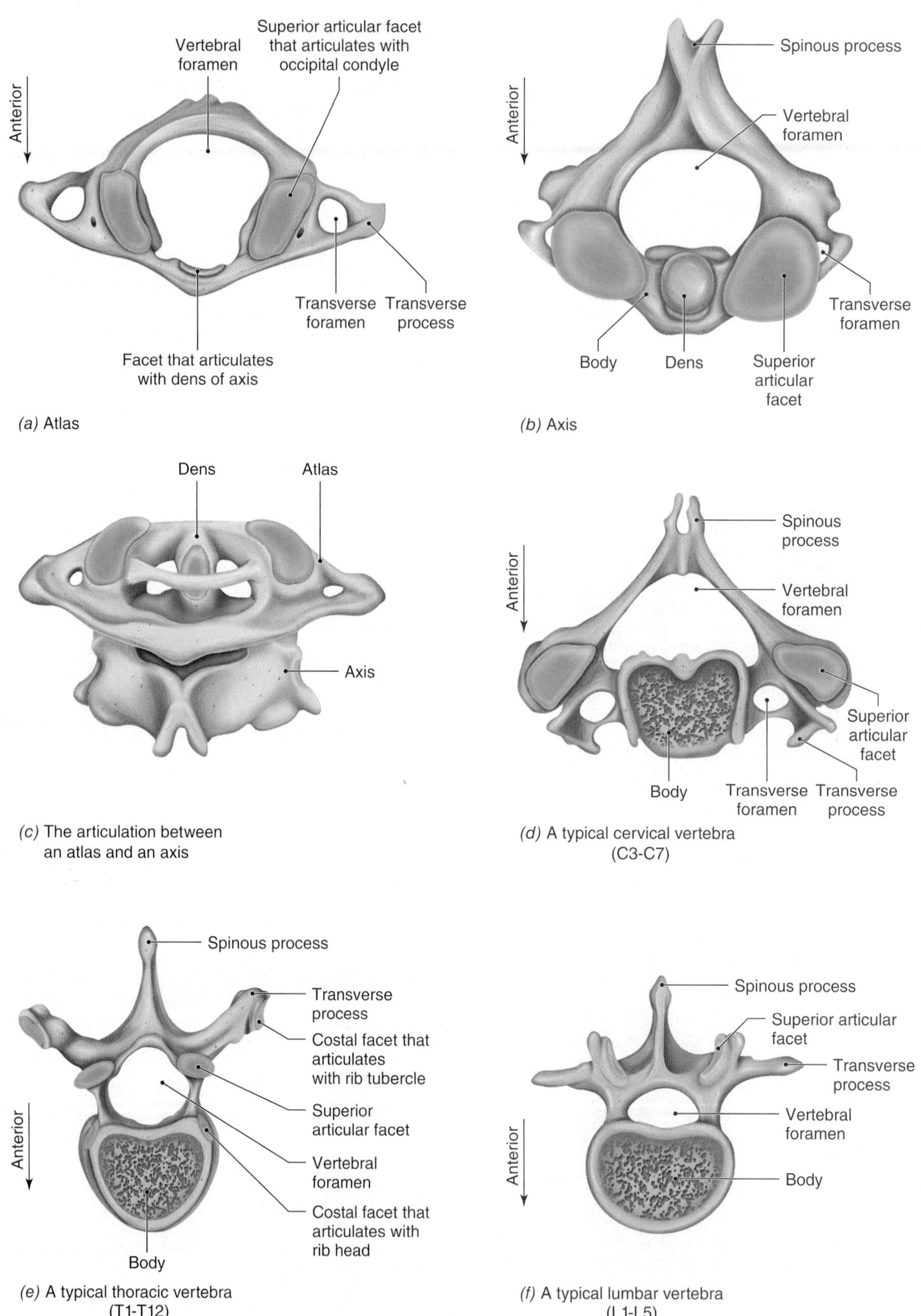

Figure 6.17 The Structures of Vertebrae.
(a), *(b)*, *(d)*, *(e)* and *(f)* are superior views. *(c)* is a posterior view. Bones are not shown to scale. APR

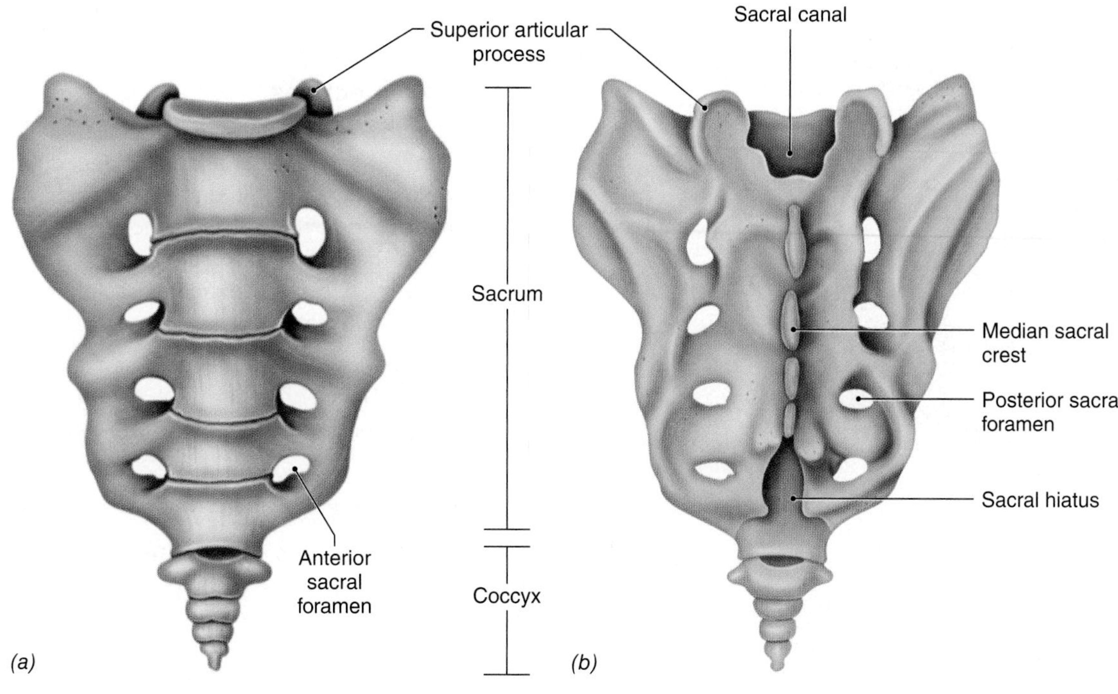

Figure 6.18 *(a)* Anterior view and *(b)* posterior view of the sacrum and coccyx. 

Sacrum

The **sacrum** (sā-k'rum) is composed of five fused sacral vertebrae (S1-S5) (figure 6.18). It articulates with the fifth lumbar vertebra and forms the posterior wall of the pelvis. The spinous processes of the fused vertebrae form the *median sacral crest* on the posterior midline. On either side of the median sacral crest are the *posterior sacral foramina,* passageways for blood vessels and nerves. *Anterior sacral foramina* on the anterior surface serve a similar function. The *sacral canal* is a continuation of the vertebral canal that carries spinal nerve roots to the sacral foramina and the *sacral hiatus,* an inferior opening proximal to the coccyx.

Coccyx

The most inferior part of the vertebral column is the **coccyx** (kok'-six), or tailbone, which is formed of three to five fused coccygeal vertebrae.

Thoracic Cage

The thoracic vertebrae, ribs, costal cartilages, and sternum form the **thoracic cage.** It provides protection for the internal organs of the thoracic cavity and supports the superior trunk, pectoral girdle, and upper limbs (figure 6.19).

Ribs

Twelve pairs of **ribs** are attached to the thoracic vertebrae. The *head* of each rib articulates with the costal facet on the body of its own vertebra, and a *tubercle* near the head articulates with the costal facet on the transverse process. The head also articulates with the costal facet on the body of the vertebra superior to it. The *shaft* of each rib curves around the thoracic cage and slopes slightly inferiorly.

The superior seven pairs of ribs are attached directly to the sternum by the **costal** (kos'-tal) **cartilages,** which extend medially from the ends of the ribs. These ribs are the *true ribs.* The remaining five pairs are the *false ribs.* The first three pairs of false ribs are attached by cartilages to the costal cartilages of the ribs just superior to them. The last two pairs of false ribs are called *floating ribs* because they lack cartilages and are not attached anteriorly. The costal cartilages give some flexibility to the thoracic cage.

Sternum

The **sternum,** or breastbone, is a flat, elongated bone located at the midline in the anterior portion of the

Clinical Insight

A biopsy of red bone marrow may be made by a *sternal puncture* because the sternum is covered only by skin and connective tissue. Under local anesthetic, a large-bore hypodermic needle is inserted into the sternum, and red bone marrow is drawn into a syringe.

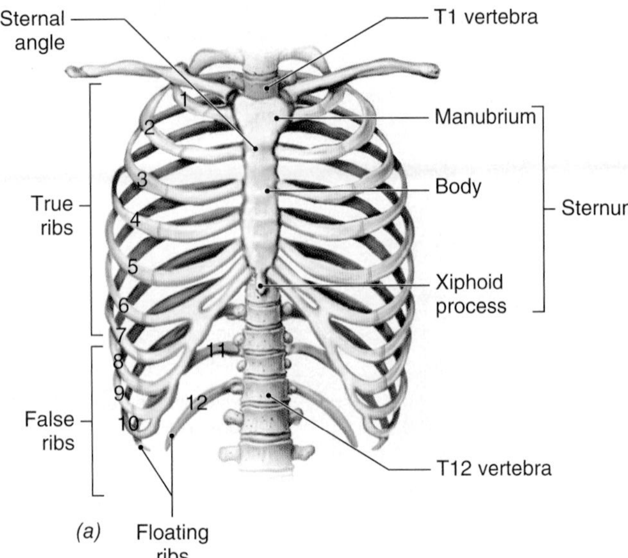

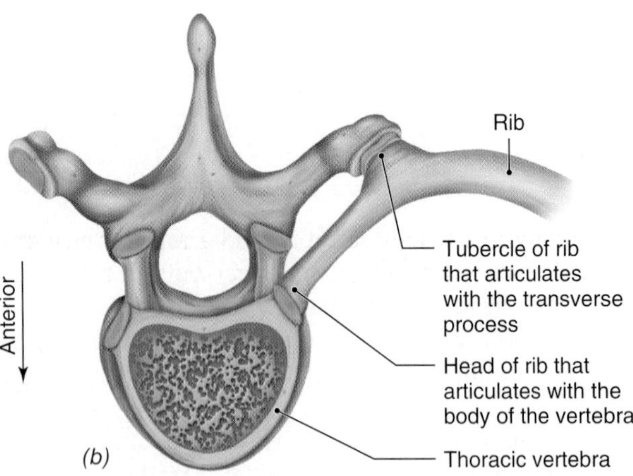

Figure 6.19 Thoracic Cage.
(a) The thoracic cage is formed by thoracic vertebrae, ribs, costal cartilages (colored light blue), and the sternum. *(b)* The articulation between a rib and a thoracic vertebra. AP|R

thoracic cage. It consists of three bones that are fused together. The *manubrium* (mah-nū′-brē-um) is the superior portion that articulates with the first two pairs of ribs; the *body* is the larger middle segment; and the *xiphoid* (zīf′-oyd) *process* is the small inferior portion.

Check My Understanding

8. How do cervical, thoracic, and lumbar vertebrae differ in structure and location?
9. How does the axial skeleton protect vital organs?

6.6 Appendicular Skeleton

Learning Objectives

12. Identify the bones of the appendicular skeleton.
13. Compare the structural and functional differences between pectoral girdle and pelvic girdle.
14. Compare the structural and functional differences between the male and female pelves.
15. Describe how the appendicular skeleton is connected to the axial skeleton.

The appendicular skeleton consists of (1) the pectoral girdle and the bones of the upper limbs, and (2) the pelvic girdle and the bones of the lower limbs (see figure 6.6).

Pectoral Girdle

The **pectoral** (pek′-to-ral) **girdle,** or shoulder girdle, consists of two clavicles (collarbones) and two scapulae (shoulder blades) (figure 6.20). Each S-shaped **clavicle** (klav′-i-cul) articulates with the acromion of a scapula laterally and with the sternum medially. The **scapulae** (skap′-ū-le, singular, *scapula*) are flat, triangular bones located on each side of the vertebral column, but they do not articulate with the axial skeleton. Instead, they are held in place by muscles, an arrangement that enables freedom of movement for the shoulder joints.

The anterior surface of each scapula is flat and smooth where it moves over the ribs. The scapular *spine* runs diagonally across the posterior surface from the *acromion* (ah-krōm′-ē-on) to the medial margin. On its lateral margin is the shallow *glenoid cavity,* which articulates with the head of the humerus. The *coracoid* (kor′-ah-koyd) *process* projects anteriorly from the superior margin of the glenoid cavity and extends inferior to the clavicle.

Upper Limb

The skeleton of each **upper limb** is composed of a humerus, an ulna, a radius, carpal bones, metacarpals, and phalanges (figure 6.21).

Humerus

The **humerus** (hū′-mer-us) articulates with the scapula at the shoulder joint, and the ulna and radius at the elbow joint. The rounded *head* of the humerus fits into the glenoid cavity of the scapula. Just inferior to the head are two large tubercles where muscles attach. The *greater tubercle* (tū′-ber-cul) is on the lateral surface, and the *lesser tubercle* is on the anterior surface. An *intertubercular sulcus* lies between them. Just distal to these tubercles is the *surgical neck,* which gets its name from the frequent fractures that occur in this area. Near the midpoint on the lateral surface is the *deltoid tuberosity* (tū-be-ros′-i-tē), a rough, elevated area where the deltoid attaches.

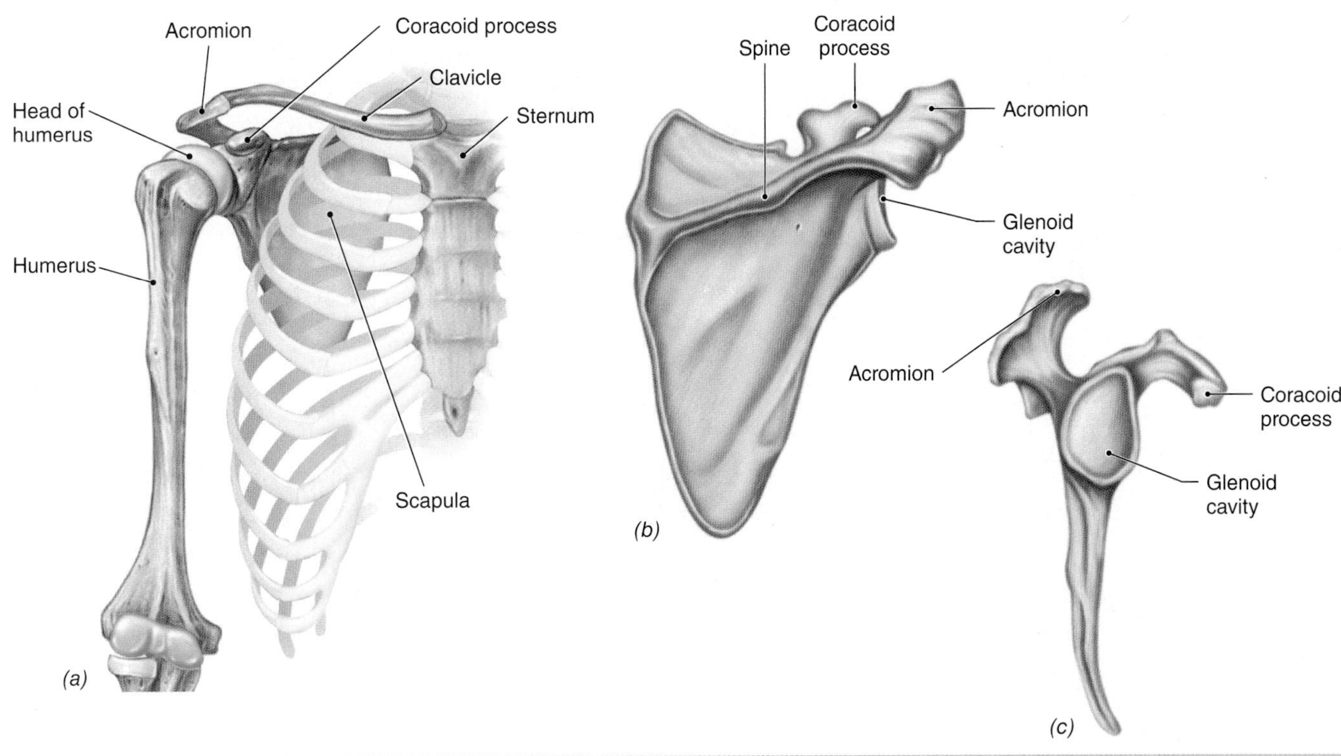

Figure 6.20 The Right Side of the Pectoral Girdle.
(a) The pectoral girdle consists of two scapulae and two clavicles. Note how the head of the humerus articulates with the glenoid cavity of the scapula. The posterior view *(b)* and the lateral view *(c)* of the right scapula. AP|R

The distal end of the humerus has two condyles. The *trochlea* (trok′-lē-ah) is the medial condyle, which articulates with the trochlear notch of the ulna. The *capitulum* (kah-pit′-ū-lum) is the lateral condyle, which articulates with the head of the radius. Just proximal to these condyles are two enlargements that project laterally and medially: the *lateral epicondyle* (ep-i-kon′-dīl) and the *medial epicondyle.* On the anterior surface between the epicondyles is a depression, the *coronoid* (kor′-o-noyd) *fossa,* that receives the coronoid process of the ulna whenever the upper limb is flexed at the elbow. The *olecranon* (o-lek′-rah-non) *fossa* is in a similar location on the posterior surface of the humerus, and it receives the olecranon of the ulna when the upper limb is extended at the elbow.

Ulna

The **ulna** (ul′-na) is the medial bone of the forearm. The proximal end of the ulna forms the *olecranon,* the bony point of the elbow. The large, half-circle depression just distal to the olecranon is the *trochlear notch,* which articulates with the trochlea of the humerus. This articulation is secured by the *coronoid process* on the distal lip of the notch.

At the distal end, the knoblike *head* of the ulna articulates with the medial surface of the radius and with the wrist bones. The *styloid process* is a small medial projection to which ligaments of the wrist are attached.

Radius

The **radius** (rā′-dē-us) is the lateral bone of the forearm. The disclike *head* of the radius articulates with the lateral proximal surface of the ulna in a way that enables the head to rotate freely when the forearm is rotated. A short distance distally from the head is the *radial tuberosity,* a elevated, roughened area where the biceps brachii attaches. At its distal end, the radius articulates with the carpal bones. A small lateral *styloid process* serves as an attachment site for ligaments of the wrist.

Carpal Bones, Metacarpals, and Phalanges

The skeleton of the hand consists of the carpal bones, metacarpals, and phalanges (figure 6.21*d*). The **carpal** (kar′-pul) **bones,** or wrist bones, consist of eight small bones that are arranged in two transverse rows of four bones each. They are joined by ligaments that allow limited gliding movement.

The **metacarpals,** bones of the palm, consist of five metacarpal bones that are numbered I to V starting with the metacarpal adjacent to the thumb. The bones of the fingers are the **phalanges** (fah-lan′-jēz, singular, *phalanx*).

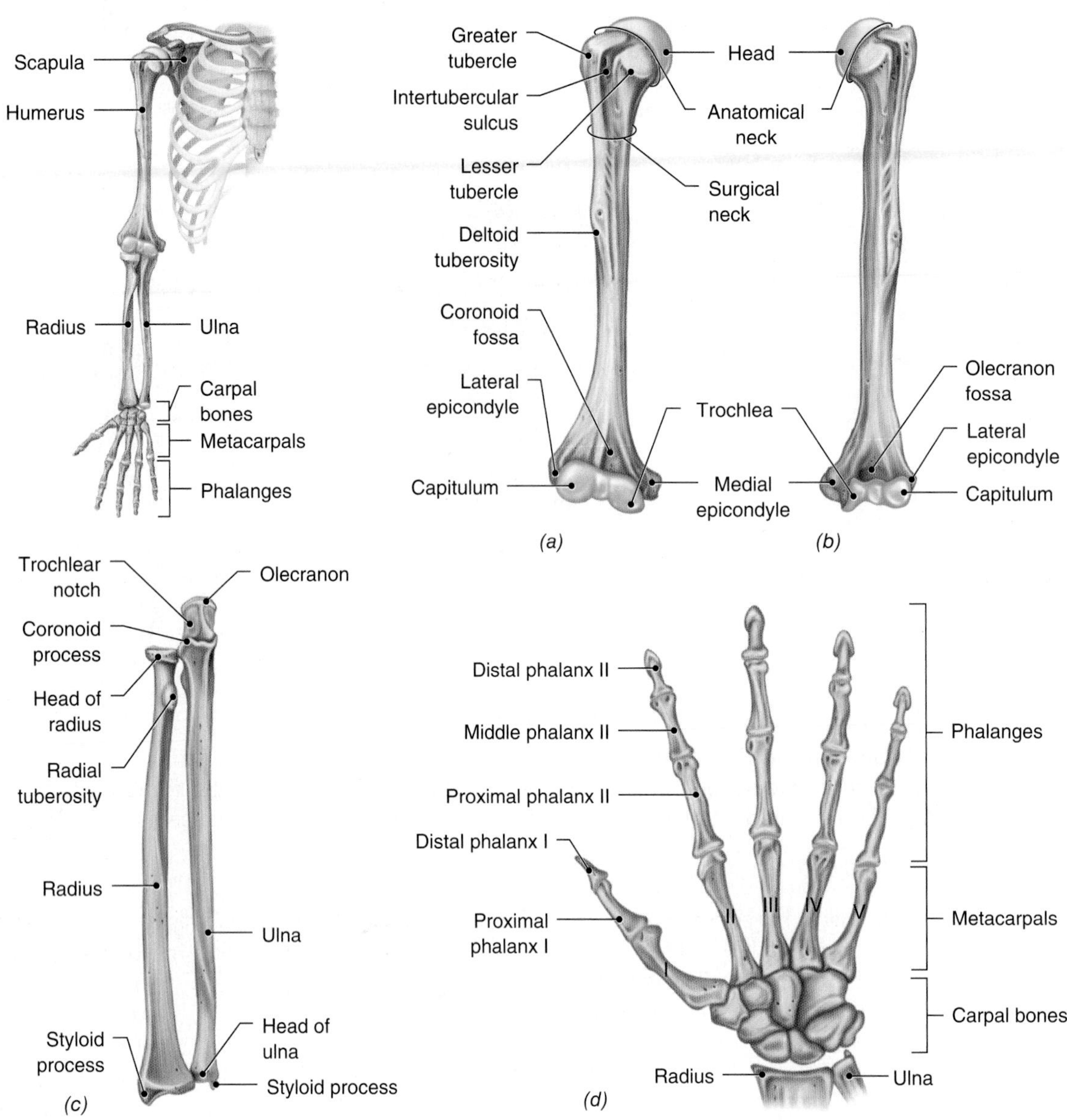

Figure 6.21 The Right Upper Limb.
(a) Anterior view of humerus; *(b)* Posterior view of humerus; *(c)* Anterior view of ulna and radius; *(d)* Posterior view of hand. AP|R

Each finger consists of three phalanges (proximal, middle, and distal), except for the thumb, which has only two (proximal and distal).

Pelvic Girdle

The **pelvic** (pel′-vik) **girdle** consists of two **coxal** (kok′sal) **bones,** or hip bones, that support the attachment of the lower limbs. The coxal bones articulate with the sacrum posteriorly and with each other anteriorly to form an almost rigid, bony **pelvis** (plural, *pelves*), as shown in figure 6.22.

Coxal Bones

Each coxal bone is formed by three fused bones–ilium, ischium, and pubis–that join at the *acetabulum* (as-e-tab′-ū-lum), the cup-shaped socket on the lateral surface. The **ilium** is the broad superior portion whose superior margin forms the *iliac crest,* the prominence of the hip. Inferior to the *posterior inferior iliac spine* is the *greater sciatic* (sī-at′-ik) *notch,* which allows the passage of blood vessels and sciatic nerve from the pelvis to the thigh. The *auricular surface* of each ilium joins with the sacrum to form a *sacroiliac joint.*

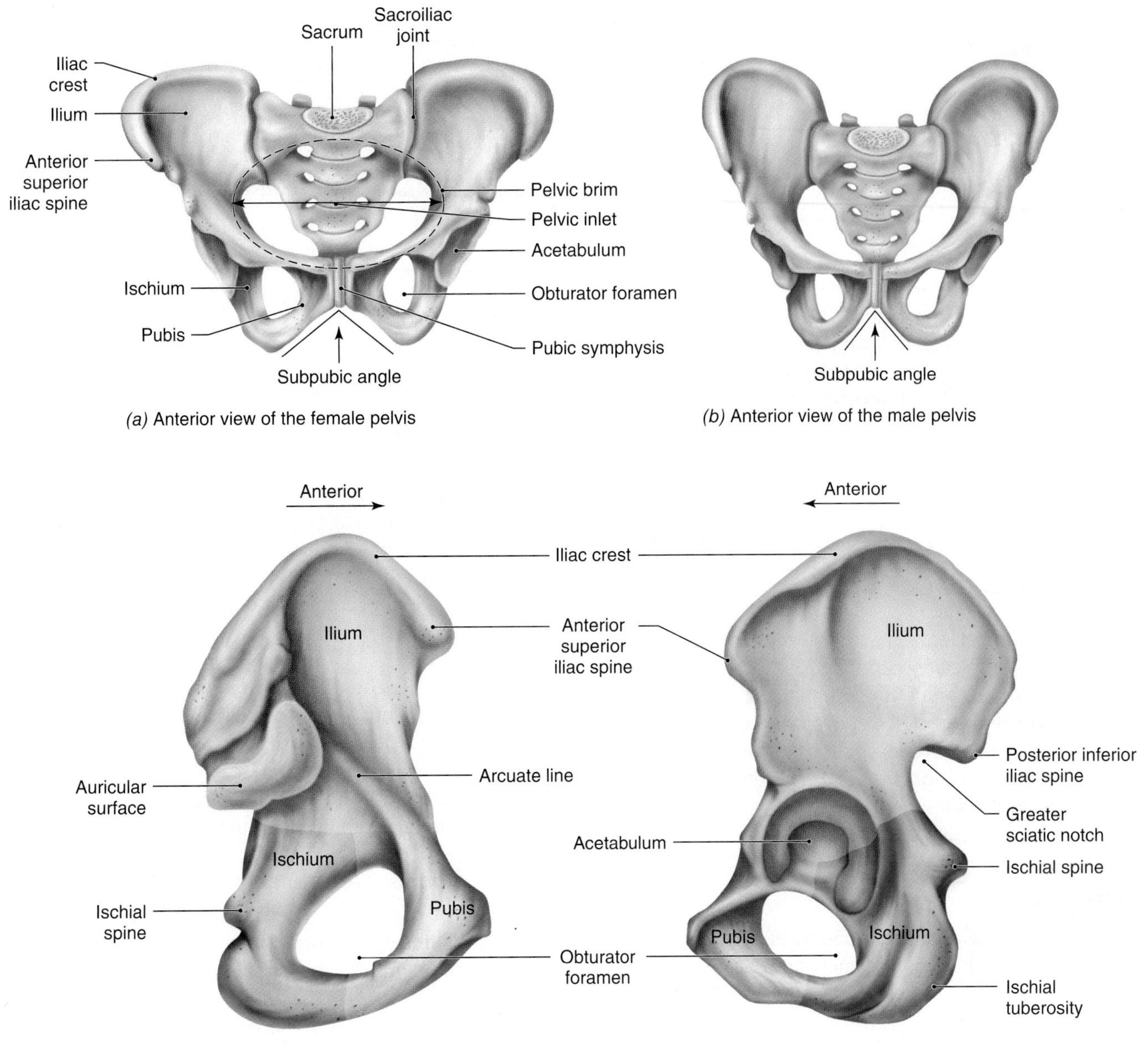

(a) Anterior view of the female pelvis

(b) Anterior view of the male pelvis

(c) Medial view of the left coxal bone

(d) Lateral view of the left coxal bone

Figure 6.22 Pelves and Coxal Bones. APR

The **ischium** forms the inferior, posterior portion of a coxal bone and supports the body when sitting. The roughened projection at the posterior, inferior angle of the ischium is the *ischial tuberosity*. Just superior to this tuberosity is the *ischial spine*, which projects medially. The distance between the left and right ischial spines in females is important during childbirth because it determines the diameter of the pelvic opening.

The **pubis** (plural, *pubes*) is the inferior, anterior portion of a coxal bone. A portion of the pubis extends posteriorly to fuse with the anterior extension of the ischium. The large opening created by this junction is the *obturator* (ob-tū-rā′-ter) *foramen*, through which blood vessels and nerves pass into the thigh. The pubes unite anteriorly to form the **pubic symphysis,** where the bones are joined by a pad of fibrocartilage.

When giving intramuscular injections in the hip, the region near the greater sciatic notch must be avoided to prevent possible injury to the large blood vessels and nerves in this area.

Table 6.3 Sexual Differences of the Pelves

Characteristic	Male	Female
General structure	Heavier; processes prominent	Lighter; processes not so prominent
Pelvic inlet	Narrower and heart-shaped	Wider and oval-shaped
Subpubic angle	Less than 90°	More than 90°
Relative width	Narrower	Wider
Acetabulum	Faces laterally	Faces laterally but more anteriorly

Table 6.3 lists the major differences between the male and the female pelves. Compare them with the male and female pelves in figure 6.22 and note the adaptations of the female pelvis for childbirth. The **pelvic inlet,** an opening superior to the pelvic cavity, is encircled by the *pelvic brim,* a circular line passing through the *arcuate line* and the superior border of pubis. Its size and shape in females are critical to the success of the birth process.

Clinical Insight

The fetus must pass through the pelvic inlet during birth. Physicians carefully measure this opening before delivery to be sure that it is of adequate size. If not, the baby is delivered via a *cesarean section.* In a cesarean section, a transverse incision is made through the pelvic and uterine walls to remove the infant.

Lower Limb

The bones of each **lower limb** consist of a femur, a patella, a tibia, a fibula, tarsal bones, metatarsals, and phalanges (figure 6.23).

Femur

The **femur,** or thigh bone, is the largest and strongest bone of the body (figure 6.23*a, b*). Structures at the proximal end include the rounded *head,* a short *neck,* and two large processes that are sites of muscle attachment: a superior, lateral *greater trochanter* (trō-kan′-ter) and an inferior, medial *lesser trochanter.* The head of the femur fits into the acetabulum of the coxal bone. The neck is a common site of fractures in older people. At the enlarged distal end are the *lateral* and *medial condyles,* surfaces that articulate with the tibia.

Patella

The **patella,** or kneecap, is located anterior to the knee joint. It is embedded in the tendon of the quadriceps femoris, which extends over the anterior of the knee to insert on the tibia. The patella offers protection to the structures within the knee joint during movement.

Tibia

The **tibia,** or shinbone, is the larger of the two bones of the leg (figure 6.23*c*). It bears the weight of the body. Its enlarged proximal portion consists of the *lateral* and *medial condyles,* which articulate with the femur to form the knee joint. The *tibial tuberosity,* a roughened area on the anterior surface just distal to the condyles, is the attachment site for the patellar ligament. The distal end of the tibia articulates with the talus, a tarsal bone, and laterally with the fibula. The *medial malleolus* (mah-lē-ō′-lus) forms the medial prominence of the ankle.

Fibula

The **fibula** is the slender, lateral bone in the leg (figure 6.23*c*). Both ends of the bone are enlarged. The proximal *head* articulates with the lateral surface of the tibia but is not involved in forming the knee joint. The distal end articulates with the tibia and talus. The *lateral malleolus* forms the lateral prominence of the ankle.

Tarsal Bones, Metatarsals, and Phalanges

The skeleton of the foot consists of the tarsal bones (ankle), metatarsals (instep), and phalanges (toes) (figure 6.23*d, e*). Seven bones compose the **tarsal bones.** The most prominent tarsal bones are the *talus,* which articulates with the tibia and fibula, and the *calcaneus* (kal-kā′n-ē -us), or

Clinical Insight

Total hip replacement (THR) has become commonplace among older persons as a way to overcome the pain and immobility caused by osteoarthritis of the hip joint. This procedure utilizes two prostheses. A polyurethane cup replaces the damaged acetabulum, and a metal shaft and ball replace the diseased head of the femur. Surfaces of the prostheses in contact with bone are porous, allowing bone to grow into them to ensure a firm attachment. Patient recovery involves stabilization of the prostheses while bone grows into them as well as normal healing from the surgery.

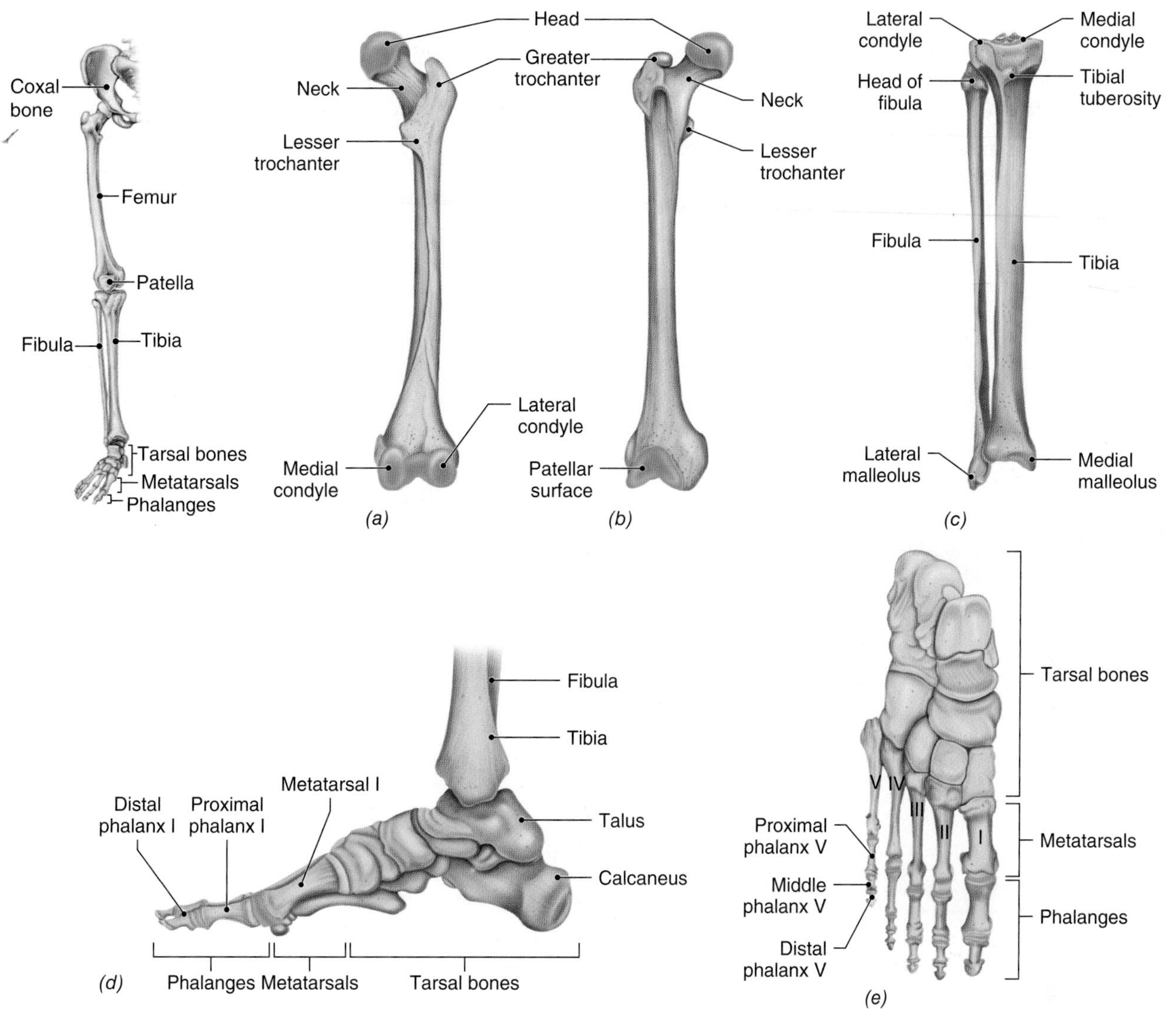

Figure 6.23 The Right Lower Limb.
(a) Posterior view of femur; *(b)* Anterior view of femur; *(c)* Anterior view of tibia and fibula; *(d)* Medial view of foot; *(e)* Superior view of foot. APR

heel bone. Five **metatarsals** support the instep. They are numbered I to V, starting with the metatarsal adjacent to the great toe. The tarsal bones and metatarsals are bound together by ligaments to form strong, resilient arches of the foot. Each toe consists of three phalanges (proximal, middle, and distal), except for the great toe, which has only two (proximal and distal).

Check My Understanding

10. What bones form the pectoral girdle and upper limbs?
11. What bones form the pelvic girdle and lower limbs?

6.7 Articulations

Learning Objectives

16. Compare the structures, functions, and locations of immovable, slightly movable, and freely movable joints.
17. Compare the types of movements allowed by freely movable joints.
18. Compare the six types of freely movable joints.

The junction between two bones or between a bone and a tooth forms an **articulation**, or **joint.** Joints allow varying degrees of movement and are categorized as immovable, slightly movable, or freely movable. As you read the following descriptions, locate the different types of joints on the corresponding illustrations of skeletal parts in figures presented earlier in the chapter.

Immovable Joints

Bones forming an immovable joint, or **synarthrosis** (sin-ar-thrō′-sis), are tightly joined and are separated by a thin band of dense connective tissue or a thin layer of hyaline cartilage. For example, skull bones, except the mandible, are joined by dense connective tissue called sutures because they resemble stitches (figure 6.24*a*). The joints between bones and teeth are also immovable joints separated by dense connective tissue. The epiphysial plates in growing bones (see figure 6.5), composed of hyaline cartilage, are also immovable joints.

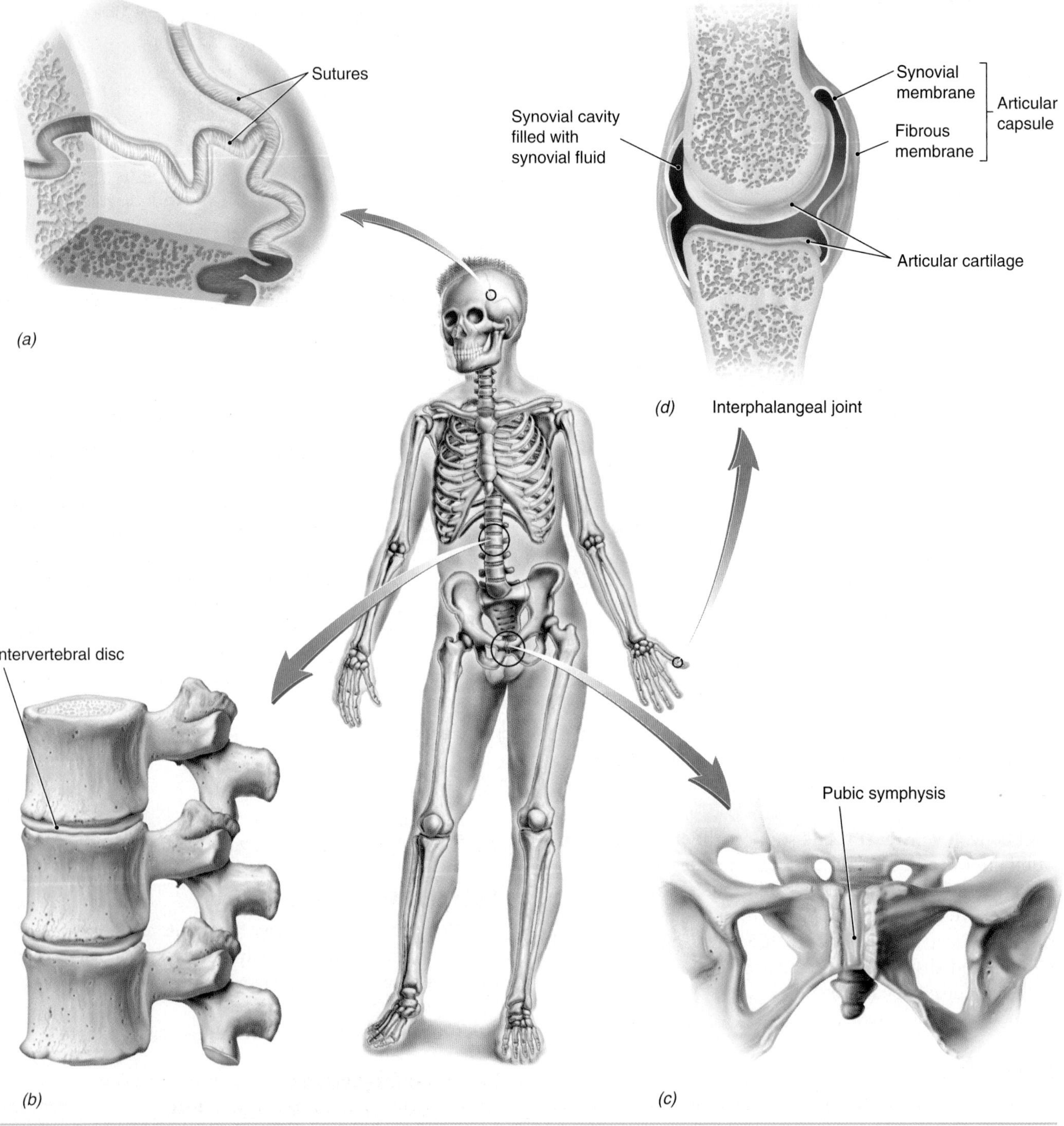

Figure 6.24 Types of Joints.
(a) Suture (synarthrosis); *(b)* Intervertebral disc (amphiarthrosis); *(c)* Pubic symphysis (amphiarthrosis); *(d)* Interphalangeal joint (diarthrosis).

Slightly Movable Joints

Bones forming a slightly movable joint, or **amphiarthrosis** (am-fē-ar-th-rō′-sis), are separated by a layer of cartilage or dense connective tissue. For example, the joints formed by adjacent vertebrae contain intervertebral discs formed of fibrocartilage (figure 6.24*b*). The limited flexibility of the discs allows slight movement between adjacent vertebrae. Other examples include the pubic symphysis (figure 6.24*c*) and sacroiliac joints.

Freely Movable Joints

Most articulations are freely movable. The structure of a freely movable joint, or **diarthrosis** (di-ar-thro′-sis), is more complex. These joints are also called **synovial** (si-no′-ve-al) **joints**. The ends of the bones forming the joint are bound together by an *articular,* or *joint, capsule.* The thick external layer of the capsule, called *fibrous membrane,* is composed of dense irregular connective tissue. The thin internal layer of the capsule, called *synovial* (si-nō′-vē-al) *membrane,* secretes *synovial fluid* that lubricates the joint. The ends of the bones are covered with *articular cartilage,* which protects bones and reduces friction (figure 6.24*d*). **Ligaments,** the cords or bands of dense regular connective tissue that connect bones together, reinforce the joints. Freely movable joints are categorized into several types based on their structure and types of movements.

Plane Joints

A *plane joint* occurs between two flat articular surfaces that slide over each other and allows for movement in one plane. Some examples of plane joints are the joints between carpal bones (figure 6.25*a*), between tarsal bones, and between clavicle and scapula.

Condylar Joints

A *condylar joint* is formed between an oval articular surface and an oval socket and allows for movements in two planes. The joints between carpal bones and radius and between metacarpals and proximal phalanges (figure 6.25*b*) are examples of condylar joints.

Saddle Joint

A *saddle joint* occurs where a saddle-like articular surface fits into a complementary depression, allowing movement in two planes. This type of joint occurs between the trapezium (a carpal bone) and metacarpal I (figure 6.25*c*).

Hinge Joints

A *hinge joint* involves a cylindrical articular surface and a complementary depression. It allows for movement similar to opening and closing a door. The elbow (figure 6.25*d*), knee, and joints between phalanges are all hinge joints.

Pivot Joints

A *pivot joint* involves a cylindrical articular surface and a complementary depression. It allows for rotation movements along a longitudinal axis. Examples of a pivot joint are the joint between atlas and axis (figure 6.25*e*) and the joint between the radius head and the ulna.

Ball-and-Socket Joints

In a *ball-and-socket joint,* a rounded head fits into a rounded socket. It allows for movements in all planes and provides the greatest range of movement of all types of freely movable joints. The ball-and-socket joints in the human body are the shoulder and hip joints (figure 6.25*f*).

Movements at Freely Movable Joints

Movement at a joint results from the contraction of skeletal muscles that span across the joint. The type of movement that occurs is determined by the type of joint and the location of the muscle or muscles involved. The more common types of movements are listed in table 6.4 and illustrated in figure 6.26.

Clinical Insight

Older persons are prone to "breaking a hip," which means that a weakened femur breaks at the neck. This usually is a consequence of osteoporosis, the excessive loss of matrix from bones. Osteoporosis is caused by a combination of factors: insufficient calcium in the diet, lack of minimal exercise, and a decline in sex hormones, especially in postmenopausal women. Not only are older persons more prone to fractures, but healing of fractures takes much longer than in younger persons.

Check My Understanding

12. Where are immovable, slightly movable, and freely movable joints found in the skeleton?
13. What types of freely movable joints occur in the body, and where are they located?

6.8 Disorders of the Skeletal System

Learning Objectives

19. Describe common disorders of bones.
20. Describe common disorders of joints.

Common disorders of the skeletal system may be categorized as disorders of bones or disorders of joints.

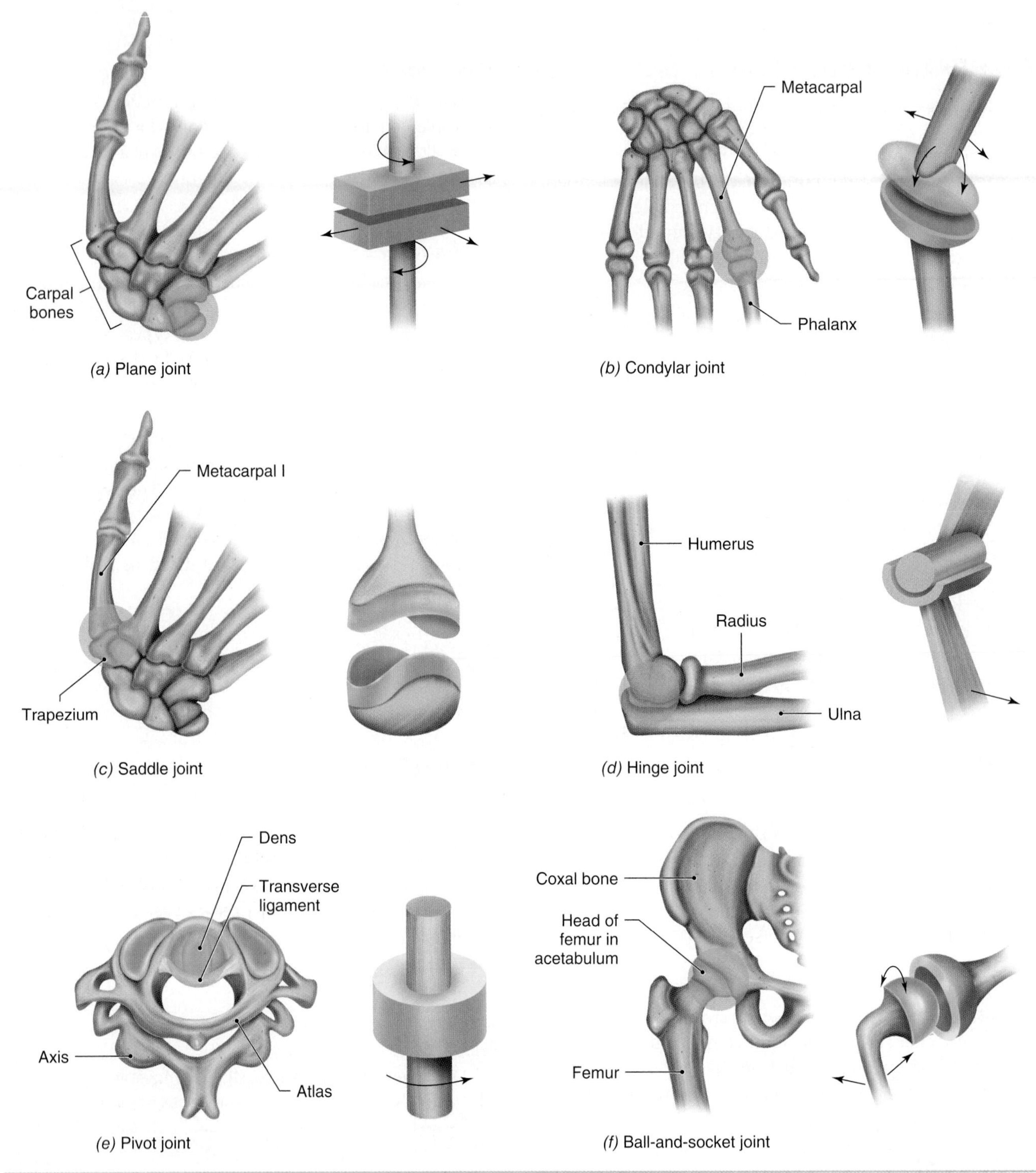

Figure 6.25 Types of Freely Movable Joints.

Orthopedics (or-thō-pē-diks) is the branch of medicine that specializes in treating diseases and abnormalities of the skeletal system.

Disorders of Bones

Fractures are broken bones. Fractures are the most common type of bone injury. Fractures are categorized as either complete or incomplete. There are also several specific subtypes, such as the examples noted here and in figure 6.27.

- **Complete:** The break is completely through the bone.
- **Compound:** A broken bone pierces the skin.
- **Simple:** A bone does not pierce the skin.
- **Comminuted:** The bone is broken into several pieces.

Table 6.4 Movements at Freely Movable Joints

Movements	Description
Flexion	Decrease in the angle of bones forming the joint
Extension	Increase in the angle of bones forming the joint
Hyperextension	Increase in the angle of bones forming the joint beyond the anatomical position
Dorsiflexion	Flexion of the foot at the ankle
Plantar flexion	Extension of the foot at the ankle
Abduction	Movement of a bone away from the midline
Adduction	Movement of a bone toward the midline
Rotation	Movement of a bone around its longitudinal axis
Medial rotation	Rotation of a limb so its anterior surface turns medially
Lateral rotation	Rotation of a limb so its anterior surface turns laterally
Circumduction	Movement of the distal end of a bone in a circle while the proximal end forms the pivot joint
Eversion	Movement of the sole of the foot laterally
Inversion	Movement of the sole of the foot medially
Pronation	Rotation of the forearm when the palm is turned inferiorly or posteriorly
Supination	Rotation of the forearm when the palm is turned superiorly or anteriorly
Protraction	Movement of a body part anteriorly
Retraction	Movement of a body part posteriorly
Elevation	Movement of a body part superiorly
Depression	Movement of a body part inferiorly
Opposition	Movement of the thumb to touch the other four fingers
Reposition	Movement of the thumb back to the anatomical position

- **Segmental:** Only one piece is broken out of the bone.
- **Spiral:** The fracture line spirals around the bone.
- **Oblique:** The break angles across the bone.
- **Transverse:** The break is at right angles to the long axis of the bone.
- **Incomplete:** The bone is not broken completely through.
- **Greenstick:** The break is only on one side of the bone, and the other side of the bone is bowed.
- **Fissured:** The break is a lengthwise split in the bone.

Osteomyelitis is an inflammation of bone and bone marrow caused by bacterial infection. It is treatable with antibiotics but not easily cured.

Osteoporosis (os-tē-ō-pō-rō′-sis) is a weakening of bones due to the removal of bone matrix, which increases the risk of fractures. This is a common problem in older persons due to inactivity and a decrease in hormone production. It is more common in postmenopausal women because of the lack of estrogens. Exercise and calcium supplements retard the decline in bone density. Therapy includes drugs that reduce bone loss or those that promote bone formation. However, such drugs must be used with caution because they can have serious side effects.

Rickets is a disease of children that is characterized by a deficiency of calcium salts in the bones. Affected children have a bowlegged appearance due to the bending of weakened femurs, tibiae, and fibulae. Rickets results from a dietary deficiency of vitamin D and/or calcium. It is rare in industrialized nations.

Disorders of Joints

Arthritis (ar-thrī′-tis) is the general term for many different diseases of joints that are characterized by inflammation, swelling (edema), and pain. Rheumatoid arthritis and osteoarthritis are the most common types.

Rheumatoid (rū′-mah-toid) *arthritis* is the most painful and crippling type. It is an autoimmune disorder, in which the joint tissues are attacked by the patient's own defenses. The synovial membrane thickens, synovial fluid accumulates causing swelling, and articular cartilages are destroyed. The joint is invaded by dense irregular connective tissue that ultimately ossifies, making the joint immovable.

Osteoarthritis, the most common type, is a degenerative disease that results from aging and wear. The articular cartilages and the bone deep to the cartilages gradually disintegrate, which causes pain and restricts movement.

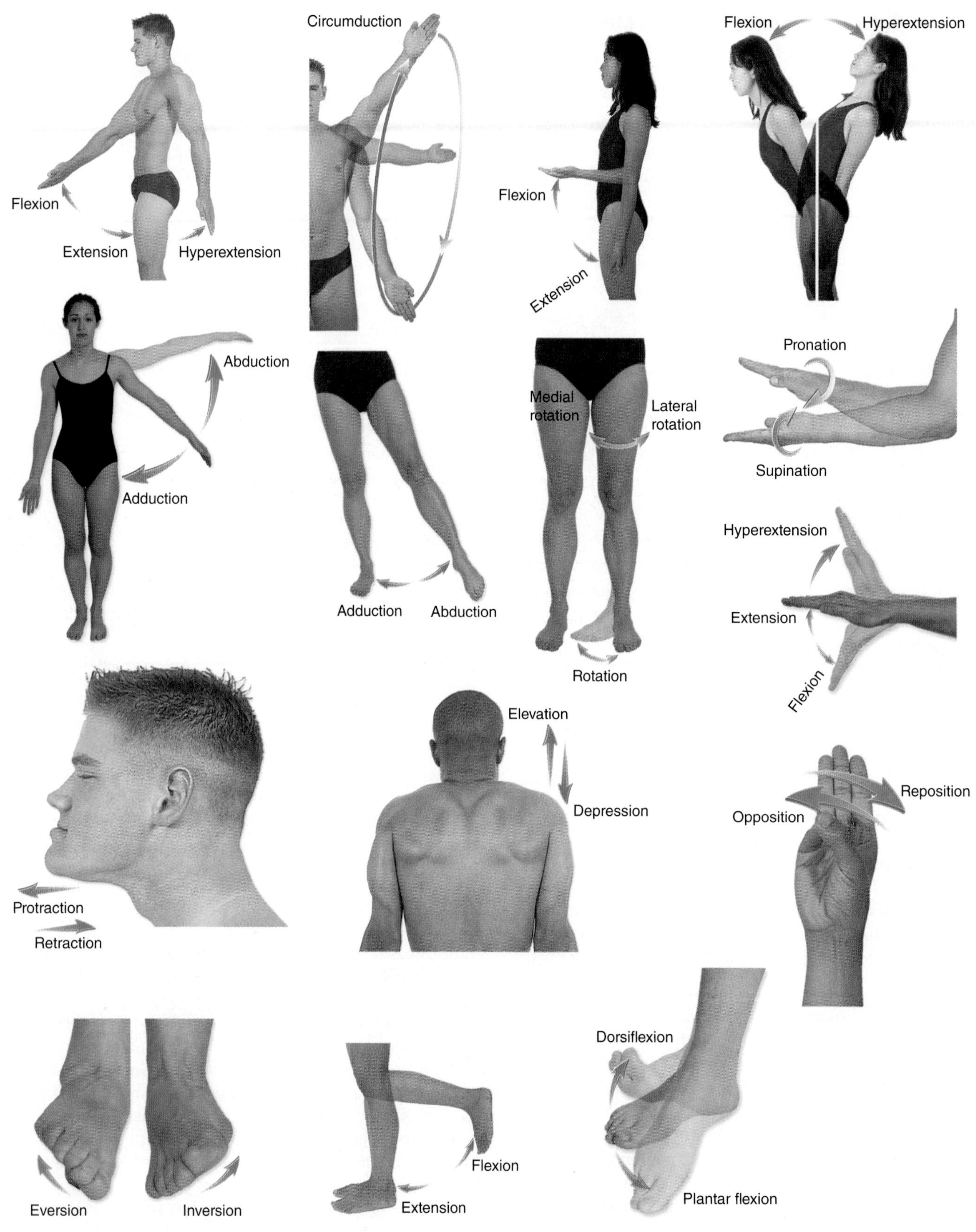

Figure 6.26 Common Movements at Freely Movable Joints.

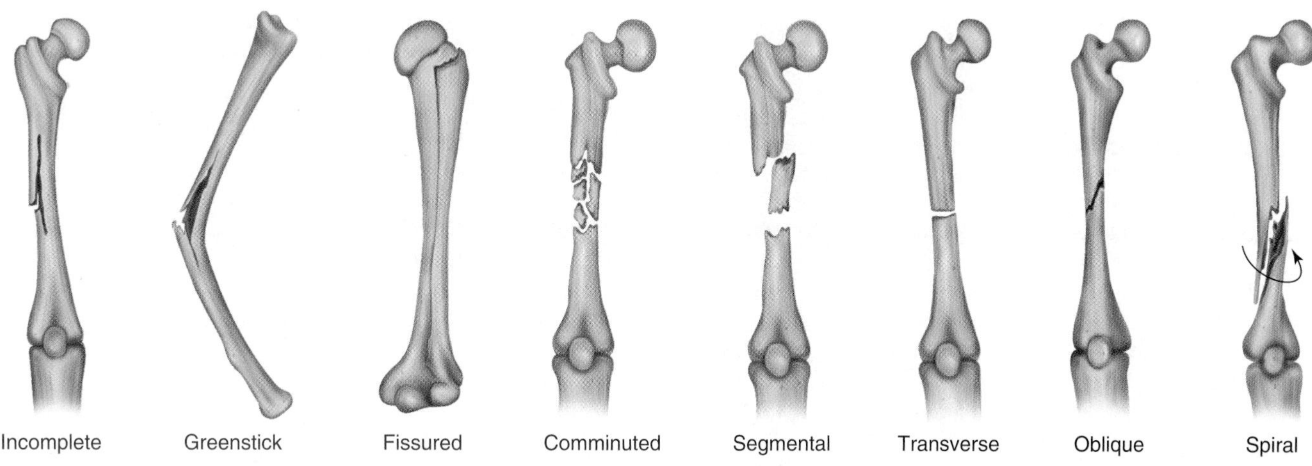

Figure 6.27 Some Types of Bone Fractures.

Dislocation is the displacement of bones forming a joint. Pain, swelling, and reduced movement are associated with a dislocation.

Herniated disc is a condition in which an intervertebral disc protrudes beyond the edge of a vertebra. A ruptured, or slipped, disc refers to the same problem. It is caused by excessive pressure on the vertebral column, which causes the *nucleus pulposus,* the centrally located gelatinous region of the disc, to protrude into the *anulus fibrosus,* the perimeter of the disc. The protruding disc may place pressure on a spinal nerve and cause considerable pain (figure 6.28).

Sprains result from tearing or excessive stretching of the ligaments and tendons at a joint without a dislocation.

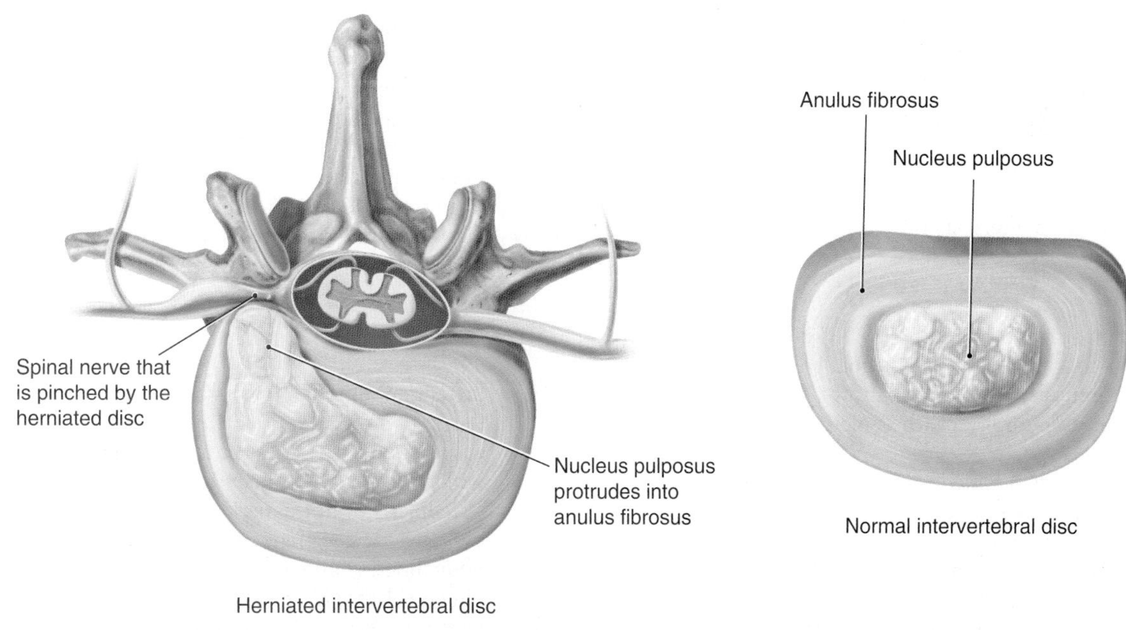

Figure 6.28 Herniated Disc.

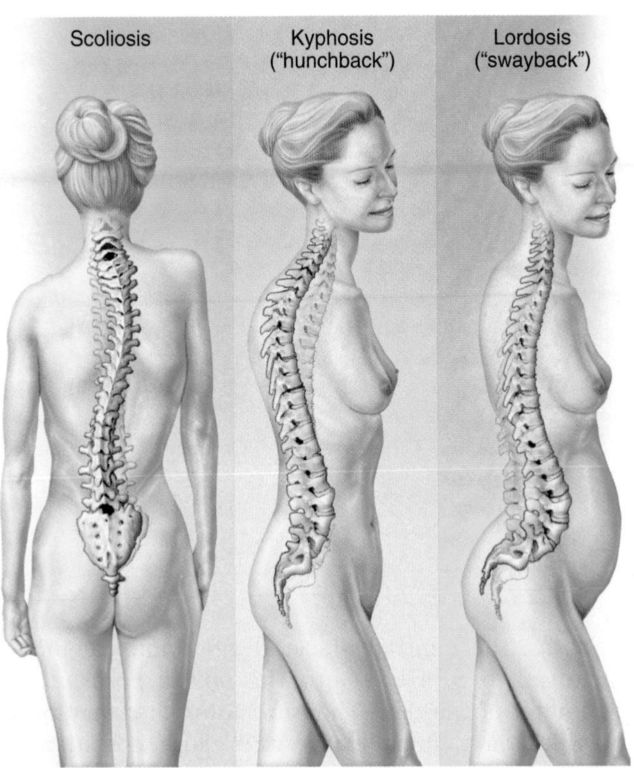

Figure 6.29 Abnormal Spinal Curvatures.

Abnormal spinal curvatures are usually congenital disorders. There are three major types (figure 6.29):

1. *Scoliosis* is an abnormal lateral curvature of the vertebral column. For some reason, it is more common in adolescent girls.
2. *Kyphosis* (kī-fō-sis) is an excessive thoracic curvature of the vertebral column, which produces a hunchback condition.
3. *Lordosis* is an excessive lumbar curvature of the vertebral column, which produces a swayback condition.

Check My Understanding

14. What are some common types of fractures?
15. How do osteoarthritis and rheumatoid arthritis differ?

Chapter Summary

6.1 Functions of the Skeletal System

- The skeletal system provides support for the body and protection for internal organs.
- The bones of the skeleton serve as sites for the attachment of skeletal muscles.
- Formed elements are produced by red bone marrow.
- Bones serve as reservoirs for calcium salts.

6.2 Bone Structure

- Based on shapes, bones are classified into short, long, sutural, flat, irregular, and sesamoid bones.
- The diaphysis is the long shaft of a long bone that lies between the epiphyses, the enlarged ends of the bone.
- Each epiphysis is joined to the diaphysis by an epiphysial plate in immature bones, or by fusion at the epiphysial line in mature bones.
- Articular cartilages protect and cushion the articular surfaces of the epiphyses.
- The periosteum covers the bone surface except for the articular cartilages.
- Compact bone forms the wall of the diaphysis and the thin superficial layer of the epiphyses.
- Spongy bone forms the internal structure of the epiphyses and the internal thin layer of the diaphysis wall.
- The diaphysis contains a medullary cavity filled with yellow marrow.
- Compact bone is formed of numerous osteons.
- Central canals contain blood vessels and nerves.
- Spongy bone is composed of interconnected bony plates called trabeculae. The spaces between trabeculae are filled with red or yellow bone marrow.
- Flat, short, and irregular bones are composed of spongy bone covered by a thin layer of compact bone.

6.3 Bone Formation

- Intramembranous bones are first formed by connective tissue membranes, which are replaced by bone.
- Connective tissue cells are transformed into osteoblasts, which deposit the spongy bone within the membrane.
- Osteoblasts from the periosteum form a layer of compact bone over the spongy bone.
- Endochondral bones are first formed of hyaline cartilage, which is later replaced by bone.
- In long bones, a primary ossification center forms in the center of the diaphysis and extends toward the epiphyses.
- Secondary ossification centers form in the epiphyses.
- An epiphysial plate of cartilage remains between the epiphyses and the diaphysis in immature bones.

- Growth in length occurs at the epiphysial plate, which is gradually replaced by bone.
- Compact bone is deposited by osteoblasts from the periosteum, and they are responsible for growth in the diameter of a bone.
- Osteoclasts hollow out the medullary cavity and reshape the bone.
- Bones are dynamic, living organs that are reshaped throughout life by the actions of osteoclasts and osteoblasts.
- Bone matrix may be removed from bones for other body needs and redeposited in bones later on.
- The number of protein fibers decreases with age. The bones of older persons tend to be brittle and weak due to the loss of fibers and calcium salts, respectively.

6.4 Divisions of the Skeleton

- The skeleton is divided into the axial and appendicular divisions.
- The axial skeleton includes the bones that support the head, neck, and trunk.
- The appendicular skeleton includes the bones of the pectoral girdle and upper limbs and the bones of the pelvic girdle and the lower limbs.

6.5 Axial Skeleton

- The axial skeleton consists of the skull, vertebral column, and thoracic cage.
- The skull consists of cranial and facial bones; all are joined by immovable joints except the mandible.
- The cranial bones are the frontal bone (1), parietal bones (2), sphenoid (1), temporal bones (2), occipital bone (1), and ethmoid (1).
- The facial bones are the maxillae (2), palatine bones (2), zygomatic bones (2), lacrimal bones (2), nasal bones (2), inferior nasal conchae (2), vomer (1), and mandible (1).
- The frontal bone, sphenoid, ethmoid, and maxillae contain paranasal sinuses.
- Cranial bones of an infant skull are separated by membranes and several fontanelles, which allow some flexibility of the skull during birth.
- Associated bones to the skull include a hyoid bone and six auditory ossicles.
- The vertebral column consists of 24 vertebrae, the sacrum, and the coccyx.
- Vertebrae are separated by intervertebral discs and are categorized as cervical (7), thoracic (12), and lumbar (5) vertebrae.
- The first two cervical vertebrae are unique. The atlas rotates on the axis when the head is turned.
- Thoracic vertebrae have costal facets on the body and transverse processes for articulation with the ribs.
- The bodies of lumbar vertebrae are heavy and strong.
- The sacrum is formed of five fused vertebrae and forms the posterior portion of the pelvis.
- The coccyx is formed of three to five fused vertebrae and forms the inferior end of the vertebral column.
- The thoracic cage consists of thoracic vertebrae, ribs, and sternum. It supports the superior trunk and protects internal thoracic organs.
- There are seven pairs of true ribs and five pairs of false ribs. The inferior two pairs of false ribs are floating ribs.
- The sternum is formed of three fused bones: manubrium, body, and xiphoid process.

6.6 Appendicular Skeleton

- The appendicular skeleton consists of the pectoral and pelvic girdles and of the bones of the limbs.
- The pectoral girdle consists of clavicles (2) and scapulae (2), and it supports the upper limbs.
- The bones of the upper limb are the humerus, the ulna, the radius, carpal bones, metacarpals, and phalanges.
- The humerus articulates with the glenoid cavity of the scapula to form the shoulder joint and with the ulna and radius to form the elbow joint.
- The ulna is the medial bone of the forearm. It articulates with the humerus at the elbow and with the radius and carpal bones at the wrist.
- The radius is the lateral bone of the forearm. It articulates with the humerus at the elbow and with the ulna and carpal bones at the wrist.
- The bones of the hand are the carpal bones (8), metacarpals (5), and phalanges (14).
- The carpal bones are joined by ligaments to form the wrist; metacarpal bones support the palm of the hand; and the phalanges are the bones of the fingers.
- The pelvic girdle consists of two coxal bones that are joined to each other anteriorly. It supports the lower limbs.
- Each coxal bone is formed by the fusion of three bones: the ilium, ischium, and pubis.
- The ilium forms the superior portion of a coxal bone and joins with the sacrum to form a sacroiliac joint.
- The ischium forms the inferior, posterior portion of a coxal bone and supports the body when sitting.
- The pubis forms the inferior, anterior part of a coxal bone. The two pubes unite anteriorly at the pubic symphysis.
- A pelvis is formed by two coxal bones and a sacrum. There are structural and functional differences between male and female pelves.
- Each lower limb consists of a femur, a patella, a tibia, a fibula, tarsal bones, metatarsals, and phalanges.
- The head of the femur is inserted into the acetabulum of a coxal bone to form a hip joint. Distally, it articulates with the tibia at the knee joint.
- The patella is a sesamoid bone in the anterior portion of the knee joint.
- The tibia articulates with the femur at the knee joint and with the talus to form the ankle joint.
- The fibula lies lateral to the tibia. It articulates proximally with the tibia and distally with the talus.
- The skeleton of the foot consists of tarsal bones (7), metatarsals (5), and phalanges (14).
- Tarsal bones form the ankle, metatarsal bones support the instep, and phalanges are the bones of the toes.

6.7 Articulations

- There are three types of joints: immovable, slightly movable, and freely movable.
- Bones forming immovable joints are closely joined by a thin layer of dense connective tissue or hyaline cartilage. Sutures in skull and epiphysial plates in growing bones are examples.
- Bones forming slightly movable joints are separated by fibrocartilage or dense connective tissue. Joints between vertebral bodies are examples.
- Bones forming freely movable joints are bound together by an articular capsule. The articular surfaces of the bones are covered by articular cartilages. The joint cavity is lubricated by synovial fluid secreted by the synovial membrane, the internal layer of articular capsule.
- There are six types of freely movable joints: plane, condylar, hinge, saddle, pivot, and ball-and-socket.
- Movements at freely movable joints include flexion, extension, hyperextension, dorsiflexion, plantar flexion, abduction, adduction, rotation, circumduction, inversion, eversion, protraction, retraction, elevation, depression, pronation, supination, opposition, and reposition.

6.8 Disorders of the Skeletal System

- Disorders of bones include fractures, osteomyelitis, osteoporosis, and rickets.
- Disorders of joints include arthritis, dislocation, herniated disc, abnormal spinal curvatures, and sprains.

Self-Review

Answers are located in appendix B.

1. The skeletal system provides ______ for the body and ______ for many internal organs.
2. The enlarged ends of a long bone are the ______, which are composed of ______ bone that is coated with a thin layer of compact bone.
3. Blood vessels and nerves enter a bone through a ______.
4. Cranial bones are formed by ______ ossification.
5. Growth in diameter of a long bone occurs by deposition of bone by osteoblasts from the ______.
6. The skull, vertebral column, and thoracic cage are part of the ______ skeleton.
7. The bone forming the lower jaw is the ______, and it articulates with the ______.
8. The first vertebra, the ______, articulates with the ______ bone of the skull.
9. True ribs are attached directly to the sternum by the ______.
10. The clavicles and scapulae form the ______.
11. The arm bone, the ______, articulates with two forearm bones, the ______ and the ______.
12. Each coxal bone is formed of three fused bones: the ______, ______, and ______.
13. The thigh bone is the ______, and it articulates distally with the ______ and ______.
14. Among freely movable joints, the elbow is an example of a ______ joint, and the shoulder is an example of a ______ joint.
15. ______ is a weakening of bones due to removal of bone matrix.

Critical Thinking

1. Explain why both osteoclasts and osteoblasts are required for proper bone development.
2. Bone repairs itself faster than cartilage. Explain why.
3. Why is osteomyelitis more likely to occur after a compound fracture than after a greenstick fracture?
4. Explain how bones may become weakened if the diet is deficient in calcium.

ADDITIONAL RESOURCES

7 CHAPTER

Muscular System

Melanie and a few of her friends head out early one morning for a short hike up a nearby mountain to a scenic overlook. As the wind gusts, forcing the temperature below freezing, they study a map and debate what trail to take. Melanie wonders if they made a good decision to hike today as her hands and feet begin to go numb despite her gloves and lined winter boots. Shivering violently, Melanie follows her friends up the mountain. The hike is strenuous because the trail they chose is both steep and rocky. The heat being created through the vigorous contractions of her skeletal muscles begins to gradually warm her body. In a short while, Melanie notices that she is no longer shivering or feeling the cold around her. By the time Melanie reaches the overlook, she is actually so warm that she begins to sweat. The friends sit on the edge of the overlook enjoying the view and each other's company. As the effects of the cold settle in once more, Melanie happily leads the way down the mountain to where a mug of hot chocolate and a roaring fire are waiting.

CHAPTER OUTLINE

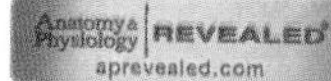

Module 6
Muscular System

SELECTED KEY TERMS

Agonist (agogos = leader) A muscle whose contraction leads an action.
Antagonist (anti = against) A muscle whose contraction opposes the action of the agonist.
Aponeurosis (apo = from; neur = cord) A broad sheet of dense regular connective tissue that attaches a muscle to another muscle or connective tissue.
Creatine phosphate An energy storage molecule found in muscle cells.
Insertion The attachment of a muscle that moves when the muscle contracts.
Motor unit A somatic motor neuron and the muscle fibers that it controls.
Muscle fiber A single skeletal muscle cell.
Muscle tone The state of slight contraction in a skeletal muscle.
Myoglobin (myo = muscle) An oxygen-storage molecule in muscle cells.
Neurotransmitter (neuro = nerve; transmit = to send across) A chemical released by terminal boutons of neurons that activates a muscle cell, gland, or another neuron.
Origin The attachment of a muscle that remains fixed when the muscle contracts.
Tendon A narrow band of dense regular connective tissue that attaches a muscle to a bone.
Tetany (tetan = rigid, stiff) A sustained muscle contraction.

MUSCLE TISSUE is the only tissue in the body that is specialized for contraction (shortening). The body contains three types of muscle tissue: skeletal, smooth, and cardiac. Each type of muscle tissue exhibits unique structural and functional characteristics. Contraction of skeletal muscle tissue produces locomotion, movement of body parts, and movement of the skin, as in making facial expressions. Cardiac muscle tissue produces the driving force responsible for pumping blood through the cardiovascular system, as you will see in chapter 12. Smooth muscle tissue is responsible for various internal functions, such as controlling the movement of blood through blood vessels and air through respiratory passageways. It is also directly involved in vision and moving contents through hollow internal organs as described in future chapters. Refresh your understanding of these tissues by referring to the discussion of muscle tissue in chapter 4. Table 7.1 summarizes the characteristics of muscle tissues.

Table 7.1 Types of Muscle Tissue

Characteristic	Skeletal	Smooth	Cardiac
Striations	Present	Absent	Present
Nucleus	Many peripherally located nuclei	Single centrally located nucleus	Usually a single centrally located nucleus
Cells	Long and parallel, called fibers	Short; tapered ends; parallel	Short and branching; intercalated discs join cells end to end to form network
Neural control	Voluntary	Involuntary	Involuntary
Contractions	Fast, variable fatigability; slow, resistant to fatigue	Slow; resistant to fatigue	Rhythmic; resistant to fatigue
Location	Attached to bones, dermis, ligaments, and other muscles	Walls of hollow visceral organs and blood and lymphatic vessels, skin, and inside eyes	Wall of the heart
Micrograph			

Contraction of a Single Fiber

It is possible to remove a single muscle fiber in order to study its contraction in the laboratory. By using electrical stimuli to initiate contraction and by gradually increasing the strength (voltage) of each stimulus, it has been shown that the fiber will not contract until the stimulus reaches a certain minimal strength. This minimal stimulus is called the **threshold stimulus.**

Whenever a muscle fiber is stimulated by a threshold stimulus or by a stimulus of greater strength, it always contracts *completely*. Thus, a muscle fiber either contracts completely or not at all—contraction is *not* proportional to the strength of the stimulus. This characteristic of individual muscle fibers is known as the **all-or-none response.**

Contraction of Whole Muscles

Much information has been gained by studying the contraction of a whole muscle of an experimental animal. In such studies, electrical stimulation is used to cause contraction, and the contraction is recorded to produce a tracing called a *myogram.*

If a single threshold stimulus is applied, some of the muscle fibers will contract to produce a single, weak contraction (a muscle twitch) and then relax, all within a fraction of a second. The myogram will look like the one shown in figure 7.8. After the stimulus is applied, there is a brief interval before the muscle starts to contract. This interval is known as the *latent phase*. Then, the muscle contracts (shortens) during the *contraction phase* and relaxes (returns to its former length) during the *relaxation phase*. If a muscle is stimulated again after it has relaxed completely, it will contract and produce a similar myogram. A series of single stimuli applied in this manner will yield a myogram like the one in figure 7.9*a*.

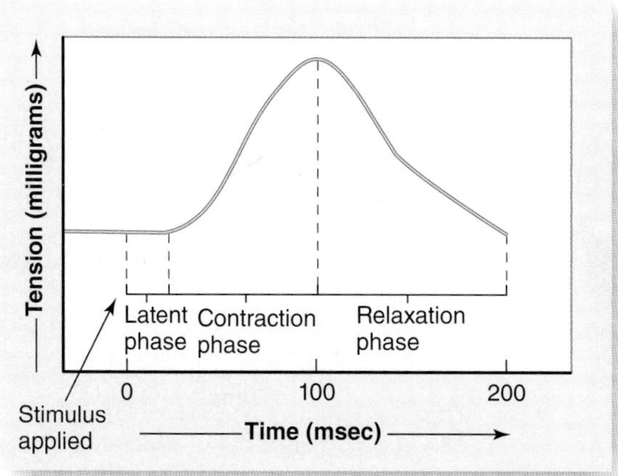

Figure 7.8 A myogram of a single muscle twitch. Note the brief latent phase, contraction phase, and longer relaxation phase.

If the interval between stimuli is shortened so that the muscle fibers cannot completely relax, the force of individual twitches combines by *summation*, which increases the force of contraction. Rapid summation produces incomplete tetany, a fluttering contraction (figure 7.9*b*). If stimuli are so frequent that relaxation is not possible, **tetany** results (figure 7.9*c*). Tetany is a state of sustained contraction without relaxation. In the body, tetany results from a rapid series of nerve impulses carried by somatic motor neurons to the muscle fibers that results in a prolonged state of contraction. Tetany for short time periods is the usual way in which muscles contract to produce body movements.

Graded Responses Unlike individual muscle fibers that exhibit all-or-none responses, whole muscles exhibit *graded responses*—that is, varying degrees of contraction. Graded responses enable the degree of muscle contraction to fit the task being performed. Obviously, more muscle fibers are required to lift a 14 kg (30 lb) weight than to lift a feather. Yet both activities can be performed by the same muscles.

Graded responses are possible because a muscle is composed of many different *motor units*, each responding to different thresholds of stimulation. In the laboratory, a weak stimulus that activates only low-threshold motor units produces a minimal contraction. As the strength of the stimulus is increased, the contractions get stronger as more motor units are activated until a **maximal stimulus** (one that activates all motor units) is applied, which produces a maximal contraction. Further increases in the strength of the stimulus (supramaximal) cannot produce a greater contraction. The same results occur in a normally functioning body. The nervous system provides the stimulation and controls the number of motor units activated in each muscle contraction. The activation of more and more motor units is known as *motor unit recruitment* (figure 7.9*d*).

Muscle Tone Even when a muscle is relaxed, some of its muscle fibers are contracting. At any given time, some of the muscle fibers in a muscle are involved in a sustained contraction that produces a constant partial, but slight, contraction of the muscle. This state of constant partial contraction, called **muscle tone,** keeps a muscle ready to respond. Muscle tone results from the alternating activation of different motor units by the nervous system so that some muscle fibers are always in sustained contraction, as seen in figure 7.10. Muscle tone of postural muscles plays an important role in maintaining erect posture.

Check My Understanding

7. What is meant by the all-or-none response?
8. How are muscles able to make graded responses?

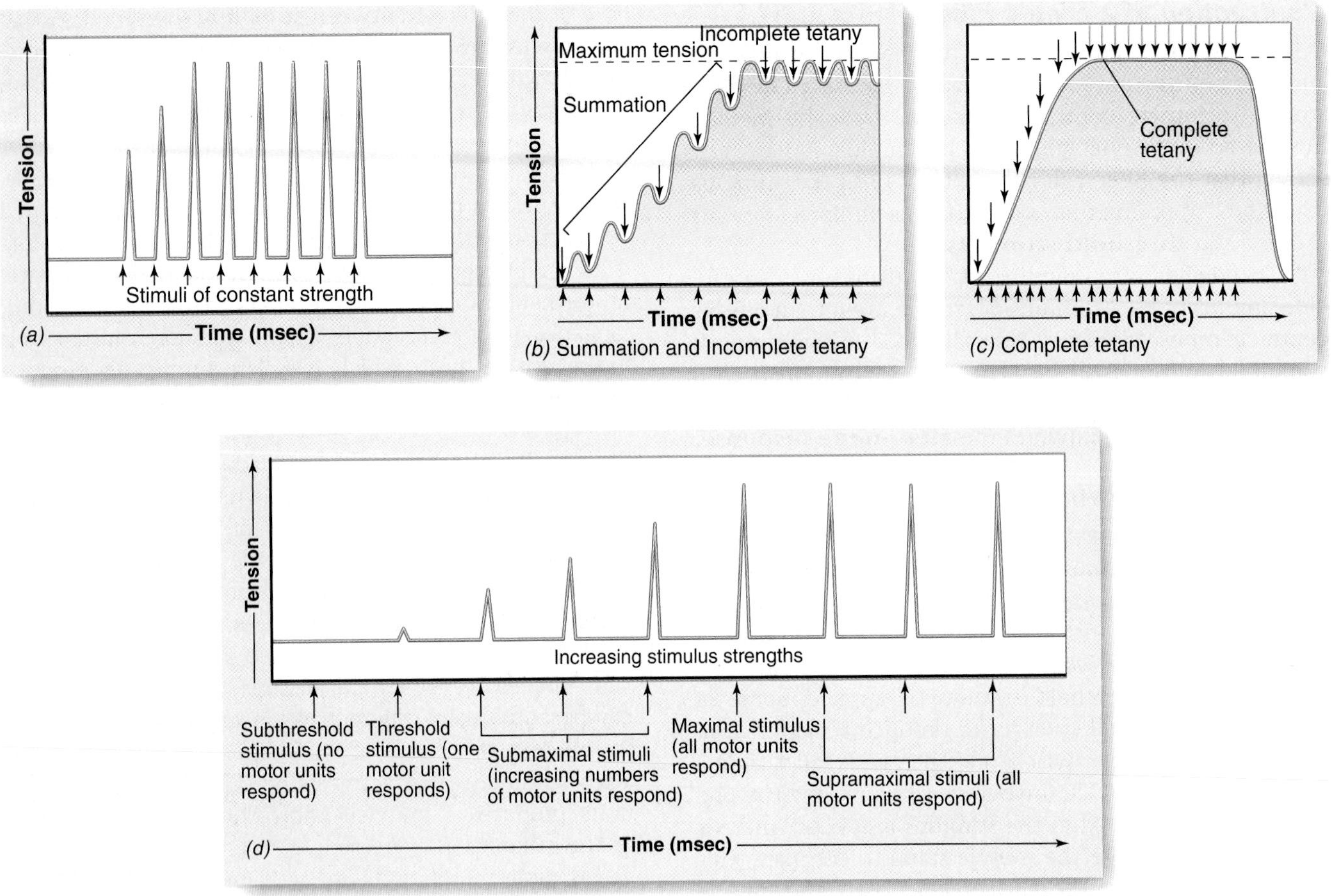

Tension
Stimuli of constant strength
(a)
Time (msec)
Maximum tension
Incomplete tetany
Summation
Tension
Time (msec)
(b) Summation and Incomplete tetany
Complete
tetany
Time (msec)
(c) Complete tetany
Tension
Increasing stimulus strengths
Subthreshold stimulus (no motor units respond)
Threshold stimulus (one motor unit responds)
Submaximal stimuli (increasing numbers of motor units respond)
Maximal stimulus (all motor units respond)
Supramaximal stimuli (all motor units respond)
(d)
Time (msec)

Figure 7.9 Myograms of *(a)* a series of simple twitches, *(b)* summation caused by incomplete relaxation between stimuli, *(c)* tetany, and *(d)* motor unit recruitment.

Key:
Motor unit 1
Motor unit 2
Motor unit 3
Somatic motor nerve
Spinal cord
Muscle fibers
Axons of Somatic motor neurons
(a)
Tension in tendon
Motor unit 1
Motor unit 2
Motor unit 3
Tension
Time
(b)

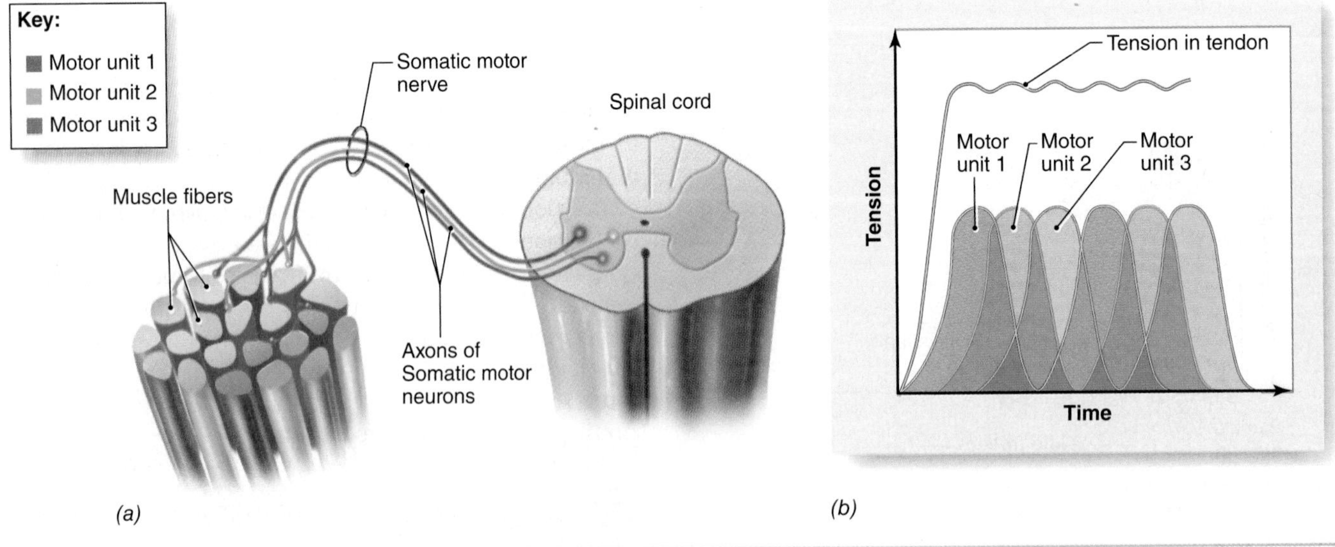

Figure 7.10 *(a)* Anatomy of motor units in a skeletal muscle. *(b)* Myogram showing mechanism of motor unit alternation in muscle tone.

7.3 Actions of Skeletal Muscles

Learning Objectives

10. Explain the relationship between a muscle's origin and insertion and its action.
11. Explain how agonists and antagonists function in the production of body movements.

Skeletal muscles are usually arranged so that the ends of a muscle are attached to bones on each side of a joint. Thus, a muscle usually extends across a joint. The type of movement produced depends upon the type of joint and the locations of the muscle attachments. Common movements at joints were discussed in chapter 6.

Origin and Insertion

During contraction, a bone to which one end of the muscle is attached moves, but the bone to which the other end is attached does not. The movable attachment of a muscle is called the **insertion,** and the immovable attachment is called the **origin.** When a muscle contracts, the insertion is pulled toward the origin.

Consider the *biceps brachii* in figure 7.11. It has two origins, and both are attached to the scapula. The insertion is on the radius, and the muscle lies along the anterior surface of the humerus. When the biceps brachii contracts, the insertion is pulled toward the origin, which results in the flexion of the forearm at the elbow.

Most muscle contractions are *isotonic contractions,* which cause movement at a joint. Walking and breathing are examples. However, some contractions may not produce movement but only increase tension within a muscle. Contractions that maintain body posture are good examples. Such contractions are *isometric contractions.*

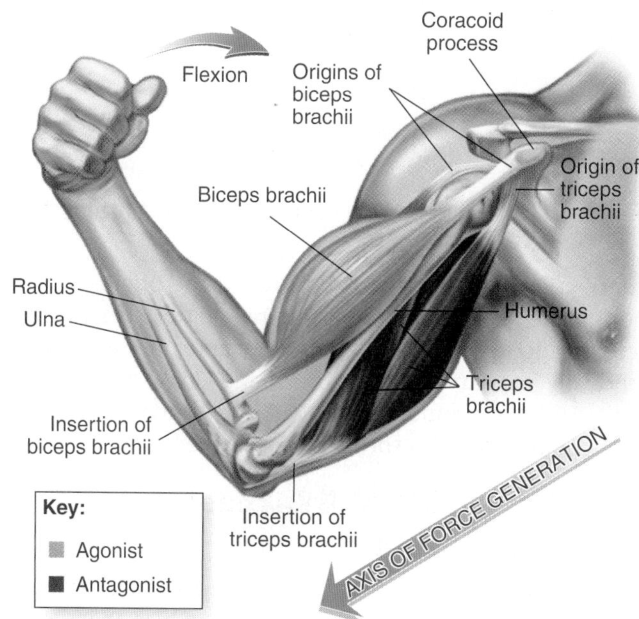

Figure 7.11 Demonstration of the actions of agonists and antagonists with origins and insertions labeled.

Muscle Interactions

Muscles function in groups rather than singly, and the groups are arranged to provide opposing movements. For example, if one group of muscles produces flexion, the opposing group produces extension. A group of muscles producing an action are called **agonists,** and the opposing group of muscles are called **antagonists.** When agonists contract, antagonists must relax, and vice versa, for movement to occur. If both groups contract simultaneously, the movable body part remains rigid. Figure 7.11 illustrates how the biceps brachii is the agonist of forearm flexion, while the triceps brachii is the antagonist.

7.4 Naming of Muscles

Learning Objective

12. List the criteria used for naming muscles.

Learning the complex names and functions of muscles can be confusing. However, the names of muscles are informative if their meaning is known. A few of the criteria used in naming muscles and examples of terms found in the names of muscles are listed below:

- **Function:** extensor, flexor, adductor, and pronator.
- **Shape:** trapezius (trapezoid), rhomboid (rhombus), deltoid (delta-shaped or triangular), biceps (two heads).
- **Relative position:** external, internal, abdominal, medial, lateral.
- **Location:** intercostal (between ribs), pectoralis (chest).
- **Site of attachment:** temporalis (temporal bone), zygomaticus (zygomatic bone).
- **Origin and insertion:** sternohyoid (sternum = origin; hyoid = insertion), sternocleidomastoid (sternum and clavicle = origins; mastoid process = insertion).
- **Size:** maximus (larger or largest), minimus (smaller or smallest), brevis (short), longus (long).
- **Orientation of fibers:** oblique (diagonal), rectus (straight), transversus (across).

7.5 Major Skeletal Muscles

Learning Objectives

13. Describe the location and action of the major muscles of the body.
14. Identify the major muscles on a diagram.

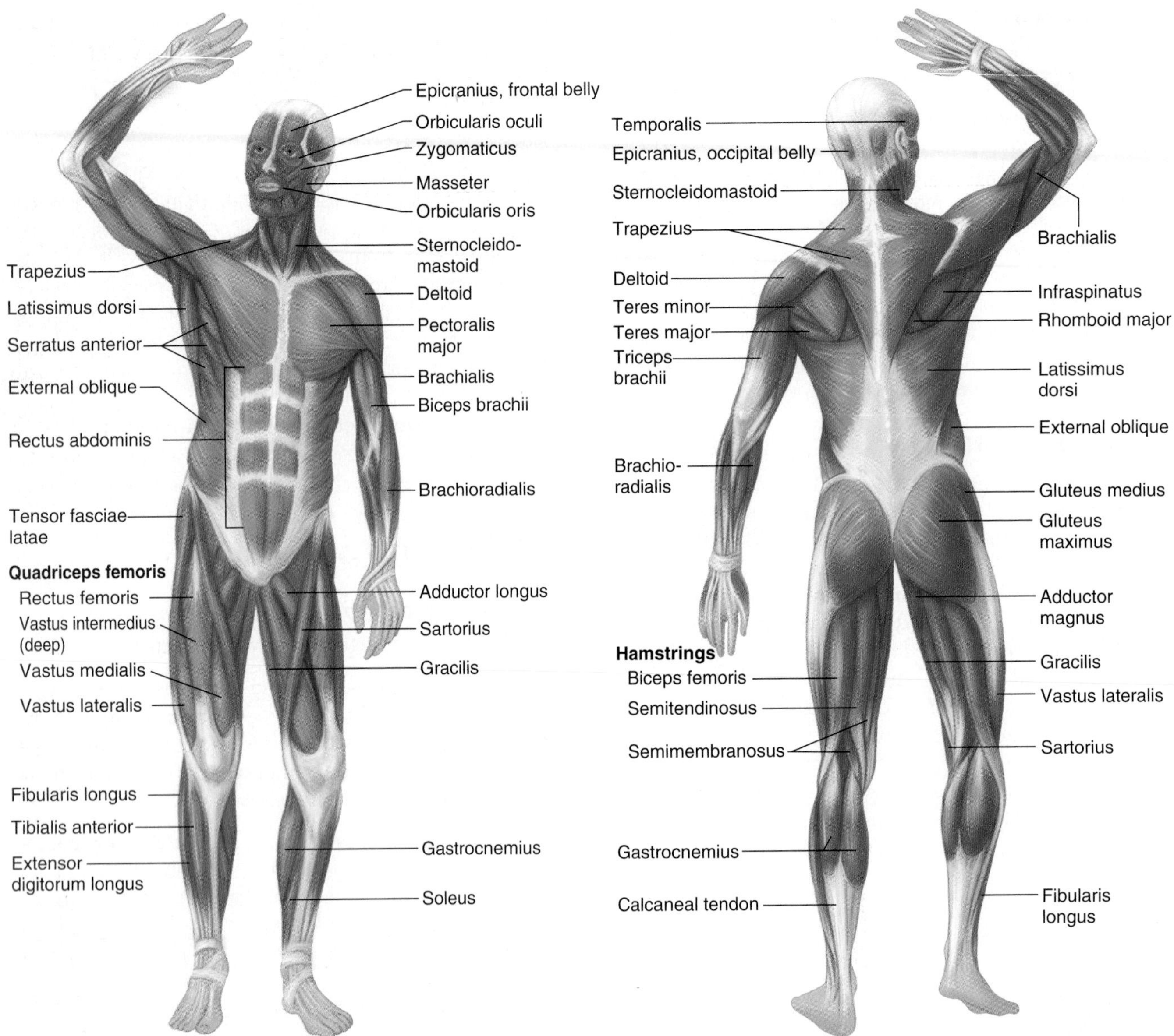

Figure 7.12 Anterior view of superficial skeletal muscles.

Figure 7.13 Posterior view of superficial skeletal muscles.

This section is concerned with the name, location, attachment, and action of the major skeletal muscles. There are more than 600 muscles in the body, but only a few of the major muscles are considered here. Most of this information is presented in tables and figures to aid your learning. The tables are organized according to the primary actions of the muscles. The pronunciation of each muscle is included, because being able to pronounce the names correctly will help you learn the names of the muscles.

As you study this section, locate each muscle listed in the tables on the related figures 7.12 to 7.25. This will help you visualize the location and action of each muscle. Also, if you visualize the locations of the origin and insertion of a muscle, its action can be determined because contraction pulls the insertion toward the origin. It may help to refresh your understanding of the skeleton by referring to appropriate figures in chapter 6. Begin your study by examining figures 7.12 and 7.13 to learn the major superficial muscles that will be considered in more detail as you progress through the chapter.

Muscles of Facial Expression and Mastication

Muscles of the face and scalp produce the facial expressions that help communicate feelings, such as anger, sadness, happiness, fear, disgust, pain, and surprise. Most have origins on skull bones and insertions on the dermis of the skin (table 7.3 and figure 7.14).

Table 7.3 Muscles of Facial Expression

Muscle	Origin	Insertion	Action
Buccinator (buk′-si-nā-tor)	Lateral surfaces of maxilla and mandible	Orbicularis oris	Compresses cheeks inward
Epicranius (ep-i-krā′-nē-us)	This muscle consists of two parts: the frontal belly and the occipital belly. They are joined by the epicranial aponeurosis, which covers the top of the skull.		
Frontal belly	Epicranial aponeurosis	Skin and muscles superior to the eyes	Elevates eyebrows and wrinkles forehead
Occipital belly	Base of occipital bone	Epicranial aponeurosis	Pulls scalp posteriorly
Orbicularis oculi (or-bik′-ū-lar-is ok′-ū-li)	Frontal bone and maxillae	Skin around eye	Closes eye
Orbicularis oris (or-bik′-ū-lar-is- o′-ris)	Muscles around mouth	Skin around lips	Closes and puckers lips; shapes lips during speech
Platysma (plah-tiz′-mah)	Fascia of superior chest	Mandible and muscles around mouth	Draws angle of mouth inferiorly
Zygomaticus (zī-gō-mat′-ik-us)	Zygomatic bone	Orbicularis oris at angle of the mouth	Elevates corners of mouth (smiling)

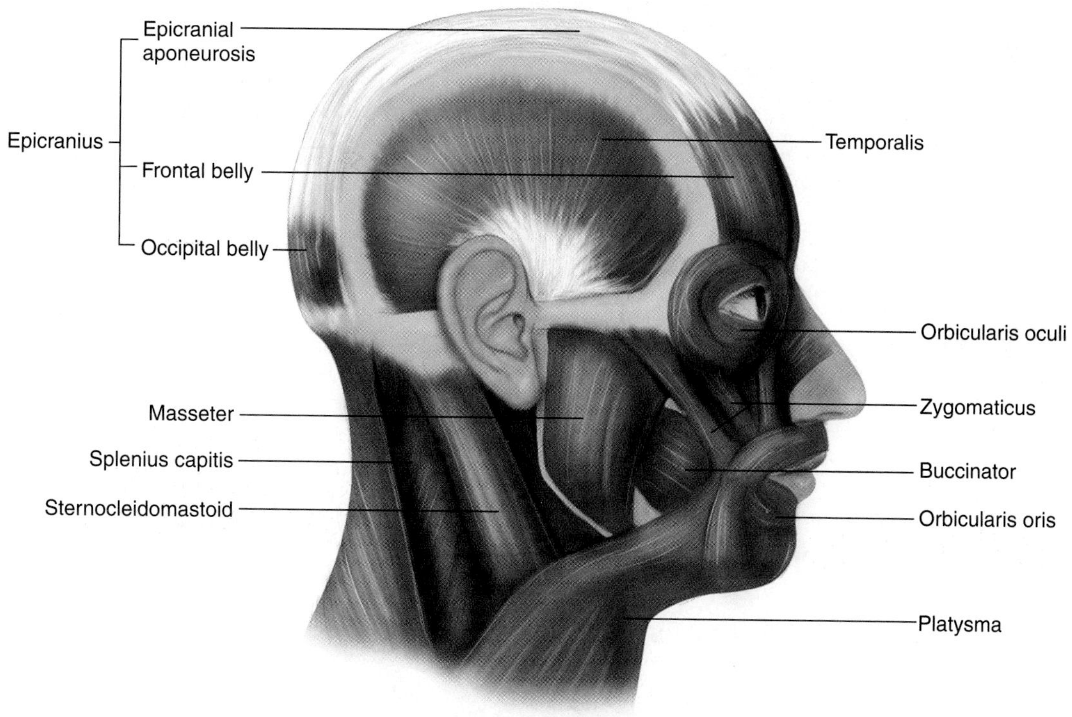

Figure 7.14 Muscles of facial expression and mastication. AP|R

The *epicranius* is an unusual muscle. It has a large *epicranial aponeurosis* that covers the top of the skull and two contractile portions: the *frontal belly* over the frontal bone and the *occipital belly* over the occipital bone.

Two major pairs of muscles elevate the mandible in the process of mastication (chewing): the *masseter* and the *temporalis* (table 7.4 and figure 7.14).

Muscles That Move the Head

Several pairs of neck muscles are responsible for flexing, extending, and rotating the head. Table 7.5 lists two of the major muscles that perform this function: the *sternocleidomastoid* and the *splenius capitis*. As noted in table 7.8, the *trapezius* can also extend the head, although this is not its major function (figures 7.14, 7.15, and 7.16).

Table 7.4 Muscles of Mastication

Muscle	Origin	Insertion	Action
Masseter (mas-se′-ter)	Zygomatic arch	Lateral surface of mandible	Elevates mandible
Temporalis (tem-po-ra′-lis)	Temporal bone	Coronoid process of mandible	Elevates mandible

Table 7.5 Muscles That Move the Head

Muscle	Origin	Insertion	Action
Sternocleidomastoid (ster-nō-klī-dō-mas′-toid)	Clavicle and sternum	Mastoid process of temporal bone	Contraction of both muscles flexes head toward chest; contraction of one muscle turns head away from contracting muscle
Splenius capitis (splē′-nē-us kap′-i-tis)	Inferior cervical and superior thoracic vertebrae	Mastoid process of temporal bone	Contraction of both muscles extends head; contraction of one muscle turns head toward same side as contracting muscle

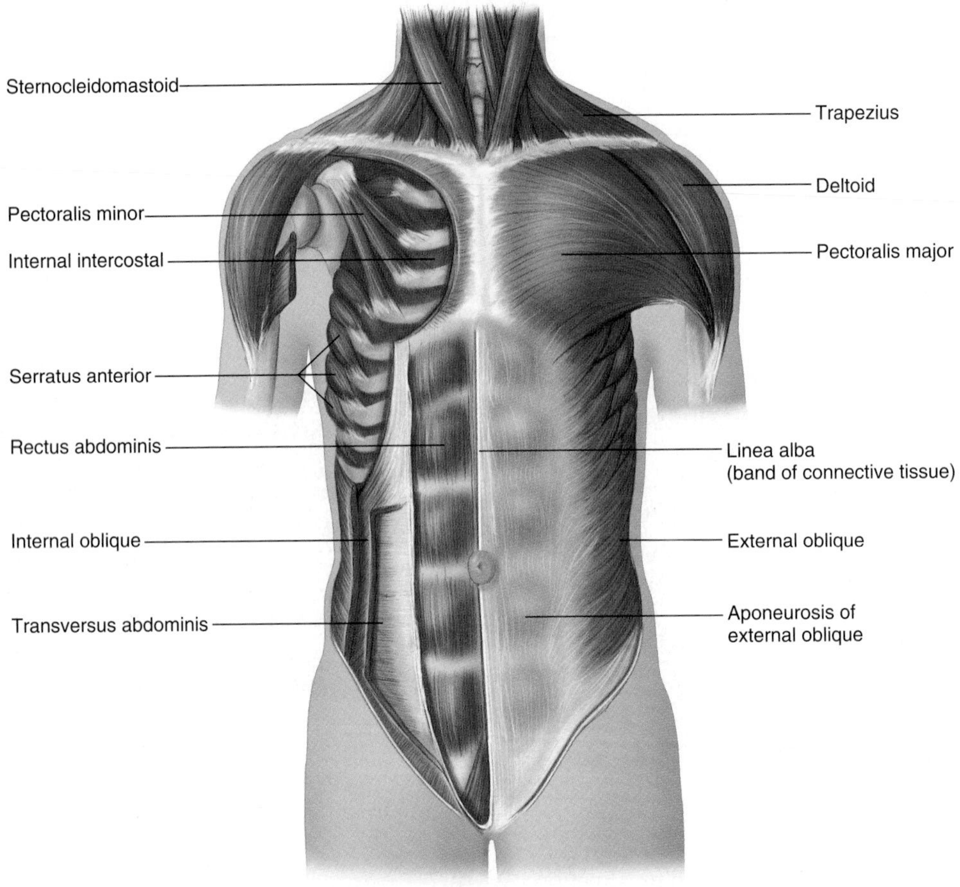

Figure 7.15 Muscles of the anterior chest and abdominal wall. The right pectoralis major is removed to show the deep muscles. APIR

Muscle of the Abdominal Wall

The abdominal muscles are paired muscles that provide support for the anterior and lateral portions of the abdominal and pelvic regions, including support for the internal organs. The muscles are named for the direction of their muscle fibers: *rectus abdominis, external oblique, internal oblique,* and *transversus abdominis.* They are arranged in overlapping layers and are attached by larger

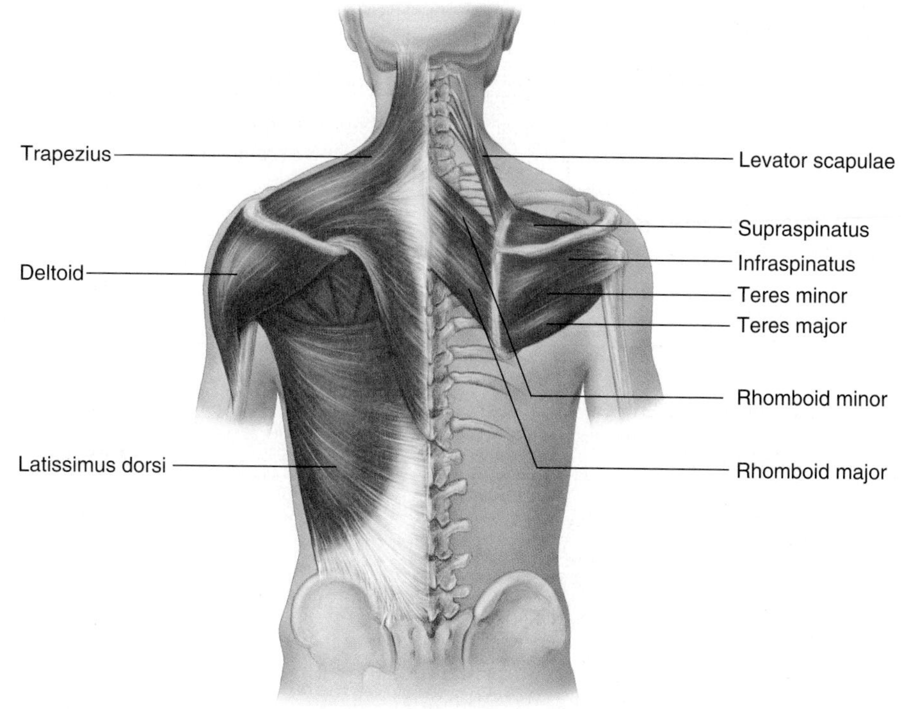

Figure 7.16 Muscles of the posterior shoulder. The right trapezius is removed to show deep muscles. APR

aponeuroses that merge at the anterior midline to form the *linea alba,* or white line (table 7.6 and figure 7.15).

Muscles of Breathing

Movement of the ribs occurs during breathing and is brought about by the contraction of two sets of muscles that are located between the ribs. The *external intercostals* elevate and protract the ribs during inspiration, and the *internal intercostals* depress and retract the ribs during expiration (table 7.7 and figure 7.15). The primary breathing muscle is the *diaphragm,* a thin sheet of muscle that separates the thoracic and abdominal cavities.

Muscles That Move the Pectoral Girdle

Pectoral girdle muscles originate on bones of the axial skeleton and insert on the scapula or clavicle. Because the scapula is supported mainly by muscles, it can be moved more freely than the clavicle. The *trapezius* is a superficial trapezoid-shaped muscle that covers much of the superior back. The *rhomboid major* and *minor* and the *levator scapulae* lie deep to the trapezius. Each *serratus anterior* is located on the lateral surface of the superior ribs near the axillary region. The *pectoralis minor* lies deep to the pectoralis major. It protracts and depresses the scapula (table 7.8 and figures 7.15 to 7.18).

Table 7.6 Muscles of the Abdominal Wall

Muscle	Origin	Insertion	Action
Rectus abdominis (rek′-tus ab-dom′-i-nis)	Pubic symphysis and pubis	Xiphoid process of sternum and costal cartilages of ribs 5 to 7	Tightens abdominal wall; flexes the vertebral column
External oblique (eks-ter′-nal o-blēk′)	Anterior surface of inferior eight ribs	Iliac crest and linea alba	Tightens abdominal wall; rotation and lateral flexion of the vertebral column
Internal oblique (in-ter′-nal o-blēk′)	Iliac crest and inguinal ligament	Cartilage of inferior four ribs, pubis, and linea alba	Same as above
Transversus abdominis (trans-ver′-sus ab-dom′-i-nis)	Iliac crest, cartilages of inferior six ribs, processes of lumbar vertebrae	Pubis and linea alba	Tightens abdominal wall

Table 7.7 Muscles of Breathing

Muscle	Origin	Insertion	Action
Diaphragm (dī-a-fram)	Lumbar vertebrae, costal cartilages of inferior ribs, xiphoid process	Central tendon located at midpoint of muscle	Forms floor of thoracic cavity; depresses during contraction, causing inspiration
External intercostals (eks-ter′-nal in-ter-kos′-tals)	Inferior border of rib above	Superior border of rib below	Elevates and protracts ribs during inspiration
Internal intercostals (in-ter′-nal in-ter-kos′-tals)	Superior border of rib below	Inferior border of rib above	Depresses and retracts ribs during expiration

Table 7.8 Muscles That Move the Pectoral Girdle

Muscle	Origin	Insertion	Action
Trapezius (trah-pē-zē′-us)	Occipital bone; cervical and thoracic vertebrae	Clavicle; spine and acromion of scapula	Elevates clavicle; adducts and elevates scapula; extends head
Rhomboid major and minor (rom-boid)	Superior thoracic vertebrae	Medial border of scapula	Adducts and elevates scapula
Levator scapulae (le-va′-tor skap′-ū-lē)	Cervical vertebrae	Superior medial margin of scapula	Elevates scapula
Serratus anterior (ser-ra′-tus)	Superior eight to nine ribs	Medial border of scapula	Depresses, protracts, and rotates scapula
Pectoralis minor (pek-to-rah′-lis)	Anterior surface of superior ribs	Coracoid process of scapula	Depresses and protracts scapula

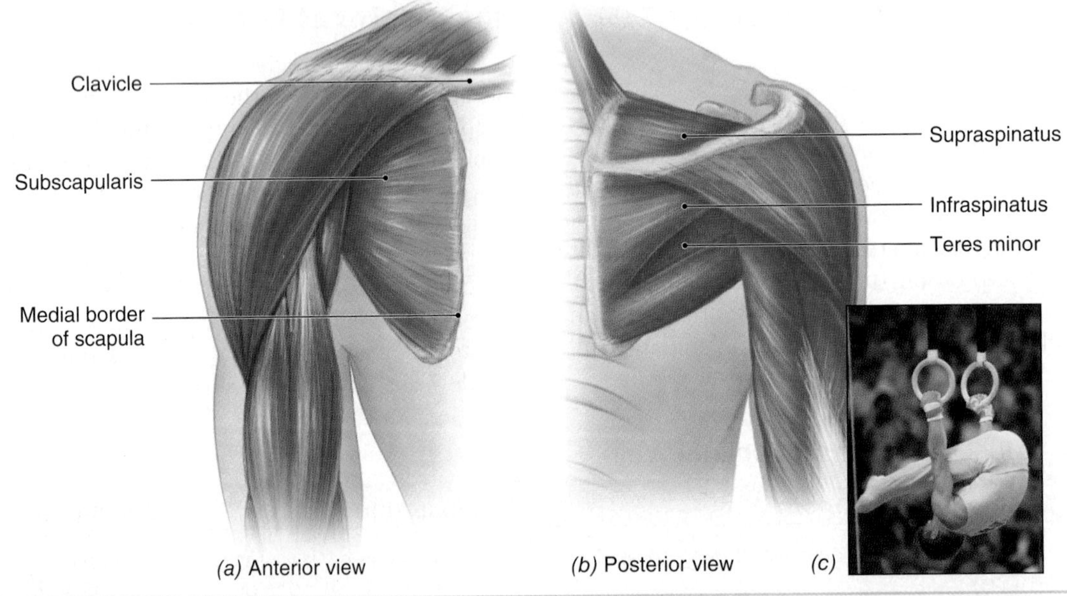

Figure 7.17 Muscles of the rotator cuff. *(a)* Anterior view showing subscapularis. *(b)* Posterior view showing supraspinatus, infraspinatus, and teres minor. *(c)* A gymnast on the rings must have a strong rotator cuff. AP|R

Muscles That Move the Arm and Forearm

Movement of the humerus is enabled by the muscles that originate on the pectoral girdle, ribs, or vertebrae and insert on the humerus. The arrangement of these muscles and the ball-and-socket joint between the humerus and scapula enable great freedom of movement for the arm. The *pectoralis major* is the large superficial muscle of the chest. The *deltoid* is the thick muscle that caps the shoulder joint. The *supraspinatus, infraspinatus,* and *teres minor* cover the posterior surface of the scapula. The anterior surface of each scapula is covered by the *subscapularis.* These four muscles and their tendons surround the head of the humerus at the shoulder joint, making up the **rotator cuff** (figure 7.17). The muscles and tendons of the rotator cuff are the only structures stabilizing the shoulder joint; thus the joint is fairly unstable compared to other joints. However, this relative lack of stability is what allows the shoulder's mobility. The *latissimus dorsi* is a broad, sheetlike muscle that covers the inferior back. The *teres major* assists the latissimus dorsi and is located just superior to it. (table 7.9 and figures 7.15 to 7.17).

Muscles moving the forearm originate on either the humerus or the scapula and insert on either the radius or the ulna. Three flexors occur on the anterior surface of the arm: the *biceps brachii, brachialis,* and *brachioradialis.* One extensor, the *triceps brachii,* is located on the posterior surface of the arm (table 7.10 and figures 7.15, 7.18, and 7.19).

Check My Understanding

9. What are the names and locations of the two parts of the epicranius muscle?
10. What muscles are involved in chewing your food?
11. What muscles turn your head to the side?
12. What muscle separates the abdominal and thoracic cavities?
13. What are the names of the abdominal muscles from deep to superficial?
14. What three muscles elevate the scapula?

Table 7.9 Muscles That Move the Arm

Muscle	Origin	Insertion	Action
Pectoralis major (pek-tō-rah′-lis)	Clavicle, sternum, and cartilages of superior ribs	Greater tubercle of humerus	Adducts, flexes, and medially rotates arm
Deltoid (del′-toid)	Clavicle and spine, and acromion of scapula	Deltoid tuberosity of humerus	Abducts, flexes, and extends arm
Latissimus dorsi (lah-tis′-i-mus dor′sī)	Inferior thoracic and lumbar vertebrae; sacrum; inferior ribs; iliac crest	Intertubercular sulcus of humerus	Adducts, extends, and medially rotates arm
Teres major (te′r-ez)	Inferior angle of scapula	Distal to lesser tubercle of humerus	Same as above
Rotator cuff muscles	These four muscles stabilize the shoulder joint		
Supraspinatus (su-prah-spī′-na-tus)	Superior to spine of scapula	Greater tubercle of humerus	Abducts arm
Infraspinatus (in-frah-spī′-na-tus)	Inferior to spine of scapula	Greater tubercle of humerus	Laterally rotates arm
Teres minor	Lateral border of scapula	Greater tubercle of humerus	Laterally rotates arm
Subscapularis (sŭ-skap-ŭ-lār′ris)	Anterior surface of scapula	Lesser tubercle of humerus	Medially rotates arm

Table 7.10 Muscles That Move the Forearm

Muscle	Origin	Insertion	Action
Biceps brachii (bi′-seps brā′-kē-i)	Coracoid process and tubercle superior to glenoid cavity of scapula	Radial tuberosity of radius	Flexes forearm and supination, also flexes arm
Brachialis (brā′-kē-al-is)	Distal, anterior surface of humerus	Coronoid process of ulna	Flexes forearm
Brachioradialis (brā-kē-ō-rā-dē-a′-lis)	Lateral surface of distal end of humerus	Lateral surface of radius superior to styloid process	Flexes forearm
Triceps brachii (trī′-seps brā′-kē-ī)	Lateral and medial surfaces of humerus and tubercle inferior to glenoid cavity of scapula	Olecranon of ulna	Extends forearm, also extends arm

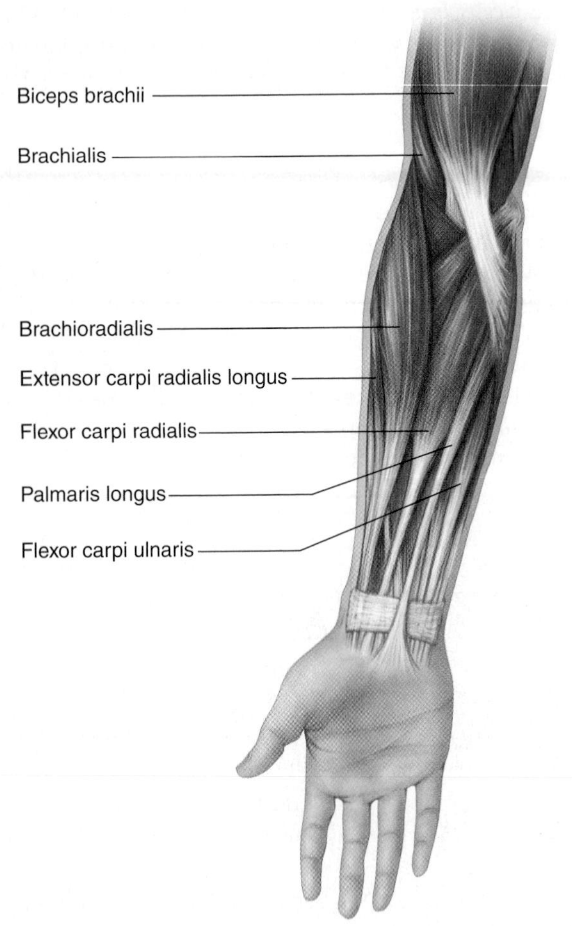

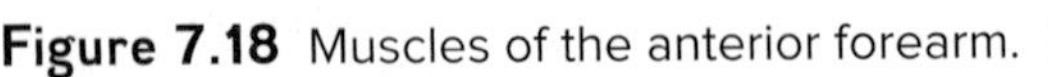

Figure 7.18 Muscles of the anterior forearm. APIR

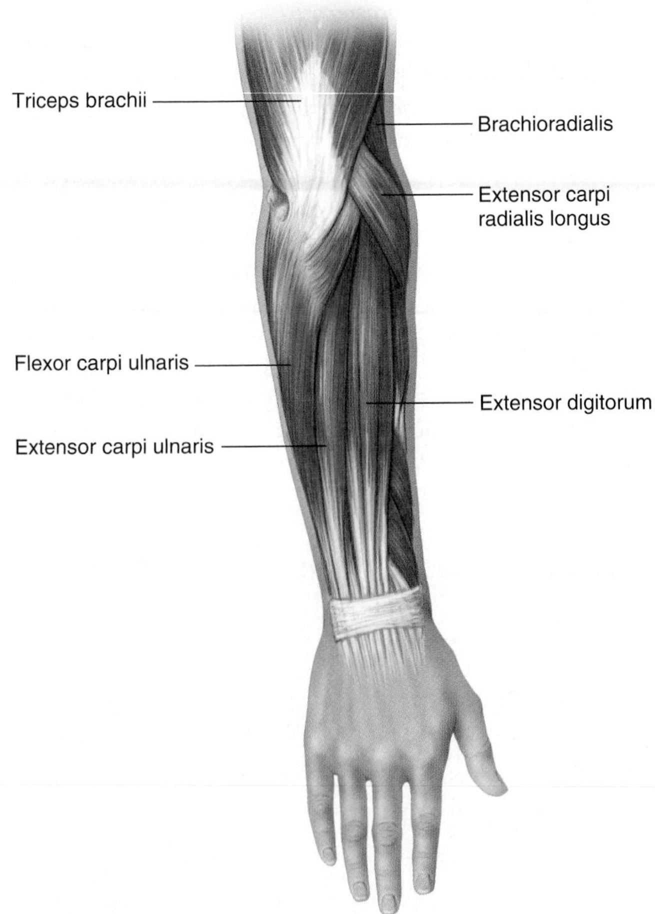

Figure 7.19 Muscles of the posterior forearm. APIR

Muscles That Move the Wrist and Fingers

Many muscles that produce the various movements of the wrist and fingers are located in the forearm. Only a few of the larger superficial muscles are considered here. They originate from the distal end of the humerus and insert on carpal bones, metacarpals, or phalanges. Flexors on the anterior surface include the *flexor carpi radialis, flexor carpi ulnaris,* and *palmaris longus.* Extensors on the posterior surface include the *extensor carpi radialis longus, extensor carpi ulnaris,* and *extensor digitorum* (table 7.11; and figures 7.18 and 7.19). Note that the tendons of these muscles are held in position by a circular ligament at the wrist.

Check My Understanding

15. What muscle abducts and extends your arm?
16. What muscle extends your forearm?
17. What muscle extends your fingers?

Muscles That Move the Thigh and Leg

Muscles moving the thigh span the hip joint. They insert on the femur, and most originate on the pelvic girdle. The *iliacus* and *psoas major* are located anteriorly, the *gluteus maximus* is located posteriorly and forms the buttocks, the *gluteus medius* is located deep to the gluteus maximus posteriorly and extends laterally, and the *tensor fasciae latae* is located laterally. The *adductor longus* and *adductor magnus* are both located medially (table 7.12 and figures 7.20, 7.21, and 7.22).

The leg is moved by muscles located in the thigh. They span the knee joint and originate on the pelvic girdle or femur and insert on the tibia or fibula. The **quadriceps femoris** is composed of four muscles that have a common tendon that inserts on the patella. However, this tendon continues as the patellar ligament, which attaches to the tibial tuberosity–the functional insertion for these muscles. The *biceps femoris, semitendinosus,* and *semimembranosus* on the posterior surface of the thigh are often collectively called the **hamstrings.** The medially

Table 7.11 Muscles That Move the Wrist and Fingers

Muscle	Origin	Insertion	Action
Flexor carpi radialis (flek′-sor kar′-pī rā-dē-a′-lis)	Medial epicondyle of humerus	Metacarpals II and III	Flexes and abducts wrist
Flexor carpi ulnaris (flek′-sor kar′-pī ul-na′-ris)	Medial epicondyle of humerus and olecranon of ulna	Carpal bones and metacarpal V	Flexes and adducts wrist
Palmaris longus (pal-ma′-ris long′-gus)	Medial epicondyle of humerus	Fascia of palm	Flexes wrist
Extensor carpi radialis longus (eks-ten′-sor kar′-pī rā-dē-a′-lis long′-gus)	Lateral epicondyle of humerus	Metacarpal II	Extends and abducts wrist
Extensor carpi ulnaris (eks-ten′-sor kar′-pī ul-na′-ris)	Lateral epicondyle of humerus	Metacarpal V	Extends and adducts wrist
Extensor digitorum (eks-ten′-sor dij-i-to′-rum)	Lateral epicondyle of humerus	Posterior surfaces of phalanges II–V	Extends fingers

Table 7.12 Muscles That Move the Thigh

Muscle	Origin	Insertion	Action
Iliacus (il′-ē-ak-us)	Fossa of ilium	Lesser trochanter of femur	Flexes thigh
Psoas major (so′-as)	Lumbar vertebrae	Lesser trochanter of femur	Flexes thigh
Gluteus maximus (glū′-tē-us mak′-si-mus)	Posterior surfaces of ilium, sacrum, and coccyx	Posterior surface of femur and iliotibial tract	Extends and laterally rotates thigh
Gluteus medius (glū′-tē-us mē′-dē-us)	Lateral surface of ilium	Greater trochanter of femur	Abducts and medially rotates thigh
Tensor fasciae latae (ten′-sor fash′-ē-ē lah-tē′)	Anterior iliac crest	Iliotibial tract	Flexes and abducts thigh
Adductor longus (ad-duk′-tor long′-gus)	Pubis near pubic symphysis	Posterior surface of femur	Adducts, flexes, and laterally rotates thigh
Adductor magnus (ad-duk′-tor mag′-nus)	Inferior portion of ischium and pubis	Same as above	Same as above

Clinical Insight

Intramuscular injections are commonly used when quick absorption is desired. Such injections are given in three sites: (1) the lateral surface of the deltoid; (2) the gluteus medius in the superior, lateral portion of the buttock; and (3) the vastus lateralis near the midpoint of the lateral surface of the thigh. These injection sites are chosen because there are no major nerves or blood vessels present that could be damaged, and the muscles have a good blood supply to aid absorption. The site chosen may vary with the age and condition of the patient.

located *gracilis* has two insertions that give it dual actions. The long, straplike *sartorius* extends diagonally across the anterior surface of the thigh and spans both the hip and knee joints. Its contraction enables the legs to cross (tables 7.12, and 7.13 and figures 7.20, 7.21, and 7.22).

Muscles That Move the Foot and Toes

Many muscles are involved in the movement of the foot and toes. They are located in the leg and originate on the femur, tibia, or fibula and insert on the tarsal bones, metatarsals, or phalanges. The posterior leg muscles include the *gastrocnemius* and *soleus,* which insert through a common tendon, the calcaneal (Achilles) tendon, which attaches to the calcaneus. The *tibialis*

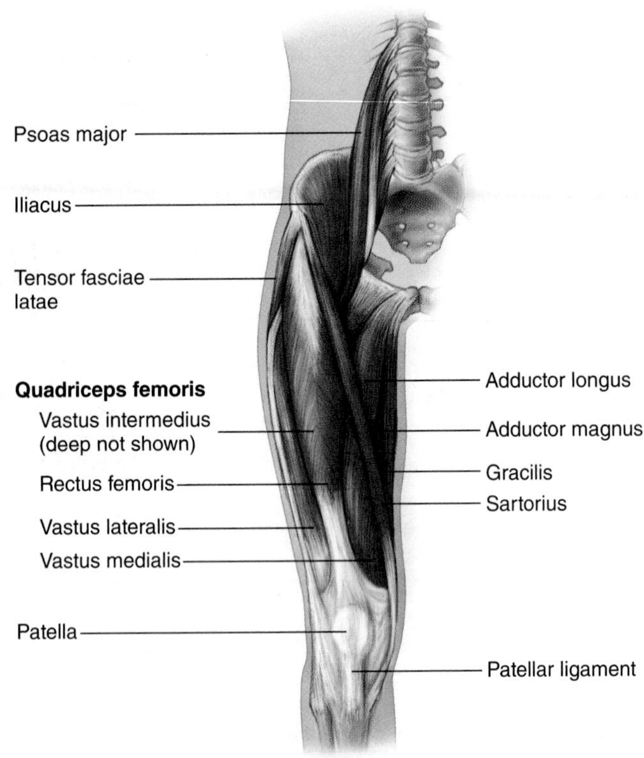

Figure 7.20 Muscles of the anterior right thigh. (Note that the vastus intermedius is deep to the rectus femoris and is not visible in this view.) APIR

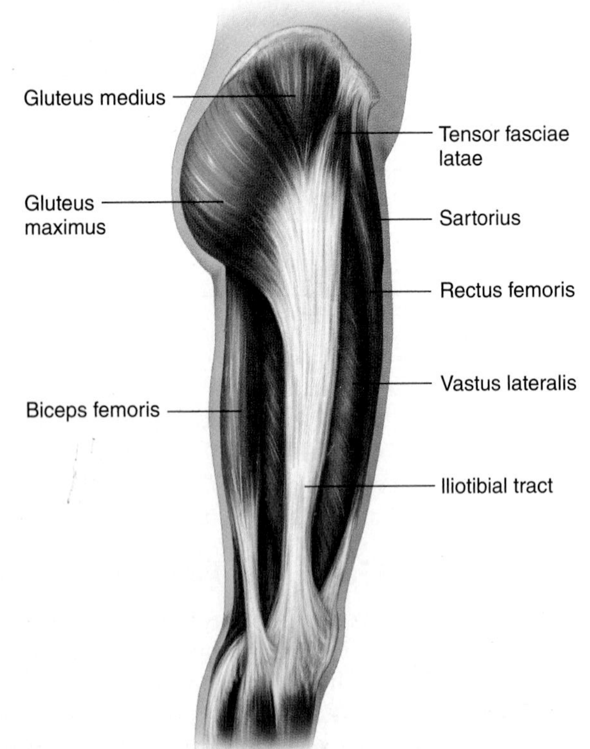

Figure 7.21 Muscles of the lateral right thigh.

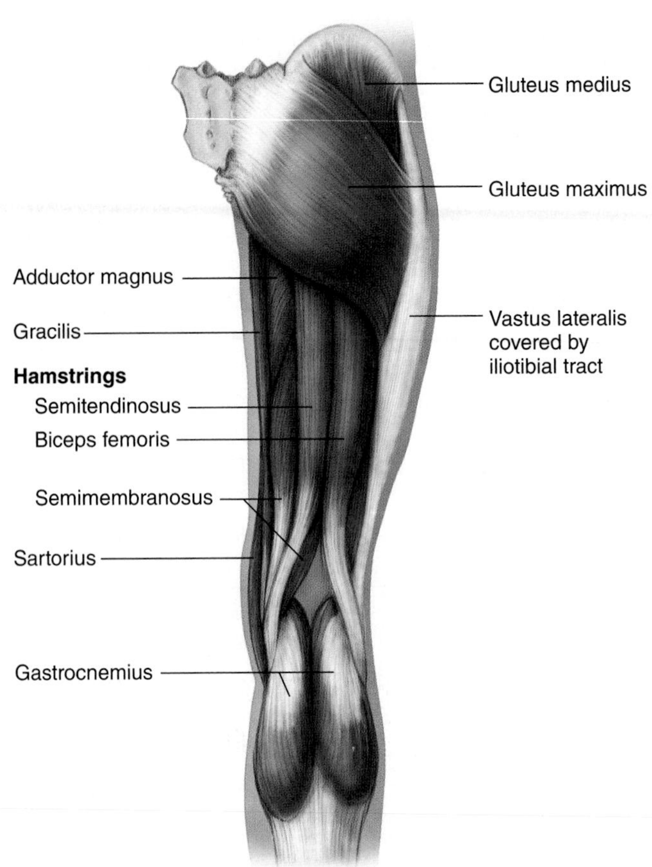

Figure 7.22 Muscles of the posterior right thigh.

anterior is anteriorly located, and the *extensor digitorum longus* lies lateral to it. Note that although the extensor digitorum extends the toes, as its name implies, it also dorsiflexes the foot. The *fibularis longus* is located on the lateral surface of the leg (table 7.14 and figures 7.23 to 7.25).

Note how the tendons are held in position by the bands of ligaments at the ankle.

Clinical Insight

Repeated stress from athletic activities may cause inflammation of a tendon, a condition known as *tendonitis*. Tendons associated with the shoulder, elbow, hip, and knee joints are most commonly affected.

Check My Understanding

18. Name the muscles that flex the thigh.
19. What are the four parts of the quadriceps femoris?
20. What is the action of muscles inserting on the calcaneus?

Table 7.13 Muscles That Move the Leg

Muscle	Origin	Insertion	Action
Quadriceps femoris (quad′-ri-seps fem′-or-is)	Four muscles of the anterior thigh that extend the leg.		
Rectus femoris (rek′-tus fem′-or-is)	Anterior inferior iliac spine and superior margin of acetabulum	Patella; tendon continues as patellar ligament, which attaches to tibial tuberosity	Extends leg and flexes thigh
Vastus lateralis (vas′-tus lat-er-a′lis)	Greater trochanter and posterior surface of femur	Same as above	Extends leg
Vastus medialis (vas′-tus me-de-a′lis)	Medial and posterior surfaces of femur	Same as above	Extends leg
Vastus intermedius (vas′-tus in-ter-mē′dē-us)	Anterior and lateral surfaces of femur	Same as above	Extends leg
Hamstrings	Three distinct muscles of the posterior thigh that flex leg and extend thigh.		
Biceps femoris (bi′-seps fem′-or-is)	Ischial tuberosity and posterior surface of femur	Head of fibula and lateral condyle of tibia	Flexes and laterally rotates leg; extends thigh
Semitendinosus (sem-ē-ten-di-nō′-sus)	Ischial tuberosity	Medial surface of tibia	Flexes and medially rotates leg; extends thigh
Semimembranosus (sem-ē-mem-brah-nō′-sus)	Ischial tuberosity	Medial condyle of tibia	Flexes and medially rotates leg; extends thigh
Gracilis (gras′-il-is)	Pubis near pubic symphysis	Medial surface of tibia	Adducts thigh; flexes leg and locks knee
Sartorius (sar-to′r-ē-us)	Anterior superior iliac spine	Medial surface of tibia	Flexes thigh and leg; abducts and laterally rotates thigh

Table 7.14 Muscles That Move the Foot and Toes

Muscle	Origin	Insertion	Action
Gastrocnemius (gas-trōk-nē′m-ē-us)	Medial and lateral condyles of femur	Calcaneus by the calcaneal tendon	Plantar flexes foot and flexes leg
Soleus (sō′l-ē-us)	Posterior surface of tibia and fibula	Calcaneus by the calcaneal tendon	Plantar flexes foot
Fibularis longus (fib-yu-lar-ris long′-gus)	Lateral condyle of tibia and head and body of fibula	Metatarsal I and tarsal bones	Plantar flexes and everts foot; supports arch
Tibialis anterior (tib-ē-a′l-is an-te′rē-or)	Lateral condyle and surface of tibia	Metatarsal I and tarsal bones	Dorsiflexes and inverts foot
Extensor digitorum longus (eks-ten′-sor dig-i-tor′-um long′-gus)	Lateral condyle of tibia and anterior surface of fibula	Phalanges of toes II-V	Dorsiflexes and everts foot; extends toes

7.6 Disorders of the Muscular System

Learning Objective

15. Describe the major disorders of the muscular system.

Some disorders of the muscle system may result from factors associated only with muscles, while others are caused by disorders of the nervous system. Certain neurological disorders are included here because of their obvious effect on muscle action.

Muscular Disorders

Cramps involve involuntary, painful tetany. The precise cause is unknown, but a cramp seems to result from chemical changes in the muscle, such as ionic imbalances or ATP deficiencies. Sometimes a severe blow to a muscle can produce a cramp.

Fibrosis (fī -brō′-sis) is an abnormal increase of connective tissue in a muscle. Usually, it results from connective tissue replacing dead muscle fibers following an injury.

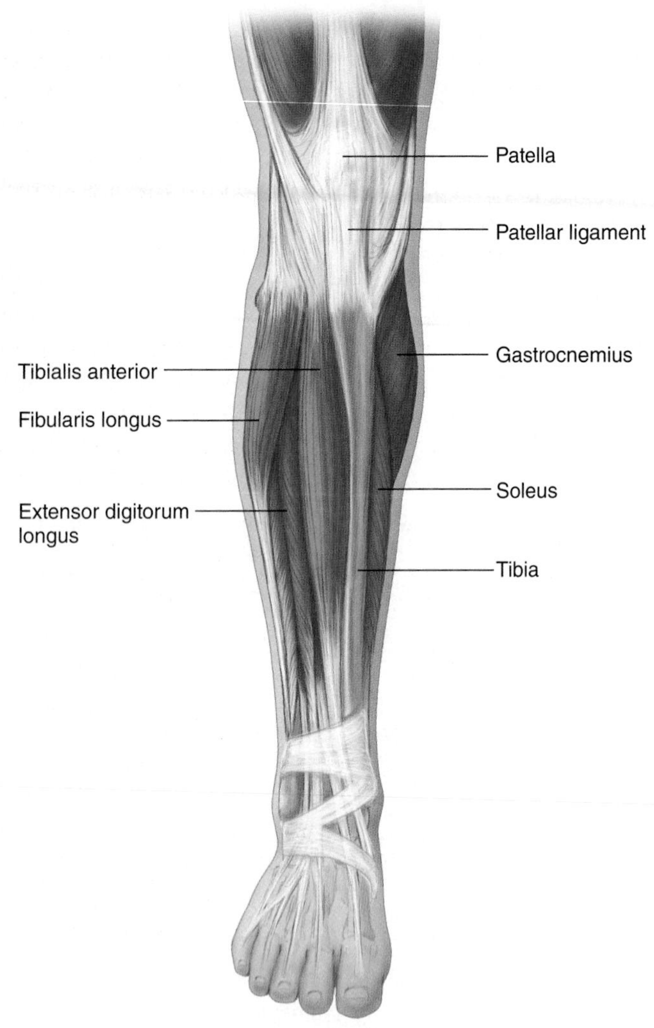

Figure 7.23 Muscles of the anterior right leg. AP|R

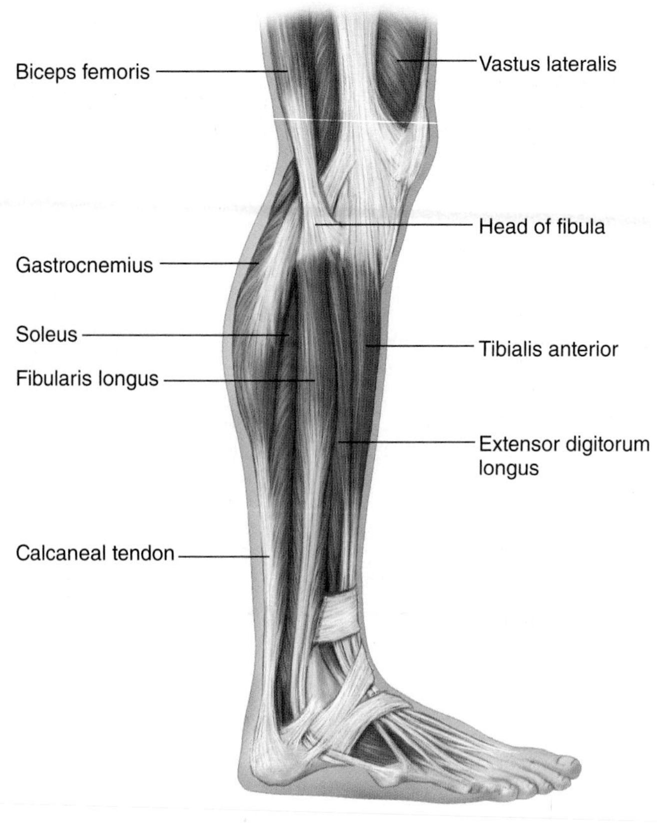

Figure 7.24 Lateral view of muscles of the right leg.

Fibromyalgia (fi-brō-mi-alj-a) is a painful condition of the muscles and joints with no known cause. Once thought to be a mental disorder, this is actually a musculoskeletal disorder that often leads to depression due to the helpless nature of the chronic symptoms.

Muscular dystrophy (dis′-trō-fē) is a general term for a number of inherited muscular disorders that are characterized by the progressive degeneration of muscles. The affected muscles gradually weaken and atrophy, producing a progressive crippling of the patient. There is no specific drug cure, but patients are encouraged to keep active and are given muscle-strengthening exercises.

Strains, or "pulled muscles," result when a muscle is stretched excessively. This usually occurs when an antagonist has not relaxed quickly enough as an agonist contracts. The hamstrings are a common site of muscle strains. In mild strains, only a few muscle fibers are damaged. In severe strains, both connective and muscle tissues are torn, and muscle function may be severely impaired.

Neurological Disorders Affecting Muscles

Botulism (boch′-ū-lizm) poisoning is caused by a neurotoxin produced by the bacterium *Clostridium botulinum.* The toxin prevents release of ACh from the terminal boutons of somatic motor axons. Without prompt treatment with an antitoxin, death may result from paralysis of breathing muscles. Poisoning results from eating improperly canned vegetables or meats that contain *C. botulinum* and the accumulated toxins.

Myasthenia gravis (mī-as-thē′-nē-ah grav′-i-is) is characterized by extreme muscular weakness caused by improper functioning of the neuromuscular junctions. It is an autoimmune disease in which antibodies are produced that attach to the ACh receptors on the motor end plate and reduce or block the stimulatory effect of ACh. Myasthenia gravis occurs most frequently in women between 20 and 40 years of age. Usually, it first affects ocular muscles and other muscles of the face and neck, which may lead to difficulty in chewing, swallowing, and talking. Other muscles of the body may be involved later. Treatment typically involves the use of acetylcholinesterase inhibitors and immunosuppressive drugs, such as the steroid prednisone.

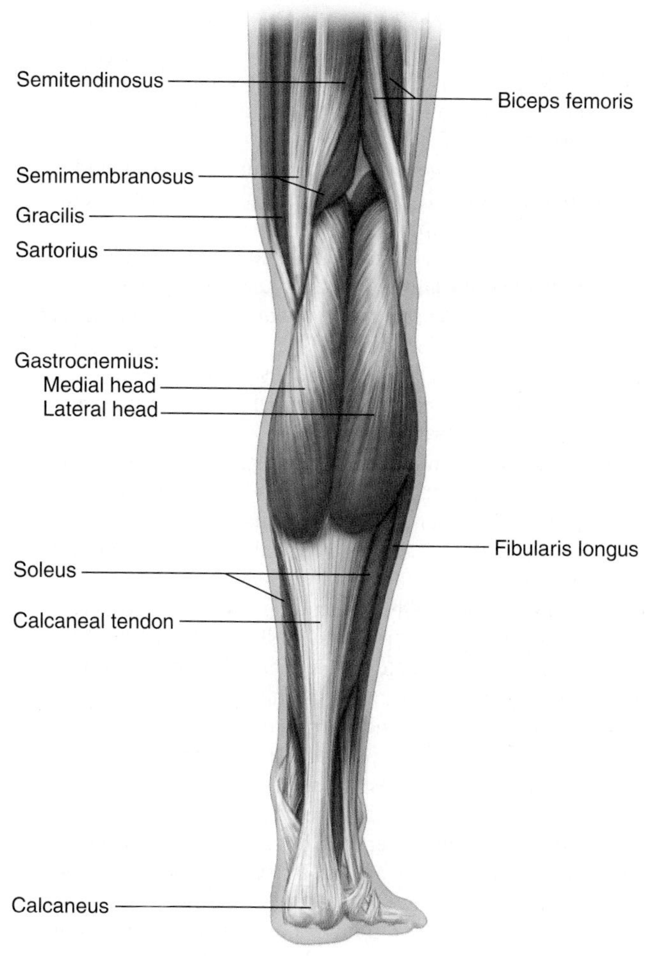

Figure 7.25 Muscles of the posterior right leg. APR

Poliomyelitis (pō-lē-ō-mī-e-lī′-tis) is a viral disease of somatic motor neurons in the spinal cord. Destruction of the somatic motor neurons leads to paralysis of skeletal muscles. It is now rare in industrialized countries due to the availability of a polio vaccine. Virtually all children in the United States receive this vaccine, which protects them from polio.

Spasms are sudden, involuntary contractions of a muscle or a group of muscles. They may vary from simple twitches to severe convulsions and may be accompanied by pain. Spasms may be caused by irritation of the motor neurons supplying the muscle, emotional stress, or neurological disorders. Spasms of smooth muscle in the walls of the digestive and respiratory tracts, or certain blood vessels can be hazardous. Hiccupping is a spasm of the diaphragm.

Tetanus (tet′-ah-nus) is a disease caused by the anaerobic bacterium *Clostridium tetani,* which is common in soil. Infection usually results from puncture wounds. *C. tetani* produces a neurotoxin that affects somatic motor neurons in the spinal cord, resulting in continuous stimulation and tetany of certain muscles. Because the first muscles affected are those that move the mandible, this disease is often called "lockjaw." Without prompt treatment, mortality is high. Young children usually receive vaccinations of tetanus toxoid to stimulate production of antibodies against the neurotoxin. Booster injections are given at regular intervals to keep the concentration of antibodies at a high level in order to prevent the disease.

Chapter Summary

- The three types of muscle tissue in the body are skeletal, smooth, and cardiac.
- Each type of muscle tissue has unique structural and functional characteristics.

7.1 Structure of Skeletal Muscle

- Each skeletal muscle is formed of many muscle fibers that are arranged in fascicles.
- Connective tissue envelops each muscle fiber, each fascicle, and the entire muscle.
- Muscles are attached to bones or other tissues by either tendons or aponeuroses.
- The sarcolemma is the plasma membrane of a muscle fiber, and the sarcoplasm (cytoplasm) contains the myofibrils, the contractile elements.
- Myofibrils consist of thick and thin myofilaments. The arrangement of the myofilaments produces the striations that are characteristic of muscle fibers.
- Each myofibril consists of many sarcomeres joined end-to-end. A sarcomere is bounded by a Z line at each end.
- I bands are light areas in a muscle tissue micrograph, and A bands are dark areas.
- The H band is the center of a sarcomere and contains only thick myofilaments.
- The terminal bouton of a somatic motor neuron is adjacent to each muscle fiber at the neuromuscular junction. The terminal bouton fits into depressions in the sarcolemma, called motor end plates. The synaptic cleft is the small space between the terminal bouton and motor end plate. The neurotransmitter ACh is contained in tiny vesicles in the terminal bouton.
- Each muscle fiber is innervated and controlled by a somatic motor neuron.
- A motor unit consists of a somatic motor neuron and all muscle fibers it innervates.

7.2 Physiology of Skeletal Muscle Contraction

- An activated terminal bouton releases ACh into the synaptic cleft. ACh attaches to ACh receptors of the motor end plate, which leads to the release of Ca^{2+} within the sarcoplasm. This, in turn, leads to the formation of cross-bridges between the heads of myosin molecules and the myosin binding sites on actin molecules. A series of ratchetlike movements pulls the thin myofilaments toward the center of the sarcomere, producing contraction.

- Acetylcholinesterase quickly breaks down ACh to prevent continued stimulation and to prepare the muscle fiber for the next stimulus.
- Energy for contraction comes from high-energy phosphate bonds in ATP.
- After cellular respiration has formed a muscle fiber's normal supply of ATP, excess energy is transferred to creatine to form creatine phosphate, which serves as a reserve supply of energy.
- Small amounts of oxygen are stored in combination with myoglobin, which gives muscle fibers a reserve of oxygen for aerobic respiration.
- Vigorous muscular activity quickly exhausts available oxygen, leading to the accumulation of lactic acid and causing excess post-exercise oxygen consumption. Heavy breathing after exercise provides the oxygen required to metabolize lactic acid and restore the pre-exercise state within the muscle fiber.
- Fatigue most likely results primarily from the lack of raw fuel in a muscle fiber.
- Large amounts of heat are produced by the chemical and physical processes of muscle contraction.
- When stimulated by a threshold stimulus, individual muscle fibers exhibit an all-or-none contraction response.
- A simple contraction consists of a latent phase, contraction phase, and relaxation phase.
- Whole muscles provide graded contraction responses, which are enabled by the number of motor units that are recruited.
- A sustained contraction of all motor units is tetany.
- Muscle tone is a state of partial contraction that results from alternating contractions of a few motor units.

7.3 Actions of Skeletal Muscles

- The origin is the immovable attachment, and the insertion is the movable attachment.
- Muscles are arranged in groups with opposing actions: agonists and antagonists.

7.4 Naming of Muscles

- Several criteria are used in naming muscles.
- These criteria include function, shape, relative position, location, site of attachment, origin and insertion, size, and orientation of fibers.

7.5 Major Skeletal Muscles

- Muscles of facial expression originate on skull bones and insert on the dermis of the skin. They include the epicranius, orbicularis oculi, orbicularis oris, buccinator, zygomaticus, and platysma.
- Muscles of mastication originate on fixed skull bones and insert on the mandible. They include the masseter and the temporalis.
- Muscles that move the head occur in the neck and superior back. They include the sternocleidomastoid and splenius capitis.
- Muscles of the abdominal wall connect the pelvic girdle, thoracic cage, and vertebral column. They include the rectus abdominis, external oblique, internal oblique, and transversus abdominis.
- The diaphragm is the major muscle of breathing.
- External intercostals and internal intercostals move the ribs, helping breathing.
- Muscles that move the pectoral girdle originate on the thoracic cage or vertebrae and insert on the pectoral girdle. They include the trapezius, rhomboid major and minor, levator scapulae, pectoralis minor, and serratus anterior.
- Muscles that move the arm originate on the thoracic cage, vertebrae, or pectoral girdle and insert on the humerus. They include the pectoralis major, deltoid, subscapularis, supraspinatus, infraspinatus, latissimus dorsi, teres major, and teres minor.
- Supraspinatus, infraspinatus, teres minor, and subscapularis make up the rotator cuff.
- Muscles that move the forearm originate on the scapula or humerus and insert on the radius or ulna. They include the biceps brachii, brachialis, brachioradialis, and triceps brachii.
- Muscles that move the wrist and fingers are the muscles of the forearm. They include the flexor carpi radialis, flexor carpi ulnaris, palmaris longus, extensor carpi radialis longus, extensor carpi ulnaris, and extensor digitorum.
- Muscles that move the thigh originate on the pelvic girdle and insert on the femur. They include the iliacus, psoas major, gluteus maximus, gluteus medius, tensor fasciae latae, adductor longus, and adductor magnus.
- Muscles that move the leg originate on the pelvic girdle or femur and insert on the tibia or fibula. They include the quadriceps femoris, biceps femoris, semitendinosus, semimembranosus, gracilis, and sartorius.
- Muscles that move the foot and toes are the muscles of the leg. They include the gastrocnemius, soleus, fibularis longus, tibialis anterior, and extensor digitorum longus.

7.6 Disorders of the Muscular System

- Disorders of muscles include cramps, fibrosis, fibromyalgia, muscular dystrophy, and strains.
- Neurological disorders that directly affect muscle action include botulism, myasthenia gravis, poliomyelitis, spasms, and tetanus.

Self-Review

Answers are located in appendix B.

1. A skeletal muscle consists of many ______, which are arranged in fascicles.
2. Muscles are attached to bones by ______.
3. A contractile unit of a myofibril is a ______.
4. A muscle contraction is triggered by ______ binding to its receptors on the motor end plate.
5. Contraction occurs when thick myofilaments pull ______ myofilaments toward the center of a sarcomere.
6. The movable attachment of a muscle is its ______.
7. The mandible is elevated by the contraction of the temporalis and the ______.
8. The abdominal muscle extending from the sternum to the pubis is the ______.
9. The broad muscle of the inferior back is the ______.
10. The shoulder muscle that abducts the arm is the ______.
11. The arm muscle that extends the forearm is the ______.
12. The large muscle that extends and laterally rotates the thigh is the ______.
13. The four-part thigh muscle that extends the leg is the ______.
14. The large superficial calf muscle that plantar flexes the foot is the ______.

Critical Thinking

1. Using what you have learned in chapters 6 and 7, predict what would happen if calcium ions were not sequestered in the sarcoplasmic reticulum and were allowed to mingle with the high levels of Pi in the sarcoplasm.
2. Predict the clinical symptoms of a person with damage to the nerve that supplies the triceps brachii. How would the agonist-antagonist relationship be disturbed?
3. Can the origin and insertion of some muscles be interchanged? Explain.
4. As a cosmetic procedure, Botox is injected in very small doses into specific facial muscles to reduce wrinkles. It is derived from a neurotoxin that prevents the release of ACh at the neuromuscular junction. Explain how Botox works.

ADDITIONAL RESOURCES

output. In about 90% of the population, the left cerebral hemisphere controls analytical and verbal skills, such as mathematics, reading, writing, and speech. In these persons, the right hemisphere controls musical, artistic and spatial awareness, imagination, and insight. In some persons, this pattern is reversed; in a few, there seems to be no specialization. Men also have greater lateralization than women, which is why damage to a hemisphere can have greater effects in men.

Diencephalon

The **diencephalon** (di-en-sef′-a-lon) is a small but important part of the brain. It lies between the brainstem and the cerebrum of the brain and consists of three major components: the thalamus, hypothalamus, and epithalamus (see figure 8.13).

Thalamus

The **thalamus** (thal′-ah-mus) consists of two lateral masses of nervous tissue that are joined by a narrow isthmus of nervous tissue called the *interthalamic adhesion.* Sensory nerve impulses (except those for smell) coming from lower regions of the brain and the spinal cord are first received by the thalamus before being relayed to the cerebral cortex. The thalamus provides a general but nonspecific awareness of sensations such as pain, pressure, touch, and temperature. It seems to associate sensations with emotions but it is the cerebral cortex that interprets the precise sensation. The thalamus also serves as a relay station for communication between motor areas of the brain.

Hypothalamus

The **hypothalamus** (hī-pō-thal′-ah-mus) is located inferior to the thalamus and anterior to the midbrain. It communicates with the thalamus, cerebrum, and other parts of the brain. The hypothalamus is the major integration center for the autonomic nervous system. In this role, it controls virtually all internal organs. The hypothalamus also is the connecting link between the brain and the endocrine system, which produces chemicals (hormones) that affect most cells in the body. This link results from hypothalamic control of the hypophysis, or pituitary gland, which is suspended from its inferior surface. Although it is small, the hypothalamus exerts a tremendous impact on body functions.

The primary function of the hypothalamus is the maintenance of homeostasis, and this is accomplished through its regulation of

- body temperature;
- mineral and water balance;
- appetite and digestive processes;
- heart rate and blood pressure;
- sleep and wakefulness;
- emotions; and
- secretion of hormones by the pituitary gland.

Epithalamus

The **epithalamus** (ep-i-thal′-ah-mus; epi = above) is a small mass of tissue located superior and posterior to the thalamus forming part of the roof of the third ventricle. The major structure within the epithalamus is the *pineal gland.* The pineal gland is stimulated to produce a hormone called *melatonin* when sunlight levels become low during the evening and overnight hours. This hormone induces sleepiness to initiate the night component of a person's day-night cycle and may assist in regulating the onset of puberty. This hormone will be discussed further in Chapter 10.

Limbic System

The thalamus and hypothalamus are associated with parts of the cerebral cortex and nuclei deep within the cerebrum to form a complex known as the **limbic system.** The limbic system is involved in memory and in emotions such as sadness, happiness, anger, and fear. It seems to regulate emotional behavior, especially behavior that enhances survival. Mood disorders, such as depression, are usually a result of malfunctions of the limbic system. It also is referred to as the "motivational system" because it provides our desire to carry out the commands created by the cerebrum.

Check My Understanding

6. How is the CNS protected from mechanical injuries?
7. What are roles of the functional areas of the cerebrum?
8. What are the functions of the thalamus, hypothalamus, and epithalamus?

Brainstem

The **brainstem** is the stalklike portion of the brain that joins higher brain centers to the spinal cord. It contains several nuclei that are surrounded by white matter. Ascending (sensory) and descending (motor) axons between higher brain centers and the spinal cord pass through the brainstem. The components of the brainstem include the midbrain, pons, and medulla oblongata (see figure 8.13).

Midbrain

The **midbrain** is the most superior portion of the brainstem. It is located posterior to the hypothalamus and superior to the pons. It contains reflex centers for head, eye, and body movements in response to visual and auditory stimuli. For example, reflexively turning the head to enable better vision or better hearing is activated by the midbrain.

Pons

The **pons** lies between the midbrain and the medulla oblongata and is recognizable by its bulblike anterior portion. It consists primarily of axons. Longitudinal axons connect lower and higher brain centers, and transverse axons connect with the cerebellum. The pons also works with the medulla oblongata by controlling the rate and depth of breathing (see chapter 14).

Medulla Oblongata

The **medulla oblongata** (me-dŭl′-ah ob-lon-ga′-ta) is the most inferior portion of the brain, and it is the connecting link with the spinal cord. Descending (motor) axons extending between the brain and the spinal cord cross over to the opposite side of the brain within the medulla oblongata. The medulla oblongata contains three integration centers that are vital for homeostasis:

1. The **respiratory rhythmicity center** controls the basic rhythm of breathing by triggering each cycle of inhale and exhale. It is also involved in associated reflexes such as coughing and sneezing.
2. The **cardiac control center** regulates the rate and force of heart contractions.
3. The **vasomotor center** regulates blood pressure and blood flow by controlling the diameter of blood vessels.

Reticular Formation

The **reticular** (re-tik′-ū-lar) **formation** is a network of axons and small nuclei of gray matter that extends from the superior spinal cord, through the brainstem, into the diencephalon. This network generates and transmits nerve impulses that arouse the cerebrum to wakefulness. A decrease in activity results in sleep. Damage to the reticular formation may cause unconsciousness or a coma.

Cerebellum

The **cerebellum** (ser-e-bel′-um) is the second largest portion of the brain. The **transverse cerebral fissure** separates it superiorly from the occipital and temporal lobes of the cerebrum. It is also positioned posterior to the pons and medulla oblongata. It is divided into two lateral hemispheres by a medial constriction, the *vermis* (ver′-mis). Gray matter forms a thin superficial layer covering the deep white matter, which forms most of the cerebellum (see figure 8.13).

The cerebellum is a reflex center that controls and coordinates the interaction of skeletal muscles. It controls posture, balance, and muscle coordination during movement. Damage to the cerebellum may result in a loss of equilibrium, muscle coordination, and muscle tone.

Table 8.2 summarizes the major brain functions.

Table 8.2 Summary of Brain Functions

Part	Function
Cerebrum	Sensory areas interpret nerve impulses as sensations. Motor areas control voluntary skeletal muscle actions. Association areas interrelate various sensory and motor areas and are involved in intellectual processes, will, memory, emotions, and personality traits. The limbic system is involved with motivation and with emotions as they relate to survival behavior.
Diencephalon	
Thalamus	Receives and relays sensory nerve impulses (except smell) to the cerebrum and motor nerve impulses to lower brain centers. Provides a general awareness of pain, touch, pressure, and temperature.
Hypothalamus	Serves as the major integration center the autonomic nervous system. Controls water and mineral balance, heart rate and blood pressure, appetite and digestive activity, body temperature, and sexual response. Is involved in sleep and wakefulness and in emotions of anger and fear. Regulates functions of the pituitary gland.
Epithalamus	Production of the hormone melatonin
Brainstem	
Midbrain	Relays sensory nerve impulses from the spinal cord to the thalamus and motor nerve impulses from the cerebrum to the spinal cord. Contains reflex centers that move the eyeballs, head, and neck in response to visual and auditory stimuli.
Pons	Relays nerve impulses between the midbrain and the medulla oblongata and between the cerebellar hemispheres. Helps medulla oblongata control breathing.
Medulla oblongata	Relays nerve impulses between the brain and spinal cord. Reflex centers control heart rate and contraction force, blood vessel diameter, breathing, swallowing, vomiting, coughing, sneezing, and hiccupping. Motor axons cross over to the opposite side.
Cerebellum	Controls posture, balance, and the coordination of skeletal muscle contractions.

and is surrounded by white matter. The *central canal* extends the length of the spinal cord and contains CSF.

The pointed projections of the gray matter, as seen in cross section, are called horns. The *anterior horns* contain the cell bodies of somatic motor neurons whose axons enter spinal nerves and carry nerve impulses to skeletal muscles. The *posterior horns* contain interneurons that receive nerve impulses from sensory axons in the spinal nerves and carry them to sites within the CNS. *Lateral horns,* found only in the thoracic and lumbar segments of the spinal cord, contain the cell bodies of autonomic motor neurons whose axons follow ANS pathways as they carry nerve impulses to cardiac and smooth muscle, glands, and adipose tissue. Interneurons form most of the gray matter in the CNS.

The horns of the gray matter divide the white matter into three regions: the *anterior, posterior,* and *lateral funiculi* (singular, funiculus). These funiculi contain **nerve tracts,** which are bundles of myelinated and unmyelinated axons of interneurons that extend superiorly and inferiorly within the spinal cord.

Functions

The spinal cord has two basic functions. It transmits nerve impulses to and from the brain, and it serves as a reflex center for spinal reflexes. Nerve impulses are transmitted to and from the brain by axons composing the nerve tracts. *Ascending* (sensory) *tracts* carry sensory nerve impulses to the brain; *descending* (motor) *tracts* carry motor nerve impulses from the brain.

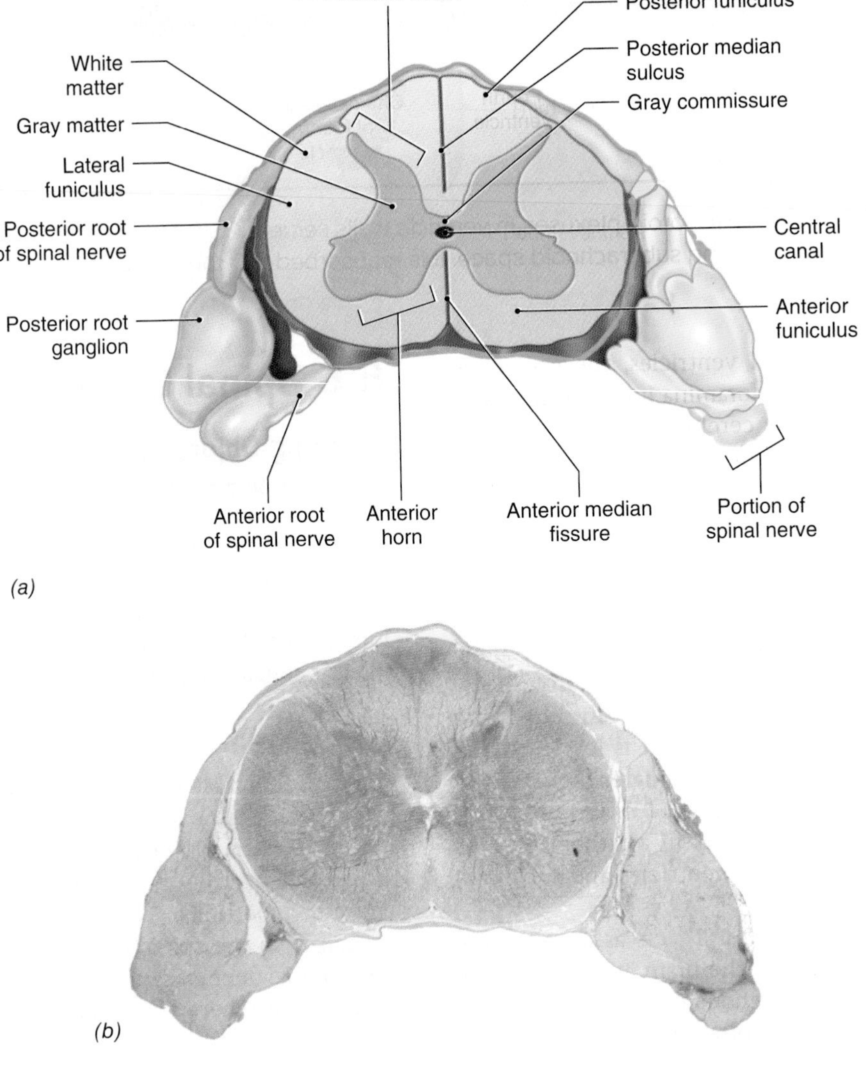

Figure 8.17 A drawing (*a*) and a photomicrograph (*b*) of the spinal cord in cross section show its basic structure. APR

Clinical Insight

In *hydrocephalus,* a congenital defect restricts the movement of CSF from the ventricles into the subarachnoid space. In severe cases, the buildup of hydrostatic pressure within an infant's brain causes a marked enlargement of the ventricles and brain and widens the fontanelles of the cranium. Without treatment, death usually results within two to three years. Treatment involves surgical insertion of a small tube to drain the excess CSF from a ventricle into the peritoneal cavity, where it is reabsorbed.

Check My Understanding

11. What is the relationship between the ventricles, the meninges, and the cerebrospinal fluid?
12. What are the functions of the spinal cord?

8.7 Peripheral Nervous System (PNS)

Learning Objectives

13. Recall the name, type, and functions for each of the 12 pairs of cranial nerves.
14. Describe the classification of the spinal nerves and the plexuses they form.
15. Explain the functions of the components involved in a reflex.

The *peripheral nervous system (PNS)* consists of cranial and spinal nerves that connect the CNS to other portions of the body, along with sensory receptors and ganglia. A **nerve** consists of axons that are bound together by connective tissue. **Motor nerves** contain mostly axons of motor neurons; **sensory nerves** contain only axons of sensory neurons; and **mixed nerves** contain both motor axons and sensory axons. Most nerves are mixed. Nerves may contain axons of both the somatic nervous system, which is involved with voluntary responses, and the autonomic nervous system, which controls involuntary (automatic) responses.

Cranial Nerves

Twelve pairs of **cranial nerves** arise from the brain and connect the brain with organs and tissues that are primarily located in the head and neck (table 8.3). Most cranial nerves arise from the brainstem. Cranial nerves are identified by both roman numerals and names. The numerals indicate the order in which the nerves arise from the inferior surface of the brain: CN I is most anterior; CN XII is most posterior (figure 8.18).

Five cranial nerves are primarily motor, three are sensory, and four are mixed.

Spinal Nerves

Arising from the spinal cord, there are thirty-one pairs of mixed nerves called **spinal nerves.** Each pair of spinal nerves is named based upon where it exits the vertebral column. The first pair of spinal nerves emerges from the spinal cord between the atlas and the occipital bone. The remaining thirty pairs of spinal nerves emerge through the *intervertebral foramina* between adjacent vertebrae, the *sacral foramina,* and the *sacral hiatus.* There are eight pairs of *cervical nerves* (C1–C8), twelve pairs of *thoracic nerves* (T1–T12), five pairs of *lumbar nerves* (L1–L5), five pairs of *sacral nerves* (S1–S5), and one pair of *coccygeal nerves* (Co) (figure 8.19). Recall from Chapter 6 that there are seven cervical vertebrae. Because the first pair of spinal nerves emerges superior to the atlas, there are eight pairs of cervical nerves instead of seven.

Spinal nerves branch from the spinal cord by two short roots that merge a short distance from the spinal cord to form a spinal nerve. The **anterior root** contains axons of motor neurons whose cell bodies are located within the spinal cord. These neurons carry motor nerve impulses from the spinal cord to effectors. The **posterior root** contains axons of sensory neurons. The swollen region in a posterior root is a **posterior root ganglion,** which contains cell bodies of sensory neurons. The long axons of these neurons carry sensory nerve impulses to the spinal cord. Observe these structures and their relationships in figures 8.12, 8.17, and 8.20.

As shown in figure 8.19, the spinal cord ends at the second lumbar vertebra. The roots of lumbar, sacral, and coccygeal spinal nerves continue inferiorly within the vertebral canal to exit between the appropriate vertebrae. These roots form the *cauda equina,* or horse's tail, in the inferior portion of the vertebral canal.

Spinal Plexuses

After a spinal nerve exits the vertebral canal, it divides into four major parts: the *anterior ramus* (plural, rami), *posterior ramus, meningeal branch,* and *ramus communicans.* The posterior ramus innervates the deep muscles and skin of the posterior trunk. The meningeal branch innervates the vertebrae, meninges, and vertebral ligaments. The ramus communicans passes to the sympathetic chain ganglia and is part of the autonomic system. The anterior rami of many spinal nerves merge to form **spinal plexuses,** networks of nerves, before continuing to the innervated structures. The anterior rami of most thoracic nerves do

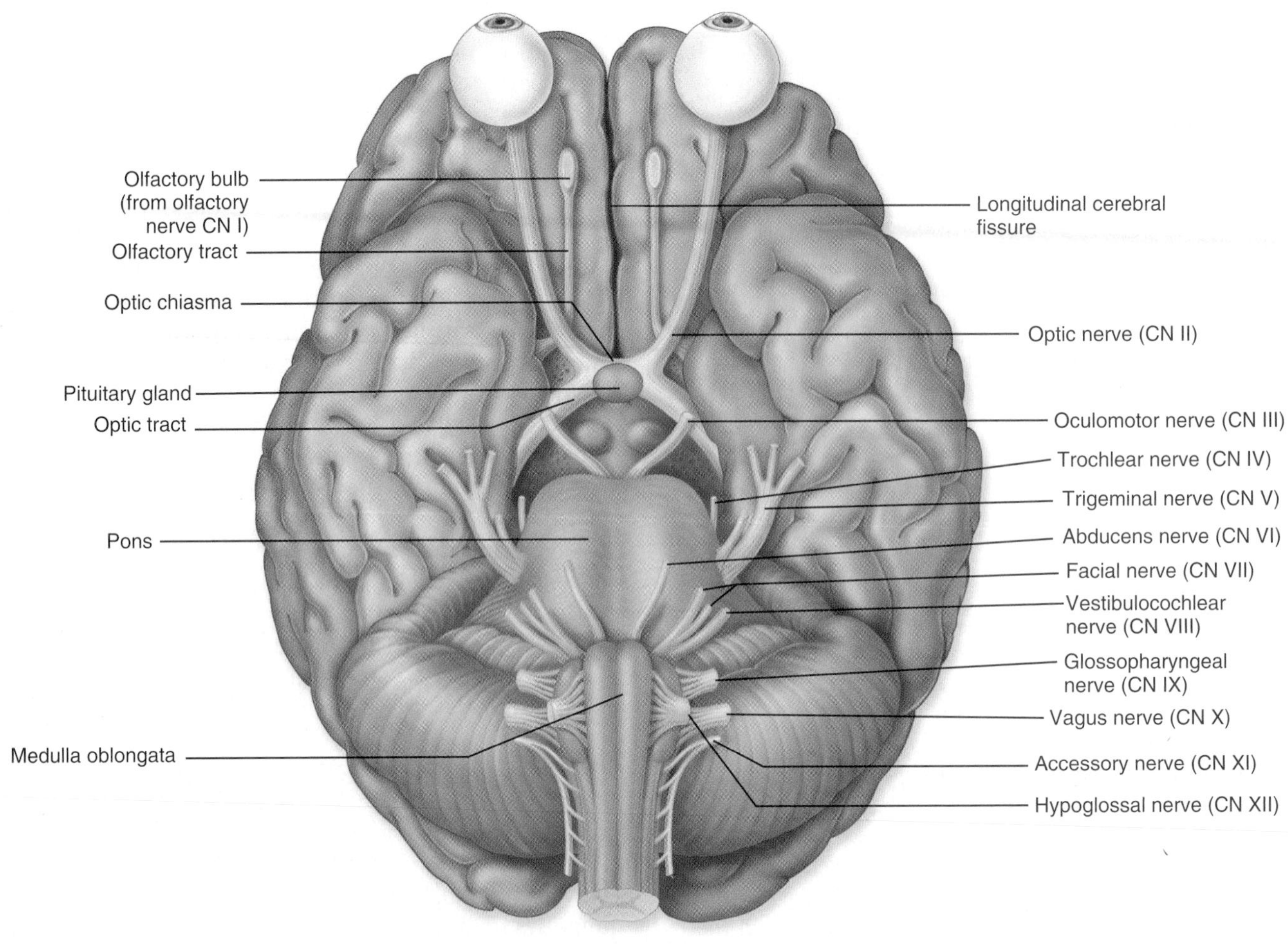

Figure 8.18 Inferior view of the brain showing the roots of the 12 pairs of cranial nerves. Cranial nerves are identified by both roman numerals and names. Most cranial nerves arise from the brainstem. AP|R

not form plexuses; rather, they form intercostal nerves. The intercostal nerves innervate the intercostal and abdominal muscles, in addition to overlying skin.

In a plexus, the axons in the anterior rami are sorted and recombined so that axons going to a specific body part are carried in the same peripheral nerve, although they may originate in several different spinal nerves. There are four pairs of plexuses: cervical, brachial, lumbar, and sacral. Because many axons from the lumbar plexus contribute to the sacral plexus, these two plexuses are sometimes called the *lumbosacral plexus* (figure 8.19).

Cervical Plexus The superior cervical nerves merge to form a *cervical plexus* on each side of the neck. The nerves from these plexuses supply the muscles and skin of the neck and portions of the head and shoulders. The paired *phrenic* (fren′-ik) *nerves,* which stimulate the diaphragm to contract and begin inspiration, also arise from the cervical plexus.

Brachial Plexus The inferior cervical nerves and perhaps nerves T1–T2 join to form a *brachial plexus* on each side of the vertebral column in the shoulder region. Nerves that serve skin and muscles of the pectoral girdle and upper limb emerge from the brachial plexuses. The *musculocutaneous, axillary, radial, median,* and *ulnar nerves* arise here.

Lumbar Plexus The last thoracic nerve (T12) and the superior lumbar nerves unite to form a *lumbar plexus* on each side of the vertebral column just superior to the coxal bones. Nerves from the lumbar plexuses supply the skin and muscles of the inferior trunk, external genitalia, and the anterior and medial thighs. The *femoral* and *obturator nerves* arise here.

Sacral Plexus The inferior lumbar nerves and the sacral nerves merge to form a *sacral plexus* on each side of the sacrum within the pelvis. Nerves from the sacral plexuses

Table 8.3 Summary of the Cranial Nerves

Nerve	Type	Function
CN I Olfactory	Sensory	Transmits sensory nerve impulses from olfactory receptors in olfactory epithelium to the brain.
CN II Optic	Sensory	Transmits sensory nerve impulses for vision from the retina of the eye to the brain.
CN III Oculomotor	Motor	Transmits motor nerve impulses to muscles that move the eyes superiorly, inferiorly, and medially; control the eyelids; adjust pupil size; and control the shape of the lens.
CN IV Trochlear	Motor	Transmits motor nerve impulses to muscles that rotate the eyes.
CN V Trigeminal	Mixed	Transmits sensory nerve impulses from scalp, forehead, face, teeth, and gums to the brain. Transmits motor nerve impulses to chewing muscles and muscles in floor of mouth.
CN VI Abducens	Motor	Transmits motor nerve impulses to muscles that move the eyes laterally.
CN VII Facial	Mixed	Transmits sensory nerve impulses from the anterior part of the tongue to the brain. Transmits motor nerve impulses to facial muscles, salivary glands, and tear glands.
CN VIII Vestibulocochlear	Sensory	Transmits sensory nerve impulses from the internal ear associated with hearing and equilibrium.
CN IX Glossopharyngeal	Mixed	Transmits sensory nerve impulses from posterior portion of the tongue, tonsils, pharynx, and carotid arteries to the brain. Transmits motor nerve impulses to salivary glands and pharyngeal muscles used in swallowing.
CN X Vagus	Mixed	Transmits sensory nerve impulses from thoracic and abdominal organs, esophagus, larynx, and pharynx to the brain. Transmits motor nerve impulses to these organs and to muscles of speech and swallowing.
CN XI Accessory	Motor	Transmits motor nerve impulses to muscles of the palate, pharynx, and larynx and to the trapezius and sternocleidomastoid muscles.
CN XII Hypoglossal	Motor	Transmits motor nerve impulses to the muscles of the tongue.

supply the skin and muscles of the buttocks and lower limbs. The *sciatic nerves*, which emerge from the sacral plexuses, are the largest nerves in the body.

Check My Understanding

13. What composes the peripheral nervous system?
14. Identify and describe the functions of the twelve cranial nerves.
15. Name and locate the major spinal plexuses.

Reflexes

Reflexes are rapid, involuntary, and predictable responses to internal and external stimuli. Reflexes maintain homeostasis and enhance chances of survival. A reflex involves either the brain or the spinal cord, a sensory receptor, sensory and motor neurons, and an effector.

Most pathways of nerve impulse transmission within the nervous system are complex and involve many neurons. In contrast, reflexes require few neurons in their pathways and therefore produce very rapid responses to stimuli. Reflex pathways are called **reflex arcs.**

Reflexes are divided into two types–autonomic and somatic–based on the effector(s) involved in the reflex. *Autonomic reflexes* act on smooth muscle, cardiac muscle, adipose tissue, and glands. They are involved in controlling homeostatic processes such as heart rate, blood pressure, and digestion. Autonomic reflexes maintain homeostasis and normal body functions at the unconscious level, which frees the mind to deal with those actions that require conscious decisions. *Somatic reflexes* act on skeletal muscles. They enable quick movements such as moving the hand away from a painful stimulus. A person is usually unaware of autonomic reflexes but is aware of somatic reflexes. Reflexes are also divided into *cranial reflexes* and *spinal reflexes*, depending upon whether the brain or the spinal cord is involved in the reflex.

Figure 8.20 illustrates a somatic spinal reflex, which withdraws the hand after sticking a finger with a tack. Three neurons are involved in this reflex. Pain receptors are stimulated by the sharp pin and form nerve impulses that are carried by a sensory neuron to an interneuron in the spinal cord. Nerve impulses pass along the interneuron to a motor neuron, which carries the nerve impulses to a muscle that contracts to move the hand. Although the brain is not involved in this reflex, it does receive

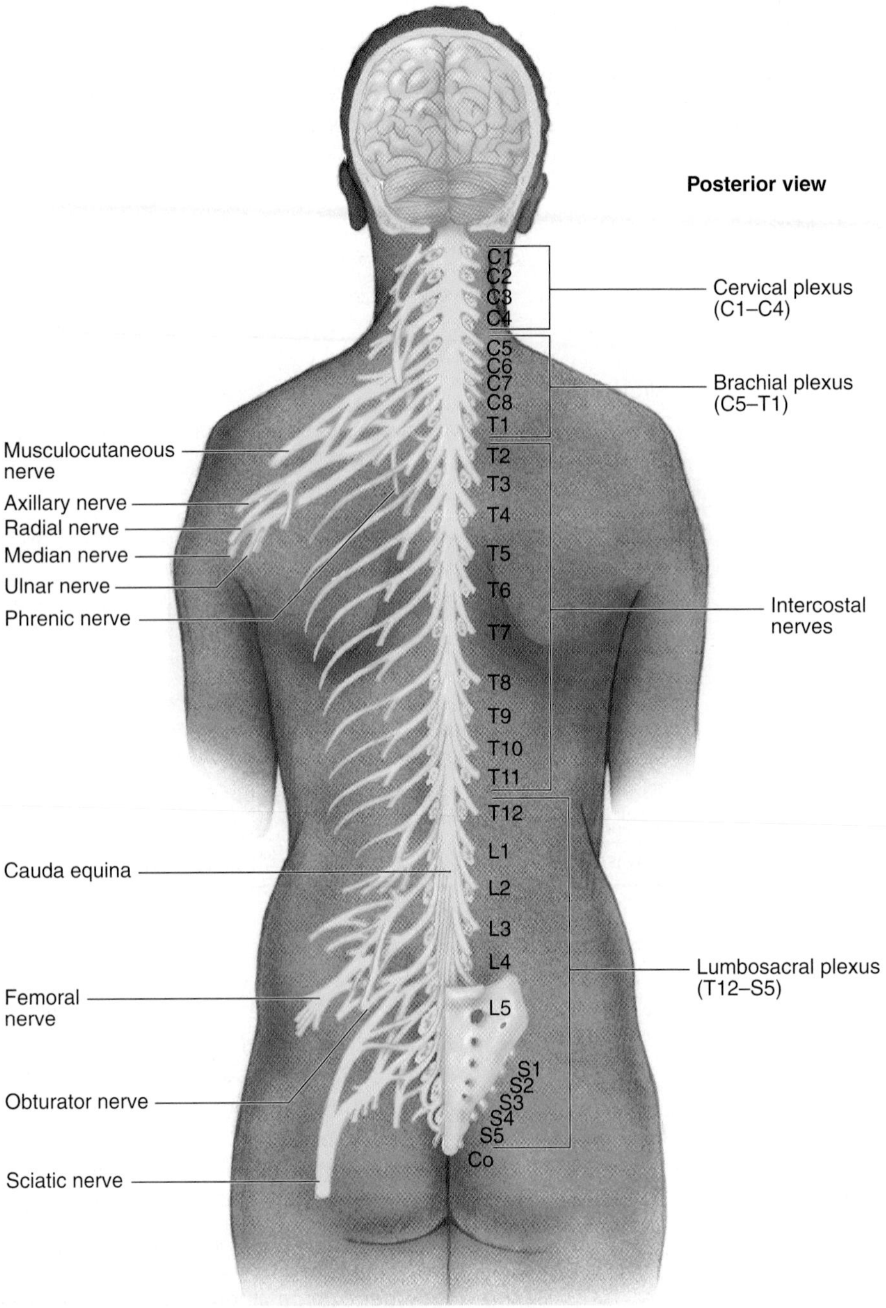

Figure 8.19 Thirty-one pairs of spinal nerves arise from the spinal cord. Anterior rami of spinal nerves in the thoracic region form the intercostal nerves. Those in other segments form nerve networks called spinal plexuses before continuing on to their target tissues. APR

sensory nerve impulses that make a person aware of a painful stimulus.

Check My Understanding

16. What is a reflex?
17. What are the components of a spinal reflex?

Clinical Insight

Because the responses of reflexes are predictable, physicians usually test a patient's reflexes in order to determine the health of the nervous system. Exaggerated, diminished, or distorted reflexes may indicate a neurological disorder.

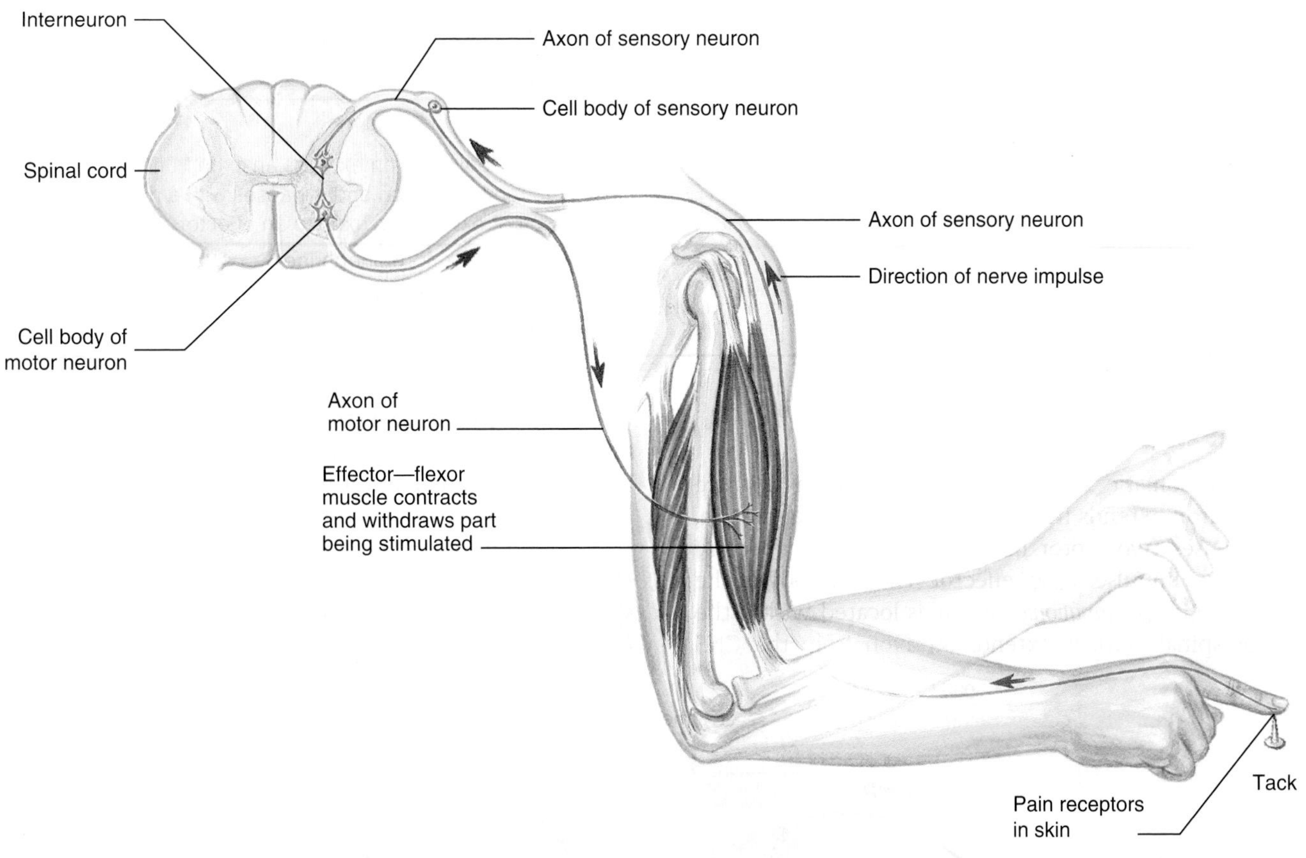

Figure 8.20 A somatic spinal reflex involving a sensory neuron, an interneuron, and a motor neuron. APIR

Clinical Insight

Because the spinal cord ends at the level of the second lumbar vertebra, spinal taps and epidural anesthetics are administered inferior to this point. For these procedures, a patient is placed in a fetal position in order to open the spaces between the posterior margins of the vertebrae. A hypodermic needle is inserted into the vertebral canal either between the third and fourth lumbar vertebrae or between the fourth and fifth lumbar vertebrae. In a *spinal tap* (lumbar puncture), a hypodermic needle is inserted into the subarachnoid space to remove cerebrospinal fluid for diagnostic purposes. An *epidural anesthetic* is given by injecting an anesthetic into the epidural space with a hypodermic syringe. The anesthetic prevents sensory nerve impulses from reaching the spinal cord via posterior roots inferior to the injection. Epidurals are sometimes used to ease pain during childbirth.

8.8 Autonomic Nervous System (ANS)

Learning Objective

16. Compare the structure and functions of the sympathetic and parasympathetic divisions.

The **autonomic** (aw-to-nom′-ik) **nervous system (ANS)** consists of portions of the central and peripheral nervous systems and functions without conscious control. Its role is to maintain homeostasis in response to changing internal conditions. The effectors under autonomic control are cardiac muscle, smooth muscle, adipose tissue, and glands. The ANS functions mostly by involuntary reflexes. Visceral sensory nerve impulses carried to the autonomic reflex centers in the hypothalamus, brainstem, or spinal cord cause visceral motor nerve impulses to be carried to effectors via cranial or spinal nerves. Higher brain centers, such as the limbic system and cerebral cortex, influence the ANS during times of emotional stress.

Table 8.4 compares the somatic and autonomic nervous systems.

- The epithalamus possesses the pineal gland, which produces the hormone melatonin. Melatonin induces sleepiness in the evenings.
- The limbic system is associated with emotional behavior, memory, and motivation.
- The brainstem consists of the midbrain, pons, and medulla oblongata. Ascending and descending axons between higher brain centers and the spinal cord pass through the brainstem.
- The midbrain is a small, superior portion of the brainstem. It contains reflex centers for movements associated with visual and auditory stimuli.
- The pons is the middle portion of the brainstem. It works with the medulla oblongata to control breathing.
- The medulla oblongata is the most inferior portion of the brainstem and is continuous with the spinal cord. It contains reflexive integration centers that control breathing, heart rate and force of contraction, and blood pressure.
- The reticular formation consists of nuclei and axons that extend from the superior spinal cord and into the diencephalon. It is involved with wakefulness.
- The cerebellum lies posterior to the fourth ventricle. It is composed of two hemispheres separated by the vermis and coordinates skeletal muscle contractions.
- The ventricles of the brain, the central canal of the spinal cord, and the subarachnoid space around the brain and spinal cord are filled with cerebrospinal fluid. Cerebrospinal fluid is secreted by a choroid plexus in each ventricle.
- Cerebrospinal fluid is absorbed into blood of the dural venous sinus in the dura mater.

8.6 Spinal Cord

- The spinal cord extends from the medulla oblongata inferiorly through the vertebral canal to the second lumbar vertebra.
- Gray matter is located internally and is surrounded by white matter. Anterior horns of gray matter contain cell bodies of somatic motor neurons; posterior horns contain interneuron cell bodies that receive incoming sensory nerve impulses; lateral horns contain cell bodies of autonomic motor neurons. White matter contains ascending and descending tracts of myelinated and unmyelinated axons.
- The spinal cord serves as a reflex center and conducting pathway for nerve impulses between the brain and spinal nerves.

8.7 Peripheral Nervous System (PNS)

- The PNS consists of cranial and spinal nerves, in addition to sensory receptors and ganglia. Most nerves are mixed nerves; a few cranial nerves are motor or sensory only. A nerve contains bundles of axons supported by connective tissue.
- The 12 pairs of cranial nerves are identified by roman numeral and name. The 31 pairs of spinal nerves are divided into 8 cervical, 12 thoracic, 5 lumbar, 5 sacral, and 1 coccygeal nerve.
- Anterior rami of many spinal nerves form spinal plexuses where axons are sorted and recombined so that all axons to a particular organ are carried in the same nerve. The four pairs of spinal plexuses are cervical, brachial, lumbar, and sacral plexuses.
- Reflexes are rapid, involuntary, and predictable responses to internal and external stimuli.
- Autonomic reflexes involve smooth muscle, cardiac muscle, adipose tissue, and glands. Somatic reflexes involve skeletal muscles.
- Cranial reflexes involve the brain, while spinal reflexes involve the spinal cord.

8.8 Autonomic Nervous System (ANS)

- The ANS involves portions of the central and peripheral nervous systems that are involved in involuntary maintenance of homeostasis.
- Two ANS motor neurons are used to activate an effector. The axon of the preganglionic neuron arises from the CNS and ends in an autonomic ganglion, where it synapses with a postganglionic neuron. The axon of the postganglionic neuron extends from the ganglion to an effector.
- The ANS is divided into two subdivisions that generally have antagonistic effects. Nerves of the sympathetic division arise from the thoracic and lumbar segments of the spinal cord and prepare the body to meet emergencies. Nerves of the parasympathetic division arise from the brain and the sacral segment of the spinal cord and function mainly in nonstressful situations.

8.9 Disorders of the Nervous System

- Disorders may result from infectious diseases, degeneration from unknown causes, malfunctions, and physical injury.
- Inflammatory neurological disorders include meningitis, neuritis, sciatica, and shingles.
- Noninflammatory neurological disorders include Alzheimer disease, cerebral palsy, CVAs, comas, concussion, dyslexia, epilepsy, fainting, headaches, mental illness, multiple sclerosis, neuralgia, paralysis, and Parkinson disease.

Self-Review

Answers are located in appendix B.

1. Nerve impulses are carried away from the cell body of a neuron by the ______ of the neuron.
2. Neurons that conduct nerve impulses from place to place within the CNS are ______.
3. A nerve impulse is formed by the sudden flow of ______ ions across the plasma membrane into a neuron.
4. Synaptic transmission is dependent upon the secretion of a ______ by an axon's terminal bouton.
5. The ______ is the only lobe of the cerebrum that cannot be seen superficially.
6. Voluntary muscle contractions are controlled by the ______ lobe of the cerebrum.
7. The ______ area of the cerebrum is involved with decision making, conscience, and personality.
8. The ______, a component of the diencephalon, regulates appetite, water balance, and body temperature.
9. The ______, a component of the brainstem, regulates heart and breathing rates.
10. Coordination of body movements is a function of the ______.
11. Cerebrospinal fluid fills the ventricles of the brain and the ______ space of the meninges.
12. The ______ horns of the spinal cord receive incoming sensory nerve impulses.
13. The ______ roots of spinal nerves consist of axons of somatic motor neurons.
14. The ______ nervous system is involved in involuntary responses that maintain homeostasis.
15. The ______ division prepares the body for physical responses to emergencies.

Critical Thinking

1. Predict the cognitive changes that will occur following physical trauma to the anterior portion of the frontal lobe.
2. Explain why damage to the medulla oblongata is life-threatening.
3. Explain the effect of an abnormally high level of potassium ions in the ECF on the ability of a neuron to create a nerve impulse.
4. If you touch a hot stove, a reflexive withdrawal of your hand is triggered at about the same time that you feel the pain. Describe the roles of the PNS and CNS in your response and sensation.
5. Explain how the ANS can both increase and decrease heart rate.

ADDITIONAL RESOURCES

9 CHAPTER

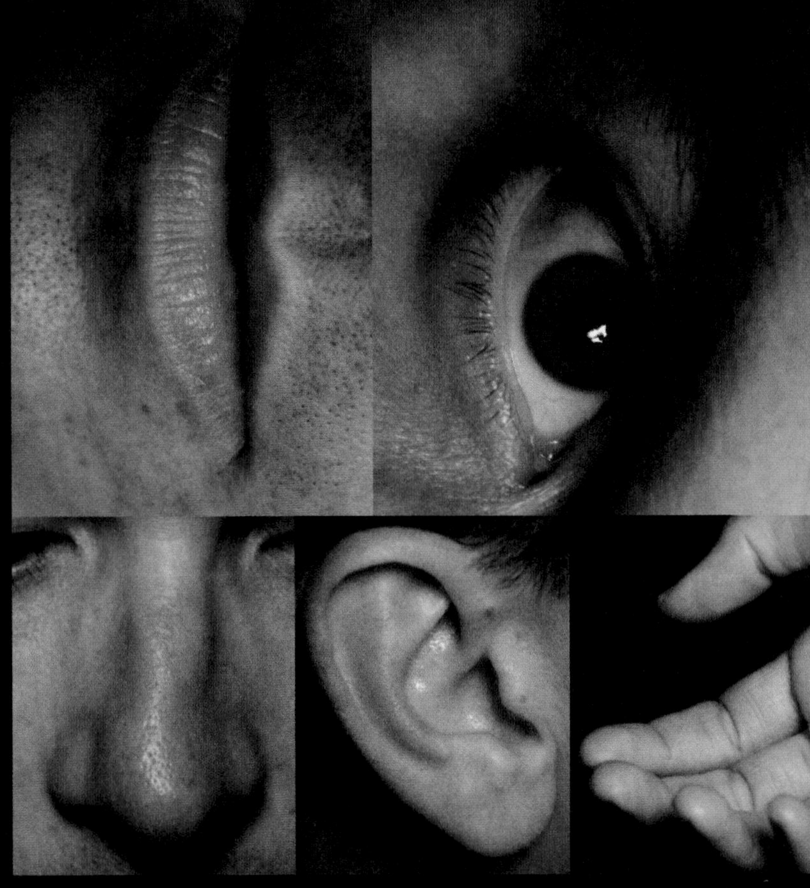

Senses

Jeremy, age 14, was born blind and deaf. He is unable to see the world around him. He cannot see the sky, the earth, or his family. He cannot see a car heading towards him, so that he can get out of the way. Jeremy cannot hear warning alarms or people yelling at him when there is danger around him. He cannot hear the spoken words used for quick, easy communication between people. Because he cannot hear words, he did not develop the auditory memories needed to produce speech. He cannot verbally express his thoughts, opinions, or desires to those around him. To survive in the world and communicate with those around him, Jeremy has had to learn to use his other senses. He uses his sense of touch to identify people and objects around him and to learn about the world by reading in braille. By feeling vibrations through his skin, he can detect the rhythm in music that is playing. His sense of smell is heightened, which allows him to detect certain types of hazards and aid in the identification of people and objects. Jeremy's life is the perfect example of just how important the senses are in maintaining health and wellness for each of us.

CHAPTER OUTLINE

Module 7
Nervous System

SELECTED KEY TERMS

Accommodation The focusing of light rays on the retina by the lens.
Chemoreceptor Sensory receptor stimulated by certain chemicals.
Cochlear hair cells Sensory receptors used in hearing.
Cones Photoreceptors for color vision.
Dynamic equilibrium Maintenance of balance when the head is in motion.
Mechanoreceptor Sensory receptor stimulated by mechanical forces such as pressure or touch.
Nociceptor Sensory receptor stimulated by tissue damage.
Olfactory receptor Sensory receptor used to detect odors in inhaled air.
Photoreceptor (photo = light) Sensory receptor stimulated by light.
Projection The process by which the brain makes a sensation seem to come from the body part being stimulated.
Proprioceptor Sensory receptor stimulated by changes in body position or movements of the body or its parts.
Retina (retin = net) The internal layer of the eye, which contains the photoreceptors.
Rods Photoreceptors for black and white vision.
Semicircular canals The portion of the internal ear containing the sensory receptors for dynamic equilibrium.
Sensory adaptation The decrease in the formation of nerve impulses by a sensory receptor when repeatedly stimulated by the same stimulus.
Spiral organ (Organ of Corti) The sense organ in the internal ear containing the sensory receptors for hearing.
Static equilibrium The maintenance of balance when the head is not in motion.
Taste bud Tongue organ that contains taste receptors.
Thermoreceptor Sensory receptor stimulated by changes in temperature.

OUR SENSES CONSTANTLY inform us of what is going on in our internal and external environments so that our body can take appropriate voluntary or involuntary action and maintain homeostasis. Several different types of **sensory receptors** are involved in sending nerve impulses to the CNS, which then initiates the appropriate response.

The senses may be subdivided into two broad categories: general senses and special senses. *General senses* include pain, touch, pressure, stretching, chemical changes, cold, and heat. *Special senses* are taste, smell, vision, hearing, and equilibrium. Each of the senses depends upon (1) sensory receptors, which detect environmental changes and form nerve impulses; (2) sensory neurons, which carry the nerve impulses to the CNS; and (3) the brain, which interprets the nerve impulses.

9.1 Sensations

Learning Objectives

1. Differentiate between sense, sensation, and perception.
2. Recall the five basic types of sensory receptors.
3. Compare the mechanisms of projection and adaptation of sensations.

Each type of sensory receptor is sensitive to a particular type of stimulus that causes the sensory receptor to form nerve impulses. The five types of sensory receptors, based on the specific stimuli to which they respond, are listed in table 9.1. These nerve impulses are carried by cranial or spinal nerves to the CNS. A **sensation** is a conscious or subconscious awareness of a change in the internal or external environment. The conscious awareness of a sensation, or **perception,** results from the interpretation of nerve impulses reaching sensory areas of the cerebral cortex. The sensation that is created is determined by the area of the brain receiving the nerve impulses rather than by the type of sensory receptor forming the nerve impulses. For example, all nerve impulses reaching the visual area of the occipital lobe are interpreted as visual sensations. A strong blow to an eye or to the back of the head may produce a visual sensation (flashes of light), although the stimulus is mechanical.

The perceived intensity of a sensation is dependent upon the frequency of nerve impulses reaching the cerebral cortex. The greater the frequency of nerve impulses, the greater is the intensity of the sensation. The frequency of nerve impulses sent to the brain is, in turn, dependent upon the action of sensory receptors. The greater the

Table 9.1 Types of Sensory Receptors

Type	Stimulus Detected
Thermoreceptors	Temperature changes
Mechanoreceptors	Mechanical forces
Nociceptors	Tissue damage
Chemoreceptors	Concentration of chemicals
Photoreceptors	Light energy

intensity of a stimulus, the greater the frequency of nerve impulse formation by sensory receptors.

Projection

Whenever a sensation occurs, the cerebral cortex projects the sensation back to the body region where the nerve impulses originated so that the sensation seems to come from that region. This phenomenon is called **projection.** For example, if your thumb is injured, the pain is projected back to your thumb so that you are aware that your thumb hurts. Similarly, projection of visual and auditory sensations gives the feeling that eyes see and ears hear. The projection of sensations has obvious survival value in pinpointing the source of a sensation because it allows for corrective action to remove harmful stimuli.

Adaptation

If a sensory receptor is repeatedly stimulated by the same stimulus, the rate of nerve impulse formation may decline until nerve impulses may not be formed at all. This phenomenon is called **sensory adaptation.** For example, when the odor of perfume is first encountered, it is very noticeable. But as the olfactory receptors become adapted to the stimulus, the strength of the sensation rapidly declines until the odor is hardly noticeable. Adaptation occurs within most sensory receptors, with the exception of those involved in pain and proprioception. Its purpose is to prevent overloading the nervous system with unimportant stimuli, such as clothes touching the body. Once a sensory receptor is adapted, a stronger stimulus is needed to form nerve impulses.

9.2 General Senses

Learning Objectives

4. Contrast the structures, locations, and functions of the sensory receptors involved in sensations of warm, cold, touch, pressure, stretch, chemical change, and pain.
5. Explain the mechanism of referred pain.

Sensory receptors for the general senses are widely distributed in the skin, muscles, tendons, ligaments, and visceral organs.

Temperature

Two types of **thermoreceptors** are located in the skin. *Warm receptors* are **free nerve endings,** which are sensory neuron dendrites, in the deep dermis that are most sensitive to temperatures above 25°C (77°F). *Cold receptors* are free nerve endings in the superficial dermis that are most sensitive to temperatures below 20°C (68°F). Temperatures below 10°C (50°F) or above 45°C (113°F) stimulate pain receptors, which results in painful sensations. Thermoreceptors adapt very quickly to constant stimulation.

Pressure, Touch, and Stretch

Pressure, touch, and stretch receptors are **mechanoreceptors** (mek-ah-no-re-cep′tors), which are sensitive to mechanical stimuli displacing the tissue in which they are located. *Lamellated (Pacinian) corpuscles* are rapidly adapting receptors used to detect deep pressure and stretch. They are located deep in the dermis, as well as in the ligaments and tendons associated with joints. (figure 9.1).

There are several types of receptors that function in the skin as touch receptors (figure 9.1). Free nerve endings extend from the dermis superficially to the spaces between the epidermal cells. These endings function primarily as pain receptors but also serve to detect touch, itch, and temperature. The free nerve endings of the *hair root plexus* functions to detect hair displacement, such as when a bug lands on the forearm. *Tactile (Meissner) corpuscles* in the superficial dermis are most abundant in hairless areas such as fingertips, palms, and lips. These rapidly adapting receptors are useful in detecting the onset of light touch to the skin. *Tactile discs* in the superficial dermis are associated with *tactile cells* in the stratum basale of the epidermis in areas such as the fingertips, hands, lips, and external genitalia. Together these slowly adapting structures function in detecting light touch and pressure, such as when reading braille.

Baroreceptors are free nerve endings that monitor stretching within distensible internal organs such as blood vessels, the stomach, and the bladder. Signals from these receptors are used to help regulate visceral reflexes such as those used to regulate blood pressure, digestion, and urination. For example, baroreceptors within the urinary bladder will trigger the urination reflex as the bladder fills and stretches. These receptors do not exhibit sensory adaptation owing to their role in regulating visceral reflexes.

Proprioceptors, such as *muscle spindles* and *tendon organs,* are used to monitor changes in skeletal muscle stretching and tendon tension during skeletal muscle contraction and relaxation (figure 9.2). These receptors keep us informed about the positioning of our body or body parts while stationary or moving. These receptors do not exhibit sensory adaptation owing to their role in maintaining posture, equilibrium, and muscle tone.

Chemoreceptors

The **chemoreceptors** that are part of the general senses are specialized neurons used to monitor body fluids for chemical changes. For example, chemoreceptors monitor changes in ion concentrations, pH, blood glucose levels, and dissolved gases. The signals created by these chemoreceptors are not processed within the cerebral cortex; this means that the sensation created within the brain cannot be consciously detected.

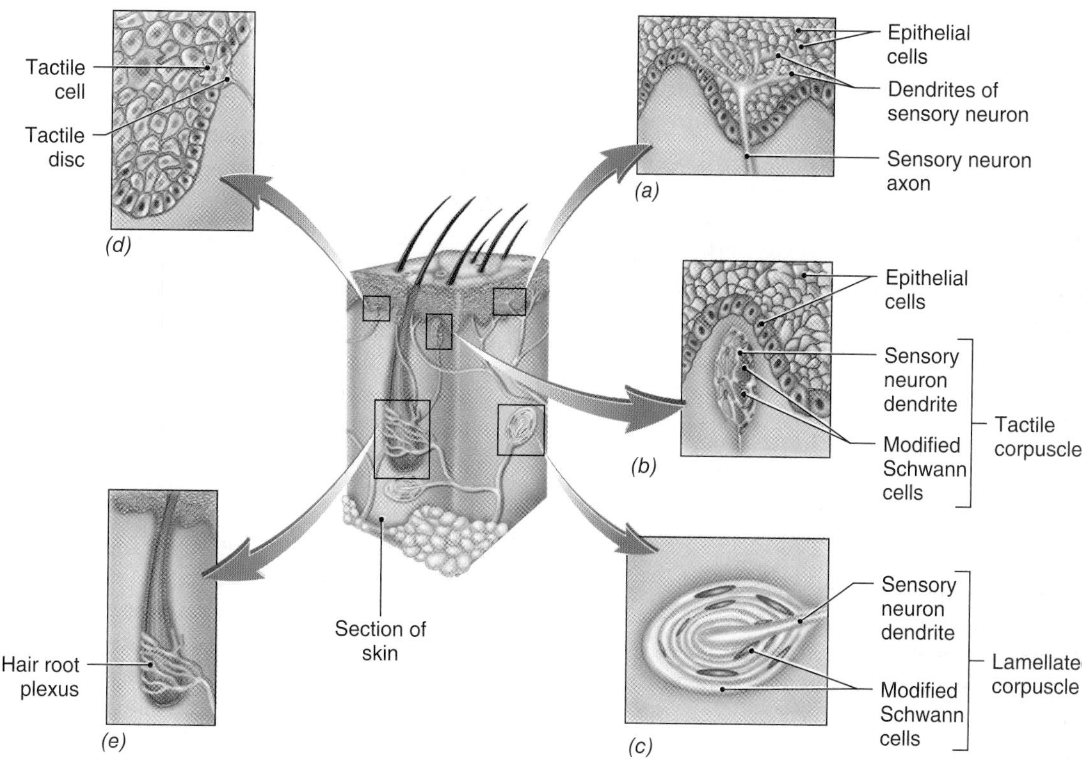

Figure 9.1 Touch and Pressure Receptors.
(a) Free nerve endings between epidermal cells detect touch, itch, pain, and temperature sensations. *(b)* Tactile corpuscles are light touch receptors located in the superifical dermis. *(c)* Lamellated corpuscles are pressure receptors located deep in the dermis, in addition to certain ligaments and tendons. *(d)* Tactile cell in stratum basale of the epidermis and tactile disc in the adjacent dermis detect light touch and pressure. *(e)* Hair root plexus around hair follicle detects movement of the hair shaft.

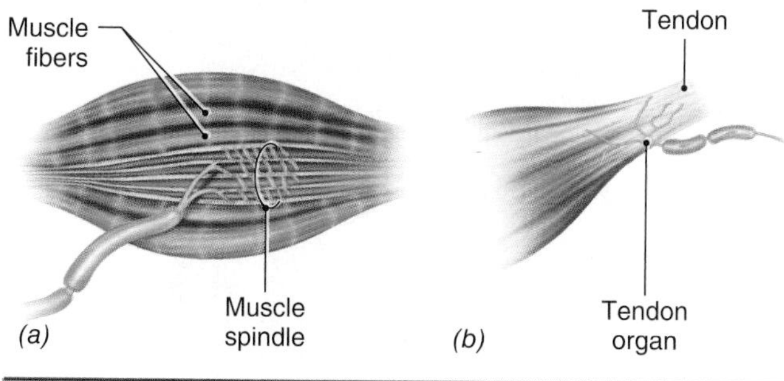

Figure 9.2 Proprioceptors. *(a)* Muscle spindle. *(b)* Tendon organ.

Pain

Nociceptors, also referred to as pain receptors, are free nerve endings, which are widespread in body tissues, except within the nervous tissue of the brain. They are especially abundant in the skin, the organ that is in direct contact with the external environment. Nociceptors are stimulated whenever tissues are damaged, and the pain sensation initiates actions by the CNS to remove the source of the stimulation. Further, nociceptors do not easily adapt like many other sensory receptors. The lack of adaptation is a protective mechanism that allows the person to be aware of a harmful stimulus until it is removed.

Referred Pain

Projection by the cerebral cortex is not always accurate when the nerve impulses originate from nociceptors in visceral organs. When damage to visceral organs occurs, pain sensations are often projected or referred to an undamaged part of the body wall or limb. This type of pain is called **referred pain.**

Pain management in the United States costs billions of dollars each year. *Analgesia* (pain reduction) and *anesthesia* (complete loss of sensation) control pain by decreasing nociceptor sensitivity, blocking nerve impulse formation, preventing nerve impulse transmission to the CNS, or interfering with pain perception within the brain.

Referred pain is consistent from person to person and is important in the diagnosis of many disorders. For example, pain caused by a heart attack is referred to the left anterior chest wall, left shoulder, and left upper limb in both genders, while women also commonly experience pain in the abdomen and jaw and between the scapulae. Referred pain is due to communication between neurons within the same nerve that are carrying nerve impulses from both visceral organs and the body wall or a limb. For example, neurons carrying nerve impulses from the heart use the same nerves as those from the left shoulder and upper limb (figure 9.3).

Check My Understanding

1. What parts of the nervous system are involved in the development of a sensation?
2. What sensory receptors are involved in the general senses?
3. What are the roles of these sensory receptors in monitoring the external and internal environments?

9.3 Special Senses

Learning Objectives

6. Contrast the location, structure, and function of olfactory and taste receptors.
7. Recall the location, structure, and function of the sensory receptors involved in hearing.
8. Distinguish the location, structure, and function of the sensory receptors involved in static equilibrium and dynamic equilibrium.
9. Identify the structures of the eye and the functions of these structures.
10. Describe the location, structure, and function of the sensory receptors involved in vision.

The sensory receptors for special senses are localized rather than widely distributed, and they, like all sensory receptors, are specialized to respond to only certain types of stimuli. There are three different kinds of sensory receptors for the special senses. Taste and olfactory receptors are chemoreceptors, which are sensitive to chemical substances. Sensory receptors for hearing and equilibrium

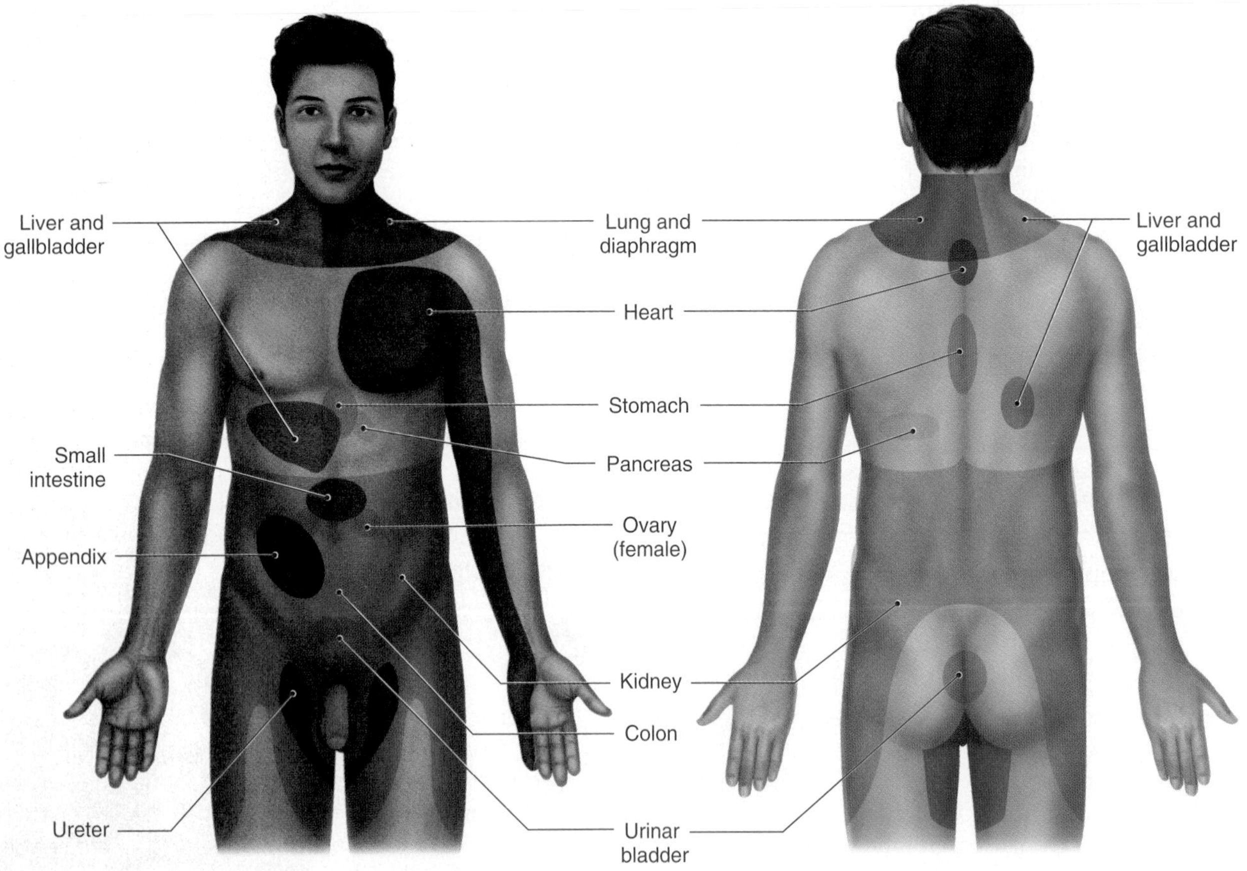

Figure 9.3 Superficial regions to which visceral pain originating from various internal organs may be referred.

are mechanoreceptors, which are sensitive to vibrations formed by sound waves and movement of the head. Sensory receptors for vision are **photoreceptors,** which are sensitive to light energy.

Taste

The chemoreceptors for taste are located in specialized microscopic organs called **taste buds.** Most taste buds are located on the tongue in small, raised structures called *lingual papillae* (figure 9.4), though some can be found in areas such as the soft palate, pharynx, and esophagus.

A taste bud consists of a bulblike arrangement of rapidly adapting **taste receptors,** called *gustatory epithelial cells,* located within the epithelium of the lingual papillae. The taste bud possesses an opening called a *taste pore.* Taste receptors have hairlike projections called *gustatory hairs* that extend through the pore and are exposed to chemicals on the tongue. Sensory axons leading to the brain are connected to the opposite end of the taste receptors. In order to activate the taste receptors, a substance must be dissolved in a liquid such as saliva.

There are five confirmed basic tastes that can be detected by the tongue: sweet, sour, salty, bitter, and umami (savory). The receptors for each basic taste are located across the tongue surface, which disproves the earlier belief that the basic tastes were mapped to specific regions of the tongue. It is probable that other substances, such as fats and Ca^{2+}, will be added as basic tastes in the near future as a result of ongoing taste research. It has been suggested that water is also a basic taste; however, not enough experimental data has been produced to support this claim. The many flavor sensations of food result from the stimulation of one or more taste receptors and, more importantly, the activation of olfactory receptors discussed in the next section.

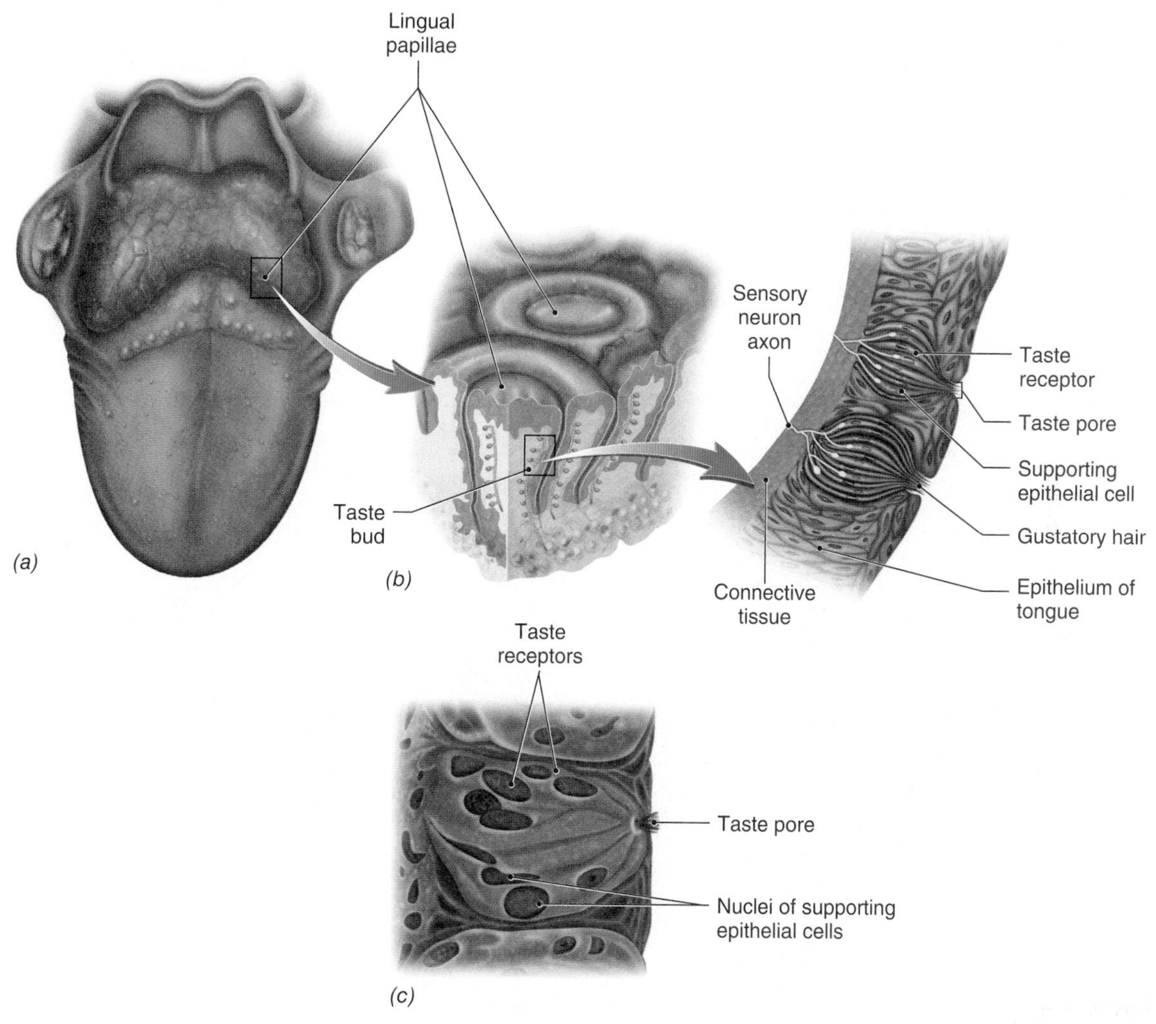

Figure 9.4 *(a)* Taste buds are located on lingual papillae of the tongue. *(b)* A taste bud contains taste receptors whose gustatory hairs protrude through the taste pore. *(c)* Photomicrograph of a taste bud. APR

The pathway of nerve impulses from taste receptors to the brain depends on where the taste receptors are located. Nerve impulses created by taste receptors on the anterior two-thirds of the tongue are carried by the facial nerve (CN VII), while those created on the posterior one-third travel over the glossopharyngeal nerve (CN IX). Nerve impulses created at the base of the tongue are carried by the vagus nerve (CN X). These cranial nerves carry the nerve impulses to the medulla oblongata, from which the nerve impulses travel to the thalamus and on to the taste areas in the parietal lobes of the cerebrum. AP|R

Smell

The **olfactory** (ō-l-fak′-tō-rē) **receptors** are located in the superior portion of the nasal cavity, including the superior nasal conchae and nasal septum. The olfactory receptors, also called *olfactory sensory neurons,* are surrounded by the supporting epithelial cells of the olfactory epithelium. The distal ends of the olfactory receptors are covered with cilia that project into the nasal cavity, where they can contact airborne molecules. Chemicals in inhaled air are in a gaseous state and must dissolve in the mucus layer covering the olfactory epithelium in order to stimulate nerve impulse formation (figure 9.5). The nerve impulses are carried by axons of the olfactory receptors, which form the olfactory nerves (CN I), to the olfactory bulbs. Here they synapse with neurons that form the olfactory tract and relay the nerve impulses to the olfactory areas deep within the temporal lobes and at the bases of the frontal lobes of the cerebrum.

It is common for a person to sniff the air when trying to detect faint odors. This is because the olfactory receptors are located superior to the usual path of inhaled air and additional force is needed to send larger amounts of air over the olfactory epithelium. Like taste receptors, olfactory receptors rapidly adapt to a particular stimulus.

The human olfactory epithelium possesses approximately 350 functional types of olfactory receptors. However, the average person can distinguish between 2,000 and 4,000 different odors. The ability to detect so many types of odors largely depends upon how the temporal lobes process the nerve impulses from various combinations of olfactory receptors. Studies have shown that women can detect, discern, and identify a wider range of odors than men. It is also possible with training to enhance your olfactory ability and potentially discern up

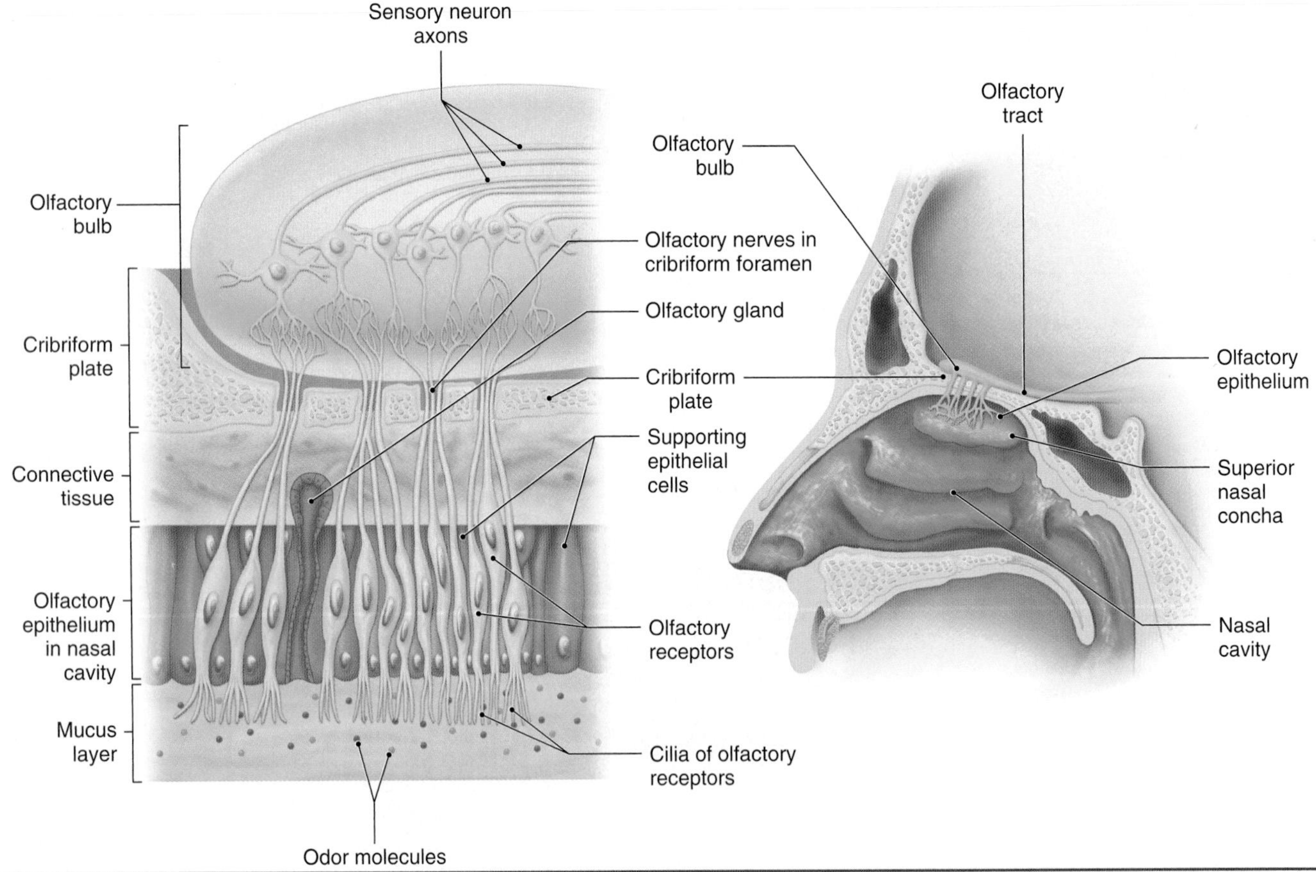

Figure 9.5 Olfactory receptors are located between supporting epithelial cells in the superior portion of the nasal cavity. AP|R

Clinical Insight

The ability to distinguish various foods relies predominantly on the sense of smell. This explains why foods seem to have little taste for a person who is suffering from a head cold. The taste and smell of appetizing foods prepare the digestive tract for digestion by stimulating the flow of saliva in the mouth and gastric juice in the stomach.

to 10,000 different odors, an ability important for those in the wine industry. The decrease in odor detection that occurs with age, which is why the elderly tend to use more cologne and perfume, is a result of receptor loss and desensitization rather than temporal lobe dysfunction. Research suggests that the olfactory epithelium is capable of detecting human pheromones. Human pheromones, which have been found in apocrine sweat and vaginal secretions, have been shown to have influence over reproductive functions. For example, pheromones from one female have been shown to lengthen or shorten the menstrual cycle of exposed females. The olfactory epithelium is also highly regenerative owing to its direct exposure to the external environment. On average, an olfactory receptor lives only approximately 60 days before being replaced.

Check My Understanding

4. Where are taste and olfactory receptors located?
5. How are the five basic taste sensations produced?

Hearing

The *ear* is the organ of hearing. It is also the organ of equilibrium. The ear is subdivided into three major parts: the external ear, middle ear, and internal ear (figure 9.6). Table 9.2 summarizes the structures of the ear and their functions.

External Ear

The external ear consists of two parts: the auricle and the external acoustic meatus. The *auricle* (pinna) is the funnel-like structure composed primarily of cartilage and skin that is attached to the side of the head. The **external acoustic meatus** is a short tube that extends from the auricle through the temporal bone to the eardrum. Sound waves striking the auricle are channeled into the external acoustic meatus. **Cerumen** (earwax) and hairs in the external acoustic meatus help to prevent foreign particles from reaching the eardrum.

Middle Ear APIR

The middle ear, or **tympanic** (tim-pan′-ik) **cavity,** is an air-filled space within the temporal bone. The tympanic

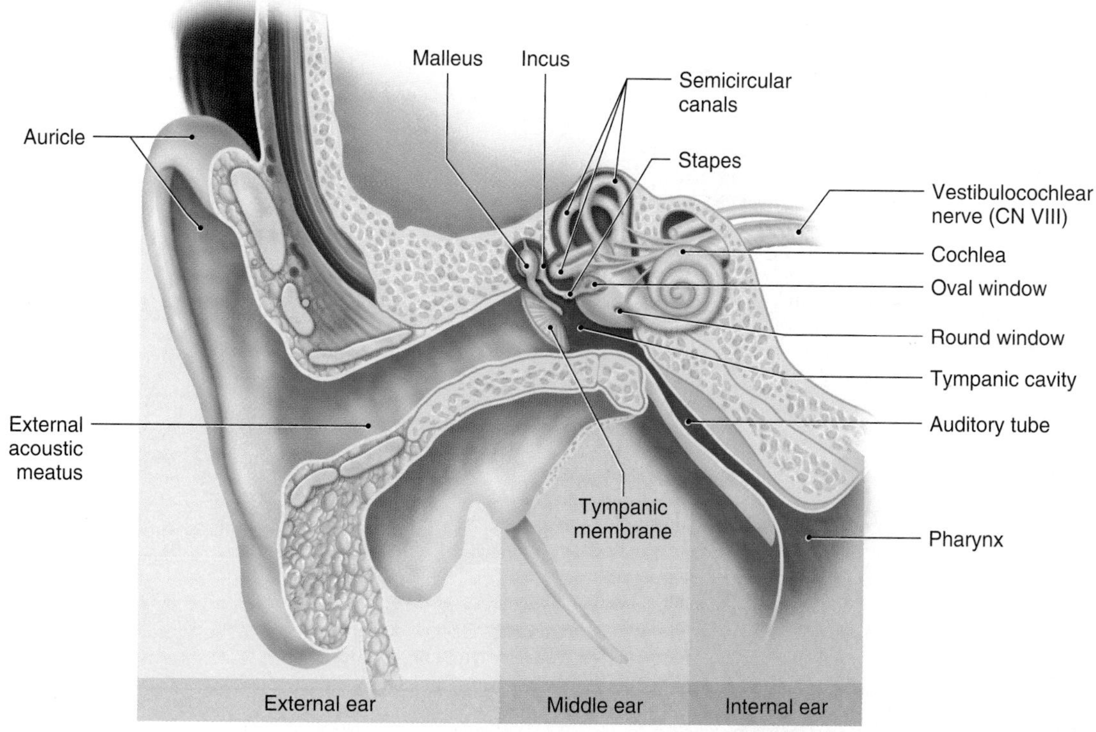

Figure 9.6 Anatomy of the Ear. APIR

membrane, auditory tube, and auditory ossicles are parts of the middle ear. The **tympanic membrane,** or eardrum, separates the tympanic cavity from the external acoustic meatus. The tympanic membrane is covered with skin externally and by a mucous membrane internally. Sound waves, or air pressure waves, entering the external acoustic meatus cause the tympanic membrane to vibrate in and out at the same frequency as the sound waves.

The **auditory** (eustachian) **tube** connects the tympanic cavity with the pharynx. Its function is to keep the air pressure within the tympanic cavity the same as the external air pressure by allowing air to enter or exit the tympanic cavity. Equal air pressure on each side of the tympanic membrane is essential for the tympanic membrane to function properly. A valve at the pharyngeal end of the tube is usually closed but it opens when a person swallows or yawns to allow air pressure to equalize. If you have experienced a rapid change in air pressure, you probably have noticed your ears "popping" as the air pressure is equalized and the tympanic membrane snaps back into place.

The **auditory ossicles** (os′-si-kulz) are three tiny bones that articulate to form a lever system from the tympanic membrane, across the tympanic cavity, to the internal ear. Each ossicle is named for its shape. The tip of the "handle" of the club-shaped *malleus* (mal′-ē-us), or hammer, is attached to the tympanic membrane and its head articulates with the *incus* (ing′-kus), or anvil. The base of the incus articulates with the *stapes* (stā′-pēz), or stirrup, whose foot plate is inserted into the oval window of the internal ear.

The vibrations of the tympanic membrane cause corresponding movements of the ossicles, which result in the stapes vibrating in the oval window. In this way, vibrations of the tympanic membrane are transmitted to the fluid-filled internal ear. Due to the size difference between the larger tympanic membrane and the smaller oval window, vibrations are amplified by the ossicles.

Internal Ear

The internal ear is embedded in the temporal bone. It consists of two series of connecting tubes and chambers, one within the other: an external *bony labyrinth* (lab′-i-rinth) and an internal *membranous labyrinth.* The two **labyrinths** are similar in shape (figure 9.7). The space between the bony and membranous labyrinths is filled with **perilymph,** whereas the membranous labyrinth contains **endolymph.** These fluids play important roles in the functions of the internal ear. The internal ear has three major parts: the cochlea, vestibule, and semicircular canals.

The **cochlea** (kok′-lē-ah) is the coiled portion of the internal ear. When viewed in cross section, as in figure 9.8, it can be seen that the cochlea is composed of three chambers that are separated from each other by membranes. The *scala vestibuli* (skā-la ves-tib′-ū-lī) and the *scala tympani,* both components of the bony labyrinth, extend the length of the cochlea and are continuous with each other at the apex of the cochlea. The scala vestibuli continues into the vestibule, which houses the membrane-covered *oval window.* The scala tympani extends toward the vestibule, ending at the membrane-covered *round window.*

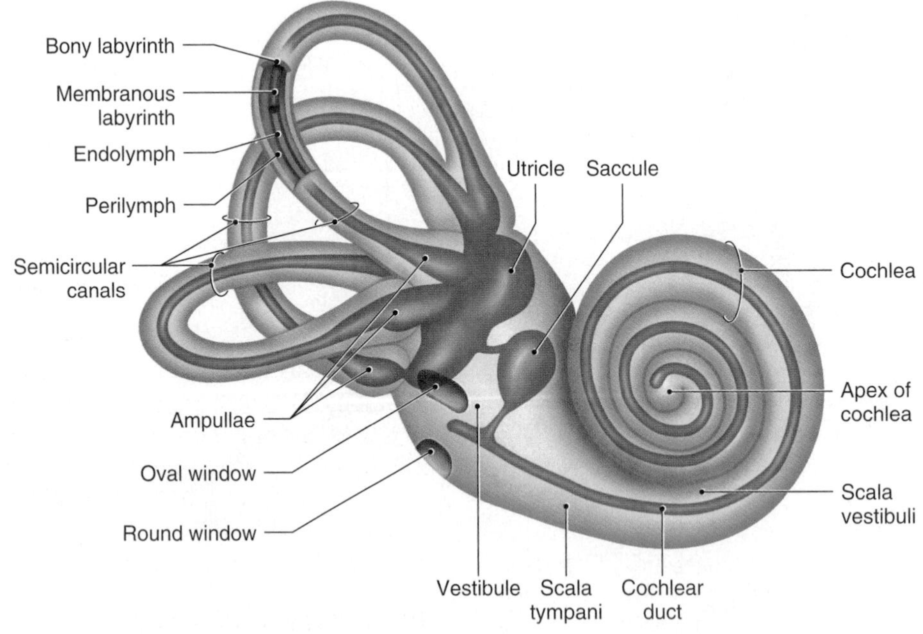

Figure 9.7 The bony (orange) and membranous (purple) labyrinths of the internal ear. Perilymph fills the space between the membranous labyrinth and the bony labyrinth. Endolymph fills the membranous labyrinth. Note that the ampullae of the semicircular canals, utricle, saccule, and cochlear duct are portions of the membranous labyrinth.

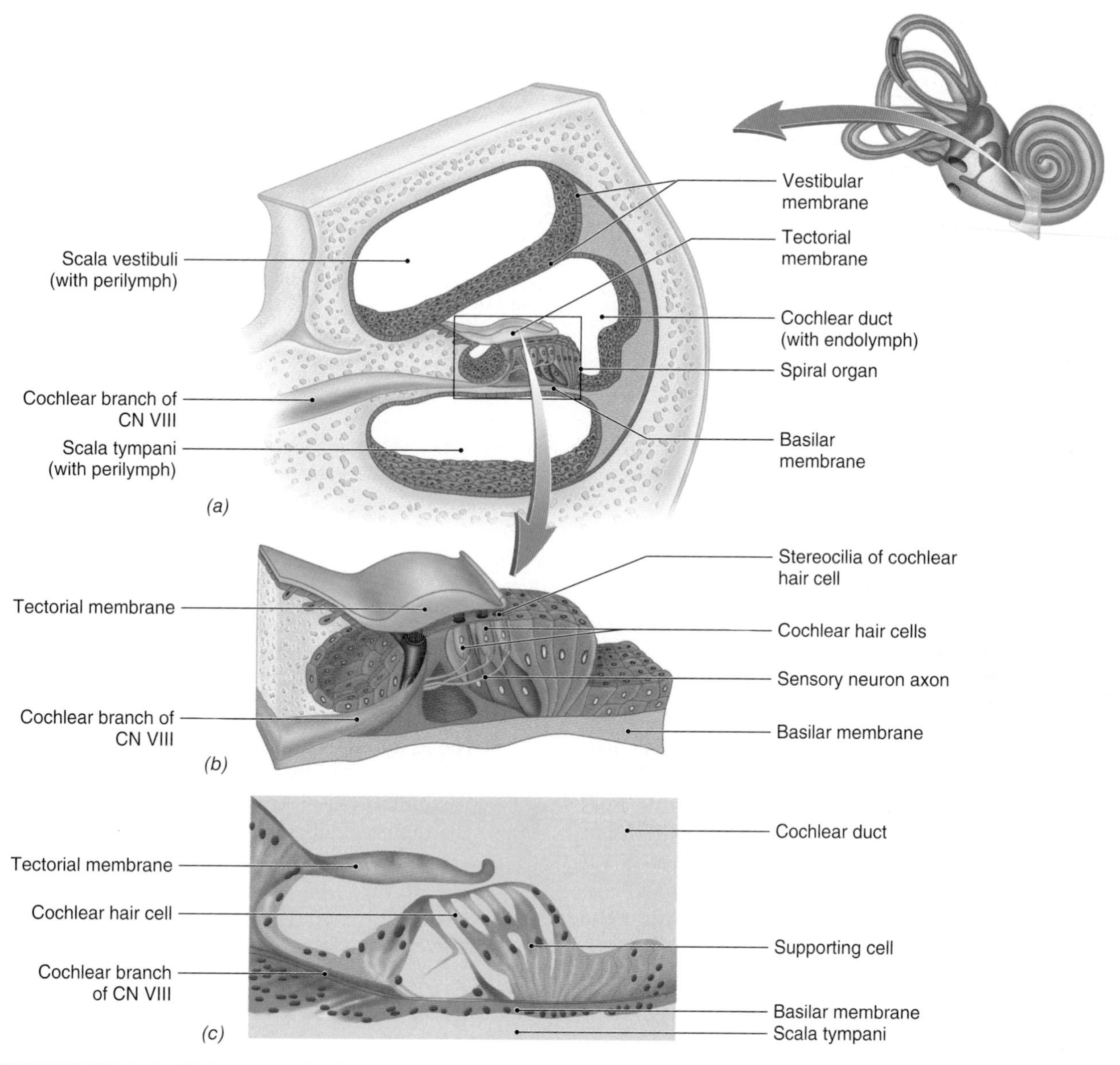

Figure 9.8 *(a)* A cross section of the cochlea shows the cochlear duct located between the scala vestibuli and the scala tympani and the spiral organ resting on the basilar membrane. *(b)* Detail of the spiral organ shows the tectorial membrane overlying the cochlear hair cells. *(c)* Photomicrograph of spiral organ. APR

The *cochlear duct,* which is part of the membranous labyrinth, extends nearly to the apex of the cochlea (see figure 9.7). As shown in figure 9.8, it is separated from the scala vestibuli by the *vestibular membrane* and from the scala tympani by the **basilar membrane.** The basilar membrane contains about 20,000 cross fibers that gradually increase in length from the base to the apex of the cochlea. The attachment of the basilar membrane to the bony center of the cochlea allows it to vibrate like the reeds of a harmonica when activated by vibrations generated by sound.

The **spiral organ** (organ of Corti), which contains the sensory receptors for sound stimuli, supported by the basilar membrane within the cochlear duct. The sensory receptors are called **cochlear hair cells,** and they have hairlike stereocilia extending from their free surfaces toward the overlying *tectorial* (tek-to′-rē-al) *membrane.* Axons of the cochlear branch of the vestibulocochlear nerve (CN VIII) exit the cochlear hair cells and lead to the brain.

Physiology of Hearing

The human ear is able to detect sound waves with frequencies ranging from near 20 to 20,000 Hertz (Hz; vibrations per second), but hearing is most acute between 2,000 and 3,000 Hz. For hearing to occur, vibrations formed by

Table 9.2 Summary of Ear Function

Structure	Function
External Ear	
Auricle	Channels sound waves into external acoustic meatus
External acoustic meatus	Directs sound waves to tympanic membrane
Tympanic membrane	Vibrates when struck by sound waves
Middle Ear	
Tympanic cavity	Air-filled space that allows tympanic membrane to vibrate freely when struck by sound waves
Auditory ossicles	Transmit and amplify vibrations produced by sound waves from the tympanic membrane to the perilymph within the cochlea
Auditory tube	Equalizes air pressure on each side of tympanic membrane
Internal Ear	
Cochlea	Fluids and membranes transmit vibrations initiated by sound waves to the spiral organ, whose cochlear hair cells generate nerve impulses associated with hearing
Saccule	Vestibular hair cells of the macula form nerve impulses associated with static and dynamic equilibrium
Utricle	Vestibular hair cells of the macula form nerve impulses associated with static and dynamic equilibrium
Semicircular canals	Vestibular hair cells of the crista ampullaris form nerve impulses associated with dynamic equilibrium

sound waves must be transmitted to the cochlear hair cells of the spiral organ. Then, the cochlear hair cells form nerve impulses that are transmitted to the hearing areas of the cerebrum for interpretation as sound sensations.

Figure 9.9 shows the structure of the internal ear with the cochlea uncoiled to show more clearly the relationships of its parts. Refer to this figure as you study the following outline of hearing physiology.

1. Sound waves enter the external acoustic meatus and strike the tympanic membrane, causing it to vibrate in and out at the same frequency and comparable intensity to the sound waves. Loud sounds cause a greater displacement of the tympanic membrane than do soft sounds.
2. Vibration of the tympanic membrane causes movement of the auditory ossicles, resulting in the in-and-out vibration of the stapes in the oval window.
3. The vibration of the stapes causes a corresponding oscillatory (back-and-forth) movement of the perilymph in the scala vestibuli and scala tympani and a corresponding movement of the membrane over the round window. This movement of the perilymph causes vibrations in the vestibular and basilar membranes.
4. The vibration of the basilar membrane causes the stereocilia of the cochlear hair cells to contact the tectorial membrane, which stimulates the formation of nerve impulses by the cochlear hair cells.
5. Nerve impulses formed by the cochlear hair cells are carried by the cochlear branch of the vestibulocochlear nerve to the hearing areas of the temporal lobes of the cerebrum, where the sensation is interpreted. Some of the axons cross over to the opposite side of the brain so that the hearing areas in each temporal lobe interprets nerve impulses originating in each ear. **APR**

Pitch and Loudness

Because of the gradually increasing length of the fibers in the basilar membrane, different portions of the basilar membrane vibrate in accordance with the different frequencies (pitch) of sound waves. Low-pitched sounds cause the longer fibers of the membrane near the apex of the cochlea to vibrate, and high-pitched sounds activate the shorter fibers of the membrane near the base of the cochlea. The pitch of a sound sensation is determined by the portion of the basilar membrane and the spiral organ that are activated by the specific sound frequency and by the parts of the hearing areas that receive the nerve impulses. Nerve impulses from different regions of the spiral organ go to slightly different portions of the hearing areas in the brain, which causes them to be interpreted as different pitches.

The loudness of the sound is dependent upon the intensity of the vibration of the basilar membrane and spiral organ, which, in turn, determines the frequency of nerve impulse formation. The greater the frequency of nerve impulses sent to the brain, the louder the sound sensation.

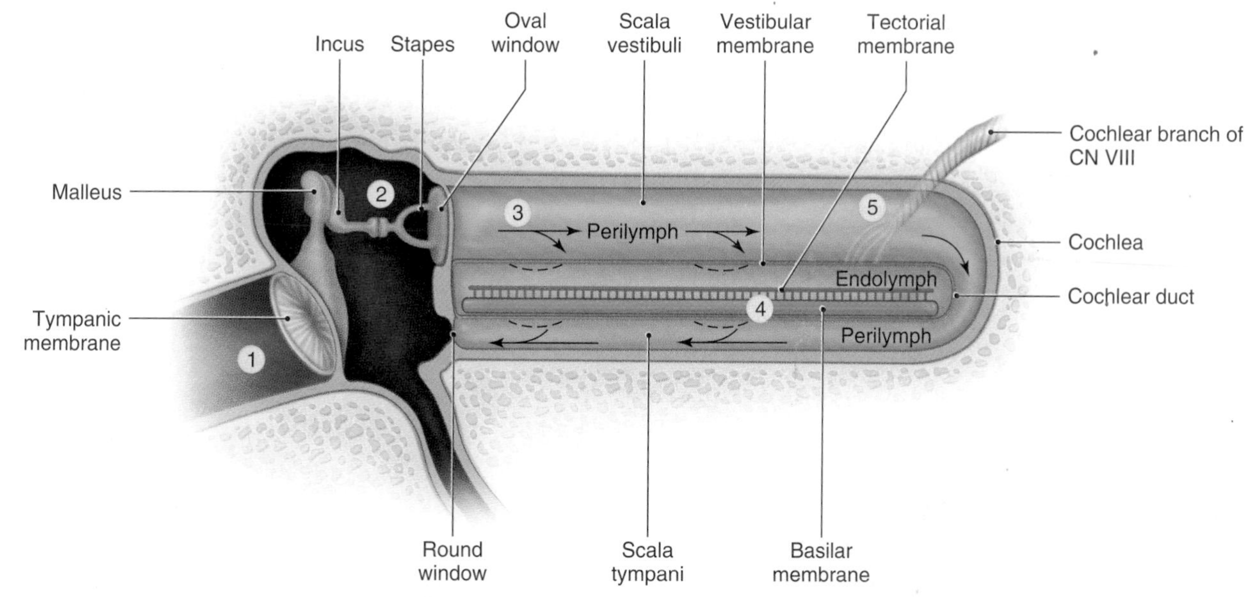

Figure 9.9 Transmission of vibrations produced by sound waves. Vibrations of the tympanic membrane are carried by the auditory ossicles to the perilymph. Oscillating movements of the perilymph cause the vibration of the basilar membrane and spiral organ, which, in turn, results in the formation of nerve impulses by the cochlear hair cells. APIR

Check My Understanding

6. How do sound waves stimulate the formation of nerve impulses?
7. How are pitch and loudness of a sound determined?

Equilibrium

Several types of sensory receptors provide information to the brain for the maintenance of equilibrium. The eyes and proprioceptors in joints, tendons, and muscles are important in informing the brain about equilibrium and the position and movement of body parts. However, unique receptors in the internal ear are crucial in monitoring two types of equilibrium. **Static equilibrium** involves the movement of the head with respect to gravitational force. **Dynamic equilibrium** involves linear acceleration in both horizontal and vertical directions, in addition to the rotational movement of the head.

Static Equilibrium

The **macula** (mak′-u-lah; plural *maculae*), an organ of static equilibrium, is located within the **utricle** (u′-tri-kul) and the **saccule** (sak′-ul), enlarged portions of the membranous labyrinth within the vestibule (see figure 9.7). Each macula contains thousands of sensory receptors called *vestibular hair cells* that possess hairlike stereocilia embedded in a gelatinous material. *Otoliths* (ō′-tō-liths), crystals of calcium carbonate, are also embedded in the gelatinous mass. The otoliths increase the weight of the gelatinous mass and make it more responsive to the pull of gravity (figure 9.10).

The mechanism of static equilibrium may be summarized as follows:

1. Changes in head position cause gravity to pull on the gelatinous mass, which bends the stereocilia. This change stimulates the vestibular hair cells to form nerve impulses that are carried by the vestibular branch of the vestibulocochlear nerve to the brain. No matter the position of the head, nerve impulses are formed that inform the brain of the head's position.
2. The cerebellum uses this information to maintain static equilibrium subconsciously.
3. Our awareness of static equilibrium results when the nerve impulses are interpreted by the cerebrum.

Dynamic Equilibrium

The maculae in the utricle and saccule also sense linear acceleration in both the horizontal and vertical directions. The mechanism is similar to that used to detect changes in static equilibrium. When the head accelerates either vertically or horizontally, inertia of the gelatinous mass causes the stereocilia of the vestibular hair cells to bend. When motion ends, the gelatinous mass continues to move for a moment, which bends the stereocilia in the opposite direction. The changes in stereocilia movement alert the brain to changes in velocity.

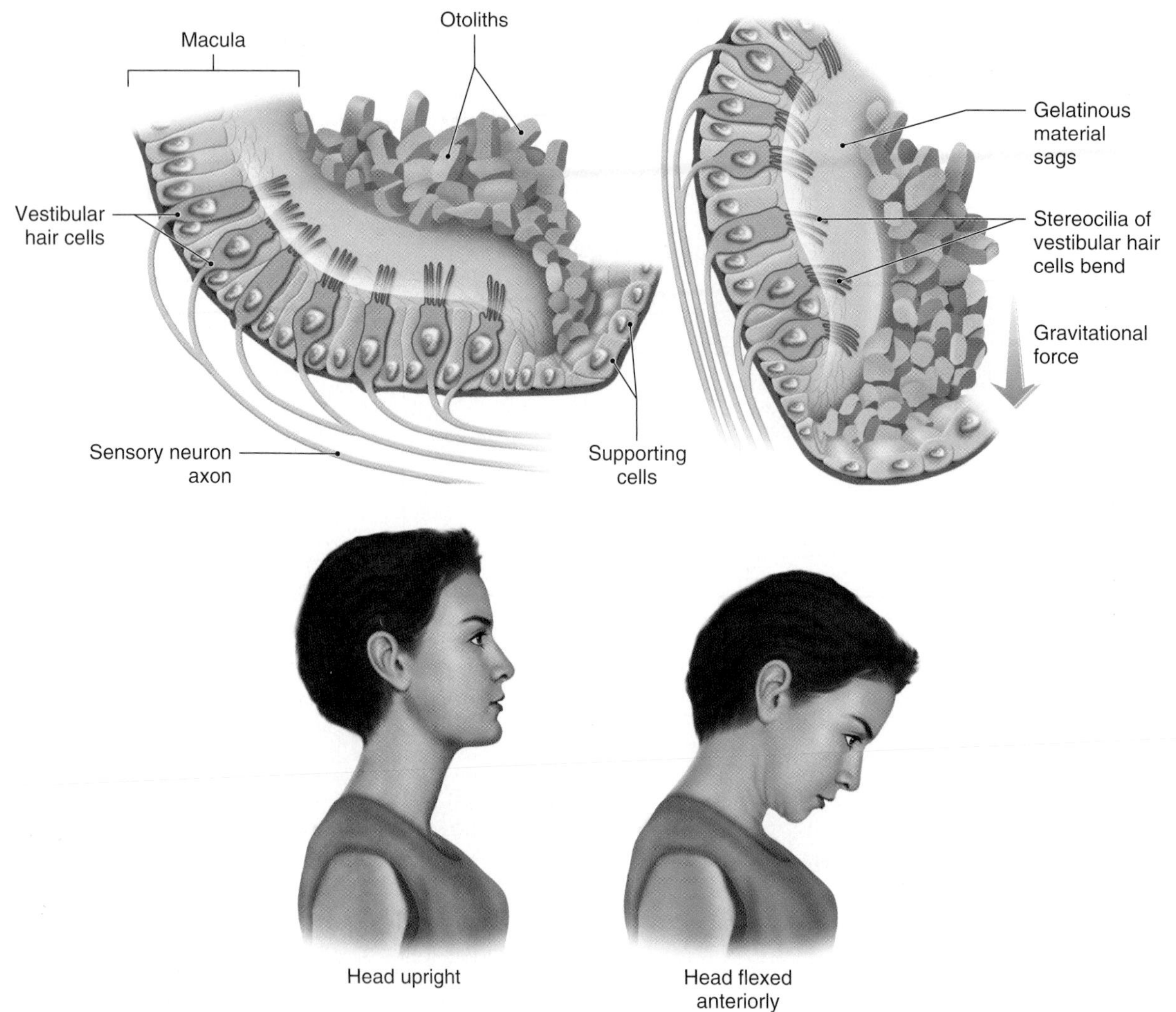

Figure 9.10 A macula is a sensory receptor for static equilibrium. Note how flexion of the head causes bending of the stereocilia of vestibular hair cells.

The membranous labyrinths of the **semicircular canals** contain the sensory receptors that detect rotational motion of the head. Examine figure 9.7 and note that the semicircular canals are arranged at right angles to each other so that each occupies a different plane in space, roughly equal to the frontal, sagittal, and transverse planes.

Near the attachment of each membranous canal to the utricle is an enlarged region called the **ampulla** (am-pˉul′-lah). Each ampulla contains a sensory organ for dynamic equilibrium called the **crista ampullaris** (kris′-ta am-pul-lar′-is). Each crista ampullaris contains a number of vestibular hair cells, whose stereocilia extend into a dome-shaped gelatinous mass called the *ampullary cupula.* Axons of the vestibular branch of the vestibulocochlear nerve lead from the vestibular hair cells to the brain (figure 9.11).

The mechanism for detecting rotational movement may be described as follows:

1. When the head is turned, the endolymph pushes against the ampullary cupula, bending the stereocilia of vestibular hair cells, which stimulates the formation of nerve impulses. The nerve impulses are carried to the brain via the vestibular branch of the vestibulocochlear nerve.
2. Because each semicircular canal is oriented in a different plane, the vestibular hair cells of the cristae are not stimulated equally with a given head movement. Thus, the brain receives a different pattern of nerve impulses for each type of head movement.
3. The cerebellum uses the nerve impulses to make adjustments below the conscious level to maintain dynamic equilibrium.
4. Awareness of rotational movement, or lack of it, results from the cerebrum interpreting the pattern of nerve impulses it receives (figure 9.11).

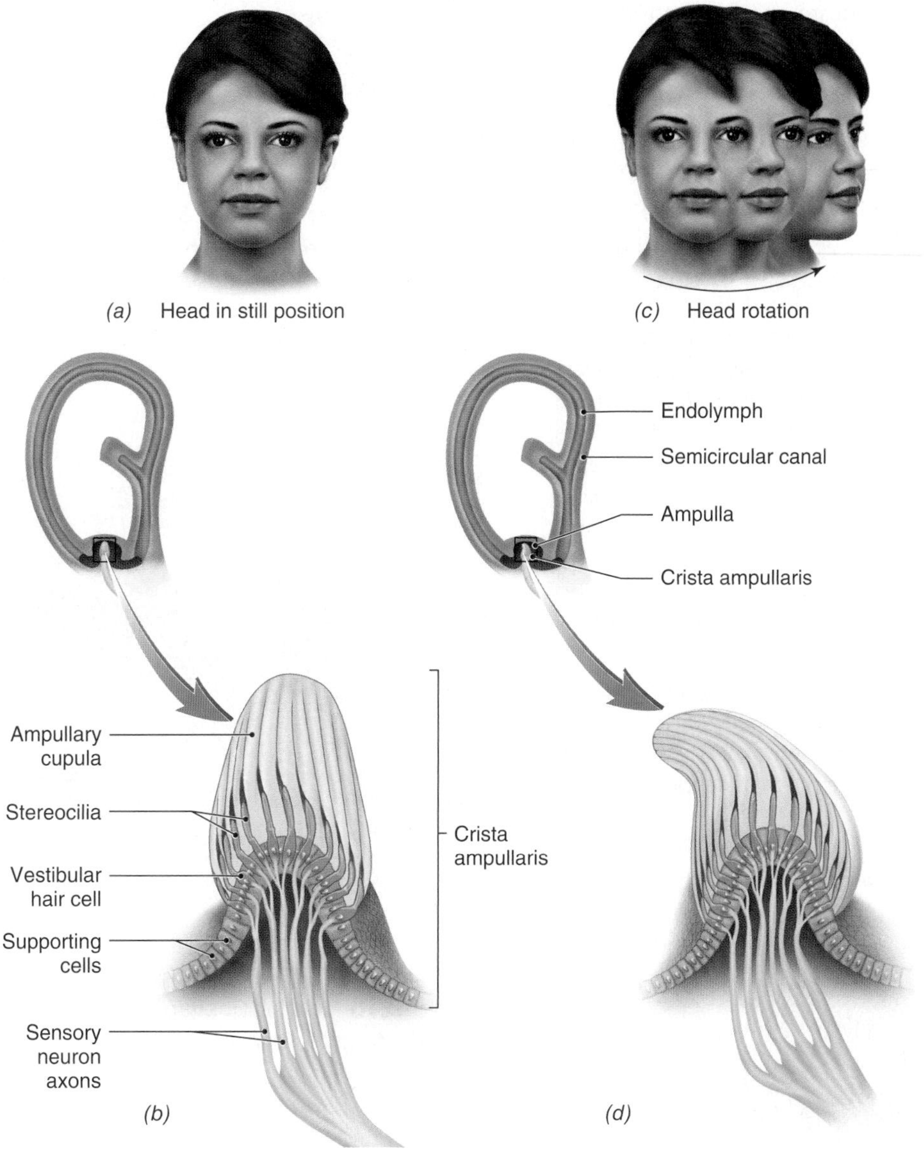

Figure 9.11 Detection of Rotational Movement.
(a) When the head is upright and stationary, *(b)* the crista ampullaris is upright. *(c)* When the head is rotated, *(d)* the endolymph bends the cupula in the opposite direction, stimulating the vestibular hair cells to form nerve impulses.

 Check My Understanding

8. What structures are involved in static and dynamic equilibrium?
9. What are the mechanisms of static and dynamic equilibrium?

Vision

Vision is one of the most important senses supplying information to the brain. The sensory receptors for light stimuli are located within the *eyes* (or *eyeballs*), the organs of vision. The eyes are located within the *orbits*, where they are protected by seven skull bones (see chapter 6). Connective tissues provide support and protective cushioning for the eyes.

Eyelids, Eyelashes, and Eyebrows

The exposed anterior surface of the eye is protected by the *eyelids*. Blinking spreads tears and mucus over the anterior eye surface to keep it moist. The internal surface of each eyelid is lined with a mucous membrane called the **conjunctiva** (kon-junk-tī′-vah), which continues across the anterior surface of the eye. Only its transparent superficial epithelium covers the cornea. Mucus from the conjunctiva helps to lubricate the eye and keep it moist. The conjunctiva also contains many blood vessels and nociceptors.

Eyelashes help to keep airborne particles from reaching the eye surface and provide some protection from excessive light. *Eyebrows,* located on the brow ridges, also shield the eyes from overhead light and divert sweat from the eyes. Observe the accessory structures in figure 9.12.

Lacrimal Apparatus

The **lacrimal** (lak′-ri-mal) **apparatus,** shown in figure 9.13, is involved in the production and removal of tears. Tears are secreted continuously by the **lacrimal gland,** which is located in the superior, lateral part of each orbit. Tears are carried to the surface of the eye by a series of tiny *excretory ducts.* The tears flow inferiorly and medially across the eye surface as they are spread over the eye surface by blinking. Once collected at the medial corner of the eye by the *lacrimal canaliculi,* tears flow into the *lacrimal sac,* and flow on through the *nasolacrimal duct* into the nasal cavity.

Tears perform an important function in keeping the anterior surface of the eye moist and in washing away foreign particles. An antibacterial enzyme (lysozyme) in tears helps to reduce the chance of eye infections.

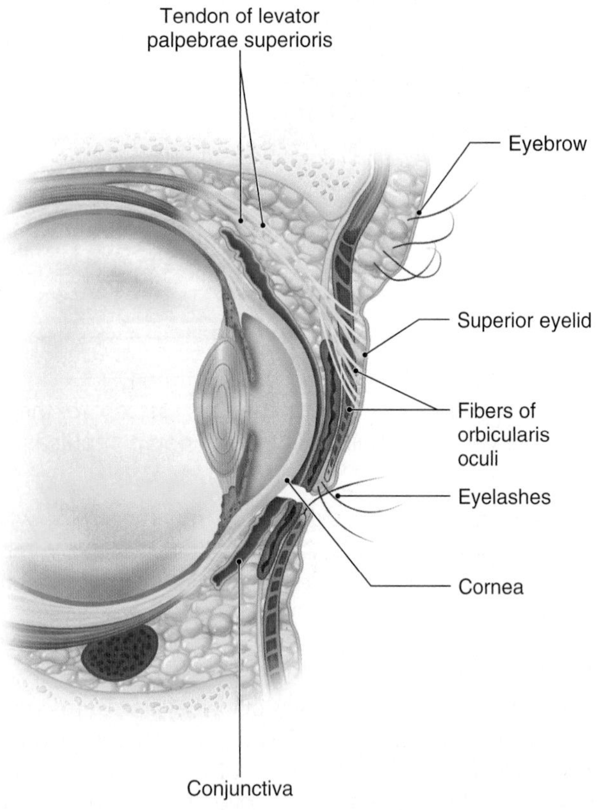

Figure 9.12 Accessory structures and the anterior portion of the eye as shown in a sagittal section. APIR

Extrinsic Muscles

Movement of the eyes must be precise and in unison to enable good vision. Each eye is moved by six **extrinsic muscles of the eyeball** that originate from the posterior of the orbit and insert on the surface of the eye. Four muscles exert a direct pull on the eye, but two muscles pass through cartilaginous loops, enabling them to exert an oblique pull on the eyeball. Although each muscle has its own action, these muscles function as a coordinated group to enable eye movements. The locations and functions of these muscles are shown in figure 9.14.

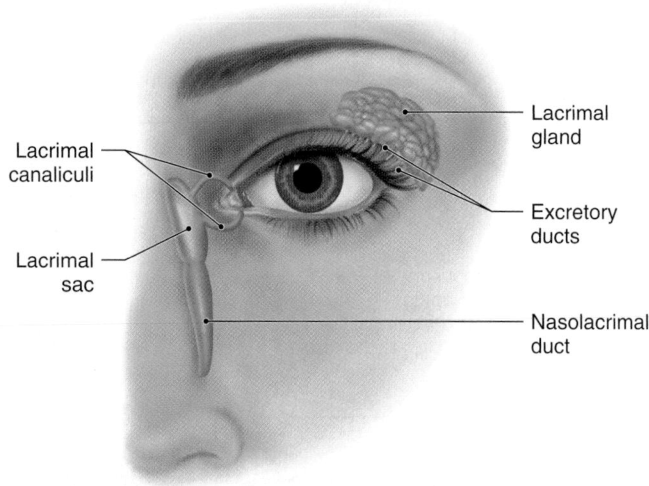

Figure 9.13 The lacrimal apparatus consists of a tear-secreting lacrimal gland and a series of ducts.

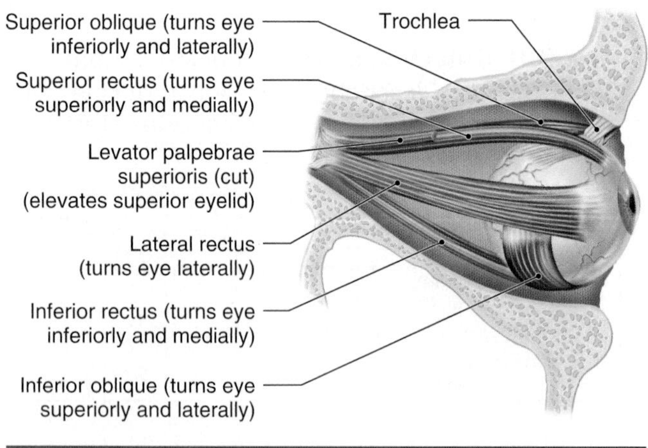

Figure 9.14 Six extrinsic muscles move the eyeball, and the levator palpebrae superioris raises the superior eyelid. The medial rectus, which is not shown in the image, turns the eye medially. APIR

Check My Understanding

10. What are the functions of the accessory structures of the eye?
11. Why does crying leads to a runny nose?

Structure of the Eye

The eye is a hollow, spherical organ about 2.5 cm (1 in) in diameter. It has a wall composed of three layers and internal spaces filled with fluids that support the walls and maintain the shape of the eye. The major parts of the eye are shown in figure 9.15.

External Layer The external layer of the eye consists of two parts: the sclera and the cornea. The **sclera** (skle′-rah) is the opaque, white portion of the eye that forms most of the external layer. The sclera is a tough, fibrous layer that provides protection for the delicate internal portions of the eye and for the optic nerve (CN II), which emerges from the posterior portion of the eye. The anterior portion of the sclera is covered by the conjunctiva. The **cornea** (kor′-nē-ah) is the anterior clear window of the eye. It has a greater convex curvature than the rest of the eyeball so that it can bend light rays as they pass through it. It lacks blood vessels and nerves that would block light rays from entering the eye.

Middle Layer The middle layer includes the choroid, ciliary body, and iris. The **choroid** (kō′-roid), which is found in all but the anteriormost portion of the layer, contains blood vessels that nourish the eye and large amounts of melanin. The absorption of light by melanin prevents back-scattering of light, which would impair vision. The **ciliary** (sil′-ē-ar-ē) **body** contains the ciliary muscles and forms a ring around the **lens** just anterior to the choroid. The *ciliary zonule* contains fibrous strands that extend from the ciliary body to the lens and hold the lens in place. Contraction and relaxation of ciliary muscles change the shape of the lens.

Although entering light rays are bent by the cornea, it is the lens that focuses light rays precisely on the retina. The transparent, somewhat elastic lens is composed of protein fibers and lacks blood vessels and nerves that would

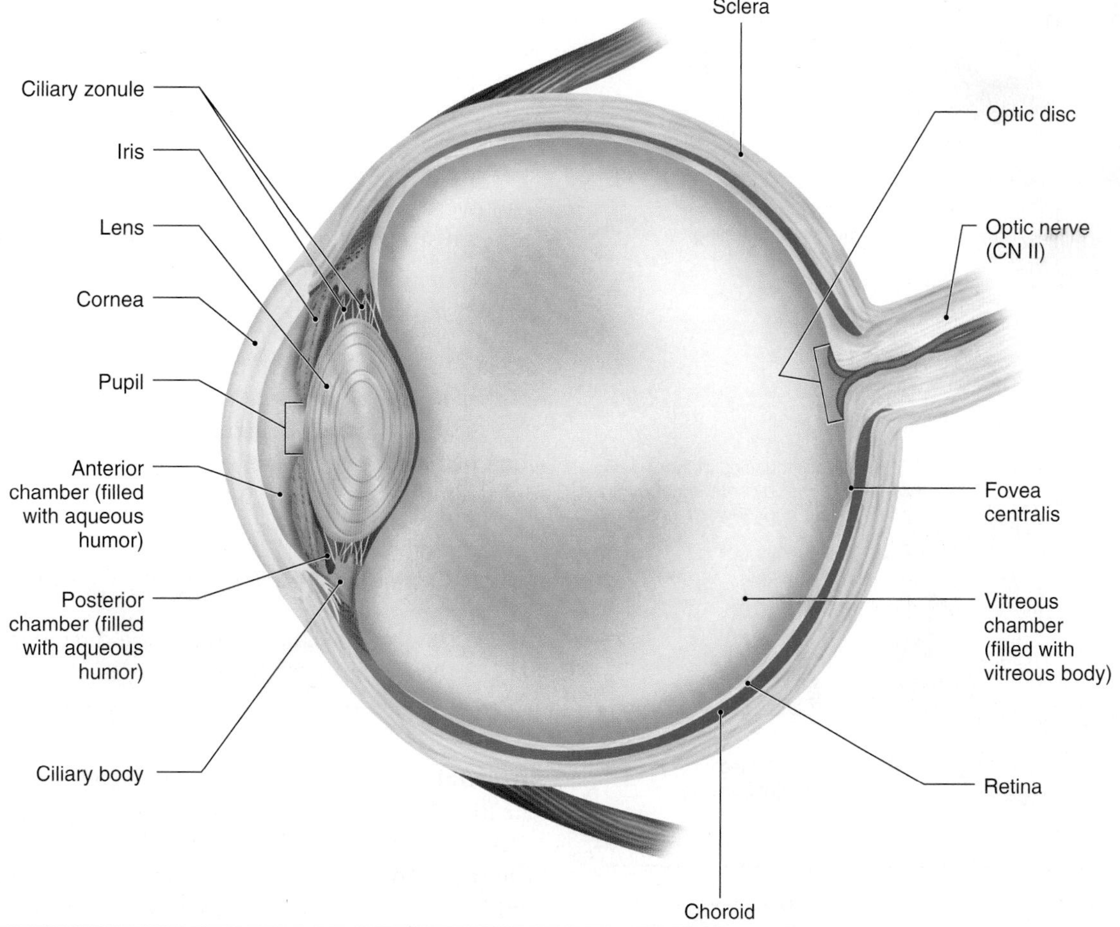

Figure 9.15 The structure of the left eye in transverse section. APR

block the passage of light rays. Contraction and relaxation of the ciliary muscles change the shape of the lens in a process called **accommodation** (figure 9.16). Contraction of the ciliary muscles relaxes the fibrous strands in the ciliary zonule and allows the lens to become more spherical in shape. The relaxation of the ciliary muscles increases tension on the fibrous strands of the ciliary zonule and causes the lens to take on a more flattened shape. In this way, the shape of the lens is adjusted for distant, intermediate, and near vision so that the image is focused precisely on the retina.

The colored portion of the eye is the **iris,** a thin disc of connective tissue and smooth muscle that extends from the ciliary body anterior to the lens. The iris controls the amount of light entering the eye by controlling the size of the pupil. The **pupil** is the opening in the center of the iris through which light passes to the lens. Its size is constantly adjusted by the iris as lighting conditions change. The pupil is constricted in bright light and is dilated in dim light.

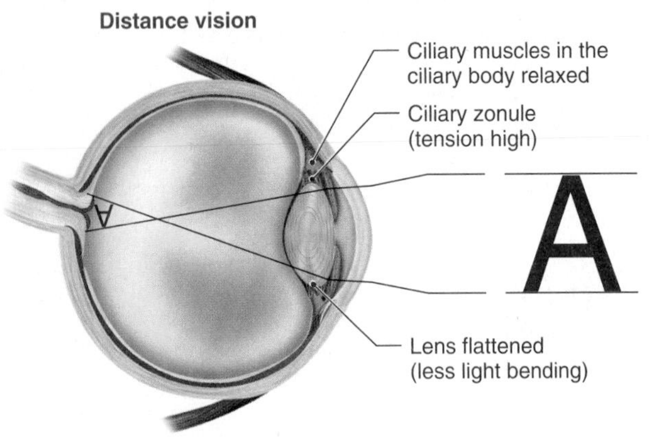

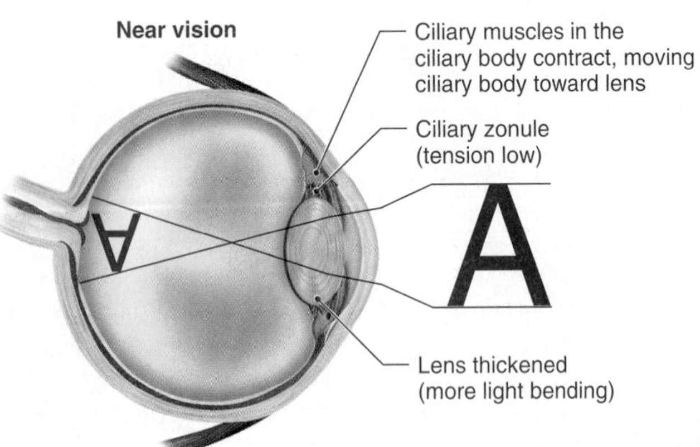

Figure 9.16 Focusing of light rays on the retina by the lens in distance vision and near vision. Note how the lens changes shape in accommodation.

Internal Layer The filmlike **retina** (ret′-i-nah) lines the internal surface of the eye posterior to the ciliary body. The retina contains two types of photoreceptor cells: rods and cones. Rod and cone anatomy can be seen in figure 9.17. The thin, elongate **rods** are photoreceptors for black and white vision because they are sensitive only to the presence of light. The shorter and thicker **cones** are photoreceptors for color vision. Because cones require bright light to function, only rods allow us to see in dim light.

The **macula** (mak′-u-lah) is a yellowish disc on the retina directly posterior to the lens. In the center of the macula, there is a small depression called the **fovea centralis** (fo′-ve-ah sen-trah′-lis). The fovea centralis contains densely packed cones, making it the area for the sharpest color vision. The density of the cones decreases with increased distance from the fovea. Rods, which are absent from the fovea, increase in density with increased distance from the fovea. Therefore, dim-light vision is best at the edge of the visual field (figure 9.18; see figure 9.15). It is important to remember that the macula on the retina is structurally and functionally different from the maculae within the internal ear.

The retina contains neurons in addition to rods and cones (see figure 9.17). Nerve impulses formed by rods and cones are transmitted to *retinal ganglion cells,* whose axons converge at the **optic disc** to form the optic nerve. The optic disc is located medial to the fovea. Because the optic disc lacks photoreceptors, it is also known as the "blind spot." However, we usually do not notice a blind spot in our field of vision because the visual fields of our eyes overlap.

An artery enters the eye and a vein exits the eye via the optic disc. These blood vessels are continuous with capillaries that nourish the internal tissues of the eye and are the only blood vessels in the body that can be viewed directly. A special instrument called an *ophthalmoscope* (of-thal′-mō-skōp) is used to look through the lens and observe these vessels. Figure 9.18 shows the appearance of blood vessels and the retina as viewed with an ophthalmoscope.

Check My Understanding

12. How are the components of the three layers of the eye involved in vision?

Internal Cavities The space between the cornea and the iris is known as the **anterior chamber,** which is filled with a watery fluid called **aqueous** (ā′-kwē-us) **humor.** The small **posterior chamber,** located between the iris and lens, is also filled with aqueous humor. The aqueous humor is filtered out of capillaries in the ciliary body, flows through the posterior chamber into the anterior

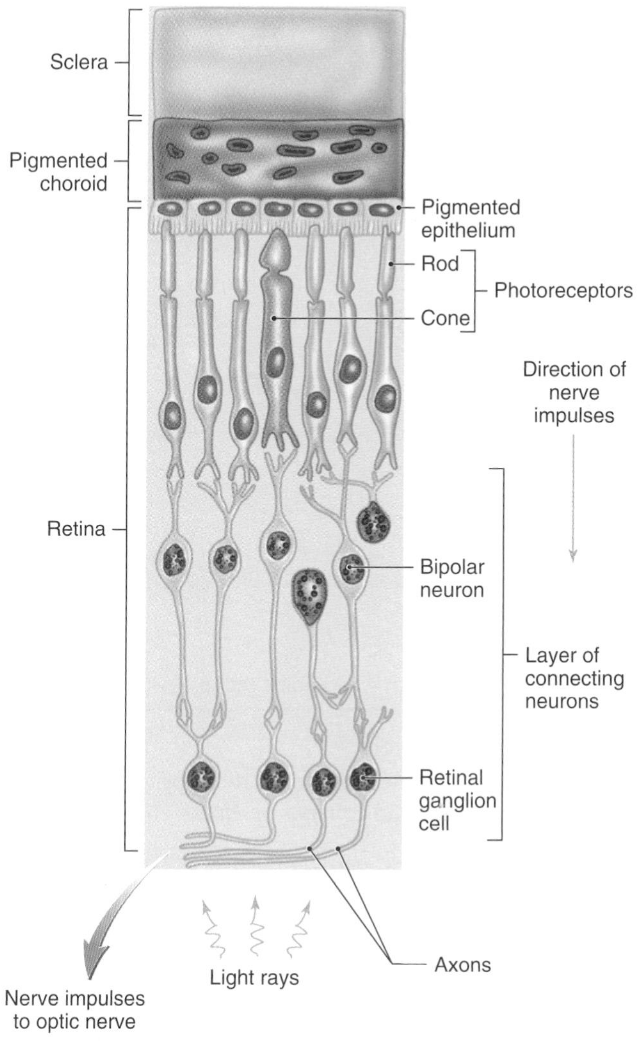

Figure 9.17 The retina consists of several cell layers. APIR

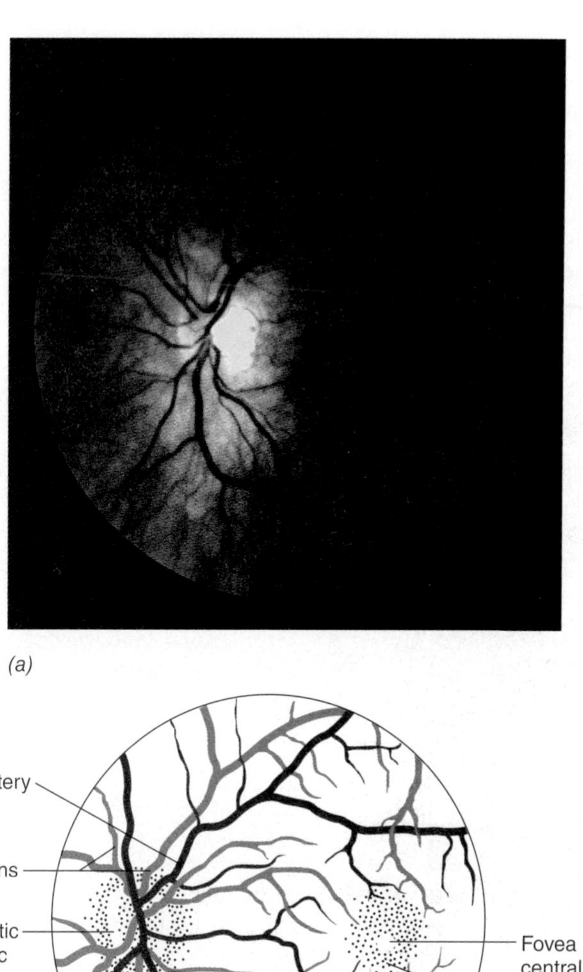

Figure 9.18 *(a)* A photo of the retina and *(b)* a diagram of the retina showing the optic disc and fovea centralis. Blood vessels enter and exit the eye at the optic disc. Axons exit the eye at the optic disc to form the optic nerve. The fovea centralis contains densely packed cones for direct color vision. APIR

chamber, and is reabsorbed into blood vessels located at the junction of the sclera and cornea. Aqueous humor is largely responsible for the internal pressure within the eye and the normal shape of the cornea. The aqueous humor also provides nourishment to the cornea and lens. Normally, it is secreted and absorbed at the same rate so that intraocular pressure is maintained at a constant level.

The large **vitreous chamber** is located posterior to the lens. It is filled with a clear, gel-like substance called the **vitreous** (vit′-rē-us) **body.** The vitreous body, which forms during embryonic development, is not reabsorbed or regenerated. The vitreous body presses the retina firmly against the wall of the eye and helps to maintain the shape of the eye.

Table 9.3 summarizes the functions of eye structures.

Physiology of Vision

Light rays coming to the eye must be precisely bent so that they are focused on the retina. This bending of the light rays is called *refraction* (rē-frak′-shun) and it is produced by the cornea and lens. The convex surface of the cornea produces the greatest refraction of light rays, while further bending (accommodation) by the lens provides a "fine adjustment" so that the image is focused precisely on the retina.

The optics of the eye cause the image to be inverted on the retina, as shown in figure 9.16. However, the visual

Table 9.3 Functions of Eye Structures

Structure	Function
External Layer	
Sclera	Provides protection and shape for eye
Cornea	Allows entrance of light and bends light rays
Middle Layer	
Choroid	Contains blood vessels that nourish deep structures and melanin that absorbs excessive light
Ciliary body	Supports and changes shape of lens in accommodation; secretes aqueous humor
Iris	Regulates amount of light entering eye by controlling the size of the pupil
Internal Layer	
Retina	Contains photoreceptors that convert light rays into nerve impulses; nerve impulses transmitted to brain via optic nerve (CN II)
Other Structures	
Lens	Bends light rays and focuses them on the retina
Anterior chamber	Contains the aqueous humor that controls intraocular pressure, maintains the shape of the cornea, provides nourishment to cornea and lens
Posterior chamber	Receives the aqueous humor produced by the ciliary body
Vitreous chamber	Contains the vitreous body that maintains shape of the eye and holds retina against choroid

Clinical Insight

Glaucoma results when the rate of absorption of aqueous humor is less than its rate of secretion. This causes a buildup of intraocular pressure that, without treatment, can compress and close the blood vessels nourishing the photoreceptors of the retina. If this occurs, the photoreceptors die and permanent blindness results.

areas of the cerebral cortex correct for this inversion so that objects are seen in their correct orientation. When images are incorrectly focused on the retina, poor vision results. Figure 9.19 shows common optical disorders and how they may be corrected with glasses, contact lenses, or Lasik surgery.

When light rays strike the retina, the light stimuli must be converted into nerve impulses that are sent to the brain. Both rods and cones contain light-sensitive pigments that break down into simpler substances when light is absorbed. The breakdown of these pigments results in the formation of nerve impulses.

Rods contain a light-sensitive pigment called **rhodopsin** that breaks down into *opsin,* a protein, and *retinal,* which is derived from vitamin A. This breakdown triggers the formation of nerve impulses that are carried via the optic nerve to the brain. Rhodopsin is resynthesized from opsin and retinal to prepare the rods for receiving subsequent stimuli. A deficiency of vitamin A may result in an insufficient amount of rhodopsin in the rods, which, in turn, may lead to *night blindness,* the inability to see in dim light.

Although the light-sensitive pigments are different in cones, they function in a similar way to rhodopsin. There are three different types of cones, and each has a pigment that responds best to a different color (wavelength) of light. One type responds best to red light, another type responds best to green light, and the third type responds to blue light. The perceived color of objects results from the combination of the cones that are stimulated and the interpretation of the nerve impulses that they form by the cerebral cortex.

Nerve Pathway Nerve impulses formed by the photoreceptors are transmitted via axons of the optic nerve to the brain. The optic nerves merge just anterior to the pituitary gland to form an X-shaped pattern called the **optic chiasma** (kī′-as-mah) (figure 9.20). Within the optic chiasma, the axons from the medial half of the retina in each eye cross over to the opposite side. Thus, the medial axons of the left eye and the lateral axons of the right eye form the right optic tract leaving the optic chiasma. Similarly, the medial axons of the right eye and the lateral axons of the left eye form the left optic tract leaving the optic chiasma. The axons of the optic tracts enter the thalamus, where they synapse with neurons that carry the nerve impulses on to the visual areas of the occipital lobes.

The crossing of the medial axons results in each visual area receiving images of the entire object but from

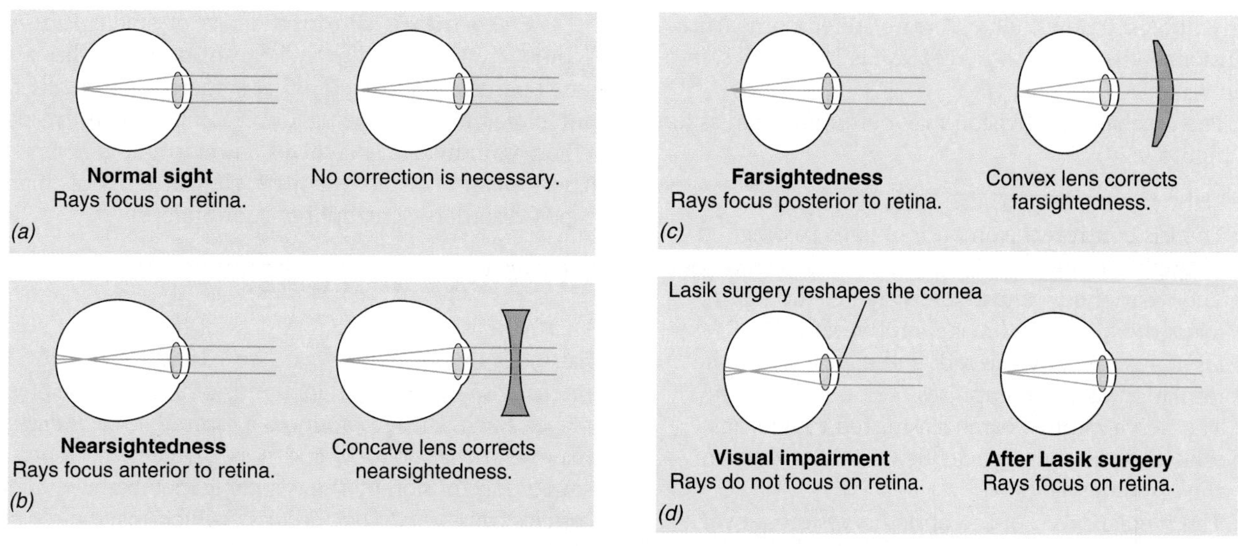

Figure 9.19 Comparison of *(a)* normal sight, *(b)* nearsightedness *(c)* farsightedness, and *(d)* eyesight with Lasik surgery.

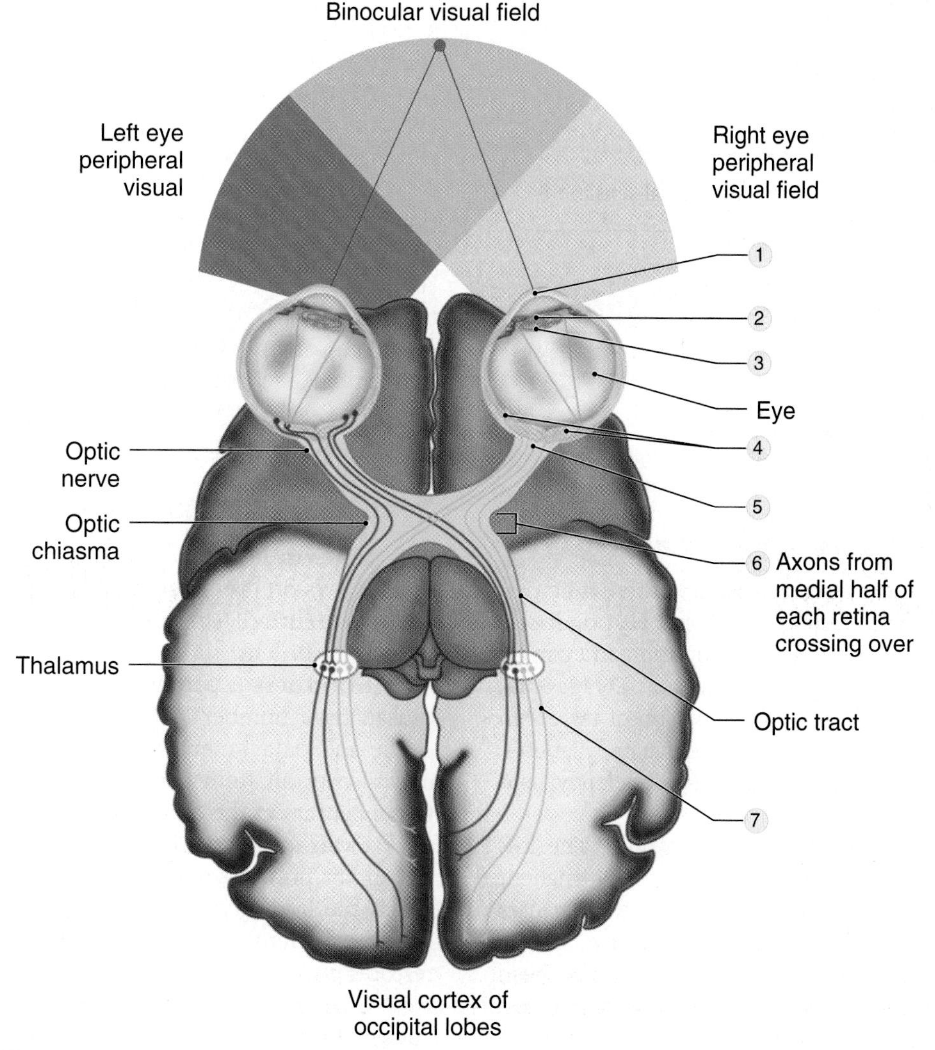

Figure 9.20 The Optic Nerve Pathway. AP|R

slightly different views, which create stereoscopic (three-dimensional) vision. Depth perception is a result of stereoscopic vision.

The mechanism of vision may be summarized as follows (figure 9.20):

1. Light rays are bent as they pass through the cornea.
2. The iris controls the amount of light passing through the pupil.
3. The ciliary body adjusts the shape of the lens to focus the light rays (image) on the retina.
4. Light absorbed by the rods and cones causes the formation of nerve impulses.
5. The nerve impulses are transmitted to neurons whose axons converge at the optic disc to form the optic nerve.
6. The medial axons of the optic nerves cross over at the optic chiasma and merge with the lateral axons on that side to form the optic tracts, which continue to the thalamus.
7. The nerve impulses are then carried to the vision areas in the occipital lobes of the cerebrum, where they are interpreted as visual images.

Check My Understanding

13. How are light rays converted into visual sensations?

9.4 Disorders of The Special Senses

Learning Objectives

11. Describe the common disorders of taste, smell, hearing, and vision.

Disorders of Taste and Smell

Ageusia is a loss of taste function, meaning there is no perception of the five basic tastes, and is rare. **Hypogeusia,** or a reduced ability to taste, is more common and can be caused by zinc deficiency and chemotherapy. **Dysgeusia,** which is a distortion or impaired perception of taste, can be caused by taste bud distortion, pregnancy, diabetes, allergy medications like albuterol, zinc deficiency, and chemotherapy.

Anosmia is the inability to detect odor. The loss can be for one odor or all odors. It may also be permanent or temporary depending upon the cause. Typical causes are inflammation of the nasal mucosa, blockage of the nasal pathways, damage to the olfactory nerve, or head trauma leading to temporal lobe damage. **Hyposmia** is a decrease in the ability to detect odors. Hyposmia is common with advanced age due to a decrease in olfactory epithelium regeneration or smoking.

Dysosmia is distorted sense of smell. *Parosmia,* a type of dysosmia, occurs when an individual has altered smell perception, meaning that something normally pleasant is perceived as being unpleasant. *Phantosmia* occurs when an individual perceives an odor that is not present. These phantom smells can be clinical signs of migraine, mood disorders, schizophrenia, or epilepsy.

Disorders of the Ear

Deafness is a partial or total loss of hearing. The cochlear hair cells of the spiral organ are easily damaged by high-intensity sounds, such as loud music and the noise of jet airplanes. Such damage produces a form of *nerve deafness* that may be partial or total, and it is permanent. Disorders of sound transmission by the tympanic membrane or auditory ossicles cause *conduction deafness,* which may be repairable by surgical means or overcome by the use of hearing aids.

Labyrinthine disease is a term applied to disorders of the internal ear that produce symptoms of dizziness, nausea, ringing in the ears (tinnitis), and hearing loss. It may be caused by an excess of endolymph, infection, allergy, trauma, circulation disorders, or aging.

Motion sickness is a functional disorder that is characterized by nausea and is produced by repetitive stimulation of the equilibrium receptors in the internal ear.

Otitis media (ō-tī′-tis mē′-dē-ah) is an acute infection of the tympanic cavity. It may cause severe pain and an outward bulging of the tympanic membrane due to accumulated fluids. Pathogens enter the middle ear from the pharynx via the auditory tube or through a perforated tympanic membrane. Young children are especially susceptible because their auditory tubes are short and horizontal, which aids the spread of bacteria from the pharynx to the tympanic cavity. AP|R

Disorders of the Eye

Astigmatism (a-stig′-mah-tizm) is the unequal focusing of light rays on the retina, which causes part of an image to appear blurred. It results from an unequal curvature of the cornea or lens.

Blindness is partial loss or lack of vision. It may be caused by a number of disorders such as cataract, glaucoma, and detachment or deterioration of the retina. It may also result from damage to the optic nerves or the visual centers in the occipital lobes of the cerebrum.

Cataract is cloudiness or opacity of the lens, which impairs or prevents vision. It is common in older people and is the leading cause of blindness. Surgical removal of the clouded lens and implantation of a plastic lens usually restores good vision.

Color blindness is the inability to perceive certain colors or, more rarely, all colors. Red-green color blindness, the most common type, is characterized by difficulty distinguishing reds and greens due to the absence

Clinical Insight

A Caucasian male, age 63, presented for a routine eye exam with complaints of a blurry spot in his central vision and difficulty reading in dim light. Visual observation of his retinas with pupil dilation showed several medium-sized spots of drusen (yellow-white fatty protein deposits) in the macula of both eyes. The patient's tentative diagnosis was intermediate stage dry **age-related macular degeneration (AMD).** AMD manifests in two forms: *wet (neovascular) macular degeneration* and *dry (non-neovascular) macular degeneration.* Dry AMD in the early stages involves the formation of small drusen in the maculae and is usually asymptomatic. Drusen number and size increases during the intermediate stage, which often leads to blurry central vision and a need for brighter light to perform visual tasks (figure 9A). Wet AMD involves the growth of new vessels into the maculae that break easily and allow for fluid leakage. The resulting inflammation damages the maculae and causes loss of central vision. A fluorescein angiogram, using a fluorescent dye injected into the bloodstream to photograph retinal blood vessels, detected no abnormal vessel growth in the maculae. Because dry AMD can spontaneously become wet AMD, the man was instructed to monitor his condition daily with an *Amsler Grid.* The presence of wavy or blurred lines when viewed is an indication of the development of wet AMD. There is no cure currently for dry AMD. However, the man was placed on an AREDS Formula Eye Vitamin regime. The high levels of vitamin A, vitamin C, vitamin E, zinc (zinc oxide), and copper (cupric oxide) in the supplement have been shown to decrease the risk of developing advanced AMP (dry or wet) by as much as 25%.

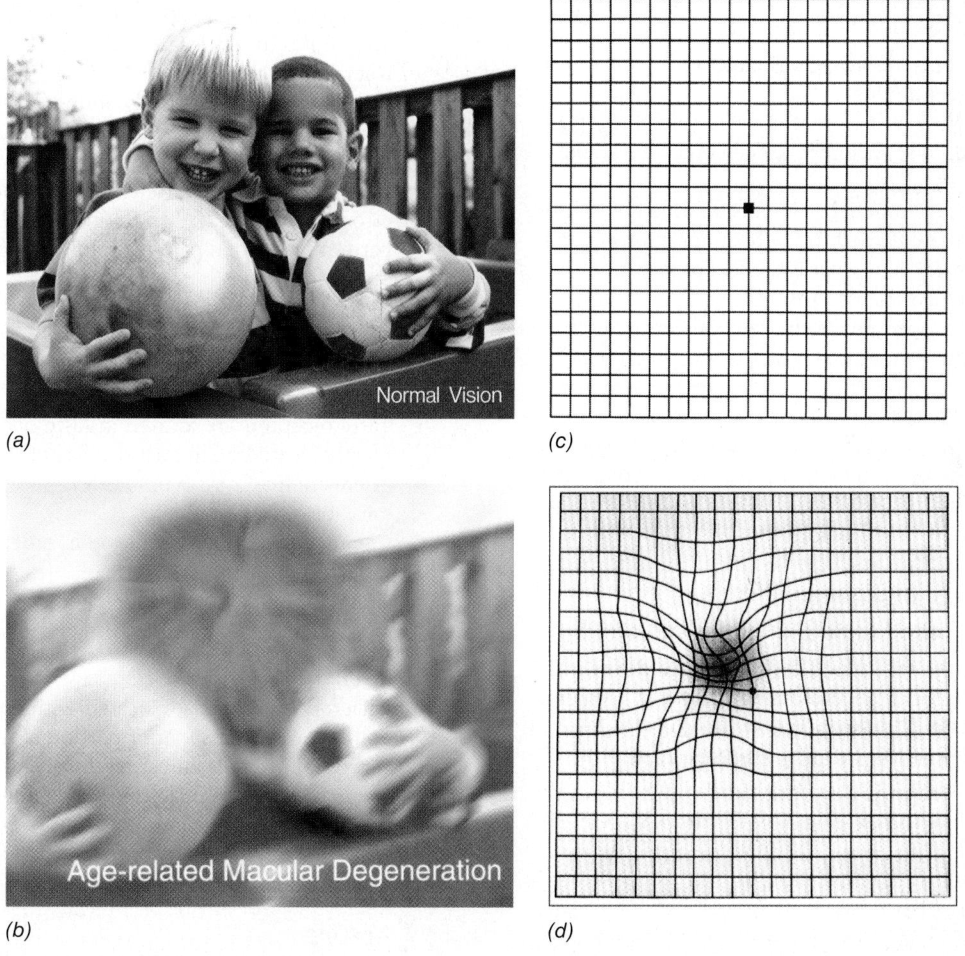

Figure 9A Macular Degeneration.
(a) Healthy vision. *(b)* Vision with macular degeneration. Appearance of Amsler grid with *(c)* healthy vision and *(d)* macular degeneration.

of either red or green cones. Color blindness is inherited, and it occurs more often in males than in females because it is a sex-linked hereditary trait.

Conjunctivitis (con-junk-ti-vī′-tis) is inflammation of the conjunctiva. It may be caused by allergic reactions, physical or chemical causes, or infections. Inflammation that results from a bacterial or viral infection is commonly called *pink eye*. Such bacterial infections are highly contagious.

Farsightedness (hyperopia) is blurred vision caused by light rays being incorrectly focused posterior to the retina. Its causes include lens abnormalities and the eye being shorter than normal.

Nearsightedness (myopia) is blurred vision caused by light rays being incorrectly focused anterior to the retina. Its causes include lens abnormalities and the eye being longer than normal.

Presbyopia (prez-bē-ō′pē-ah) is the diminished ability of the lens to accommodate for near vision due to a decrease in its elasticity. It is a natural result of aging. At age 20, an object can be clearly observed about 10 cm (4 in) from the eye. At age 60, an object must be about 75 cm (30 in) from the eye to be clearly observed.

Retinoblastoma (ret-i-nō-blas-tō′-mah) is a cancer of immature retinal cells. It constitutes about 2% of the cancers in children.

Strabismus (strah-biz′-mus) is a disorder of the extrinsic eye muscles in which the eyes are not directed toward the same object simultaneously. Treatment may include eye exercises, corrective lenses, or corrective surgery.

Chapter Summary

9.1 Sensations

- Each type of sensory receptor is most sensitive to a particular type of stimulus.
- The five types of sensory receptors are thermoreceptors, mechanoreceptors, nociceptors, chemoreceptors, and photoreceptors.
- Sensations result from nerve impulses formed by sensory receptors that are carried to the brain for interpretation.
- Perception is the conscious awareness of sensation that is created by the cerebral cortex.
- A particular sensory area of the cerebral cortex interprets all nerve impulses that it receives as the same type of sensation.
- The brain projects the sensation back to the body region from which the nerve impulses seem to have originated.
- Sensory receptors for touch, pressure, warm, cold, taste, and smell adapt to repetitious stimulaton by decreasing the rate of nerve impulse formation.

9.2 General Senses

- Thermoreceptors are located in the dermis of the skin. Warm receptors are located deeper than cold receptors.
- Pressure receptors are located in tendons and ligaments of joints and deep in the dermis of the skin.
- There are four types of touch receptors. Tactile corpuscles in the superficial dermis are important for detecting the onset of light touch in hairless areas. The hair root plexus detects movement of a hair follicle. Tactile discs and tactile cells work together to detect light touch, which is necessary for reading braille. Free nerve endings have a secondary role in detecting touch sensations.
- Baroreceptors detect stretching in distensible internal organs to regulate activities such as blood pressure and urination.
- Proprioceptors in skeletal muscles and tendons are involved in the maintenance of erect posture and muscle tone.
- Chemoreceptors are specialized neurons that monitor body fluids for chemical changes.
- Nociceptors are free nerve endings that detect painful stimuli. They are especially abundant within the skin and visceral organs and exhibit no adaptation as a protective mechanism.
- Referred pain is pain from visceral organs that is erroneously projected to the body wall or limbs. The pattern of referred pain is useful in diagnosing disorders of visceral organs.

9.3 Special Senses

Taste

- Taste receptors are located in taste buds, which are mostly located on lingual papillae of the tongue.
- There are five basic types of taste receptors: sour, sweet, salty, bitter, and umami.
- Chemicals must be in solution in order to stimulate the taste receptors.
- Nerve impulses from taste receptors are carried by the facial, glossopharyngeal, and vagus nerves to the brain.

Smell

- Olfactory receptors are located within the olfactory epithelium of the superior portion of the nasal cavity.
- Airborne molecules must be dissolved in the mucus layer covering the olfactory epithelium to stimulate the olfactory receptors.
- Nerve impulses are carried by the olfactory nerves to the olfactory bulbs and then by the olfactory tracts to the brain.

Hearing

- The ear is subdivided into the external, middle, and internal ear.
- The external ear consists of (a) the auricle, which directs sound waves into the external acoustic meatus; and (b) the external acoustic meatus, which channels sound waves to the tympanic membrane.

- The middle ear, or tympanic cavity, consists of (a) the tympanic membrane, which vibrates when struck by sound waves; (b) the auditory tube, which enables the equalization of air pressure on each side of the tympanic membrane; and (c) the auditory ossicles, which transmit vibrations of the tympanic membrane to the oval window of the internal ear.
- The internal ear is embedded in the temporal bone. It consists of (a) the membranous labyrinth, which is filled with endolymph, and lies within (b) the bony labyrinth, which is filled with perilymph. The major portions of the internal ear are the cochlea, vestibule, and semicircular canals.
- The cochlear duct, part of the membranous labyrinth, is bordered by the vestibular and basilar membranes. The spiral organ on the basilar membrane within the cochlear duct contains the cochlear hair cells, which are used to detect the vibrations produced by sound waves.
- Sound waves vibrate the tympanic membrane. These vibrations are transmitted by the auditory ossicles to the oval window of the internal ear, which then sets up oscillatory movements of the perilymph in the cochlea. The movements of the perilymph vibrate the basilar membrane and spiral organ, resulting in the formation of nerve impulses.
- Nerve impulses are carried to the brain by the cochlear branch of the vestibulocochlear nerve (CN VIII).

Equilibrium

- Sensory receptors in joints, muscles, eyes, and the internal ear send nerve impulses to the brain that are associated with equilibrium.
- The maculae within the saccule and utricle contain vestibular hair cells that are sensory receptors for static equilibrium and linear acceleration, which is a type of dynamic equilibrium.
- The cristae ampullaries within the ampullae of the semicircular canals contain vestibular hair cells that are the sensory receptors for rotational movement of the head, which is a type of dynamic equilibrium.

Vision

- The eyes contain the sensory receptors for vision. Accessory organs include the extrinsic muscles of the eyeball, eyelids, eyelashes, eyebrows, and lacrimal apparatus.
- The wall of the eye is composed of three layers. The external layer consists of (a) the sclera, which supports and protects internal structures; and (b) the cornea, which allows light to enter the eye. The middle layer consists of (a) the darkly pigmented choroid, which contains blood vessels to nourish internal structures and melanin to absorb excess light; (b) the ciliary body, which changes the shape of the lens to focus light rays on the retina; and (c) the iris, which controls the amount of light entering the eye via the pupil. The internal layer consists of the retina, which contains the photoreceptors.
- Fluids fill the cavities of the eye and give it shape. Aqueous humor fills the anterior and posterior chambers and is primarily responsible for maintaining intraocular pressure. The vitreous body fills the vitreous chamber and helps to hold the retina against the choroid.
- Light rays pass through the cornea, aqueous humor, lens, and vitreous body to reach the retina. Light rays are primarily refracted by the cornea, but the lens focuses them on the retina.
- The photoreceptors are rods and cones. Rods are adapted for dim-light, black and white vision. Cones are adapted for bright-light, color vision. There are three types of cones based on the color of light that they primarily absorb: red, green, and blue.
- Nerve impulses formed by the photoreceptors are carried to the brain by the optic nerve (CN II). Axons from the medial half of each eye cross over to the opposite side at the optic chiasma. The optic tracts continue from the optic chiasma and carry nerve impulses to the thalamus, which then relays the nerve impulses to the occipital lobes of the cerebrum.
- The slightly different retinal images of each eye enable stereoscopic vision when the two images are superimposed by the brain.

9.4 Disorders of the Special Senses

- Disorders of taste and smell include ageusia, hypogeusia, dysgeusia, anosmia, hypoosmia, dysosmia, parosmia, and phantosmia.
- Disorders of the ears include deafness, labyrinthine disease, motion sickness, and otitis media.
- Disorders of the eyes include age-related macular degeneration, astigmatism, blindness, cataract, color blindness, conjunctivitis, farsightedness, nearsightedness, presbyopia, retinoblastoma, and strabismus.

Self-Review

Answers are located in appendix B.

1. Movement of a hair follicle is detected by a ______.
2. Pain originating from visceral organs but projected to parts of the body wall and limbs is called ______.
3. Sensory receptors for taste are localized in bulblike aggregations called ______.
4. ______ receptors are located in the epithelium of the superior nasal cavity.
5. The decline in nerve impulse formation when a sensory receptor is repeatedly exposed to the same stimulus is called ______.
6. The cochlea of the internal ear is involved in the sense of ______, while the ______, ______, and ______ are involved in the sense of equilibrium.
7. The membranous labyrinth is filled with a fluid called ______.

8. The sensory receptors for hearing are located in the ______, which is supported by the ______ membrane within the cochlear duct.
9. Vibrations of the tympanic membrane are transmitted to the internal ear by the ______.
10. The sensory receptors for static equilibrium are located in maculae of the ______ and ______.
11. When light rays enter the eyes, they first are refracted by the ______ and then are focused on the retina by the ______.
12. The fovea centralis contains only ______, which are sensory receptors for ______ vision.
13. Muscles within the ______ are responsible for changing the shape of the lens.
14. The retina is pressed against the choroid by the ______ in the vitreous chamber.
15. The medial axons of the optic nerve cross over at the ______.

Critical Thinking

1. Why can loud noises, such as music in a car or a jackhammer, lead to hearing loss over time?
2. Explain the benefits of sensory adaptation.
3. Free nerve endings are the most abundant sensory receptors in the body. How is this beneficial?
4. Will a hearing aid improve hearing in both neural and conduction deafness? Explain.
5. Lasik surgery, which is used to eliminate nearsightedness, changes the shape of the cornea. How does this improve vision when the problem is caused by the shape of the lens or eyeball?

ADDITIONAL RESOURCES

10 CHAPTER

Endocrine System

Katherine, an endocrinologist in Los Angeles, has just finished with her last patient of the day and is headed off to her daily yoga class. An endocrinologist is a doctor who specializes in treating patients with hormonal imbalances. As she drives through traffic, she finds it amusing that the practice of yoga is a good metaphor for the endocrine system. That must be why she enjoys it so much. The ability to successfully maintain and change yoga positions requires focused control and coordination over muscle contraction and relaxation throughout every area of the body. If balance is lost at any time, the yoga position is lost and the person will fall, even possibly become injured. The endocrine system functions in a similar fashion to maintain the body's homeostasis. Many glands work in concert, releasing hormones in precise amounts and with perfect timing, to maintain the health and balance of a human being. If even one of those hormones is produced incorrectly, the entire body can be thrust out of balance. Loss of balance within the body can be debilitating, which is why Katherine knows her medical practice provides such a valuable service.

CHAPTER OUTLINE

Module 8
Endocrine System

SELECTED KEY TERMS

Endocrine gland (endo = within; crin = secrete) A ductless gland whose secretions diffuse into the blood for distribution.
Gene expression The use of DNA to promote protein synthesis.
Hormone (hormon = to set in motion) A chemical messenger secreted by an endocrine gland.
Hypersecretion (hyper = above) Production of an excessive amount of a secretion.
Hyposecretion (hypo = below) Production of an insufficient amount of secretion.
Negative-feedback mechanism A mechanism that returns a condition to its healthy state, thereby maintaining homeostasis.
Paracrine signal (para = near) Local chemical signal that affects targets cells within the same tissue from which it is produced.
Positive-feedback mechanism A mechanism that amplifies a condition and moves it away from homeostasis.
Prostaglandin A class of chemicals that produce a response in nearby cells.
Second messenger An intracellular substance, activated by a nonsteroid hormone, that produces the specific cellular effect associated with the hormone.
Target cell A cell whose functions are affected by a specific hormone.

TWO INTERRELATED REGULATORY SYSTEMS coordinate body functions and maintain homeostasis: the nervous system and the *endocrine* (en′-do-krin) *system.* Unlike the almost instantaneous coordination by the nervous system, the endocrine system provides slower but longer-lasting coordination. The endocrine system consists of cells, tissues, and organs, collectively called **endocrine glands,** that secrete **hormones** (chemical messengers) into the interstitial fluid. The hormones then pass into the blood for transport to other tissues and organs, where they alter cellular functions (figure 10.1). In contrast, exocrine gland secretions are carried from the gland by a duct (tube) to an internal or external surface.

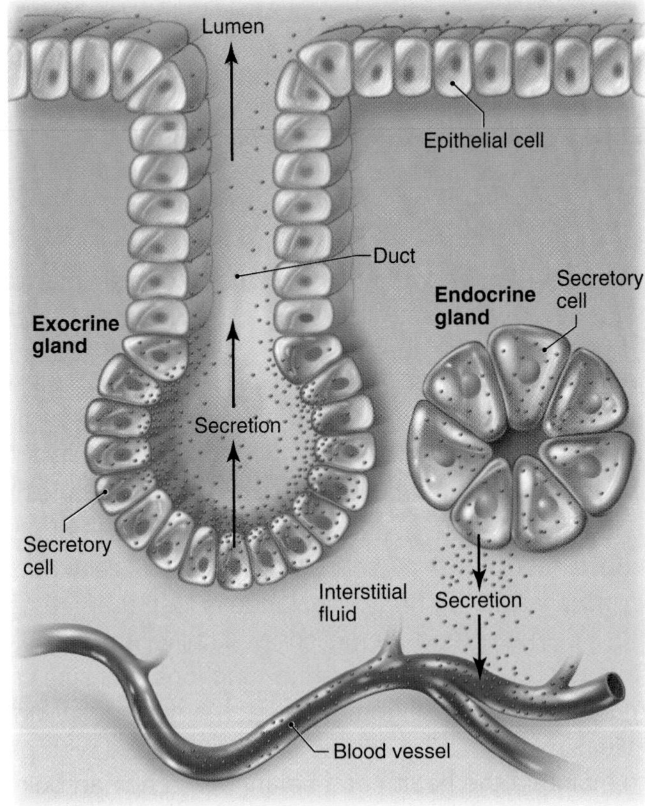

Figure 10.1 Exocrine gland and endocrine gland compared.

10.1 The Chemical Nature of Hormones

Learning Objectives

1. Distinguish between endocrine and exocrine glands.
2. Distinguish between neurotransmitters, paracrine signals, and hormones.
3. Explain the three negative-feedback mechanisms that control hormone secretion.
4. Compare the mechanisms of action of steroid and nonsteroid hormones.

There are various modes of communication utilized within the human body (figure 10.2). Neural communication, which was described in chapter 8, uses the release of neurotransmitters at a synapse to transmit a signal from a neuron to another cell. Paracrine communication involves the release of **paracrine signals,** also referred to as "local hormones." Paracrine signals are released within a tissue and are used to affect the function of neighboring cells within that tissue.

Endocrine communication involves the release of hormones into the blood for distribution throughout the body. Because hormones are transported within the blood supply, virtually all body cells are exposed to them. However, a hormone will create a response only in its **target cells,**

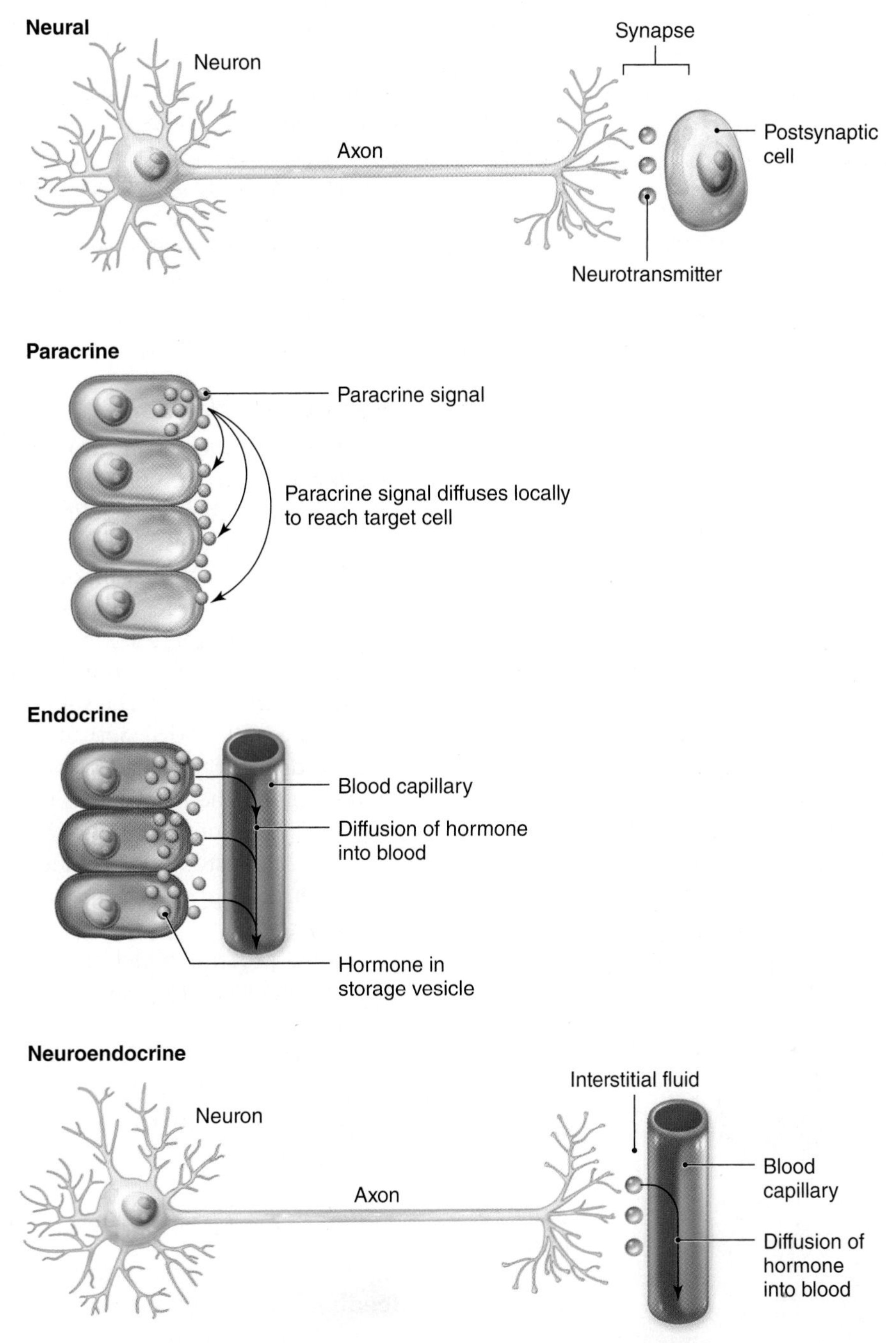

Figure 10.2 Comparison of modes of communication in the human body.

which are cells that possess receptors specific for that hormone. Non-target cells lack these hormone-specific receptors and are unaffected by the hormone. Neuroendocrine communication is a hybrid mechanism in which a neuron releases a hormone that enters a blood vessel. The majority of this chapter focuses on the role of endocrine and neuroendocrine communication in homeostasis.

Hormones are secreted in very small amounts, so their concentrations in the blood are extremely low. However, because they act on cells that have specific receptors for particular hormones, large quantities are not necessary to produce effects. Chemically, hormones may be classified in two broad groups: **steroids,** which are derived from cholesterol, and **nonsteroids,** which are derived from amino acids, peptides, or proteins.

Eicosanoids (i-ko′-sa-noyds) are another group of molecules secreted by cells that cause specific actions in other cells. These lipid molecules act as paracrine signals because they are released into the interstitial fluid and typically affect only nearby cells. Prostaglandins and

Clinical Insight

Aspirin and acetaminophen are widely used pain relievers. They function by inhibiting the synthesis of prostaglandins involved in the inflammatory response, which often is the basis of pain and fever.

leukotrienes are examples. **Prostaglandins** produce a variety of effects ranging from promoting inflammation and blood clotting to increasing uterine contraction in childbirth and raising blood pressure. **Leukotrienes** help regulate the immune response and promote inflammation and some allergic reactions.

Mechanisms of Hormone Action

A hormone produces its effect by binding to a target cell's receptors for that hormone. The more receptors it binds to, the greater is the effect on the target cell. All hormones affect target cells by altering their metabolic activities. For example, they may change the rate of cellular processes in general, or they may promote or inhibit specific cellular processes. The end result is that homeostasis is maintained. Figure 10.3 shows the major endocrine glands.

Steroid and Thyroid Hormones

Steroid hormones and thyroid hormones act on DNA in a cell's nucleus and affect **gene expression** (figure 10.4). (1) Because they are lipid-soluble (see chapter 3), they can easily move through the phospholipid bilayers of plasma membranes to (2) enter the nucleus. (3) After a hormone enters the nucleus, it combines with an intracellular receptor to form a hormone-receptor complex. (4) The hormone-receptor complex interacts with DNA, activating specific genes that synthesize messenger RNA (mRNA). (5) The mRNA exits the nucleus and interacts with ribosomes, which results in the synthesis of specific proteins, usually enzymes. Then the newly formed proteins produce the specific effect that is characteristic of the particular hormone.

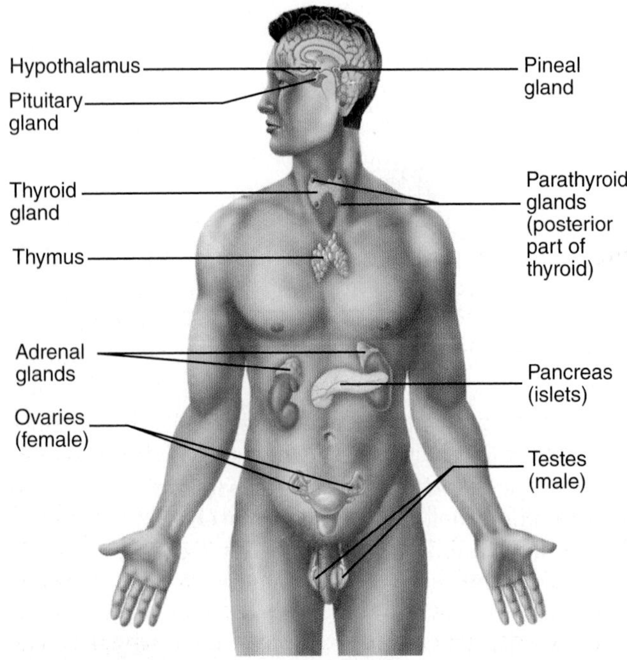

Figure 10.3 The major endocrine glands.

Nonsteroid Hormones

Nonsteroid hormones are proteins, peptides, or modified amino acids that are not lipid-soluble, meaning they cannot pass across the phospholipid bilayer. Two messengers are required for these hormones to produce their effect on a target cell. The *first messenger* is the nonsteroid hormone bound to a receptor on the plasma membrane. The first messenger leads to the formation of a *second messenger* that is often, but not always, *cyclic adenosine monophosphate (cAMP)*. The **second messenger** is formed within the cell, and it activates or inactivates enzymes that produce the characteristic effect for the hormone (figure 10.4). When a cAMP is the second messenger, the sequence of events is as follows.

(1) A nonsteroid hormone binds to a receptor on the target cell's plasma membrane to (2) form a hormone-receptor complex. (3) This complex activates a membrane protein (G protein), which, in turn, activates a membrane enzyme (adenylate cyclase), (4) which catalyzes the formation of cyclic adenosine monophosphate (cAMP) from ATP within the cytosol. (5) The cAMP activates enzymes that catalyze the activation or inactivation of cellular enzymes, which produce the cellular changes associated with the specific hormone.

Control of Hormone Production

Most hormone secretion is usually regulated by a **negative-feedback mechanism** that works to maintain homeostasis. When the blood concentration of a regulated substance begins to decrease, the endocrine gland is stimulated to increase the secretion of its hormone. The increased hormone concentration stimulates target cells to raise the blood level of the substance back to normal. When the substance returns to normal levels, the endocrine gland is no longer stimulated to secrete the hormone, and the secretion and concentration of the hormone decrease. Negative feedback keeps hormone levels in the blood relatively stable (figure 10.5). However, there are a few body processes that are hormonally regulated through **positive-feedback mechanisms**. An example that you will see later in this chapter and in chapter 18 is the production of oxytocin during labor and delivery.

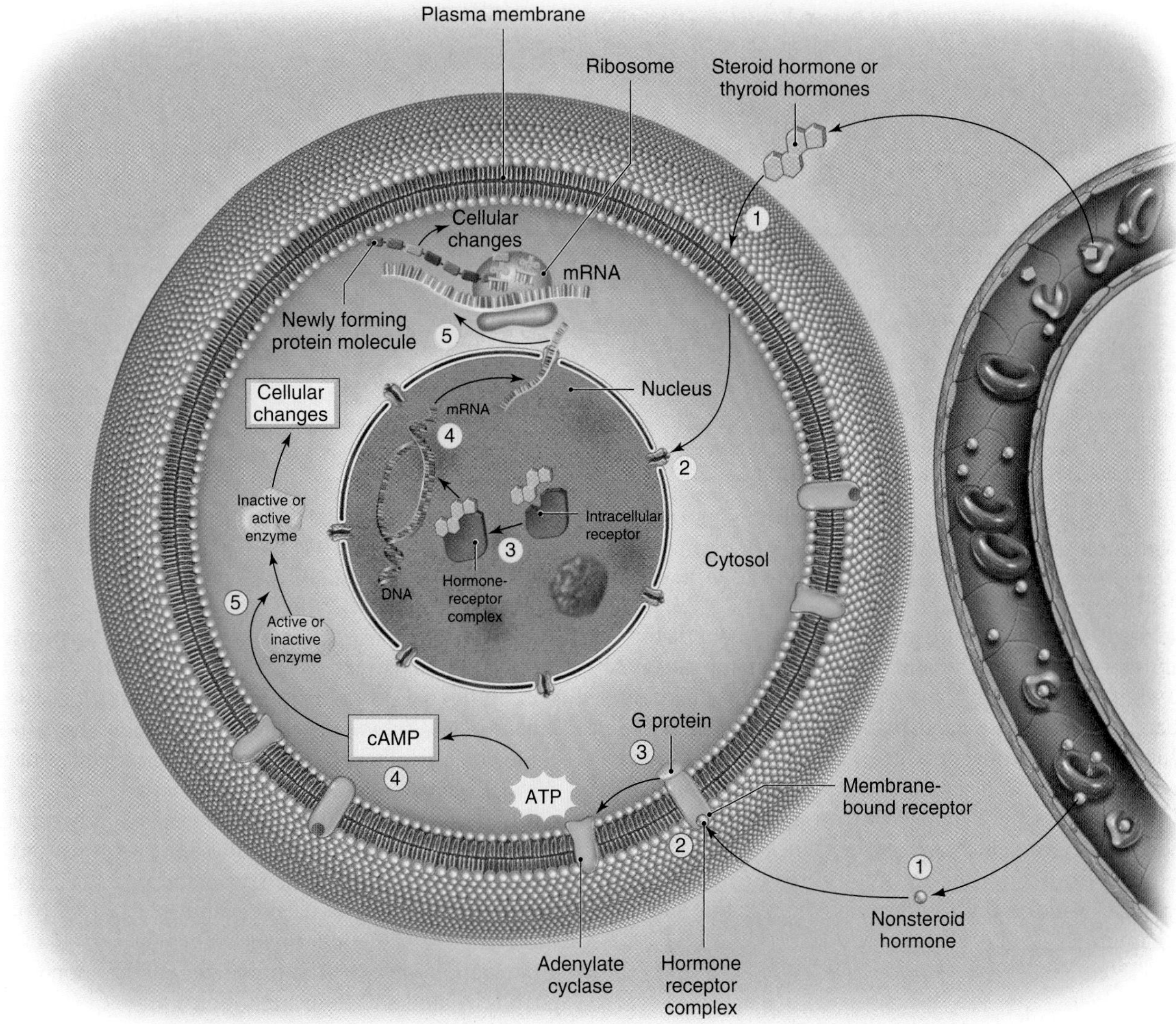

Figure 10.4 Comparison of mechanisms of hormone action. Mechanisms are numbered to match descriptions within text. APR

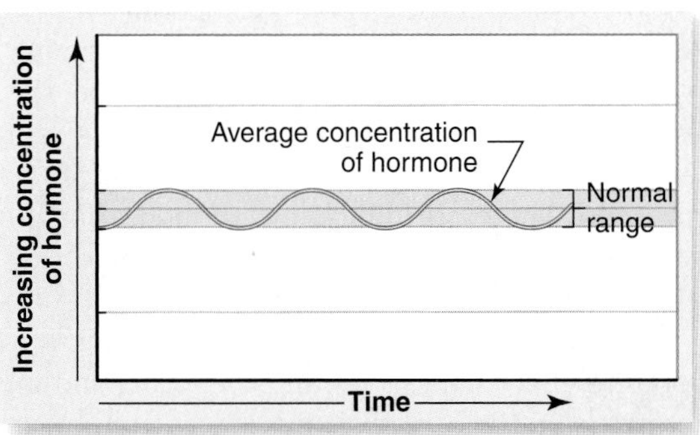

Figure 10.5 Negative-feedback mechanism controls the concentration of a hormone in the blood. The concentration may fluctuate slightly above and below the hormone's average concentration.

As shown in figure 10.6, endocrine glands are controlled by these negative-feedback mechanisms in three ways. (1) In *hormonal control* (figure 10.6*a*), the hypothalamus and anterior lobe of the pituitary gland release hormones that stimulate other endocrine glands to produce hormones. These hormones feedback and affect the function of the hypothalamus and anterior lobe. (2) In *neural control* (figure 10.6*b*), the nervous system stimulates an endocrine gland to produce a hormone, which affects target cells in the body. The actions of the target cells feedback on the nervous system to alter its activity. (3) In *humoral control* (figure 10.6*c*), a chemical change in the blood stimulates an endocrine gland to produce a hormone, which in turn affects target cells. The actions of the target cells then create a change in blood levels of the chemical, which feeds back and alters the activity of the endocrine gland.

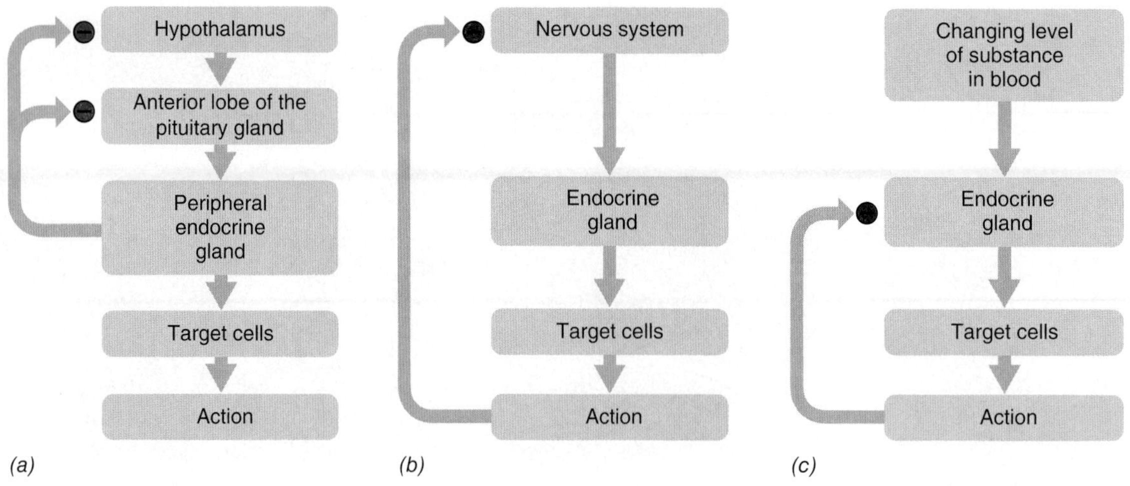

Figure 10.6 Negative-feedback mechanisms used to control the release of various hormones. *(a)* Hormonal control. *(b)* Neural control. *(c)* Humoral control. APIR

These feedback mechanisms may have either stimulatory or inhibitory effects on the hormone production pathway.

The production of hormones is normally precisely regulated so that there is no **hypersecretion** (excessive production) or **hyposecretion** (deficient production). However, hormonal disorders do occur, and they usually result from severe hypersecretion or hyposecretion. Because endocrine disorders are specifically related to individual glands, disorders in this chapter are considered when each gland is discussed rather than at the end of the chapter.

Check My Understanding

1. How do steroid and nonsteroid hormones produce their effects on target cells?
2. How are hormones and prostaglandins similar but different?
3. How is the secretion of hormones regulated?

10.2 Pituitary Gland

Learning Objectives

5. Describe how the production of each of the anterior lobe hormones is controlled.
6. List the actions of hormones of the anterior lobe of the pituitary gland.
7. Describe how the production of each of the posterior lobe hormones is controlled.
8. List the actions of hormones of the posterior lobe of the pituitary gland.
9. Describe the major pituitary gland disorders.

The **pituitary** (pi-tū′-i-tar-ē) **gland,** or **hypophysis** (hī-pof′-i-sis), is attached to the hypothalamus by a short stalk. It rests in a depression of the sphenoid bone, the sella turcica, which provides protection. The pituitary gland consists of two major parts that have different functions: an *anterior lobe* and a *posterior lobe.* Although the pituitary gland is small, it regulates many body functions. The pituitary gland is controlled by neurons and hormones that originate in the **hypothalamus,** as shown in figure 10.7. The hypothalamus serves as a link between the brain and the endocrine system and is itself an endocrine gland. Table 10.1 summarizes the hormones of the pituitary gland and their functions.

Control of the Anterior Lobe

Special neurons (neurosecretory cells) in the hypothalamus regulate the secretion of hormones from the anterior lobe by secreting *releasing* and *inhibiting hormones.* The hypothalamic hormones enter the hypophyseal portal veins, which carry them directly into the anterior lobe without circulating throughout the body. In the anterior lobe, the hormones exert their effects on specific groups of cells. There is a releasing hormone for each hormone produced by the anterior lobe. There are inhibiting hormones for growth hormone and prolactin. As the names imply, releasing hormones stimulate the production and release of hormones from the anterior lobe, while inhibiting hormones have the opposite effect. The secretion of releasing and inhibiting hormones by the hypothalamus is regulated by various hormonal negative-feedback mechanisms.

Control of the Posterior Lobe

The posterior lobe is controlled by the neural negative-feedback mechanism described previously and shown in

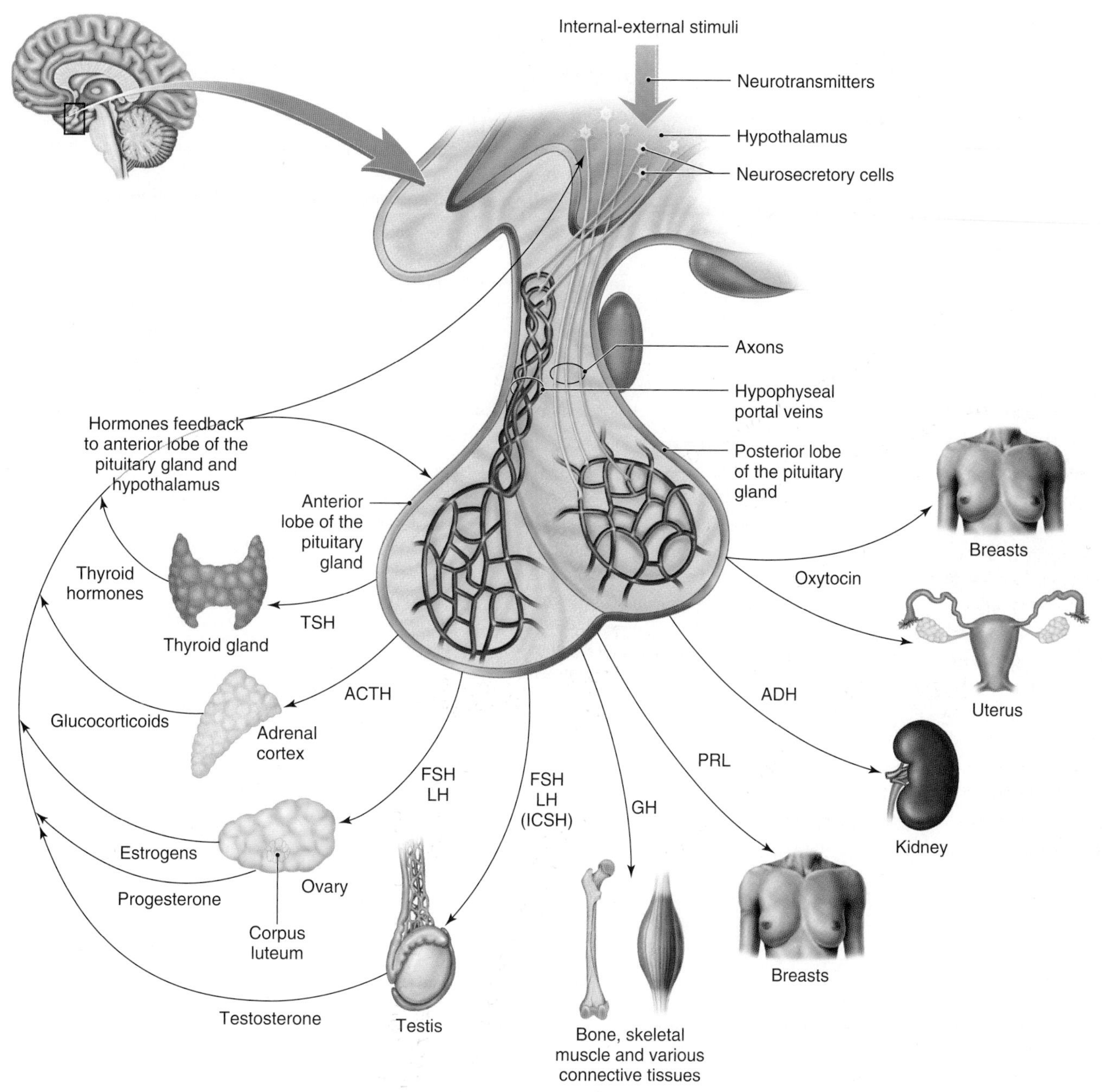

Figure 10.7 Control of pituitary gland secretions. Hypothalamic hormones are secreted by modified neurons and carried by the hypophyseal portal veins to the anterior lobe of the pituitary gland, where they either stimulate or inhibit the secretion of anterior lobe hormones. Nerve impulses stimulate modified neurons in the hypothalamus to secrete hormones that are released from their terminal boutons within the posterior lobe of the pituitary gland. APR

figure 10.6*b*. Special neurons that originate in the hypothalamus have axons that extend into the posterior lobe of the pituitary gland. Nerve impulses passed along these neurosecretory axons cause the release of hormones from their terminal boutons within the posterior lobe, where they diffuse into the blood. Note that the posterior lobe hormones are formed by neurosecretory cells originating in the hypothalamus and not by cells of the posterior lobe of the pituitary gland. They are only released within the posterior lobe.

Anterior Lobe Hormones

The anterior lobe of the pituitary gland is sometimes called the "master gland" because it affects so many body functions. It produces and secretes six hormones: growth hormone (GH), thyroid-stimulating hormone (TSH), adrenocorticotropic hormone (ACTH), follicle-stimulating hormone (FSH), luteinizing hormone (LH), and prolactin (PRL).

Table 10.1 Hormones of the Pituitary Gland

Hormone	Control	Action	Disorders
Anterior Lobe Hormones			
Growth hormone (GH)	Growth-hormone-releasing hormone (GHRH); growth-hormone-inhibiting hormone (GHIH)	Promotes growth of body cells and cell division; promotes protein synthesis; increases the use of fat and glucose for ATP	Hyposecretion in childhood causes pituitary dwarfism. Hypersecretion in childhood causes gigantism; in adults, it causes acromegaly.
Thyroid-stimulating hormone (TSH)	Thyrotropin-releasing hormone (TRH)	Stimulates thyroid gland to produce thyroid hormones	Hyposecretion leads to secondary hypothyroidism. Hypersecretion leads to secondary hyperthyroidism.
Adrenocorticotropic hormone (ACTH)	Corticotropin-releasing hormone (CRH)	Stimulates adrenal cortex to secrete glucocorticoids and androgens	
Follicle-stimulating hormone (FSH)	Gonadotropin-releasing hormone (GnRH)	In ovaries, stimulates development of ovarian follicles and secretion of estrogens; in testes, stimulates the production of sperm	
Luteinizing hormone (LH)	Gonadotropin-releasing hormone (GnRH)	In females, promotes ovulation, development of the corpus luteum, which leads to the production and secretion of progesterone, preparation of uterus to receive embryo, and preparation of mammary glands for milk secretion; in males, stimulates testes to secrete testosterone	
Prolactin (PRL)	Prolactin-releasing hormone (PRH); prolactin-inhibiting hormone (PIH)	Stimulates milk secretion and maintains milk production by mammary glands	
Posterior Lobe Hormones			
Antidiuretic hormone (ADH)	Concentration of water in body fluids	Promotes retention of water by kidneys	Hyposecretion causes diabetes insipidus.
Oxytocin (OT)	Stretching of uterus; stimulation of nipples	Stimulates contractions of uterus in childbirth and contraction of milk glands when nursing infant In both sexes, promotes parental caretaking and involved in feeling of pleasure associated with sexual experiences	

Growth Hormone

As the name implies, **growth hormone (GH)** stimulates the division and growth of body cells. Increased growth results because GH promotes the synthesis of proteins and other complex organic compounds. GH also increases available energy for these synthesis reactions by promoting the release of fat from adipose tissue, the use of fat in cellular respiration, and the conversion of glycogen to glucose. Although GH is more abundant during childhood and puberty, it is secreted throughout life.

Regulation of growth hormone secretion is by two hypothalamic hormones with antagonistic functions. GH-releasing hormone (GHRH) stimulates GH secretion, and GH-inhibiting hormone (GHIH) inhibits GH secretion. Whether the hypothalamus releases GHRH or GHIH depends upon changes in blood chemistry. For example, following strenuous exercise, a low level of blood sugar (hypoglycemia), and an excess of amino acids in the blood trigger the secretion of GHRH. Conversely, high levels of blood sugar (hyperglycemia) stimulate the secretion of GHIH.

Disorders If hypersecretion of GH occurs during the growing years, the individual becomes very tall—sometimes nearly 2.5 m (8 ft) in height. This condition is known as **gigantism.** If the hypersecretion of GH occurs in an adult after full growth in height has been attained, it produces a condition known as **acromegaly** (ak-rō-meg′-ah-lē). Because the growth of long bones has been completed, only the bones of the face, hands, and feet continue to grow. Over time, the individual develops heavy, protruding brow ridges, a jutting mandible, and enlarged hands and feet. Both gigantism and acromegaly may result from tumors of the anterior lobe. Affected persons may have other health problems due to hypersecretion of other anterior lobe hormones.

If hyposecretion of GH occurs during childhood, body growth is limited. In extreme cases, this results in **pituitary dwarfism.** Affected persons have well-proportioned body parts but may be less than 1 m (3 ft) in height. They may suffer from other maladies due to a deficient supply of other anterior lobe hormones.

Thyroid-Stimulating Hormone

Thyroid-stimulating hormone (TSH) stimulates the thyroid gland to produce thyroid hormones. Blood concentrations of thyroid hormones control the negative-feedback mechanism for TSH production. Low levels of thyroid hormones activate the hypothalamus to secrete *thyrotropin-releasing hormone (TRH),* which stimulates release of TSH by the anterior lobe. Conversely, high concentrations of thyroid hormones inhibit the secretion of TRH, which decreases production of TSH. Because TSH controls the thyroid gland, disorders of TSH secretion lead to thyroid disorders.

Adrenocorticotropic Hormone

Adrenocorticotropic (ad-re-nō-kor-ti-kō-trō-p′-ik) **hormone (ACTH)** controls the secretion of hormones produced by the adrenal cortex (the superficial portion of the adrenal gland). ACTH production is controlled by *corticotropin-releasing hormone (CRH)* from the hypothalamus. CRH release is controlled by blood levels of ACTH and glucocorticoids from the adrenal cortex through negative-feedback mechanisms. Low levels of ACTH in the blood trigger the production and release of CRH. High blood levels of ACTH inhibit the production of CRH. Low levels of glucocorticoids from the adrenal cortex activate the hypothalamus to secrete CRH, which stimulates the release of ACTH from the anterior lobe. High levels of glucocorticoids inhibit CRH secretion, and thus inhibit the production and secretion of ACTH. Excessive stress may stimulate the production of excessive amounts of ACTH by overriding the negative-feedback control.

Gonadotropins

The **follicle-stimulating hormone (FSH)** and **luteinizing** (lū-tē-in-īz-ing) **hormone (LH)** affect the gonads (testes and ovaries). Their release is stimulated by *gonadotropin-releasing hormone (GnRH)* from the hypothalamus. The onset of puberty in both sexes is caused by the start of FSH secretion. In females, FSH acts on the ovaries to promote the development of ovarian follicles, which contain ova and produce estrogens, the primary female sex hormones. In males, FSH acts on testes to promote sperm production. In females, LH stimulates ovulation and the development of the corpus luteum, a temporary gland in the ovary that produces progesterone, another female sex hormone. In males LH is often referred to as *interstitial cell stimulating hormone (ICSH)* because it affects the interstitial cells of the testes, where it stimulates the secretion of testosterone. Further discussion of FSH and LH can be found in chapter 17.

Prolactin

Prolactin (prō-lak′-tin) **(PRL)** helps to initiate and maintain milk production by the mammary glands after the birth of an infant. Prolactin stimulates milk secretion after the mammary glands have been prepared for milk production by other hormones, including female sex hormones. In males, PRL increases the activity of LH in the testes, thus increasing testosterone production. Prolactin secretion is regulated by the antagonistic actions of *prolactin-releasing hormone (PRH)* and *prolactin-inhibiting hormone (PIH)* produced by the hypothalamus.

Posterior Lobe Hormones

Posterior lobe hormones are good examples of neuroendocrine secretion. The posterior lobe stores and releases two hormones: the antidiuretic hormone and oxytocin. Both of these hormones are secreted by neurons that originate in the hypothalamus and extend into the posterior lobe. The hormones are released into the blood within the posterior lobe and are distributed throughout the body (see figure 10.7).

Antidiuretic Hormone

The **antidiuretic** (an-ti-dī-ū-ret′-ik) **hormone (ADH)** promotes water retention by the kidneys to reduce the volume of water that is excreted in urine. ADH secretion is regulated by special neurons that detect changes in the water concentration of the blood. If water concentration decreases, secretion of ADH increases to promote water retention by the kidneys. If water concentration increases, secretion of ADH decreases, causing more water to be excreted in urine. By controlling the water concentration of blood, ADH helps to control blood volume and blood pressure. Further discussion of ADH can be found in chapter 16.

Disorders A severe hyposecretion of ADH results in the production of excessive quantities (20–30 liters per day) of dilute urine, a condition called **diabetes insipidus** (dī-ah-bē′-tēz in-sip′-i-dus). Diabetes means "overflow," and insipidus means "tasteless." Thus, diabetes insipidus essentially means to have overflow of tasteless urine. Conversely, mellitus means "sweet," so diabetes mellitus is an overflow of sweet urine. In diabetes insipidus, the affected person is always thirsty and must drink water almost constantly.

Clinical Insight

Pitocin, a synthetic oxytocin, is one of several drugs that is used to clinically induce labor. After delivery, these drugs may also be used to increase the muscle tone of the uterus and to control uterine bleeding.

This condition may be caused by injuries or tumors that affect any part of the ADH regulatory mechanism, such as the hypothalamus or posterior lobe of the pituitary gland, or nonfunctional ADH receptors in the kidneys.

Oxytocin

Oxytocin (ok-sē-tō′-sin) **(OT)** is released in large amounts during childbirth. It stimulates and strengthens contraction of the smooth muscles of the uterus, which culminates in the birth of the infant. It also has an effect on the mammary glands. Stimulation of a nipple by a suckling infant causes the release of OT, which, in turn, contracts the milk glands of the breast, forcing milk into the milk ducts, where it can be removed by the suckling infant.

Unlike other hormones, oxytocin secretion is controlled by a positive-feedback mechanism. For example, the greater the nipple stimulation by a suckling infant, the more OT released and the more milk available for the infant. When suckling ceases, OT production ceases.

OT is also produced in males and nonpregnant females, where it plays a role in creating parental care-taking behaviors and feelings of pleasure associated with sexual intercourse.

Check My Understanding

4. How does the hypothalamus control the secretions of the pituitary gland?
5. What are the functions of anterior lobe and posterior lobe hormones?

10.3 Thyroid Gland APR

Learning Objectives

10. Describe how the production of thyroid hormones is controlled.
11. List the actions of thyroid hormones.
12. Describe how the production of calcitonin is controlled.
13. List the actions of calcitonin.
14. Describe the major thyroid disorders.

The **thyroid gland** is located just inferior to the larynx. It consists of two lobes, each one lateral to the trachea, that are connected by an anterior isthmus (figure 10.8). Table 10.2 summarizes the control, action, and disorders of the thyroid gland.

Thyroxine and Triiodothyronine

Iodine atoms are essential for the formation and functioning of two similar thyroid hormones, produced by groups of cells forming *thyroid follicles* that respond to TSH. **Thyroxine** is the primary hormone. It is also known as $\mathbf{T_4}$ because each molecule contains four iodine atoms. The other hormone, **triiodothyronine** (tri″i-o″do-thi′ro-nen) or $\mathbf{T_3}$, contains three iodine atoms in each molecule. Both T_4 and T_3 exert their effect on body cells, and they have similar functions. They increase the metabolic rate, promote protein synthesis, and enhance neuron function. T_3 and T_4 are the primary factors that determine the basal metabolic rate (BMR), the number of calories required at rest to maintain life. Thyroid hormones are also important during infancy and childhood for normal development of the nervous, skeletal, and muscular systems. Secretion of these hormones is stimulated by TSH from the anterior lobe of the pituitary gland, and TSH, in turn, is regulated by a negative-feedback mechanism as described in the discussion of the anterior lobe of the pituitary gland.

Disorders Hypersecretion, hyposecretion, and iodine deficiencies are involved in the thyroid disorders: Graves disease, simple goiter, cretinism, and myxedema.

Graves disease results from the hypersecretion of thyroid hormones. It is thought to be an autoimmune disorder in which antibodies bind to TSH receptors, stimulating excessive hormone production. It is characterized by restlessness and increased metabolic rate with possible weight loss. Usually, the thyroid gland is somewhat enlarged, which is called a **goiter** (goy-ter), and eyes bulge due to the swelling of tissues posterior to the eyes, producing what is called an *exophthalmic* (ek-sof-thal-mik) *goiter*.

Simple goiter is an enlargement of the thyroid gland that results from a deficiency of iodine in the diet. Without adequate iodine, hyposecretion of thyroid hormones occurs and the thyroid gland enlarges due to overstimulation with TSH in an attempt to produce more thyroid hormones. In some cases, the thyroid gland may become the size of an orange. Goiter can be prevented by including very small amounts of iodine in the diet. For this reason, salt manufacturers produce "iodized salt," which contains sufficient iodine to prevent simple goiter.

Cretinism (kre′-tin-izm) is caused by a severe hyposecretion of thyroid hormones in infants. Without treatment, it produces severe mental and physical retardation. Cretinism is characterized by stunted growth, abnormal bone formation, mental retardation, sluggishness, and goiter.

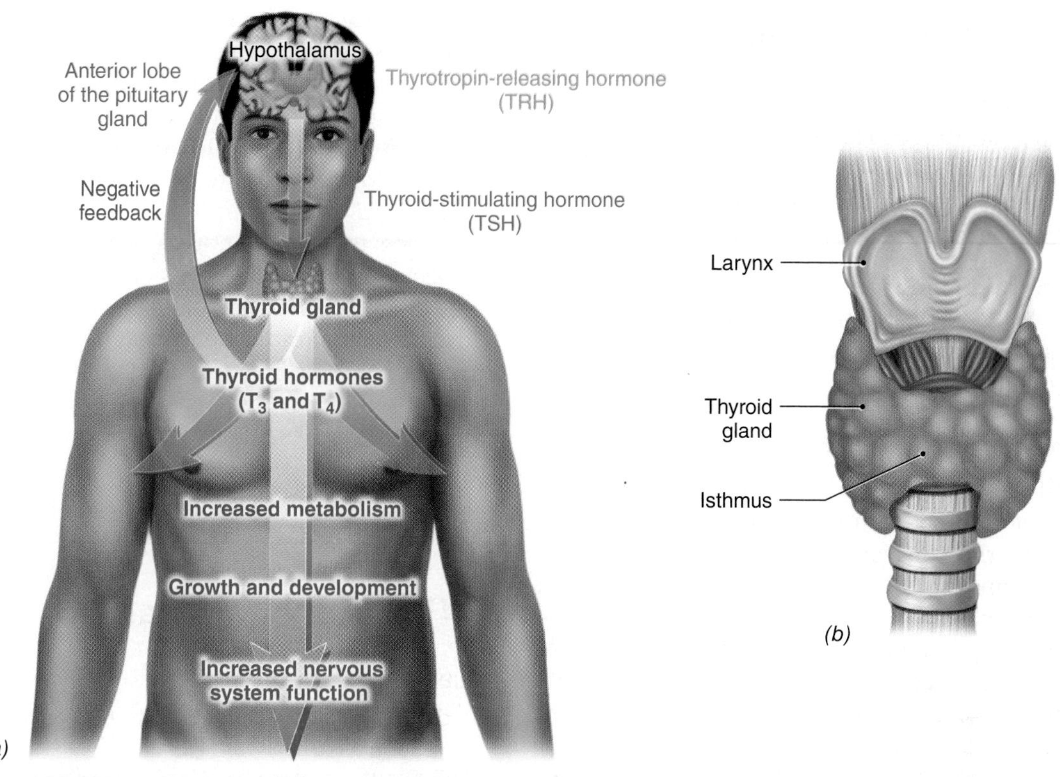

Figure 10.8 Anatomy and Physiology of the Thyroid Gland.
(a) Negative-feedback mechanism of thyroid control. *(b)* The thyroid gland consists of two lobes connected anteriorly at the isthmus. APIR

Table 10.2 Hormones of the Thyroid Gland

Hormone	Control	Action	Disorders
Thyroxine (T_4) and triiodothyronine (T_3)	TSH from anterior lobe of the pituitary gland	Increase metabolic rate; accelerate growth; stimulate neural activity	Hyposecretion in infants and children causes cretinism; in adults, it causes myxedema. Hypersecretion causes Graves disease. Iodine deficiency causes simple goiter.
Calcitonin (CT)	Blood Ca^{2+} level	Decreases blood Ca^{2+} levels by promoting Ca^{2+} deposition in bones, inhibiting removal of Ca^{2+} from bones, promoting excretion of Ca^{2+} by kidneys	

Myxedema (mik-se-de′-mah) is caused by severe hyposecretion of thyroid hormones in adults. It is characterized by sluggishness, weight gain, weakness, dry skin, goiter, and puffiness of the face.

Calcitonin

The thyroid gland produces a third hormone, **calcitonin** (kal-si-to′-nin) **(CT)**, from cells called *C cells* that are located between thyroid follicles. C cells do not respond to the hormonal mechanism the same as thyroid follicles do but respond to a humoral negative-feedback mechanism linked to blood Ca^{2+} levels. Calcitonin decreases blood Ca^{2+} by inhibiting the bone-resorbing action of osteoclasts, increasing the rate of Ca^{2+} deposition by osteoblasts, and promoting Ca^{2+} excretion by the kidneys. An excess of Ca^{2+} in the blood stimulates the thyroid gland to secrete calcitonin. The concentration of Ca^{2+} in the blood is important because it plays vital roles in metabolism, including maintenance of healthy bones, conduction of nerve impulses, muscle contraction, and clotting of blood. The function of calcitonin is antagonistic to parathyroid hormone, which is discussed in the next section.

10.4 Parathyroid Glands

Learning Objectives

15. Describe how the production of parathyroid hormone is controlled.
16. List the actions of parathyroid hormone.
17. Describe the major parathyroid disorders.

The **parathyroid glands** are small glands that are located on the posterior surface of the thyroid gland. There are usually four parathyroid glands, two glands on each lobe of the thyroid (figure 10.9).

Parathyroid Hormone

Parathyroid glands secrete **parathyroid hormone (PTH),** the most important regulator of blood Ca^{2+} levels. PTH increases the concentration of blood Ca^{2+} by promoting the removal of Ca^{2+} from bones by osteoclasts and by inhibiting Ca^{2+} deposition by osteoblasts. PTH acts in the kidneys to inhibit excretion of Ca^{2+} into urine and trigger the activation of vitamin D (also a hormone). Both PTH and vitamin D increase Ca^{2+} absorption by the small intestine. The antagonistic actions of PTH and calcitonin maintain blood Ca^{2+} homeostasis (figure 10.10 and table 10.3).

Disorders **Hypoparathyroidism,** the hyposecretion of PTH, can produce devastating effects. Without treatment, the concentration of blood Ca^{2+} may drop to levels that impair neural and muscular activity. The effect on cardiac muscle may result in cardiac arrest and sudden death. Tetany of skeletal muscles may occur, and death may result from a lack of oxygen due to the inability of breathing muscles to function normally.

Hyperparathyroidism, the hypersecretion of PTH, causes too much Ca^{2+} to be removed from bones and raises blood Ca^{2+} to abnormally high levels. Without treatment, Ca^{2+} loss results in soft, weak bones that are prone to spontaneous fractures. The excess Ca^{2+} in the blood may lead to the formation of kidney stones or may be deposited in abnormal locations creating bone spurs (abnormal bony growths).

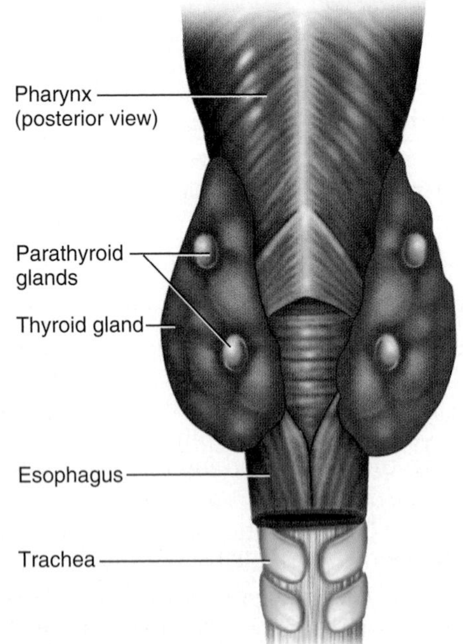

Figure 10.9 Two small parathyroid glands are located on the posterior surface of each lobe of the thyroid gland.

 Check My Understanding

6. What are the actions of thyroid hormones?
7. How is the level of blood Ca^{2+} regulated?

10.5 Adrenal Glands

Learning Objectives

18. Describe how the production of adrenal hormones is controlled.
19. List the actions of adrenal hormones.
20. Describe the major adrenal disorders.

There are two **adrenal glands;** one is located on top of each kidney. Each adrenal gland consists of two portions that are distinct endocrine glands: the deep adrenal medulla and the superficial adrenal cortex (figure 10.11). Table 10.4 summarizes the control, action, and disorders of the adrenal gland.

Hormones of the Adrenal Medulla

The **adrenal medulla** secretes **epinephrine** (adrenaline) and **norepinephrine** (noradrenaline), two closely related hormones that have very similar actions on target cells. Epinephrine forms about 80% of the secretions.

The sympathetic division of the autonomic nervous system regulates the secretion of adrenal medullary hormones. They are secreted whenever the body is under stress, and they duplicate the action of the sympathetic division on a bodywide scale. The medullary hormones have a stronger and longer-lasting effect in preparing the body for "fight or flight." The effects of epinephrine and norepinephrine include (1) a decrease in blood flow to the viscera and skin; (2) an increase in blood flow to the skeletal muscles, lungs, and nervous system; (3) conversion of glycogen to glucose to raise the glucose level in the blood; and (4) an increase in the rate of cellular respiration. Epinephrine and norepinephrine are particularly important in short-term stress situations. In times of chronic stress the adrenal cortex makes further adjustment as will be discussed in the next section.

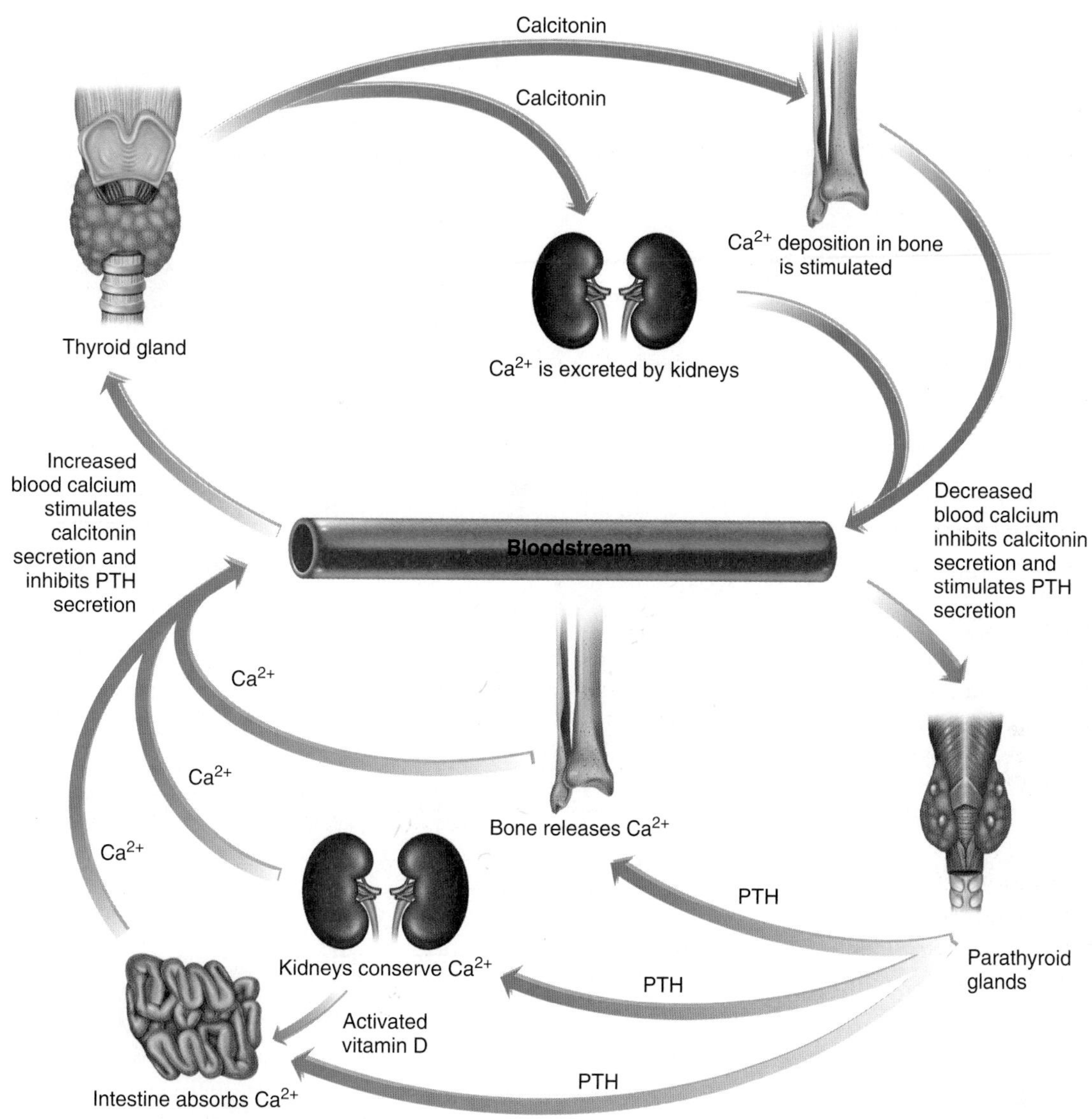

Figure 10.10 Calcium Homeostasis.
The concentration of Ca^{2+} in the blood controls the secretion of calcitonin and PTH. APR

Table 10.3 Parathyroid Hormone

Hormone	Control	Action	Disorders
Parathyroid hormone (PTH)	Blood Ca^{2+} level	Increases blood Ca^{2+} level by promoting Ca^{2+} removal from bones and Ca^{2+} reabsorption by kidneys	Hyposecretion causes tetany, which may result in death. Hypersecretion causes weak, deformed bones that may fracture spontaneously.

Hormones of the Adrenal Cortex

Several different steroid hormones are produced by the **adrenal cortex,** but the most important ones are aldosterone, cortisol, and the sex hormones.

Aldosterone (al-dō-ster′-ōn) is the most important mineralocorticoid secreted by the adrenal cortex. **Mineralocorticoids** regulate the concentration of electrolytes (mineral ions) in body fluids. Aldosterone stimulates the kidneys to retain sodium ions (Na^+) and to excrete potassium ions (K^+). This action not only maintains the normal balance of Na^+ and K^+ in body fluids but also maintains blood volume and blood pressure. The reabsorption of Na^+ into the blood causes anions, such as chloride (Cl^-) and bicarbonate (HCO_3^-), to be reabsorbed due to their opposing

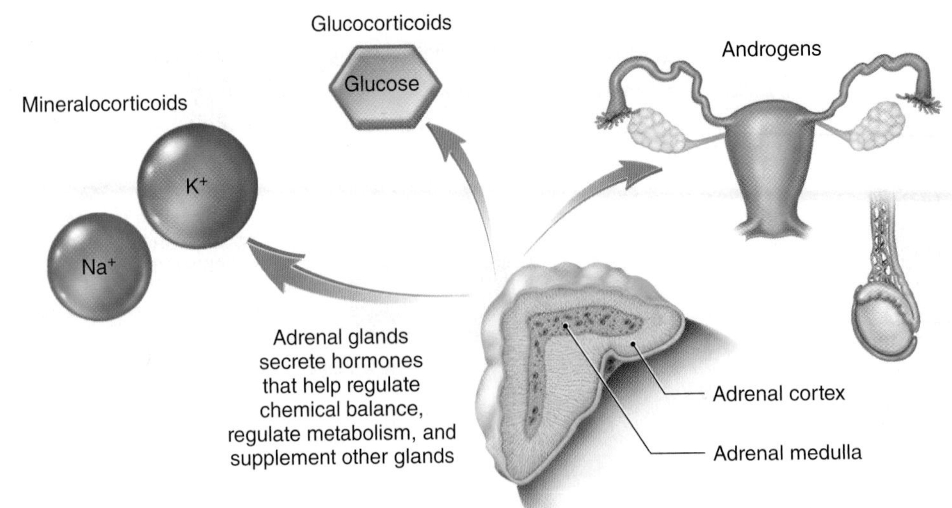

Figure 10.11 An adrenal gland consists of a superficial adrenal cortex and a deep adrenal medulla. AP|R

Table 10.4 Hormones of the Adrenal Glands

Hormone	Control	Action	Disorders
Adrenal Medulla			
Epinephrine and norepinephrine	Sympathetic division of the autonomic nervous system	Prepare body to meet emergencies; increase heart rate, cardiac output, blood pressure, and metabolic rate; increase blood sugar by converting glycogen to glucose; dilate respiratory passages	Hypersecretion causes prolonged responses. Hyposecretion causes no major disorders.
Adrenal Cortex			
Aldosterone	Blood electrolyte levels, angiotensin II	Increases blood levels of sodium and water, which decreases blood levels of potassium; increases blood pressure	Hypersecretion inhibits neural and muscular activity, and also causes edema.
Cortisol	ACTH from anterior lobe of the pituitary gland	Promotes formation of glucose from noncarbohydrate nutrients; provides resistance to stress and inhibits inflammation	Hyposecretion causes Addison disease. Hypersecretion causes Cushing syndrome.
Androgens	ACTH from anterior lobe of the pituitary gland	Effects are insignificant in normal adult males; contribute to the sex drive in females.	Hypersecretion as a result of tumors; causes masculinization in females.

charges. And it causes water to be reabsorbed by osmosis, which maintains blood volume and blood pressure. Aldosterone secretion is stimulated by several factors, including (1) a decrease in blood level of Na^+, (2) an increase in blood level of K^+, or (3) a decrease in blood pressure, which leads to angiotensin II production, as will be discussed in chapter 16.

Glucocorticoids are so named because they affect glucose metabolism. There are three major actions of glucocorticoids. (1) In response to chronic stress, glucocorticoids ensure a constant fuel supply by promoting the conversion of noncarbohydrate nutrients into glucose. This is important because carbohydrate sources, such as glycogen, may be exhausted after several hours without food or strenuous exercise. (2) They facilitate the utilization of glucose by cells. (3) They reduce inflammation.

Cortisol (kor′-ti-sol) is the most important of several glucocorticoids that are secreted by the adrenal cortex under the stimulation of ACTH. The blood levels of glucocorticoids are kept in balance because they exert a negative-feedback control on the secretion of CRH and ACTH, as described in the section of this chapter discussing the anterior lobe of the pituitary gland.

The adrenal cortex also secretes small amounts of **androgens** (male sex hormones) and estrogens in response to ACTH from the anterior lobe of the pituitary gland. The estrogens have little significant function. The

Clinical Insight

Everyone experiences stressful situations. Stress may be caused by physical or psychological stimuli that are perceived as threatening. Whereas mild stress can stimulate creativity and productivity, severe and prolonged stress can have serious consequences.

The hypothalamus is the initiator of the stress response. When stress occurs, the hypothalamus activates the sympathetic division of the autonomic nervous system and the secretion of epinephrine and norepinephrine by the adrenal medulla. Thus, both neural and hormonal activity prepare the body to meet the stressful situation by increasing blood glucose, heart rate, breathing rate, blood pressure, and blood flow to the muscular and nervous systems.

Simultaneously, the hypothalamus stimulates the release of ACTH from the anterior lobe of the pituitary gland. ACTH, in turn, causes the secretion of glucocorticoids by the adrenal cortex. Glucocorticoids increase the levels of amino acids and fatty acids in the blood and promote the formation of additional glucose from noncarbohydrate nutrients.

All of these responses prepare the body for an immediate response to cope with a stressful situation.

Prolonged stress may cause several undesirable side effects from the constant secretion of large amounts of epinephrine and glucocorticoids, such as decreased immunity and high blood pressure—problems that are common in our society.

androgens promote the early development of male reproductive organs, but in adult males their effects are masked by sex hormones produced by testes. In females, adrenal androgens contribute to the female sex drive. In both sexes, excessive production results in exaggerated male characteristics. AP|R

Disorders **Cushing syndrome** results from hypersecretion by the adrenal cortex. It may be caused by an adrenal tumor or by excessive production of ACTH by the anterior lobe of the pituitary gland. This syndrome is characterized by high blood pressure, an abnormally high blood glucose level, protein loss, osteoporosis, fat accumulation on the trunk, fatigue, edema, and decreased immunity. A person with this condition tends to have a full, round face and an enlarged abdomen.

Addison disease results from a severe hyposecretion by the adrenal cortex. It is characterized by low blood pressure, low blood glucose and sodium levels, an increase in the blood potassium level, dehydration, muscle weakness, and increased skin pigmentation. Without treatment to control blood electrolytes, death may occur in a few days.

Check My Understanding

8. How do secretions of the adrenal medulla prepare the body to react in emergencies?
9. How does the adrenal cortex help to maintain blood pressure?

10.6 Pancreas

Learning Objectives

21. Describe the control of pancreatic hormones.
22. List the actions of pancreatic hormones.
23. Describe the major pancreatic disorders.

The **pancreas** (pan′-krē-as) is an elongate organ that is located posterior to the stomach (figure 10.12). It is both an exocrine gland and an endocrine gland. Its exocrine functions are performed by secretory cells that secrete digestive enzymes into tiny ducts within the gland. These ducts merge to form the pancreatic duct, which carries the secretions into the small intestine. Its endocrine functions are performed by secretory cells that are arranged in clusters or clumps called the **pancreatic islets.** Their secretions diffuse into the blood. The islets contain alpha cells and beta cells. Alpha cells produce the hormone glucagon; beta cells form the hormone insulin. Table 10.5 summarizes the control, action, and disorders of the pancreas.

Glucagon

Glucagon (glū′-kah-gon) increases the concentration of glucose in the blood. It does this by activating the liver to convert glycogen and certain noncarbohydrates, such as amino acids, into glucose. Glucagon helps to maintain the blood level of glucose within normal limits even when carbohydrates are depleted due to long intervals between meals. Epinephrine stimulates a similar action, but glucagon is more effective. Glucagon secretion is controlled by the blood level of glucose via a negative-feedback mechanism. A low level of blood glucose stimulates glucagon secretion, and a high level of blood glucose inhibits glucagon secretion.

Clinical Insight

Persons with inflamed joints often receive injections of *cortisone,* a glucocorticoid, to temporarily reduce inflammation and the associated pain. Such a procedure is fairly common in sports medicine.

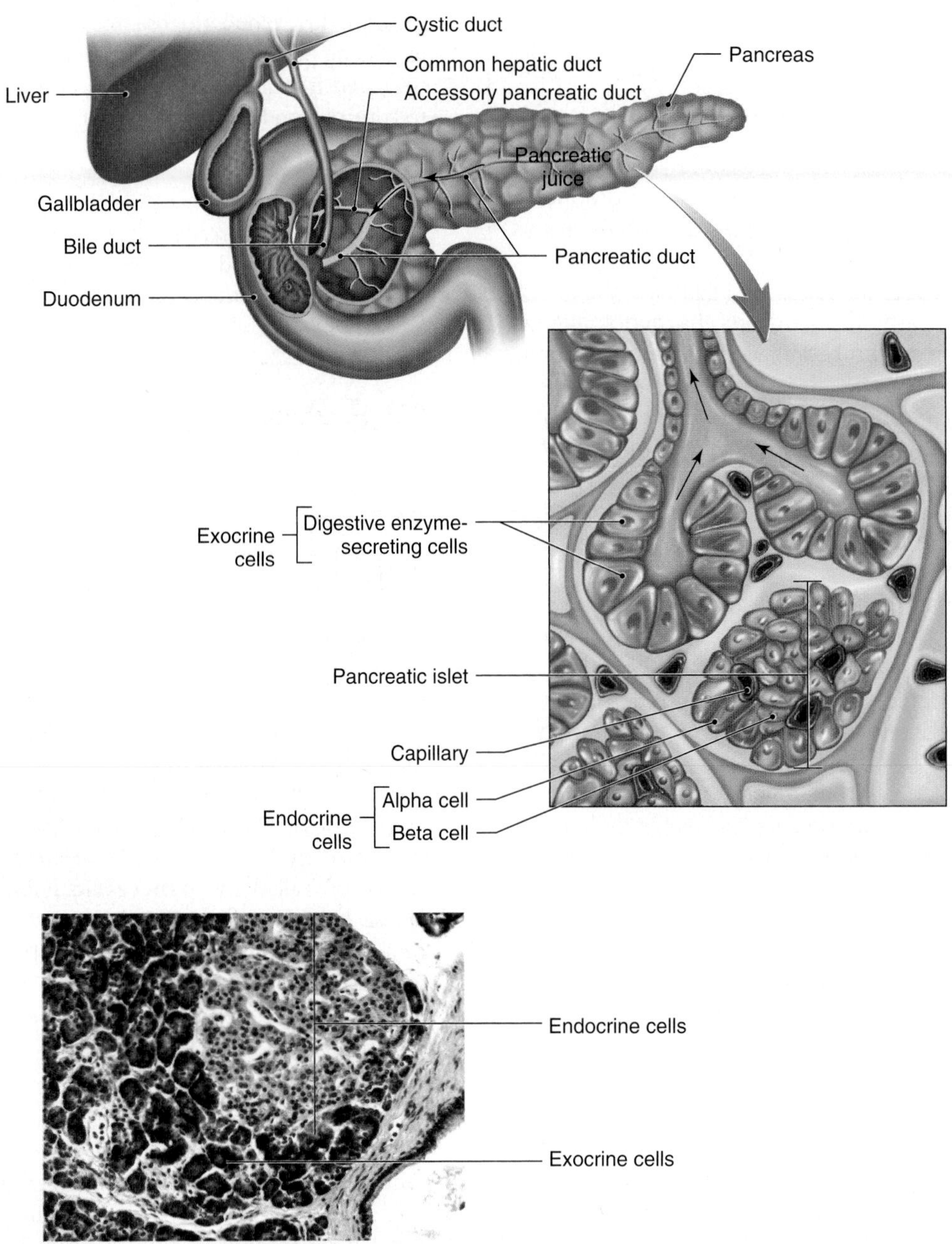

Figure 10.12 The pancreas is both an endocrine and an exocrine gland. The hormone-secreting alpha and beta cells are grouped in clusters, called pancreatic islets. Other pancreatic cells secrete digestive enzymes. APR

Table 10.5 Hormones of the Pancreas

Hormone	Control	Action	Disorders
Glucagon	Blood glucose level	Increases blood glucose by stimulating the liver to convert glycogen and other nutrients into glucose	
Insulin	Blood glucose level	Decreases blood glucose by aiding movement of glucose into cells and promoting the conversion of glucose into glycogen	Hyposecretion causes type I diabetes mellitus. Hypersecretion may cause hypoglycemia.

Insulin

The effect of **insulin** on the level of blood glucose is opposite that of glucagon. Insulin decreases blood glucose by aiding the movement of glucose into body cells, where it can be used as a source of energy. Without insulin, glucose is not readily available to most cells for cellular respiration. Insulin also stimulates the liver to convert glucose into glycogen for storage. Figure 10.13 shows how the antagonistic functions of glucagon and insulin maintain the concentration of glucose in the blood within normal limits. Like glucagon, the level of blood glucose regulates the secretion of insulin. High blood glucose levels stimulate insulin secretion; low levels inhibit insulin secretion.

Disorders **Diabetes mellitus** (dī-ah-bē′-tēz mel-lī′-tus) is caused by the hyposecretion of insulin or the inability of target cells to recognize it due to a loss of insulin receptors. *Type I* or *insulin-dependent diabetes* is an autoimmune metabolic disorder that usually appears in persons less than 20 years of age. For this reason, it is sometimes called juvenile diabetes, although the condition persists for life. Type I diabetes results when the immune response destroys the beta cells in pancreatic islets. Because the metabolism of carbohydrates, fats, and proteins is affected, persons with type I diabetes must follow a restrictive diet. They must also check their blood glucose level several times a day and inject themselves with insulin, or receive insulin from an implanted insulin pump, to keep their blood glucose concentration within normal limits.

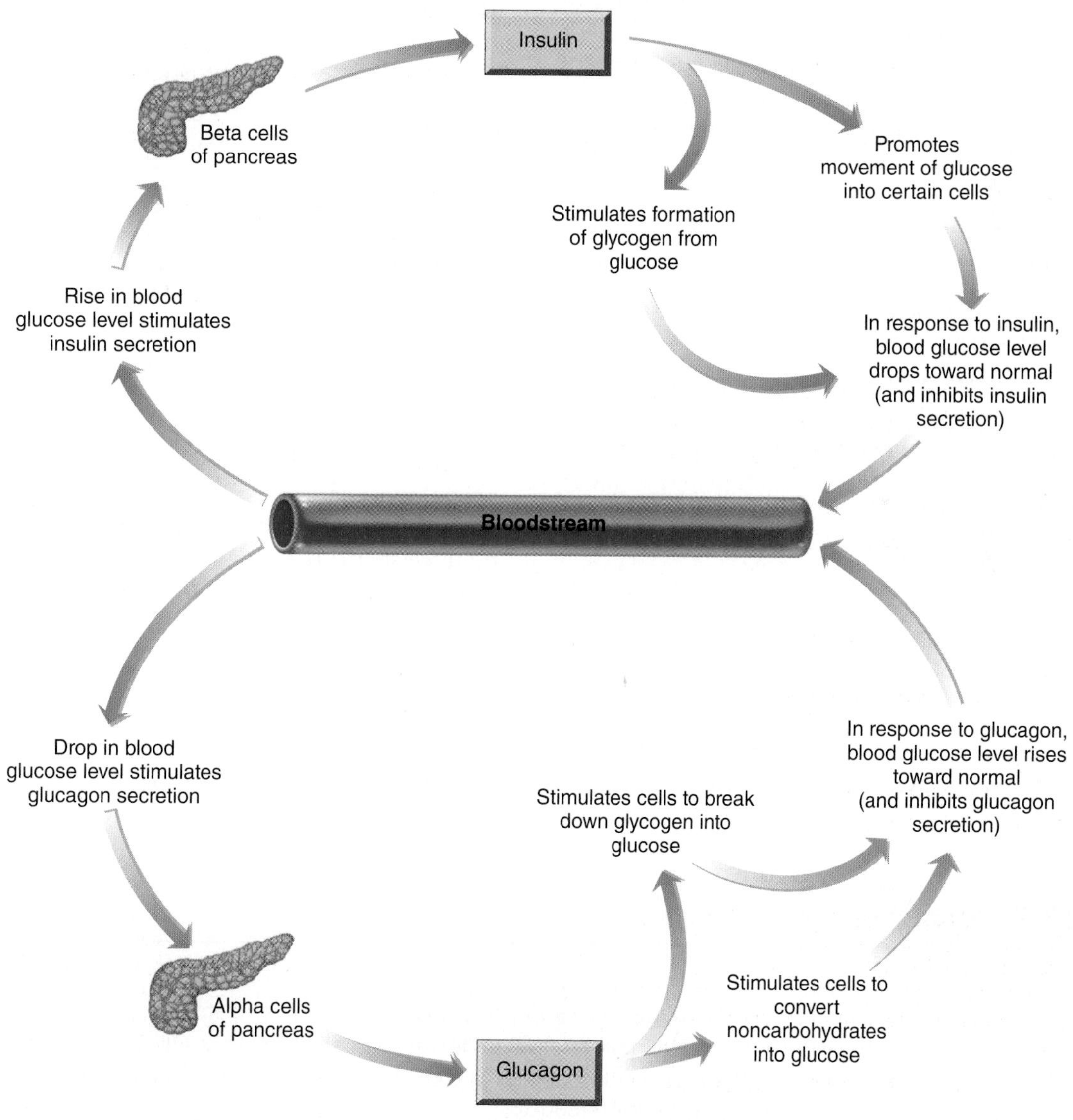

Figure 10.13 Insulin and glucagon function together to help maintain a relatively stable blood glucose level. Negative-feedback mechanism responding to blood glucose level controls the secretion of both hormones. APR

The vast majority of diabetics have *type II* or *insulin-independent diabetes,* which is caused by a reduction of the insulin receptors on target cells. This form of diabetes, also called adult-onset diabetes, usually appears after 40 years of age in persons who are overweight. The symptoms are less severe than in type I diabetes and can be controlled by a careful diet and oral medications that help regulate blood levels of glucose. The current increase in obesity among children and young adults is of concern because it may lead to an increase in type II diabetes. In either case, the result is **hyperglycemia,** excessively high levels of glucose in the blood. With insufficient insulin or a reduction in target insulin receptors, glucose cannot get into cells easily, and cells must rely more heavily on triglycerides as an energy source for cellular respiration. The products of this reaction tend to decrease blood pH (acidosis), which can inactivate vital enzymes and may lead to death.

An excessive production of insulin, or overdose of insulin, may lead to **hypoglycemia,** a condition characterized by excessively low blood glucose levels. Symptoms include acute fatigue, weakness, increased irritability, and restlessness. In extreme conditions, it may lead to an insulin-triggered coma.

Check My Understanding

10. How does the pancreas regulate the level of blood glucose?

10.7 Gonads

Learning Objectives

24. Describe how the production of female sex hormones is controlled.
25. List the actions of female sex hormones.
26. Describe how the production of male sex hormones is controlled.
27. List the actions of male sex hormones.

The gonads are the sex glands: the ovaries and testes. They not only produce oocytes and sperm, respectively, but also secrete the sex hormones. Table 10.6 summarizes the actions of the sex hormones. The gonads and their hormones are covered in more detail in chapter 17.

Female Sex Hormones

The **ovaries** are the female gonads. They are small, almond-shaped organs located in the pelvic cavity. The ovaries begin to function at the onset of puberty when the gonadotropins (FSH and LH) are released from the anterior lobe of the pituitary gland. Subsequently, ovarian hormones, FSH, and LH interact in an approximately 28-day *ovarian cycle* in which their concentrations increase and decrease in a rhythmic pattern.

Estrogens (es′-trō-jens), the primary female sex hormones, are several related compounds that are secreted by developing ovarian follicles that also contain an oocyte (developing egg). Estrogens stimulate the development and maturation of the female reproductive organs and the secondary sex characteristics (e.g., female fat distribution, breasts, and broad hips). They also help to grow and maintain the uterine lining (endometrium) to support a pregnancy.

Progesterone (prō-jes′-te-rōn) is secreted by the corpus luteum, a gland that forms from the empty ovarian follicle after the oocyte has been released by ovulation. It helps prepare the uterus for receiving a preembryo and maintains the pregnancy. It also helps to prepare the mammary glands for milk production.

Male Sex Hormone

The **testes** are paired, ovoid organs located inferior to the pelvic cavity in the scrotum, a sac of skin located posterior to the penis. The seminiferous tubules of the testes produce sperm, the male sex cell; and the interstitial cells (cells between the tubules) secrete the male hormone **testosterone** (tes-tos′-te-rōn). Testosterone stimulates the development and maturation of the male reproductive organs, the secondary sex characteristics (e.g., growth of facial and body hair, low voice, narrow hips, and heavy muscles and bones), the male sex drive, and helps stimulate sperm production.

Table 10.6 Hormones of Ovaries and Testes

Hormone	Control	Action
Ovaries		
Estrogens	FSH	Development of female reproductive organs, secondary sex characteristics, and sex drive; prepares uterus to receive a preembryo and helps maintain pregnancy
Progesterone	LH	Prepares uterus to receive a preembryo and maintains pregnancy; prepares mammary glands for milk production
Testes		
Testosterone	LH (ICSH)	Development of male reproductive organs, secondary sex characteristics, and sex drive

10.8 Other Endocrine Glands and Tissues

Learning Objectives

28. Describe the actions of melatonin.
29. Describe the action of the thymus.

There are a few other glands and tissues of the body that secrete hormones and are part of the endocrine system. These include the pineal gland, the thymus, the kidneys, the heart, and certain small glands in the lining of the stomach and small intestine. Hormones released from the kidneys, heart, and digestive system will be covered in their respective chapters. In addition, the placenta is an important temporary endocrine organ during pregnancy. It is considered in chapter 18.

Pineal Gland

The **pineal** (pin′-ē-al) **gland** is a small, cone-shaped nodule of endocrine tissue that is located in the epithalamus of the brain near the roof of the third ventricle. It secretes the hormone **melatonin** (mel-ah-tō′-nin), which seems to inhibit the secretion of gonadotropins and may help control the onset of puberty. Melatonin seems to regulate wake-sleep cycles and other biorhythms associated with the cycling of day and night. The secretion of melatonin is regulated by exposure to light and darkness. When exposed to light, nerve impulses from the retinas of the eyes are sent to the pineal gland, causing a decrease in melatonin production. During darkness, these nerve impulses decrease, and melatonin secretion is increased. Secretion is greatest at night and lowest in the day, which keeps our sleep-wakefulness cycle in harmony with the day-night cycle.

As frequent fliers know, jet lag results when the sleep-wakefulness cycles are out of sync with the day-night cycle. Jet lag can be more quickly reversed by exposure to bright light with wavelengths similar to sunlight, because the melatonin cycle is resynchronized to the new day-night cycle.

Thymus

The **thymus** is located in the mediastinum superior to the heart. It is large in infants and children but it shrinks with age and is greatly reduced in adults. It plays a crucial role in the development of immunity, which is discussed in chapter 13. The thymus produces several hormones, collectively called **thymosins** (thi-mo′-sins), which are involved in the maturation of T lymphocytes, a type of white blood cell. Thymosins also seem to have some anti-aging effects. Hence, after the thymus shrinks, we age.

Chapter Summary

- The endocrine system is composed of hormone-secreting cells, tissues, and organs.
- Exocrine glands have a duct; endocrine glands are ductless.
- Hormones are chemical messengers that are carried by the blood throughout the body, where they modify cellular functions of target cells.

10.1 The Chemical Nature of Hormones

- There are four major types of communication in the body: 1) neural, 2) paracrine, 3) endocrine, 4) neuroendocrine. All target cells have receptors for chemical messengers that affect them.
- Prostaglandins are not secreted by endocrine glands. They are formed by most body cells and have a distinctly local (paracrine) effect.
- The major endocrine glands are the adrenal glands, gonads, pancreas, parathyroid glands, pineal gland, pituitary gland, thymus, and thyroid gland. In addition, the hypothalamus functions like an endocrine gland in some ways.
- Hormones may be classified chemically as either steroid hormones or nonsteroid hormones.
- Steroid hormones and thyroid hormones combine with a receptor within the target cell and interact with DNA to affect production of mRNA. All other nonsteroid hormones combine with a receptor in the plasma membrane of the target cell, which activates a membrane enzyme that promotes synthesis of cyclic AMP (cAMP), a second messenger. Cyclic AMP, in turn, activates other enzymes that bring about cellular changes.
- Production of most hormones is controlled by a negative-feedback mechanism.
- The negative-feedback mechanisms of hormone production work one of three ways: (1) hormonal, (2) neural, and (3) humoral.
- Endocrine disorders are associated with severe hyposecretion or hypersecretion of various hormones. Hyposecretion may result from injury. Hypersecretion is sometimes caused by a tumor.

10.2 Pituitary Gland

- The pituitary gland is attached to the hypothalamus by a short stalk. It consists of an anterior lobe and a posterior lobe.
- The hypothalamus secretes releasing hormones and inhibiting hormones that are carried to the anterior lobe by the hypophyseal portal veins. The releasing and inhibiting hormones regulate the secretion of anterior lobe hormones.
- Anterior lobe hormones are
  a. growth hormone (GH), which stimulates growth and division of body cells;
  b. thyroid-stimulating hormone (TSH), which activates the thyroid gland to secrete thyroid hormones;
  c. adrenocorticotropic hormone (ACTH), which stimulates the secretion of hormones by the adrenal cortex;

  d. follicle-stimulating hormone (FSH) and luteinizing hormone (LH), which affect the gonads (in females, FSH stimulates production of estrogens by the ovaries, and the development of the ovarian follicles, leading to oocyte production; in males, it activates sperm production by the testes; in females, LH promotes ovulation and stimulates development of the corpus luteum, which produces progesterone; in males, it stimulates testosterone production); and
  e. prolactin (PRL), which initiates and maintains milk production by the mammary glands.
- Hyposecretion of GH in childhood causes pituitary dwarfism. Hypersecretion of GH in childhood causes gigantism, while during adulthood it causes acromegaly.
- Hyposecretion and hypersecretion of TSH leads to secondary thyroid disorders.
- Hormones of the posterior lobe are formed by neurons in the hypothalamus and are released within the posterior lobe.
- There are two posterior lobe hormones:
  a. antidiuretic hormone (ADH) promotes retention of water by the kidneys;
  b. oxytocin stimulates contraction of the uterus during childbirth, contractions of mammary glands in breast-feeding, and parental caretaking behaviors and sexual pleasure in both genders.
- Hyposecretion of ADH causes diabetes insipidus.

10.3 Thyroid Gland

- The thyroid gland is located just inferior to the larynx, with two lobes lateral to the trachea.
- TSH stimulates the secretion of thyroxine (T_4) and triiodothyronine (T_3), which increase cellular metabolism, protein synthesis, and neural activity.
- Iodine is an essential component of the T_4 and T_3 molecules.
- Calcitonin decreases the level of blood Ca^{2+} by promoting Ca^{2+} deposition in bones. It also promotes the excretion of Ca^{2+} by the kidneys. Its secretion is controlled humorally by the level of Ca^{2+} in the blood.
- Hypersecretion of thyroid hormones causes Graves disease. Iodine deficiency causes simple goiter.
- Hyposecretion of thyroid hormones in infants and children causes cretinism; in adults, it causes myxedema.

10.4 Parathyroid Glands

- The parathyroid glands are embedded in the posterior surface of the thyroid gland.
- Parathyroid hormone increases the level of blood Ca^{2+} by promoting Ca^{2+} removal from bones, Ca^{2+} absorption from the intestine, and Ca^{2+} retention by the kidneys. PTH also activates vitamin D, which helps stimulate Ca^{2+} absorption by intestine.
- Parathyroid secretion is controlled humorally by the level of blood Ca^{2+}.
- Parathyroid hormone and calcitonin work antagonistically to regulate blood Ca^{2+} levels.
- Hyposecretion of PTH causes tetany, which may result in death. Hypersecretion causes weak, soft, deformed bones that may fracture spontaneously.

10.5 Adrenal Glands

- An adrenal gland is located superior to each kidney. Each gland consists of two parts: a deep adrenal medulla and a superficial adrenal cortex.
- The adrenal medulla secretes epinephrine and norepinephrine, which prepare the body to deal with emergency situations. They increase the heart rate, circulation to nervous and muscular systems, and glucose level in the blood.
- The adrenal cortex secretes a number of hormones that are classified as mineralocorticoids, glucocorticoids, and androgens.
- Aldosterone is the most important mineralocorticoid. It helps to regulate the concentration of electrolytes in the blood, especially sodium and potassium ions, which increases blood pressure.
- Cortisol is the most important glucocorticoid. It promotes the formation of glucose from noncarbohydrate sources and inhibits inflammation. Its secretion is regulated by ACTH.
- Cortisol is involved in the response to chronic stress.
- Small amounts of androgens are secreted. They have little effect in adult males but contribute to the sex drive in adult females.
- Hyposecretion of cortisol causes Addison disease. Hypersecretion causes Cushing syndrome.

10.6 Pancreas

- The pancreas is both an exocrine and an endocrine gland. Its hormones are formed by the pancreatic islets, and their secretions are controlled by the level of blood glucose.
- Glucagon, from the alpha cells, increases the level of blood glucose by stimulating the liver to form glucose from glycogen and some noncarbohydrate sources.
- Insulin, from the beta cells, decreases the level of blood glucose by aiding the movement of glucose into cells.
- The antagonistic functions of glucagon and insulin keep the level of blood glucose within normal limits.
- Hyposecretion of insulin or a decrease in the number of insulin receptors causes diabetes mellitus. Hypersecretion may cause hypoglycemia.

10.7 Gonads

- Gonads are the sex glands: the ovaries in females and the testes in males. They secrete sex hormones, in addition to producing sex cells. The secretion of these hormones is controlled by FSH and LH.
- Estrogens are secreted by ovarian follicles and they stimulate development of female reproductive organs and secondary sex characteristics. Estrogens also help to prepare the uterus for a preembryo and help to maintain pregnancy.
- Progesterone is secreted mostly by the corpus luteum of the ovary after ovulation. It prepares the uterus for the preembryo, maintains pregnancy, and prepares the mammary glands for milk production.
- The testes secrete testosterone, the male sex hormone that stimulates the development of the male reproductive organs and secondary sex characteristics.

10.8 Other Endocrine Glands and Tissues

- The pineal gland is located near the roof of the third ventricle of the brain. It secretes melatonin, which seems to lead to the inhibition of secretion of FSH and LH by the anterior lobe of the pituitary gland. The pineal gland also seems to be involved in biorhythms.
- The thymus is located in the thoracic cavity superior to the heart. It secretes thymosins, which are involved in the maturation of white blood cells called T lymphocytes.
- Thymosins also seem to have anti-aging effects.

Self-Review

Answers are located in appendix B.

1. Chemical coordination of body functions is the function of the ______ system, whose glands secrete ______ that serve as chemical messengers.
2. A particular hormone affects only those cells that have ______ for that hormone.
3. ______ hormones use a second messenger to produce their characteristic effects on cells.
4. The secretion of most hormones is regulated by a ______ mechanism.
5. The secretion of pituitary hormones is regulated by a part of the brain called the ______.
6. The pituitary gland secretes four hormones that regulate secretion of other endocrine glands. ______ acts on the thyroid gland; ACTH acts on the ______; ______ and ______ act on the gonads.
7. Metabolic rate is regulated by ______ secreted by the ______.
8. The concentration of Ca^{2+} in the blood is regulated by two hormones with antagonistic actions: ______ promotes Ca^{2+} deposition in bones; ______ promotes Ca^{2+} removal from bones.
9. Secretions of the adrenal ______ prepare the body to react in emergencies.
10. The primary hormone regulating the concentration of mineral ions in the blood is ______.
11. The pancreatic hormone that increases the concentration of blood glucose is ______.
12. The primary sex hormones in females are ______ and ______; the male sex hormone is ______.

Critical Thinking

1. Some hormones affect many widely distributed cells in the body but others affect relatively few, localized cells. Explain how this occurs.
2. A blood test indicates that a patient has a low level of thyroxine. What are three possible causes of this condition? Explain.
3. A tumor in the parathyroid gland causes hypersecretion of PTH. Predict (1) the effects of this hormone on the skeletal system and (2) the effects on calcitonin production.
4. Using what you have learned about the endocrine system, explain why individuals who work the "night shift" have such a hard time staying awake.

ADDITIONAL RESOURCES

11 CHAPTER

Blood

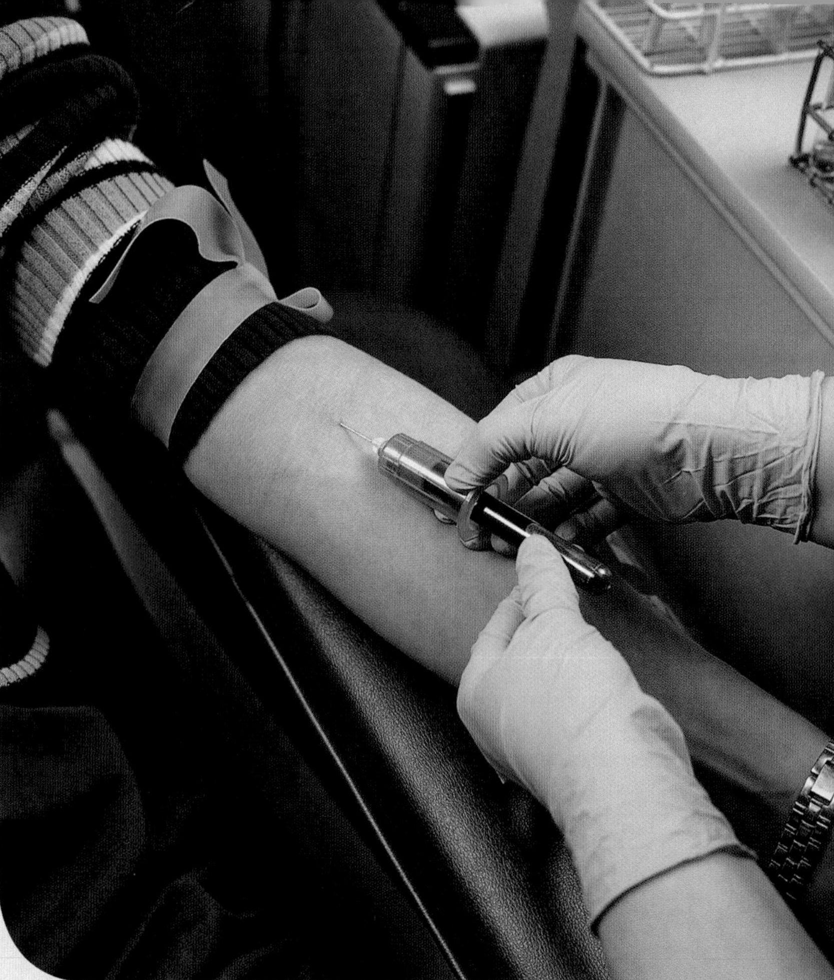

Phillip, at the age of 35, has been actively donating blood at the local Red Cross chapter for ten years. Since he is type AB+, his whole blood donations can be used to help only type AB+ patients in need. However, at his last visit, Phillip learned that he had the ability to help more people by donating his platelets and plasma specifically. Cancer patients undergoing chemotherapy can suffer from platelet deficiency, which results in an increased risk of bleeding. These patients usually benefit from platelet transfusions to supplement what their own bodies cannot produce. Plasma, specifically the proteins within it, is frequently used to treat many rare diseases, such as bleeding disorders, immune deficiency disorders, and rabies. Because Phillip has type AB+ blood, his plasma lacks antibodies that are capable of creating adverse reactions in people with other blood types. Since his plasma can be transfused into anyone with need safely, Phillip is considered a "universal plasma donor." Phillip's next appointment is in a few weeks and he is excited that, by donating specific blood components, he will be able to do so much for so many.

CHAPTER OUTLINE

Module 9
Cardiovascular System

SELECTED KEY TERMS

Agglutination (agglutin = to stick together) The clumping of red blood cells in an antigen-antibody reaction.
Coagulation The formation of a blood clot.
Embolus A moving blood clot or foreign body in the blood.
Formed elements The solid components of blood: red blood cells, white blood cells, and platelets.
Hematopoiesis (hemato = blood; poiesis = to make) The formation of formed elements.
Hemoglobin (hemo = blood) The pigmented protein in red blood cells, involved in transporting oxygen and carbon dioxide.
Hemostasis (hemo = blood; stasis = standing still) The stoppage of bleeding.
Plasma The liquid portion of blood.
Platelet A cellular fragment in blood, involved in blood clot formation.
Red blood cell A hemoglobin-containing blood cell that transports respiratory gases; an erythrocyte.
Thrombus A stationary blood clot or foreign body in a blood vessel.
White blood cell A blood cell that has defensive and immune functions; a leukocyte.

BLOOD IS USUALLY CONFINED WITHIN THE HEART AND BLOOD VESSELS as it transports materials from place to place within the body. Substances carried by blood include oxygen, carbon dioxide, nutrients, waste products, hormones, electrolytes, and water. Blood also has several regulatory and protective functions that will be described in this chapter.

11.1 General Characteristics of Blood

Learning Objective

1. Describe the general characteristics and functions of blood.

Blood is classified as a connective tissue that is composed of **formed elements** (the solid components, including blood cells and platelets) suspended in **plasma,** the liquid portion (matrix) of the blood. It is one of the two fluid connective tissues in the body. Blood is heavier and about four times more viscous than water. It is slightly alkaline, with a pH between 7.35 and 7.45. The volume of blood varies with the size of the individual, but it averages 5 to 6 liters in males and 4 to 5 liters in females. Blood comprises about 8% of the body weight.

About 55% of the blood volume consists of plasma, and 45% is made up of formed elements. Because the majority of the formed elements are red blood cells (RBCs), it can be said that almost 45% of the blood volume consists of red blood cells. White blood cells (WBCs) and platelets combined form less than 1% of the blood volume (figure 11.1).

The great number of formed elements in blood is hard to imagine. There are approximately 5 million RBCs, 7,500 WBCs, and 300,000 platelets in one single microliter (µl). A single drop of blood due to a finger stick (approximately 50 ul) contains 250 million RBCs!

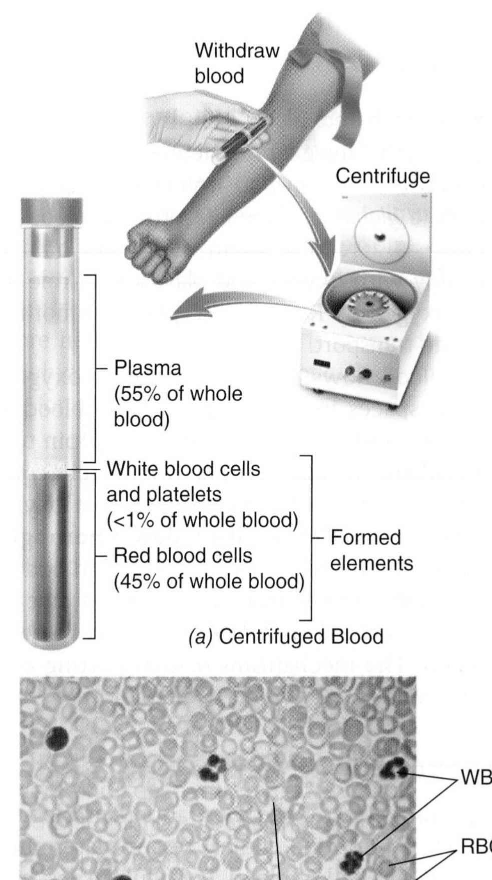

Figure 11.1 Blood Consists of Plasma and Formed Elements.
(a) If blood is centrifuged, the RBCs sink to the bottom of the tube and the liquid plasma forms the top layer. WBCs and platelets form a thin layer between the two. *(b)* The microscopic appearance of formed elements in a smear of blood. APR

if they plug an artery and deprive vital tissues of blood. Blood clots form more frequently in veins than in arteries, causing a condition known as *thrombophlebitis*, which is inflammation of the veins due to a blood clot.

Sometimes, a clot formed in a vein breaks free and is carried by the blood only to lodge in an artery, often a branch of a pulmonary artery. A moving blood clot or foreign body in the blood is called an **embolus,** and when it blocks a blood vessel, the resulting condition is known as an **embolism.** An embolism can produce very serious and sometimes fatal results if it lodges in a vital organ and blocks the flow of blood.

Chapter Summary

11.1 General Characteristics of Blood

- Blood is composed of plasma (55%) and formed elements (45%). Red blood cells constitute nearly all of the formed elements.
- Blood is heavier and about four times more viscous than water, and it is slightly alkaline.
- About 8% of the body weight consists of blood. Blood volume ranges between 4 and 6 liters.

11.2 Red Blood Cells

- Red blood cells are biconcave discs that lack nuclei and other organelles, and contain a large amount of hemoglobin. Their primary function is the transport of respiratory gases.
- Hemoglobin is composed of heme, an iron-containing pigment, and globin, a protein. It plays a vital role in oxygen transport and participates in carbon dioxide transport.
- RBCs are very abundant in the blood. They number 4.7 to 6.1 million per µl in males and 4.2 to 5.4 million per µl in females.
- RBCs are formed from hemocytoblasts in the red bone marrow. The rate of production is controlled by the oxygen concentration of the blood via a negative-feedback mechanism. A decreased oxygen concentration stimulates kidney and liver cells to release erythropoietin, which stimulates increased production of RBCs by red bone marrow.
- Iron, amino acids, vitamin B12, and folic acid are essential for RBC production.
- RBCs live about 120 days before they are destroyed by macrophages in the spleen and liver. In hemoglobin breakdown, the iron ions are recycled for use in forming more hemoglobin. Bilirubin, a yellow pigment, is a waste product of hemoglobin breakdown. Amino acids from globin are recycled for use in making new proteins.

11.3 White Blood Cells

- White blood cells are also formed from hemocytoblasts in the red bone marrow. They retain their nuclei and other organelles, and number 4,500 to 10,000 per µl of blood.
- WBCs help to defend the body, and most of their activities occur within body tissues.
- The five types of WBCs are categorized into two groups. Granulocytes have visible cytoplasmic granules and include neutrophils, eosinophils, and basophils. Agranulocytes lack visible cytoplasmic granules and include lymphocytes and monocytes.
- Neutrophils and monocytes are phagocytes that destroy bacteria and clean up cellular debris.
- Eosinophils help to reduce inflammation and destroy parasitic worms.
- Basophils promote inflammation.
- Lymphocytes play vital roles in immunity.

11.4 Platelets

- Platelets are fragments of megakaryocytes in the red bone marrow. They number 150,000 to 400,000 per µl of blood.
- Platelets play a crucial role in hemostasis by forming platelet plugs and starting coagulation.

11.5 Plasma

- Plasma, the liquid portion of the blood, consists of over 90% water along with a variety of solutes, including nutrients, nitrogenous wastes, proteins, electrolytes, and respiratory gases.
- There are three major types of plasma proteins. Albumins are most numerous. Their major functions include the transport of hydrophobic substances, and helping to maintain the osmotic pressure and pH of the blood. Alpha and beta globulins transport lipids and lipid-soluble vitamins. Gamma globulins are antibodies that are involved in immunity. Fibrinogen is a soluble protein that is converted into insoluble fibrin during coagulation.
- Less than 1% of plasma proteins are enzymes and hormones.
- Nitrogenous wastes in plasma include urea, uric acid, ammonia, and creatinine.
- Electrolytes include ions of sodium, potassium, calcium, bicarbonate, phosphate, and chloride. Electrolytes help to maintain the pH and osmotic pressure of the blood, in addition to the ionic balance between blood and interstitial fluid.

11.6 Hemostasis

- Hemostasis is a series of processes involved in the stoppage of bleeding. It consists of three processes: vascular spasm, platelet plug formation, and coagulation.
- Vascular spasm reduces blood loss until the other processes can occur.
- Platelets stick to the damaged tissue of the blood vessel wall and to each other to form a platelet plug.
- Platelets and the damaged blood vessel wall initiate clot formation by releasing platelet factors and

thromboplastin, which cause the formation of prothrombin activator. Prothrombin activator converts prothrombin into thrombin, which, in turn, converts fibrinogen into fibrin. Fibrin strands form the clot.
- After clot formation, fibroblasts invade the clot and gradually replace it with dense irregular connective tissue as the clot is dissolved by enzymes.

11.7 Human Blood Types

- Blood types are determined by the presence or absence of specific antigens on the plasma membranes of red blood cells.
- The four ABO blood types, A, B, AB, and O, are based on the presence or absence of A antigen and B antigen.
- Anti-A and anti-B antibodies are spontaneously formed against the antigen(s) that is (are) not present on a person's RBCs.
- Blood with RBCs containing the Rh antigen is typed as Rh+. Blood without the Rh antigen is typed as Rh−.
- Anti-Rh antibodies are produced only after Rh+ RBCs are introduced into a person with Rh− blood. Once a person is sensitized in this way, a subsequent transfusion of Rh+ blood results in agglutination of the transfused RBCs.
- If incompatible blood is transferred, agglutination of the transfused RBCs occurs. The clumped RBCs plug small blood vessels, depriving tissues of nutrients and oxygen. The result may be fatal.
- Transfusions must be made using only compatible blood types. Types A, B, AB, and O blood recipients can only receive RBCs with antigens that will not trigger an agglutination reaction with antibodies present in plasma. Type Rh+ blood recipients can receive the RBCs of types Rh− and Rh+ blood. Type Rh− blood recipients can receive the RBCs of type Rh− blood only.
- Hemolytic disease of the newborn occurs in newborn infants when a sensitized Rh− woman is pregnant with an Rh+ fetus. Her anti-Rh antibodies pass through the placenta into the fetus and agglutinate the fetal RBCs, producing anemia and jaundice.

11.8 Disorders of the Blood

- Anemia is the most common disorder, and it may result from a variety of causes.
- Other disorders include polycythemia, infectious mononucleosis, leukemia, hemophilia, thrombocytopenia, thrombosis, and embolism.

Self-Review

Answers are located in appendix B.

1. About ______% of blood consists of RBCs.
2. The red color of blood results from the presence of _____ in ______.
3. All formed elements are derived from stem cells, the ______, within red bone marrow.
4. A decreased blood concentration of ______ promotes the formation of the hormone ______, which stimulates RBC production.
5. RBCs are destroyed in the spleen and ______.
6. Fighting against invasion of pathogens is the function of nucleated formed elements called ______.
7. The two major phagocytic WBCs are ______ and ______.
8. The release of histamine by ______ helps to promote inflammation.
9. WBCs that destroy parasitic worms and fight inflammation are the ______.
10. Immunity is the prime function of ______.
11. The fluid carrier of solutes and formed elements in blood is the ______.
12. Damaged blood vessel walls and ______ start coagulation by releasing thromboplastin and platelet factors.
13. Blood clot formation involves converting ______, a soluble plasma protein, into an insoluble protein called ______.
14. ABO blood types are named for the ______ on the surface of RBCs.
15. Blood type B+ can receive the RBCs of blood types _____ safely in a transfusion.

Critical Thinking

1. In the days before RhoGAM, some Rh− women had more than one Rh+ baby and never had a problem with hemolytic disease of the newborn. How do you explain this?
2. What are the differences between coagulation and agglutination?
3. Why can persons with type O blood donate blood to any other blood type?
4. Why is a CBC a useful test in monitoring the homeostasis of the human body?

ADDITIONAL RESOURCES

12 CHAPTER

The Cardiovascular System

A two-alarm fire is called in and the alarm begins to sound in the local fire station. Charlie, a veteran firefighter, begins shout directions as he and the others in his unit don their gear. As they travel to the site of the blaze, Charlie is so focused on the task at hand that he is barely aware of the cardiovascular changes occurring within his body. His heart rate increases in order to increase his blood pressure, which in turn increases blood flow through his body. Changes within his blood vessels allow blood flow to be prioritized to organs that will be called upon once he arrives at the scene. Increasing activity in his skeletal muscle tissue, cardiac muscle tissue, and nervous tissue requires elevated rates of ATP production, which in turn require an increase in the delivery of oxygen, glucose, and fatty acids. Increased blood flow to the lungs, liver, and adipose tissue is needed to maintain sufficient levels of these vital chemicals. By the time the fire truck reaches the scene, Charlie is physically prepared to rush into the burning building to rescue trapped inhabitants, thanks in part to the actions of his cardiovascular system.

CHAPTER OUTLINE

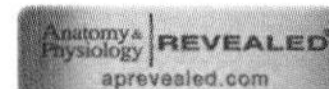

Module 9
Cardiovascular System

SELECTED KEY TERMS

Arteries Blood vessels that carry blood away from the heart.
Atrium (atrium = vestibule) A heart chamber that receives blood returned to the heart by veins.
Capillaries Tiny blood vessels in tissues where exchange of materials between the blood and interstitial fluid occurs.
Cardiac output The volume of blood pumped from each ventricle in one minute.
Cardiac cycle The sequence of events that occur during one heartbeat.
Diastole The relaxation phase of the cardiac cycle.
Pulmonary circuit (pulmo = lung) The blood pathway that transports blood to and from the lungs.
Stroke volume The volume of blood pumped from each ventricle per heartbeat.
Systemic circuit The blood pathway that transports blood to and from all parts of the body except the lungs.
Systole The contraction phase of the cardiac cycle.
Vasoconstriction (vas = vessel) Contraction of vessel smooth muscle to decrease the diameter of the blood vessel.
Vasodilation Relaxation of vessel smooth muscle to increase the diameter of the blood vessel.
Veins Blood vessels that carry blood toward the heart.
Ventricle (ventr = underside) A heart chamber that pumps blood into an artery.

THE HEART AND BLOOD VESSELS form the *cardiovascular* (kar-dē-ō-vas′-kū-lar) *system.* The heart pumps blood through a closed system of blood vessels. Figure 12.1 shows the general scheme of circulation of blood in the body. Blood vessels colored blue carry deoxygenated (oxygen-poor) blood; those colored red carry oxygenated (oxygen-rich) blood. Large arteries carry blood away from the heart and branch into smaller and smaller arteries that open into capillaries, the smallest blood vessels, where materials are exchanged with body tissues. Capillaries open into small veins that merge to form larger and larger veins, and the largest veins return blood to the heart. APIR

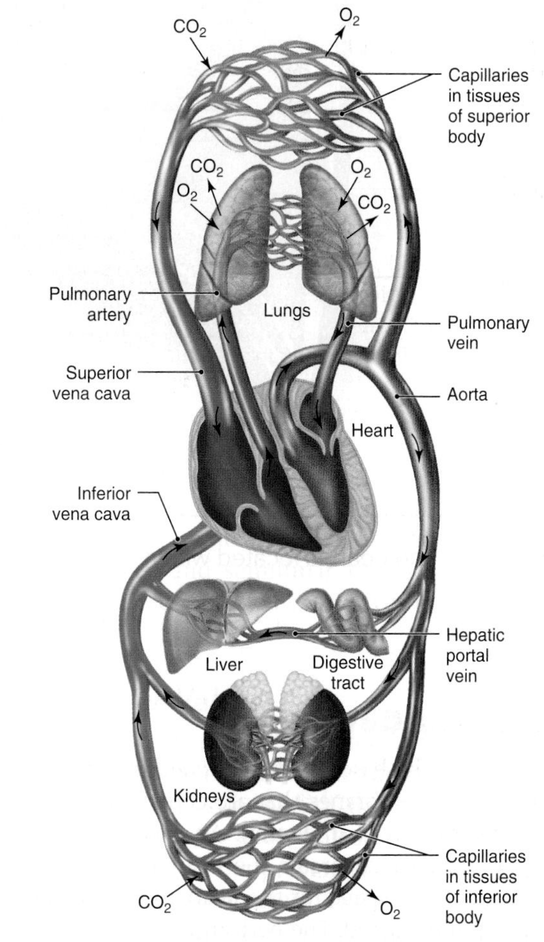

Figure 12.1 The general scheme of the cardiovascular system. Blood vessels carrying oxygenated blood are colored red; those carrying deoxygenated blood are colored blue. APIR

12.1 Anatomy of the Heart

Learning Objectives

1. Identify the protective coverings of the heart.
2. Describe the parts of the heart and their functions.
3. Trace the flow of blood through the heart.
4. Describe the blood supply to the heart.

The heart is a four-chambered muscular pump that is located within the mediastinum in the thoracic cavity. It lies between the lungs and just superior to the diaphragm. The *apex* of the heart is the inferior pointed end, which extends toward the left side of the thoracic cavity at the level of the fifth rib. The *base* of the heart is the superior portion, which is attached to several large blood vessels at the level of the second rib. The heart is about the size of a closed fist. Note the relationship of the heart with the surrounding organs in figure 12.2.

Flow of Blood Through the Heart

Figure 12.7 diagrammatically shows the flow of blood through the heart and the major vessels attached to the heart. Blood is oxygenated as it flows through the lungs and becomes deoxygenated as it releases oxygen to body tissues. Trace the flow of blood through the heart and major vessels in figure 12.7 as you read the following description.

The right atrium receives deoxygenated blood from all parts of the body except the lungs via three veins: the superior and inferior venae cavae and the coronary sinus. The **superior vena cava** (vē′-nah kā′-vah) returns blood from the head, neck, shoulders, upper limbs, and thoracic and abdominal walls. The **inferior vena cava** returns blood from the inferior trunk and lower limbs. The **coronary sinus** drains deoxygenated blood from cardiac muscle tissue. Simultaneously, the left atrium receives oxygenated blood returning to the heart from the lungs via the **pulmonary veins.** Blood flows from the left and right atria into the corresponding ventricles. About 70% of the blood flow into the ventricles is passive, and about 30% results from atrial contraction.

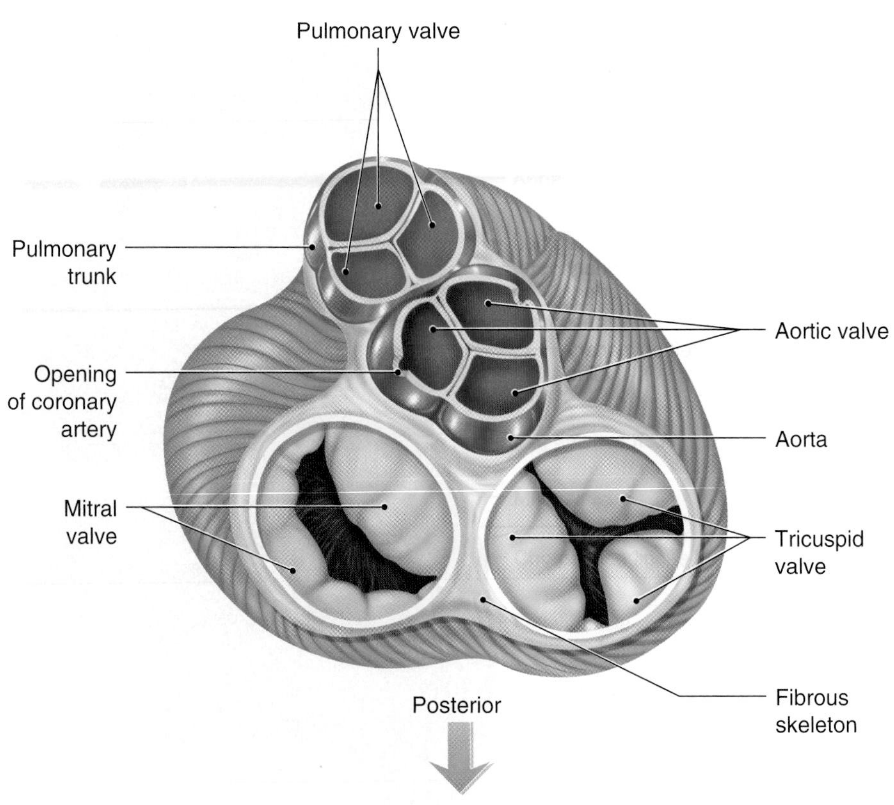

Figure 12.6 A superior view of the heart valves. Note the fibrous skeleton of the heart.

After blood has flowed from the atria into their respective ventricles, the ventricles contract. The right ventricle pumps deoxygenated blood into the **pulmonary trunk.** The pulmonary trunk branches to form the **left** and **right pulmonary arteries,** which carry blood to the lungs. The left ventricle pumps oxygenated blood into the **aorta** (ā-or′-tah). The aorta branches to form smaller arteries that carry blood to all parts of the body except the lungs. Locate these major blood vessels associated with the heart in figures 12.2, 12.4, 12.5, and 12.7.

Because the heart is a double pump, there are two basic pathways, or circuits, of blood flow as shown in figure 12.7. The **pulmonary circuit** carries deoxygenated blood from the right ventricle to the lungs and returns oxygenated blood from the lungs to the left atrium. The **systemic circuit** carries oxygenated blood from the left ventricle to all parts of the body except the lungs and returns deoxygenated blood to the right atrium.

Blood Supply to the Heart

The heart requires a constant supply of blood to nourish its own tissues. Blood is supplied by **left** and **right coronary** (kor′-ō-na-rē) **arteries**, which branch from the aorta just distal to the aortic valve (figures 12.6 and 12.18a). Blockage of a coronary artery may result in a heart attack. After passing through capillaries in cardiac muscle tissue, blood is returned via **cardiac** (kar′-dē-ak) **veins**, which lie next to the coronary arteries. These veins empty into the **coronary sinus,** which drains into the right atrium. Locate these blood vessels in figures 12.2 and 12.5 and note the adipose tissue that lies alongside the vessels. Also, study the relationships of the atria, ventricles, and large blood vessels associated with the heart.

Check My Understanding

1. What are the names and functions of the heart chambers?
2. What are the names and functions of the heart valves?
3. Trace a drop of blood as it flows through the heart and the pulmonary and systemic circuits.
4. Describe the flow of blood throughout the myocardium.

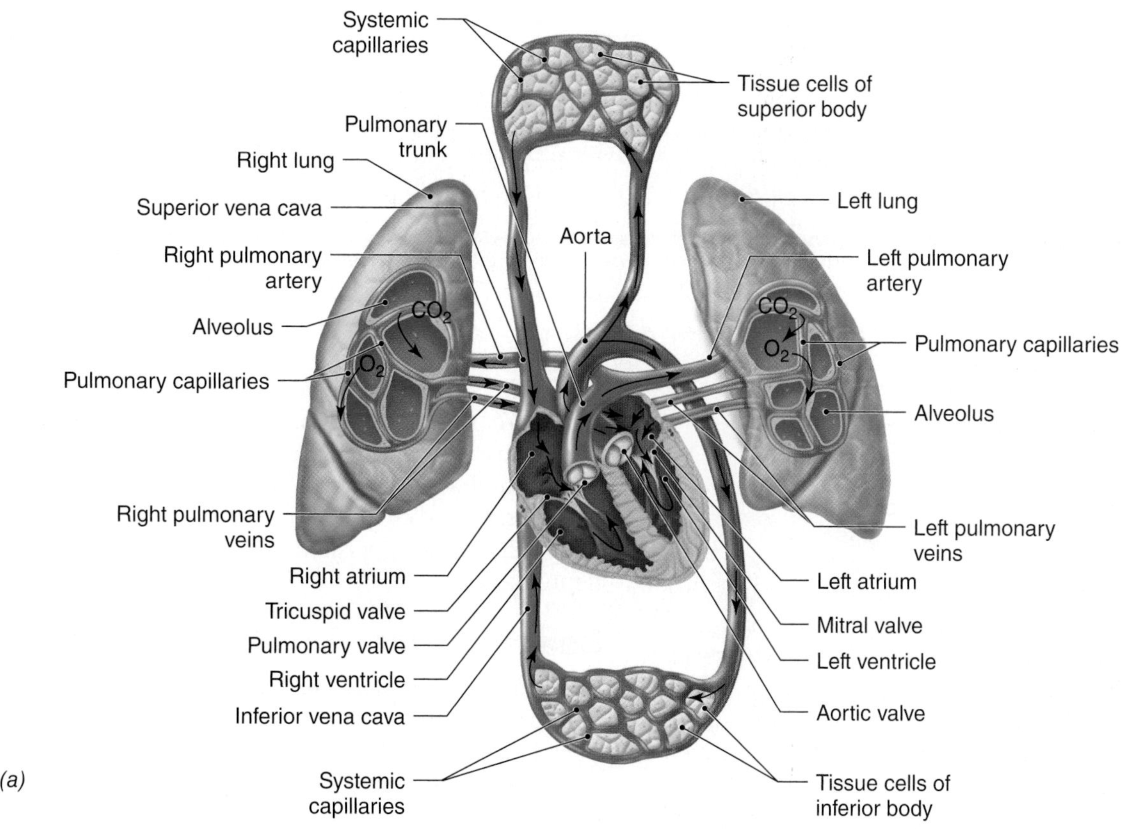

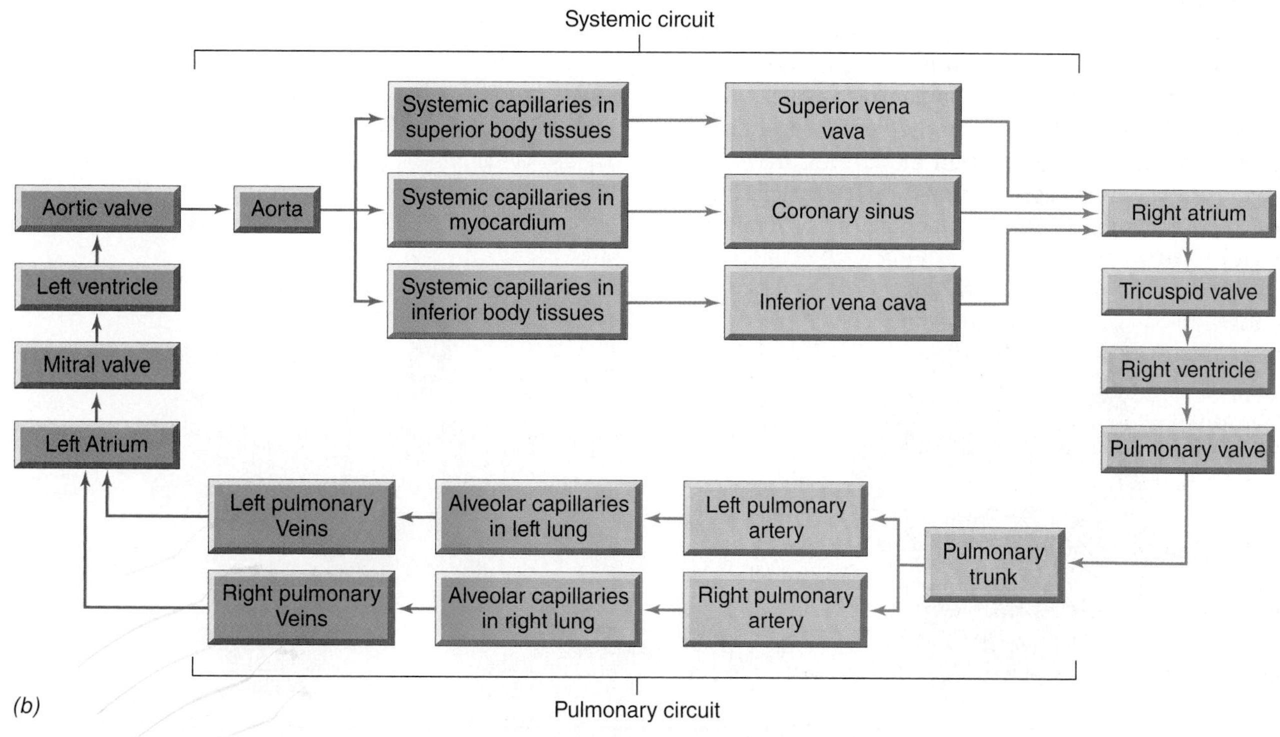

Figure 12.7 Blood flow through the heart and the systemic and pulmonary circuits. Heart chambers and vessels colored red carry oxygenated blood. Those colored blue carry deoxygenated blood.

Clinical Insight

If cusps of an AV valve collapse and open into the atrium, some blood may regurgitate (backflow) into the atrium during ventricular contractions. This is what happens in a disorder known as *mitral valve prolapse (MVP).* In some cases, it causes no serious dysfunction. In others, fatigue and shortness of breath may occur. Persons with MVP are susceptible to *endocarditis,* inflammation of the endocardium, caused by some species of *Streptococcus* bacteria. Endocarditis can result in scarring of the valve cusps, which further decreases valve function. Persons with MVP are often advised to take antibiotics prior to dental work to prevent bacteria from entering the blood and being carried to the heart.

12.2 Cardiac Cycle

Learning Objectives

5. Describe the events of the cardiac cycle.
6. Describe the sounds of the heartbeat.

The **cardiac cycle** refers to the sequence of events that occur during one heartbeat. The contraction phase of a cardiac cycle is known as **systole** (sis′-to-lē); the relaxation phase is called **diastole** (dī′-as-to-lē). These phases are illustrated in figure 12.8. Note that the ventricles are relaxed when the atria contract, and the atria are relaxed when the ventricles contract. Systole increases blood pressure within a chamber, while diastole decreases blood pressure within a chamber.

When both the atria and ventricles are relaxed between beats, blood flows passively into the atria from the large veins leading to the heart and then passively into the ventricles. Then, the atria contract (atrial systole), forcing more blood into the ventricles so that they are filled. Immediately thereafter, the ventricles contract. Ventricular systole produces high blood pressure within the ventricles, which causes both AV valves to close and both semilunar valves to open. Opening of the semilunar valves allows blood to move into the arteries leading from the heart. Ventricular diastole immediately follows and the decrease in ventricle pressure allows the AV valves to open. Simultaneously, the semilunar valves close because of the greater blood pressure within the arteries. The cardiac cycle is then repeated. Study these relationships in figure 12.8.

Heart Sounds

The sounds of the heartbeat are usually described as *lub-dup* (pause) *lub-dup,* and so forth. These sounds are produced by the closing of the heart valves. The first sound results from the closing of the AV valves in the beginning of ventricular systole. The second sound results from the closing of the semilunar valves in the beginning of ventricular diastole. If any of the heart valves are defective and do not close properly, an additional sound, known as a heart murmur, may be heard.

Check My Understanding

5. What are the events of a cardiac cycle?
6. What produces the heart sounds?

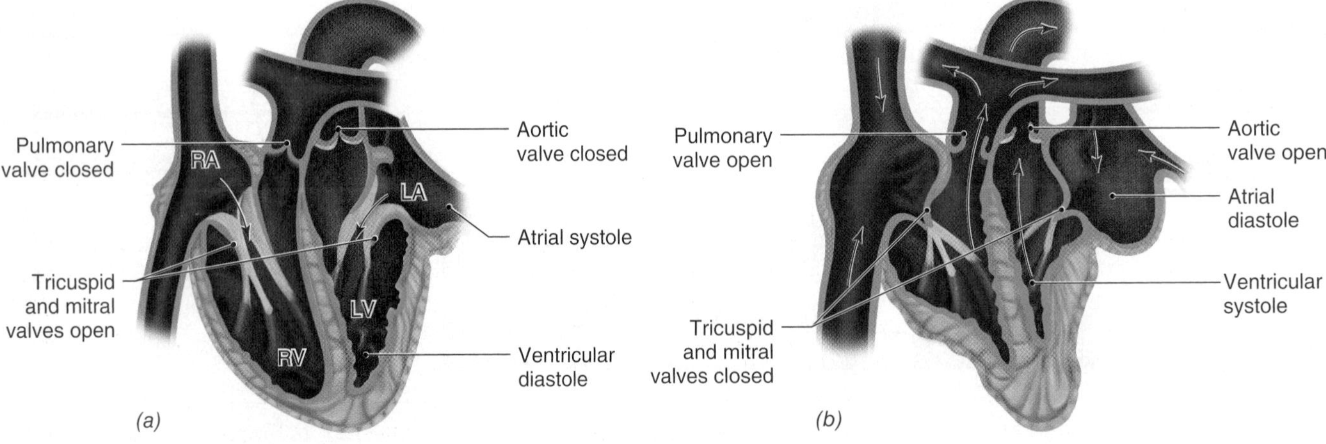

Figure 12.8 The Cardiac Cycle.
(a) Blood flows from the atria into the ventricles during ventricular diastole. *(b)* Blood is pumped from the ventricles during ventricular systole. APR

12.3 Heart Conduction System

Learning Objective

7. Describe the parts of the heart conduction system and their functions.

The heart is able to contract on its own because it contains specialized cardiac muscle tissue that spontaneously forms impulses and transmits them to the myocardium to initiate contraction. This specialized tissue forms the *conduction system* of the heart, which consists of the sinoatrial node, atrioventricular node, AV bundle, bundle branches, and ventricular fibers. Observe the location of the conduction system and its parts in figure 12.9.

The **sinoatrial** (sī-nō-ā′-trē-al) **node (SA node)** is located in the right atrium at the junction of the superior vena cava. It is known as the pacemaker of the heart because it rhythmically forms electrical impulses to initiate each heartbeat. The impulses are transmitted to the myocardium of the atria, where they produce a simultaneous contraction of the atria. The flow of impulses causes contraction of the atria from superior to inferior, forcing blood into the ventricles. At the same time, the impulses are carried to the **atrioventricular node (AV node),** which is located in the right atrium near the junction with the interventricular septum.

There is a brief time delay as the impulses pass slowly through the AV node, which allows time for the ventricles to fill with blood. From the AV node, the impulses pass along the **AV bundle** *(bundle of His),* a group of large fibers that divide into **left** and **right bundle branches** extending inferiorly to the interventricular septum and superior to the lateral walls of the ventricles. The smaller **ventricular** *(Purkinje)* **fibers** arise from the bundle branches and carry the impulses to the myocardium of the ventricles, where they stimulate ventricular contraction. The distribution of the ventricular fibers causes the ventricles to contract from the apex superiorly so that blood is forced into the pulmonary trunk and aorta.

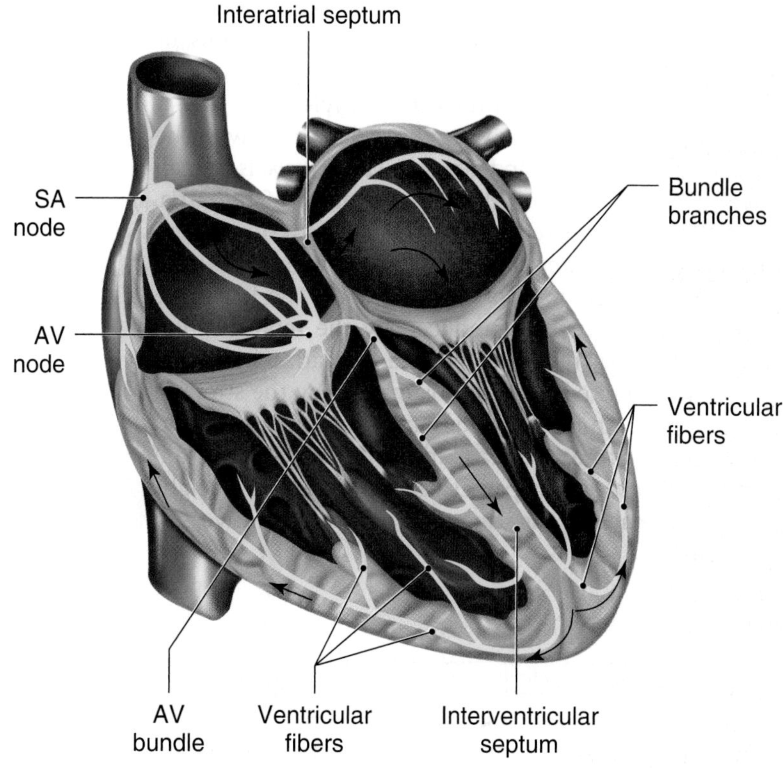

Figure 12.9 The heart conduction system. Arrows indicate the flow of impulses from the SA node. AP|R

Clinical Insight

If a coronary artery is partially obstructed by the fatty deposits of *atherosclerosis* (see the disorders section in this chapter for details), portions of the myocardium may be deprived of adequate blood. This produces chest pain known as *angina pectoris.* In severe cases, treatment may involve one of two approaches: coronary angioplasty or coronary bypass surgery.

In *coronary angioplasty,* a catheter that contains a balloon at its tip is inserted into an artery of an upper or lower limb and is threaded into the affected coronary artery. The balloon is positioned at the obstruction and is inflated for a few seconds to compress the fatty deposit and enlarge the lumen of the affected coronary artery. A meshlike metal tube called a stent is then inserted and positioned at the site of the obstruction to hold open the artery. The stent may be coated with a chemical that inhibits the growth of cells to minimize the chances that the artery will become obstructed again.

In *coronary bypass surgery,* a portion of an artery or a vein from elsewhere in the body is removed and is surgically grafted, providing a bypass around the obstruction to supply blood to the distal portion of the affected coronary artery.

Neurons of the sympathetic division extend axons from the cardiac control center down the spinal cord to the thoracic region. There the sympathetic axons exit the spinal cord to innervate the SA node, AV node, and portions of the myocardium. The transmission of nerve impulses causes the sympathetic axons to secrete *norepinephrine* at synapses in the heart. Norepinephrine increases the heart rate and strengthens the force of myocardial contraction. Physical and emotional stresses, such as exercise, excitement, anxiety, and fear, stimulate the sympathetic division to increase heart rate and contraction strength.

Parasympathetic axons arise from the cardiac control center and exit in the vagus nerve (CN X) to innervate the SA and AV nodes. The transmission of nerve impulses causes the parasympathetic axons to secrete *acetylcholine* at the heart synapses, which decreases the heart rate. The greater the frequency of parasympathetic nerve impulses sent to the heart, the slower the heart rate. Excessive blood pressure and emotional factors, such as grief and depression, stimulate the parasympathetic division to decrease the heart rate.

When the heart is at rest, more parasympathetic nerve impulses than sympathetic nerve impulses are sent to the heart. As cellular needs for blood increase, a decrease in the frequency of parasympathetic nerve impulses and an increase in sympathetic nerve impulses cause heart rate to increase.

Other Factors Affecting Heart Function

Age, sex, physical condition, temperature, epinephrine, thyroxine, and the blood levels of calcium and potassium ions also affect the heart rate and contraction strength.

The resting heart rate gradually declines with age, and it is slightly faster in females than in males. Average resting heart rates in females are 72 to 80 beats per minute, as opposed to 64 to 72 beats per minute in males. People who are in good physical condition have a slower resting heart rate than those in poor condition. Athletes may have a resting heart rate of only 40 to 60 beats per minute. An increase in body temperature, which occurs during exercise or when feverish, increases the heart rate.

Epinephrine, which is secreted by the adrenal glands during stress or excitement, affects the heart like norepinephrine–it increases the rate and strength of heart contractions. An excess of thyroxine produces a lesser, but longer-lasting, increase in heart rate.

Reduced levels of blood Ca^{2+} decrease the rate and strength of heart contraction, while increased levels of blood Ca^{2+} increase heart rate and contraction strength, and prolong contraction. In extreme cases, an excessively prolonged contraction may result in death. Excessive levels of blood K^+ decrease both heart rate and contraction strength. A high dose of K^+ is often used in lethal injections, in which the abnormally high levels of blood K^+ cause the heart to stop contracting. Abnormally low levels of blood K^+ may cause potentially life-threatening abnormal heart rhythms.

Check My Understanding

9. How are the heart rate and contraction strength regulated?
10. What other factors affect the heart rate and contraction strength?

12.5 Types of Blood Vessels

Learning Objectives

9. Describe the structure and function of arteries, arterioles, capillaries, venules, and veins.
10. Describe how materials are exchanged between capillary blood and interstitial fluid.

There are three basic types of blood vessels: arteries, capillaries, and veins. They form a closed system of tubes that carry blood from the heart to the tissue cells and back to the heart. Table 12.3 compares these three types.

Structure of Arteries and Veins

The walls of arteries and veins are composed of three distinct layers. The *tunica externa*, the most superficial layer, is formed of dense irregular connective tissue that includes both collagen and elastic fibers. These fibers provide support and elasticity for the vessel. The *tunica media*, the middle layer, usually is the thickest layer. It consists of smooth muscle cells that encircle the blood vessel. The smooth muscle cells not only provide support but also produce changes in the diameter of the blood vessel by contraction or relaxation. The *tunica intima*, the deepest layer, forms the internal lining of blood vessels. It consists of a simple squamous epithelium, called the *endothelium*, supported by thin layers of areolar connective tissue containing elastic and collagen fibers.

The walls of arteries and veins have the same basic structure. However, arterial walls are thicker because their tunica media contains more smooth muscle and elastic connective tissues as an adaptation to the higher blood pressure found in them. The tunica media of veins possesses very little smooth muscle, which leads to a much thinner wall. Veins possess larger lumens than arteries; as a result, they can hold a larger volume of blood. Another difference is that large veins, but not arteries, contain valves formed of endothelium. Venous valves prevent a backflow of blood. Compare the structure of arteries and veins in figure 12.12.

Table 12.3 Comparison of Arteries, Capillaries, and Veins

Type of Vessel	Function	Structure
Arteries	Carry blood from the heart to the capillaries Control blood flow and blood pressure	Composed of tunica intima, tunica media, and tunica externa Contain more smooth muscle and elastic connective tissues than veins
Capillaries	Enable exchange of materials between blood and interstitial fluid	Microscopic vessels composed of endothelium supported by areolar connective tissue
Veins	Return blood from capillaries to the heart Serve as storage areas for blood	Composed of tunica intima, tunica media, and tunica externa Have thinner walls and larger lumens than arteries Large veins have venous valves.

Arteries

Arteries carry blood away from the heart. They branch repeatedly into smaller and smaller arteries and ultimately form microscopic arteries called **arterioles** (ar-te′-rē-ōls). As arterioles branch and form smaller arterioles, the thickness of the tunica media decreases. The walls of the smallest arterioles consist of only the tunica intima and a few encircling smooth muscle cells. Arteries, especially the arterioles, play an important role in the control of blood flow and blood pressure.

Capillaries

Arterioles connect with **capillaries,** the most numerous and the smallest blood vessels. A capillary's diameter is so small that RBCs must pass through it in single file. The walls of capillaries consist of an endothelium supported by a layer of areolar connective tissue. These extremely thin walls facilitate the exchange of materials between blood in capillaries and tissue cells.

The distribution of capillaries in body tissues varies with the metabolic activity of each tissue. Capillaries are especially abundant in active tissues, such as muscle and nervous tissues, where nearly every cell is near a capillary. Capillaries are less abundant in connective tissues and are absent in some tissues, such as cartilage, epidermis, and the lens and cornea of the eye.

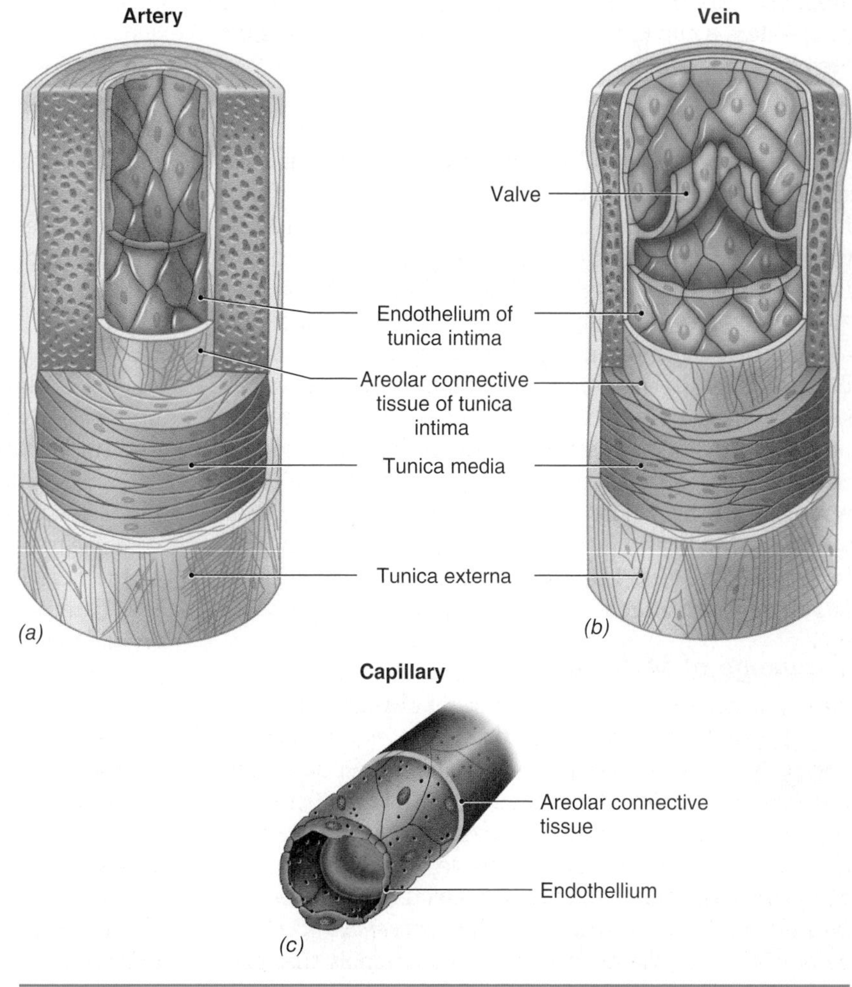

Figure 12.12 *(a)* The wall of an artery. *(b)* The wall of a vein. *(c)* The wall of a capillary. APR

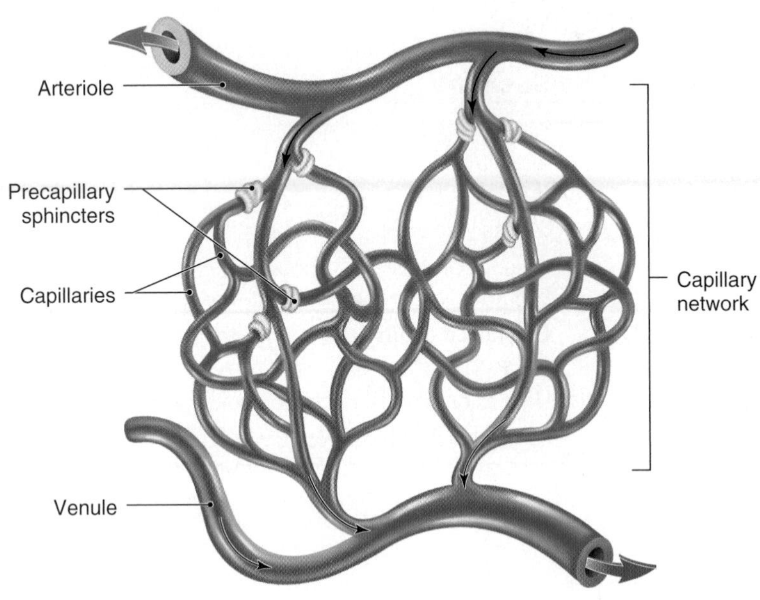

Figure 12.13 A capillary network. Precapillary sphincters regulate the blood flow from an arteriole into a capillary. Oxygenated blood (red) enters a capillary network. Deoxygenated blood (blue) exits the capillaries and enters a venule.

Blood flow in capillaries is controlled by *precapillary sphincters,* smooth muscle cells encircling the bases of capillaries at the arteriole-capillary junctions (figure 12.13). Contraction of a precapillary sphincter inhibits blood flow to its capillary network. Relaxation of the sphincter allows blood to flow into its capillary network to provide oxygen and nutrients for the tissue cells. The flow of blood in capillary networks occurs intermittently. When some capillary networks are filled with blood, others are not. Capillary networks receive blood according to the needs of the cells that they serve. For example, during physical exercise blood is diverted from capillary networks in the digestive tract to fill the capillary networks in skeletal muscles. This pattern of blood distribution is largely reversed after a meal.

Exchange of Materials

The continual exchange of materials between the blood and tissue cells is essential for life. Cells require oxygen and nutrients to perform their metabolic functions, and they produce carbon dioxide and other metabolic wastes that must be removed by the blood.

The cells of tissues are enveloped in a thin film of extracellular fluid called **interstitial fluid,** or *tissue fluid,* that fills tissue spaces and lies between the tissue cells and the capillaries. Therefore, all materials that pass between the blood and tissue cells must pass through the interstitial fluid. Dissolved substances such as oxygen and nutrients diffuse from blood in the capillary into the interstitial fluid and from the interstitial fluid into tissue cells. Carbon dioxide and metabolic wastes diffuse in the opposite direction.

Recall that the capillary walls are so thin that materials can readily diffuse through them, and the junctions between these cells are not tight so fluid is able to move between the cells. Two opposing forces determine the movement of fluid between capillary blood and interstitial fluid: osmotic pressure and blood pressure. Osmotic pressure of the blood results from plasma proteins. Osmotic pressure tends to "pull" fluid from interstitial fluid into the capillaries by osmosis. Blood pressure against the capillary walls results from the force of ventricular contractions. It tends to push fluid out of the capillaries into the interstitial fluid. This type of transport, forcing substances through a membrane due to greater hydrostatic pressure on one side of the membrane, is known as **filtration.**

At the arteriolar end of a capillary, blood pressure exceeds osmotic pressure, so fluid moves out of the capillary into the interstitial fluid. In contrast, at the venular end of the capillary, osmotic pressure exceeds blood pressure, so fluid moves from the interstitial fluid into the capillary by osmosis (figure 12.14). About nine-tenths of the fluid that moves from the arteriolar end of a capillary into the interstitial fluid returns into the venular end of the capillary. The remainder is picked up by the lymphoid system and ultimately is returned to the blood (see chapter 13).

Veins

After blood flows through the capillaries, it enters the **venules,** the smallest **veins.** Several capillaries merge to form a venule. The smallest venules consist only of endothelium and areolar connective tissue, but larger venules also contain smooth muscle tissue. Venules unite to form small veins. Small veins combine to form progressively larger veins as blood is returned to the heart. Larger veins, especially those in the upper and lower limbs, contain valves that prevent a backflow of blood and aid the return of blood to the heart.

Because nearly 60% of the blood volume is in veins at any instant, veins may be considered as storage areas for blood that can be carried to other parts of the body in times of need. Venous sinusoids in the liver and spleen are especially important reservoirs. If blood is lost by hemorrhage, both blood volume and pressure decline. In response, the sympathetic division sends nerve impulses to constrict the muscular walls of the veins, which reduces the venous volume while increasing blood volume and pressure in the heart, arteries and capillaries. This effect compensates for the blood loss. A similar response occurs during strenuous muscular activity in order to increase the blood flow to skeletal muscles.

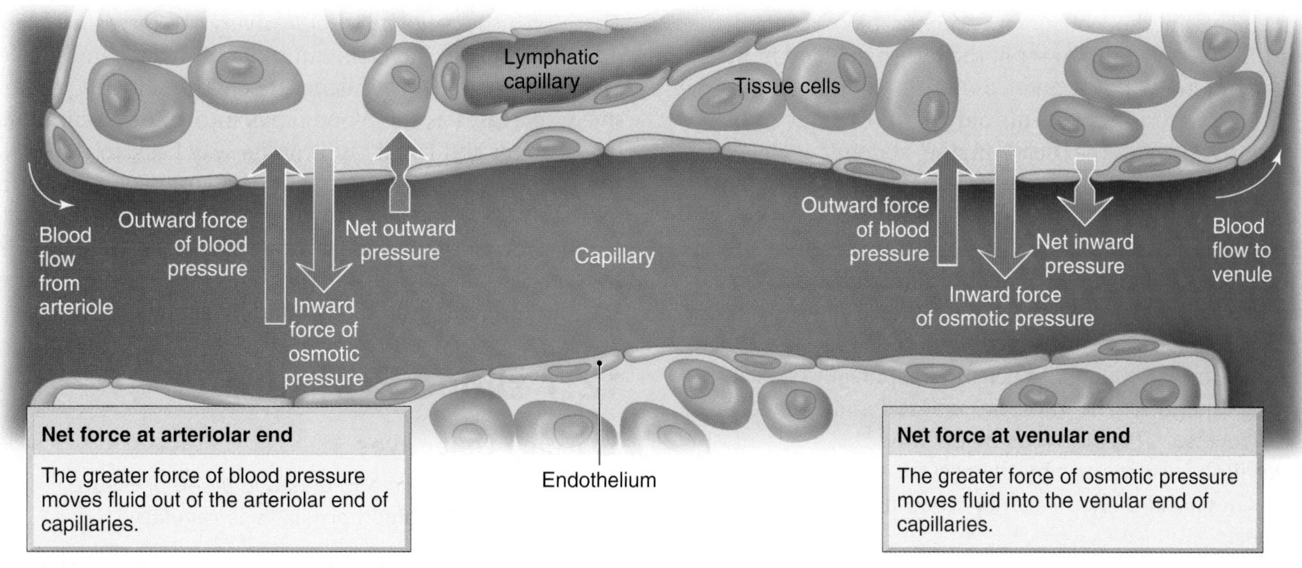

Figure 12.14 Fluid exchange across capillary walls. Fluid moves out of or into capillaries according to the net difference between blood pressure and osmotic pressure. Solutes diffuse out of or into capillaries according to each solute's concentration gradient. APR

Check My Understanding

11. Compare the structure and function of arteries, capillaries, and veins.
12. How does the exchange of materials occur between blood in capillaries and tissue cells?

12.6 Blood Flow

Learning Objective

11. Describe the mechanism of blood circulation.

Blood circulates because of differences in blood pressure. Blood flows from areas of higher pressure to areas of lower pressure. Blood pressure is greatest in the ventricles and lowest in the atria. Figure 12.15 shows the decline of blood pressure in the systemic circuit with increased distance from the left ventricle.

Contraction of the ventricles creates the blood pressure that propels the blood through the arteries. However, the pressure declines as the arteries branch into an increasing number of smaller and smaller arteries and finally connect with the capillaries. The decline in blood pressure occurs because of the increased distance from the ventricle. By the time blood has left the capillaries and entered the veins, there is very little blood pressure remaining to return the blood to the heart. The return of venous blood is assisted by three additional forces: *skeletal muscle contractions, respiratory movements*, and *gravity*.

Contractions of skeletal muscles compress the veins, forcing blood from one valved segment to another and on toward the heart because the valves prevent a backflow of

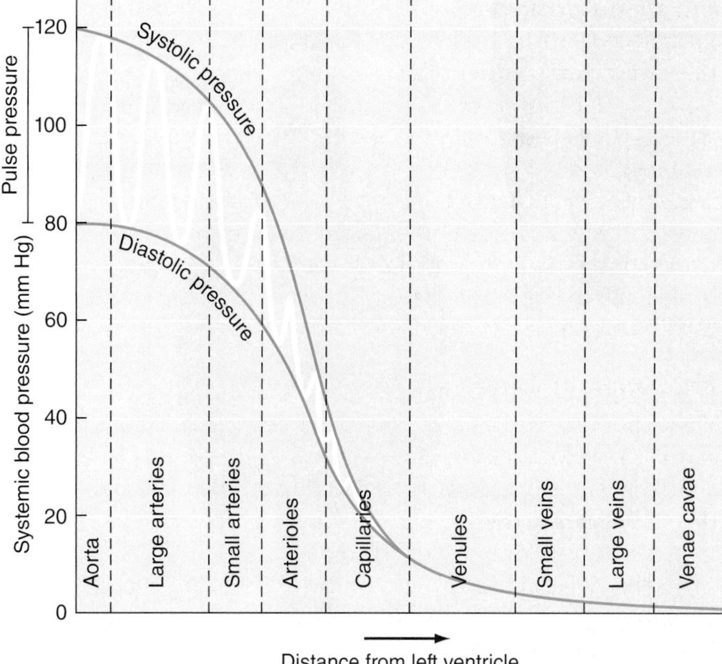

Figure 12.15 Blood pressure decreases as distance from the left ventricle increases.

blood. This method of moving venous blood toward the heart is especially important in the return of blood from the upper and lower limbs, and it is illustrated in figure 12.16.

Respiratory movements aid the movement of blood superiorly toward the heart in the abdominopelvic and thoracic cavities. The inferior movement of the diaphragm as it contracts during inspiration decreases the pressure within the thoracic cavity and increases the pressure within the abdominopelvic cavity. The higher pressure in the abdominopelvic cavity forces blood to move from the abdominopelvic veins superiorly into thoracic veins, where the pressure is reduced. When the diaphragm relaxes and moves superiorly, the thoracic and abdominopelvic pressures reverse. Backflow of blood into the veins of the lower limb is prevented by the presence of venous valves.

Gravity aids the return of blood in veins superior to the heart.

Velocity of Blood Flow

The velocity of blood flow varies inversely with the overall cross-sectional area of the *combined* blood vessels.

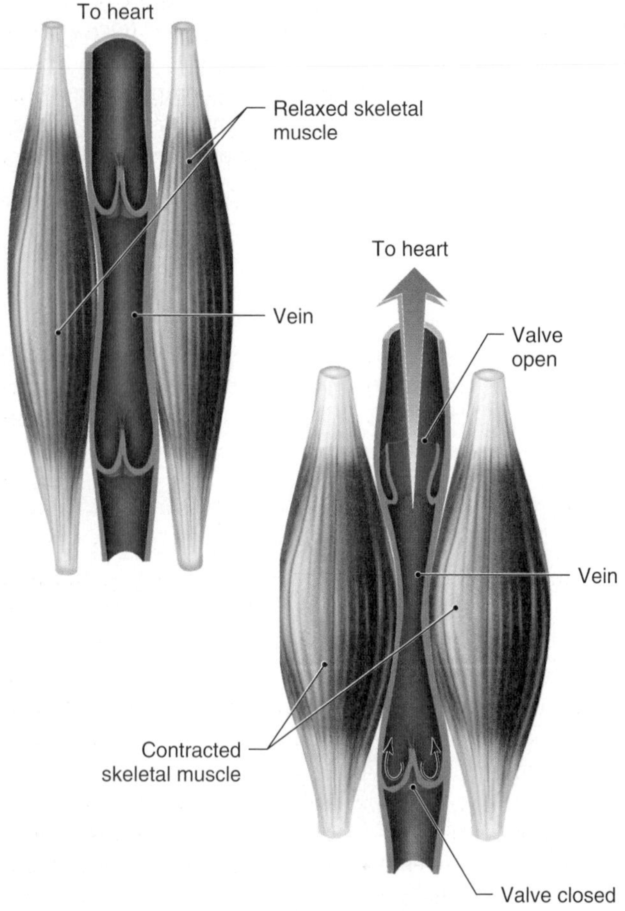

Figure 12.16 Contraction of skeletal muscles compresses veins and aids the movement of blood toward the heart.

Therefore, the velocity progressively decreases as blood flows through an increasing number of smaller and smaller arteries and into the capillaries. Then, the velocity progressively increases as the blood flows into a decreasing number of larger and larger veins on its way back to the heart.

Blood velocity is fastest in the aorta and slowest in the capillaries, an ideal situation providing for the rapid circulation of the blood and yet sufficient time for the exchange of materials between blood in the capillaries and the interstitial fluid surrounding tissue cells.

12.7 Blood Pressure

Learning Objectives

12. Compare systolic and diastolic blood pressure.
13. Describe how blood pressure is regulated.

The term *blood pressure*, the force of blood against the wall of the blood vessels, usually refers to arterial blood pressure in the systemic circuit—in the aorta and its branches. Arterial blood pressure is greatest during ventricular contraction (systole) as blood is pumped into the aorta and its branches. This pressure is called the **systolic blood pressure,** and it optimally averages 110 millimeters of mercury (mm Hg) when measured in the brachial artery. The lowest arterial pressure occurs during ventricular relaxation (diastole). This pressure is called the **diastolic blood pressure,** and it optimally averages 70 mm Hg (figure 12.15).

The difference between the systolic and diastolic blood pressures is known as the *pulse pressure* (figure 12.15). The alternating increase and decrease in arterial blood pressure during ventricular systole and diastole causes a comparable expansion and contraction of the elastic arterial walls. This pulsating expansion of the arterial walls follows each ventricular contraction, and it may be detected as the *pulse* by placing the fingers on a superficial artery. Figure 12.17 identifies the name and location of superficial arteries where the pulse may be detected.

Factors Affecting Blood Pressure

Three major factors affect blood pressure: cardiac output, blood volume, and peripheral resistance. An increase in any of these factors causes an increase in blood pressure, while a decrease in any of these causes a decrease in blood pressure.

Clinical Insight

A blood pressure of 110/70 mm Hg is optimal. Each 20 mm Hg of systolic pressure over 115, and each 10 mm Hg of diastolic pressure over 75 doubles the risk of heart attack, stroke, and kidney disease.

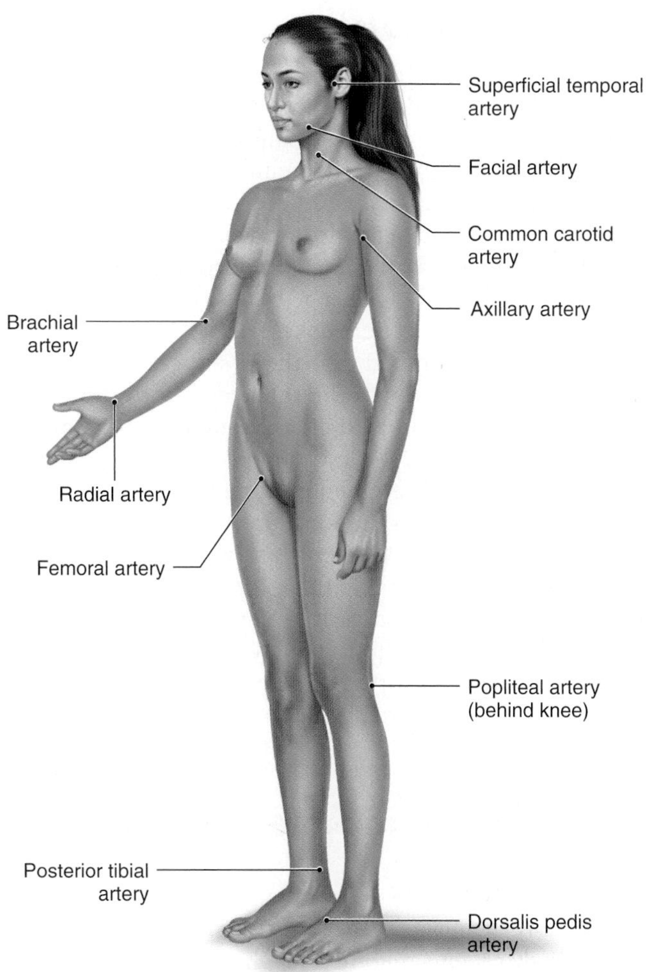

Figure 12.17 Locations and arteries where the pulse may be detected. See figures 12.19 and 12.20 for specific locations of these arteries.

Recall that cardiac output is determined by the heart rate and the stroke volume. An increase or decrease in cardiac output causes a comparable change in blood pressure.

Blood volume may be decreased by severe hemorrhage, vomiting, diarrhea, or reduced water intake. The decrease in blood volume causes a decrease in blood pressure. Many drugs used to treat hypertension (abnormally high blood pressure) act as diuretics, meaning they increase urine volume and as a result decrease blood volume. As soon as the lost fluid is replaced, blood pressure returns to normal. Conversely, if the body retains too much fluid, blood volume and blood pressure increase. A high-salt diet is a risk factor for hypertension because it causes the blood to retain more water as a result of osmosis, leading to an increase in blood volume.

Peripheral resistance is the opposition to blood flow created by friction of blood against the walls of blood vessels. Increasing peripheral resistance will increase blood pressure, while decreasing peripheral resistance decreases blood pressure. Peripheral resistance is determined by *vessel diameters, total vessel length,* and *blood viscosity.* Arterioles play a critical role in controlling blood pressure by changing their diameters. As arterioles constrict, peripheral resistance increases and blood pressure increases accordingly. As arterioles dilate, peripheral resistance and blood pressure decrease. Peripheral resistance is directly proportional to the total length of the blood vessels in the body: the longer the total length of the vessels, the greater their resistance to flow. Obese people tend to have hypertension partly because their bodies contain more blood vessels to serve the extra adipose tissue. *Viscosity* is the resistance of a liquid to flow. For example, water has a low viscosity, while honey has a high viscosity. Blood viscosity is determined by the ratio of plasma to formed elements and plasma proteins. Increasing viscosity, or shifting the ratio in favor of the formed elements and plasma proteins, increases peripheral resistance and blood pressure. Both dehydration (loss of water from plasma) and polycythemia (elevated RBC count) can increase viscosity. Abnormally high levels of blood lipids and sugar are also risk factors for hypertension because they increase blood viscosity, in addition to promoting the formation of plaque on the vessel walls. Decreasing viscosity through over-hydration or certain types of anemia (see chapter 11) will decrease peripheral resistance and blood pressure.

Control of Peripheral Resistance

The sympathetic division of the ANS controls peripheral resistance primarily by regulating the diameter of blood vessels, especially arterioles. The integration center is the **vasomotor center** in the medulla oblongata. An increase in the frequency of sympathetic nerve impulses to the smooth muscle of blood vessels produces **vasoconstriction**, which increases resistance. The increase in resistance increases blood pressure and blood velocity. This response accelerates the rate of oxygen transport to cells and the removal of carbon dioxide from blood by the lungs. A decrease in sympathetic nerve impulse frequency results in **vasodilation**, which decreases resistance. The decrease in resistance decreases blood pressure and blood velocity.

Like the cardiac control center, the activity of the vasomotor center is modified by nerve impulses from higher brain areas, and sensory nerve impulses from baroreceptors and chemoreceptors in the aortic arch and the internal and external carotid arteries. For example, a decrease in pressure, pH, or oxygen concentration of the blood stimulates vasoconstriction. Conversely, an increase in these values promotes vasodilation.

In addition, arterioles and precapillary sphincters are affected by localized changes in blood concentrations of oxygen, carbon dioxide, and pH. These local effects override the control by the vasomotor center, through a process called *autoregulation,* and increase the rate of exchange of materials between tissue cells and the capillaries. For example, if a particular muscle group is active for an extended period, a localized decrease in oxygen concentration and an increase in carbon dioxide concentration result. These chemical changes stimulate the vasodilation of local arterioles and precapillary sphincters, which increases the flow of blood into capillary networks of the affected muscles to provide more oxygen and to remove more carbon dioxide.

 Check My Understanding

13. How does blood pressure affect the flow of blood through blood vessels?
14. How are systolic and diastolic blood pressure different?
15. How do cardiac output, blood volume, and peripheral resistance affect blood pressure?

12.8 Circulation Pathways

Learning Objective

14. Compare the systemic and pulmonary circuits.

As noted earlier, the heart is a double pump that serves two distinct circulation pathways: the pulmonary and systemic circuits. These circuits were shown earlier in figure 12.7.

Pulmonary Circuit

The **pulmonary circuit** carries deoxygenated blood to the lungs, where oxygen and carbon dioxide are exchanged between the blood and the air in the lungs. The right ventricle pumps deoxygenated blood into the pulmonary trunk, a short, thick artery that divides to form the left and right pulmonary arteries. Each pulmonary artery enters a lung and divides repeatedly to form arterioles, which continue into the alveolar capillaries that surround the air sacs (alveoli) of the lungs (see chapter 14). Oxygen diffuses from the air in the alveoli into the capillary blood, and carbon dioxide diffuses from the blood into the air in the alveoli. Blood then flows from the capillaries into venules, which merge to form small veins, which, in turn, join to form progressively larger veins. Two pulmonary veins emerge from each lung to carry oxygenated blood back to the left atrium of the heart.

Systemic Circuit

The systemic circuit carries oxygenated blood to the tissue cells of the body and returns deoxygenated blood to the heart. The left ventricle pumps the freshly oxygenated blood, received from the pulmonary circuit, into the aorta for circulation to all parts of the body except the lungs. The aorta branches to form many major arteries, which continually branch to form arterioles leading to capillaries, where the exchange of materials between the blood and interstitial fluid takes place. Oxygen diffuses from the capillary blood into the tissue cells, while carbon dioxide diffuses from the tissue cells into the blood. From the capillaries, blood enters venules, which merge to form small veins, which join to form progressively larger veins. Ultimately, veins from the superior body (head, neck, shoulders, upper limbs, and superior trunk) join to form the superior vena cava, which returns blood from these regions back to the right atrium. Similarly, veins from the inferior body (inferior trunk and lower limbs) enter the inferior vena cava, which also returns blood into the right atrium. The **coronary sinus** drains the blood from the myocardium into the right atrium (see figure 12.5).

12.9 Systemic Arteries

Learning Objective

15. Identify the major systemic arteries and the organs or body regions that they supply.

Major Branches of the Aorta

The aorta ascends from the heart, arches to the left and posterior to the heart, and descends through the thoracic and abdominal cavities just anterior to the vertebral column. Because of its size, the aorta is divided into four regions: the ascending aorta, the aortic arch, the thoracic aorta, and the abdominal aorta. Figure 12.18 shows the major branches of the aorta and their relationships to the internal organs. Tables 12.4 and 12.5 list the major branches of the aorta and the organs and body regions that they supply.

The first arteries to branch from the aorta are the left and right coronary arteries, which supply blood to the heart. They branch from the aorta just distal to the aortic valve in the base of the *ascending aorta.*

Three major arteries branch from the *aortic arch.* In order of branching, they are the **brachiocephalic** (brăk-ē-ō-se-fal′-ik) **trunk,** the **left common carotid** (kah-rot′-id) **artery,** and the **left subclavian** (sub-klā′-vē-an) **artery.**

Pairs of **posterior intercostal** (in-ter-kos′-tal) **arteries** branch from the *thoracic aorta* to supply the intercostal

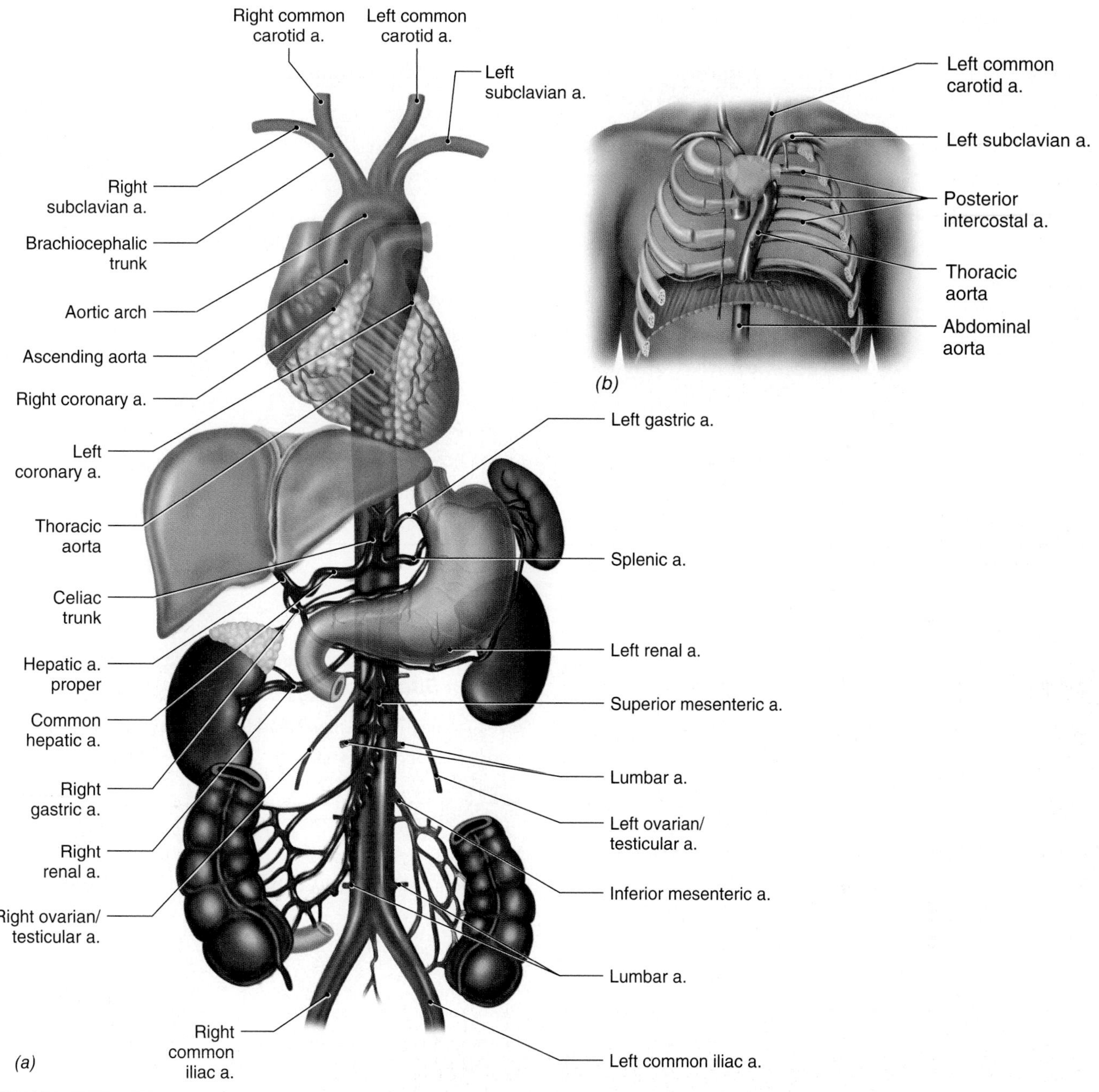

Figure 12.18 *(a)* The major arteries that branch from the aorta. *(b)* Major arteries supplying the thoracic cage. (a. = artery) APIR

muscles between the ribs and other organs of the thoracic wall. A number of other small arteries supply the organs of the thoracic cavity.

Once the aorta descends through the diaphragm, it is called the *abdominal aorta,* and it gives off several branch arteries to the abdominal wall and visceral organs. The **celiac** (sē′-lē-ak) **trunk** is a short artery that divides to form three branch arteries: (1) the **left gastric artery** supplies the stomach and esophagus, (2) the **splenic artery** supplies the spleen, stomach, and pancreas, and (3) the **common hepatic artery** supplies the liver, gallbladder, stomach, duodenum, and pancreas.

The **superior mesenteric** (mes-en-ter′-ik) **artery** supplies the pancreas, most of the small intestine, and the proximal portion of the large intestine. The left and right **renal arteries** supply the kidneys. The left and right **ovarian arteries** supply the ovaries in females. The left and right **testicular arteries** supply the testes in males.

Table 12.4 Major Arteries Branching from the Ascending Aorta, Aortic Arch, and Thoracic Aorta

Artery	Origin	Region Supplied
Coronary	Ascending aorta	Myocardium
Brachiocephalic trunk	Aortic arch	Branches as below
Right common carotid	Brachiocephalic trunk	Right side of head and neck
Right subclavian	Brachiocephalic trunk	Right shoulder and upper limb, thoracic wall
Left common carotid	Aortic arch	Left side of head and neck
External carotid	Common carotid	Scalp, face, and neck
Internal carotid	Common carotid	Brain
Left subclavian	Aortic arch	Left shoulder and upper limb, thoracic wall
Vertebral	Subclavian	Neck and brain
Axillary	Subclavian	Axilla and shoulder
Brachial	Axillary	Arm
Radial	Brachial	Forearm and hand
Ulnar	Brachial	Forearm and hand
Posterior intercostal	Thoracic aorta	Thoracic wall

Table 12.5 Major Arteries Branching from the Abdominal Aorta

Artery	Origin	Region Supplied
Celiac trunk	Abdominal aorta	Liver, stomach, spleen, gallbladder, esophagus, and pancreas
Common hepatic	Celiac trunk	Liver, gallbladder, stomach, duodenum, and pancreas
Left gastric	Celiac trunk	Stomach and esophagus
Splenic	Celiac trunk	Spleen, stomach, and pancreas
Renal	Abdominal aorta	Kidney
Superior mesenteric	Abdominal aorta	Pancreas, small intestine, and proximal part of large intestine
Ovarian, testicular	Abdominal aorta	Ovaries or testes
Lumbar	Abdominal aorta	Lumbar region of back
Inferior mesenteric	Abdominal aorta	Distal part of large intestine
Common iliac	Abdominal aorta	Pelvic region and lower limb
Internal iliac	Common iliac	Pelvic wall, pelvic viscera, external genitalia, and medial thigh
External iliac	Common iliac	Pelvic wall and lower limb
Femoral	External iliac	Thigh
Popliteal	Femoral	Knee
Anterior tibial	Popliteal	Leg (anterior) and foot
Posterior tibial	Popliteal	Leg (posterior) and foot

Several pairs of **lumbar arteries** supply the walls of the abdomen and back. The **inferior mesenteric artery** supplies the distal portion of the large intestine.

At the level of the iliac crests, the aorta divides to form two large arteries, the left and right **common iliac** (il′-ē-ak) **arteries,** which carry blood to the inferior portions of the trunk and to the lower limbs.

Arteries Supplying the Head and Neck

The head and neck receive blood from several arteries that branch from the common carotid and subclavian arteries. Note in figures 12.18 and 12.19 that the brachiocephalic trunk branches to form the **right common carotid**

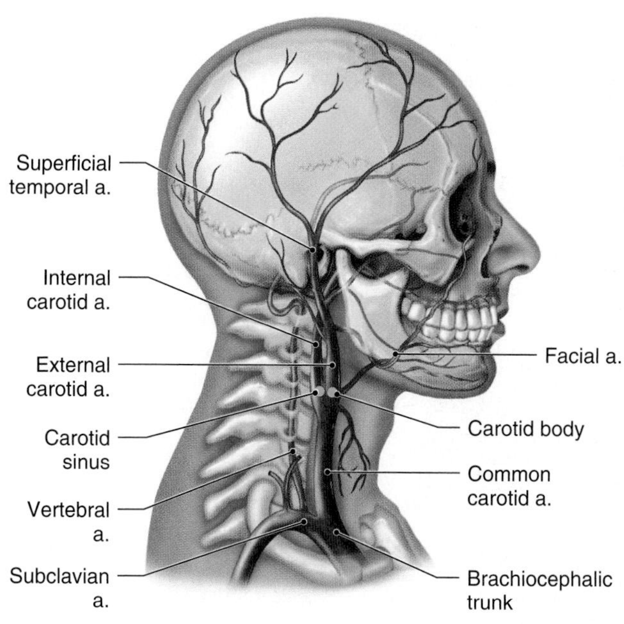

Figure 12.19 Major arteries supplying the head and neck. (a. = artery) APIR

artery and the **right subclavian artery.** The left common carotid and left subclavian arteries branch directly from the aortic arch.

Each common carotid artery divides in the neck to form an **external carotid artery** and the **internal carotid artery.** Near the junction of external and internal carotid arteries are the *carotid body* (the site of chemoreceptors) and *carotid sinus* (the site of baroreceptors), which send sensory nerve impulses to the cardiac control and vasomotor centers in the medulla oblongata. The external carotid arteries give rise to a number of smaller arteries that carry blood to the neck, face, and scalp. The internal carotid arteries enter the cranium and provide the major supply of blood to the brain.

The neck and brain are also supplied by the **vertebral arteries.** They branch from the subclavian arteries and pass superiorly through the transverse foramina of cervical vertebrae to enter the cranium.

Arteries Supplying the Shoulders and Upper Limbs APIR

The subclavian artery provides branches to the shoulder and passes inferior to the clavicle to become the **axillary artery,** which supplies branches to the thoracic wall and axillary region. The axillary artery continues into the arm to become the **brachial artery,** which provides branches to serve the arm. At the elbow, the brachial artery divides to form a **radial artery** and an **ulnar artery,** which supply the forearm and wrist and merge to form a network of arteries supplying the hand (figure 12.20 and table 12.4).

Arteries Supplying the Pelvis and Lower Limbs APIR

As noted earlier, the left and right common iliac arteries branch from the inferior end of the aorta. Each common iliac branches within the pelvis to form internal and external iliac arteries. The **internal iliac artery** is the smaller branch that supplies the pelvic wall, pelvic organs, external genitalia, and medial thigh muscles. The **external iliac artery** is the larger branch, and it supplies the anterior pelvic wall and continues into the thigh, where it becomes the femoral artery (figure 12.20).

The **femoral artery** gives off branches that supply the anterior and medial muscles of the thigh. The largest branch is the **deep femoral artery,** which serves the posterior and lateral thigh muscles. As the femoral artery descends, it passes posterior to the knee and becomes the **popliteal** (pop-li-té′-al) **artery,** which supplies certain muscles of the thigh and leg, as well as the knee. The popliteal artery branches just inferior to the knee to form the anterior and posterior tibial arteries.

The **anterior tibial artery** descends between the tibia and fibula to supply the anterior and lateral portions of the leg, and it continues to become the **dorsalis pedis,** which supplies the ankle and foot. The **posterior tibial artery** lies posterior to the tibia and supplies the posterior portion of the leg, and it continues to supply the ankle and the plantar surface of the foot. Its largest branch is the **fibular artery,** which serves the lateral leg muscles (table 12.5).

Clinical Insight

The pulse may be taken at any superficial artery, but the radial artery at the wrist and the common carotid artery in the neck are the most commonly used sites. Blood pressure is usually taken in the brachial artery of the arm. The radial artery at the wrist and the femoral artery at the groin are the common entry sites for angioplasty, a procedure in which a wire is fed into the arteries for widening narrowed or obstructed coronary or other systemic arteries.

Check My Understanding

16. What is the arterial pathway of blood from the left ventricle to the right side of the brain?
17. What is the arterial pathway of blood from the left ventricle to the small intestine?
18. What is the arterial pathway of blood from the left ventricle to the superior surface of the foot?

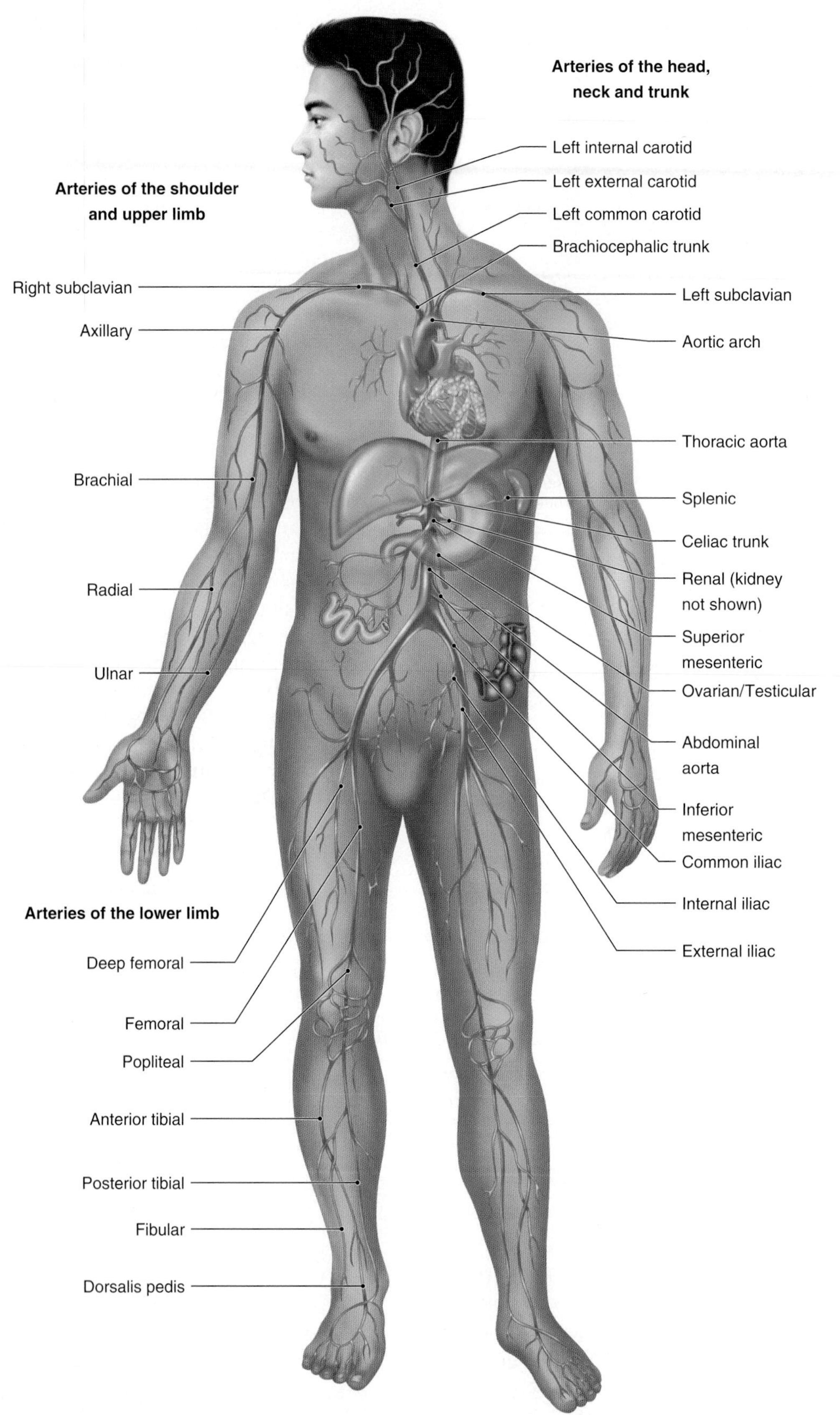

Figure 12.20 Major systemic arteries.

12.10 Systemic Veins

Learning Objectives

16. Identify the major systemic veins and the organs or body regions that they drain.

The systemic veins receive deoxygenated blood from capillaries and return the blood to the heart. Ultimately, all systemic veins merge to form two major veins, the superior and inferior venae cavae, that empty into the right atrium of the heart.

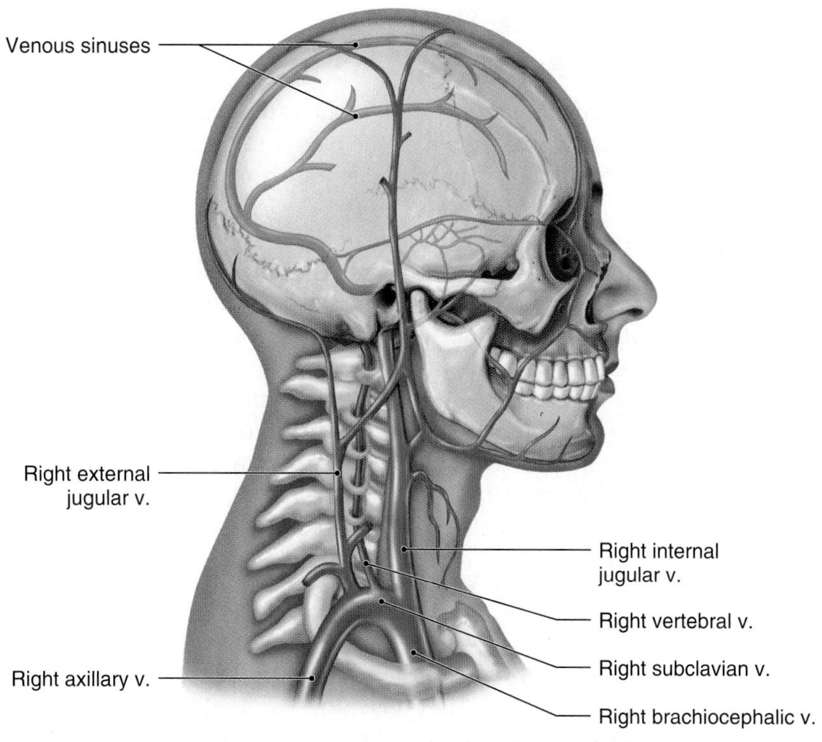

Figure 12.21 Major veins draining the head and neck. (v. = vein) APR

Veins Draining the Head and Neck

As shown in figure 12.21, superficial areas of the head and neck are drained by the left and right **external jugular** (jug′-ū-lar) **veins,** which lead into the left and right **subclavian veins,** respectively. The left and right **vertebral veins** carry blood from the cervical spinal cord and deep neck regions into the subclavian veins as well.

Most of the blood from the brain, face, and neck is carried by the left and right **internal jugular veins.** Each internal jugular vein merges with a subclavian vein to form a **brachiocephalic vein.** The left and right brachiocephalic veins join to form the **superior vena cava,** which returns blood to the right atrium of the heart (table 12.6).

Table 12.6 Major Veins Draining to the Superior Vena Cava

Vein	Region Drained and Location	Receiving Vein
Head and Neck		
External jugular	Face, scalp, and neck	Subclavian
Vertebral	Cervical spinal cord and deep neck	Subclavian
Internal jugular	Brain, face, and neck	Brachiocephalic
Upper Limb and Shoulder		
Radial	Hand and forearm (deep)	Brachial
Ulnar	Hand and forearm (deep)	Brachial
Basilic	Medial upper limb (superficial)	Axillary
Cephalic	Lateral upper limb (superficial)	Axillary
Brachial	Arm (deep)	Axillary
Axillary	Axilla and shoulder	Subclavian
Subclavian	Shoulder and thoracic wall	Brachiocephalic
Trunk		
Brachiocephalic	Head, neck, shoulder, upper limb, and thoracic wall	Superior vena cava
Azygos	Thoracic and abdominal walls	Superior vena cava
Posterior intercostal	Thoracic wall	Azygos
Ascending lumbar	Abdominal wall	Azygos

Veins Draining the Shoulders and Upper Limbs AP|R

Deep regions of the forearm are drained by the **radial** and **ulnar veins.** These two veins join at the elbow to form the **brachial vein,** which drains the deep areas of the arm (figure 12.22).

Superficial regions of the hand, forearm, and arm are drained by the laterally located **cephalic** (se-fal′-ik) **vein** and the medially located **basilic** (bah-sil′-ik) **vein.** Note the **median cubital** (kyū-bi-tal) **vein,** which connects the basilic and cephalic veins.

The basilic and brachial veins merge in the axilla to form the **axillary vein,** which, in turn, joins with the cephalic vein to form the **subclavian vein.** As noted earlier, the subclavian vein joins with the internal jugular vein to form the brachiocephalic vein (table 12.6).

Veins Draining the Pelvis and Lower Limbs AP|R

The **anterior** and **posterior tibial veins** drain the foot and deep regions of the leg. They join inferior to the knee to form the **popliteal vein.** The **small saphenous** (sah-fē′-nus) **vein** drains the superficial posterior part of the leg and merges with the popliteal vein. The **fibular vein** drains the lateral portion of the leg and joins with the popliteal vein at the knee to form the **femoral vein,** which drains the deep regions of the thigh and hip.

The **great saphenous vein** originates from the venous arches in the foot, and it drains the medial and superficial portions of the foot, leg, and thigh. It merges with the femoral vein to form the **external iliac vein.** The external iliac vein and the **internal iliac vein** receive branches that drain the superior thigh and pelvic areas, and they merge to form the **common iliac vein.** The left and right common iliac veins merge to form the **inferior vena cava,** which returns blood to the right atrium of the heart (see figure 12.22).

Clinical Insight

The median cubital vein is the vein of choice when drawing a sample of blood for clinical tests. It is easily located just deep to the skin on the anterior surface of the elbow joint. In coronary bypass surgery, a segment of the internal thoracic artery, saphenous vein, or radial artery is grafted to the afflicted coronary artery on each side of the blockage. The subclavian vein is a common site for implanting the central line, a long-term catheter for administering medications and taking blood samples.

Veins Draining the Abdominal and Thoracic Walls AP|R

The **azygos** (az′-i-gō-s) **vein** drains most of the thoracic and abdominal walls, and it empties into the superior vena cava near the right atrium. The azygos vein receives blood from a number of smaller veins, including the **posterior intercostal veins** and the **ascending lumbar vein,** which drains the wall of the abdomen (figure 12.23).

Veins Draining the Abdominal Viscera AP|R

The **hepatic portal vein** carries blood from the stomach, intestines, spleen, and pancreas to the liver instead of the inferior vena cava. The hepatic portal vein is formed by the union of the **superior mesenteric vein,** which drains the small intestine and proximal large intestine, and the **splenic vein,** which drains the spleen. The splenic vein receives blood from the **inferior mesenteric vein,** which drains the distal large intestine, and the **pancreatic vein,** which drains the pancreas. The **gastric veins,** from the stomach, drain directly into the hepatic portal vein. All of these veins compose the **hepatic portal system.**

After entering the liver, the blood flows through the venous sinusoids, where materials are either removed or added before the blood enters the **hepatic veins,** which empty into the inferior vena cava (figure 12.24a). Note that 75% of the blood supply to the liver comes from the hepatic portal vein; the rest comes from the hepatic artery proper (see figure 12.18*a*). The hepatic portal system allows the liver to monitor and adjust the concentrations of substances in blood coming from the digestive tract before it enters the general circulation.

The left and right **renal veins** carry blood from the kidneys, and the left and right **ovarian** or **testicular veins** return blood from the ovaries in females or the testes in males, respectively. Both renal veins and the right ovarian or testicular vein drain into the inferior vena cava. The left ovarian or testicular vein empties into the left renal vein (figure 12.24*b* and table 12.7).

Check My Understanding

19. What is the venous pathway of blood from the left side of the head to the right atrium?
20. What is the venous pathway of blood from the small intestine to the right atrium?
21. What is the venous pathway of blood from the posterior portion of the ankle to the right atrium?

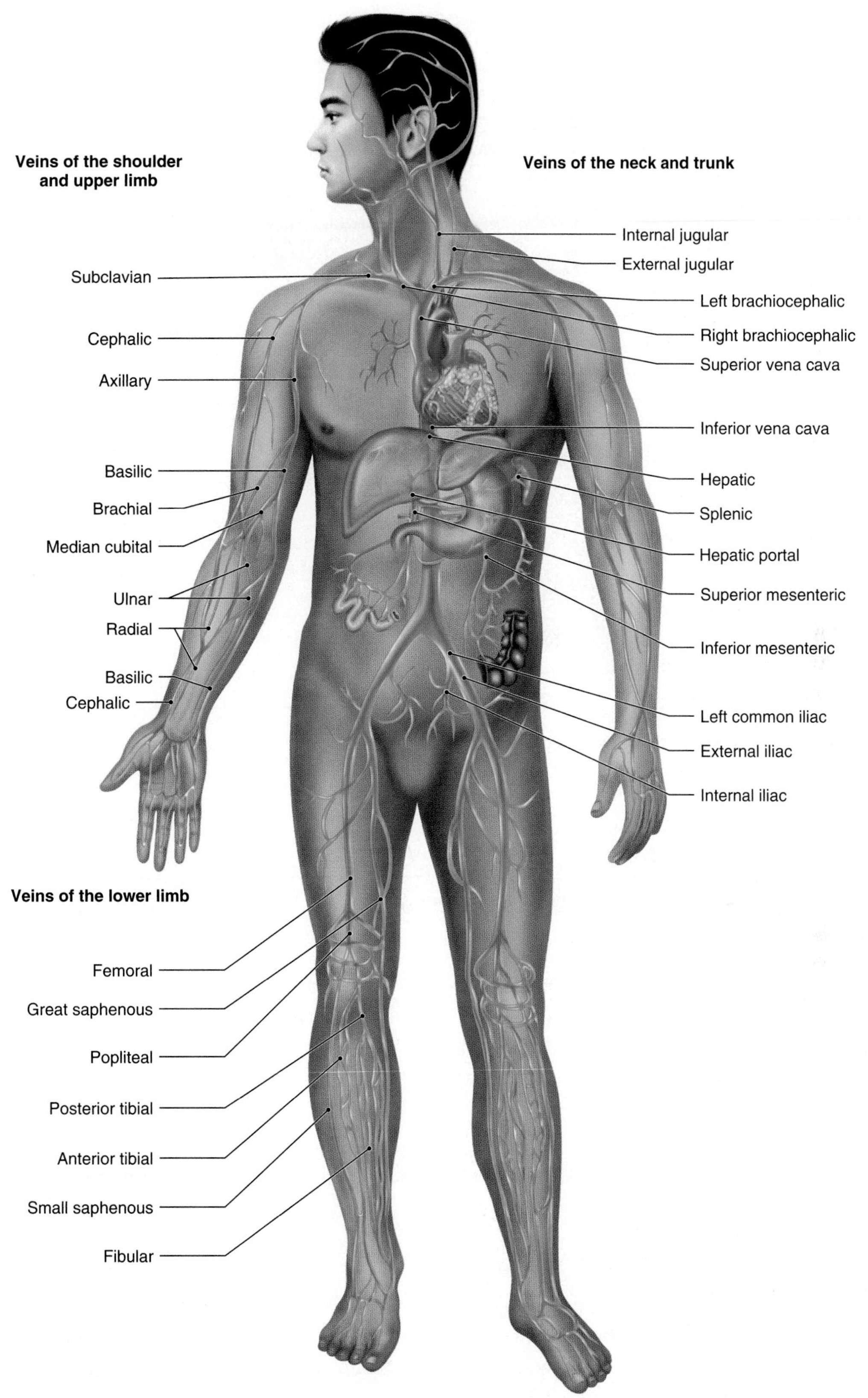

Figure 12.22 Major systemic veins.

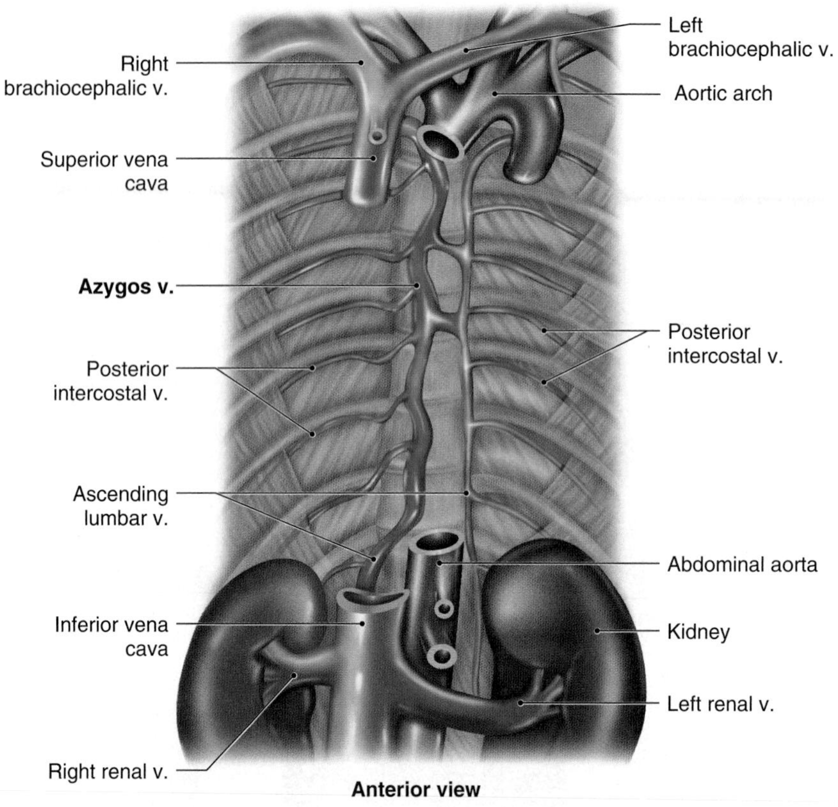

Figure 12.23 Major veins of the thoracic cavity.

Table 12.7 Major Veins Draining to the Inferior Vena Cava

Vein	Region Drained and Location	Receiving Vein
Lower Limb		
Anterior and posterior tibial	Foot and leg	Popliteal
Popliteal	Knee	Femoral
Small saphenous	Posterior leg (superficial)	Popliteal
Great saphenous	Lower limb (superficial)	Femoral
Femoral	Thigh	External iliac
Trunk		
Hepatic portal	Intestines, stomach, spleen, and pancreas	Liver sinusoids
Hepatic	Liver	Inferior vena cava
Renal	Kidneys	Inferior vena cava
Right ovarian or testicular	Right ovary or testis	Inferior vena cava
Left ovarian or testicular	Left ovary or testis	Left renal
External iliac	Pelvic wall and lower limb	Common iliac
Internal iliac	Viscera in pelvis and inferior abdominal cavity, pelvic wall, and superior thigh	Common iliac
Common iliac	Pelvic and inferior abdominal regions and lower limb	Inferior vena cava

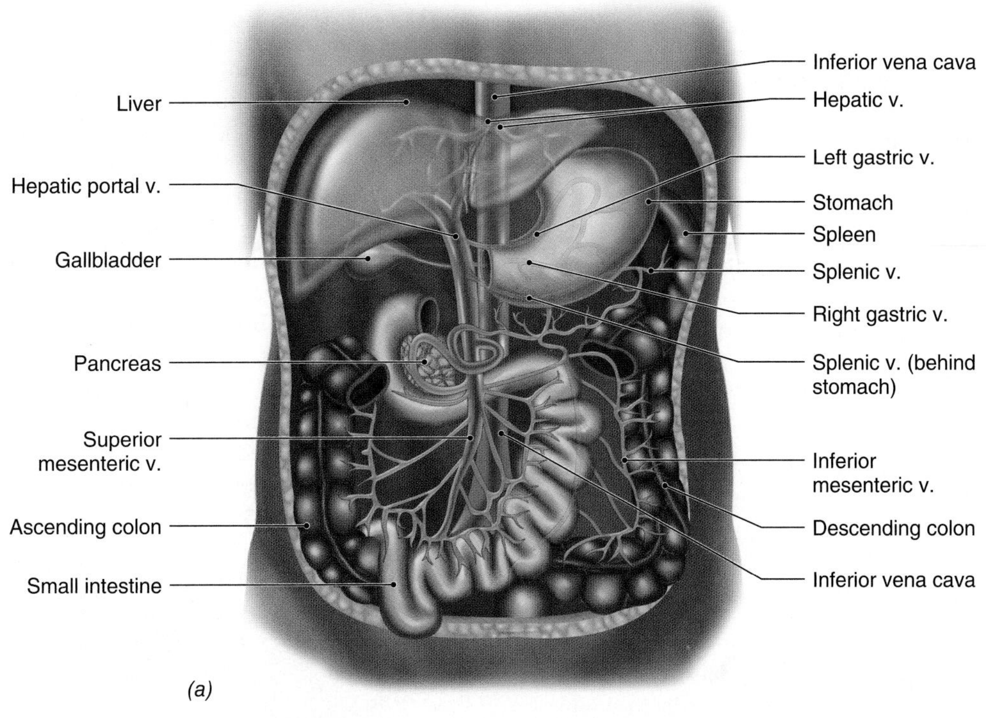

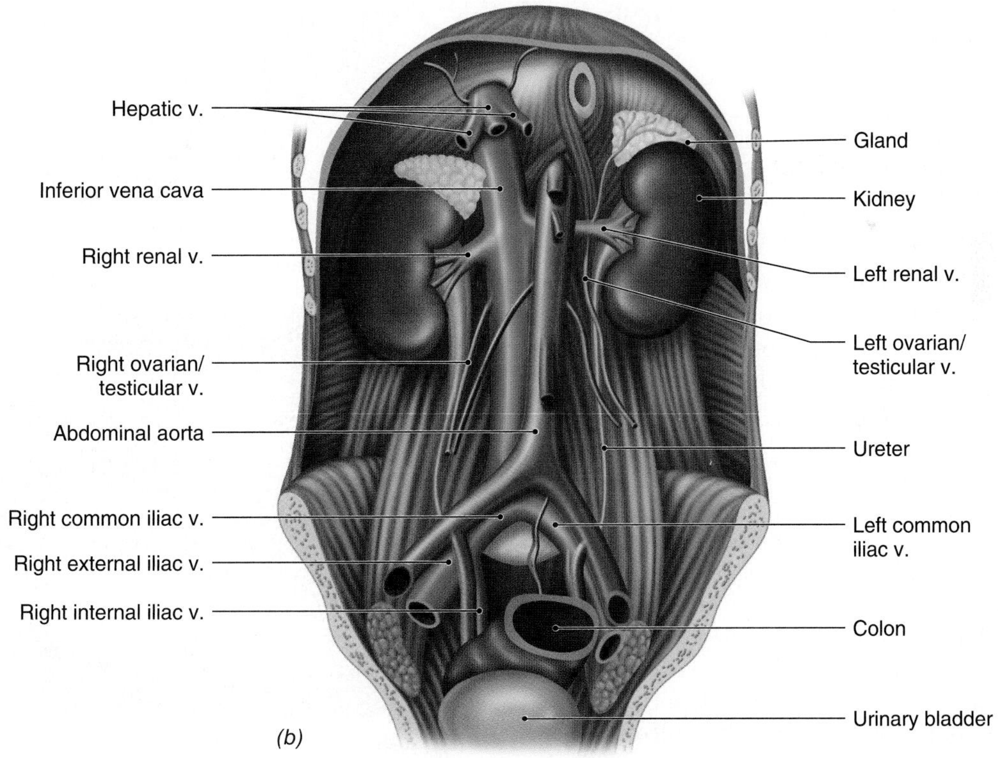

Figure 12.24 *(a)* An anterior view of the veins of the hepatic portal system. *(b)* An anterior view of the major veins draining into the inferior vena cava. AP|R

12.11 Disorders of the Heart and Blood Vessels

Learning Objective

17. Describe the common disorders of the heart and blood vessels.

These disorders are grouped according to whether they affect primarily the heart or the blood vessels. In some cases, the underlying cause of a heart ailment is a blood vessel disorder.

Heart Disorders

Arrhythmia (ah-rith'-mē-ah), or dysrhythmia, refers to an abnormal heartbeat. It may be caused by a number of factors, including damage to the heart conduction system, drugs, electrolyte imbalance, or a diminished supply of blood via the coronary arteries. In addition to irregular heartbeats, arrhythmia includes

- Bradycardia–a slow heart rate of less than 60 beats per minute. Note that the bradycardia in well-trained athletes is a healthy condition because it saves energy during resting heart contraction and has a greater potential to increase cardiac output.
- Tachycardia–a fast heart rate of over 100 beats per minute.
- Heart flutter–a very rapid heart rate of 200 to 300 beats per minute.
- Fibrillation–a very rapid heart rate in which the contractions are uncoordinated so that blood is not pumped from the ventricles. Ventricular fibrillation is usually fatal without prompt treatment.

Congestive heart failure (CHF) is the acute or chronic inability of the heart to pump out the blood returned to it by the veins. Symptoms include fatigue; edema (accumulation of fluid) of the lungs, feet, and legs; and excess accumulation of blood in internal organs. CHF may result from atherosclerosis of the coronary arteries, which deprives the myocardium of adequate blood.

Heart murmurs are unusual heart sounds. They are usually associated with defective heart valves, which allow a backflow of blood. Unless there are complications, heart murmurs have little clinical significance.

Myocardial infarction (mī-ō-kar'-dē-al in-fark'-shun) is the death of a portion of the myocardium due to an obstruction in a coronary artery. The obstruction is usually a blood clot that has formed as a result of atherosclerosis. This event is commonly called a "heart attack," and it may be fatal if a large portion of the myocardium is deprived of blood.

Pericarditis is the inflammation of the pericardium and is usually caused by a viral or bacterial infection. It may be quite painful as the inflamed membranes rub together during each heart cycle.

Blood Vessel Disorders

An **aneurysm** (an'-yū-rizm) is a weakened portion of a blood vessel that bulges externally, forming a balloonlike sac filled with blood. Rupture of an aneurysm in a major artery may produce a fatal hemorrhage.

Arteriosclerosis (ar-te"-rē-ō-skle-rō'-sis) is hardening of the arteries. It results from calcium deposits that accumulate in the tunica media of arterial walls and is usually associated with atherosclerosis.

Atherosclerosis is the formation of fatty deposits (cholesterol and triglycerides) along the tunica intima of arterial walls. The atherosclerotic plaques reduce the lumen of the arteries and increase the probability of blood clots being formed. Such deposits in the coronary, carotid, or cerebral arteries may lead to serious circulatory problems (figure 12.25).

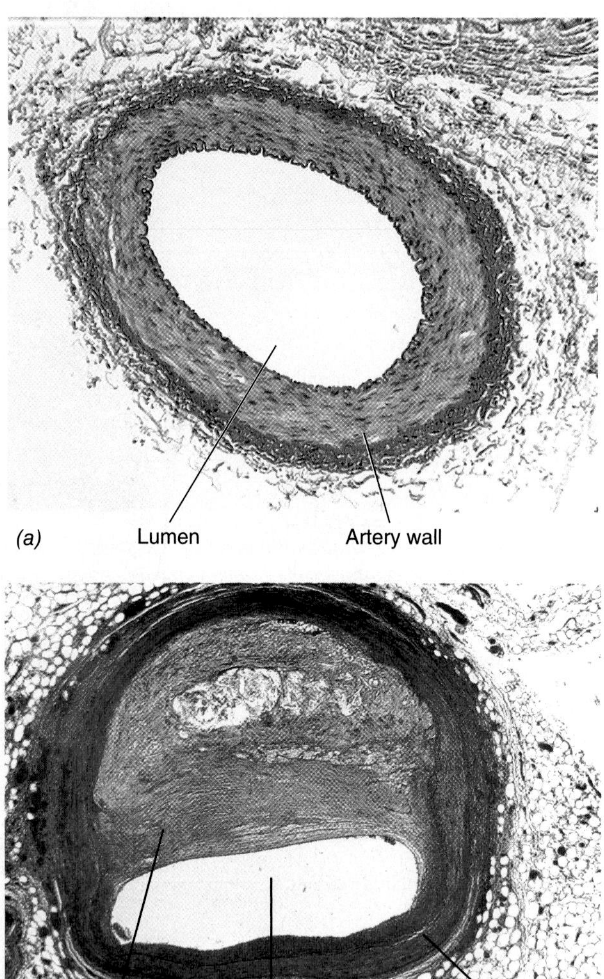

Figure 12.25 Cross sections of *(a)* a normal artery and *(b)* an atherosclerotic artery whose lumen is diminished by fatty deposits. Atherosclerosis promotes the formation of a blood clot within the artery.

Hypertension refers to chronic high blood pressure. It is the most common disease affecting the heart and blood vessels. Blood pressure that exceeds 140/90 mm Hg is indicative of hypertension. A systolic blood pressure of 120 to 139 mm Hg and a diastolic blood pressure of 80 to 89 mm Hg is considered to be *prehypertension*. Hypertension may be caused by a variety of factors, but persistent stress and smoking are commonly involved.

Phlebitis (flē-bī′-tis) is inflammation of a vein, and it most often occurs in a lower limb. If it is complicated by the formation of a blood clot, it is called *thrombophlebitis*.

Varicose veins are veins that have become dilated and swollen because their valves are not functioning properly. Heredity seems to play a role in their occurrence. Pregnancy, standing for prolonged periods, and lack of physical activity reduce venous return and promote varicose veins in the lower limbs. Chronic constipation promotes their occurrence in the anal canal, where they are called **hemorrhoids** (hem′o-royds).

Chapter Summary

12.1 Anatomy of the Heart

- The heart wall consists primarily of the myocardium, a thick layer of cardiac muscle tissue. It is lined internally by the thin endocardium and externally by the thin epicardium.
- The pericardial sac is composed of the parietal layer of the serous pericardium and fibrous pericardium. The pericardial cavity is filled with pericardial fluid.
- The heart contains four chambers. The superior chambers are the left and right atria, which receive blood returning to the heart. The inferior chambers are the left and right ventricles, which pump blood out of the heart. There are no openings between the atria or between the ventricles.
- Atrioventricular valves allow blood to flow between each atrium and its corresponding ventricle but prevent a backflow of blood. The mitral valve lies between the left atrium and left ventricle. The tricuspid valve lies between the right atrium and right ventricle.
- Semilunar valves allow blood to be pumped from the ventricles into their associated arteries but prevent a backflow of blood. The aortic valve is located in the base of the aorta. The pulmonary valve is located in the base of the pulmonary trunk.
- The right atrium receives deoxygenated blood from the superior and inferior venae cavae and the coronary sinus, and the left atrium receives oxygenated blood from the pulmonary veins.
- The right ventricle pumps deoxygenated blood into the pulmonary trunk, which divides into the left and right pulmonary arteries, which lead to the lungs. At the same time, the left ventricle pumps oxygenated blood into the aorta, which leads to all parts of the body except the lungs.
- The myocardium receives blood from the coronary arteries, which branch from the ascending aorta. Blood is returned from the myocardium by the cardiac veins, which open into the coronary sinus, which leads to the right atrium.

12.2 Cardiac Cycle

- The cardiac cycle includes both contraction (systole) and relaxation (diastole) phases.
- During atrial diastole, blood returns to the atria and flows on into the ventricles. Atrial systole forces more blood into the ventricles to fill them.
- During ventricular diastole, blood flows into the ventricles. Ventricular systole pumps blood from the ventricles into their associated arteries.
- The normal lub-dup heart sound is caused by the closure of the heart valves. The first sound results from the closure of the atrioventricular valves. The second sound results from the closure of the semilunar valves.

12.3 Heart Conduction System

- The SA node is the pacemaker, which rhythmically initiates impulses that cause the heart contractions.
- Impulses pass through the atria, causing atrial systole, and simultaneously reach the AV node.
- Impulses pass from the AV node along the AV bundle and bundle branches to the ventricular fibers, which transmit the impulses to the myocardium, causing ventricular systole.
- An electrocardiogram is a recording of the formation and transmission of impulses through the heart conduction system.
- An electrocardiogram consists of a P wave, a QRS complex, and a T wave, and it is used in the diagnosis of heart ailments.

12.4 Regulation of Heart Function

- Cardiac output is a measure of heart function. It is determined by stroke volume and heart rate:

$$CO = SV \times HR$$

- Cardiac output is regulated by factors internal and external to the heart.
- Heart rate and stroke volume are controlled by the autonomic nervous system. The cardiac control center is in the medulla oblongata. It receives sensory nerve impulses from baroreceptors and chemoreceptors, and is also affected by nerve impulses from the cerebrum and hypothalamus.

- Sympathetic axons release norepinephrine at heart synapses, which causes an increase in the heart rate and contraction strength. Parasympathetic axons release acetylcholine at heart synapses, which causes a decrease in heart rate.
- The dynamic balance in the frequency of sympathetic and parasympathetic nerve impulses reaching the heart adjusts the heart rate and stroke volume to meet body needs.
- Heart rate and stroke volume are also affected by age, sex, physical condition, temperature, epinephrine, thyroid hormones, and the blood concentration of Ca^{2+} and K^{+}.

12.5 Types of Blood Vessels

- The three basic types of blood vessels are arteries, capillaries, and veins. Large arteries and veins are formed of a superficial tunica externa of dense irregular connective tissue, a middle tunica media of smooth muscle, and a deep tunica intima of endothelium supported by areolar connective tissues.
- Arteries have thick, muscular walls and carry blood from the heart. Large arteries divide repeatedly to form the smallest arteries, arterioles, which connect with capillaries.
- Capillaries are the smallest and most numerous blood vessels. They are composed of an endothelium supported by a layer of areolar connective tissue. Their thin walls allow an exchange of materials between the blood and the interstitial fluid. Dissolved substances are exchanged by diffusion. Fluid exits the arteriolar end of a capillary because blood pressure is greater than osmotic pressure, and it reenters at the venular end of the capillary because osmotic pressure is greater than blood pressure.
- Veins have thinner walls than arteries and carry blood from capillaries toward the heart. The smallest veins are venules, which lead from capillaries and merge to form small veins. Large veins contain valves that prevent a backflow of blood.

12.6 Blood Flow

- Blood circulates from areas of higher pressure to areas of lower pressure. Blood pressure is highest in the ventricles and lowest in the atria.
- Systemic blood pressure declines as blood is carried from the arteries through the capillaries and through the veins. Skeletal muscle contractions and respiratory movements are important forces that aid the return of venous blood.
- Blood velocity varies inversely with the cross-sectional area of the combined blood vessels. Blood velocity is fastest in the aorta and slowest in the capillaries. The velocity progressively increases as the blood flows from capillaries to the larger veins.

12.7 Blood Pressure

- Optimal systolic blood pressure is 115 mm Hg. Optimal diastolic blood pressure is 75 mm Hg.
- The difference between systolic and diastolic pressures is the pulse pressure. The pulse may be detected by palpating superficial arteries.
- Blood pressure is determined by three factors: cardiac output, blood volume, and peripheral resistance.
- Peripheral resistance is determined by vessel diameters, total vessel length, and blood viscosity.
- The vasomotor center in the medulla oblongata provides the autonomic control of blood vessel diameter. In this way, the autonomic nervous system controls peripheral resistance and blood pressure.
- Local autoregulation of arterioles overrides autonomic control and regulates blood flow in capillaries according to the needs of the local tissues.

12.8 Circulation Pathways

- The pulmonary circuit carries blood from the heart to the lungs and back again.
- The systemic circuit carries blood from the heart to all parts of the body, except the lungs, and back again.

12.9 Systemic Arteries

- The aorta is divided into the ascending aorta, aortic arch, thoracic aorta, and abdominal aorta.
- The major branch arteries of the aorta are the coronary, brachiocephalic trunk, left common carotid, left subclavian, posterior intercostals, celiac trunk, superior mesenteric, renal, ovarian/testicular, lumbar, inferior mesenteric, and common iliac arteries.
- The major arteries supplying the head and neck are paired arteries. Each common carotid artery branches to form the external and internal carotid arteries. The external carotid supplies the neck, face, and scalp. The internal carotid is the major artery supplying the brain. The vertebral arteries supply the neck and brain.
- Each shoulder and upper limb is supplied by a subclavian artery, which becomes the axillary artery, which becomes the brachial artery of the arm. The brachial artery branches to form the radial and ulnar arteries of the forearm.
- Each common iliac artery branches to form internal and external iliac arteries. The external iliac enters the thigh to become the femoral artery, which becomes the popliteal artery near the knee. The popliteal branches inferior to the knee to form the anterior and posterior tibial arteries.

12.10 Systemic Veins

- Veins draining the head and neck are paired veins. On each side, the external jugular and vertebral veins empty into the subclavian vein. The internal jugular vein and subclavian merge to form the brachiocephalic vein. The left and right brachiocephalic veins join to form the superior vena cava.
- The ascending lumbar veins and the posterior intercostal veins enter the azygos vein, which opens into the superior vena cava.

- Radial and ulnar veins of the forearm join to form the brachial vein of the arm. The basilic vein joins the brachial vein to form the axillary vein, which, in turn, receives the cephalic vein to form the subclavian vein.
- Anterior and posterior tibial veins merge inferior to the knee to form the popliteal vein. The popliteal vein receives the small saphenous and fibular veins to form the femoral vein. The great saphenous vein extends from the foot to join with the femoral vein near the hip, which forms the external iliac vein. The external iliac vein joins with the internal iliac vein to form the common iliac vein. The left and right common iliac veins merge to form the inferior vena cava.
- The splenic vein receives the inferior mesenteric vein and the pancreatic vein, and merges with the superior mesenteric vein to form the hepatic portal vein. The gastric vein drains into the hepatic portal vein. The hepatic portal vein empties into the liver sinusoids. The hepatic vein carries blood from the liver to the inferior vena cava.
- The right ovarian or testicular vein and the paired renal veins empty into the inferior vena cava. The left ovarian or testicular vein drains into the left renal vein.

12.11 Disorders of the Heart and Blood Vessels

- Disorders of the heart include arrhythmia, congestive heart failure, heart murmurs, myocardial infarction, and pericarditis.
- Disorders of blood vessels include aneurysm, arteriosclerosis, atherosclerosis, hypertension, phlebitis, and varicose veins.

Self-Review

Answers are located in appendix B.

1. The membranous sac around the heart is the ______.
2. The ______ is the heart chamber receiving oxygenated blood from the lungs.
3. The ______ valve prevents a backflow of blood from the right ventricle into the ______.
4. During ______ diastole blood fills the atria; during ventricular ______ blood is pumped into arteries leading from the heart.
5. The heartbeat originates in the ______ node: the ______ node relays impulses along the AV bundle and ventricular fibers to the ventricular myocardium.
6. Nerve impulses from ______ axons increase the heart rate; nerve impulses from the ______ axons decrease the heart rate.
7. Blood is carried from the heart in ______ and returned to the heart in ______.
8. Vessels with the thickest walls are ______; those with the thinnest walls are ______.
9. The exchange of materials between capillary blood and interstitial fluid occurs by ______ and ______.
10. The heart chambers and vessels in the pulmonary circuit are ______ ventricle, pulmonary ______, pulmonary ______, alveolar capillaries, pulmonary ______, ______ atrium.
11. The arterial pathway of blood from the heart to the right side of the brain is ascending aorta, aortic arch, ______, common carotid, and ______.
12. The arterial pathway of blood from the heart to the liver is ascending aorta, aortic arch, ______ aorta, abdominal aorta, ______, ______, and ______.
13. The venous pathway returning blood from the digestive tract to the heart is ______, liver ______, ______, and ______ vena cava.
14. The venous pathway returning blood from the posterior of the knee is ______, femoral, ______, ______, and ______ vena cava.
15. The venous pathway from the little finger to the heart is basilic, ______, ______, ______, and ______ vena cava.

Critical Thinking

1. How do the heart valves keep blood flowing in one direction?
2. What factors are involved in creating blood pressure?
3. What enables the heart to change its rate and strength of contraction as needed?
4. What is the advantage of the hepatic portal system?
5. Why are there more superficial veins than superficial arteries?

ADDITIONAL RESOURCES

ANATOMY & PHYSIOLOGY

13
CHAPTER

Lymphoid System and Defenses Against Disease

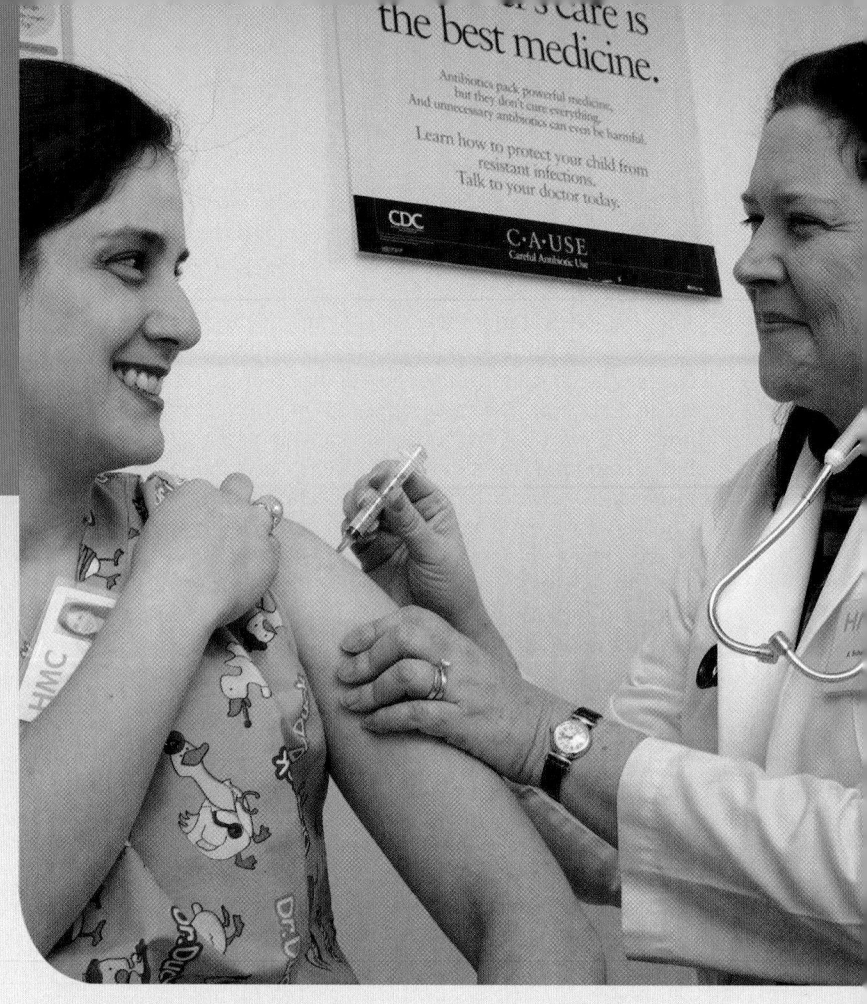

Adele, a 32-year-old receptionist, woke up this morning not feeling well and, while watching the morning news, heard another report about a new strain of influenza sweeping across the country. Recently, several contagious diseases have begun to circulate through the people in her office. She and her twelve coworkers share printers, a water cooler, and a small kitchen space. It is easy to pass a viral or bacterial infection to another person if you are not careful. After suffering for almost two weeks with the flu last year, Adele was proactive regarding her health this year. She went to the local pharmacy a few months ago and received her flu and pneumonia vaccinations. Thanks to the vaccinations, her body now possesses defensive cells, called lymphocytes, to defend her against both diseases if she encounters them. However, these new lymphocytes will defend her against only the strains of flu and bacterial pneumonia in the vaccinations. As she begins to shiver slightly from a fever that has begun to develop, she now hopes that she has not been exposed to a new strain of flu or bacterial pneumonia, against which she will have no defense.

CHAPTER OUTLINE

Module 10
Lymphatic System

SELECTED KEY TERMS

Allergen A foreign substance that stimulates an allergic reaction.
Antibody (anti = against) A protein produced by plasma cells in response to a specific antigen.
Antigen A foreign substance capable of stimulating the production of antibodies.
Complement A group of plasma proteins that destroy pathogens.
Immunity (immun = free) Resistance to specific antigens.
Immunocompetent (im-mu-no-kom′-pe-tent) Capable of responding to a foreign antigen.
Inflammation (inflam = to set on fire) A localized response to damaged or infected tissues that is characterized by swelling, redness, pain, and heat.
Lymph (lymph = clear water) The fluid connective tissue transported in lymphatic vessels.
Lymph node A lymph-filtering secondary lymphoid organ.
Lymphatic vessel A vessel that transports lymph.
Pathogen A disease-causing organism or substance.
Primary lymphoid organ An organ where lymphocytes become immunocompetent.
Red bone marrow Primary lymphoid organ responsible for the production of all formed elements.
Secondary lymphoid organ An organ where immunocompetent lymphocytes reside.
Spleen The largest lymphoid organ. Stores formed elements and clears pathogens from the blood.
Thymus Primary lymphoid organ responsible for T cell maturation.

THE *LYMPHOID (LYMPHATIC) SYSTEM* is closely related to the cardiovascular system, both structurally and functionally (figure 13.1). A network of lymphatic vessels drains excess interstitial fluid (the approximate 10–15% that has not been returned directly to the blood capillaries) and returns it to the bloodstream in a one-way flow that moves slowly toward the subclavian veins. Additionally, the lymphatic vessels in the small intestine function in lipid absorption and lymphocytes aid in the body's defense against **pathogens** (disease-causing organisms or substance).

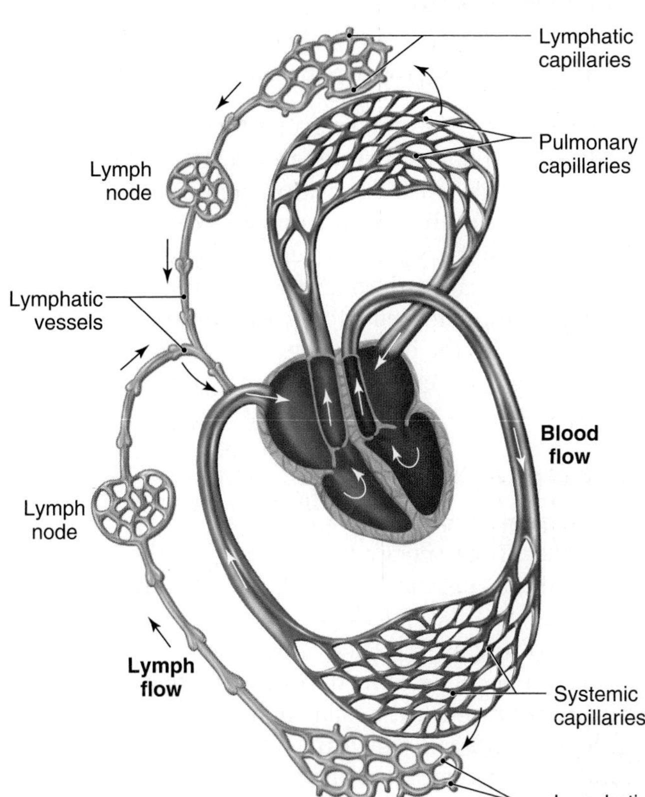

Figure 13.1 Lymphatic vessels transport fluid from interstitial spaces to the bloodstream. AP|R

13.1 Lymph and Lymphatic Vessels

Learning Objectives

1. Describe the formation of lymph.
2. Describe the pathway of lymph in its return to the blood.
3. Describe how lymph is propelled through lymphatic vessels.

The lymphatic network of vessels begins with the microscopic **lymphatic capillaries.** Lymphatic capillaries are closed-ended tubes that form vast networks in the interstitial spaces within most vascular tissues (figure 13.2). Notably, these capillaries are not found in the CNS. Instead the CNS relies on the flow of CSF to removes excess fluid from nervous tissue. Because the walls of lymphatic capillaries are composed of endothelial cells with unique junctions, interstitial fluid, proteins, and microorganisms can easily enter the vessels but cannot leave and reenter the interstitial space. Once fluid enters the lymphatic capillaries, it becomes a fluid connective tissue referred to as **lymph** (limf). Adequate lymphatic drainage is needed to prevent the accumulation of interstitial fluid, a condition called *edema* (ĕ-de′mă). Additionally, within the villi of the small intestine, lymphatic capillaries called *lacteals* (lak′te-alz) transport absorbed lipids and lipid-soluble vitamins away from the digestive tract.

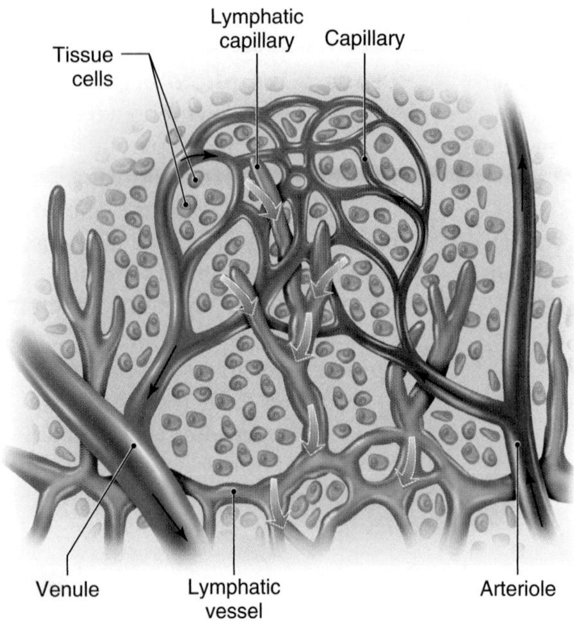

Figure 13.2 Lymphatic capillaries (green) are microscopic, closed-ended tubes that begin in the interstitial spaces of most tissues.

From merging lymphatic capillaries, the lymph is carried into **lymphatic vessels.** These lymphatic vessels merge into even larger vessels called **lymphatic trunks** that are named after large body regions (figure 13.3). The walls of lymphatic vessels and trunks are much like those of veins. They have the same three layers and also contain valves to prevent backflow. The pressure that keeps the lymph moving comes from the massaging action produced by skeletal muscle contractions, intestinal movements, respiratory pressure changes (the same venous return mechanisms described in chapter 12), and from peristaltic contractions of some lymphatic vessels. Interconnecting lymphatic trunks eventually empty into one of the two principal vessels: the **thoracic duct** and the **right lymphatic duct** (figure 13.3). The larger thoracic duct drains lymph from the left thoracic region, left upper limb, left side of the head and neck, and all areas inferior to the diaphragm. The thoracic duct begins in the abdominal cavity as a saclike enlargement called the **cisterna chyli** (sis-ter′nă ki′le), which collects lymph from the lower limbs and the intestinal region. The thoracic duct then ascends along the vertebral column and drains into the left subclavian vein near the left internal jugular vein. The smaller right lymphatic duct receives lymph from the right upper limb, right thoracic region, and right side of the head and neck. The right lymphatic duct empties into the right subclavian vein near the right internal jugular vein (figure 13.3).

Clinical Insight

Edema is the swelling of localized tissues due to the accumulation of excess interstitial fluid. It results from either too much fluid exiting the blood in capillaries or insufficient removal of fluid by lymphatic vessels. There are a variety of causes for edema. For example, a sedentary lifestyle, which leads to the breakdown of valves in lower limb veins, can result in edema of the lower limbs. Also, removal of lymphatic vessels and lymph nodes during cancer surgery often leads to edema of the affected area.

Check My Understanding

1. What is lymph and how is it returned to the blood?
2. Why is the return of lymph to the blood important?

13.2 Lymphoid Organs

Learning Objective

4. Describe the locations and functions of the red bone marrow, thymus, lymph nodes, and spleen.

Lymphoid structures can be found throughout the body. While all lymphoid structures are capable of lymphocyte production, the red bone marrow and thymus are considered **primary lymphoid organs** because all WBCs, especially lymphocytes, originate in these organs. After production in the red bone marrow most lymphocytes and other immune cells go to **secondary lymphoid organs,** such as the lymph nodes and spleen that become the sites of proliferation of lymphocytes and immune responses.

Red Bone Marrow

Red bone marrow is hematopoietic (blood-forming) tissue found in the spongy bone of most of the axial skeleton and the proximal epiphyses of the humerus and femur. As described in chapter 11, red bone marrow is the site of origin of all formed elements in the blood. Not all lymphocytes formed in the red bone marrow are **immunocompetent,** capable of recognizing and attacking foreign antigens, when they exit the marrow. To be immunocompetent, a lymphocyte must be able to elicit an immune response. **B cells,** or *B lymphocytes,* stay in the bone marrow until they are immunocompetent before moving on to secondary lymphoid organs. Other lymphocytes, however, must first move to the thymus for maturation before moving to secondary lymphoid organs.

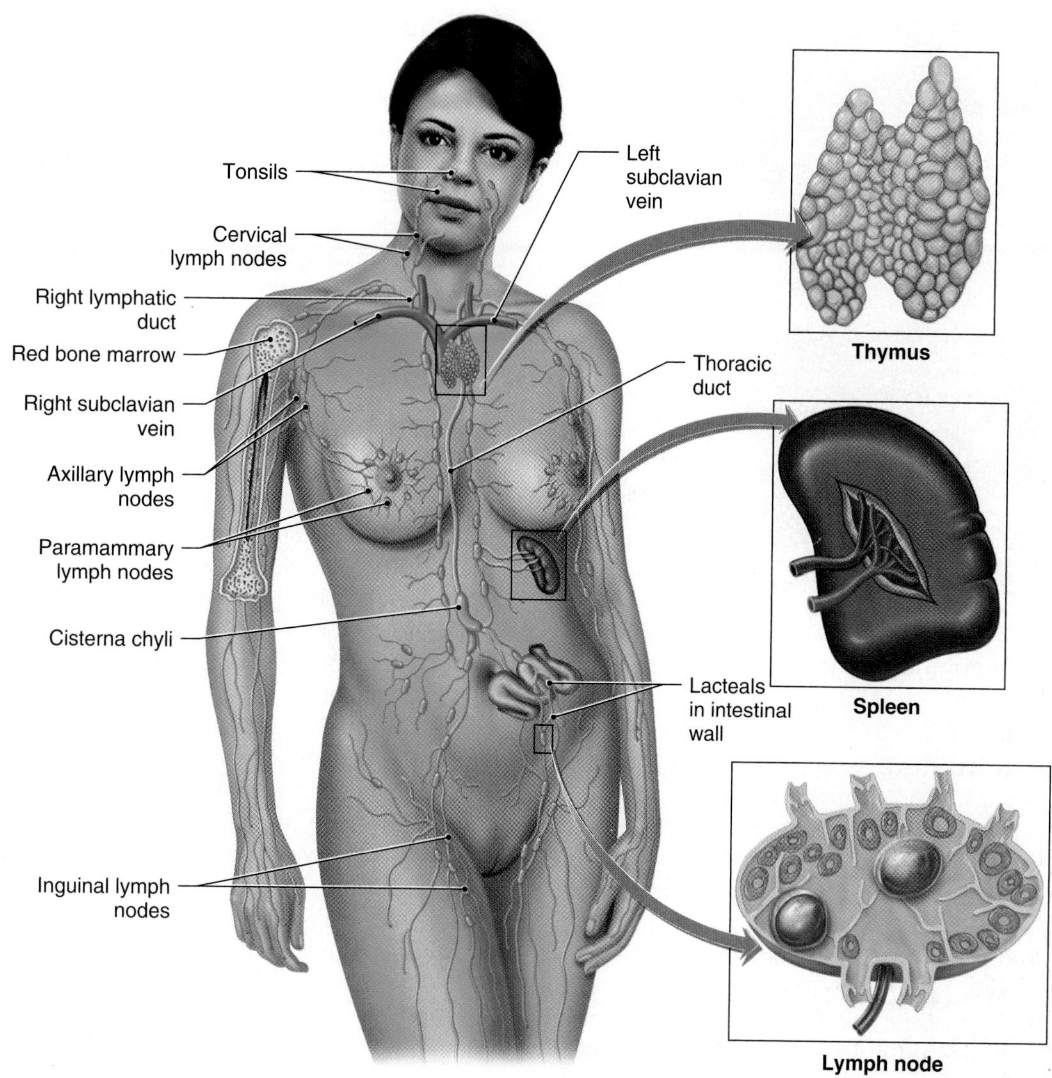

Figure 13.3 The lymphoid system showing the principal lymph nodes and other lymphoid organs. Lymph from the right upper limb, the right side of the head and neck, and the right thoracic region drains through the right lymphatic duct into the right subclavian vein. Lymph from the remainder of the body drains through the thoracic duct into the left subclavian vein. AP|R

Thymus AP|R

The **thymus** is a soft, bilobed gland located in the mediastinum superior to the heart (figure 13.3). It is large (40 g) in infants and children, but after puberty it begins to atrophy and becomes quite small (12 g) in adults. The thymus plays a key role in the development of the lymphoid system before birth and during early childhood. Until the lymphoid system matures at about two years of age, an infant is more susceptible to disease than older children. The major function of the thymus is the differentiation of a class of lymphocytes called **T cells,** or *T lymphocytes,* into immunocompetent cells. The thymus produces hormones called *thymosins* that promote the differentiation and division of T cells, making them immunocompetent. After maturation, T cells are distributed by the blood to secondary lymphoid organs and lymphoid tissues throughout the body.

Lymph Nodes

Lymph nodes usually occur in groups along the larger lymphatic vessels. They are widely distributed in the body, but they do occur in large collections in the inguinal, axillary, and cervical regions of the body, as well as within the ventral cavity. There are no lymph nodes in the CNS.

Structure of Lymph Nodes

Lymph nodes are roughly bean-shaped and 1.0 to 2.5 cm in length. Figure 13.4 is a photograph of a lymph node

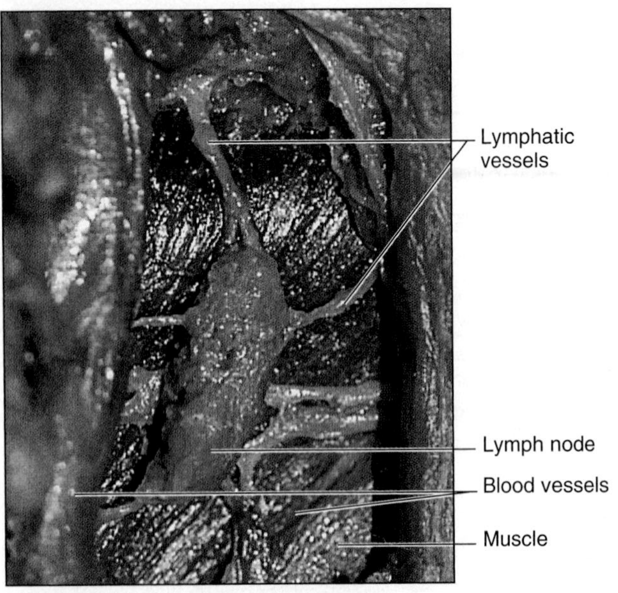

Figure 13.4 A lymph node and its associated vessels.

in situ that shows lymphatic vessels leading to and from the node. Figure 13.5 shows the structure of a section of a lymph node. Note that the lymph node consists of a number of small subunits called **lymphoid nodules.** The lymphoid nodules are collections of lymphocytes and macrophages within reticular tissue and are the sites of activation and proliferation of lymphocytes.

Lymph enters a lymph node through several *afferent* (af′-er-ent) *lymphatic vessels* and flows through the *lymphatic sinuses,* which surround the lymphoid nodules. Lymph is collected from the sinuses and enters *efferent* (ef′-er-ent) *lymphatic vessels,* which carry lymph away from the lymph node. The indentation of the node where efferent lymphatic vessels emerge is called the *hilum.*

Function of Lymph Nodes

A major function of lymph nodes is the filtration and cleansing of the lymph as it passes through a node. They are the only lymphoid organs that filter the lymph. Damaged cells, cancerous cells, cellular debris, bacteria, and viruses become trapped in the reticular tissue of the lymph node and are destroyed by the action of lymphocytes and macrophages. Lymphocytes act against cancerous cells and pathogens, such as bacteria and viruses. Macrophages engulf cellular debris, immobilized or dead bacteria, and viruses.

Spleen

The **spleen** is the largest lymphoid organ. It is located posterior to the stomach near the diaphragm in the left upper quadrant of the abdominopelvic cavity (see figure 13.3). The false ribs provide protection against physical injury. The spleen is a soft, purplish organ 5 to 7 cm (2–3 in) wide and 13 to 16 cm (5–6 in) long. It contains numerous centers for lymphocyte proliferation and a large venous sinus filled with blood.

The spleen resembles a large lymph node. Like lymph nodes, it is enveloped by a thin capsule of dense irregular connective tissue and is subdivided by reticular tissue into many compartments. The compartments contain two basic types of tissues that are named for their appearance in fresh, unstained tissue. *White pulp* consists of large numbers of lymphocytes that cluster around tiny branches of the splenic artery. This tissue is mostly concerned with the immune functions of the spleen. *Red pulp* occupies the rest of a compartment, surrounding the white pulp and the venous sinuses. It is a storage area for formed elements and a site where worn-out red blood cells and pathogens are removed from the blood (figure 13.6). Before birth, the spleen and liver are the major blood-forming organs, but this function is later taken over by

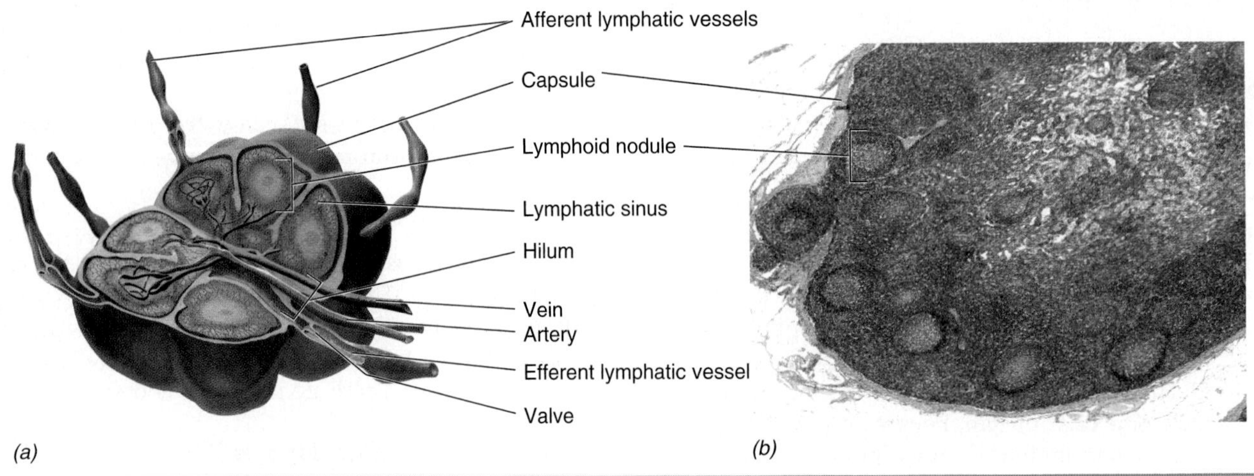

Figure 13.5 Internal Structure of a Lymph Node.
(a) Cut away showing internal structure. *(b)* Photomicrograph of a lymph node. APR

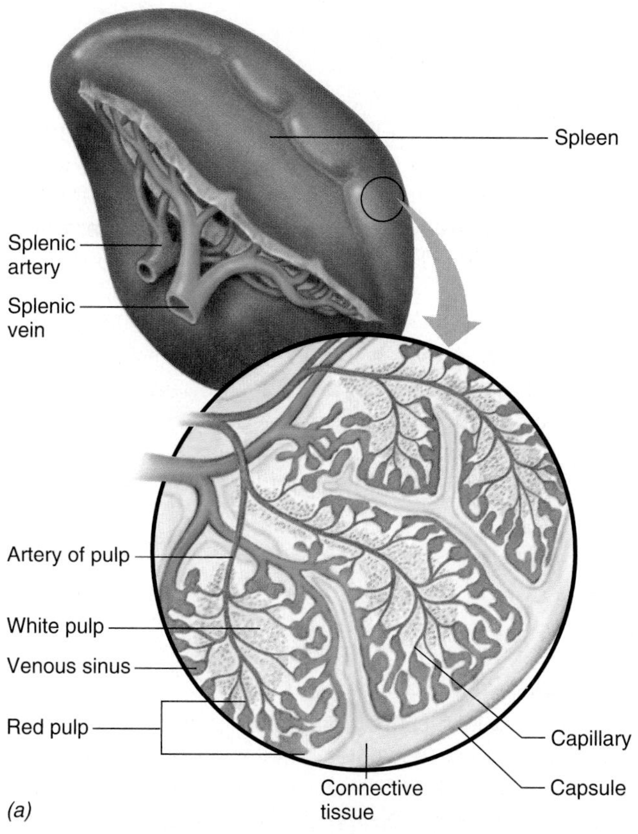

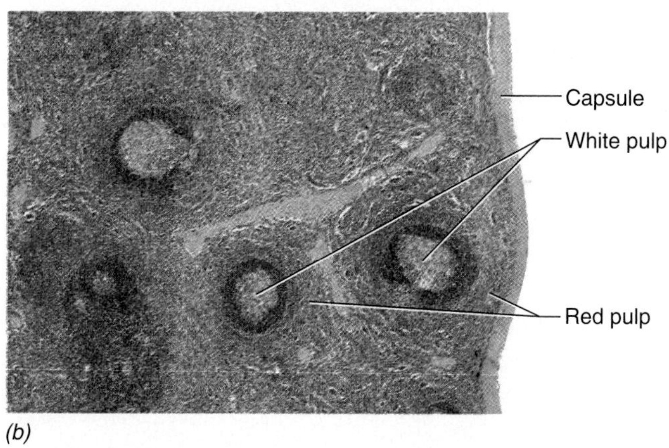

Figure 13.6 The Spleen.
(*a*) The spleen is subdivided into compartments by reticular tissue. (*b*) A photomicrograph of spleen tissue showing white pulp and red pulp. APR

red bone marrow. After birth, the spleen's role is related to both lymphoid and cardiovascular functions:

- It cleanses and filters the blood much like lymph nodes cleanse lymph. Lymphocytes and macrophages destroy pathogens, and macrophages clean up the debris.
- It stores a reserve supply of RBCs and platelets, which can be released into the blood in times of need, such as after a hemorrhage.
- It is a major site for RBC destruction and recycling as described in chapter 11.
- It is a major site of lymphocyte activation and proliferation.

In spite of these important functions, the spleen is not essential for life. But following a splenectomy, a person may be more susceptible to potential pathogens and the effects of hemorrhage.

13.3 Lymphoid Tissues

Learning Objective

5. Describe the locations and functions of the tonsils and mucosa associated lymphoid tissues.

The tonsils and mucosa associated lymphoid tissues are not structurally organs; however, they function as secondary lymphoid organs because they are sites of immune responses.

Tonsils APR

Tonsils (ton′-sils) are clusters of lymphoid tissue located just deep to the mucous membrane in the pharynx (fayr′-inks), or throat, and oral cavity. Like all lymphoid tissue, they contain both lymphocytes and macrophages to fight pathogens. Their function is to intercept and destroy pathogens that enter through the nose and mouth before they can reach the blood.

There are three kinds of tonsils that are strategically located to carry out this function:

1. The *palatine tonsils* are located bilaterally at the junction of the oral cavity and the pharynx.
2. The *pharyngeal* (fah-rin′-je-al) *tonsil* is located posterior to the nasal cavity in the superior portion of the pharynx. This tonsil is commonly called the *adenoid.*
3. The lingual (ling′-gwal) tonsils are located on the base of the tongue in the posterior oral cavity.

Tonsils are larger in young children and play an important role in their defense against pathogens. Sometimes the palatine tonsils and pharyngeal tonsil become so overloaded and swollen with pathogens, a condition called *tonsillitis,* that they must be surgically removed by a *tonsillectomy.*

Mucosa Associated Lymphoid Tissue

Individual lymphoid nodules, like those found within lymph nodes, are located throughout the body, especially in the areolar connective tissues of mucous membranes.

These collections of numerous macrophages and lymphocytes trapped in reticular tissue provide additional barriers to invasion by pathogens. Large clusters of many lymphoid nodules located in the mucous membranes of the respiratory, digestive, urinary, and reproductive tracts are referred to collectively as **MALT (mucosa associated lymphoid tissue).** The *appendix*, an extension of the large intestine located in the right lower quadrant of the abdominopelvic cavity, is part of the MALT that helps control bacterial growth in the large intestine. Table 13.1 outlines the components of the lymphoid system.

Check My Understanding

3. What are the functions of the red bone marrow, thymus, lymph nodes, spleen, tonsils, and MALT?

Clinical Insight

The spread of cancerous cells occurs when cells break away from the primary tumor and are carried to other sites via the lymphoid system or the blood. This process, called *metastasis* (me-tas′-ta-sis), generates secondary cancerous growths called *metastases* wherever the cancerous cells lodge. Because metastatic cells are frequently carried in the lymph, secondary cancerous growths often occur in lymph nodes on lymphatic vessels draining the region of the primary cancer. Knowledge of the location of lymphatic vessels and lymph nodes, and the direction of lymph flow, is important in the detection and treatment of metastases.

Table 13.1 Components of the Lymphoid System

Component	Characteristics	Function
Lymphatic Capillaries	Microscopic closed-ended tubes in interstitial spaces	Collect interstitial fluid from interstitial spaces; collect and transport dietary lipids and lipid-soluble vitamins; once in a lymphatic capillary, fluid is called lymph
Lymphatic Vessels	Formed by merging of lymphatic capillaries; structure similar to veins; contain valves; merge to form lymphatic trunks that drain into either the right lymphatic duct or the thoracic duct	Transport lymph and empty it into the subclavian veins
Lymphoid Organs		Sites of lymphocyte production or proliferation, and immune responses
Primary Lymphoid Organs		
Red Bone Marrow	Located mostly in spongy bone of the skeleton	Site of origination of all lymphocytes
Thymus	Bilobed gland located superior to the heart; size decreases with age	Site of T cells maturation; secretes hormones called thymosins, which stimulate maturation of T cells
Secondary Lymphoid Organs		
Lymph nodes	Small, bean-shaped organs arranged in groups along lymphatic vessels	Sites of lymphocyte proliferation; house T cells and B cells that are responsible for immunity; macrophages phagocytose pathogens and cellular debris from lymph
Spleen	Large lymphoid organ containing venous sinuses	RBC and platelet reservoir; macrophages phagocytose pathogens, cellular debris, and worn formed elements from the blood; house lymphocytes
Lymphoid Tissues		
Tonsils	Masses of lymphoid tissue within the mucosae of pharynx and oral cavity	Protect against invasion of pathogens that are ingested or inhaled
MALT (mucosa associated lymphoid tissue)	Masses of lymphoid tissue within the mucosae of respiratory, digestive, urinary, and reproductive tracts	Guards against pathogens that penetrate the epithelium of mucosa

13.4 Nonspecific Resistance

Learning Objective

6. Identify the components of nonspecific resistance.

Nonspecific resistance provides protection against all pathogens and foreign substances, but it is not directed against a specific pathogen. Nonspecific defense mechanisms include mechanical barriers, chemical actions, phagocytosis, inflammation, and fever.

Mechanical Barriers

The most obvious *mechanical barriers* against pathogens are the skin and the mucous membranes. The closely packed epidermal cells of the skin make penetration by pathogens very difficult and the acidic pH of the skin discourages bacterial growth. Mucous membranes are less effective barriers than the skin. However, they secrete mucus that entraps pathogens and airborne particles and usually prevents their contact with the underlying membranes. The continuous flow of tears over the eyes, the production and swallowing of saliva, the movement of vaginal secretions, and the passage of urine through the urethra are examples of fluid mechanical barriers that help to flush away pathogens before they can attack body tissues.

Chemical Actions

Various body chemicals, including certain enzymes, provide a nonspecific defense against pathogens. A few examples will illustrate the effect of these chemicals.

Tears, saliva, nasal secretions, and perspiration contain the enzyme *lysozyme,* which destroys certain types of bacteria and helps to protect underlying tissues.

Mucus is continuously produced by epithelia lining the respiratory and digestive tracts. Pathogens entering the nose and mouth tend to be trapped in the mucus on the surface of the tonsils and are destroyed there. Pathogens the tonsils miss are swallowed at frequent intervals. Upon reaching the stomach, most pathogens are destroyed by gastric juice, either by its acidic pH or by the enzyme *pepsin.* Pepsin acts by digesting the proteins composing the pathogens.

Cells that are infected with a virus produce *interferon,* a substance that stimulates uninfected cells to synthesize special proteins that inhibit the replication of viruses within them. In this way, the rapid growth of viruses may be inhibited.

The blood contains a group of plasma proteins known as **complement,** which are named because their actions complement the actions of antibodies. Complement proteins can bind to certain pathogens initiating a chain of events that leads to the destruction of the pathogen. The binding of complement is known as *complement fixation.* The fixed complement punches holes in the pathogen's plasma membrane, causing the cell to burst and the pathogen to be destroyed. Subsequently, the resulting debris is cleaned up by phagocytes (neutrophils and macrophages). Complement proteins also enhance phagocytosis and inflammation.

Phagocytosis

Phagocytosis (fag″-ō-sī-tō′-sis) is the engulfing and destruction (by digestion) of pathogens, damaged or cancerous cells, and cellular debris by neutrophils and macrophages. When an infection occurs, neutrophils and monocytes are quickly attracted to the infected tissues. Monocytes entering the tissues become transformed into macrophages, large cells that are especially active in phagocytosis.

Some macrophages wander among the tissues, searching out and phagocytizing pathogens and cellular debris. Others become fixed (stationary) in particular locations in the body, where they phagocytize pathogens that are passing by. Fixed macrophages are especially abundant along the internal walls of blood and lymphatic vessels and in the spleen, lymph nodes, liver, and red bone

The most common pathogens affecting humans are bacteria and viruses. Bacteria are very small, single-celled organisms that lack a true nucleus and other complex cellular organelles. Their DNA is concentrated, but it is not enclosed in a nuclear envelope as in higher organisms. Bacteria are simple organisms but they have the necessary metabolic machinery required for life and reproduction. Bacterial pathogens may cause disease by releasing toxins (poisons), releasing enzymes that damage cells, or entering and destroying cells. Antibiotics are effective in treating most bacterial infections.

In contrast, a virus is composed of nucleic acids, either DNA or RNA, enveloped by a protein coat. Viruses are thousands of times smaller than bacteria. A virus attaches to a cell's surface receptor and penetrates into the cell. Once inside, it takes over the cell's DNA and metabolic machinery, causing the cell to replicate hundreds or thousands of viruses, which burst forth as the cell is destroyed. The released viruses then move on to attack other cells. Antibiotics are not effective in treating viral infections.

marrow. The wandering and fixed macrophages compose the **tissue macrophage system,** which plays a major role in the destruction of potential pathogens.

Inflammation

Inflammation is a localized response to infection or injury that promotes the destruction of pathogens and the healing process. It is characterized by redness, pain, heat, and swelling of the affected tissues.

When injury or infection occurs, several mechanisms produce chemicals, such as complement proteins and histamine, that cause dilation of the arterioles and increase the permeability of blood capillaries in the affected area. The increased blood flow to the local area produces redness and heat. The increased movement of fluids out of the blood capillaries produces swelling (edema) of the tissues. Pain results from irritation of nociceptors by pathogens, swelling, or chemicals released by infected cells.

Some of the chemicals of the inflammatory response attract WBCs to the affected area. In bacterial infections, neutrophils and macrophages actively phagocytize the pathogens and damaged cells. The accumulated mass of living and dead WBCs, tissue cells, and bacteria may form a thick, whitish fluid called **pus.**

Fluids from blood capillaries that enter the affected area contain both fibrinogen and fibroblasts. Fibrinogen may be converted into fibrin to form a clot that is subsequently penetrated and enveloped by fibers formed by the fibroblasts. This action tends to seal off the infected area and prevent the spread of pathogens to neighboring tissues.

The continued action of WBCs usually brings the infection under control. Then, the dead pathogens and cells are cleaned up by phagocytes, and new cells are produced by cell division to repair any damage to the tissues.

Fever

Fever is a high body temperature that accompanies infections and is a normal part of the immune response. It serves a useful purpose as long as the body temperature does not get too high. The increased body temperature inhibits growth of certain pathogens and increases the rate of body processes, including those that fight infection.

Table 13.2 summarizes the major components of nonspecific resistance.

 Check My Understanding

4. What is the method of action of mechanical barriers, chemical actions, phagocytosis, inflammation, and fever?

Table 13.2 Summary of Major Components of Nonspecific Resistance

Component	Function
Mechanical Barriers	
Intact skin	Closely packed cells and multiple cell layers prevent entrance of pathogens
Intact mucous membranes	Closely arranged cells retard entrance of pathogens; not as effective as intact skin
Mucus	Traps pathogens in digestive, respiratory, urinary, and reproductive tracts
Saliva	Washes pathogens from oral surfaces
Tears	Wash pathogens from surface of eye
Urine	Washes pathogens from urethra
Vaginal secretions	Wash pathogens from vaginal canal
Chemical Actions	
Acidic pH of skin	Retards growth of many bacteria
Gastric juice	Kills pathogens that are swallowed
Interferon	Helps to prevent viral infections
Lysozyme	Antimicrobial enzyme in nasal secretions, perspiration, saliva, and tears that kills some pathogens
Complement	Group of plasma proteins that enhance inflammation and phagocytosis. Also cause direct death of pathogens by puncturing plasma membranes.
Other Mechanisms	
Fever	Speeds up body processes and inhibits growth of pathogens
Inflammation	Promotes nonspecific resistance; confines infection; attracts WBCs
Phagocytosis	Phagocytes engulf and destroy pathogens

13.5 Immunity

Learning Objectives

7. Compare nonspecific resistance and specific resistance.
8. Explain the mechanism of cell-mediated immunity.
9. Explain the mechanism of antibody-mediated immunity.

In contrast to nonspecific resistance, **immunity** (i-mū′-ni-tē), or specific resistance, is directed at specific antigens. An immune response involves the production of specific cells and substances to attack a specific antigen. Immunity has "memory," that is, if the same pathogen should reenter the body at a later date, the immune response is quicker and stronger than during the first encounter. Lymphocytes play several important roles in immunity.

An immune response involves one or both of two distinct processes: **cell-mediated immunity** and **antibody-mediated immunity** (figure 13.7).

Specialization of Lymphocytes

During fetal development all formed elements are initially produced by the yolk sac. Later in fetal development the liver and spleen take over the production of formed elements. Around the time of birth the red bone marrow takes over the production of most formed elements, including lymphocytes. However, lymphocytes must mature and become specialized within primary lymphoid organs before they can participate in immunity. About half of the unspecialized lymphocytes pass to the thymus, where they become immunocompetent T cells. The other half of the lymphocytes become specialized in the bone marrow to become immunocompetent B cells. T and B cells are carried by blood to secondary lymphoid organs, such as the lymph nodes, spleen, and tonsils, where they proliferate to form large populations of T and B cells, and some are released into the blood. About 75% of circulating lymphocytes are T cells, and about 25% are B cells (figure 13.8).

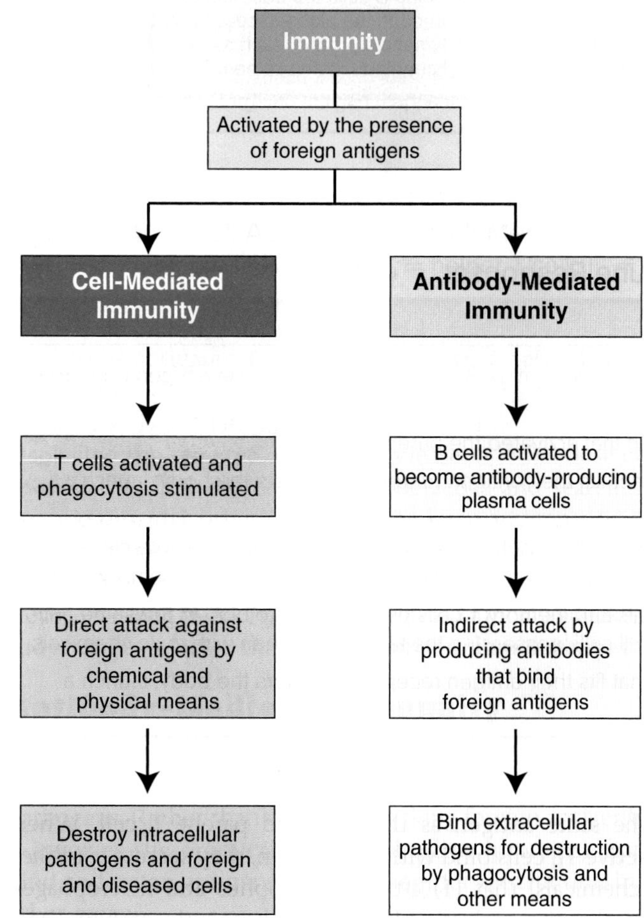

Figure 13.7 Overview of Immunity.

Recognizing Pathogens

All cells, including human cells, have surface recognition molecules called **antigens.** (Recall that antigens are used in typing blood.) In fact, the cells of each person have a unique set of antigens. Antigens are usually large molecules, such as proteins and glycoproteins.

During the specialization process, lymphocytes "learn" to distinguish "self" antigens from "nonself," or foreign, antigens. Thereafter, lymphocytes can recognize an invading pathogen or self cells with abnormal antigens, such as cancerous cells and cells infected by a virus, and launch an attack. Unfortunately, lymphocytes also recognize a transplanted organ as foreign, as in graft rejection, and in some cases fail to recognize certain body tissues as "self" and attack the body's own tissues, as in autoimmune diseases.

The maturation process produces thousands (and perhaps millions) of different kinds of T and B cells, each with specific receptors capable of binding with (recognizing) a specific antigen. Therefore, when an antigen of a pathogen or foreign cell enters the body, only the T or B cells that have a receptor that can bind with the specific antigen are involved in the immune response. Table 13.3 summarizes the roles of B and T cells in immunity.

Cell-Mediated Immunity

In cell-mediated (cellular) immunity, T cells directly attack and destroy foreign cells or diseased body cells, such as cancerous cells, and develop a memory of their antigens in case they should reappear in the future. Cell-mediated immunity also destroys *intracellular* pathogens, especially viruses.

A cell-mediated immune response begins when an **antigen-presenting cell (APC),** often a macrophage, engulfs a foreign (nonself) antigen and displays the antigen, in combination with its own self proteins, on its plasma membrane. When a T cell, which has been programmed during maturation to recognize this particular antigen, binds to the antigen and the APC self proteins, it becomes activated. The activated T cell undergoes repeated mitotic

Table 13.5 Types of Immunity

Type	Mechanism	Result
Naturally acquired active immunity	Infection by live pathogens	Person is ill with the disease; immune response destroys pathogens and leaves memory T and B cells to prevent later infection
Artificially acquired active immunity	Receives vaccine of weakened, dead, or inactivated pathogens or their antigenic parts	Immune response occurs without the person becoming ill; memory T and B cells remain to prevent later infection
Naturally acquired passive immunity	Antibodies passed from mother to fetus in utero or to newborn via breast milk	Enables short-term immunity for newborn infant without stimulating an immune response
Artificially acquired passive immunity	Receives injection of antibodies against a specific antigen	Enables short-term immunity without stimulating an immune response

13.7 Rejection of Organ Transplants

Learning Objective

12. Explain how rejection of a transplanted organ occurs.

Organ transplants are viable treatment options for persons with terminal disease of certain organs such as the heart, kidneys, and liver. Except for the surgery, the major problem encountered is that the patient's lymphoid system recognizes the transplanted organ as foreign and launches an attack against it. The problem is reduced by carefully determining compatibility of the tissues of both donor and recipient. This is usually done by comparing the antigens on the surfaces of leukocytes, called the *human leukocyte antigens (HLAs) group A*, of donor and recipient. A match does not have to be 100% to make a transplant, but the closer to 100%, the better chance of avoiding rejection.

To overcome the normal immune response and organ rejection, immunosuppressive therapy is administered following transplant surgery. The lymphoid system must be suppressed sufficiently to prevent rejection of the organ but not enough to eliminate immunity against pathogens. Achieving this delicate balance has been aided by the use of *cyclosporine*, a selective immunosuppressive drug derived from fungi. Cyclosporine inhibits T cell functions but has minimal effects on B cells. Because T cells are primarily responsible for organ rejection, rejection is minimized, and B cells are able to provide antibody-mediated immunity against pathogens. In spite of advances in immunosuppressive therapy, bacterial and viral infections are the primary causes of death among organ transplant recipients. Immunosuppression also increases the risk of cancer.

13.8 Disorders of the Lymphoid System

Learning Objective

13. Describe the common disorders of the lymphoid system.

Infectious Disorders

Acquired immunodeficiency syndrome (AIDS) is a viral disease that is approaching epidemic proportions. It is caused by the *human immunodeficiency virus (HIV)*, which attacks and kills helper T cells and invades macrophages, which serve as a reservoir for the virus. In time, the immune defenses of the victim are greatly reduced, and the patient becomes susceptible to opportunistic diseases that ultimately lead to death. These secondary diseases include pneumonia caused by *Pneumocystis carinii* and a cancer known as *Kaposi sarcoma*, disorders that are rarely encountered except in AIDS patients. At present, there is no cure for AIDS, although major research efforts have provided drugs to slow its progress.

Although HIV has been found in most body fluids, it appears that sufficient HIV concentrations for transmission to other persons are not present in tears or saliva. Transmission does not occur through routine, nonintimate contact; instead, it primarily occurs through blood exchanges and sexual intercourse. Transmission occurs through exchanges of blood, most commonly by the use of contaminated hypodermic needles and by exposure of open wounds or mucous membranes to infected blood. Vaginal fluids and semen of infected persons are effective transmitting agents in sexual intercourse. Also, infected mothers may transmit HIV to infants during childbirth.

Elephantiasis (el-e-fan-tī′-ah-sis) is a chronic condition characterized by greatly swollen (edematous) lower limbs or other body parts, which become "elephantlike" in appearance. This occurs because the lymphatic vessels are blocked by masses of microscopic roundworms, which causes fluid to accumulate excessively in the tissues drained by the plugged lymphatic vessels. The microscopic worms are transmitted by the bites of certain species of mosquitoes found in tropical regions.

Tonsilitis is the inflammation of the tonsils. It usually results from bacterial infections that cause the tonsils to become sore and swollen. If this condition becomes chronic and interferes with breathing or swallowing, or becomes a persistent focal point for spreading infections, the tonsils may be surgically removed. Tonsillectomies are much less common now that effective antibiotics are available and the role of tonsils in immunity is better understood.

Lymphadenitis (lim-fad-en-ahy′-tes) is inflammation of the lymph nodes. It is a common complication of bacterial infections and is often described as "swollen glands." Doctors often feel the cervical region for swollen lymph nodes to see if they are inflamed due to fighting an infection in the head or neck, which drains into the cervical lymph nodes.

A severely swollen lymph node is called a *bubo*. Buboes are named after the characteristic swelling of the inguinal and axillary lymph nodes in individuals infected with the bubonic plague, a bacterial infection of the lymphoid system that contributed to the Black Death. The Black Death is the largest known pandemic (large infectious outbreak), which killed about half of the population of Europe (estimated at up to 200 million people worldwide) from 1347 to 1351.

Noninfectious Disorders

Allergy (al′-er-jē), or *hypersensitivity*, refers to an abnormally intense immune response to an antigen that is harmless to most people. Such antigens are called **allergens** to distinguish them from antigens associated with disease. Once a person is sensitized to an allergen, an allergic reaction results whenever subsequent exposure to that allergen occurs. Allergic reactions may be either immediate or delayed.

Immediate reactions result when allergens bind with IgE on the surface of mast cells. This interaction causes these cells to secrete substances—such as histamine—that stimulate an inflammatory response. Immediate allergic reactions may be either localized or systemic (whole body). Localized reactions, such as hay fever, hives, allergy-based asthma, and digestive disorders, are unpleasant but rarely life threatening. In contrast, systemic allergic reactions, also known as *anaphylaxis* (an-ah-fi-lak′-sis), are often life threatening. They quickly impair breathing and may cause circulatory failure due to a sudden drop in blood pressure as blood vessels dilate and fluid moves into the tissues. Allergic reactions to penicillin and bee stings can cause systemic allergic responses. APIR

Delayed allergic reactions appear one to three days after exposure to the antigen. Delayed allergic reactions result from cytokines released by T cells. The dermatitis that occurs following contact with poison ivy and some cosmetic chemicals is a common delayed allergic reaction.

Autoimmune diseases result when T and B cells, for unknown reasons, recognize certain body tissues as foreign antigens and produce an immune response against them. This problem may result because certain body molecules have changed slightly and are no longer recognizable as self. Some of the most common autoimmune diseases are

- *rheumatoid arthritis (RA),* which destroys joints;
- *type I diabetes,* which destroys beta cells in the pancreas;
- *multiple sclerosis (MS),* which destroys the myelin sheath in the CNS;
- *Graves disease,* which stimulates the thyroid gland to produce excessive amounts of thyroid hormones;
- *myasthenia gravis,* which impairs nerve impulse transmission at neuromuscular junctions; and
- *Hashimoto disease,* which destroys the thyroid gland, creating hypothyroidism. It is the most common cause of hypothyroidism in the United States.

Lymphoma (lim-fō′-mah) is a general term referring to any tumor of lymphoid tissue. There are several types of lymphomas. One type of malignant lymphoma is *Hodgkin lymphoma,* which is cancer of lymphoid tissue involving the production of B cells. It is characterized by lymphadenitis, fatigue, and sometimes fever and night sweats. Early treatment with chemotherapy or radiation yields a high cure rate.

Severe combined immunodeficiency (SCID) is a group of disorders resulting from several different genetic defects. They are characterized by a marked deficit or absence of both T cells and B cells. Thus, the lymphoid system of affected individuals is basically nonfunctional. Infants have little or no protection against pathogens and usually die within a year without treatment. Transplants of normal stem cells from red bone marrow or umbilical cord blood have proved to be successful in some cases, and gene therapy trials look promising for treating this devastating disease.

Check My Understanding

9. What is so serious about the AIDS virus destroying helper T cells?
10. What is the cause of autoimmune diseases?

Chapter Summary

13.1 Lymph and Lymphatic Vessels

- As interstitial fluid accumulates in interstitial spaces, it is picked up by lymphatic capillaries. Once the fluid is within a lymphatic vessel, it is called lymph.
- Lymph is carried by lymphatic vessels and is ultimately returned to the blood. Lymphatic vessels contain valves that keep the lymph moving in one direction.
- Lymphatic vessels merge to form lymphatic trunks, which drain major regions of the body. Lymphatic trunks ultimately join one of two collecting ducts.
- The right lymphatic duct receives lymphatic trunks that drain the superior right portion of the body; it empties into the right subclavian vein.
- The thoracic duct receives lymphatic trunks from the rest of the body; it empties into the left subclavian vein.
- The propulsive forces that move lymph through the vessels are skeletal muscle contractions, intestinal movements, respiratory pressure changes (the same venous return mechanisms described in chapter 12), and peristaltic contractions of some lymphatic vessels.

13.2 Lymphoid Organs

- Lymphoid organs include red bone marrow, the thymus, lymph nodes, and the spleen.
- Primary lymphoid organs are sites of lymphocyte production.
- Secondary lymphoid organs are sites of lymphocyte proliferation and immune responses.
- Red bone marrow is the site of origination for all lymphocytes.
- B cells become immunocompetent in the red bone marrow, but T cells migrate to the thymus to become immunocompetent.
- Immunocompetence is the ability to recognize and respond to foreign antigens.
- The thymus is located superior to the heart within the mediastinum.
- Lymph nodes are lymphoid organs occurring in groups along lymphatic vessels, where they filter lymph and serve as a site of lymphocyte activation and proliferation.
- The spleen filters and cleanses the blood and contains a reservoir of RBCs and platelets that are squeezed into circulation if more blood is needed. It destroys old and damaged RBCs as well as activates an immune response to foreign antigens in the blood.

13.3 Lymphoid Tissues

- Tonsils are groups of lymphatic tissues within the mucosae of pharynx and oral cavity. They intercept pathogens entering the pharynx from the mouth and nose.
- Mucosa associated lymphoid tissue (MALT) consists of collections of WBCs within the mucosae of respiratory, digestive, urinary, and reproductive tracts.

13.4 Nonspecific Resistance

- Nonspecific resistance provides general protection against all pathogens, but it is not directed at any particular pathogen.
- Mechanical barriers include the skin, mucous membranes, mucus, tears, saliva, and urine.
- Protective chemicals include gastric juice, interferon, enzymes such as lysozyme and pepsin, and fluids with a low pH.
- Complement is a group of plasma proteins that poke holes in the plasma membranes of pathogens. Complement also activates phagocytosis and inflammation.
- Phagocytosis is a major mechanism of nonspecific resistance. Neutrophils, monocytes, and wandering and fixed macrophages actively engulf and destroy pathogens. The wandering and fixed macrophages form the tissue macrophage system, which plays a major role in protection against pathogens.
- Inflammation helps control infection by attracting WBCs and macrophages and by increasing the blood supply to the affected area.
- Fever increases the rate of defense processes and inhibits the growth of certain pathogens.

13.5 Immunity

- Immunity provides protective mechanisms against specific antigens. Lymphocytes and macrophages play key roles in immunity.
- Maturing lymphocytes learn to distinguish molecules composing the body (self) from foreign (nonself) molecules. Large nonself molecules that can stimulate an immune response are called antigens.
- Undifferentiated lymphocytes are first formed in the fetal yolk sac with all formed elements. The liver and spleen take over formed element production in the fetal life as well before eventually moving to the red bone marrow in a newborn infant. About half of the lymphocytes, the B cells, stay in the red bone marrow to become immunocompetent. The other half are carried to the thymus, where they become immunocompetent T cells. Immunocompetent T cells and B cells are dispersed to secondary lymphoid organs and tissues throughout the body and also circulate in the blood.
- T cells provide cell-mediated immunity. T_H cells and T_C cells are activated when they bind to an antigen and self protein presented by an APC. Activated T_H cells form a clone of active T_H cells and T_M cells; activated T_C cells form a clone of active T_C cells and T_M cells.
- T_H cells release cytokines that stimulate cell division and immune action of T and B cells and phagocytosis. T_C cells destroy foreign and diseased cells with chemicals. After the antigen is destroyed, T_M cells remain to attack the antigen if it reenters the body.
- B cells provide antibody-mediated immunity. When an antigen binds to receptors on a B cell, it is engulfed and displayed on the cell surface along with self antigens.

A T_H cell binds to the antigen-self protein complex and secretes cytokines, which activate the B cell. The B cell forms a clone of identical B cells. Most of the cloned B cells become plasma cells, and a few become memory B cells. Plasma cells produce antibodies that bind to the antigen, enabling easier destruction of the antigen by phagocytosis and other means. Memory cells remain to attack the antigen if it reenters the body.
- Antibodies are specific proteins produced by plasma cells against specific antigens. There are five classes of antibodies: IgG, IgA, IgM, IgD, IgE. Each performs a special role.

13.6 Immune Responses

- The first contact with a specific antigen produces the primary immune response. Subsequent contacts with the same antigen produce secondary immune responses that are more rapid and intense.
- There are two basic types of immune responses: active and passive. Active immunity results when the body produces its own memory cells and antibodies. Passive immunity results from antibodies that have been produced by another person, an animal, or synthetically.

13.7 Rejection of Organ Transplants

- The normal response of the lymphoid system is to attack and destroy organ transplants because they are foreign. Rejection is minimized by a good match between donor and recipient and by the administration of immunosuppressive drugs.
- The use of cyclosporine inhibits functions of T cells but leaves functions of B cells essentially intact.

13.8 Disorders of the Lymphoid System

- Infectious disorders include AIDS, elephantiasis, lymphadenitis, and tonsilitis.
- Noninfectious disorders of the lymphoid system include allergy, autoimmune diseases, and lymphoma.

Self-Review

Answers are located in appendix B.

1. The lymphoid system removes excess ______ from interstitial spaces.
2. Lymphatic vessels carry ______ from body tissues to the ______ veins.
3. Filtering and cleansing lymph is the function of ______, which also are sites of ______ production.
4. Filtering and cleansing blood is a function of the largest lymphoid organ, the ______, which also removes worn-out ______ cells.
5. Mucous membranes, tears, and gastric juice provide ______ resistance against disease.
6. The major WBC phagocytes involved in nonspecific resistance are ______ and ______.
7. Fighting infection is aided by ______, which increases blood flow and attracts phagocytes to the affected area.
8. Lymphocytes that become immunocompetent in the ______ become T cells.
9. To start an immune response, part of an antigen must be displayed by an ______ cell where a specific ______ cell can bind to it.
10. Subtype cells formed in a clone of T cells are ______ T cells, ______ T cells, and ______ T cells.
11. B cells produce ______-mediated immunity, and T cells produce ______-mediated immunity.
12. Cytokines released by ______ cells promote rapid cell division in activated ______ cells and ______ cells.
13. An activated B cell forms a clone consisting of ______ cells, which produce ______, and memory B cells.
14. Immunity acquired by contact with a pathogen and producing antibodies against it is known as ______ immunity.
15. ______ artificially introduces an antigen into the body and produces a primary immune response, so if the real pathogen subsequently enters the body, a faster, more intense ______ immune response destroys it.

Critical Thinking

1. How would a nonfunctional thymus affect immunity?
2. How can T cells and B cells make receptors for so many different antigens?
3. Why would an inability to remove excess fluid from a tissue lead to a potentially life-threatening situation?
4. How is vaccination good for both the person receiving the vaccination and also people that person comes in contact with?
5. Why is it difficult to cure AIDS?

ADDITIONAL RESOURCES

14 CHAPTER

Respiratory System

One fall weekend, Jesse drives home from college to attend the homecoming football game at his old high school. When Jesse arrives at the game, it is as if the entire town has come out to watch their team battle the rival high school. For three hours, Jesse cheers loudly for every great play and yells at the referee for every bad call. In fact, the stadium is so loud that he has to shout to talk to some old friends who have sat down nearby. When the game ends, Jesse heads home feeling triumphant over his alma mater's victory. However, when he wakes up in the morning, his neck is sore and his voice is very raspy and barely audible. His mother, a registered nurse at the local hospital, diagnoses Jesse as having acute laryngitis. Jesse's loud cheering has inflamed his larynx, or "voice box," which is causing the soreness and difficulty speaking. As he rests his voice and drinks some hot tea with honey, Jesse thinks to himself that supporting his school was completely worth a little discomfort.

CHAPTER OUTLINE

Module 11
Respiratory System

SELECTED KEY TERMS

Alveolar gas exchange The exchange of oxygen and carbon dioxide between the air in alveoli and the blood in alveolar capillaries.
Alveolus (alveol = small cavity) A microscopic air sac within a lung.
Breathing The movement of air into and out of the lungs.
Bronchial tree (bronch = windpipe) The branching bronchi.
Expiration (ex = from; spirat = breathe) Movement of air out of the lungs; exhalation.
Glottis The opening between the vocal folds within the larynx.
Inspiration Movement of air into the lungs: inhalation.
Larynx (laryn = gullet) The cartilaginous box located between the pharynx and the trachea that contains vocal folds.
Pharynx (pharyn = throat) The cavity between the mouth and the esophagus or larynx, used in both breathing and swallowing; the throat.
Surfactant A chemical in alveoli that reduces surface tension and prevents alveolar collapse.
Systemic gas exchange The exchange of oxygen and carbon dioxide between the blood in systemic capillaries and the tissue cells.
Trachea (trache = windpipe) The tube carrying air between the larynx and the bronchi.

THE PRIMARY ROLE of the respiratory system is to make oxygen available to cells for cellular respiration and to remove carbon dioxide, the main byproduct of that metabolism. The entire process of respiration encompasses five unique and sequential processes:

1. **Breathing** (pulmonary ventilation)–the movement of air into and out of the lungs.
2. **Alveolar gas exchange**–the exchange of oxygen and carbon dioxide between the air in alveoli and the blood in alveolar capillaries.
3. **Gas transport**–transport of oxygen and carbon dioxide between the lungs and tissues, accomplished by the cardiovascular system.
4. **Systemic gas exchange**–the exchange of oxygen and carbon dioxide between the blood in systemic capillaries and the tissue cells.
5. **Cellular respiration**–the use of oxygen and production of carbon dioxide during ATP production.

The structures of the respiratory system are involved directly in only two of these processes: breathing and alveolar gas exchange, which is collectively referred to as **external respiration.** Systemic gas exchange and cellular respiration together are referred to as **internal respiration.**

The respiratory system does more than just exchange respiratory gases. It also helps to detect odors, produce sounds, regulate blood pH, trap and defend the body from airborne pathogens, and assist in the movement of venous blood and lymph.

14.1 Structures of the Respiratory System

Learning Objective

1. Describe the structures and functions of the respiratory system.

The *respiratory system* is subdivided into upper and lower respiratory tracts. The *upper respiratory tract* includes the nose and pharynx. The *lower respiratory tract* includes the larynx, trachea, bronchi, and lungs (figure 14.1*a*).

Nose

The protruding portion of the *nose* is supported by bone and nasal cartilage (figure 14.1*b*). The nasal bones form a rigid support for the bridge of the nose, and nasal cartilage supports the remaining portions and is responsible for the flexibility of the nose. The *nostrils,* or *nares,* are the two external openings in the nose that allow air to enter and leave the nasal cavity. Stiff hairs around the nostrils tend to keep out large airborne particles and insects.

The **nasal cavity** is the internal chamber of the nose that is surrounded by skull bones. It is separated from the oral cavity by the **palate** (roof of the mouth), which consists of two basic portions. The anterior *hard palate* is formed by the palatine processes of the maxillae and the palatine bones. The posterior soft palate is composed of skeletal muscle tissue. The nasal cavity is divided into left and right portions by the **nasal septum,** a vertical partition of bone and nasal cartilage that is located on the midline. Three *nasal conchae* project from each lateral wall and serve to increase the surface area of and create air turbulence in the nasal cavity.

The superior nasal concha and superior nasal septum are lined with **olfactory mucosa** containing the olfactory epithelium, the tissue containing the olfactory receptors used in detecting chemicals for the sense of smell (see chapter 9). The rest of the nasal cavity, larynx, trachea, and bronchi are covered by **respiratory mucosa** containing pseudostratified ciliated columnar epithelium. Goblet cells within the epithelium produce mucus to coat the epithelial surface. As air flows through the nasal cavity, it is warmed by the blood-rich mucosae and is moistened by the mucus. In addition, airborne particles, including microorganisms,

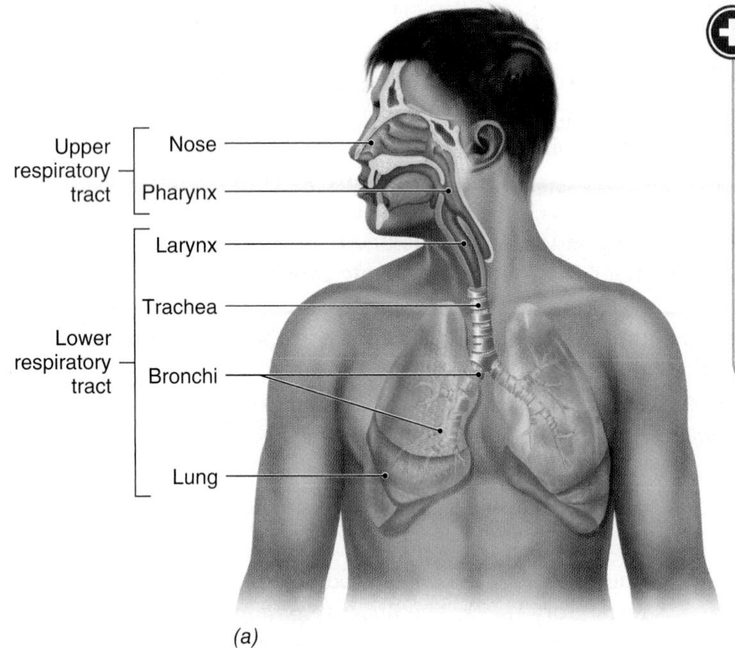

Clinical Insight

Tobacco smoke paralyzes cilia of the epithelia lining the air passages. As a result, mucus and entrapped particles are not effectively removed. Prolonged irritation by tobacco smoke causes the ciliated epithelium to be replaced with stratified squamous epithelium, which cannot clear the airways of mucus. The resulting mucus accumulations lead to smoker's cough and provide a fertile site for the growth of microorganisms.

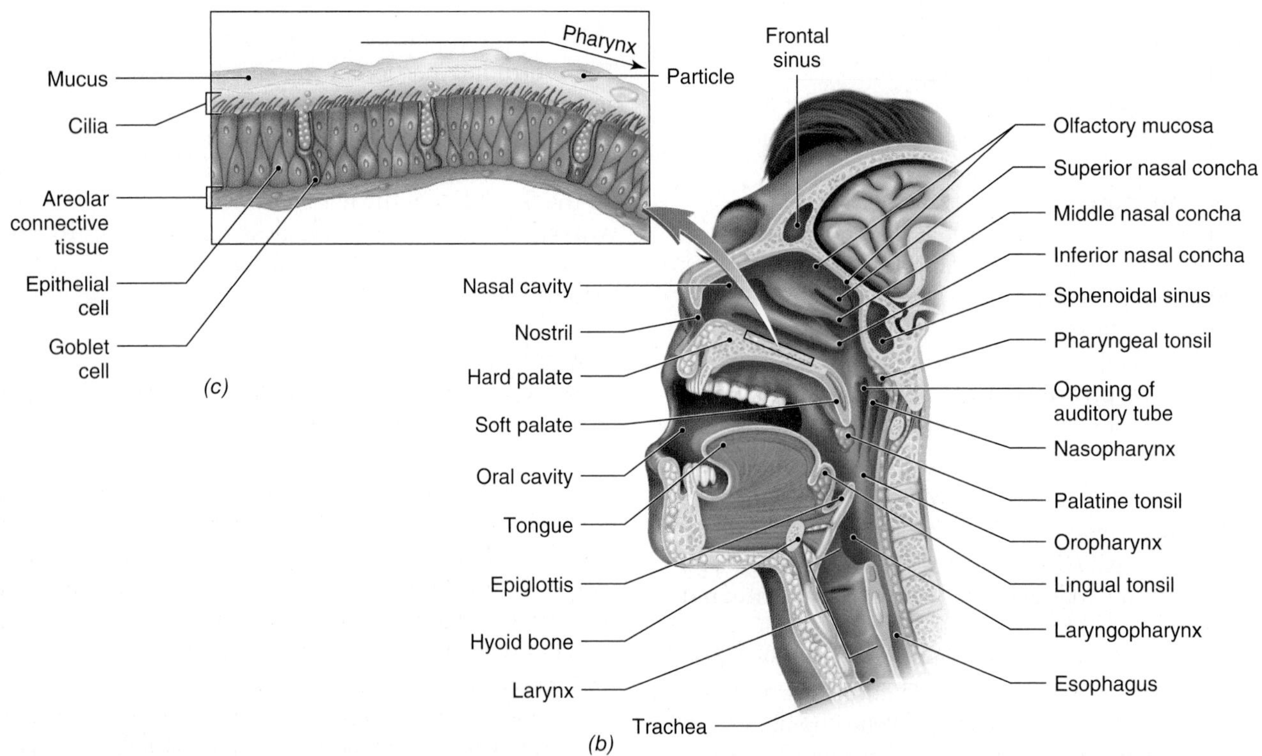

Figure 14.1 *(a)* The general organization of the respiratory system. *(b)* Major structures of the upper respiratory tract. *(c)* Mucus and entrapped particles are moved by cilia of the mucosa from the nasal cavity to the pharynx. APR

are trapped in the mucus of the respiratory mucosa. Within the larynx, trachea, and bronchi, cilia of the epithelium slowly move the layer of mucus with its entrapped particles toward the pharynx (figure 14.1c), where it is swallowed. Upon reaching the stomach, most microorganisms in the mucus are destroyed by the gastric juice.

Several bones surrounding the nasal cavity contain **paranasal sinuses,** air-filled cavities. Sinuses are located in the frontal bone, ethmoid, maxillae, and sphenoid adjacent to the nasal cavity. The sinuses lighten the skull and serve as sound-resonating chambers during speech. The sinuses open into the nasal cavity, which increases nasal cavity surface area, and are lined with ciliated mucosae that are continuous with the mucosae of the nasal cavity. The secreted mucus drains into the nasal cavity.

Pharynx

The **pharynx** (fayr′-inks), commonly called the throat, is a short passageway that lies posterior to the nasal and oral cavities and extends inferiorly to the larynx and esophagus. It has a muscular wall and it is lined with mucosae containing stratified squamous epithelium. As shown in figure 14.1*b*, the pharynx consists of three parts: the *nasopharynx* posterior to the nose; the *oropharynx* posterior to the mouth; and the *laryngopharynx* posterior to the larynx.

The *auditory* (eustachian) *tubes* (figure 14.1*b*), which extend to the middle ear, open into the nasopharynx. Air moves in or out of the auditory tubes to equalize the air pressure on each side of the tympanic membrane.

The tonsils, clumps of lymphoid tissue, occur at the openings to the pharynx. The *palatine tonsils* are located bilaterally at the junction of the oropharynx and the oral cavity. The *pharyngeal tonsil* (adenoid) is located in the superior nasopharynx, and the *lingual tonsils* are found on the posterior tongue. Tonsils are sites of immune reactions and may become sore and swollen when infected. Enlargement of the palatine tonsils tends to make swallowing painful and difficult. A swollen pharyngeal tonsil tends to block the flow of air between the nasal cavity and the pharynx, which promotes mouth breathing and can create snoring. When breathing through the mouth, air is not adequately warmed, filtered, and moistened.

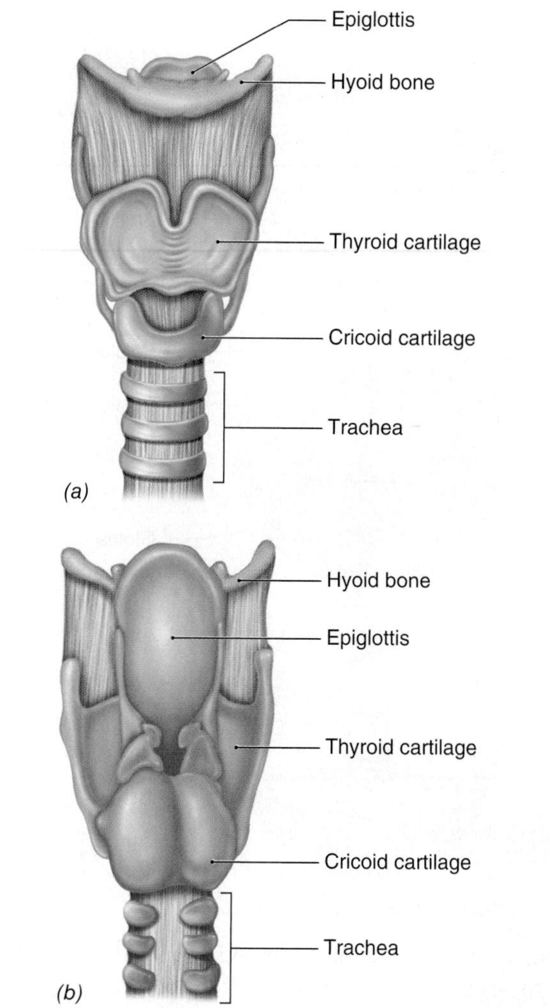

Figure 14.2 The Larynx.
(a) Anterior view. *(b)* Posterior view. APR

Larynx

The **larynx** (layr′-inks) is a boxlike structure, composed of several cartilages, that provides a passageway for air between the pharynx and the trachea. The three largest cartilages are the *thyroid cartilage,* which projects anteriorly to form the Adam's apple; the *cricoid cartilage,* which forms the attachment to the trachea; and the *epiglottis,* a cartilaginous flap that helps to keep solids (food) and liquid from entering the larynx. The larynx is supported by ligaments that extend from the hyoid bone (figure 14.2).

The **vocal folds** are two bands of elastic connective tissue covered by mucosa located within the larynx (figure 14.3). They are relaxed during normal breathing, but when contracted, they vibrate to produce vocal sounds when exhaled air passes over them. The pitch (high or low tone) of a sound is determined by the vibration frequency of the vocal folds. High frequency vibrations lead to high pitch sound and vice versa. The loudness (volume) of a sound is related to the vibration amplitude of the vocal folds. The larger the amplitude the louder the volume and vice versa. The opening between the vocal folds, the **glottis,** leads to the trachea. The **vestibular folds,** which lie superior to the vocal folds, are composed of a small amount of elastic connective tissue covered by mucosa. They prevent food from entering the glottis and are not involved in sound production.

Because the oropharynx and laryngopharynx are also passageways for food, a mechanism exists to prevent food from entering the larynx and to direct food into the esophagus (ē-sof′-ah-gus), the flexible tube that carries food to the stomach. When swallowing, muscles lift the larynx superiorly, which causes the epiglottis to fold over and cover the opening into the larynx. This action directs food into the esophagus, whose opening is located just posterior to the larynx. Sometimes this mechanism does not work perfectly and a small amount of food or drink enters the larynx, stimulating a coughing reflex that usually expels the substance.

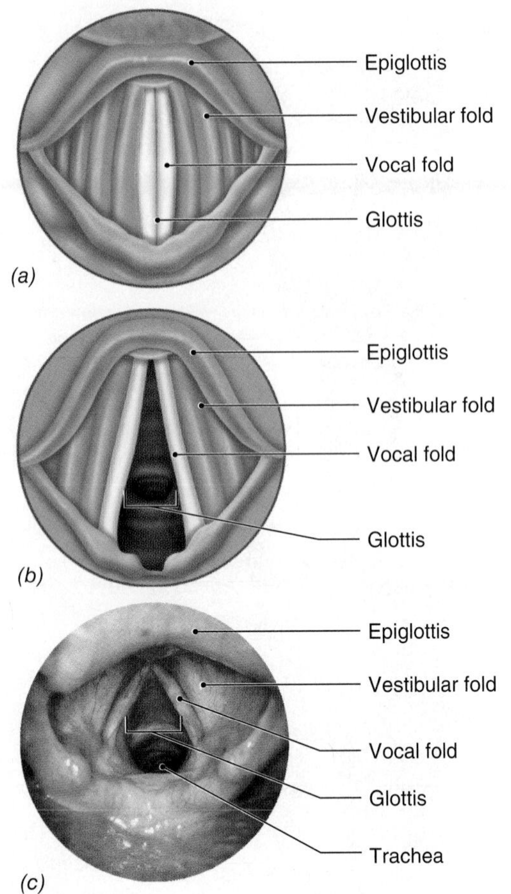

Figure 14.3 Vocal folds viewed superiorly with the glottis *(a)* closed and *(b)* open. *(c)* Photograph of the glottis and vocal folds.

Trachea

The **trachea** (trā′-kē-ah), or windpipe, is a tube that extends from the larynx into the thoracic cavity, where it branches to form the right and left main bronchi. The walls of the trachea are supported by C-shaped *tracheal cartilages* that hold the passageway open in spite of the air pressure changes that occur during breathing (figure 14.4*a*). The open portion of the tracheal cartilages is oriented posteriorly against the esophagus (see figure 14.2*b*). This orientation allows the esophagus to expand slightly as food passes down to the stomach.

The internal wall of the trachea is lined with mucosa containing pseudostratified ciliated columnar epithelium, the same type of epithelium lining most of the upper respiratory tract. Mucus produced by the goblet cells coats the surface of the epithelium and traps airborne particles, including microorganisms. The beating cilia move the mucus and entrapped particles superiorly to the pharynx where they are coughed out or swallowed. Microorganisms are usually killed by gastric juice in the stomach.

Bronchi, Bronchioles, and Alveoli

The trachea branches at about midchest into the left and right *main bronchi* (brong′-kī; singular, bronchus). Each main bronchus enters its respective lung, where it branches to form smaller *lobar bronchi,* one for each lobe of the lung. Lobar bronchi branch to form segmental bronchi that lead to different bronchopulmonary segments of each lung lobe. The bronchi continue to branch into smaller and smaller bronchi. Because the bronchi resemble tree branches, they are collectively called the **bronchial tree** (figure 14.4*a*).

The walls of the bronchi contain cartilaginous rings similar to those of the trachea, but as the branches get progressively smaller, the amount of cartilage gradually decreases and finally is absent in the very small tubes called the **bronchioles** (brong′-kē-ōls). As the amount of cartilage decreases, the amount of smooth muscle increases. The smooth muscle plays an important role in regulating the airflow through the air passageways. Contraction of the smooth muscle causes *bronchoconstriction,* which decreases airflow. Relaxation of the smooth muscle results in *bronchiodilation,* which increases airflow. Air passageways larger than bronchioles are lined with ciliated mucosae that continue to trap and remove airborne particles. Bronchioles are lined with mucosae containing simple cuboidal epithelium, so foreign particles that reach them are not effectively removed. Bronchioles branch to form smaller and smaller bronchioles that lead to microscopic **alveolar ducts,** which terminate in tiny air sacs called **alveoli** (al-vē′-ō-lī; singular, *alveolus*). Alveoli resemble tiny grapes clustered about an alveolar duct and are composed of simple squamous epithelium (figure 14.5*a*).

The primary function of the bronchial tree and bronchioles is to carry air into and out of the alveoli during breathing. The exchange of respiratory gases occurs between the air in the alveoli and the blood in the capillary networks that surround the alveoli (figure 14.5*b*). Oxygen and carbon dioxide diffuse readily through the thin **respiratory membrane,** which is composed of squamous cells of the alveolar wall and the capillary wall (figure 14.5*c*). Alveoli are extremely numerous—about 300 million in each lung. They have a combined surface area of about 75 square meters and can hold about 6,000 ml of air.

Alveoli contain very small spaces, which are coated with a watery fluid. The attraction (surface tension) between water molecules would cause the alveoli to collapse if it were not for surfactant. **Surfactant** (ser-fak′-tant) is a mixture of lipoproteins secreted by special cells in alveoli. It reduces the attraction between water molecules and keeps alveoli open so they may fill with air during inspiration. Without surfactant, alveoli would collapse and become very difficult to reinflate.

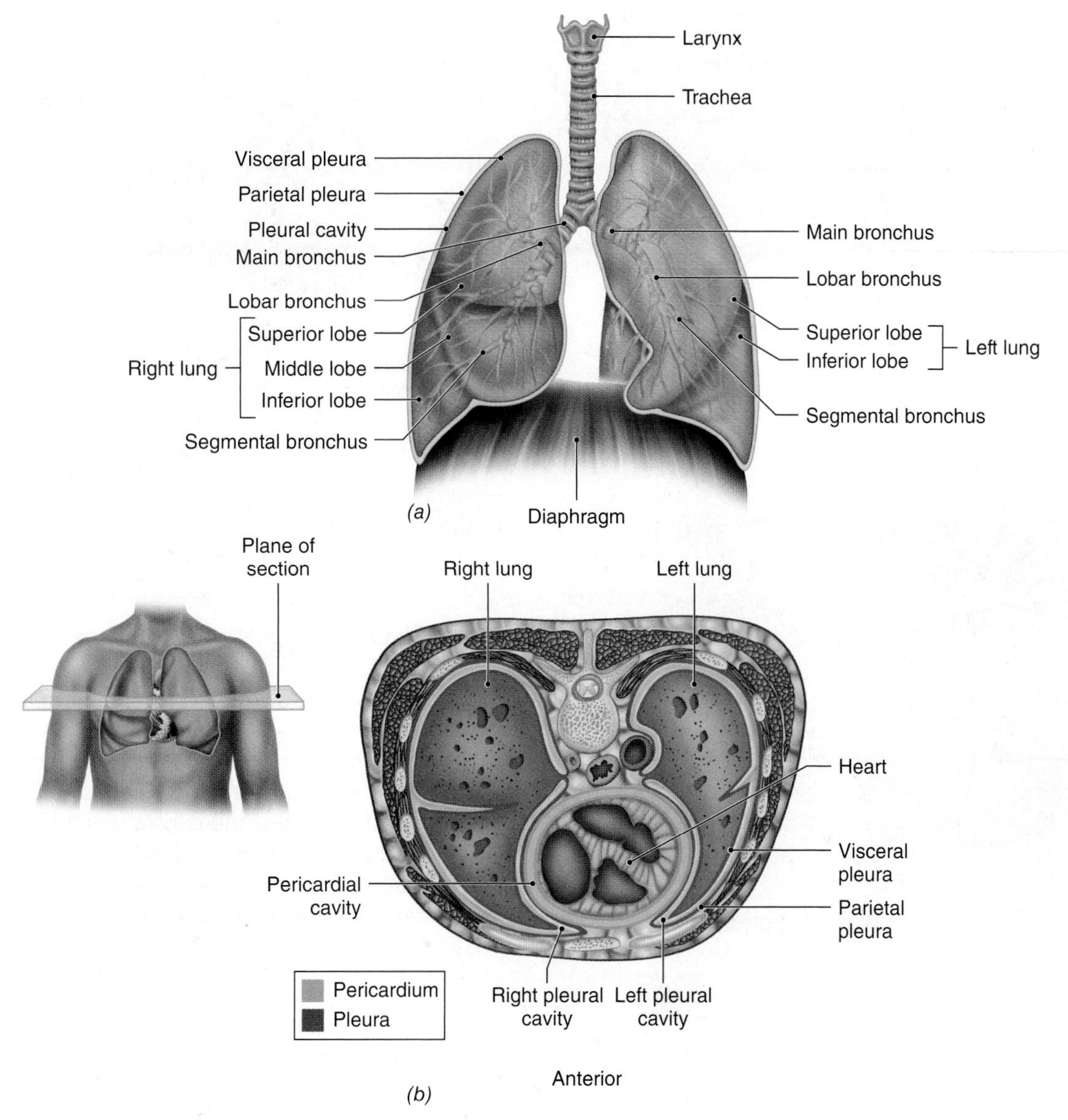

Figure 14.4 *(a)* Anterior view of the lower respiratory tract. *(b)* Transverse section of the thoracic cavity shows the pleurae and the relationship of the lungs to other thoracic organs. APIR

Lungs

The paired **lungs** are large organs that occupy much of the thoracic cavity. They are roughly cone-shaped and are separated from each other by the mediastinum. Each lung is divided into lobes. The left lung has two lobes (superior and inferior) and is somewhat smaller than the right lung, which is composed of three lobes (superior, middle, and inferior) (see figure 14.4*a*). Each lobe is supplied by a lobar bronchus, blood and lymphatic vessels, and nerves. The lungs consist primarily of air passages, alveoli, blood and lymphatic vessels, nerves, and connective tissues, giving the lungs a soft, spongy texture.

Two layers of serosae, called pleural membranes, enclose and protect each lung. The **visceral pleura** is firmly attached to the surface of each lung, and the **parietal pleura** lines the internal wall of the thoracic cage. The potential space between the visceral and parietal pleurae is known as the **pleural cavity.** A thin film of pleural fluid occupies the pleural cavity and reduces friction between the pleurae as the lungs inflate and deflate during breathing. Although lungs are elastic and tend to recoil, the attraction of water molecules in the pleural fluid within the pleural cavity keeps the visceral and parietal pleurae stuck together (figure 14.4*b*).

Table 14.1 summarizes the functions of the respiratory structures.

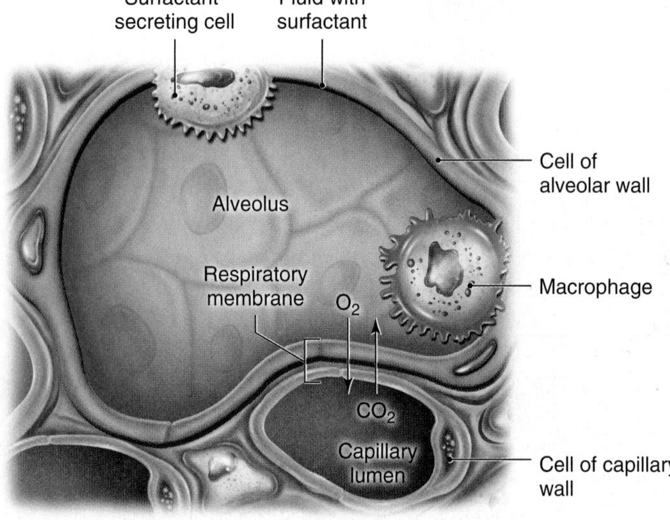

Figure 14.5 *(a)* Bronchioles branch to form smaller and smaller bronchioles, which lead to microscopic alveolar ducts that terminate in a cluster of alveoli. *(b)* Each alveolus is enveloped by a capillary network. Deoxygenated blood (blue) enters the capillary network. Oxygenated blood (red) exits the capillary network. *(c)* Cross-sectional view of an alveolus shows the respiratory membrane consisting of alveolar and capillary walls. AP|R

Check My Understanding

1. What are the functions of the nose?
2. What are the three divisions of the pharynx, and where are they located?
3. What are the functions of the glottis and epiglottis?
4. What are the functions of the vocal folds and vestibular folds?
5. What is the function of the cartilage in the walls of the trachea and bronchi?
6. What are the functions of the bronchi and bronchioles?
7. Why is surfactant important?

Table 14.1 Summary of Functions of the Respiratory Structures

Component	Function
Nose	Nostrils allow air to enter and exit the nasal cavity; the nasal cavity filters, warms, and moistens the inhaled air
Pharynx	Carries air between the nasal cavity and the larynx; filters, warms, and moistens the inhaled air; serves as passageway for food from the mouth to the esophagus; equalizes air pressure with middle ear via auditory tube
Larynx	Carries air between the pharynx and trachea; contains vocal folds for producing sounds in vocalization; prevents objects from entering the trachea
Trachea	Carries air between the larynx and the bronchi; filters, warms, and moistens the inhaled air
Bronchi	Carries air between the trachea and the bronchioles; filters, warms, and moistens the inhaled air
Bronchioles	Regulates the rate of airflow through bronchoconstriction and bronchodilation
Alveoli	Allow the gas exchange between the air in the alveoli and the blood in surrounding capillaries

14.2 Breathing

Learning Objective

2. Describe the mechanism of breathing.

Breathing, or pulmonary ventilation, is the process that exchanges air between the atmosphere and the alveoli of the lungs. Air moves into and out of the lungs along an air pressure gradient–from regions of higher pressure to regions of lower pressure. There are three pressures that are important in breathing:

1. *Atmospheric pressure* is the pressure of the air that surrounds the earth. Atmospheric pressure at sea level is 760 mm Hg, but at higher elevations it decreases because there is less air at higher elevations.
2. *Intra-alveolar (intrapulmonary) pressure* is the air pressure within the lungs. As we breathe in and out, this pressure fluctuates between being lower than atmospheric pressure and higher than atmospheric pressure.
3. *Intrapleural pressure* is the pressure within the pleural cavity. It is about 2 to 6 mm Hg below the atmospheric pressure during various phases of breathing. This lower intrapleural pressure is often described as "negative pressure," and it keeps the lungs stuck to the internal walls of the thoracic cage and helps expand the lungs, even as the thoracic cage expands and contracts during breathing. If the intrapleural pressure were to equal atmospheric pressure, the lungs would collapse and be nonfunctional.

The presence of air in the pleural cavity is called a *pneumothorax* (nū-mō-thō′-raks). This may occur due to a thoracic injury or surgery that allows air to enter the pleural cavity. It also occurs in emphysema patients when air escapes from ruptured alveoli into the pleural cavity. A pneumothorax causes the affected lung to collapse and become nonfunctional. Because each lung is in a separate pleural cavity, the collapse of one lung does not adversely affect the other lung. Treatment involves removing the intrapleural air to restore the normal pressure so that the lung may inflate.

Inspiration

The process of moving air into the lungs is called **inspiration,** or inhalation. When the lungs are at rest, the air pressure in the lungs is the same as the atmospheric pressure. In order for air to flow into the lungs, the intra-alveolar pressure must be decreased to below atmospheric pressure. This change allows for air to flow from the higher air pressure in the atmosphere towards the lower air pressure within the lungs. The contraction of the diaphragm and the external intercostals during inspiration causes an increase in lung volume, which results in a decrease in intra-alveolar pressure.

The dome-shaped **diaphragm** is a thin sheet of skeletal muscle separating the thoracic and abdominal cavities. When it contracts, the diaphragm pulls inferiorly and becomes flattened, which increases the volume of the thoracic cavity. At the same time, contraction of the *external intercostals* elevates and protracts the ribs and pushes the sternum anteriorly, which further increases the volume of the thoracic cavity (figures 14.6, 14.7*a*).

Because the negative intrapleural pressure and the surface tension of the pleural fluid keep the visceral pleura stuck to the parietal pleura, the lungs are pulled along when the thoracic cage expands. Therefore, the expansion of the thoracic cavity increases the volume of the lungs, which decreases the intra-alveolar pressure. Then, the higher atmospheric pressure forces air through the air passageways into the lungs until intra-alveolar and atmospheric pressures are equal. Quiet inspiration requires the contraction of the diaphragm and the external intercostals only. Forceful inspiration requires the involvement of additional muscles in the neck and chest, such as the sternocleidomastoid, scalenes, serratus anterior, and pectoralis minor (figure 14.6). The contraction of these muscles elevates and protracts the ribs to a greater extent, leading to a greater increase in the volume of the thoracic cavity. Through this further increase in thoracic volume, intra-alveolar pressure decreases to a greater extent, which results in greater airflow into the lungs.

Expiration

Expiration, or exhalation, occurs when the diaphragm and external intercostals relax, allowing the thoracic cage and lungs to return to their original size. This results in a decrease in the volume of the thoracic cavity and lungs. The decrease in lung volume increases intra-alveolar pressure to a level higher than atmospheric pressure. The higher intra-alveolar pressure forces air out of the lungs until intra-alveolar and atmospheric pressures are equal.

Expiration during quiet breathing is a rather passive process because the abundant elastic connective tissue in the lungs and thoracic wall causes them to return to their original size as soon as the muscles of inspiration relax. However, a forceful expiration is possible by contraction of the *internal intercostals* (figure 14.6), which depresses and retracts the ribs, and by the muscles of the abdominal wall, which move the abdominal viscera and diaphragm

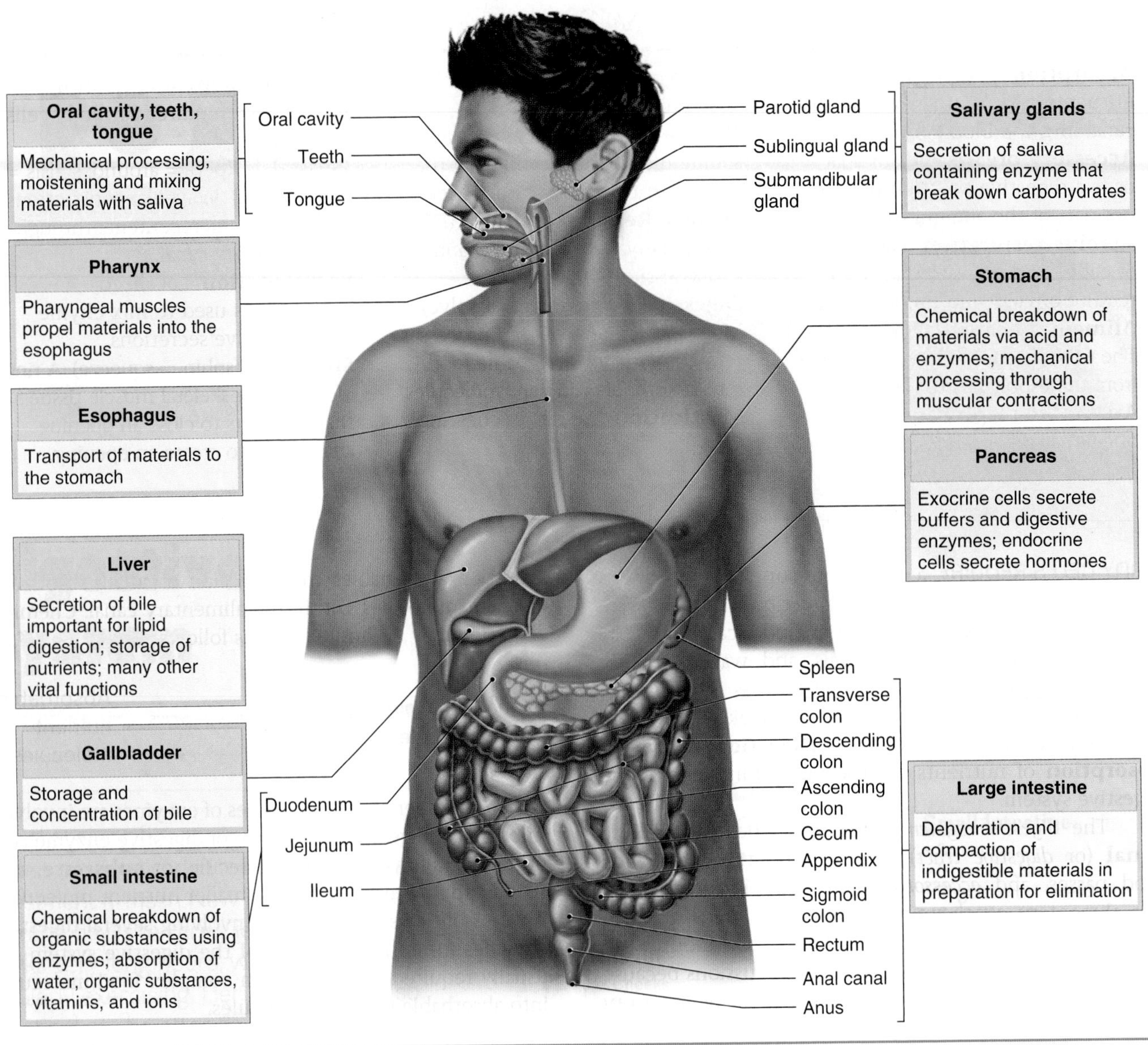

Figure 15.1 Major Organs of the Digestive System. APR

muscular layer, submucosa, and mucosa. These layers may be modified in the various regions but remain as distinct layers. Compare figure 15.2 to the discussion that follows.

The **serosa** is the most superficial layer. It consists of the *visceral peritoneum* and is continuous with the *parietal peritoneum,* which lines the internal surface of the abdominal wall. Cells of the peritoneum secrete peritoneal fluid, which keeps the membrane surfaces moist and reduces friction as parts of the alimentary canal rub against each other and the abdominal wall.

The **muscular layer** lies just deep to the serosa. It consists of two layers of smooth muscle that differ in the orientation of their muscle cells. Muscle cells of the more superficial muscular layer are arranged longitudinally. Their contractions shorten the tube. Muscle cells of the deeper muscular layer are arranged circularly around the tube. Their contractions constrict the tube. The functions of muscular layer are discussed next in the "Movements" section.

The **submucosa** lies between the muscular layer and the mucosa. It contains nerves, mucous glands, lymphatic vessels, and blood vessels embedded in areolar connective tissue.

The deepest layer is the **mucosa.** It consists of an epithelium that is in direct contact with the lumen and is supported by areolar connective tissue containing a few smooth muscle cells. The epithelium is often folded

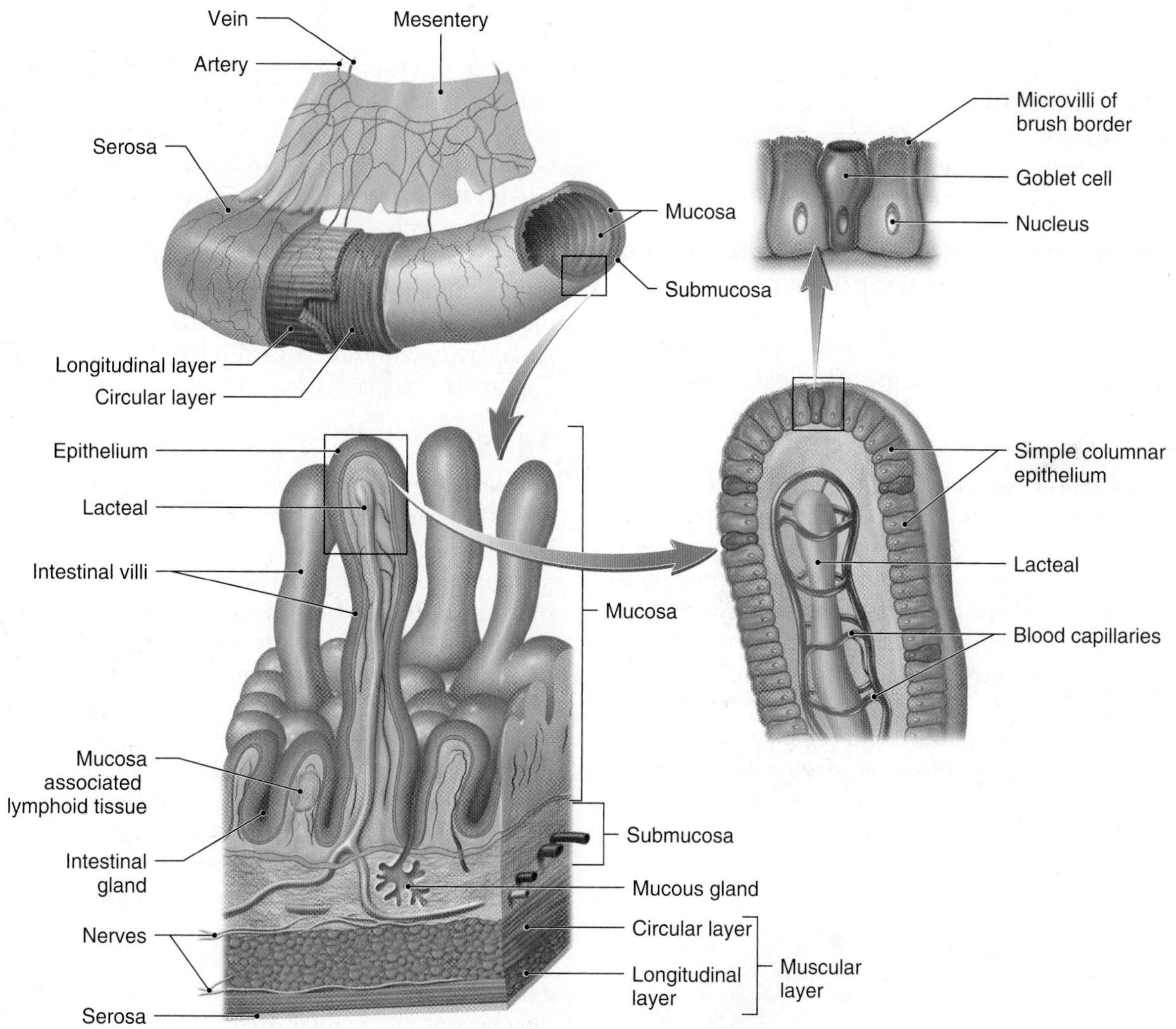

Figure 15.2 The wall of the alimentary canal consists of four layers as shown here in a section of the small intestine. From superficial to deep, they are the serosa, muscular layer, submucosa, and mucosa.

to increase the surface area that is in contact with food. The mucosa has different functions in different parts of the alimentary canal. In some regions, it secretes only mucus, which protects underlying cells. In others, it secretes mucus and digestive secretions containing enzymes and absorbs nutrients.

Movements

Contraction of the smooth muscle layers produces two different types of movement in the alimentary canal: mixing movements and propelling movements. Mixing movements are called **segmentation** and involve ringlike contractions followed by relaxation at multiple places along the alimentary canal. The alternating segmental contractions help to mix the food with the digestive secretions. These are especially frequent in the small intestine.

The movement that propels food through the alimentary canal is called **peristalsis** (per-i-stal′-sis). In peristalsis, coordinated contraction and relaxation of the circular and longitudinal muscular layers produce a wave of contraction along the alimentary canal that pushes the food in front of it. The movement resembles pushing toothpaste along a toothpaste tube towards its opening. In this way, peristaltic contractions move food from one portion of the alimentary canal to another.

Check My Understanding

1. What organs compose the alimentary canal?
2. What is chemical digestion?
3. How is food moved through the alimentary canal and what layer of the wall aids in this movement?

15.3 Mouth

Learning Objectives

4. Compare deciduous and permanent teeth.
5. Describe the basic structure of a tooth.
6. Describe digestion in the mouth.

The *mouth,* or *oral cavity,* is involved in the intake of food (ingestion), mechanically breaking it into small pieces, mixing it with saliva to begin chemical digestion, and swallowing it. The mouth is surrounded by the cheeks, palate, and tongue. Examine figures 15.3 and 15.4.

Cheeks

The *cheeks* form the lateral walls of the mouth. Skin covers their external surfaces, and nonkeratinized stratified squamous epithelium lines their internal surfaces. The cheeks help to hold food within the mouth, while contractions of muscles within them produce facial expressions. The anterior portions of the cheeks form the *lips,* which surround the opening into the mouth. Lips are sensitive, highly mobile structures with important roles in producing speech and detecting touch and temperature stimuli. Their pinkish color results from numerous blood vessels near their surfaces.

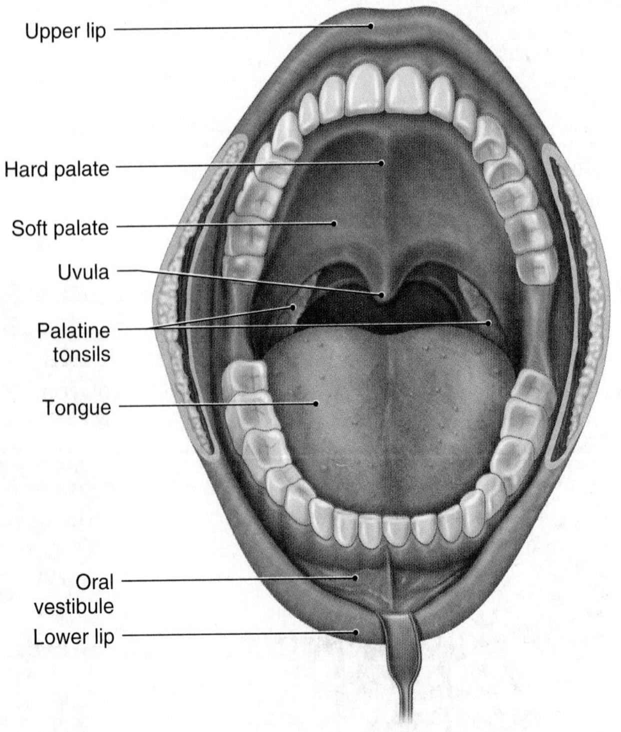

Figure 15.3 Major Structures Associated with the Mouth.

Palate

Recall from chapter 14 that the palate forms the roof of the mouth and separates the oral cavity from the nasal cavity. The hard palate provides protection to the nasal cavity while chewing, in addition to allowing simultaneous breathing and chewing. The soft palate ends posteriorly in a cone-shaped *uvula* that extends inferiorly at the back of the oral cavity. The uvula is very sensitive to touch stimuli. This sensitivity causes the soft palate to move superiorly during swallowing, which closes off the nasal cavity and directs food inferiorly into the pharynx. The uvula also plays a role in triggering the gag reflex.

Tongue

The *tongue* covers the floor of the oral cavity. It is composed primarily of skeletal muscle that is covered by a mucous membrane. An anterior fold of the mucous membrane on the inferior surface of the tongue attaches the tongue to the floor of the mouth. This membranous attachment, known as the *frenulum* (fren′-ū-lum) *of the tongue,* limits the posterior movement of the tongue.

Sometimes the frenulum of the tongue is too short and restricts tongue movements that are required for normal speech. A person with this problem is said to be "tongue-tied." Cutting the frenulum of the tongue to allow freer movement usually solves the problem.

The superior surface of the tongue contains numerous tiny projections called **lingual papillae** (pah-pil′-ē) that are the locations for taste buds, the sensory receptors for the sense of taste. Lingual papillae also give the tongue a rough texture, which aids in its manipulation of food during chewing and in mixing food with saliva. In swallowing, the tongue pushes food posteriorly into the pharynx. The tongue possesses *lingual glands* that produce an enzyme called *lingual lipase.* Lingual lipase aids in the digestion of triglycerides into fatty acids and glycerides within the acidic environment of the stomach. The movement of the tongue is also important for the production of speech.

Teeth

Teeth are important accessory digestive structures that mechanically break food into smaller pieces during **mastication** (mas-ti-kā′-shun), or chewing. Humans develop two sets of teeth: deciduous and permanent teeth.

The **deciduous teeth,** the first set, start to erupt through the gums at about six months of age. Central incisor (in-sī-zers) teeth come in first, and second molar teeth erupt last. There are 20 deciduous teeth, 10 in each jaw, and all of them are in place by three years of age. Deciduous teeth are gradually shed starting at about six years of age, and they are usually lost in the same order in which they emerged.

The **permanent teeth** begin appearing at about six years of age when the first molar teeth (six-year molar teeth) erupt. All of the permanent teeth, except the third

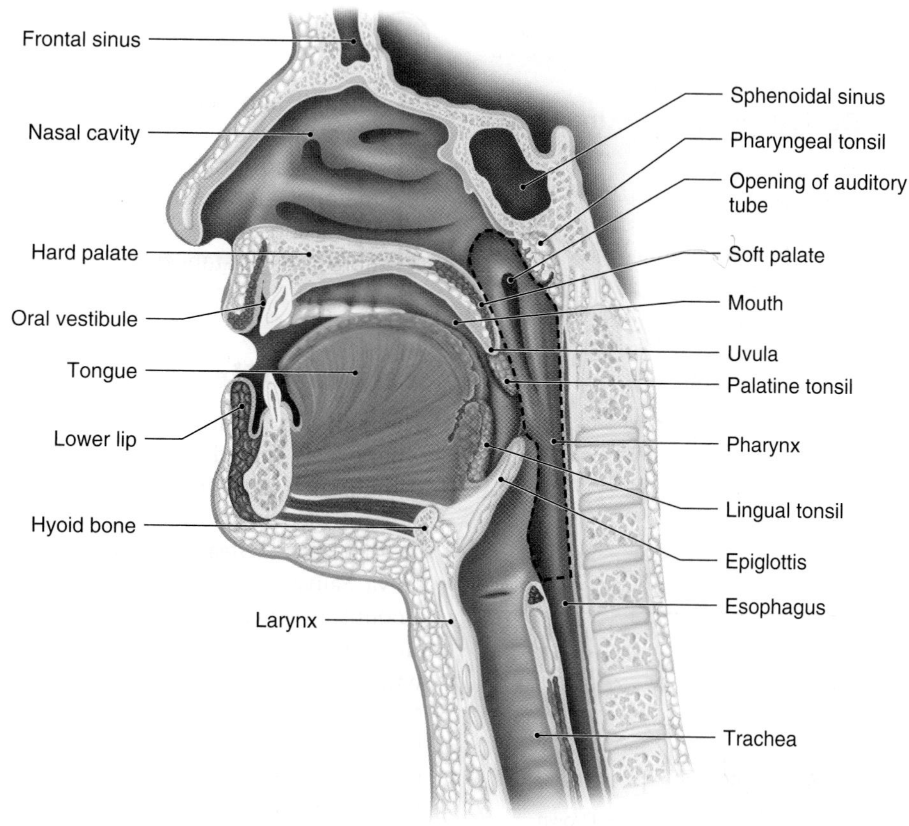

Figure 15.4 The structural relationships of the nasal cavity, mouth, pharynx, esophagus, and larynx are shown in sagittal section. The pharynx is outlined by the black dotted line. APIR

molar teeth, are in place by age 16. The third molar teeth (wisdom teeth) erupt between 17 and 21 years of age, or they may never emerge. In many persons, there is insufficient room for the third molar teeth, so they become impacted and often must be surgically removed. The 32 permanent teeth, 16 in each jaw, consist of four different types: incisor teeth, canine teeth, premolar teeth, and molar teeth (figure 15.5).

The chisel-shaped *incisor teeth* are adapted for biting off pieces of food. The *canine teeth* are used to grasp and tear tough food morsels. The somewhat flattened surfaces of the *premolar teeth* and *molar teeth* are used to crush and grind food. Examine figure 15.5 and table 15.1.

Figure 15.6 shows the basic structure of a tooth. Each tooth consists of two major parts: a root and a crown. The *crown* is the portion of the tooth that is exposed and not covered by the *gingiva* (jin-ji-vah), or gum, covering the underlying bone. The *root* is embedded in a socket, called an *alveolus*. A hard substance called *cement* attaches the *root*, through tough *periodontal ligaments*, to the alveolus. The junction of the crown and root is known as the *neck* of the tooth.

Most of a tooth is composed of **dentin,** a hard, bone-like substance. The crown of the tooth has a layer of **enamel** overlying the dentin. Enamel is the hardest substance in the

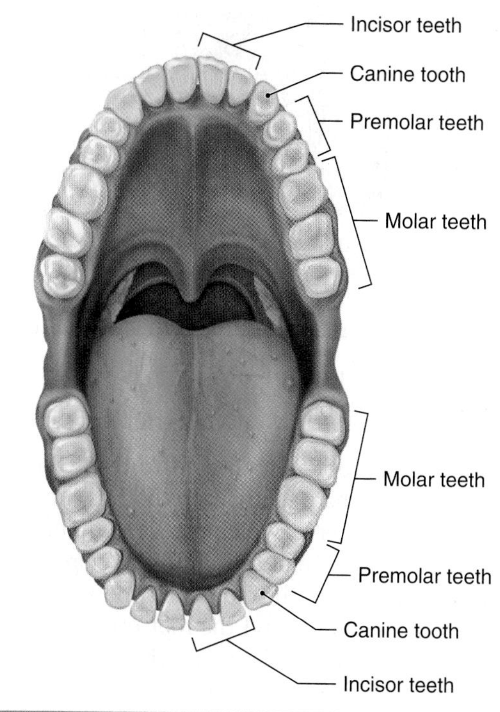

Figure 15.5 There are 32 permanent teeth, 16 in each jaw. Note the location and shape of the incisor teeth and molar teeth. APIR

Clinical Insight

Failure of the LES to close properly allows for gastric secretions to regurgitate into the inferior esophagus. The *gastroesophageal reflux* (GER) irritates the esophageal lining and usually produces a burning sensation called "heartburn." Causes of GER include malfunction of the LES, alcohol consumption, smoking, and excess acid production in the stomach. Avoiding foods that stimulate acid production in the stomach, such as caffeine and tomatoes, can reduce GER symptoms. Medications that neutralize stomach acid or reduce its production can also alleviate GER symptoms. GER that occurs chronically may indicate a more serious condition called *Gastroesophageal Reflux Disease* (GERD). The chronic esophageal irritation associated with GERD can lead to cellular changes that increase a person's risk of esophageal cancer or Barrett esophagus, which involves changes in the esophageal lining to resemble that of the intestine.

peristaltic contractions propel the food toward the stomach. The esophageal mucosa produces mucus to lubricate the esophagus and aid the passage of food.

At the junction of the esophagus and stomach, the *lower esophageal,* or *cardiac, sphincter* (sfink′-ter) prevents regurgitation of stomach contents into the esophagus. But when the peristaltic wave that is propelling food toward the stomach reaches the sphincter, it relaxes and allows food to enter the stomach. The lower esophageal sphincter (LES) is a physiological sphincter rather than an anatomical one because it is not seen in cadavers. It is believed to be caused by muscle tone within the esophagus or surrounding diaphragm. AP|R

Check My Understanding

7. How does the swallowing reflex push the food to the esophagus?
8. How does the esophagus carry food to the stomach?

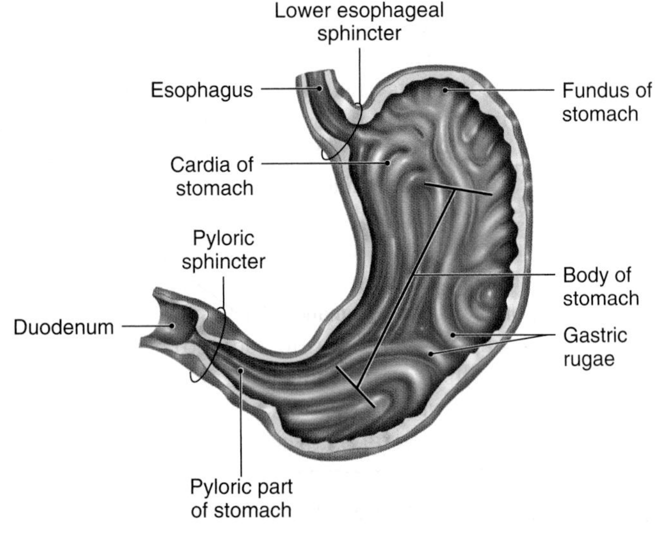

Figure 15.8 The pouchlike stomach receives food from the esophagus and releases chyme into the duodenum. AP|R

15.5 Stomach

Learning Objectives

9. Describe the structure and functions of the stomach.
10. Explain the control of gastric secretions.

As shown in figure 15.8, the J-shaped **stomach** is a pouchlike portion of the alimentary canal. It lies just inferior to the diaphragm in the left upper quadrant of the abdominopelvic cavity. The basic functions of the stomach are temporary storage of food, mixing food with gastric juice, and starting the chemical digestion of proteins.

Structure

The stomach may be subdivided into four regions: the cardia, fundus, body, and pyloric part. The *cardia* (closest to the heart) is a relatively small area that receives food from the esophagus. The *fundus* expands superior to the level of the cardia and serves as a temporary storage area. The *body* is the largest region of the stomach, and it is located between the fundus and pyloric part. The *pyloric part* is the narrow portion located near the junction with the duodenum.

The *pyloric sphincter* is a thickened ring of circular muscle cells that is located at the junction of the stomach and duodenum. This muscle usually is contracted, closing the stomach outlet, but it relaxes to let stomach contents pass into the small intestine.

The stomach possesses some specializations to accommodate its functions. The muscular layer contains a third layer of oblique muscle cells, which allows the stomach to better mix food with gastric secretions. The mucosa is quite thick, when compared to other organs of the alimentary canal. In an empty stomach, the mucosa and submucosa are organized into numerous folds called gastric *rugae* (rū-jē). These folds allow the lining to stretch as the stomach fills with food. The mucosa is dotted with numerous pores called *gastric pits*. Gastric pits receive secretions from **gastric glands** that extend deep into the mucosa.

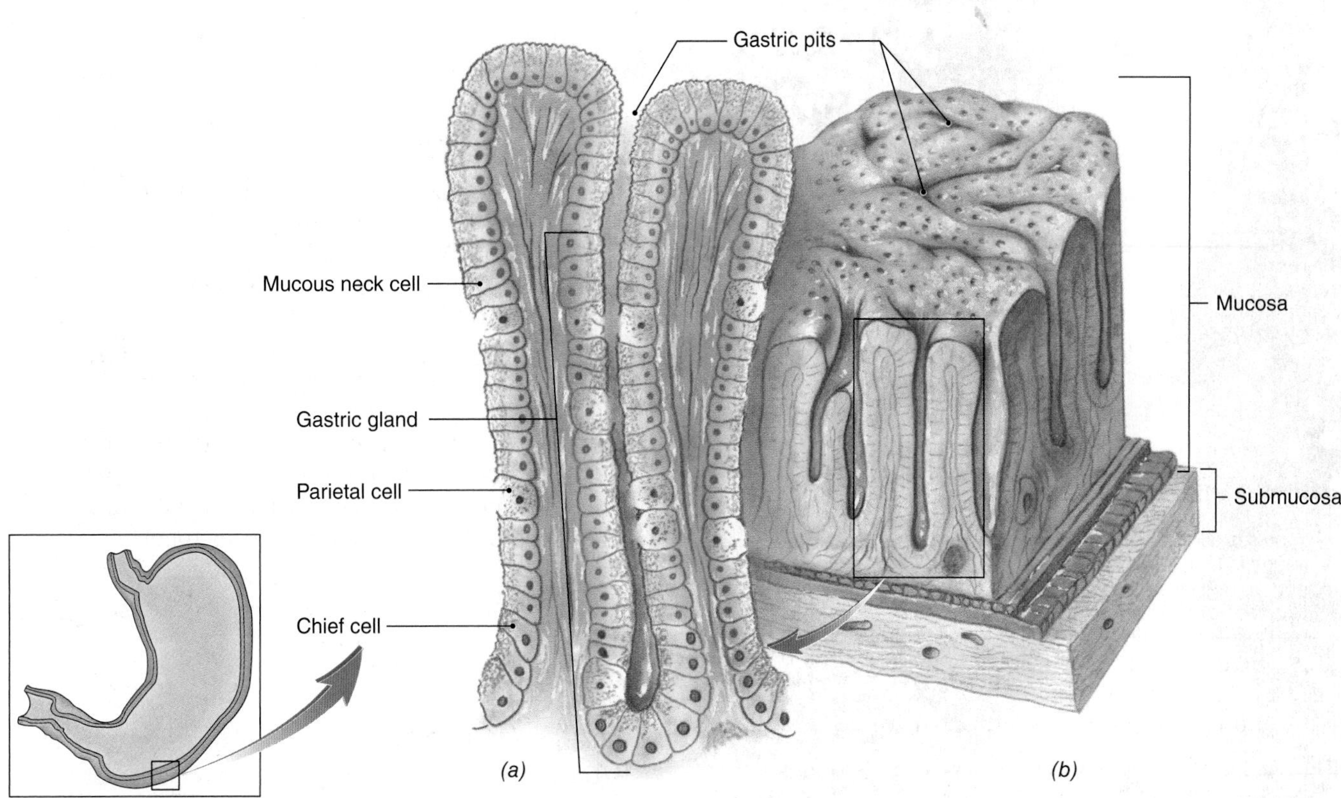

Figure 15.9 (*a*) An enlargement of the gastric glands shows the locations of mucous neck cells, parietal cells, and chief cells. (*b*) The thick stomach mucosa is dotted with gastric pits, openings that receive the secretions from the gastric glands. AP|R

Gastric Juice

The secretion of the gastric glands is known as **gastric juice.** *Mucous neck cells,* located near the opening to the gastric pit, secrete mucus to coat and protect the mucosa from the action of digestive secretions. *Chief cells,* located in the deepest portions of the gastric glands, secrete the digestive enzymes pepsinogen (inactive form of pepsin), gastric lipase, and rennin. *Parietal cells,* located in the mid-portion of the gastric glands, secrete hydrochloric acid (HCl) and intrinsic factor (figure 15.9).

As food is mixed with gastric juice and as chemical digestion occurs, it is converted into an acidic, semiliquid substance called **chyme** (kīm). Small amounts of chyme are released intermittently into the duodenum by the relaxing of the pyloric sphincter.

Control of Gastric Secretion

The rate of gastric secretion is controlled by both neural and hormonal means and is a good example of a positive-feedback mechanism. Gastric juice is produced continuously, but its secretion is greatly increased whenever food is on the way to, or already in, the stomach. The sight, smell, or thought of appetizing food, food in the mouth, or food in the stomach stimulates the transmission of parasympathetic nerve impulses that increase the secretion of gastric juice. These nerve impulses also, along with food in the stomach and stomach stretching, stimulate certain stomach cells to secrete a hormone called **gastrin.** Gastrin is absorbed into the blood and is carried to gastric glands, increasing their secretions (figure 15.10).

As stomach contents are gradually emptied into the small intestine, there is a decrease in the frequency of parasympathetic nerve impulses and an increase in the frequency of sympathetic nerve impulses to the stomach, which reduces the secretion of gastric juice. When chyme passes from the stomach into the small intestine, it stimulates the intestinal mucosa to release two hormones: **cholecystokinin** (kō-lē-sis-tō-kīn′-in) **(CCK)** and **secretin** (se′-krē′-tin), which reduce both the motility of the stomach and the secretion of gastric juice.

Digestion and Absorption

Food entering the stomach is thoroughly mixed with gastric juice by ripplelike, mixing contractions of the stomach wall. Gastric juice is very acidic (pH 2) due to an abundance of HCl. **Pepsin** is the most important digestive enzyme in gastric juice, and it is secreted in an inactive form that prevents digestion of the cells secreting it.

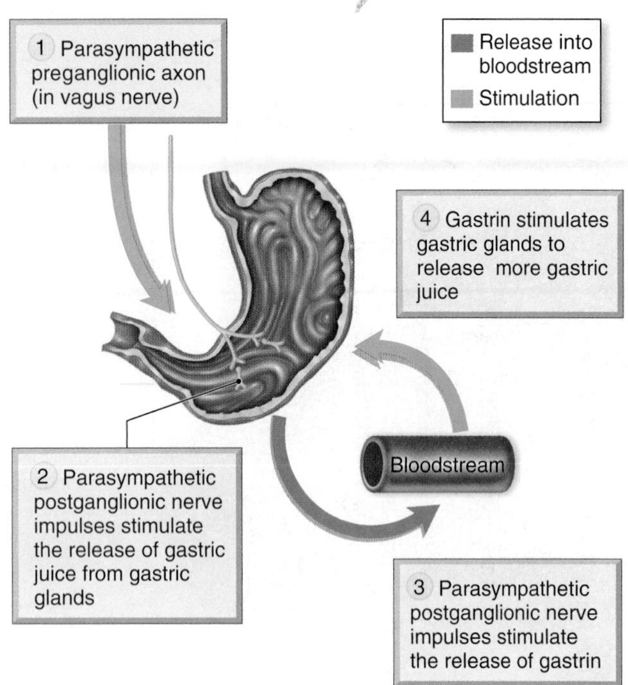

Figure 15.10 Neural and hormonal control of gastric secretions. AP|R

Once it is released into the stomach, pepsin is activated by the strong acidity of gastric juice. Pepsin acts on proteins and breaks these complex molecules into shorter amino acid chains called *peptides*. However, peptides are still much too large to be absorbed and require further digestion in the small intestine.

Gastric juice contains a substance known as **intrinsic factor** that is essential for the absorption of vitamin B12 by the small intestine.

The gastric juice of infants contains two unique enzymes that help to improve the digestion of milk proteins and lipids. **Rennin** (ren-in) curdles milk proteins, which keeps them in the stomach longer and makes them more easily digested by pepsin. **Gastric lipase** acts on triglycerides and breaks them into fatty acids and monoglycerides. Except for a few substances such as water, minerals, some drugs, and alcohol, little absorption occurs in the stomach.

When semiliquid chyme passes from the stomach into the **duodenum,** the first part of the small intestine, secretions from the pancreas and liver are emptied into the duodenum. The secretions from these **accessory organs** play important roles in digestion within the small intestine.

Check My Understanding

9. What digestive processes occur in the stomach?
10. How are gastric secretions controlled?

Clinical Insight

Gastric ulcers produce stomach pain one to three hours after eating. Without treatment, they can perforate the stomach wall, producing internal bleeding or peritonitis. Gastric ulcers result from persistent erosion of the alkaline mucus that coats the stomach lining. Most recurring gastric ulcers are caused by an acid-resistant bacterium, *Helicobacter pylori,* which erodes the protective mucosa, allowing gastric juice to attack deeper cells. Other contributing factors that promote HCl secretion or reduce mucus production include stress, smoking, alcohol, coffee, aspirin, and nonsteroidal antiinflammatory drugs. Treatment involves antibiotics to kill the bacteria and drugs to reduce gastric secretion.

15.6 Pancreas

Learning Objective

11. Describe the control and functions of pancreatic secretions.

The **pancreas** is a small, pennant-shaped gland located posterior to the pyloric part of the stomach. It is connected by a duct to the duodenum, approximately 10 cm distal to the pyloric sphincter. The pancreas has both endocrine and exocrine functions. The majority of the cells within the pancreas secrete **pancreatic juice,** which is the digestive (exocrine) function of the pancreas. Pancreatic juice is collected by tiny ducts that merge to form large ducts, which enter the **pancreatic duct.** The pancreatic duct extends the length of the pancreas and usually forms a smaller *accessory pancreatic duct.* The pancreatic duct joins with the bile duct where they both empty their secretions into the duodenum. Their common opening is controlled by the *hepatopancreatic sphincter,* which dilates to allow pancreatic juice and bile to enter the duodenum. The accessory pancreatic duct allows pancreatic juice to enter the duodenum independently of bile (figure 15.11).

Control of Pancreatic Secretion

Pancreatic secretion, like gastric secretion, is controlled by both neural and hormonal mechanisms. Neural control is via parasympathetic axons. When parasympathetic nerve impulses activate the stomach mucosa, they also stimulate the pancreas to secrete pancreatic juice.

Hormonal control of pancreatic secretion results from two hormones that stimulate different types of pancreatic cells. Acid chyme entering the duodenum stimulates the intestinal mucosa to release the hormone secretin. Secretin is carried by blood to the pancreas, where it stimulates

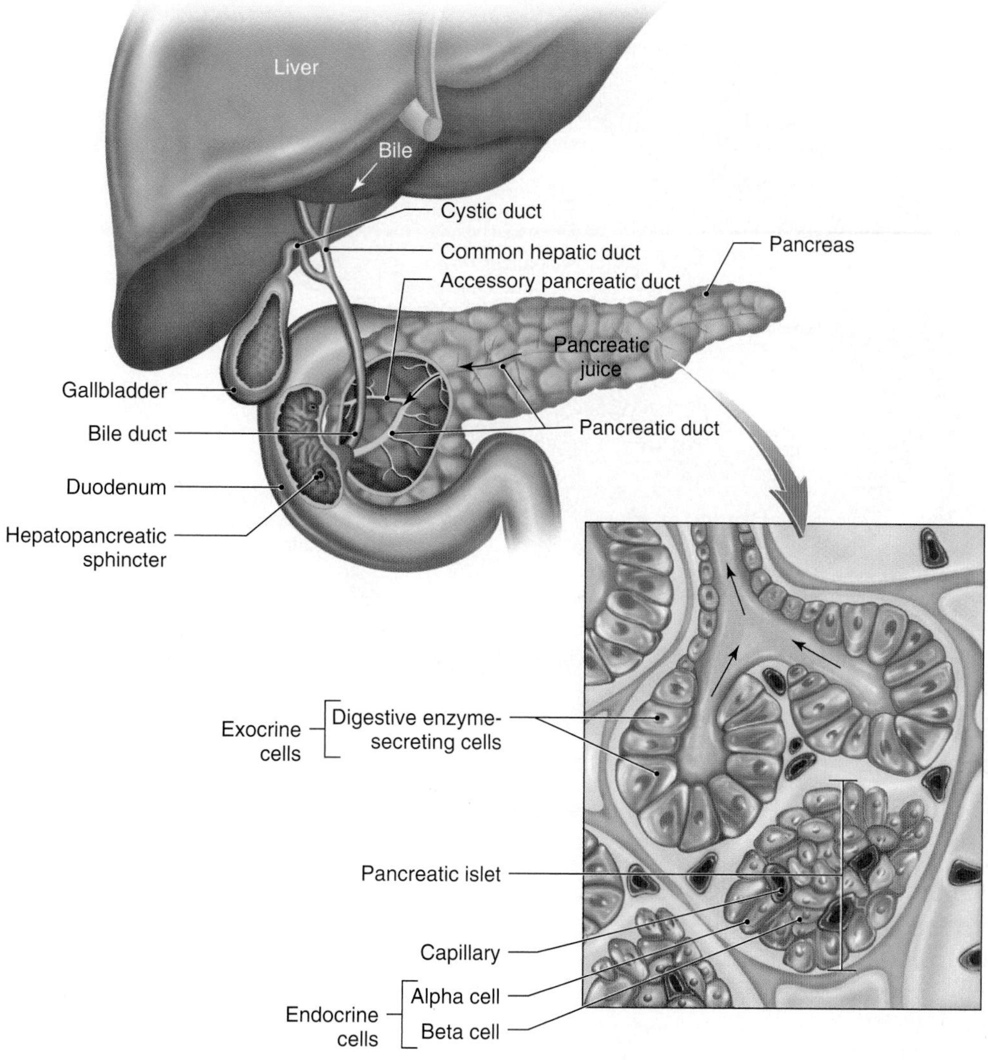

Figure 15.11 The pancreas is closely associated with the liver and duodenum. The inset shows secretory cells involved in the exocrine and endocrine functions of the pancreas. APR

secretion of pancreatic juice that is rich in bicarbonate ions (figure 15.12). Bicarbonate ions neutralize the acidity of the chyme entering the small intestine. Lipid-rich chyme stimulates production of cholecystokinin by the intestinal mucosa. CCK stimulates secretion of pancreatic juice that is rich in digestive enzymes.

Table 15.2 summarizes the major hormones that regulate digestive secretions.

Digestion by Pancreatic Enzymes

Pancreatic juice contains enzymes that act on each of the major classes of energy foods: carbohydrates, fats, and proteins. Their digestive actions occur within the small intestine.

Pancreatic amylase, like salivary amylase, acts on *starch* and *glycogen,* splitting these polysaccharides into *maltose,* a disaccharide. Maltose, however, is too big to be absorbed by the body.

Pancreatic lipase acts on *fats* (triglycerides) and splits them into *monoglycerides* and *fatty acids* that are absorbable.

Trypsin is the major pancreatic enzyme in pancreatic juice. Trypsin splits *proteins* into peptides. Like pepsin in the stomach, it is secreted in an inactive form and is activated only when mixed with intestinal secretions within the small intestine. This mode of secretion prevents the pancreatic cells from being digested by their own enzymatic secretions.

Check My Understanding

11. What are the functions of the enzymes in pancreatic juice?
12. How is pancreatic secretion controlled?

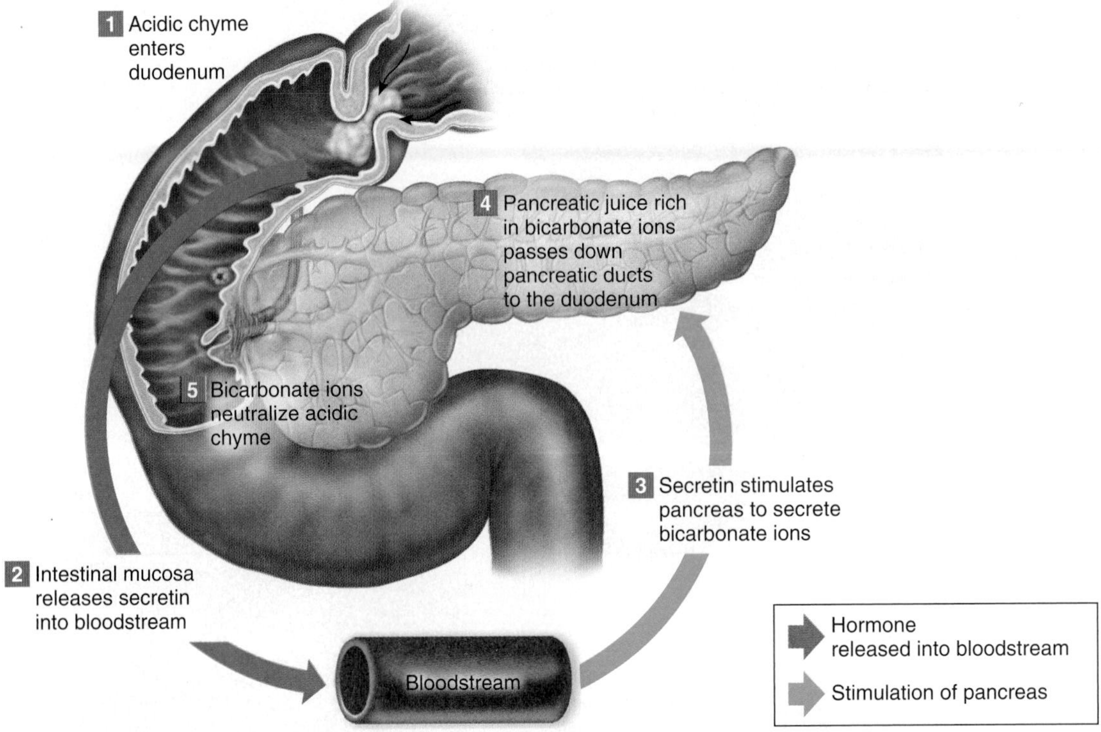

Figure 15.12 Hormonal Control of Pancreatic Secretion.

Table 15.2 Major Hormones Regulating Digestive Secretions

Hormone	Action	Source
Gastrin	Stimulates gastric juice secretion	Gastric mucosa; parasympathetic nerve impulses, stomach stretching, and food in the stomach stimulate release of gastrin
Cholecystokinin	Reduces gastric juice secretion; stimulates secretion of pancreatic juice that is rich in digestive enzymes; stimulates contraction of gallbladder and relaxes hepatopancreatic sphincter causing release of bile	Intestinal mucosa; lipid-rich chyme stimulates the release of cholecystokinin
Secretin	Stimulates secretion of pancreatic juice that is rich in bicarbonate ions; inhibits gastric secretion	Intestinal mucosa; acid chyme stimulates the release of secretin

15.7 Liver

Learning Objectives

12. Describe the location and functions of the liver.
13. Explain how bile release is stimulated.

The **liver** is the largest gland in the body. It weighs about 1.4 kg (3 lb) and is dark reddish brown in color. The liver is located mostly in the right upper quadrant of the abdominopelvic cavity just inferior to the diaphragm, where it is protected by the inferior ribs.

The liver has many important and vital functions, though most are not associated with digestion. (1) The liver produces and secretes bile, a substance that aids in the digestion and absorption of lipids, and heparin, a blood anticoagulant. It also produces and secretes plasma proteins (see chapter 11). (2) The liver plays a critical role in carbohydrate metabolism. When blood glucose levels are elevated, the liver can convert and store the excess as glycogen or triglycerides. When blood glucose levels are low, the liver can convert glycogen, glycerol, fatty acids, and amino acids into glucose. (3) As part of lipid metabolism, the liver forms lipoproteins for the transport of fatty acids, triglycerides, and cholesterol. It also can synthesize cholesterol and use it to form bile salts. (4) As a part of protein metabolism, the liver removes the amine groups from amino acids so that the remainder of the molecules can be used in cellular respiration

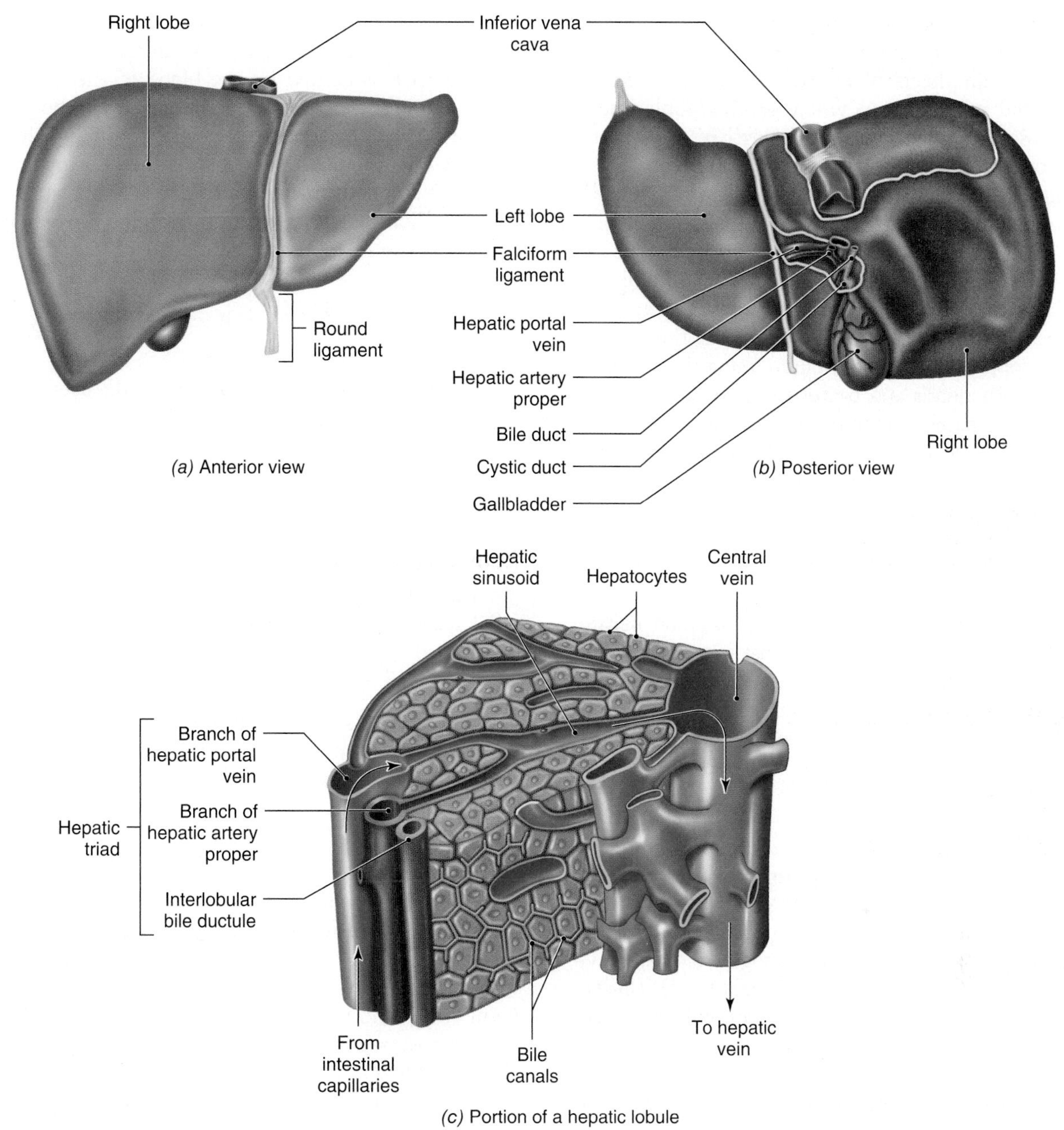

Figure 15.13 Anatomy of the Liver. AP|R

or in forming glucose and triglycerides. (5) The liver is used for the storage of triglycerides, glycogen, iron, and vitamins A, D, E, K, and B12. (6) It detoxifies the blood by modifying many drugs and toxic chemicals to form less toxic compounds. (7) Using phagocytosis, the liver removes worn-out blood cells and any bacteria present.

The liver is encased in a dense irregular connective tissue capsule that, in turn, is covered by the peritoneum for additional support. A ligament of dense regular connective tissue, called the *falciform ligament*, joins the liver to the diaphragm and the anterior abdominal wall and separates the two main lobes: a larger *right lobe* and a smaller *left lobe*. Several blood vessels and the common hepatic duct enter or exit the liver from a small area on the posterior surface of the liver (figure 15.13).

The liver receives blood from two sources. The *hepatic artery proper* brings oxygenated blood to the *hepatocytes* (liver cells). The *hepatic portal vein* brings deoxygenated, nutrient-rich blood from the digestive tract. As blood flows through the liver, hepatocytes remove, modify, or add substances to the blood before it leaves the liver via the *hepatic veins*.

Microscopically, the liver consists of multitudes of **hepatic lobules,** which serve as the structural and functional units. Each lobule is a short, roughly hexagonal cylinder with a central vein running through its core from which thin sheets of hepatocytes radiate. *Hepatic triads,* located at the corners where several lobules meet, are composed of three vessels: a branch of the hepatic artery proper, a branch of the hepatic portal vein, and a small *interlobular bile ductule* carrying bile. Between the sheets of hepatocytes are *hepatic sinusoids,* blood-filled spaces that carry blood from the hepatic artery proper and hepatic portal vein to the central vein of the lobule. As blood flows through the hepatic sinusoids, an exchange of materials occurs between the blood and the hepatocytes. Macrophages in the epithelium lining the sinusoids remove cellular debris and bacteria. The central veins of the lobules ultimately merge to form the hepatic veins.

As noted, the production of bile is the only digestive function of the liver. Bile is collected in tiny ducts that merge to form the interlobular ductules of the hepatic triads, which in turn unite to form the *right* and *left hepatic ducts* exiting the right and left lobes of the liver. The right and left hepatic ducts merge to form the *common hepatic duct,* which carries bile out of the liver. The common hepatic duct and the *cystic duct,* a short duct that extends from the gallbladder, merge to form the **bile duct,** which carries bile to the duodenum. The cystic duct carries bile to and from the **gallbladder,** a small, pear-shaped sac that stores bile temporarily between meals (figures 15.11 and 15.13).

Bile

Hepatocytes continuously produce **bile,** a yellowish green liquid. Bile consists of water, bile salts, bile pigments, cholesterol, and minerals. Bile pigments, such as yellow-colored bilirubin, are waste products of hemoglobin breakdown that are excreted through bile (see chapter 11). *Jaundice* is a medical condition in which too much bilirubin is circulating in the blood due to liver or kidney malfunction or excessive red blood cell destruction. The excess bilirubin ends up being deposited within the skin, cornea, and mucous membranes, causing yellow discoloration.

Bile salts are the only bile components that play a digestive role. When in contact with fatty substances, they break up large fat globules into very small droplets, a process called *emulsification* (ē-mul-si-fi-kā′-shun). Emulsification greatly increases the surface area of the fats exposed to water and lipases. In this way, bile salts aid the digestion of fats. Bile salts also aid the absorption of fatty acids, cholesterol, and lipid-soluble vitamins by the small intestine.

Release of Bile

Bile normally enters the duodenum only when chyme is present. When the intestine is empty, the hepatopancreatic sphincter at the base of the bile duct constricts, which forces bile to enter the gallbladder for temporary storage.

When lipid-rich chyme enters the duodenum, it stimulates the release of cholecystokinin from the intestinal mucosa. CCK is carried by the blood to the gallbladder, where it stimulates contraction of muscles in the gallbladder wall. The contractions eject bile from the gallbladder into the bile duct. CCK also relaxes the hepatopancreatic sphincter so bile is injected into the small intestine. Note that this hormonal control releases bile only when it is needed in the small intestine (figure 15.14).

Check My Understanding

13. Where is the liver located?
14. What is the digestive function of the liver?
15. How is bile release controlled?

15.8 Small Intestine

Learning Objectives

14. Describe digestion in the small intestine.
15. Explain how the end products of digestion are absorbed.

The small intestine is about 2.5 cm (1 in) in diameter and 4 to 5 m (12–15 ft) in length. It begins at the pyloric sphincter of the stomach, fills much of the abdominopelvic cavity, and empties into the large intestine. Most of the digestive processes and absorption of nutrients occur in the small intestine.

Structure

There are three sequential segments composing the small intestine. The **duodenum** (dū-o-dē′-num) is a very short section, about 25 to 30 cm long, that receives chyme from the stomach. The middle section is the **jejunum** (je-jū′-num), and it is about 160 to 200 cm long. The last and longest segment is the **ileum** (il′-ē-um), which is about 170 to 215 cm long. The ileum joins with the large intestine at the *ileal* (il-ē-al) *orifice.*

The small intestine is suspended from the posterior abdominal wall by the **mesentery** (mes′-en-ter-ē), double folds of the peritoneum that provide support but allow movement. Blood vessels, lymphatic vessels, and nerves serving the small intestine are also supported by the mesentery (figure 15.15).

The mucosa of the small intestine is modified to provide a very large surface area. The distinctive velvety appearance of the intestinal mucosa results from the presence of **intestinal villi,** tiny projections from the mucosa that are extremely abundant (see figure 15.2). Each villus is covered by simple columnar epithelium and contains a centrally located **lacteal,** a lymphatic capillary. A blood capillary network surrounds the lacteal. At the bases of the villi are tiny pits that open into **intestinal glands,** which secrete intestinal juice (figure 15.16).

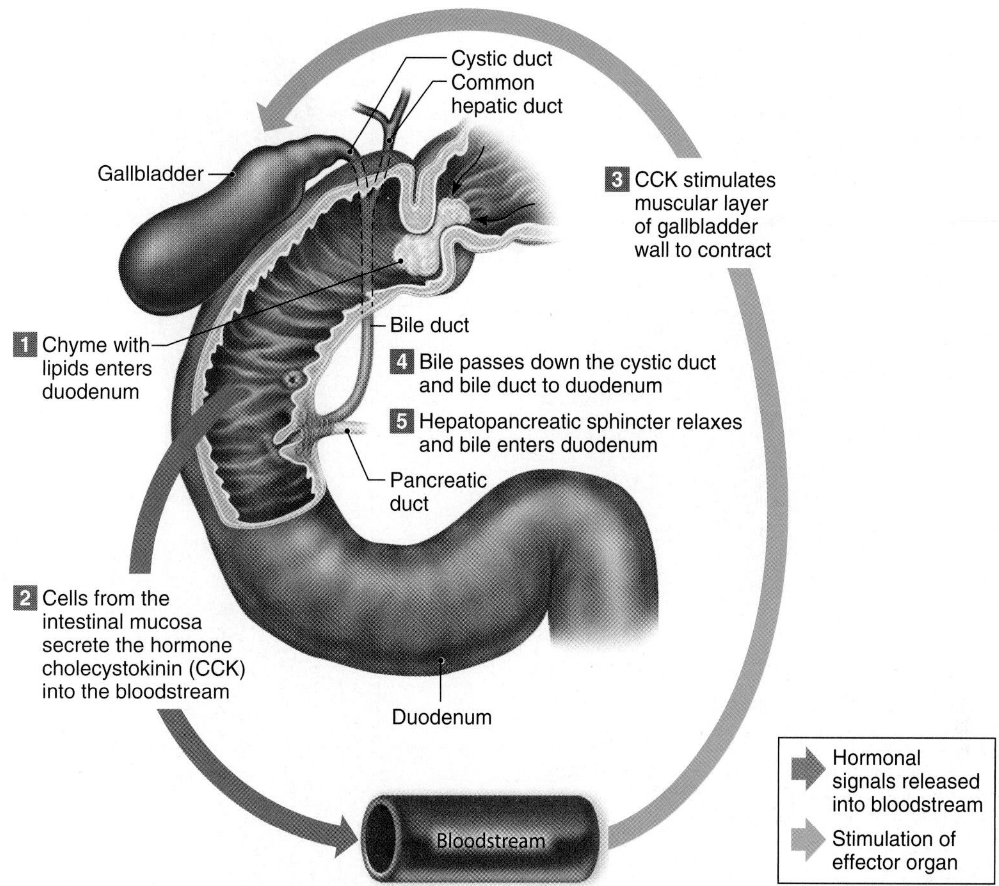

Figure 15.14 Hormonal Control of Bile Secretion.

The mucosal surface area in contact with chyme and digestive fluids is further increased by the presence of numerous microvilli forming a "brush border" (see chapter 4 and figures 15.2 and 15.17).

Intestinal Juice

The fluid secreted by the intestinal glands is known as **intestinal juice.** It is slightly alkaline and contains abundant water and mucus. Intestinal juice provides an appropriate environment for the action of bile salts and pancreatic digestive enzymes within the small intestine. Recall that trypsin in pancreatic juice is activated only after being mixed with intestinal secretions.

Regulation of Intestinal Secretion

The presence of chyme in the small intestine provides mechanical stimulation of the mucosa that activates the secretion of intestinal juice and enzymes. Chyme also causes an expansion of the intestinal wall, triggering a neural reflex that sends parasympathetic nerve impulses to the mucosa. The nerve impulses stimulate an increase in the rate of intestinal secretions.

Digestion and Absorption

Vigorous segmentation within the small intestine mixes chyme with bile, pancreatic juice, and intestinal juice. The emulsification of fats by bile and the continued digestion of carbohydrates, fats, and proteins by pancreatic and brush border enzymes occur within the small intestine. Brush border enzymes are embedded within the brush border of the small intestine mucosa. These actions complete the digestive process within the small intestine.

There are three brush border enzymes that split disaccharides into monosaccharides. (1) **Maltase** converts *maltose* into *glucose;* (2) **sucrase** converts *sucrose* into *glucose* and *fructose;* and (3) **lactase** converts *lactose* into *glucose* and *galactose.*

Brush border enzymes acting on proteins are also present. Various **peptidases** split *peptides* into *amino acids.* Table 15.3 summarizes the enzymes involved in the digestion of carbohydrates, fats, and proteins.

Carbohydrate digestion begins in the mouth and concludes in the small intestine. The end products of carbohydrate digestion are monosaccharides, the simple sugars *glucose, fructose,* and *galactose.* These sugars are absorbed

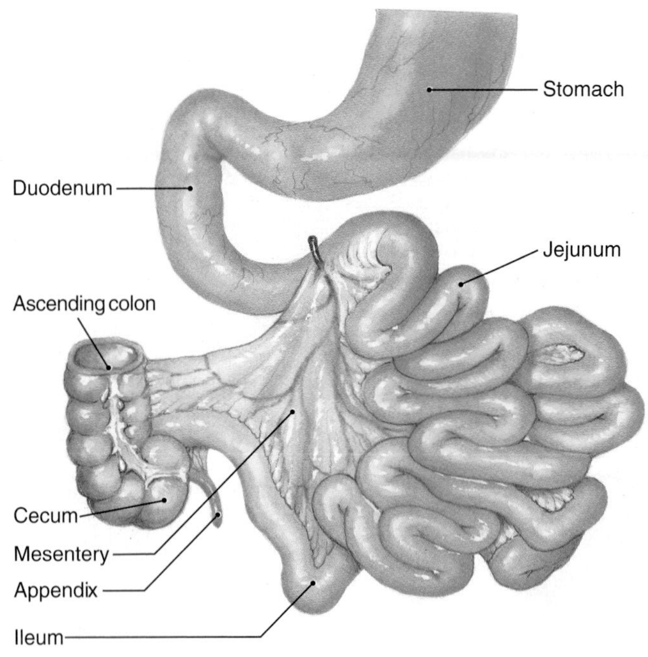

Figure 15.15 The small intestine consists of the duodenum, jejunum, and ileum. Chyme from the stomach enters the duodenum. After digestion and absorption, chyme residues pass from the ileum into the large intestine. APIR

across the epithelium and into the capillaries of the villi through both facilitated diffusion and active transport.

Fat (triglyceride) digestion primarily occurs in the small intestine. The end products are *monoglycerides* and *fatty acids*. Very small, or short-chain, fatty acids are absorbed by simple diffusion across the epithelium and into the capillaries of the villi. All other lipids require an alternate means of absorption. Bile salts interact to form structures called *micelles*, small transportation spheres that are hydrophilic on their surface and hydrophobic in their core. Micelles absorb large fatty acids, monoglycerides, cholesterol (a steroid), phospholipids, and lipid-soluble vitamins into their core and transport them to the intestinal brush border. The contents of the micelles move by simple diffusion into the epithelial cells. Once inside the epithelial cells, the fatty acids and monoglycerides recombine to form triglycerides. The triglycerides combine in small clusters with phospholipids, steroids, and lipid-soluble vitamins. These clusters are coated with protein and form structures known as **chylomicrons** (kī-lō-mī′-krons). The protein coat makes chylomicrons water-soluble. Chylomicrons move out of the epithelial cells by exocytosis and enter the lacteals of the villi, as shown in figure 15.17. They are carried by lymphatic vessels to the left subclavian vein, where lymph from the intestine enters the blood.

Protein digestion begins in the stomach and concludes in the small intestine. The end products are *amino acids*, which are actively absorbed across the epithelium and into the capillaries of villi.

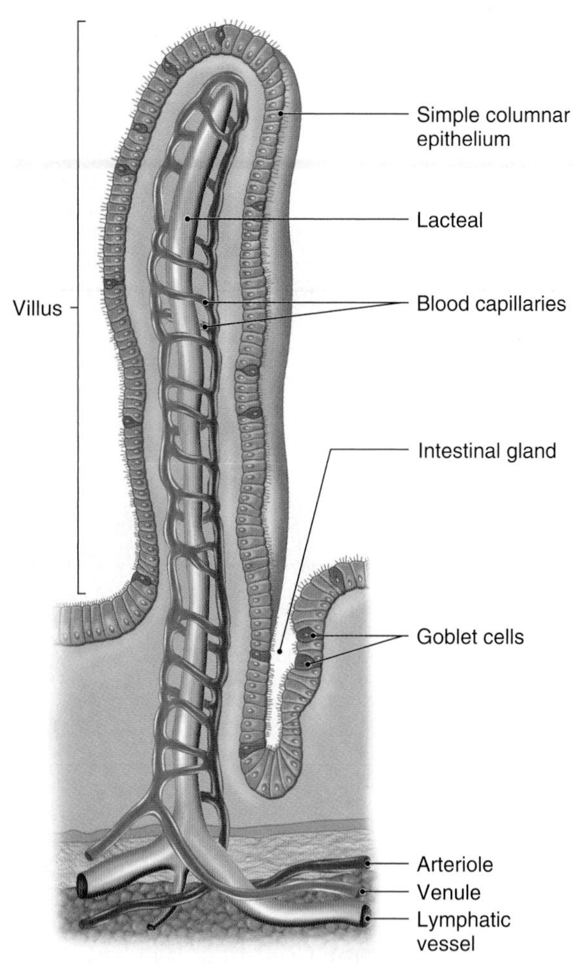

Figure 15.16 The Structure of a Villus.

In addition to the end products of digestion, other needed substances are absorbed in the small intestine. For example, water, minerals, and water-soluble vitamins are absorbed into the capillaries of villi. Materials absorbed into the blood are carried from the intestines to the liver via the hepatic portal vein. After processing by the liver, appropriate concentrations of nutrients are released into the general circulation to serve the needs of tissue cells. In this way, the liver contributes to the overall homeostasis of the body.

Clinical Insight

Lactose intolerance is caused by a deficiency or absence of lactase. The presence of undigested lactose in the intestines produces an osmotic gradient that prevents the normal reabsorption of water into the blood and, even worse, actually causes water to be drawn into the intestines from interstitial fluid. The result is diarrhea, flatulence, bloating, and intestinal cramps. Afflicted persons can avoid this problem if they take a tablet or liquid containing lactase before meals containing milk or milk products.

Table 15.3 Summary of the Major Digestive Enzymes and Their Actions

Enzyme	Substrate	Product
Saliva		
Salivary amylase	Starch and glycogen	Maltose
Gastric Juice		
Pepsin	Proteins	Peptides
Pancreatic Juice		
Pancreatic amylase	Starch and glycogen	Maltose
Pancreatic lipase	Triglycerides	Monoglycerides* and fatty acids*
Trypsin	Proteins	Peptides
Brush Border Enzymes		
Maltase	Maltose	Glucose*
Sucrase	Sucrose	Glucose* and fructose*
Lactase	Lactose	Glucose* and galactose*
Peptidases	Peptides	Amino acids*

*End products of digestion.

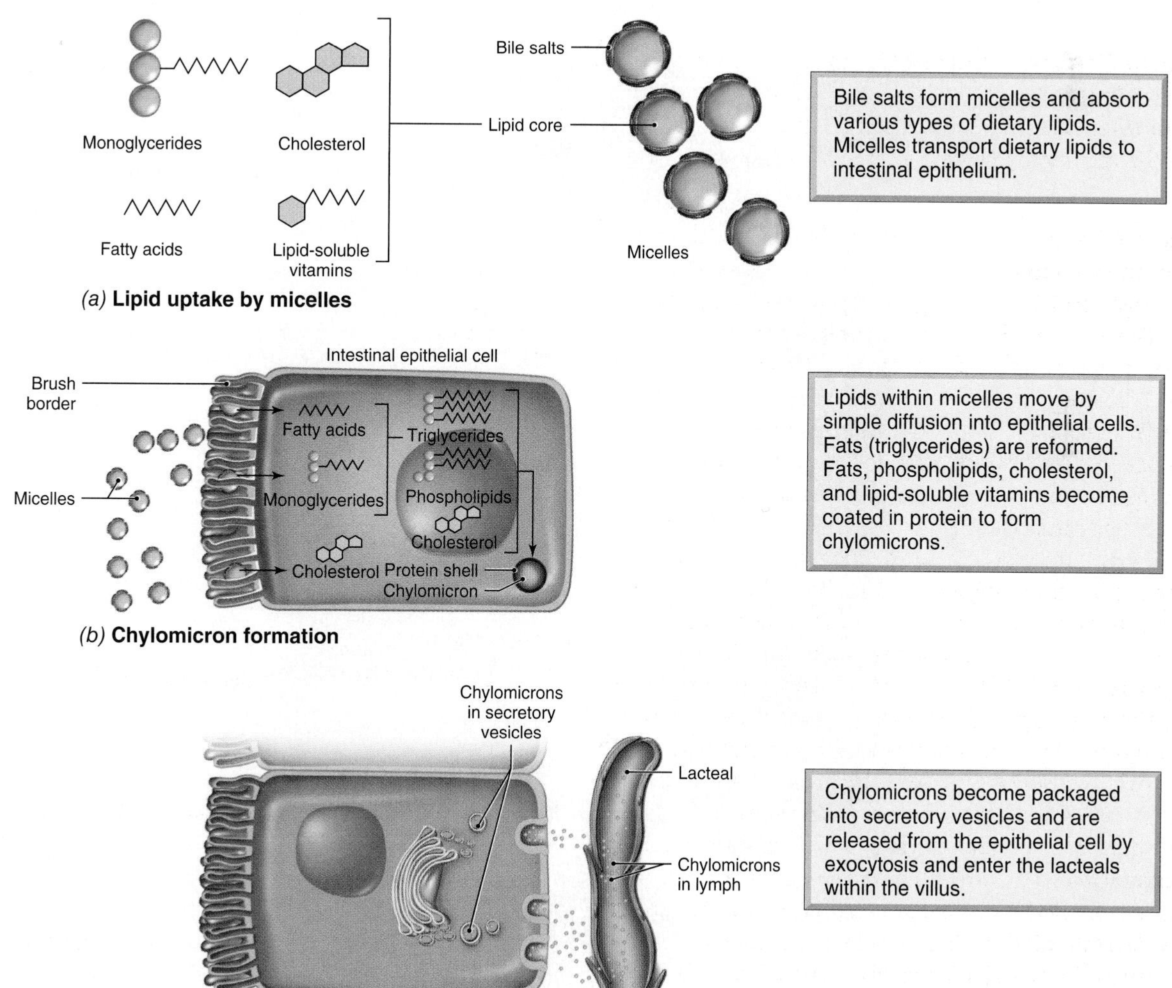

Figure 15.17 Absorption of Dietary Lipids.

Check My Understanding

16. What are the three segments of the small intestine?
17. What digestive processes occur in the small intestine?
18. How are end products of digestion absorbed?

15.9 Large Intestine

Learning Objective

16. Describe the structure and functions of the large intestine.

The small intestine joins with the large intestine at the ileal orifice. This opening is closed most of the time but opens to allow chyme to enter the large intestine.

Structure

The *large intestine* gets its name because its diameter (6.5 cm; 2.5 in) is larger than that of the small intestine, although its length (1.5 m; 5 ft) is much shorter. The large intestine consists of four segments: cecum, colon, rectum, and anal canal.

The first portion of the large intestine is the pouchlike *cecum,* which bulges inferior to the ileal orifice. The slender, wormlike **appendix** extends from the cecum and, although it has no digestive function, it contributes to the immune defense of the body.

The **colon** forms most of the large intestine and is subdivided into four segments. The *ascending colon* extends superiorly from the cecum along the right side of the abdominopelvic cavity. As it nears the liver, it turns left to become the *transverse colon.* Near the spleen, the transverse colon turns inferiorly to become the *descending*

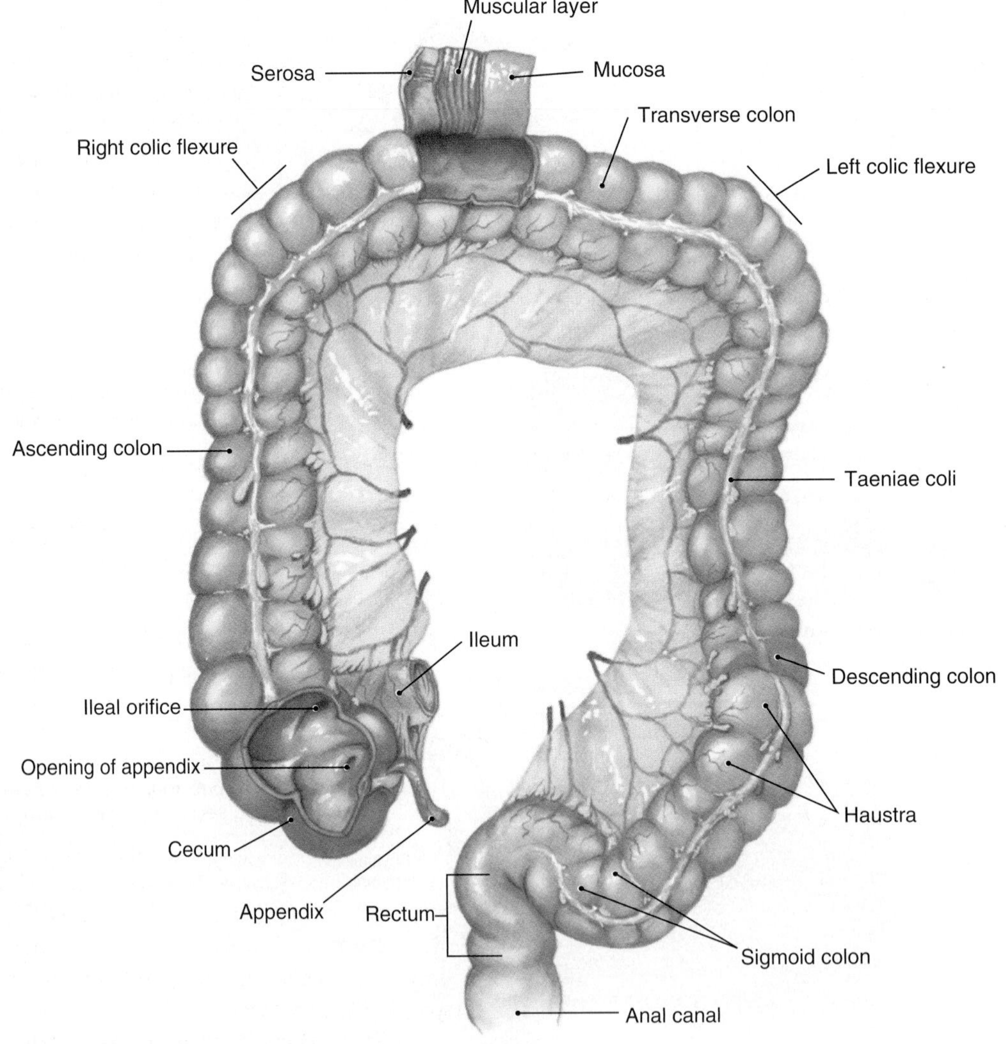

Figure 15.18 Anterior View of the Large Intestine. APR

colon along the left side of the abdominopelvic cavity. Near the pelvis, the descending colon becomes the *sigmoid colon,* which is characterized by an S-shaped curvature leading to the rectum (figure 15.18).

The **rectum** is the straight portion of the large intestine that continues inferiorly from the sigmoid colon through the pelvic cavity and ends at the **anal canal.** The anal canal is the last 3 cm of the large intestine and its external opening is the **anus.** The mucosa of the anal canal is folded to form the *anal columns,* which contain networks of arteries and veins. The anus is kept closed except during defecation by the involuntarily controlled *internal anal sphincter* and the voluntarily controlled *external anal sphincter* (figure 15.19).

The colon has a puckered appearance when viewed externally. This results because the longitudinal muscles are not uniformly layered but are reduced to three longitudinal bands, the *taeniae coli,* that run the length of the colon. Contraction of the taeniae coli gathers the colon into a series of pouches called *haustra.* Like the small intestine, the large intestine is supported by a mesentery.

The mucosa of the large intestine is also different from that of the small intestine. Villi are absent, and the simple columnar epithelium contains numerous mucus-producing goblet cells.

Functions

Chyme residue entering the large intestine contains water, minerals, bacteria, and other substances that were not digested or absorbed while in the small intestine. There are no digestive enzymes secreted by the large intestine.

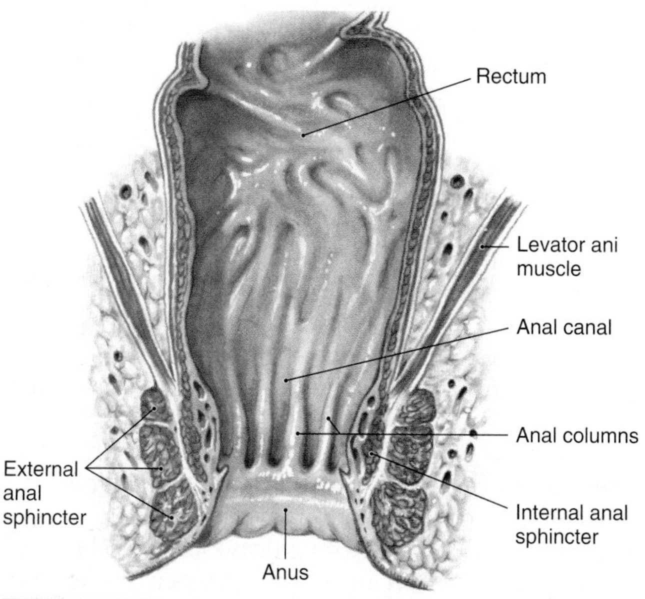

Figure 15.19 The rectum and the anal canal are located at the distal end of the alimentary canal.

Instead, intestinal bacteria decompose the undigested food molecules. This action yields certain B vitamins and vitamin K, in addition to gas (flatus). The mucosa of the large intestine secretes large quantities of mucus that lubricate the intestinal lining and reduce abrasion as materials are moved along.

A major function of the large intestine is the absorption of water, some minerals, and vitamins as the contents slowly move through the colon. Much of this absorption occurs before the chyme reaches the descending colon, where it is congealed to form the **feces** (fē-sēz). Feces contain large amounts of bacteria, mucus, and water as well as undigested food molecules.

Movements

Segmentation and peristalsis within the large intestine are more sluggish than those of the small intestine. Vigorous peristalsis occurs only two to four times a day, usually following a meal. These peristaltic contractions are called *mass movements* because they move the contents of the descending and sigmoid colons toward the rectum. The defecation reflex is activated when the rectum fills with feces and its wall is stretched. Parasympathetic nerve impulses stimulate muscular contractions that increase pressure within the rectum and relax the internal anal sphincter. Defecation, or expulsion of feces, occurs if the external anal sphincter is voluntarily relaxed. If its contraction is voluntarily maintained, defecation is postponed.

 Check My Understanding

19. What are the segments of the large intestine?
20. What are the functions of the large intestine?

15.10 Nutrients: Sources and Uses

Learning Objectives

17. Identify the sources and uses of carbohydrates, lipids, proteins, vitamins, and major minerals.
18. Explain the three major steps of cellular respiration.

Nutrients are chemicals in foods that provide energy for powering life processes; chemicals aiding or enabling life processes; or materials to construct molecules for the normal development, growth, and maintenance of the body. There are six groups of nutrients: carbohydrates, lipids, proteins, vitamins, minerals, and water. All six groups provide raw materials for constructing new molecules, but only carbohydrates, lipids, and proteins provide energy to sustain life processes.

Essential nutrients are those nutrients that the body cannot synthesize and must obtain in food in order to construct other molecules necessary for life. The essential nutrients include certain amino acids, certain fatty acids, most vitamins, minerals, and water. Because the essential nutrients are not all present in any one food, a balanced diet is required. The conversion of raw materials into molecules for life processes is a major role of the liver.

The Institute of Medicine of the U.S. National Academy of Sciences has developed nutritional recommendations, called the **Dietary Reference Intake (DRI),** to assist in the planning and assessment of nutrient intake. DRI includes the **Recommended Daily Allowance (RDA)** for each nutrient. The RDA for a nutrient is the average daily intake level that is sufficient to meet the nutritional needs of a healthy person. Most nutritional labels provide RDA but not DRI recommendations. DRI and RDA recommendations can be found through the U.S. Department of Agriculture (USDA) website (fnic.nal.usda.gov).

Energy Foods and Cellular Respiration

Carbohydrates, fats, and proteins are called "energy foods" because they are used in cellular respiration to release the energy in their chemical bonds for ATP production. Recall from the discussion in chapters 3 and 7 that cellular respiration includes both anaerobic and aerobic components. **Anaerobic respiration,** or *glycolysis,* occurs within the cytosol of a cell, while **aerobic respiration** occurs within mitochondria, where the enzymes catalyzing the reactions are located. There are two sequential, linked aerobic processes: the *citric acid cycle* and the *electron transport chain.* When oxygen is available, a nutrient, such as glucose, is completely degraded to carbon dioxide and water, in order to release energy. Approximately, 40% of the energy is captured in ATP, while the remainder is lost as heat. ATP does not store energy, but it carries it to where it is needed to power life processes. The released heat energy is important in maintaining a normal body temperature.

Follow the cellular respiration of glucose in figure 15.20. A molecule of glucose (containing 6 carbon atoms) is split during glycolysis to form 2 molecules of pyruvic acid (each containing 3 carbon atoms) and 2 ATP. Each pyruvic acid molecule is converted to acetyl-CoA (each molecule containing 2 carbon atoms), which releases CO_2. Each acetyl-CoA molecule enters the citric acid cycle. With each turn of the cycle, one acetyl-CoA molecule is broken down to release CO_2, H^+, and high-energy electrons. Substrate reactions associated with the citric acid cycle produce 2 ATP.

The higher-energy electrons pass along the molecular carriers of the electron transport chain from a higher-energy level to a lower-energy level–much like water flowing down a staircase. At each transfer (step) within the chain, energy is released to form ATP and the electrons move to the next lower energy level (next lower step in the staircase). Ultimately, all of the available energy is extracted, and the electrons, H^+, and O_2 combine to form water (H_2O). Electron transfer yields a total of 32 to 34 ATP, depending upon the cell in which cellular respiration occurs. When added to the 2 ATP from glycolysis and the 2 ATP from the citric acid cycle, one molecule of glucose can yield a total of 36 to 38 ATP.

The end products of digestion for lipids (fats) and proteins also may be used in cellular respiration, either directly or after conversion into compatible molecules. Figure 15.20 shows the major points of entry into cellular respiration for these molecules. The number of ATP produced varies with the type of molecule broken down.

Carbohydrates

Nearly all **carbohydrates** in the diet come from plant foods. Glycogen is the only carbohydrate in animal foods. While there is very little of this polysaccharide in meat, animal liver is an abundant source. Monosaccharides are in honey and fruits; disaccharides are found in table sugar and dairy products; and starch, a polysaccharide, occurs in cereals, vegetables, and legumes (e.g., beans, peas, peanuts).

Cellulose is a polysaccharide that is abundant in plant foods, but it cannot be digested by humans because they lack the necessary digestive enzymes. However, it is an important dietary component, because it provides fiber (roughage) that increases the bulk of the intestinal contents, which aids the function of the large intestine. Evidence suggests that high-fiber diets reduce the risk of certain colon disorders, such as diverticulitis and colon cancer.

Carbohydrates are used mostly as an energy source, with glucose as the primary carbohydrate molecule used in cellular respiration. Recall from chapter 10 that the hormone insulin plays a crucial role in moving glucose into cells. Most cells can live by obtaining energy from fatty acids or amino acids via cellular respiration, but some cells, notably neurons, are dependent upon a steady supply of glucose. For this reason, the functions of the nervous system decline if the concentration of blood glucose decreases.

The liver, along with the hormones insulin and glucagon, is involved in the regulation of glucose concentration in the blood. In response to insulin, excess glucose is converted into glycogen for storage primarily in the liver but also in skeletal muscles. If excess glucose still remains, it is converted into triglycerides and is stored in adipose tissues. When blood glucose levels decline, glucagon signals the liver to convert glycogen into glucose. If still more glucose is needed, triglycerides are converted into glycerol and fatty acids. Then glycerol may be converted into glucose.

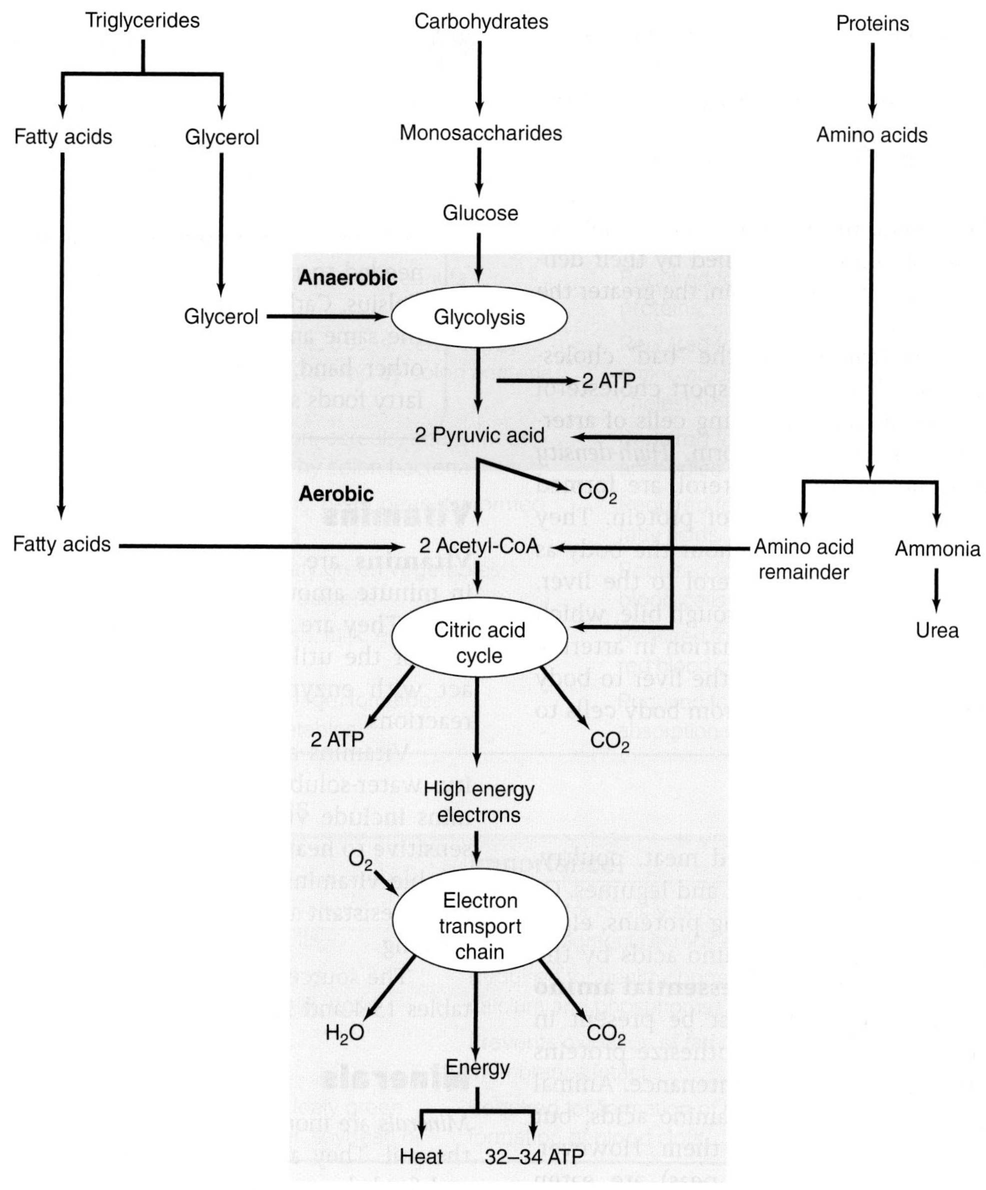

Figure 15.20 Energy foods and cellular respiration. Lipids, carbohydrates, and proteins may be broken down to release energy for ATP production. Note the three major steps of cellular respiration.

Lipids

Lipids include triglycerides, phospholipids, steroids, and lipid-soluble vitamins, (A, D, E, and K), but triglycerides are the most common lipids in the diet. Triglycerides may be either saturated or unsaturated (see chapter 2). Fats and oils contain a mixture of saturated, monounsaturated (one double bond), and polyunsaturated (more than one double bond) fatty acids. Coconut and palm oils, dairy products, and beef fat contain mostly saturated fatty acids. Peanut, olive, and canola oils consist mostly of monounsaturated fatty acids. Safflower, sunflower, and corn oils contain mostly polyunsaturated fatty acids. Cholesterol is present in dairy products, red meats, and egg yolks.

Lipids are essential components of the diet, although excessive amounts are not desirable. Phospholipids form the major portion of plasma membranes and the myelin sheaths of neurons. Triglycerides are important energy sources for many cells, including liver and skeletal muscle cells. Excess triglycerides are stored in adipose tissue, where they form the largest energy reserve in the body.

While cholesterol is not used as an energy source, it forms parts of plasma membranes and is used in the synthesis of bile salts and steroid hormones. The liver helps to regulate the concentration of triglycerides and cholesterol in the blood.

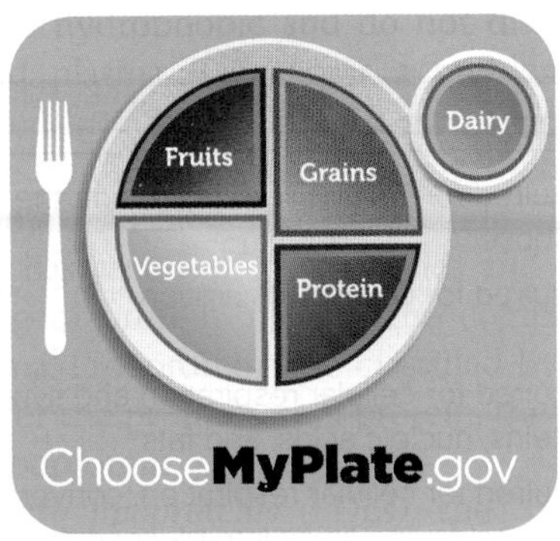

Figure 15.21 MyPlate.

to 2 cups of fruit, 1½ to 3 cups of vegetables, 4 to 8 ounces of grains, 2 to 6½ ounces of protein, 3 cups of dairy, and 5 to 7 teaspoons of fats and oils. The exact number of servings needed depends upon a person's age and gender. However, many people have trouble maintaining appropriate portion sizes when actually placing food on their plates. Measuring food servings with cups and scales requires discipline and consistency, which many find difficult to maintain for long periods of time. MyPlate creates a visual guide, based on a 9-inch plate, for the portion sizes of each food group at every meal. The visual reference removes the need to measure food servings at meals, allowing individuals to more quickly place correct portion sizes on their plate. For example, one-half of your plate needs to be filled with fruits and vegetables at each meal (figure 15.21).

 Check My Understanding

21. How is energy produced from one glucose molecule through cellular respiration?
22. How are monosaccharides, cholesterol and amino acids used in the body?

15.11 Disorders of the Digestive System

Learning Objective

19. Describe the major disorders of the digestive system.

Inflammatory Disorders

Appendicitis is an acute inflammation of the appendix. First symptoms include referred pain in the umbilical region and nausea. Later, pain is localized in the right lower quadrant of the abdominal wall. Surgical removal of the appendix is the standard treatment.

Inflammatory bowel disease (IBD) is a group of chronic disorders that cause painfully swollen, inflamed intestines, in addition to abdominal cramps and bloody diarrhea. IBDs result from an overactive immune response that continuously promotes inflammation, which causes permanent changes in the intestinal tissues and places sufferers at greater risk for developing intestinal cancer. There is a genetic predisposition for developing IBD. Treatment includes corticosteroids and immunosuppressant drugs. Surgical removal of the affected portions of the intestines is necessary when drug therapy fails to provide remission. The major IBDs are Crohn disease and ulcerative colitis. *Crohn disease* may affect any portion of the alimentary canal but most often affects the distal portion of the ileum. *Ulcerative colitis* affects only the large intestine and/or rectum.

Irritable bowel syndrome (IBS) is an intestinal disorder causing abdominal bloating, cramping pain, constipation, and diarrhea. Unlike IBDs, IBS does not permanently damage intestinal tissues, with symptoms usually controlled by dietary and lifestyle changes.

Diverticulitis (dī-ver-tik-ū-lī′-tis) is a disorder of the large intestine. Small, saclike outpockets of the colon often develop from a diet lacking sufficient fiber (bulk). Diverticulitis is the inflammation of these diverticula and may cause considerable pain, bloating, or diarrhea.

Hemorrhoids is a condition in which one or more veins in the anal canal become enlarged and inflamed. Chronic constipation contributes to the development of hemorrhoids.

Hepatitis, an inflammation of the liver, may be caused by several factors, including viruses, drugs, or alcohol. It is characterized by jaundice, fever, and liver enlargement. Hepatitis is the number one reason for liver transplants. There are three common viral types of hepatitis in the US.

Hepatitis A (infectious hepatitis) is caused by the hepatitis A virus, which is spread by sexual contact and fecal contamination of food and water. It is highly contagious, but an effective vaccine and treatment are available. Symptoms are usually mild. Recovery takes four to six weeks.

Hepatitis B (serum hepatitis) is a serious infection. It is spread through contaminated blood, saliva, vaginal secretions, and semen. Most people recover completely, but some retain live viruses for years and unknowingly serve as carriers of the disease. Chronic hepatitis may produce cirrhosis or lead to liver cancer. Hepatitis B vaccine is the best protection against hepatitis B infection. It consists of a series of three spaced injections and may be given to persons of all ages, including infants. Chronic hepatitis B patients are treated with antiviral medications.

Clinical Insight

Obesity has become a national health concern. Studies have shown that obesity is a significant health risk, perhaps second in importance only to smoking. Obesity results from consuming food containing more kilocalories than is necessary to meet the body's energy needs. Underlying causes of obesity include heredity, eating habits, inadequate exercise, and poor diet. People who are genetically predisposed toward obesity have a very difficult time controlling their weight. Excess kilocalories are stored as fats in adipose tissue. An excess of 3,500 kcal equals one pound of fat. The only way to get rid of stored fat is to consume fewer kilocalories than needed so the stored fat will be mobilized and broken down. Exercise increases the body's need for energy, which helps burn fat more rapidly.

An effective way to determine if a person is overweight or obese is to calculate the **body mass index (BMI)** and measure the waist. It is calculated by dividing a person's weight in kilograms by the square of his or her height in meters: $BMI = Wt/Ht^2$. To determine the weight in kilograms, divide the weight in pounds by 2.2. To determine the height in meters, divide the height in inches by 39.37. Someone with a BMI higher than 25 to 27 is considered overweight, and a person whose BMI has a value of 30 or more is considered obese. According to the National Institutes of Health, waist measurements over 35 inches in females and 40 inches in males indicate obesity.

Hepatitis C, the most common type of hepatitis in the US, is spread primarily by contaminated blood but also by contaminated food and water. Symptoms are usually so mild a person may not even know he or she has it until liver damage shows up, maybe decades later. It can lead to liver failure, cirrhosis, and cancer. In some cases, the symptoms are mild and short-lived so medications are usually not needed. Most cases progress to the chronic form of the disease and require antiviral medications. There is no Hepatitis C vaccine currently available.

Periodontal disease refers to a variety of conditions characterized by inflammed and bleeding gingivae, in addition to degeneration of the gingivae, cement, periodontal ligaments, and the alveoli which causes loosening of the teeth.

Peritonitis is the acute inflammation of the peritoneum that lines the abdominal cavity and covers abdominal organs. It may result from bacteria entering the peritoneal cavity due to contamination during accidents, surgery or by a ruptured intestine or appendix.

Noninflammatory Disorders

Cirrhosis of the liver is characterized by scarring, which results from dense irregular connective tissue replacing destroyed hepatocytes. It may be caused by hepatitis, alcoholism, nutritional deficiencies, or liver parasites.

Constipation is a condition in which defecation is difficult because feces are harder and drier than normal. This results from feces remaining in the colon for a longer than normal period, which allows more water to be absorbed.

Dental caries–tooth decay–result from the excess acid produced by certain microorganisms that live in the mouth and use food residues for their nutrients. Residues of carbohydrates, especially sugars, nurture these microorganisms and increase their acid production.

Diarrhea is the production of watery feces due to the abnormally rapid movement of chyme through the colon, which decreases the amount of time available to absorb water. Increased peristalsis may result from a number of causes, including inflammation and chronic stress.

Eating disorders result from an obsessive concern about weight control, especially among young adult females. There are two major types of eating disorders: anorexia and bulimia.

Anorexia nervosa is self-imposed starvation that results in malnutrition and associated physiological changes. Patients with this disorder see themselves as overweight, although others see them as very thin. Death can occur due to the complications of prolonged starvation.

Bulimia is characterized by frequent overeating and purging by self-induced vomiting. Fears of being overweight, depression, and stress are associated factors. The exact cause is unknown. Bulimia may lead to such complications as an imbalance of electrolytes, erosion of tooth enamel by stomach acids, and constipation.

Gallstones result from crystallization of cholesterol in bile. They commonly occur in the gallbladder, but they may be carried into the bile duct where they block the flow of bile. Severe pain and often jaundice accompanies such blockage. Treatment may include drugs that dissolve the gallstones, shock-wave therapy to break up the stones, or surgical removal of the gallstones and gallbladder.

Chapter Summary

15.1 Digestion: An Overview

- Digestion involves both mechanical and chemical processes. Mechanical digestion is the physical process of breaking food into smaller particles. Chemical digestion is an enzymatic process that converts nonabsorbable food molecules into absorbable nutrient molecules.
- A number of different digestive enzymes are required for the chemical digestion of food molecules.

15.2 Alimentary Canal: General Characteristics

- The alimentary canal is the long tube through which food passes from the esophagus to the anus. Digestion of food and absorption of nutrients occur in some portions of the alimentary canal.
- The wall of the alimentary canal is composed of four layers. From superficial to deep, they are the serosa, muscular layer, submucosa, and mucosa.
- Food is moved through the alimentary canal by peristaltic contractions. Segmentation mixes food with digestive secretions.

15.3 Mouth

- The mouth is surrounded by the cheeks, palate, and tongue. Teeth are embedded in the alveoli of the maxilla and mandible.
- The tongue is used to manipulate food during chewing and swallowing and to speak. It is the main location for taste buds. It produces lingual lipase, which aids in digesting triglycerides into fatty acids and glycerides within the stomach.
- Humans have two sets of teeth: deciduous and permanent. There are 32 permanent teeth divided into four types: incisor teeth, canine teeth, premolar teeth, and molar teeth.
- A tooth is composed of two major parts: a crown covered with enamel and a root embedded in an alveolus. Dentin forms most of the tooth. The centrally located pulp cavity contains blood vessels and nerves. The root is anchored to bone by cement and by periodontal ligaments.
- Three pairs of salivary glands (parotid, submandibular, and sublingual) secrete saliva into the mouth. Salivary secretion is under reflexive neural control. Saliva cleanses and lubricates the mouth and binds food together.
- Saliva contains salivary amylase, which breaks down starch and glycogen into maltose, and lysozyme, which kills certain bacterial types.

15.4 Pharynx and Esophagus

- The pharynx connects the oral and nasal cavities with the esophagus and larynx. Its digestive function is to transport food from the mouth to the esophagus.
- The swallowing reflex causes the epiglottis to cover the laryngeal opening, directing food into the esophagus.
- Pharyngeal, palatine, and lingual tonsils are located near the entrance to the pharynx.
- The esophagus carries food by peristalsis from the pharynx to the stomach.
- The lower esophageal sphincter relaxes to let food enter the stomach.

15.5 Stomach

- The stomach is a pouchlike enlargement of the alimentary canal. It is located in the left upper quadrant of the abdominopelvic cavity and consists of four regions: cardia, fundus, body, and pyloric part.
- The pyloric sphincter relaxes to allow chyme to enter the duodenum.
- The stomach mucosa contains gastric glands that secrete gastric juice, whose components include hydrochloric acid, pepsin, rennin, gastric lipase, and intrinsic factor. Gastric juice converts food into chyme.
- The secretion of gastric juice is regulated by neural reflexes and the hormones gastrin and cholecystokinin, which are produced by the gastric and intestinal mucosae, respectively.
- Pepsin acts on proteins and breaks them into peptides.
- Rennin is a gastric enzyme secreted by infants. It curdles milk, which aids digestion of milk proteins.
- In infants, gastric lipase breaks down triglycerides into fatty acids and monoglycerides.
- Gastric juice contains intrinsic factor, which is necessary for absorption of vitamin B12.
- Little absorption occurs in the stomach.

15.6 Pancreas

- The pancreas is located adjacent to the duodenum and pyloric part of the stomach. The pancreas secretes pancreatic juice, which is carried to the duodenum by the pancreatic duct. The hormones secretin and cholecystokinin from the intestinal mucosa stimulate the secretion of pancreatic juice.
- Pancreatic digestive enzymes act on each of the three types of energy food. Pancreatic amylase breaks down starch and glycogen into maltose. Pancreatic lipase breaks down triglycerides into monoglycerides and fatty acids. Trypsin breaks down proteins into peptides.

15.7 Liver

- The liver is located in the right upper quadrant of the abdominopelvic cavity.
- The liver performs many functions. It plays a critical role in the metabolism of carbohydrates, lipids, and proteins. It stores fat, glycogen, iron, and lipid-soluble vitamins. It detoxifies drugs and toxins within the blood. It phagocytizes worn-out blood cells and bacteria. It produces and secretes plasma proteins, heparin, and bile.

- Hepatic lobules are the structural and functional units of the liver.
- Bile is continuously secreted by the liver. Between meals, bile is stored in the gallbladder. It is released when the hormone cholecystokinin from the intestinal mucosa stimulates contraction of the gallbladder. Bile is carried to the duodenum by the bile duct.
- Bile emulsifies fats, which aids their digestion by lipases.

15.8 Small Intestine

- The small intestine occupies much of the abdominopelvic cavity and consists of three segments: duodenum, jejunum, and ileum. Most of the digestive processes and absorption of nutrients occur in the small intestine.
- The mucosa contains numerous intestinal glands that secrete intestinal juice and villi that absorb nutrients. Secretion of intestinal juice is activated by a neural reflex.
- The brush border enzymes maltase, sucrase, and lactase act on corresponding disaccharides to form the monosaccharides: glucose, fructose, and galactose. Brush border peptidases convert peptides into amino acids.
- End products of digestion are absorbed into the villi. Monosaccharides, amino acids, water-soluble vitamins, minerals, and very small fatty acids cross the epithelium and enter the blood capillaries of the villi. Large fatty acids, monoglycerides, cholesterol, and lipid-soluble vitamins are transported in micelles to the intestinal lining and are absorbed into the epithelial cells. Monoglycerides and fatty acids are then recombined to form triglycerides. Clusters of triglycerides, other lipids, and lipid-soluble vitamins are coated with protein to form chylomicrons, which enter the lacteals of the villi.
- Undigested and unabsorbed materials exit the small intestine and enter the large intestine through the ileal orifice.

15.9 Large Intestine

- The large intestine consists of the cecum, colon, rectum, and anal canal. The appendix is an appendage of the cecum.
- The large intestine is gathered into a series of pouches by the taeniae coli. Its mucosa lacks villi and secretes only mucus.
- The absorption of water and the formation and expulsion of feces are major functions of the large intestine. Bacteria decompose the undigested materials.
- Mass peristaltic movements propel the feces into the rectum, initiating the defecation reflex, which relaxes the internal anal sphincter. Voluntary relaxation of the external sphincter allows expulsion of the feces.

15.10 Nutrients: Sources and Uses

- Nutrients include carbohydrates, lipids, proteins, vitamins, minerals, and water. The liver is involved in the metabolism of many nutrients.
- Essential nutrients are nutrients that must be consumed because they cannot be synthesized by the body.
- Cellular respiration consists of glycolysis, the citric acid cycle, and an electron transport chain. Respiration of one molecule of glucose yields 36 to 38 ATP.
- End products of carbohydrate, lipid, and protein digestion may be used in cellular respiration either directly or after modification.
- Dietary carbohydrates come primarily from plant foods. Cellulose is a indigestible polysaccharide that provides fiber in the diet.
- The liver regulates the concentration of glucose in the blood. Excess glucose is converted to glycogen or fats for storage. These reactions may be reversed to release more glucose into the blood.
- Dietary lipids are mostly triglycerides that occur either as saturated fats, usually in animal foods, or as unsaturated fats, usually in plant foods. Cholesterol occurs in egg yolks, milk, and meats.
- Lipids form important parts of plasma membranes and myelin sheaths of neurons. Fats are an energy source for many cells. Excess fats are stored in adipose tissue. The liver helps to regulate the concentration of triglycerides and cholesterol in the blood.
- Dietary proteins occur in meats, milk, eggs, cereals, nuts, and legumes. The eight essential amino acids cannot be synthesized by the body. Only animal proteins contain all of the essential amino acids.
- Amino acids are used primarily to synthesize protein in the body. These proteins form plasma proteins, enzymes, certain hormones, and structural parts of cells. Amino acids may be deaminated by the liver and used to form glucose or fat or may be used as an energy source by cells.
- Vitamins are organic molecules required in minute amounts for normal functioning of the body. Water-soluble vitamins include the B vitamins and vitamin C. Lipid-soluble vitamins include vitamins A, D, E, and K.
- Vitamins act with enzymes to speed up essential chemical reactions, such as cellular respiration, in cells.
- Minerals are inorganic substances that are necessary for the normal functioning of the body. Minerals are obtained by plants from the soil and are passed on to animals, including humans, eating the plants.
- Many minerals are a part of organic compounds in the body. Other minerals are deposited as salts in bones and teeth, and some occur as ions in body fluids.
- There are seven major minerals that are required in moderate amounts in the diet: calcium, phosphorus, potassium, sulfur, sodium, chloride, and magnesium. Other minerals of the body are required in trace amounts.
- MyPlate is a visual guide to help people regulate the portion sizes for each food group at meals.

15.11 Disorders of the Digestive System

- Inflammatory disorders include appendicitis, inflammatory bowel disease (IBD), irritable bowel syndrome (IBS), diverticulitis, hemorrhoids, hepatitis, periodontal disease, and peritonitis.
- Other disorders include eating disorders (anorexia nervosa and bulimia), cirrhosis, constipation, dental caries, diarrhea, and gallstones.

Gross Anatomy

Each kidney is convex laterally and concave medially with a medial indentation called the *hilum*. Blood vessels, lymphatic vessels, nerves, and the ureter enter or exit at the hilum. An adult kidney is about 12 cm long, 7 cm wide, and 2.5 cm thick.

The internal macroscopic anatomy of a kidney is best observed in frontal section, as shown in figure 16.2*a*. Two functional regions of the kidney are evident: the renal cortex and the renal medulla. The **renal cortex** is the relatively thin, superficial layer. Deep to the renal cortex is the **renal medulla,** which contains the cone-shaped *renal pyramids*. The apex, or *renal papilla,* of each pyramid extends toward the renal pelvis, the most central structure of the kidney. Narrow portions of the renal cortex, the *renal columns,* extend into the renal medulla between the renal pyramids.

The work of the kidneys is performed by microscopic structures called **nephrons** (nef′-rons). Nephrons originate in the renal cortex, dip into the renal medulla, return to the renal cortex, and ultimately join a **collecting duct,** as shown in figure 16.2*b*. Nephrons produce urine, which flows into the collecting ducts of renal pyramids.

The renal papilla of each renal pyramid fits into a funnel-shaped **minor calyx** (kā′-lix), which receives urine from the collecting ducts. Two or three minor calyces (kā′-li-sēz) converge to form a **major calyx,** and two or three major calyces merge to form the funnel-like **renal pelvis.** The renal pelvis is contiguous with the ureter. Thus, the pathway of urine from nephrons to ureter is as follows: nephrons → collecting ducts → minor calyces → major calyces → renal pelvis → ureter. Urine is carried by the ureter to the urinary bladder by peristalsis.

Microscopic Anatomy

Each kidney contains about 1 million nephrons, the functional units of the kidneys. A nephron consists of two major parts: a renal corpuscle and a renal tubule. Figure 16.3 shows the structure of a nephron and its associated blood vessels.

Renal corpuscles are located in the renal cortex of the kidneys. Each renal corpuscle is composed of a **glomerulus** (glō-mer′-ū-lus, plural, *glomeruli*), a tuft of capillaries, which is enclosed in a double-walled **glomerular** *(Bowman)* **capsule.** The glomerular capsule is an expanded extension of a renal tubule.

A **renal tubule** leads away from the glomerular capsule and consists of three sequential segments. The first part of the renal tubule is the *proximal convoluted tubule* (PCT). It leads from the glomerular capsule to the *nephron loop,* the U-shaped second part of the tubule. The descending limb of the nephron loop descends into the renal medulla, and

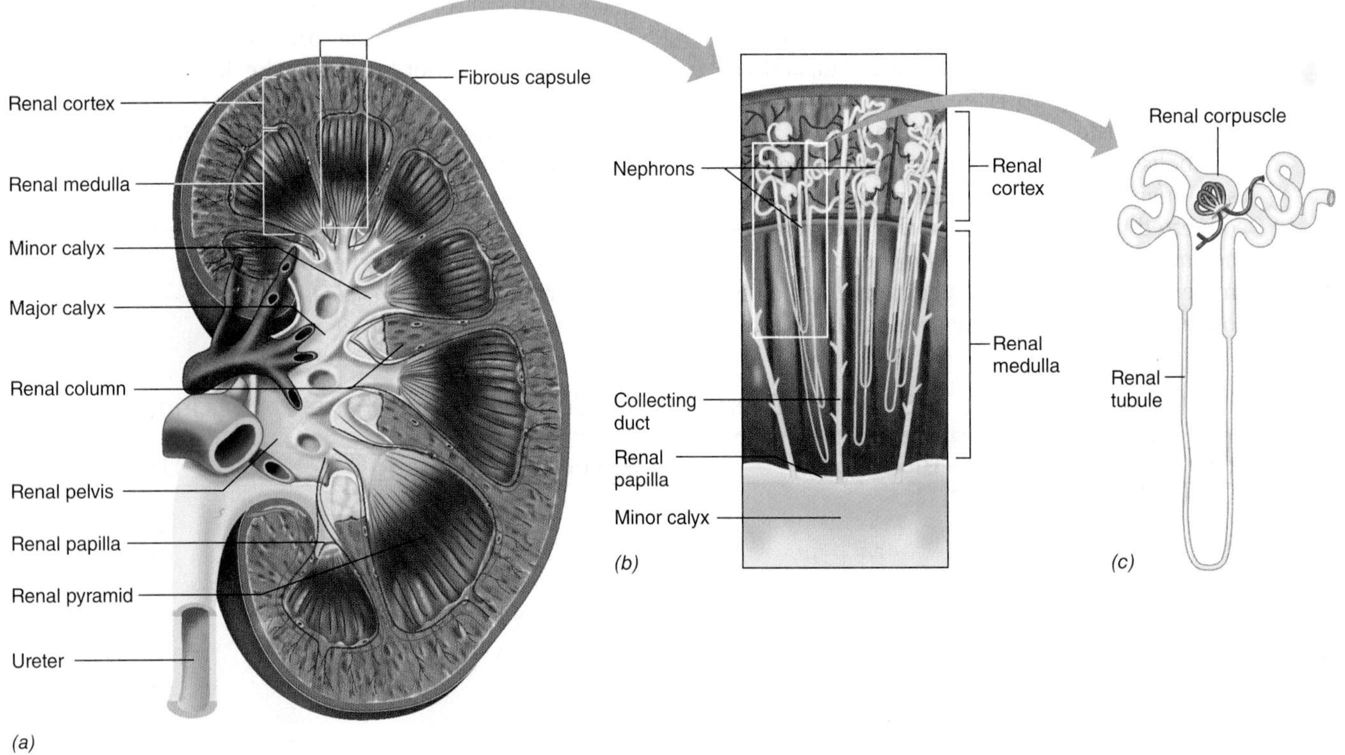

Figure 16.2 Internal Structure of the Kidney.
(a) Frontal section of a kidney. *(b)* A renal pyramid showing the orientation of nephrons and collecting ducts. *(c)* A single nephron. APIR

the ascending limb of the nephron loop ascends back into the renal cortex. The ascending limb of the nephron loop is continuous with the *distal convoluted tubule* (DCT), the third segment of the renal tubule, which unites with a collecting duct. Several nephrons unite with a single collecting duct. Collecting ducts begin in the renal cortex and extend the length of a renal pyramid to its papilla, where the collecting ducts merge before emptying into a minor calyx.

Types of Nephrons

There are two types of nephrons in the kidney: about 80% are cortical nephrons, and about 20% are juxtamedullary nephrons. The glomerular capsules of *cortical nephrons* are located superficially in the renal cortex. The nephron loops of these nephrons are located almost entirely in the renal cortex of the kidney. Cortical nephrons are important in adjusting the composition of the urine. In contrast, the glomerular capsules of *juxtamedullary nephrons* are located deep in the renal cortex near the renal medulla. The nephron loops of these nephrons penetrate deep into the medulla. Juxtamedullary nephrons play an important role in regulating water content of the blood plasma.

Renal Blood Supply

The kidneys receive a large volume of blood–1,200 ml per minute, which is about one-fourth of the total cardiac output. Each kidney receives blood via a *renal artery*, which branches from the abdominal aorta. Within each kidney, the renal artery branches to form three or four *segmental arteries*, which branch further to form several *interlobar arteries* that run along the renal columns between the renal pyramids to the renal cortex. These arteries branch to form smaller and smaller arteries and finally form arterioles.

In the renal cortex, **afferent glomerular arterioles** branch from the smallest arteries, and each afferent glomerular arteriole carries blood to a glomerulus. Blood leaves a glomerulus in an **efferent glomerular arteriole.** Note that a glomerulus is a capillary ball between two arterioles. The efferent glomerular arteriole usually leads to **peritubular capillaries,** which surround the cortical portion of the renal tubule. Sometimes the efferent glomerular arteriole leads to the **vasa recta,** which are vessels surrounding the nephron loops and collecting ducts within the renal medulla. Blood from the peritubular capillaries and vasa recta enters a venule, progressively larger veins that merge to form *interlobar veins*, which finally join to form the renal vein. A *renal vein* carries blood from each kidney to the inferior vena cava (figures 16.3 and 16.4; see figure 16.1).

Juxtaglomerular Complex

The **juxtaglomerular** (juks-tah-glo-mer′-u-lar) **complex** of each nephron is located where the ascending limb of the

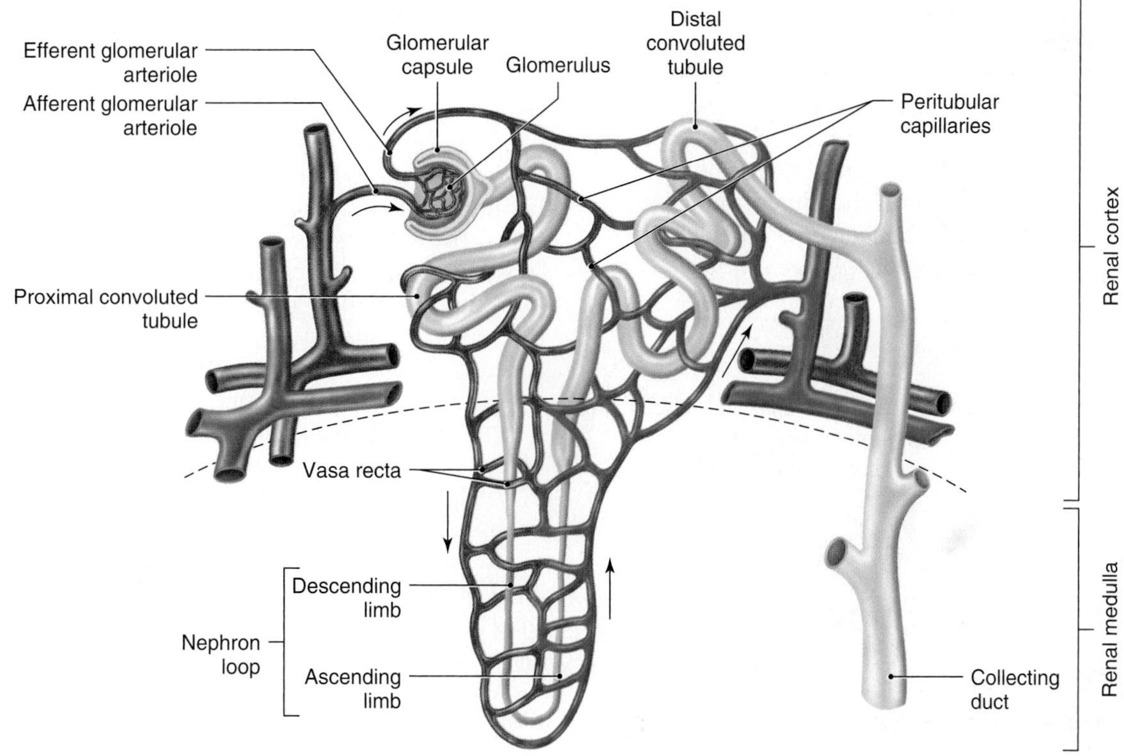

Figure 16.3 A nephron and its associated blood vessels. The nephron has been stretched out to show its parts more clearly. Arrows show the direction of blood flow. APR

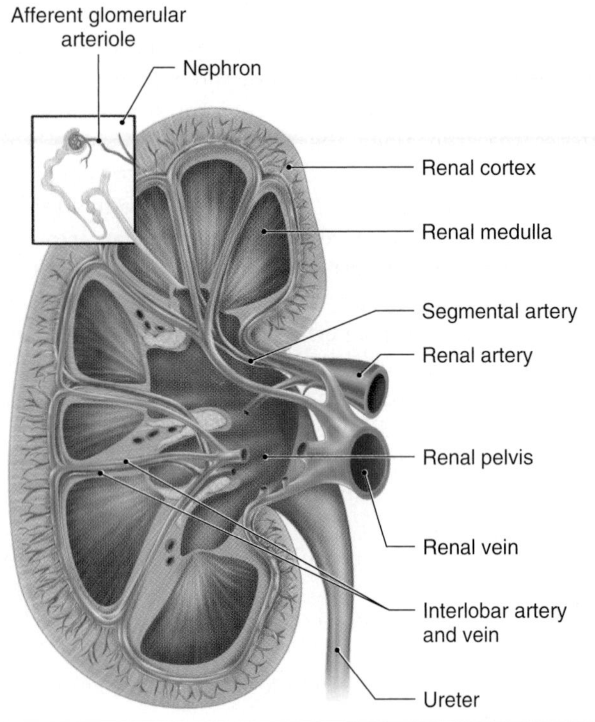

Figure 16.4 Main branches of the renal artery and renal vein. AP|R

nephron loop contacts the afferent and efferent glomerular arterioles near the glomerulus (figure 16.5). It consists of two groups of specialized cells: granular cells and the macula densa. *Granular cells,* or juxtaglomerular epithelioid cells, are large smooth muscle cells in the walls of the afferent and efferent glomerular arterioles near their attachment to the glomerulus. The *macula densa* consists of rather narrow, tightly packed cells composing the ascending limb where it contacts the afferent and efferent glomerular arterioles. The juxtaglomerular complex helps to regulate blood pressure, as you will see shortly.

Check My Understanding

3. What are the structural and functional differences between cortical and juxtamedullary nephrons?
4. How does blood flow to, through, and away from the nephron?

16.3 Urine Formation

Learning Objectives

5. Compare glomerular filtration, tubular reabsorption, and tubular secretion.
6. Explain how urine is formed.
7. Indicate the normal components of urine.

The formation of urine is a homeostatic mechanism that maintains the composition and volume of blood plasma within normal limits. In the production of urine, nephrons perform three basic functions: (1) They regulate the concentration of solutes, such as nutrients and ions, in blood plasma, and this also regulates blood pH. (2) They regulate the concentration of water in blood plasma, which in turn helps regulate blood pressure. (3) They remove metabolic wastes and excess substances from the blood plasma.

Four processes are crucial to the formation of urine: (1) **Glomerular filtration** moves water and solutes, except plasma proteins, from blood plasma into the glomerular capsule. The fluid that enters the glomerular capsule is called **glomerular filtrate.** Formed elements normally are not part of the glomerular filtrate. Once the glomerular filtrate passes from the glomerular capsule into the renal tubule it is renamed **tubular fluid.** (2) **Tubular reabsorption** removes useful substances from the tubular fluid and returns them into the blood plasma, and (3) **tubular secretion** moves additional wastes and excess substances from the blood plasma into the tubular fluid. (4) **Water conservation** removes water from the urine, returning it into the blood plasma (figure 16.6).

Glomerular Filtration

Urine formation starts with glomerular filtration, a process that forces some of the water and dissolved substances in blood plasma from the glomeruli into the glomerular capsules. Two major factors are responsible for glomerular filtration: (1) the increased permeability of glomerular capillary walls and (2) the elevated blood pressure within the glomeruli.

Glomerular capillaries are much more permeable to substances in the blood plasma than are other capillaries because their walls contain numerous pores, as shown in figure 16.5*c*. These pores allow water and most dissolved substances to easily pass through the capillary walls into the glomerular capsules. Unlike other capillaries, glomerular capillaries are enveloped by specialized cells called *podocytes,* which have numerous fingerlike cellular extensions that wrap around the capillaries. Podocytes help prevent plasma proteins, and formed elements from entering glomerular capsules.

The elevated glomerular blood pressure results because the diameter of the efferent glomerular arteriole is smaller than that of the afferent glomerular arteriole. Because blood can enter a glomerulus at a faster rate than it can leave it, the greater blood volume within the glomerulus creates an increase in blood pressure. The glomerular blood pressure provides the force for glomerular filtration.

The result of glomerular filtration is the production of glomerular filtrate that consists of the same substances that compose blood plasma, except for plasma proteins that are

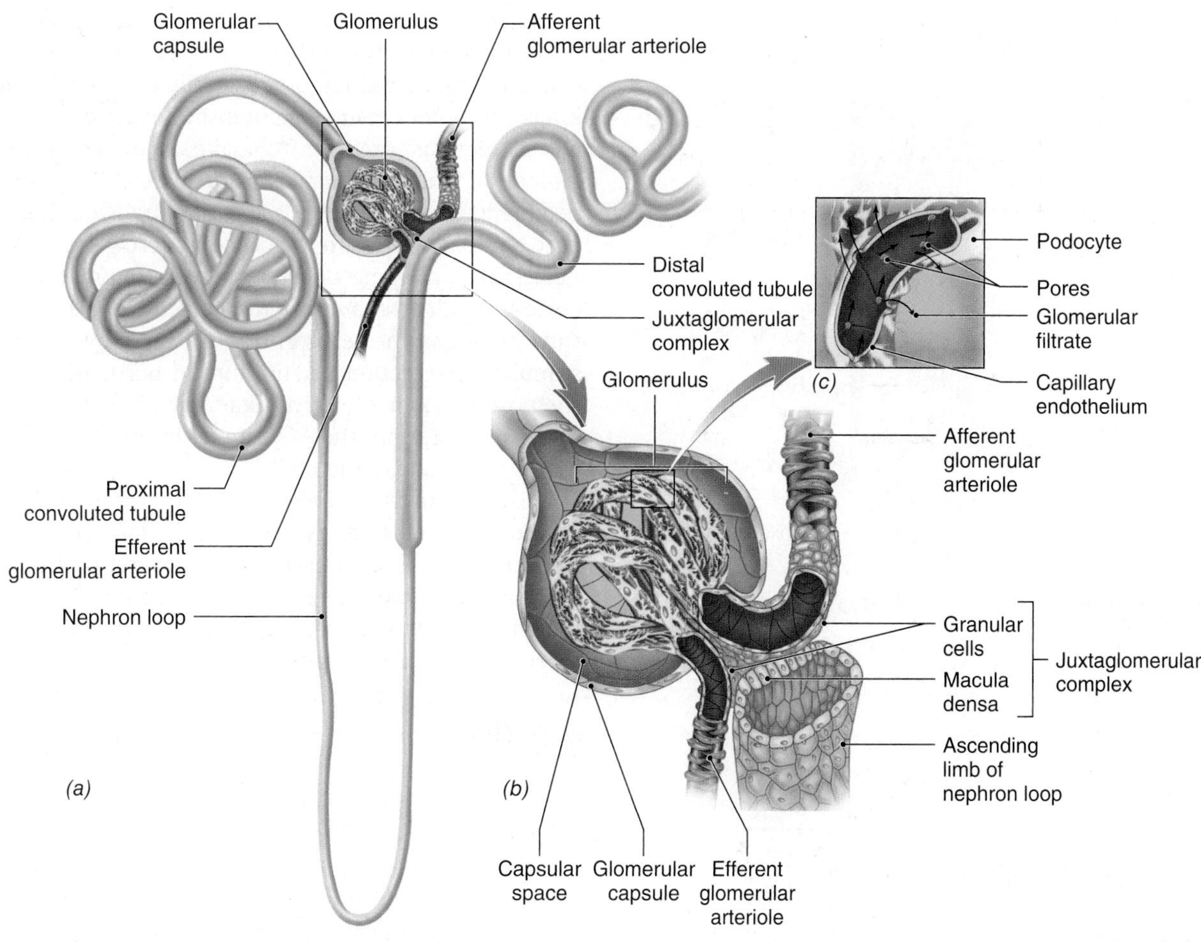

Figure 16.5 Juxtaglomerular Complex.
(*a*) The juxtaglomerular complex is located where the ascending limb of the nephron loop contacts the afferent and efferent glomerular arterioles near a glomerulus. (*b*) Enlargement showing the granular cells and macula densa, which compose the juxtaglomerular complex. (*c*) Magnification of the pores through which glomerular filtration occurs.

too large to pass through the pores of the glomerular capillaries. Because glomerular filtration is a nonselective process, the concentrations of these substances are the same in both blood plasma and glomerular filtrate (table 16.1).

Glomerular Filtration Rate

Glomerular filtration rate (GFR) is about 125 ml per minute, or 7.5 liters per hour. This means that the entire volume of blood is filtered every 40 minutes! In 24 hours, about 180 liters (nearly 45 gallons) of glomerular filtrate is produced. However, most of the glomerular filtrate is reabsorbed, as you will see shortly.

Maintenance of a relatively stable GFR is necessary for normal kidney function. The GFR varies directly with glomerular blood pressure, which, in turn, is primarily determined by systemic blood pressure. The GFR is regulated by three homeostatic processes: renal autoregulation, sympathetic control, and the renin-angiotensin mechanism. These processes operate primarily by controlling the diameter of afferent glomerular arterioles to keep the GFR within normal limits.

Renal autoregulation is the mechanism that keeps the GFR within normal limits, without extrinsic neural or hormonal control, in response to moderate variations in systemic blood pressure. One way this occurs is by the response of smooth muscle in the afferent glomerular arteriole wall to blood pressure changes. If the afferent glomerular arteriole is stretched by increased blood pressure, it contracts; the smaller diameter of the afferent glomerular arteriole means less blood volume enters the glomerulus, thus decreasing blood pressure within the glomerulus. A decline in blood pressure causes the arteriole to dilate, which keeps the GFR stable.

Another autoregulatory mechanism involves the juxtaglomerular complex. The juxtaglomerular complex monitors the GFR by using the macula densa to sense changes in the flow rate and chemical composition in the tubular fluid of the ascending limb. If the GFR increases, the

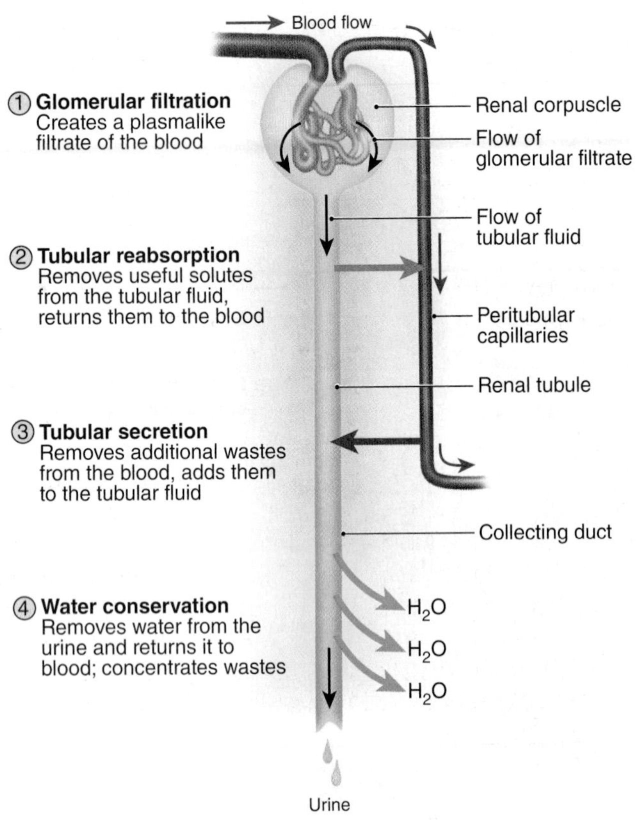

Figure 16.6 The major steps of urine formation.

granular cells constrict the afferent glomerular arteriole, keeping the GFR normal. If the GFR declines, the afferent glomerular arteriole relaxes to maintain a normal GFR.

Sympathetic control is a function of the sympathetic division of the autonomic nervous system, which overrides renal autoregulation in times of large systemic blood pressure shifts or during the "fight or flight" response. If a drop in systemic blood pressure is detected, afferent glomerular arterioles are constricted, which decreases glomerular pressure and GFR. This decreases urine formation, which conserves water to maintain normal blood pressure and volume. If an increase in blood pressure is detected, the afferent glomerular arteriole dilates as a result of the sympathetic control being removed, which increases glomerular pressure and the GFR. This results in an increase in urine production and water excretion to maintain normal blood pressure and volume.

The *renin-angiotensin mechanism* is triggered when the juxtaglomerular complex detects a reduced GFR and releases the enzyme *renin.* Renin is secreted in response to (1) sympathetic stimulation; (2) a drop in blood pressure in the afferent glomerular arteriole; and (3) detection of a reduction of Na^+, K^+, and Cl^- levels in the tubular fluid in the ascending limb of the nephron loop by the macula densa (see figure 16.5). Renin converts a plasma protein (angiotensinogen), which is formed by the liver, into angiotensin I. Angiotension I is rapidly converted into angiotensin II by the *angiotensin-converting enzyme (ACE)* released from endothelial cells of capillaries in the lungs and other organs. Angiotensin II constricts the efferent glomerular arterioles to maintain blood pressure in glomeruli, which maintains an adequate GFR in spite of a decline in blood pressure. It also acts to restore blood volume and blood pressure by (1) constricting systemic arterioles; (2) stimulating aldosterone secretion by the adrenal cortex, which promotes the reabsorption of Na^+, which in turn promotes the reabsorption of water by osmosis; (3) stimulating secretion of antidiuretic hormone (ADH) by the posterior lobe of the pituitary gland, which promotes water reabsorption; and (4) stimulating thirst, which promotes water intake (figure 16.7).

Atrial natriuretic peptide (ANP) is secreted by atria of the heart when they are stretched by an excessive blood volume. ANP promotes water excretion by increasing GFR, and inhibiting Na^+ reabsorption in the DCT, which results in a decrease in blood volume and, in turn, a decrease in blood pressure.

Clinical Insight

ACE inhibitors are a group of drugs commonly used to treat hypertension. These drugs help to reduce the blood pressure by decreasing the activity of ACE, resulting in decreased production of angiotensin II. Decreased production of angiotensin II leads to decreased blood pressure.

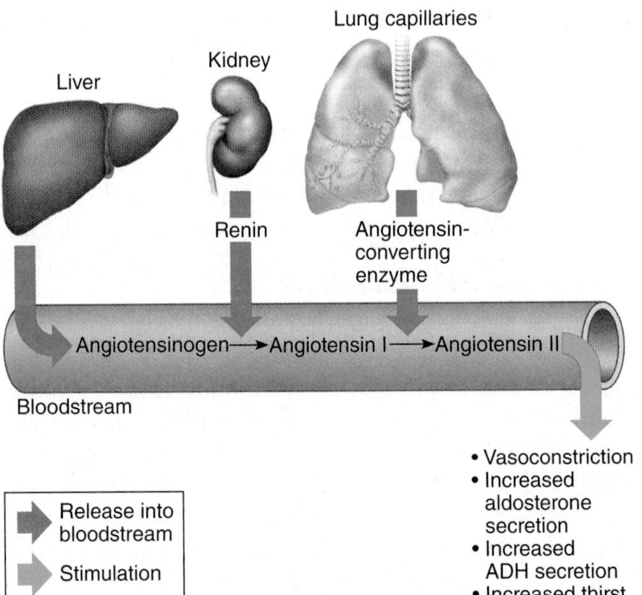

Figure 16.7 The renin–angiotensin mechanism. The multiple actions of angiotensin II help to maintain the GFR and systemic blood pressure.

Tubular Reabsorption and Tubular Secretion

Tubular reabsorption and tubular secretion involve both active and passive transport mechanisms. The effectiveness of passive transport of any substance depends upon (1) the permeability of the renal tubule, peritubular capillaries, and vasa recta to the substance, and (2) the concentration gradient of the substance.

Events in the Proximal Convoluted Tubule

About 65% of the tubular fluid is reabsorbed in the PCT. All nutrients, such as glucose and amino acids, are actively reabsorbed here. Positively charged ions, such as those of Na^+, K^+, and Ca^{2+}, are also actively reabsorbed. The active reabsorption of positively charged ions causes negatively charged ions, such as Cl^- and HCO_3^-, to be passively reabsorbed by electrochemical attraction. The reabsorption of these substances increases the osmotic pressure of the blood plasma in the peritubular capillaries and decreases the osmotic pressure of the tubular fluid. This causes water to be passively reabsorbed from the tubular fluid by osmosis (figure 16.8).

Tubular secretion is the process that extracts substances from blood plasma in the peritubular capillaries and secretes them into the tubular fluid in the renal tubule. Metabolic wastes, such as urea and uric acid, and drugs, are removed from the blood in this way. It not only removes unwanted wastes but also helps regulate the pH of body fluids by selectively removing H^+ and HCO_3^-.

Events in the Nephron Loop

Water is passively reabsorbed by osmosis from the descending limb of the nephron loop into the capillaries supplied by the vasa recta. The ascending limb is impermeable to water, but solutes are reabsorbed passively by diffusion from the proximal ascending limb (figure 16.9).

The ascending limb actively pumps Na^+ out of the tubular fluid, and K^+ and Cl^- follow passively into the interstitial fluid. Some K^+ reenter the tubule, but Na^+ and Cl^- remain in the interstitial fluid. The accumulation of ions in the interstitial fluid of the renal medulla establishes a strong osmotic gradient for the reabsorption of water from the descending limb and the collecting duct. Establishing this osmotic gradient is a major function of the nephron loop.

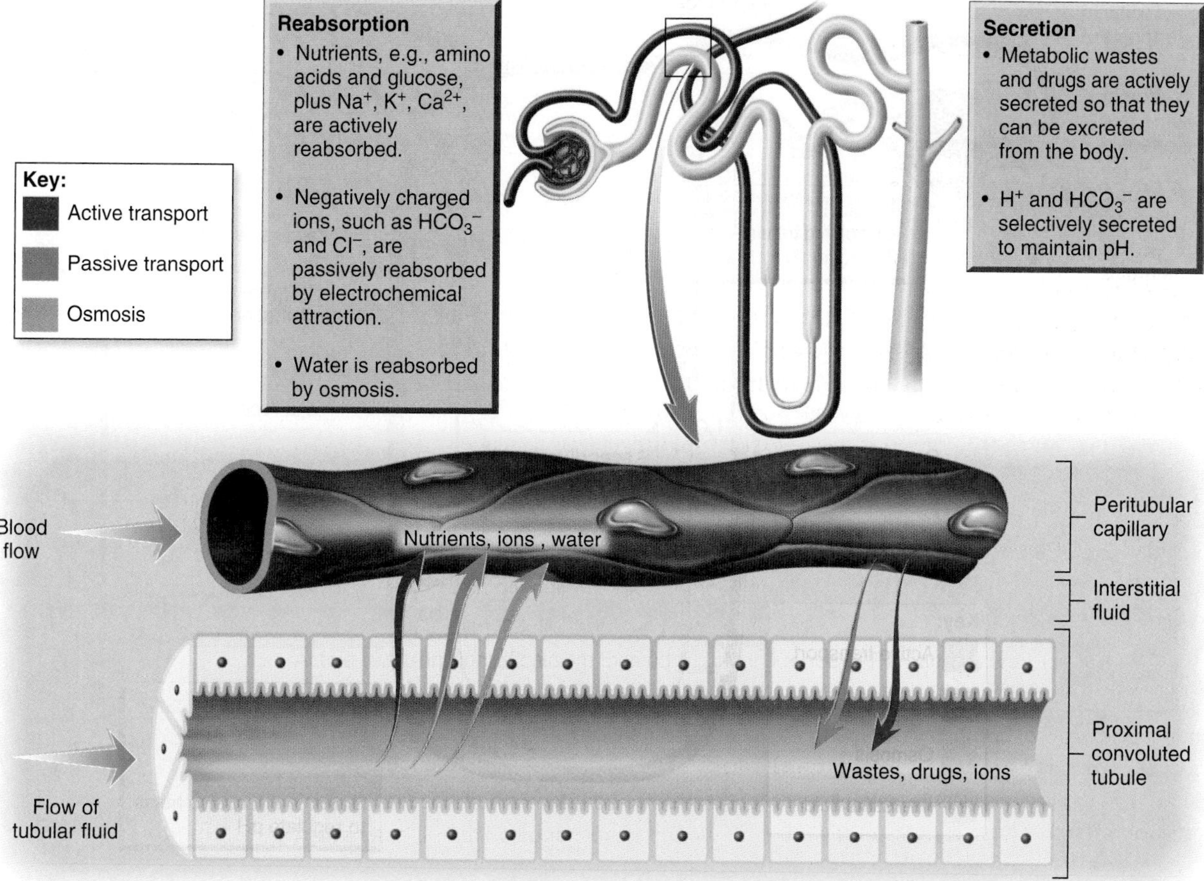

Figure 16.8 Reabsorption and secretion in the proximal convoluted tubule.

found in urine are glucose, proteins, formed elements, hemoglobin, and bile pigments. The presence of any of these substances suggests possible pathological conditions. Normal values of urine components and some indications of abnormal values are listed on the inside back cover.

Check My Understanding

5. What are the mechanisms of glomerular filtration, tubular reabsorption, tubular secretion, and water conservation?
6. For each section of a nephron and the collecting duct, what substances are reabsorbed and secreted during the formation of the urine entering the minor calyx?

16.4 Excretion of Urine

Learning Objectives

8. Describe the structure and function of the ureters, urinary bladder, and urethra.
9. Describe the control of micturition.

The term *urinary tract* refers collectively to the renal pelvis, the ureters, the urinary bladder, and the urethra. These structures function to carry urine from the kidneys to the external environment. Urine passes from the renal pelvis into the ureter and is carried by peristalsis to the urinary bladder. Urine is voided from the urinary bladder through the urethra.

Ureters

Each **ureter** is a slender tube about 25 cm (10 in) long that extends from a kidney to the urinary bladder. It begins at the kidney with the funnel-shaped renal pelvis and enters the inferior lateral margin of the urinary bladder (figure 16.10; see figure 16.1).

The wall of a ureter is formed of three layers. The external fibrous layer is composed of dense irregular connective tissue. The middle layer consists of smooth muscle cells that produce peristaltic waves for urine transport. The internal layer is a mucosa that is continuous with that of the renal pelvis and the urinary bladder. A flaplike fold of mucosa in the urinary bladder covers the opening of the ureter, and it functions as a valve that prevents backflow of urine into the ureter.

Clinical Insight

A routine urinalysis is a common clinical test that provides information about kidney function and also about general health of the body. Kidney function is also assessed by two blood tests. The *blood urea nitrogen (BUN)* test evaluates how effectively the kidney removes urea from the blood. The normal BUN value is 6 to 20 mg per 100 ml of blood. In acute kidney failure and in later stages of chronic kidney failure, BUN may range from 50 to 200 mg per 100 ml of blood. *Blood (serum) creatinine* is another test that assesses kidney effectiveness. Creatinine levels in the blood are normally stable (0.6–1.5 mg/100 ml), so an increase indicates a decrease in kidney function.

Urinary Bladder

The **urinary bladder** is a hollow, muscular organ located posterior to the pubic symphysis within the pelvic cavity. It lies inferior to the parietal peritoneum. The urinary bladder provides temporary storage of urine, and its size and shape vary with the volume of urine that it contains. When filled with urine, it is almost spherical as its superior surface expands. When empty, its superior surface collapses, giving a deflated appearance.

The internal floor of the urinary bladder contains the *trigone* (trī′-gōn), a smooth, triangular area that contains an opening at each of its angles. The openings of the ureters are located at the two laterally located posterior angles, and the opening of the urethra is located at the anterior angle (figure 16.10).

Four layers compose the wall of the urinary bladder. The most internal layer is the *mucosa*, which is composed of transitional epithelium that is adapted to the repeated stretching of the urinary bladder wall. The epithelium stretches, and its thickness decreases as the urinary bladder fills with urine.

The mucosa is supported by the underlying *submucosa* formed of areolar connective tissue containing an abundance of elastic fibers. Blood vessels and nerves supplying the urinary bladder are present in the submucosa.

Smooth muscle cells compose the third, and thickest, layer. These cells form a muscle called the *detrusor* (dē-trū′-sor). The detrusor is relaxed as the urinary bladder fills with urine, and it contracts as urine is expelled. Cells of the detrusor form an *internal urethral sphincter* at the junction of the urinary bladder and the urethra.

The external layer consists of the parietal peritoneum, but it covers only the superior portion of the urinary bladder. The remainder of the urinary bladder surface is coated with dense irregular connective tissue.

Urethra

The **urethra** is a thin-walled tube that carries urine from the urinary bladder to the external environment. The urethral wall contains smooth muscle cells and is supported by connective tissue. The internal lining is a mucosa that is continuous with the mucosa of the urinary bladder. An

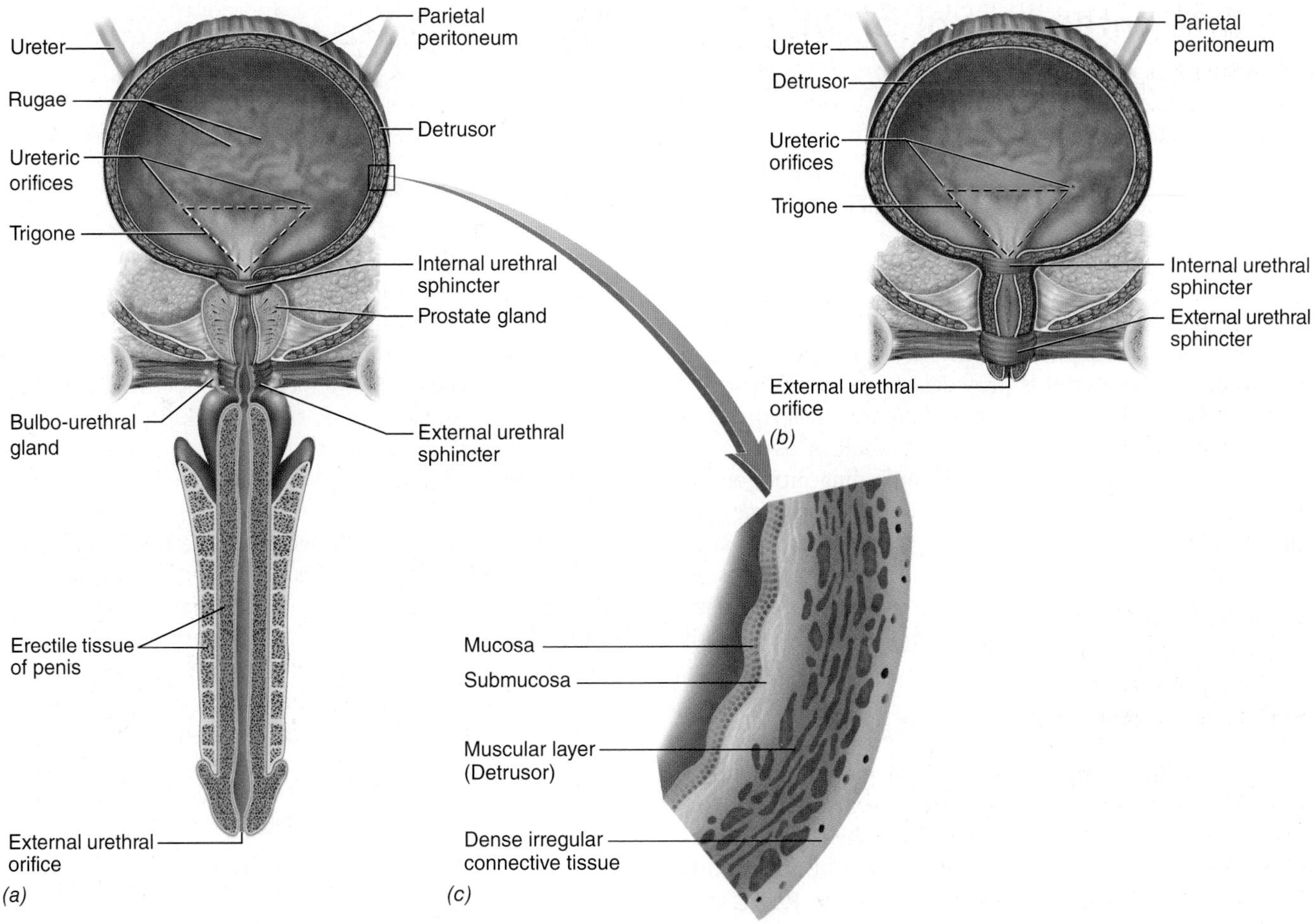

Figure 16.10 Urinary Bladder and Urethra in Frontal Section.
(a) Male. *(b)* Female. *(c)* Magnification of the wall of the urinary bladder. APR

external urethral sphincter, which is composed of skeletal muscle fibers, is located where the urethra penetrates the pelvic floor.

The female urethra is quite short, about 3 to 4 cm (1.5 in) in length. The external urethral orifice, its external opening, lies anterior to the vaginal orifice. The male urethra is much longer, about 16 to 20 cm (6–8 in) in length, because the urethra runs the length of the penis. The external urethral orifice is at the tip of the penis.

Micturition APR

Micturition (mik-tū-rish′un), or urination, is the act of expelling urine from the urinary bladder. Although the urinary bladder may hold up to 1,000 ml of urine, micturition usually occurs long before that volume is attained. When 200 to 400 ml of urine have accumulated in the urinary bladder, stretch receptors in the urinary bladder wall are stimulated and they trigger the *micturition reflex.* This reflex sends parasympathetic nerve impulses to the detrusor, causing rhythmic contractions. As this reflex continues, it causes the involuntarily controlled internal urethral sphincter to open and the person becomes aware of the desire to urinate. The act of urinating then becomes a consciously controlled process. If the voluntarily controlled external urethral sphincter is relaxed, micturition occurs; if it is not relaxed, micturition is postponed.

Micturition may be postponed by keeping the external sphincter voluntarily closed, and in a few moments the urge to urinate subsides. After more urine enters the urinary bladder, the micturition reflex is activated again, and the urge to urinate returns. Micturition cannot be postponed for long periods of time. After a while, the reflex overwhelms voluntary control and micturition occurs, ready or not.

An infant is not able to be toilet trained until neural development allows control of the external urethral sphincter muscle. Voluntary control is possible shortly after two years of age.

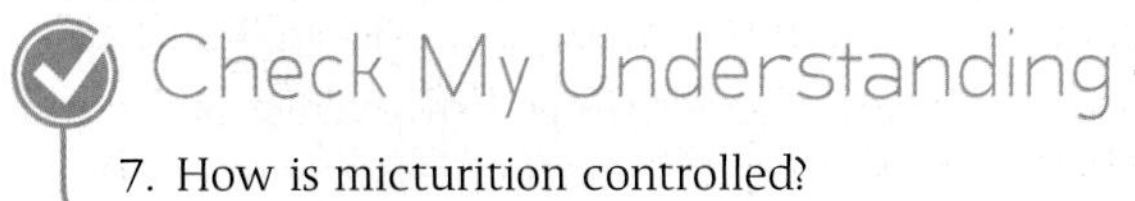

7. How is micturition controlled?

16.5 Maintenance of Blood Plasma Composition

Learning Objectives

10. Explain how water balance is maintained in body fluids.
11. Explain how electrolyte balance is maintained in body fluids.
12. Explain how pH balance is maintained in body fluids.

The composition and volume of blood plasma are affected by diet, cellular metabolism, and urine production. The intake of food and liquids provides the body with water and a variety of nutrients, including minerals, that are absorbed into the blood. Cellular metabolism uses nutrients and produces waste products, including nitrogenous wastes. Urine production retains essential nutrients and minerals in the blood plasma but removes some water along with excess substances and nitrogenous wastes. In healthy people, the kidneys are able to keep the composition and volume of the blood plasma relatively constant in spite of variations in diet and cellular activity.

Water and Electrolyte Balance

Two important components of blood plasma and other body fluids are water and electrolytes, and their concentrations in body fluids must be maintained within normal limits. Recall that water is the solvent of body fluids in which the chemical reactions of life occur. Recall that electrolytes are substances that form ions when dissolved in water, and they are so named because they can conduct an electric current when dissolved in water. For example, sodium chloride is an electrolyte that forms sodium and chloride ions when dissolved in water.

The concentrations of water and electrolytes in body fluids are interrelated because the concentration of one affects the concentration of the other. For example, the concentration of electrolytes establishes the osmotic pressure that enables water to be reabsorbed by osmosis.

Water Balance

The intake of water is largely regulated by the *thirst center* located in the hypothalamus of the brain. The thirst center is activated when it detects an increase in solute concentration in the blood. It is also activated by angiotensin II when blood pressure declines significantly. An awareness of thirst stimulates water intake to replace water lost from body fluids. Water intake must balance water loss, and this averages about 2,500 ml per day.

The body loses water in several ways, but about 60% of the total water loss occurs in urine. In addition, water is lost in the humidified air exhaled from the lungs, in feces, and in perspiration (figure 16.11). However, it is the kidneys that regulate the concentration of water in the blood plasma by controlling the volume of water lost in urine.

Clinical Insight

Substances that increase the production of urine are known as *diuretics*. Physicians often prescribe a diuretic to reduce the volume of body fluids in patients with edema or hypertension.

The volume of water lost in urine varies with both the volume of water lost by other means and the volume of water intake. These factors affect the action of the kidneys simultaneously, but we consider them separately to better understand how they influence kidney function.

In general, the more water that is lost through other means, the less water that is lost in urine. For example, if excessive water loss occurs through perspiration or diarrhea, more water is reabsorbed from the renal tubule and collecting duct. The result is a smaller volume of more concentrated urine. Conversely, if water loss through other means is minimal, water reabsorption is reduced, and a larger volume of more dilute urine is produced.

Similarly, the greater the intake of water, the less water is reabsorbed and a larger volume of more dilute urine is produced. Conversely, a lower water intake means more water is reabsorbed and a smaller volume of more concentrated urine is produced.

You can see that regulating water balance is a dynamic process and that water balance is largely controlled by the amount of water reabsorbed from renal tubules and collecting ducts into the blood plasma. Whether more or less water is reabsorbed is dependent upon ADH secreted by

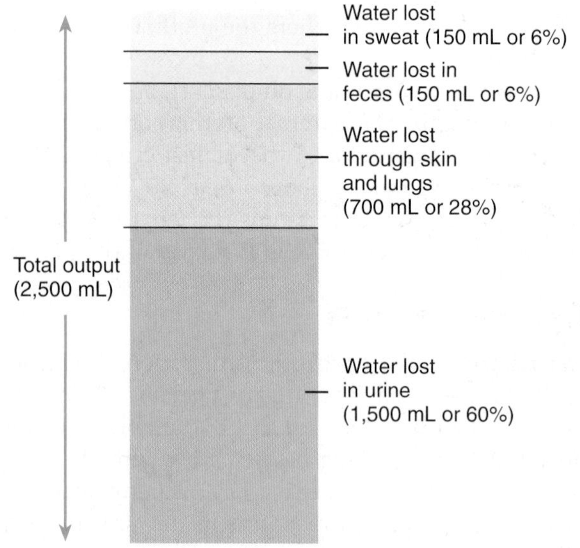

Figure 16.11 Pathways of water loss. Urine formation is the most important process that regulates water loss.

the posterior lobe of the pituitary gland. ADH promotes water reabsorption by increasing the permeability of the DCT and collecting ducts to water.

When the water concentration of blood is excessive, ANP is secreted and ADH secretion declines. The combined effect is that less water is reabsorbed and a greater volume of urine (and water) is excreted. The result is a decrease in water concentration in the blood. Conversely, when the water concentration of blood decreases, ANP is not secreted, ADH secretion is increased, more water is reabsorbed, and a smaller volume of urine is produced. ADH minimizes water loss in urine, but it cannot prevent it. Thus, water must be replenished daily by fluid intake.

Electrolyte Balance

Important electrolytes in body fluids include ions of sodium, potassium, calcium, chloride, phosphate, sulfate, and bicarbonate. Electrolytes are obtained from the intake of food and fluids. A craving for salt results when electrolytes are in low concentration in body fluids.

Electrolyte balance is regulated largely by active reabsorption of positively charged ions, which, in turn, secondarily controls the passive reabsorption of negatively charged ions by electrochemical attraction. Sodium ions are the most important ions to be regulated because they compose about 90% of the positively charged ions in extracellular fluids. Certain hormones play important roles in maintaining electrolyte balance.

Clinical Insight

If water loss significantly exceeds water intake for several days, extracellular fluids may become more concentrated, causing water to move out of the cells by osmosis. This condition, known as *cellular dehydration,* may lead to serious complications unless the water loss is quickly restored. In serious cases, dehydration may result in fever, mental confusion, or coma.

Aldosterone is a hormone that regulates the balance of sodium and potassium ions in the blood plasma by stimulating the active reabsorption of sodium ions and the active secretion of potassium ions by the DCT. Thus, aldosterone causes an exchange of sodium and potassium ions between the tubular fluid and the blood plasma until the blood concentrations of these two ions returns to normal. The adrenal cortex is stimulated to secrete aldosterone by (1) an increase of K^+ in the blood, (2) a decrease of Na^+ in the blood, and (3) angiotensin II. As long as blood concentrations of sodium and potassium ions are normal, aldosterone is not secreted. In contrast to aldosterone, ANP promotes the excretion of sodium ions and water by inhibiting sodium reabsorption, and thus osmosis, in the DCT and collecting duct when excess blood volume is detected by the atria of the heart.

The blood concentration of Ca^{2+} is regulated mainly by the actions of PTH and active vitamin D. When the blood Ca^{2+} concentration declines, the parathyroid glands are stimulated to secrete PTH. PTH promotes an increase in blood Ca^{2+} by stimulating three different processes: (1) the reabsorption of Ca^{2+} ions from the DCT, (2) the movement of Ca^{2+} from bones into the blood, and (3) the activation of vitamin D. Active vitamin D has the same actions as PTH; in addition, it increases the absorption of Ca^{2+} in foods by the small intestine. When the blood Ca^{2+} level returns to normal, PTH secretion is decreased. The lack of PTH is usually sufficient to decrease blood Ca^{2+} levels. Table 16.3 summarizes the effect of hormones that act on the kidneys.

During times of rapid bone remodeling, such as childhood or pregnancy, *calcitonin* is secreted by the thyroid gland. Calcitonin plays an antagonistic role to PTH by promoting the deposition of calcium in bones, which reduces the level of blood calcium.

Acid–Base Balance

The arterial blood pH must be maintained within rather narrow limits–pH 7.35 to pH 7.45–for body cells to function properly. Arterial blood pH below 7.35 is called **acidosis,** and arterial blood pH above 7.45 is called

Table 16.3 Hormones Acting on the Kidneys

Hormone	Source	Action
Aldosterone	Adrenal cortex	Stimulates reabsorption of Na^+ from the tubular fluid into the blood plasma; stimulates the secretion of K^+ from blood plasma into the tubular fluid
Antidiuretic hormone	Posterior lobe of pituitary	Stimulates the reabsorption of water from the tubular fluid into the blood plasma by making the DCT and collecting ducts more permeable to water; decreases the volume of urine produced
Atrial natriuretic peptide	Heart	Inhibits reabsorption of Na^+ from DCT and collecting ducts
Parathyroid hormone and active vitamin D	Parathyroid glands Liver and kidneys	Stimulates the reabsorption of Ca^{2+} from the tubular fluid into the blood plasma

alkalosis. Cellular metabolism produces products that tend to upset the acid-base balance. These products, such as lactic acid, phosphoric acid, and carbonic acid, tend to make the blood more acidic, as shown in figure 16.12.

Acids are substances that release hydrogen ions (H^+) when they are in water, which decreases the pH and increases the acidity of the liquid. Strong acids release more H^+ than weak acids. **Bases** are substances that, when placed in water, release ions that can combine with hydrogen ions, such as OH^- or HCO_3^-. Body fluids contain both acids and bases, and the balance between them determines pH. The balance of acids and bases in the body is regulated by three processes: (1) **buffers,** which act directly in the body fluids; (2) the **respiratory mechanism,** which controls carbonic acid levels; and (3) the **renal mechanism,** which regulates H^+ and HCO_3^- levels.

Buffers

The blood and other body fluids contain chemicals known as buffers (see chapter 2) that prevent significant changes in pH. Buffers are able to combine with or release H^+ ions as needed to stabilize the pH. If the H^+ concentration is excessive, buffers combine with some H^+ to reduce their concentration. Conversely, if too few H^+ are present,

Clinical Insight

The kidneys have a tremendous functional reserve. Renal insufficiency becomes evident only after about 75% of the renal functions have been lost. As the development of renal failure progresses, patients must rely on *hemodialysis* as a means of removing wastes and excessive substances from the blood. In hemodialysis, the patient's blood is pumped through selectively permeable tubes that are immersed in a dialyzing solution within a "kidney machine." Nitrogenous wastes and excessive electrolytes diffuse from the blood into the dialyzing solution, while certain needed substances, such as buffers, diffuse from the dialyzing solution into the blood. In this way, the concentration of wastes and electrolytes in the patient's blood are temporarily restored within normal limits. Hemodialysis may be required two to three times per week for patients with chronic kidney failure.

An alternative method is called *continuous ambulatory peritoneal dialysis (CAPD).* In this technique, 1 to 3 liters of dialyzing fluid are introduced into the peritoneal cavity through an opening made in the abdominal wall. Waste products and excessive substances diffuse from blood vessels in the peritoneum into the dialyzing solution, which is drained after two to three hours. This technique is less costly, may be done at home, and allows the patient to move about during the procedure. However, it must be done more frequently than dialysis using a kidney machine, and there is a greater chance of serious infection.

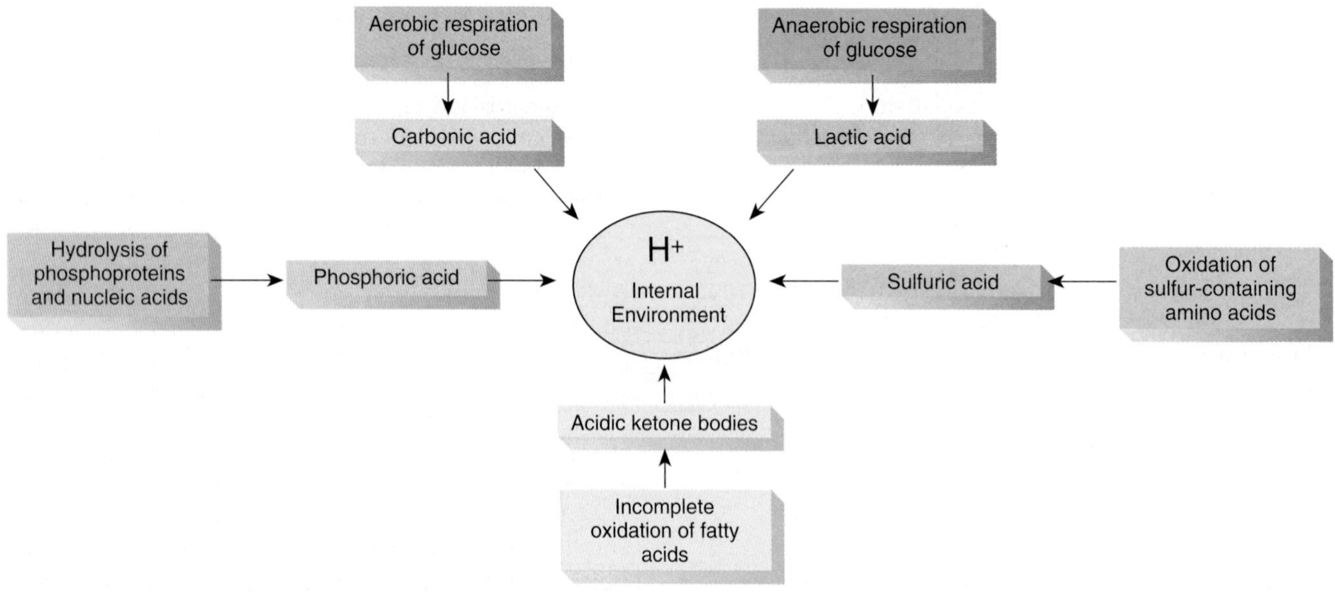

Figure 16.12 Examples of metabolic processes that increase the hydrogen ion concentration of body fluids.

buffers release some H^+ to increase their concentration to within normal limits. In this way, buffers help to keep the blood pH relatively constant.

The **bicarbonate buffer system** relies on a mixture of carbonic acid and HCO_3^-. Carbonic acid (H_2CO_3) forms by the hydration of carbon dioxide and then dissociates into HCO_3^- and H^+. The bicarbonate buffer system is particularly important in regulating the acid–base balance of extracellular fluids, such as blood. Additionally, as you will soon see, the respiratory mechanism and the renal mechanism directly influence the bicarbonate buffer system.

$$CO_2 + H_2O \leftrightarrow H_2CO_3 \leftrightarrow HCO_3^- + H^+$$

The **phosphate buffer system** relies on a mixture of HPO_4^- and $H_2PO_4^-$. The phosphate buffer system is most important in regulating the acid–base balance of intracellular fluid. The following reaction can proceed to the right to release H^+ and decrease pH, or it can proceed to the left to bind H^+ and increase pH.

$$H_2PO_4^- \leftrightarrow HPO_4^{2-} + H^+$$

The most abundant and powerful buffer system in the body is the **protein buffer system.** Proteins are able to act as buffers because amino acids have acidic (–COOH) side groups, which release H^+ when pH is elevated and decrease the pH, and amino acids also have amine (–NH_2) side groups that bind H^+ when pH is decreased and elevate the pH. Below are the reactions of the protein buffer system.

$$-COOH \rightarrow -COO^- + H^+ \quad \text{and} \quad -NH_2 + H^+ \rightarrow -NH_3$$

Respiratory Mechanism

The respiratory system also plays a significant role in regulating H^+ concentration of body fluids. The respiratory mechanism alters the bicarbonate buffer system by changing the levels of CO_2 in the body. Recall that the production of CO_2 results in the formation of carbonic acid, which dissociates to release H^+. When CO_2 production increases and blood pH decreases, the respiratory rhythmicity center in the medulla oblongata stimulates an increase in the rate and depth of breathing to remove the excess CO_2. Conversely, when the H^+ concentration of the blood is decreased, the rate and depth of breathing are decreased until the blood H^+ concentration increases to normal.

Renal Mechanism

The renal mechanism is able to control the bicarbonate buffer system. By selectively excreting excess H^+ or HCO_3^- in urine, kidneys help to maintain the normal pH of body fluids. The kidneys also have the ability to produce HCO_3^- and H^+ from carbon dioxide in times of shortage.

Check My Understanding

8. How do ADH, aldosterone, PTH, and ANP regulate water and electrolyte concentrations in blood?
9. What are the three buffer systems?
10. How does the kidney adjust blood pH?

16.6 Disorders of the Urinary System

Learning Objective

13. Describe the common disorders of the urinary system.

Inflammatory Disorders

Cystitis (sis-tī′-tis) is the inflammation of the urinary bladder. It is often caused by bacterial infection. Females are more prone to cystitis because their shorter urethra makes it easier for bacteria to reach the urinary bladder.

Glomerulonephritis (glō-mer-ū-lō-ne-frī′-tis) is the inflammation of a kidney involving the glomeruli. It may be caused by bacteria or bacterial toxins. The inflamed glomeruli become more permeable, allowing formed elements and proteins to leak into the glomerular filtrate and remain in the urine.

Pyelonephritis (pī-e-lō-ne-frī′-tis) is the inflammation of the renal pelvis and nephrons. If only the renal pelvis is involved, the condition is called *pyelitis.* These infections result from bacteria carried by blood from other places in the body or by migration of bacteria from distal portions of the urinary tract.

Urethritis is the inflammation of the urethra. It may be caused by several types of bacteria, but the bacterium *Escherichia coli* is the most common. Urethritis is more common in females.

Noninflammatory Disorders

Diuresis, or polyuria, is the excessive production of urine. It results from inadequate tubular reabsorption of water and is characteristic of diabetes insipidus and diabetes mellitus.

Renal calculi (kal′-kū-li), or kidney stones, result from crystallization of uric acid or of calcium or magnesium salts in the renal pelvis. They can cause extreme pain, especially when moving through a ureter by peristalsis. Ultrasound waves can be used to break up the stones, as an alternative to surgery.

Renal failure is characterized by a reduction in urine production and a failure to maintain the normal volume and composition of body fluids. It may occur suddenly (acute) or gradually (chronic). Renal failure leads to *uremia,* a toxic condition caused by excessive nitrogenous wastes in the blood, and ultimately to *anuria,* a cessation of urine production. Hemodialysis and/or a kidney transplant may be necessary.

Chapter Summary

16.1 Functions of the Urinary System

- The functions of the urinary system are the maintenance of blood plasma composition, the secretion of renin and erythropoietin, and the excretion of nitrogenous wastes.
- A major function of the kidneys is the removal of excess nitrogenous wastes--urea, uric acid, and creatinine–in order to keep their concentrations in the blood within normal limits.

16.2 Anatomy of the Kidneys

- The paired kidneys are located against the superior posterior abdominal wall.
- Internal structure of a kidney consists of two recognizable functional parts: a superficial renal cortex, and a deep renal medulla composed of renal pyramids.
- Nephrons are the functional units of the kidneys. Each nephron consists of a renal corpuscle and a renal tubule. A renal corpuscle is composed of a glomerulus and a glomerular capsule. A renal tubule is composed of a proximal convoluted tubule, the nephron loop, and a distal convoluted tubule.
- Each nephron joins with a collecting duct that empties into a minor calyx of the renal pelvis.
- The blood supply for each kidney is provided by a renal artery and drainage is through the renal vein.
- An afferent glomerular arteriole brings blood to a glomerulus. Blood exits the glomerulus via an efferent glomerular arteriole and flows through either the peritubular capillaries, which surround the cortical portion of the renal tubule or vasa recta, which surround the medullary portion of the nephron loop.
- The juxtaglomerular complex consists of modified cells of the afferent and efferent glomerular arterioles and the ascending limb of the nephron loop at their point of contact.

16.3 Urine Formation

- The process of urine formation regulates the composition and volume of blood plasma by removing excess nitrogenous wastes and surplus substances from the blood plasma.
- Urine is formed by four sequential processes: glomerular filtration, tubular reabsorption, tubular secretion, and water conservation.
- In glomerular filtration, water and dissolved substances (except plasma proteins and formed elements) in blood plasma are filtered from the glomerulus into the glomerular capsule. Glomerular filtration results from the increased permeability of glomerular capillaries and the elevated blood pressure within the glomerulus.
- About 180 liters of glomerular filtrate are formed in a 24-hour period.
- Glomerular filtration rate is proportional to the glomerular blood pressure. Glomerular blood pressure is maintained by mechanisms that control the diameters of the afferent and efferent glomerular arterioles.
- Glomerular blood pressure generally varies directly with systemic blood pressure.
- Glomerular filtration rate is regulated by renal autoregulation, sympathetic control, and the renin-angiotensin mechanism.
- In tubular reabsorption, needed substances are reabsorbed back into the blood plasma of the peritubular capillaries and vasa recta by either active or passive transport.
- Positively charged ions are actively reabsorbed. Negatively charged ions are passively reabsorbed by electrochemical attraction to the positively charged ions. Water is passively reabsorbed by osmosis.
- Most tubular reabsorption occurs in the PCT, especially of nutrients such as amino acids and glucose, but other portions of the renal tubule are also involved.
- In tubular secretion, certain substances are actively or passively secreted into the tubular fluid from the blood plasma. Uric acid and hydrogen ions are actively secreted. Potassium ions are secreted both actively and passively. Most tubular secretion occurs in the DCT.
- The nephron loop selectively reabsorbs water in the descending limb and Na^+ and Cl^- in the ascending limb, creating an osmotic gradient in the renal medulla.
- Urine is concentrated by water reabsorption from the collecting duct (water conservation), so it is hypertonic to blood plasma.
- The daily production of urine is 1.5 to 2.0 liters. Normal urine is a clear, pale yellow to amber fluid with a characteristic odor. The color is due to the presence of urochrome.
- Urine is usually slightly acidic but the pH may range from 4.8 to 7.5.
- Urine is heavier than water due to the dissolved substances that it contains.
- Abnormal substances that may be in urine are glucose, proteins, formed elements, hemoglobin, and bile pigments.

16.4 Excretion of Urine

- A ureter is a slender tube that carries urine by peristalsis into the urinary bladder.
- Urine is temporarily held in the urinary bladder. The urinary bladder is located posterior to the pubic symphysis in the pelvic cavity.
- The wall of the urinary bladder consists of the mucosa, submucosa, muscular layer, and dense irregular connective tissue. The parietal peritoneum covers only its superior surface. The muscular layer consists of smooth muscle cells forming the detrusor.
- A thickening of the detrusor at the urinary bladder-urethra junction forms the internal urethral sphincter. The external urethral sphincter in both genders is formed of skeletal muscle fibers in the floor of the pelvis.
- Micturition is the process of voiding urine from the urinary bladder. Urine is expelled from the urinary bladder through the urethra.
- When the urinary bladder contains 200 to 400 ml of urine, the micturition reflex is triggered, causing the detrusor to contract rhythmically. Continued contractions open the involuntarily controlled internal urethral sphincter. If the voluntarily controlled external urethral sphincter is relaxed, micturition occurs. If not, micturition is postponed.

16.5 Maintenance of Blood Plasma Composition

- The prime function of the kidneys is to maintain the volume and composition of the blood plasma in spite of variations in diet and metabolic processes.
- Water intake must equal water loss. Most water is lost in urine, but other avenues include exhaled air, perspiration, and feces.
- The volume of water lost in urine is decreased when water loss via other means is increased, and vice versa.
- Antidiuretic hormone, which is released from the posterior lobe of the pituitary gland, increases the permeability of DCT and collecting ducts to water and thereby promotes water reabsorption by osmosis.
- Electrolyte intake must replace electrolyte loss. Electrolytes are conserved largely by the active reabsorption of positively charged ions that passively pull along negatively charged ions by electrochemical attraction.
- Aldosterone regulates the blood concentration of sodium and potassium ions by stimulating the active reabsorption of sodium ions from and the active secretion of potassium ions into the DCT.
- Atrial natriuretic peptide promotes the excretion of sodium ions from the DCT and collecting duct.
- The concentration of calcium ions in the blood is regulated by the actions of two hormones acting in the DCT.
- Parathyroid hormone stimulates the reabsorption of calcium from bones into the blood and the reabsorption of calcium ions by the kidneys. Active vitamin D assists PTH and also increases the absorption of calcium by the small intestine. In contrast, calcitonin promotes the deposition of calcium in bones.
- The maintenance of the blood pH between 7.35 and 7.45 includes three major mechanisms: buffers in the blood either combine with or release hydrogen ions as needed; carbon dioxide is removed by the lungs; and renal tubules regulate the rate of hydrogen and bicarbonate ions secreted into the tubular fluid.
- Arterial blood pH less than 7.35 is called acidosis. Arterial blood pH greater than 7.45 is called alkalosis.

16.6 Disorders of the Urinary System

- Inflammatory disorders include cystitis, glomerulonephritis, pyelonephritis, and urethritis.
- Noninflammatory disorders include diuresis, renal calculi, and renal failure.

Self-Review

Answers are located in appendix B.

1. Renal corpuscles, proximal convoluted tubules, and distal convoluted tubules are located in the ______ of a kidney.
2. Renal pyramids are located in the ______ of a kidney and are composed mostly of ______, which collect urine from the nephrons.
3. The functional units of the kidneys are called ______.
4. The force powering glomerular filtration is ______ blood pressure, which in turn is primarily determined by the ______ blood pressure.
5. In glomerular filtration, water and dissolved substances are forced from blood plasma in the glomerulus into the ______, where the fluid is called ______.
6. In tubular reabsorption in the PCT, certain substances pass from the ______ in the renal tubules into blood plasma in the ______.
7. Glucose, amino acids, and positively charged ions are ______ reabsorbed, while water is ______ reabsorbed by ______.
8. Fluid in the collecting duct is called ______.
9. A severe drop in blood pressure causes the juxtaglomerular complex to secrete ______, which triggers a mechanism to raise blood pressure by chemical means.
10. ______ hormone promotes the reabsorption of water by making the DCT and ______ more permeable to water.
11. The hormone ______ promotes the reabsorption of sodium ions and secretion of ______ ions.
12. Water lost in urine composes about ______% of the total water loss from the body.
13. The ______ carry urine to the urinary bladder by ______.
14. Urine is voided from the urinary bladder through the ______.
15. The ______ urethral sphincter is involuntarily controlled, while the ______ sphincter is voluntarily controlled.

Critical Thinking

1. Predict the consequences of a problem with the granular cells.
2. How would perspiring heavily on a hot day affect a person's urine production?
3. How does kidney failure affect RBC production?
4. Why might kidney failure lead to hyperventilation?
5. Why does kidney failure usually lead to cardiovascular disorders, and vice versa?

ADDITIONAL RESOURCES

17

CHAPTER

Reproductive Systems

Out of all of the organ systems, the reproductive system is the only one that is not essential for the survival of an individual. However, it is not insignificant in its ability to influence homeostasis. Consider Debbie and Patrick, fraternal twins born in Pennsylvania. From birth through childhood, their physical attributes were quite similar aside from their external genitalia. The general shape to their torsos and limbs showed no significant differences, indicating similar skeletal and muscular development. The voices of both children exhibited the higher pitch common to young children. However, at puberty, these similarities ended. Debbie's pelvis widened and her mammary glands developed. Patrick's skeletal mass increased greatly to match the enlargement of his musculature, resulting in a broadening of his shoulders and chest. The similarities in their voices ended as Patrick's larynx enlarged, resulting in a decrease in vocal pitch. These visible and audible changes matched the internal changes designed to prepare the children for reproduction as adults. All of these changes in form and function are the product of reproductive system activity. Though not essential for personal survival, this system clearly is needed for the survival of the human race.

CHAPTER OUTLINE

Module 14
Reproductive System

SELECTED KEY TERMS

Androgens Generic term for hormones related to testosterone.
Estrogens The female sex hormones.
Gamete A sex cell, either a sperm or an ovum.
Gonads (gone = seed) The primary sex glands–the ovaries and testes.
Meiosis (mei = less) A form of cell division in which the daughter cells contain one-half the number of chromosomes as the parent cell.
Menopause (men = month; paus = stop) The cessation of monthly female reproductive cycles.
Menstrual cycle The repetitive monthly changes in the endometrium.
Oogenesis (oo = ovum, egg; genesis = origin) The process of ova formation.
Ovarian cycle The repetitive monthly changes in the ovary.
Ovulation The release of a secondary oocyte from an ovary.
Puberty (puber = grown up) The age at which reproductive organs mature.
Semen (semin = seed) Fluid composed of sperm and secretions of male accessory glands.
Spermatogenesis The process of sperm formation in the testes.

REPRODUCTION IS THE PROCESS by which life is sustained from one generation to the next. The human male and female reproductive systems are specially adapted for their roles in reproduction. The **gonads** (gō′-nads)–the ovaries and testes–form the **gametes** (sex cells). Other reproductive organs nurture male and female sex cells or transport them to sites where they may unite. After fertilization occurs, an ovum develops within the female reproductive system and culminates in the birth of an infant. Sexual maturation and the development of sex cells in both sexes and pregnancy in females are regulated by hormones secreted by the pituitary gland and the gonads.

17.1 Male Reproductive System

Learning Objectives

1. Describe the locations and functions of the male reproductive organs.
2. Describe spermatogenesis.
3. Describe the sources, contents, chemical characteristics, and functions of semen.

The primary functions of the male reproductive system are the production of male sex hormones, the formation of **sperm,** and the placement of sperm in the female reproductive tract, where one sperm can unite with a female sex cell. The organs of the male reproductive system include (1) paired testes, which produce sperm and male sex hormones; (2) accessory ducts that store and transport sperm; (3) accessory glands, whose secretions form part of the semen; and (4) external genitalia, including the scrotum and penis (figure 17.1).

Testes

The paired **testes** (tes′-tēz, singular, *testis*) are the male gonads, or sex glands. Each testis is protected and supported by a capsule of dense irregular connective tissue. Septa (partitions) of connective tissue radiate into the testis from its posterior surface, dividing the testis into internal subdivisions called *lobules.* Each lobule contains several highly coiled **seminiferous** (se-mi-nif′-er-us) **tubules.** Seminiferous tubules are lined with **spermatogenic epithelium,** which is formed of spermatogenic cells and supporting cells. *Spermatogenic cells* divide to produce sperm, while *supporting cells* support and nourish the spermatogenic cells and help regulate sperm formation. The cells that fill the spaces between the seminiferous tubules are known as **interstitial** (in-ter-stish′-al) **cells,** and they produce male sex hormones (figure 17.2*a, b*).

Spermatogenesis

Spermatogenesis (sper-mah-tō-jen′-e-sis) is the process that produces sperm by the division of the spermatogenic cells in the spermatogenic epithelium. Spermatogenesis begins at **puberty** (pū′-ber-tē), the age at which reproductive organs mature, and continues throughout the life of a male. Sexual maturity and sperm production are controlled by follicle-stimulating hormone (FSH) and luteinizing hormone (LH) from the anterior lobe of the pituitary and by testosterone from interstitial cells of the testes. LH is often called interstitial cell–stimulating hormone (ICSH) in males. Hormonal relationships are discussed later in this chapter.

The large, superficial cells near the basement membrane of a seminiferous tubule are known as **spermatogonia** (sper-mah-to-gō′-nē-ah, singular, *spermatogonium*). Each spermatogonium contains 46 chromosomes (23 pairs), the normal number of chromosomes for human body cells. Each spermatogonium divides by mitosis to produce two spermatogonia, referred to as type A and type B spermatogonia, each with 46 chromosomes. The type A spermatogonium remains next to the basement membrane of the tubule. It

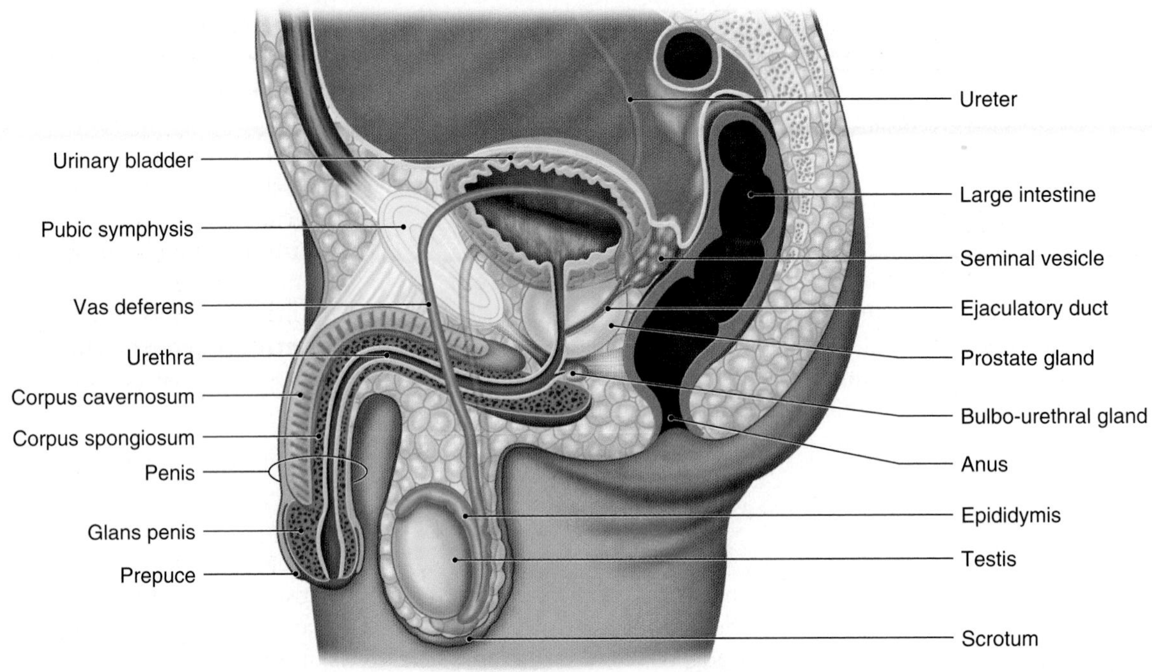

Figure 17.1 Male Reproductive System in a Median Section. APIR

will serve as the "stem" spermatogonium and will divide repeatedly by mitosis. The type B spermatogonium that is pushed toward the lumen of the tubule undergoes changes to become a **primary spermatocyte** (figure 17.2*c*).

Primary spermatocytes divide by **meiosis** (mi-ō′-sis), a special type of cell division. Meiosis requires two successive divisions and reduces the number of chromosomes in the daughter cells by one-half.

Meiotic divisions in spermatogenesis are summarized in figures 17.2*c* and 17.3. Each primary spermatocyte, containing 46 chromosomes, divides in meiosis I to form two **secondary spermatocytes,** each containing 23 chromosomes. Prior to meiosis I, the chromosomes replicate. Each replicated chromosome is composed of two *chromatids* joined together at a region called a *centromere.* During metaphase of meiosis I, the replicated chromosomes are arranged as homologous pairs. During cytokinesis, the members of each chromosome pair are separated into different daughter cells. Thus, each secondary spermatocyte contains 23 replicated chromosomes (figure 17.3).

The genetic diversity of the sperm is created by the following two mechanisms. First, *random alignment* of the paired homologous chromosomes on the cellular equator occurs during meiosis I, so that the daughter cells contain different combinations of maternal and paternal chromosomes. Second, *crossover,* the exchange of some DNA between the paired homologous chromosomes, occurs during meiosis I, so that some chromosomes contain genes from both parents.

In meiosis II, the chromatids separate into different daughter cells so each secondary spermatocyte divides to form two **spermatids,** each containing 23 chromosomes. Each spermatid attaches to a supporting cell, gradually loses much of its cytoplasm, and develops a flagellum to form a sperm containing 23 chromosomes.

Examine figure 17.2*c* and note how cells in stages of spermatogenesis are arranged in sequence from superficial to deep, with sperm located in the lumen of the tubule. Once the sperm are completely formed, they are carried into the epididymis, where they are temporarily stored while they mature.

A mature sperm consists of a head, neck, and flagellum. The flattened *head* is composed of a compact nucleus containing 23 chromosomes. The anterior portion of the head is covered by a caplike structure, the *acrosome.* The acrosome contains enzymes that help the sperm penetrate a female sex cell. The *neck* connects the head to the flagellum. The *flagellum* has a middle piece, a principal piece, and an end piece. The middle piece contains mitochondria, where ATP is formed to power the movements of the flagellum, which enables movement (figure 17.4).

Accessory Ducts

Sperm pass through a series of accessory ducts as they are carried from the testes to the external environment. These accessory ducts include the epididymis, vas deferens, ejaculatory duct, and urethra. These structures are collectively referred to as the *male reproductive tract.*

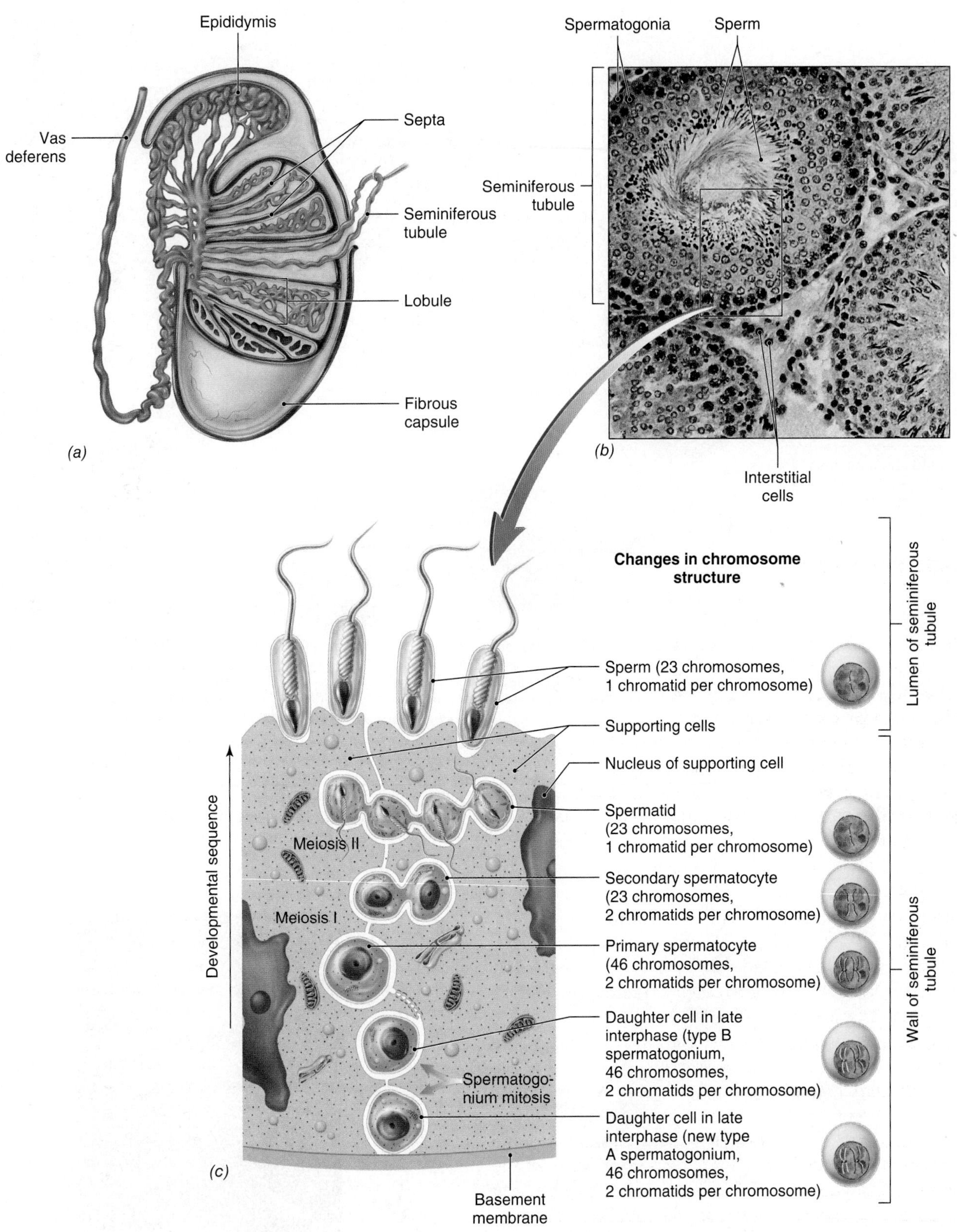

Figure 17.2 (*a*) Testis in sagittal section; (*b*) a photomicrograph of seminiferous tubules in cross section (200×); (*c*) a diagram showing the sequence of stages in spermatogenesis. Note that cells in developmental stages are embedded in supporting cells until sperm are released into the lumen of a seminiferous tubule. APR

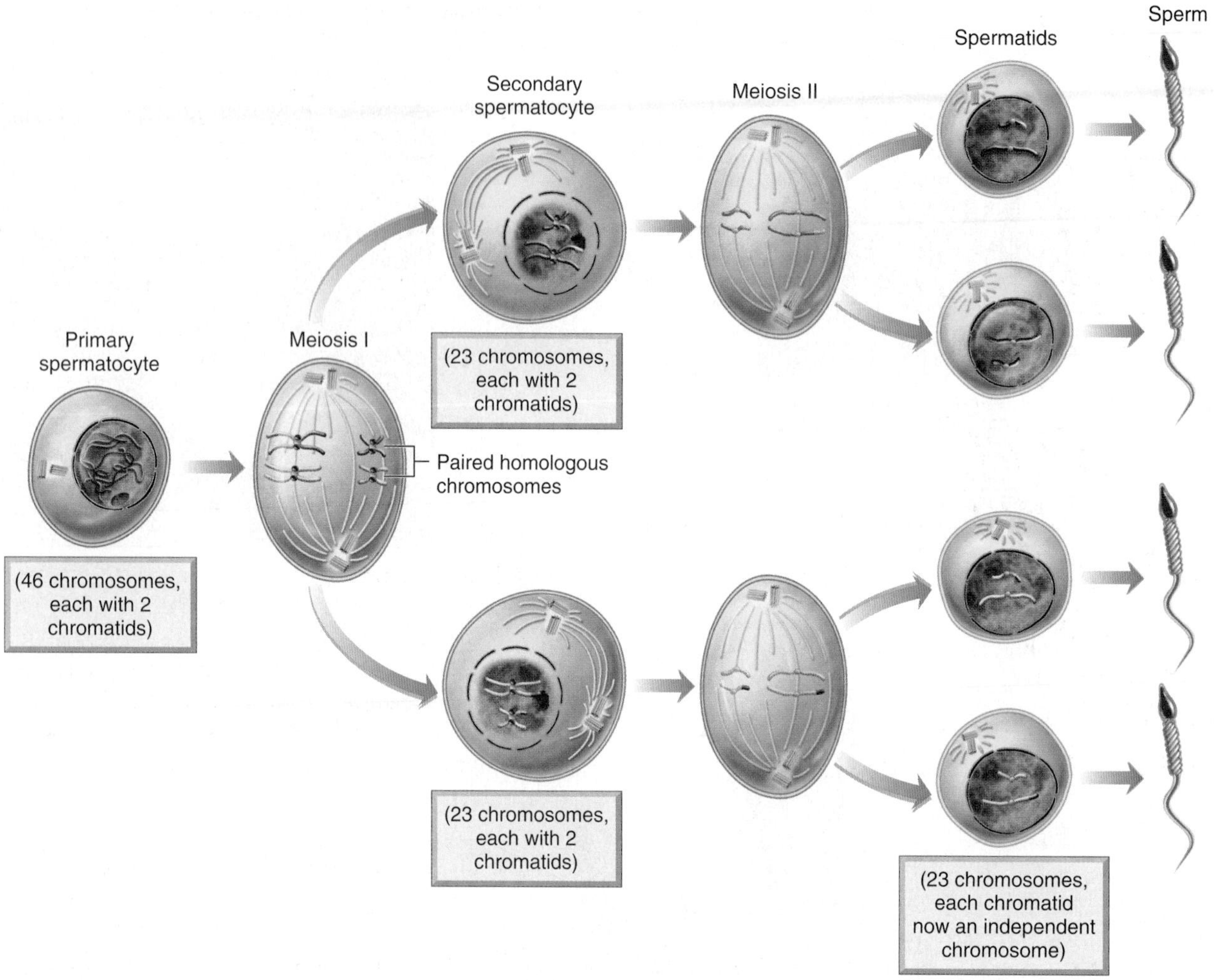

Sperm
Spermatids
Secondary spermatocyte
Meiosis II
Primary spermatocyte
Meiosis I
(23 chromosomes, each with 2 chromatids)
Paired homologous chromosomes
(46 chromosomes, each with 2 chromatids)
(23 chromosomes, each with 2 chromatids)
(23 chromosomes, each chromatid now an independent chromosome)

Figure 17.3 Spermatogenesis requires two successive meiotic divisions plus the development of spermatids into sperm. APR

Acrosome
Nucleus
Head
Neck
Middle piece
Flagellum
Principal piece
End piece
(a)
(b)

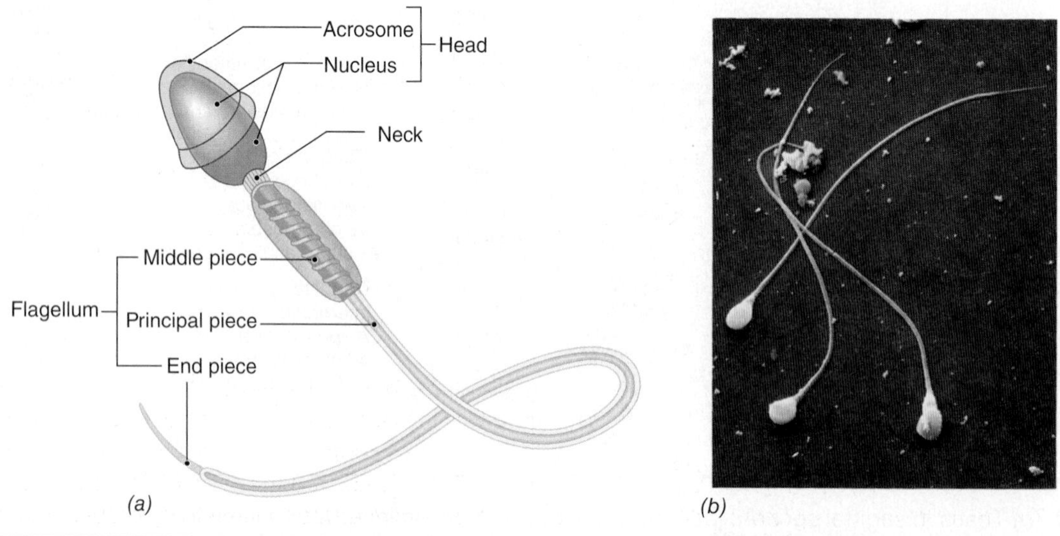

Figure 17.4 (*a*) Parts of a mature sperm. (*b*) Scanning electron photomicrograph of human sperm (1,400×).

Epididymis

The seminiferous tubules of a testis lead to a number of small ducts that open into the epididymis. The **epididymis** (ep-i-did′-i-mis, plural, *epididymides*) appears as a comma-shaped organ that lies along the superior and posterior margins of a testis (figure 17.5; see figure 17.1). Upon close examination, the epididymis is shown to be a long (6 m), tightly coiled, slender tube that is continuous with the vas deferens (see figure 17.2*a*).

Sperm mature as they are slowly moved (10–14 days) through the epididymis by weak peristaltic contractions. The mature sperm are stored in the epididymis until they are ejaculated. The sperm stored for more than two months are destroyed and absorbed by the epididymis.

Vas Deferens

As shown in figures 17.1 and 17.5, a **vas deferens** (vas def′-er-enz, plural, *vasa deferentia*) extends from the epididymis superiorly in the scrotum, passes through the inguinal canal, and enters the pelvic cavity. It runs along the lateral surface of the urinary bladder and merges with the duct from a seminal vesicle inferior to the urinary bladder. The duct formed by this merger is an ejaculatory duct. The vasa deferentia have rather thick, muscular walls that move the sperm by peristalsis.

Ejaculatory Duct

Each short **ejaculatory duct** is formed by the merger of a vas deferens and a duct from a seminal vesicle. The ejaculatory ducts enter the prostate gland and merge with the urethra within the prostate (figure 17.5; see figure 17.1). During ejaculation, muscular contractions of the ejaculatory ducts mix seminal vesicle secretions with sperm and propel them into the urethra.

Urethra

The **urethra** is a thin-walled tube that extends from the urinary bladder through the penis to the external environment (figure 17.5; see figure 17.1). The urethra serves a dual role in the male. It transports urine from the urinary bladder during micturition, as noted in chapter 16, and it also carries semen, which includes sperm, during ejaculation. Control mechanisms prevent urine and semen passing at the same time.

Accessory Glands

Three different types of exocrine glands produce secretions involved in the reproductive process. These glands are the seminal vesicles, prostate gland, and bulbo-urethral glands (figure 17.5; see figure 17.1).

Seminal Vesicles

The **seminal vesicles** are paired glands located on the posterior surface of the urinary bladder. The duct of each seminal vesicle merges with the vas deferens on the same side to form an ejaculatory duct near the posterior surface on the prostate gland. The alkaline secretions of the seminal vesicles help to keep semen alkaline and contain fructose and prostaglandins. Secretions by the seminal vesicles compose about 60% of semen.

Prostate Gland

The **prostate gland** is a pear-shaped gland that encircles the urethra where it exits the urinary bladder. The ejaculatory ducts pass through the posterior portion of the prostate to join with the urethra within the prostate. Prostatic fluid is forced through 20 to 30 tiny ducts into the urethra during ejaculation. The secretion is an alkaline, milky fluid containing substances that activate the swimming movements of sperm. It forms about 30% of semen.

Bulbo-urethral Glands

The **bulbo-urethral** (bul-bō-ū-rē′-thral) **glands** are two small, spherical glands that are located inferior to the prostate gland near the base of the penis (figure 17.5; see figure 17.1). These glands secrete an alkaline, mucuslike fluid into the urethra in response to sexual stimulation. This secretion neutralizes the acidity of the urethra and lubricates the end of the penis in preparation for sexual intercourse.

Semen

The **semen** (sē′-men) is the fluid passed from the urethra during ejaculation. It consists of the fluids secreted by the bulbo-urethral glands, seminal vesicles, and prostate gland along with sperm and fluid from the testes. The alkalinity (pH 7.5) of semen protects the sperm by neutralizing the acidity of the male's urethra and the female's vagina. Fructose from seminal vesicles provides the nutrient energy for sperm, and prostatic fluid activates their swimming movements. After semen is deposited in the vagina during sexual intercourse, prostaglandins in seminal vesicle secretions stimulate reverse peristalsis of the uterus and uterine tubes, which accelerates the movement of sperm through the female reproductive tract. The volume of semen in a single ejaculation may vary from 2 to 5 ml, with 50 to 150 million sperm per milliliter. Although only one sperm participates in fertilization, many sperm are necessary for fertilization to occur.

Clinical Insight

Male infertility may be caused by a number of factors, such as hormone imbalances, duct blockage, a low sperm count, abnormal sperm, or a low fructose concentration in the semen. The minimum sperm count for male fertility is considered to be at least 20 million sperm per milliliter of semen.

17.2 Male Sexual Response

Learning Objective

4. Describe the male sexual response.

In the absence of sexual stimulation, the vascular sinusoids in the erectile tissue of the penis contain a small amount of blood and the penis is flaccid (flak′-sid), or soft. Sexual stimulation initiates parasympathetic nerve impulses that cause the dilation of the arterioles and constriction of the venules supplying the erectile tissue. These vascular changes cause the erectile tissue to become engorged with blood, which produces **erection,** a condition in which the penis swells, lengthens, and becomes erect. The parasympathetic nerve impulses also stimulate the secretion from the bulbo-urethral glands.

Continued sexual stimulation of the glans, as in sexual intercourse, culminates in **orgasm,** which is characterized by **ejaculation** (ē-jak-ū-lā′-shun) of semen and a feeling of intense pleasure. Just prior to ejaculation, sympathetic nerve impulses stimulate peristaltic contractions of the epididymides, vasa deferentia, and ejaculatory ducts along with contractions of the seminal vesicles and prostate gland. These contractions force semen into the urethra. Then, ejaculation occurs as certain skeletal muscles contract rhythmically from the proximal penis distally, forcing semen through the urethra to the external environment. A feeling of general relaxation follows.

Immediately after ejaculation, the vascular changes that produced erection are reversed. Sympathetic nerve impulses cause the constriction of the arterioles and dilation of the venules supplying the erectile tissue, allowing the accumulated blood to leave the penis. The penis becomes flaccid as blood is carried away. After orgasm, erection is not possible for a time period that varies from less than an hour to several hours.

Check My Understanding

5. What division of the autonomic nervous system stimulates peristalsis in the male reproductive tract?

17.3 Hormonal Control of Reproduction in Males

Learning Objectives

5. Describe the actions of testosterone.
6. Describe the hormonal control of testosterone secretion.
7. Describe the hormonal control of spermatogenesis.

The onset of male sexual development begins around the ages of 11 or 12 and is completed by ages 15 to 17. The mechanisms initiating the onset of puberty are not well understood, but the sequence of events is known. Hormones of the hypothalamus, anterior lobe of the pituitary gland, and testes are involved. Melatonin from the pineal gland may also be involved.

Puberty begins when unknown stimuli trigger the hypothalamus to secrete **gonadotropin-releasing hormone (GnRH),** which is carried to the anterior lobe of the pituitary gland by the blood. GnRH activates the release of **luteinizing hormone (LH)** and **follicle-stimulating hormone (FSH)** from the anterior lobe of the pituitary gland. LH promotes growth of the interstitial cells of the testes and stimulates their secretion of **testosterone** (tes-tos′-te-rōn), the primary male sex hormone. FSH and testosterone in combination act on the seminiferous tubules, stimulating spermatogenesis (figure 17.6).

Action of Testosterone

Male sex hormones are collectively called **androgens** (an′-dro-jens), and testosterone is the most important one. They are produced primarily by the testes. Testosterone secretion starts in fetal development, continues for a brief time after birth, and then nearly ceases until puberty.

At puberty, testosterone stimulates the maturation of the male reproductive organs, the continuation of spermatogenesis, and the development of the male secondary sex characteristics. **Secondary sex characteristics** are the physical features that distinguish sexes from each other. Male secondary sex characteristics include (1) growth of body hair, especially on the face, axillary, and pubic regions; (2) increased muscular development; (3) development of heavy bones, broad shoulders, and narrow pelvis; and (4) deepening of the voice due to enlargement of the larynx and thickening of the vocal folds. Less obvious effects are increases in the rate of cellular metabolism and RBC production. Both metabolic rate and the concentration of RBCs in the blood are greater in males than in females. Testosterone production starts to decline at about 40 years of age, resulting in a gradual decline in the functions of reproductive organs and secondary sex characteristics.

Regulation of Male Sex Hormone Secretion

Like many other hormones, testosterone production is controlled by a negative-feedback mechanism, as shown in figure 17.6. Note that the secretion of GnRH starts the feedback mechanism. After puberty, this mechanism maintains the testosterone level in the blood within normal limits.

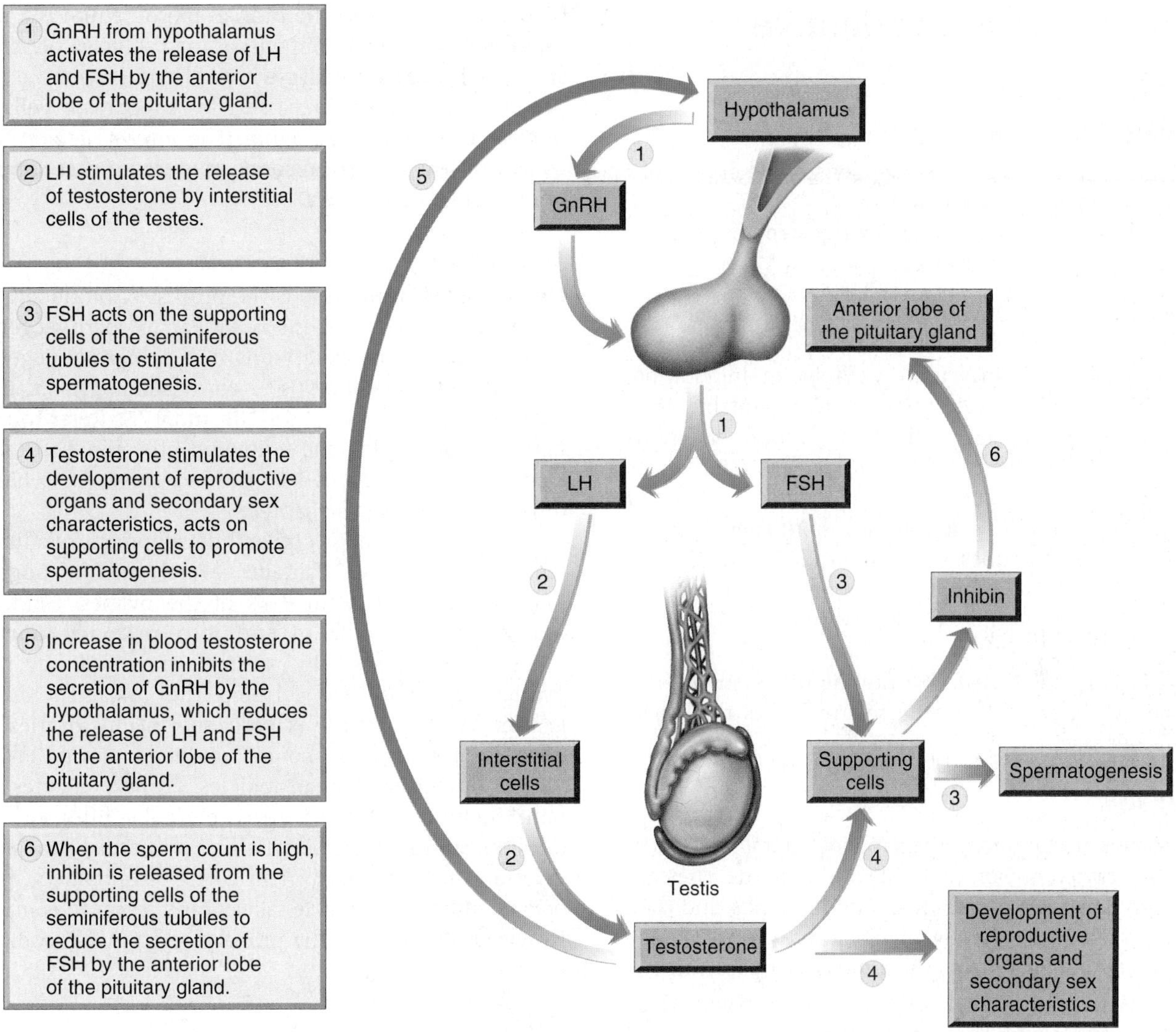

1 GnRH from hypothalamus activates the release of LH and FSH by the anterior lobe of the pituitary gland.

2 LH stimulates the release of testosterone by interstitial cells of the testes.

3 FSH acts on the supporting cells of the seminiferous tubules to stimulate spermatogenesis.

4 Testosterone stimulates the development of reproductive organs and secondary sex characteristics, acts on supporting cells to promote spermatogenesis.

5 Increase in blood testosterone concentration inhibits the secretion of GnRH by the hypothalamus, which reduces the release of LH and FSH by the anterior lobe of the pituitary gland.

6 When the sperm count is high, inhibin is released from the supporting cells of the seminiferous tubules to reduce the secretion of FSH by the anterior lobe of the pituitary gland.

Hypothalamus
GnRH
Anterior lobe of the pituitary gland
LH
FSH
Inhibin
Interstitial cells
Testis
Supporting cells
Spermatogenesis
Testosterone
Development of reproductive organs and secondary sex characteristics

Figure 17.6 Hormonal Control of Spermatogenesis and Testosterone Secretion.

As testosterone concentration increases in the blood, it inhibits the production of GnRH by the hypothalamus, which, in turn, reduces the release of LH and FSH from the anterior lobe of the pituitary gland. The decrease in LH production causes a decrease in testosterone secretion by the interstitial cells of the testes, which reduces testosterone concentration in the blood. The reduction in FSH decreases sperm production.

Conversely, as testosterone concentration in the blood declines, the hypothalamus is stimulated to secrete GnRH, which, in turn, promotes the release of LH and FSH by the anterior lobe of the pituitary. The increase in LH production causes an increase in testosterone secretion by the testes, which increases the testosterone concentration in the blood. The increase in FSH increases sperm production.

Testosterone maintains the male reproductive organs and spermatogenesis. However, sperm production is fine-tuned by the hormone **inhibin,** which is secreted by supporting cells within the seminiferous tubules. When the sperm count is high, secretion of inhibin is increased, which decreases sperm production by decreasing FSH secretion by the anterior lobe of the pituitary. When the sperm count is low, inhibin secretion decreases, and sperm production is increased by an increase in FSH secretion. Thus, inhibin works to keep sperm concentration in semen within normal limits without altering the secretion of LH and testosterone.

Check My Understanding

6. How is sperm production regulated?
7. What are the functions of testosterone?
8. How is testosterone secretion regulated?

17.4 Female Reproductive System

Learning Objectives

8. Describe the location and functions of the organs of the female reproductive system.
9. Describe oogenesis and the development of the corresponding ovarian follicles.

The female reproductive system produces female sex hormones and female sex cells and transports the sex cells to a site where they may unite with sperm. In addition, the female reproductive system provides a suitable environment for the development of the embryo and fetus and is actively involved in the birth process. The organs of the female reproductive system include the paired ovaries, which produce female sex hormones and sex cells; paired uterine tubes, which transport the female sex cells; a uterus, where internal development of the embryo and fetus occurs; a vagina, which serves as the female copulatory organ and birth canal; and the external genitalia (figure 17.7; figure 17.10). The uterine tubes, the uterus, and the vagina are collectively referred to as the *female reproductive tract*.

Ovaries

The **ovaries** are located in the superior, lateral portions of the pelvic cavity, one on each side of the uterus. They are about the same size and shape as large almonds, and they are supported by several ligaments. The external surface of an ovary is covered by the **ovarian mesothelium** (figure 17.8). Deep to the ovarian mesothelium is a more dense region, which contains ovarian follicles of various sizes. Each **ovarian follicle** is a spherical structure composed of an oocyte enveloped by supporting cells. An **oocyte** is an immature ovum. The internal region of an ovary consists of areolar connective tissue that supports nerves and blood vessels.

Oogenesis

Oogenesis (ō-ō-jen′-e-sis) is the process of producing **ova** (singular, *ovum*). Oogenesis is similar to spermatogenesis with a few notable exceptions. The process of oogenesis and the development of the corresponding ovarian follicles are shown in figures 17.8*b* and 17.9. Refer to these figures as you study the following process of oogenesis. Oogenesis is closely tied to the ovarian cycle, which is discussed later in this chapter.

By the fifth month of fetal development of a human female, the ovaries contain several million **oogonia** (ō-ō-gō-nē-ah), the stem cells of the ovaries. Gradually, most of the oogonia mature into **primary oocytes** (ō-ō-sītz) surrounded by a single layer of squamous *follicular epithelial cells* forming the *primordial ovarian follicles*. When a female is born, the infant's ovaries lack oogonia and contain about 2 million primary oocytes within primordial ovarian follicles. Primary oocytes contain 46 chromosomes, so they are capable of meiotic cell division. Starting at puberty, a few primordial ovarian follicles are activated each month. The development of the oocyte and ovarian follicle takes nearly a year to complete. However, only one of the primary oocytes will complete meiosis I each month.

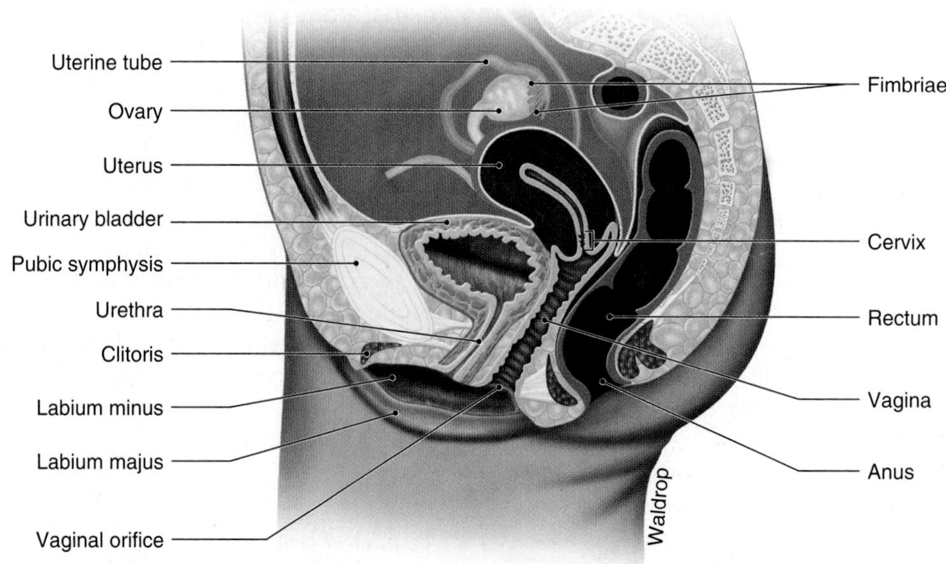

Figure 17.7 Female Reproductive System in a Median Section. APR

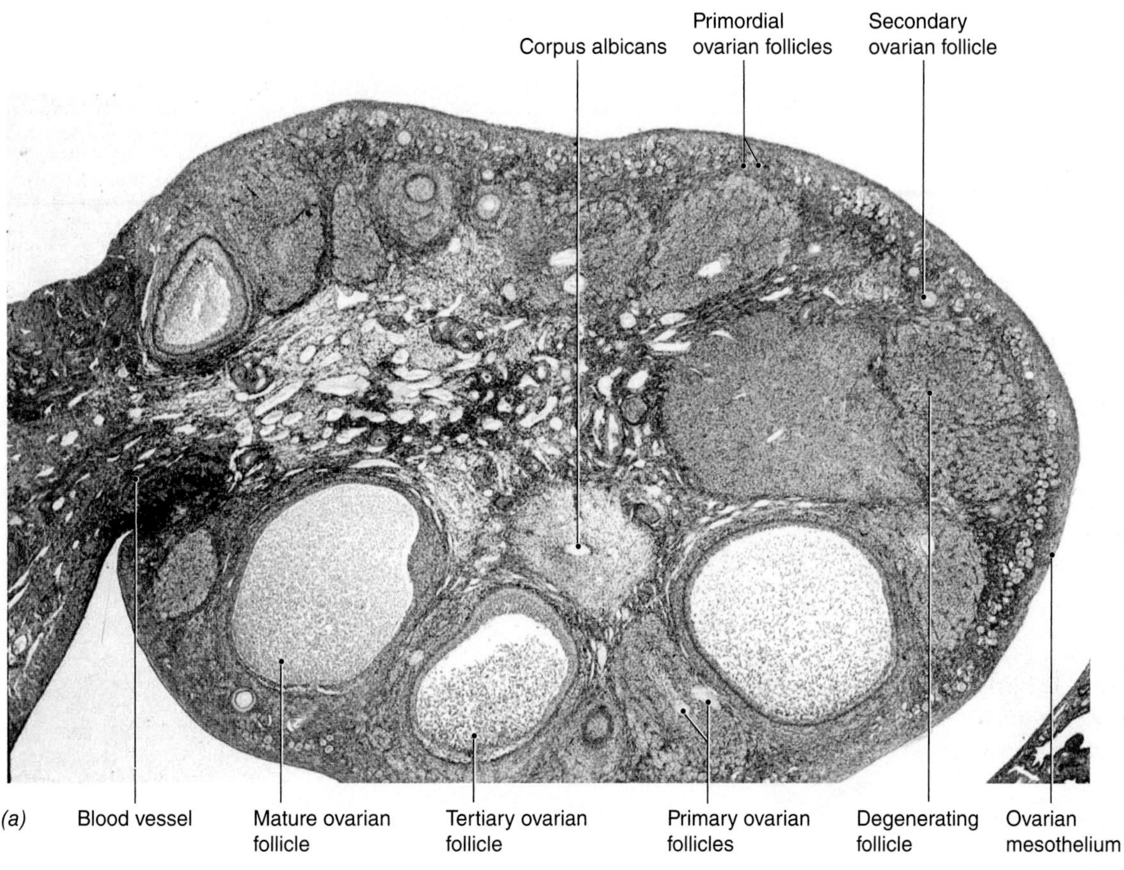

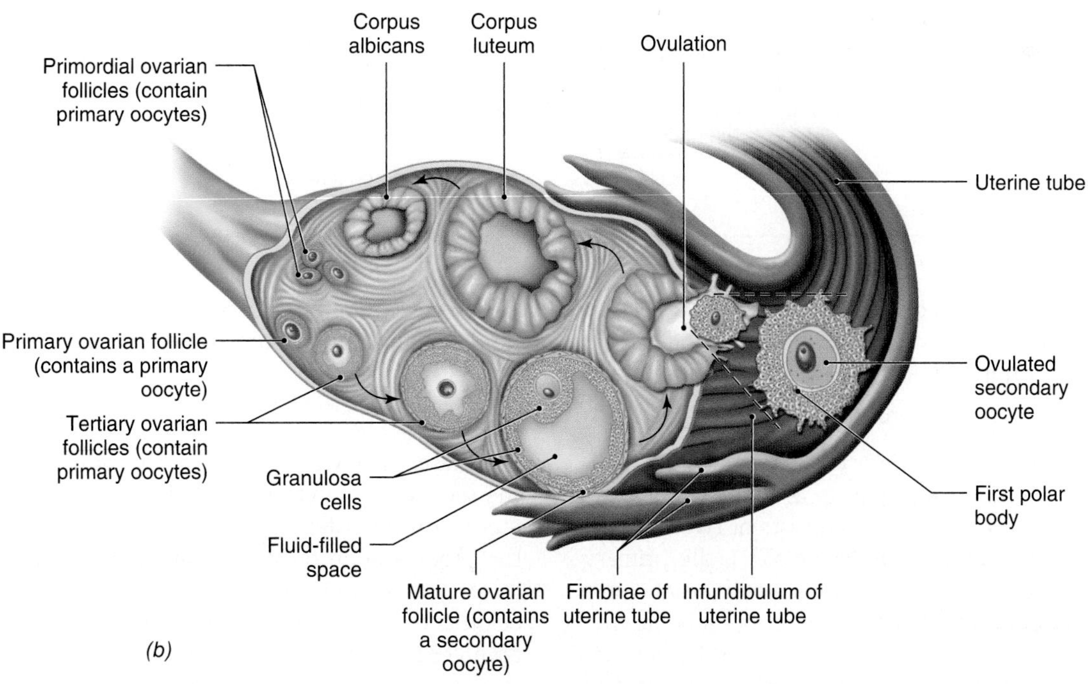

Figure 17.8 (*a*) Photomicrograph of an ovary (30×); (*b*) Stages of ovarian follicular development in an ovary, release of a secondary oocyte at ovulation, and formation of a corpus luteum. AP|R

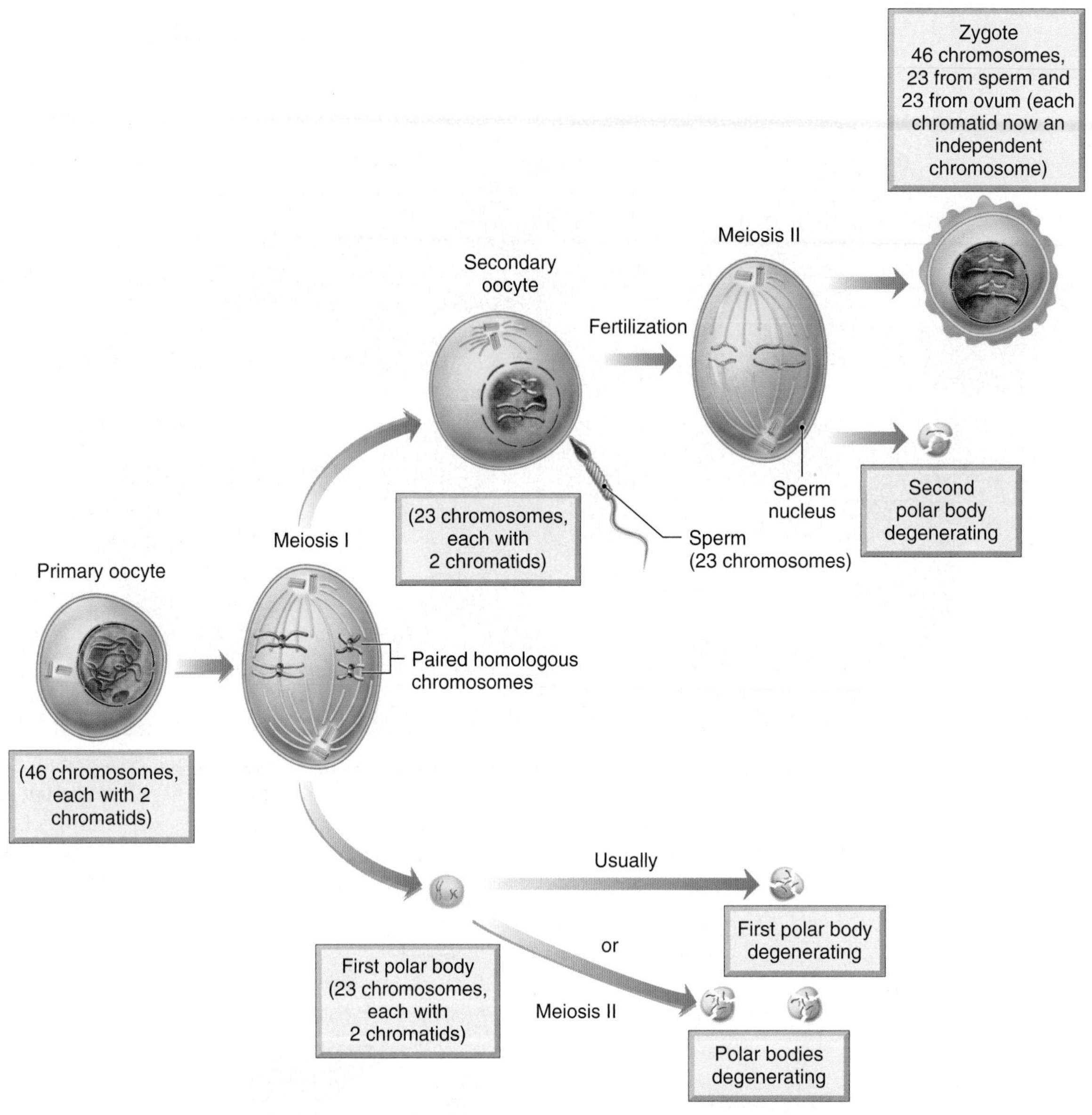

Figure 17.9 Oogenesis.
In meiosis I, the primary oocyte divides to form a secondary oocyte and a polar body. If the secondary oocyte is penetrated by a sperm, it divides in meiosis II to form an ovum and a second polar body. Union of ovum nucleus and sperm nucleus forms a zygote. Polar bodies degenerate.

As shown in Figure 17.9, chromosomes are replicated prior to meiosis I and line up as homologous pairs at metaphase. Each chromosome consists of two chromatids joined by the centromere. In meiosis I, the primary oocyte divides to form two dissimilar cells, each with only 23 chromosomes. Only one member of each chromosome pair is distributed to each daughter cell. The smaller cell, the first **polar body,** contains almost no cytoplasm and performs no further function. It is formed as a convenient way to get rid of 23 chromosomes. The larger cell is the **secondary oocyte,** and it contains nearly all of the cytoplasm that was present in the primary oocyte. It is the secondary oocyte that is released from an ovary each month during a female's reproductive life. **Ovulation** is the process of an ovary releasing a secondary oocyte. As in spermatogenesis, the genetic diversity of the secondary oocytes results from the random alignment of the paired homologous chromosomes and the crossover between the paired homologous chromosomes during meiosis I.

Meiosis II is completed only if the secondary oocyte is penetrated by a sperm. Meiosis II forms another polar body and an ovum, which receives nearly all of the cytoplasm.

The chromatids of each chromosome separate, and they are distributed into different daughter cells. Therefore, both ovum and second polar body contain 23 chromosomes. The fusion of ovum nucleus and sperm nucleus forms a **zygote,** the first cell of a preembryo. The polar bodies simply degenerate.

Uterine Tubes

The paired **uterine tubes** receive and transport the secondary oocyte, are the sites of fertilization, and transport the preembryo if fertilization occurs. Each uterine tube extends laterally from the superior lateral surface of the uterus to an ovary. The lateral end of each tube forms a funnel-shaped expansion, the *infundibulum* (in-fun-dib′-ū-lum), that partially envelops an ovary but is not connected to it. Each infundibulum is subdivided into a number of fingerlike processes called *fimbriae* (fim′-brē-ē) that may touch the ovary (figure 17.10).

The internal lining of a uterine tube consists of simple ciliated columnar epithelium and secretory cells. The beating of the cilia creates a current that helps draw the ovulated secondary oocyte into the infundibulum. Then, the beating cilia and the peristaltic contractions of the uterine tube move the oocyte slowly toward the uterus.

Uterus

The **uterus** (ū′-ter-us) is located in the pelvic cavity posterior to the urinary bladder. As shown in figure 17.7, it is located superior to the vagina and is bent anteriorly over the urinary bladder. The uterus is a hollow organ with thick, muscular walls. Its primary function is to provide an appropriate internal environment for a developing embryo and fetus.

Clinical Insight

Infections of the female reproductive tract can easily migrate through the uterine tubes into the pelvic cavity, where the infections become far more serious.

The uterus has three major regions. The *fundus,* the superior rounded region, joins with the uterine tubes laterally. The *body,* the middle portion, is enlarged and rounded. The *cervix,* the inferior tubular portion, is inserted into the superior end of the vagina.

The walls of the uterus are composed of three layers. The *endometrium* (en-dō-mē′-trē-um), the mucosal layer, is the internal layer of the uterus. The *myometrium* is the thick layer of smooth muscle that forms most of the wall thickness. The *perimetrium,* the serous layer, is the external layer of the uterus (figure 17.10).

Vagina

The **vagina** (vah-jī′-nah) is the collapsible tube that extends from the uterus to the external environment. It serves as the female copulatory organ and the birth canal. The vagina is located posterior to the urethra and anterior to the rectum (figure 17.10; see figure 17.7).

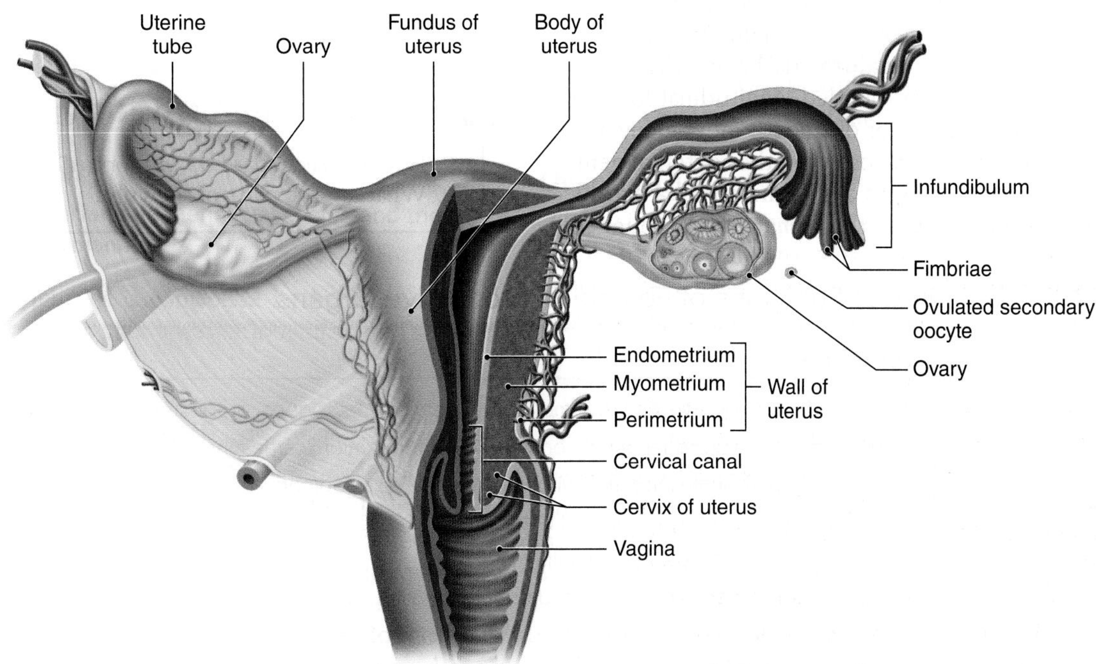

Figure 17.10 Posterior View of the Female Reproductive Organs. APIR

stimulates ovulation, which releases the secondary oocyte surrounded by granulosa cells. Ovulation occurs 14 days before the onset of the next menstruation, regardless of the length of the cycle. In a 28-day ovarian cycle, days 1-14 are known as the *follicular phase.*

After ovulation, LH stimulates the transformation of the remaining granulosa cells into the corpus luteum. The corpus luteum secretes increasing amounts of progesterone and lesser amounts of estrogens during days 15 to 25. The increasing blood levels of progesterone, aided by estrogens released from the corpus luteum, begins a negative-feedback mechanism that inhibits the secretion of GnRH by the hypothalamus. The lack of GnRH, aided by the release of inhibin from the corpus luteum, inhibits the release of LH and FSH by the anterior lobe of the pituitary gland.

If the secondary oocyte is not fertilized, the corpus luteum degenerates into the nonfunctional corpus albicans when blood LH levels decline. Loss of the corpus luteum causes a rapid decline in the blood levels of estrogens and progesterone from days 25 to 28. Days 15-28 of the ovarian cycle are known as the *luteal phase.* Once the low blood level of progesterone no longer inhibits the hypothalamus and allow the secretion of GnRH, the next reproductive cycle begins. If the secondary oocyte is fertilized, the corpus luteum continues to enlarge and produce increasing amounts of progesterone and estrogens, and degenerates by 16 to 20 weeks of development (see chapter 18 for details).

Menstrual Cycle

The menstrual, or uterine, cycle refers to the series of changes in the endometrium that occur each month unless pregnancy occurs. These changes are responses to fluctuating blood levels of estrogens and progesterone. The menstrual cycle has four phases: menstruation, proliferative phase, secretory phase, and premenstrual phase (see figure 17.12).

Menstruation starts a new menstrual cycle. It begins on the first day of the next menstrual cycle and lasts from three to five days.

The *proliferative phase* is characterized by a buildup of the endometrium under stimulation by estrogens, whose concentration in the blood increases as the dominant tertiary ovarian follicle develops. This phase begins at the end of menstruation and ends at ovulation.

Following ovulation, both progesterone and estrogens are produced by the corpus luteum, and they continue the development of the endometrium in the *secretory phase.* Estrogens promote the continued thickening of the endometrium. Progesterone stimulates the formation of blood vessels and glands in the endometrium, preparing it to receive a preembryo. If fertilization of a secondary oocyte does not occur, the *premenstrual phase* begins. The blood levels of estrogens and progesterone drop rapidly, triggering the breakdown of the endometrium. This eventually leads to menstruation on the first day of the next menstrual cycle.

Menopause, the cessation of regular menstrual cycles, usually begins around age 45-55 and can last up to ten years. The onset is usually gradual with menstrual cycles becoming irregular. During this time, a woman can still conceive, so menopause is not considered to be complete until the cycles have not occurred for one year.

Aging of the ovaries is the cause of menopause. There are fewer primary ovarian follicles to respond to FSH and LH from the anterior lobe of the pituitary, and ovulation does not occur. Therefore, the secretion of estrogens and progesterone by the ovaries is greatly curtailed.

The decline in female sex hormones often is accompanied by physical symptoms such as headaches, insomnia, and depression. Hot flashes, caused by sudden and temporary dilation of dermal blood vessels, are perhaps the most common symptom. Hormone replacement therapy (HRT), the administration of estrogens and progesterone, was previously used to treat the unpleasant symptoms of menopause. Its use has been greatly curtailed because studies have shown that it increases the risk of breast cancer, strokes, and heart disease.

Check My Understanding

12. How are FSH, LH, inhibin, estrogens, and progesterone involved in ovarian and menstrual cycles?

17.7 Mammary Glands

Learning Objective

12. Describe the structure of mammary glands and breasts.

Both males and females possess **mammary glands.** The mammary glands of males and immature females are similar. At puberty, estrogens and progesterone stimulate the development of female mammary glands. Estrogens start breast and mammary gland development, and progesterone stimulates the maturation of the mammary glands so that they are capable of secreting milk. Another hormone from the anterior lobe of the pituitary, *prolactin,* is required for milk production.

Mature mammary glands are female accessory reproductive structures that are specialized for milk production (figure 17.14). They are located in the subcutaneous tissue of female breasts. The breasts are formed superficial to the pectoralis major. Breasts contain large amounts of areolar and adipose tissues that surround and cushion the mammary glands. Dense irregular connective tissue within the breasts is attached to the dermis and to the fascia of the pectoralis major to provide support for the breasts and mammary glands. Superficially, a circle of pigmented skin, the *areola* (ah-rē′-ō-lah), is located near the anterior portion of each breast. A *nipple* containing erectile tissue is located in the center of each areola.

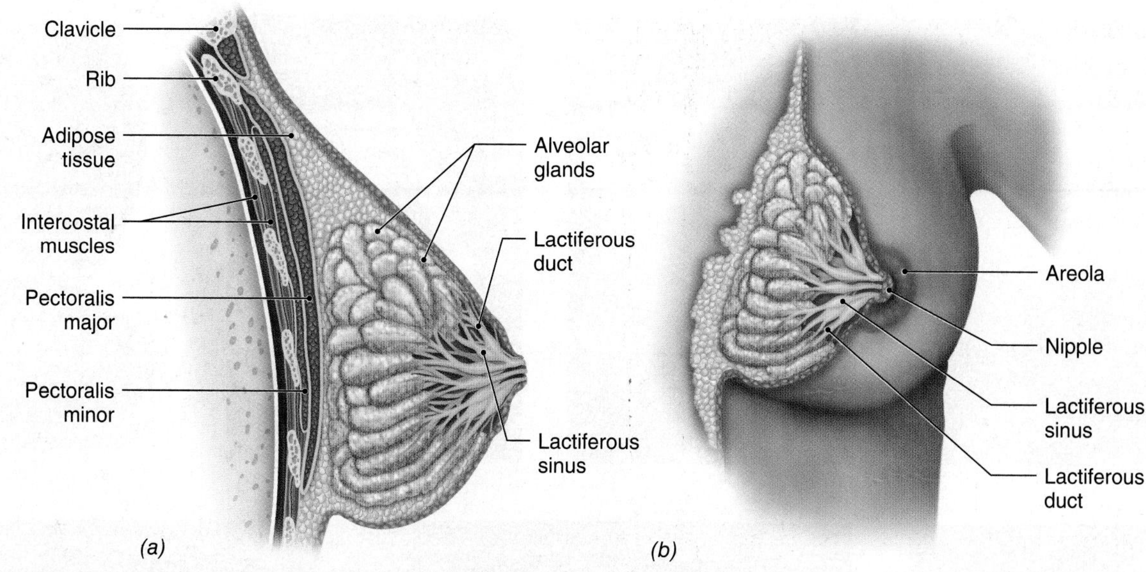

Figure 17.14 Anatomy of the Female Breast and Mammary Glands. *(a)* Sagittal section. *(b)* Anterior view. APR

Each mammary gland consists of 15 to 25 lobes containing *alveolar glands*. Alveolar glands produce milk under stimulation by prolactin after the birth of an infant. Milk is carried from the alveolar glands by lactiferous ducts, which open into *lactiferous sinuses* that lead to the nipple and the external environment. Regulation of lactation is discussed in chapter 18.

Check My Understanding

13. What hormone is required for the production of milk?

17.8 Birth Control

Learning Objective

13. Describe how various methods of birth control work.

Birth control methods may be categorized into several groups based on their mode of action: hormonal, chemical and behavioral contraceptive methods, anti-implantation devices, sterilization, and induced abortion. Figure 17.15 shows a few birth control devices, and Table 17.3 summarizes the effectiveness of the various birth control methods.

Contraception

Contraception is the prevention of conception, which is the union of sperm nucleus and ovum nucleus. Contraceptive methods are designed to prevent sperm from reaching and penetrating a secondary oocyte, which leads to the formation of an ovum. There are several types of contraceptives, including the use of hormones, barriers to sperm, spermicides (chemicals that kill sperm), and behavioral methods.

Hormonal Methods

Hormonal birth control methods use a higher concentration of synthetic progesterone and a lower concentration of synthetic estrogens to prevent ovulation. These hormones provide a negative-feedback control that inhibits the secretion of GnRH by the hypothalamus, which reduces the secretion of FSH and LH by the anterior lobe of the pituitary. This prevents maturation of ovarian follicles so ovulation does not occur. Use of hormonal birth control methods is timed to a woman's menstrual cycle.

Hormonal birth control methods are very effective, but some women may experience unpleasant side effects, such as headache, irregular menstruation, nausea, and bloating. Women who smoke or have a history of blood clots, stroke, heart disease, liver disease, or uncontrolled diabetes have an elevated health risk when using hormonal birth control methods.

There are several ways to use hormonal birth control.

The **oral contraceptive,** "the pill," was the first available method of delivery, and it is still widely used. Several types are available, but the combination pill, which contains both synthetic progesterone and estrogens, is most commonly used. One pill is taken each day for three weeks of the menstrual cycle.

Some combination birth control pills may be used in higher dosages as *morning after pills (MAPs),* for postcoital emergency contraception. When used within 72 hours of unprotected intercourse, the hormone combination disrupts normal hormonal controls and tends to prevent either fertilization or implantation.

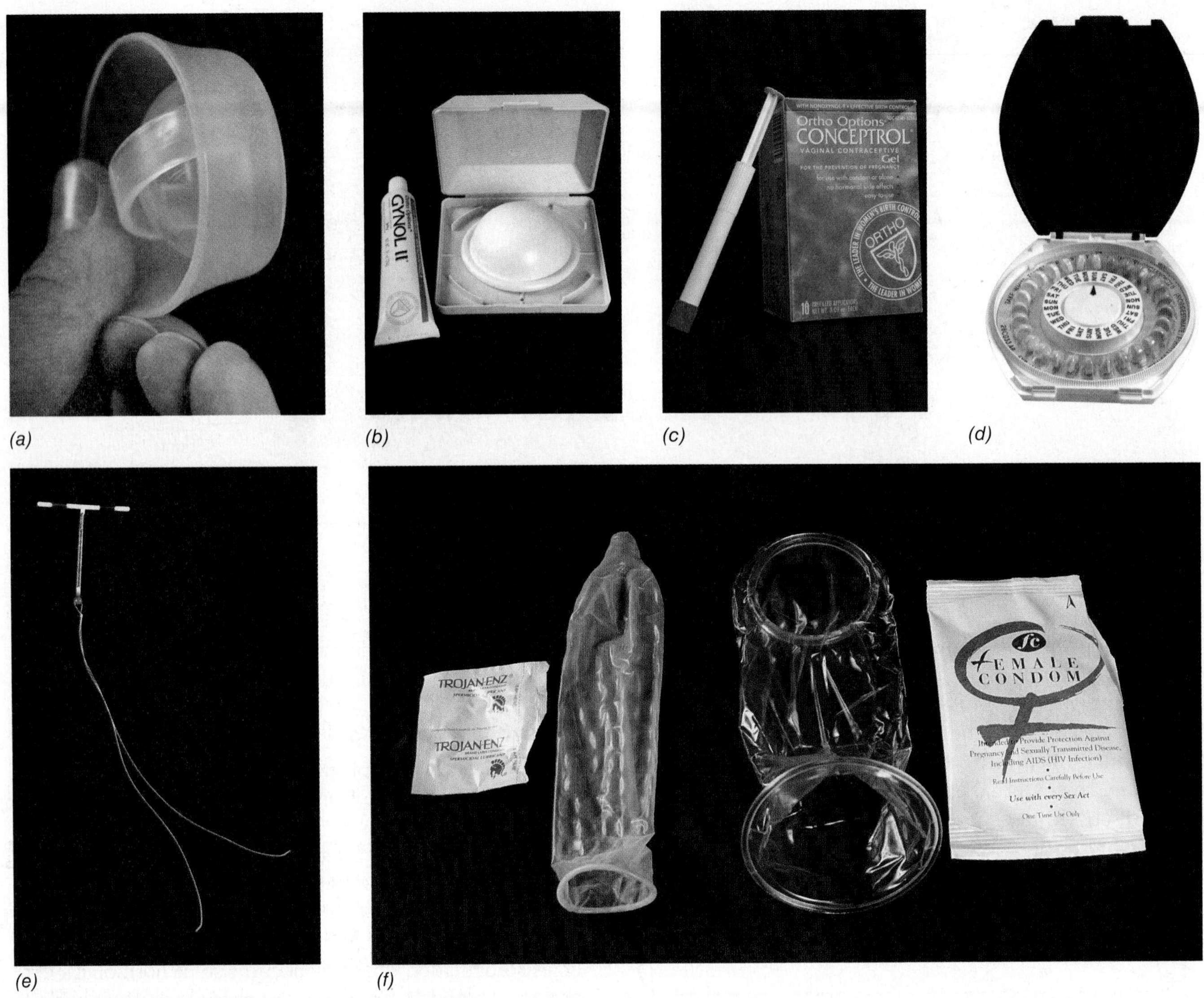

(a) (b) (c) (d) (e) (f)

Figure 17.15 Examples of Birth Control Devices and Chemicals.
(a) Cervical cap, *(b)* diaphragm and spermicide, *(c)* spermicide gel, *(d)* oral contraceptive, *(e)* intrauterine device, and *(f)* male and female condoms.

A **birth control patch** is a small adhesive skin patch containing the synthetic hormones. It is applied to the skin, and the hormones are absorbed through the skin. A new patch is used each week for three weeks.

A **vaginal ring** is a soft plastic ring that is wedged into the superior part of the vagina, where it remains and releases synthetic hormones over a three-week period. It is removed at the start of the fourth week.

Norplant is a set of tiny silicone rods containing synthetic progesterone that are surgically implanted deep to the skin of the arm or over the scapula. The released progesterone acts similarly to an oral contraceptive by inhibiting ovulation, and the implant is effective for up to five years. The implant may be removed to terminate contraception.

Depo-Provera is a synthetic progesterone that is administered by muscular injection at three-month intervals. It protects against pregnancy by preventing ovulation and altering the endometrium to inhibit implantation of a preembryo. Its use often causes weight gain as a side effect, and it can be a considerable health risk, so it must be used under a physician's care. Women with a personal or strong family history of high blood pressure, asthma, kidney disease, migraine headaches, or breast cancer are at increased risk if using Depo-Provera.

Barriers

There are three types of **barriers** that are designed to prevent sperm from entering the uterus: condom, diaphragm, and cervical cap.

Table 17.3 Effectiveness of Birth Control Methods

Effectiveness values are approximate because they vary with how the birth control methods are used.

Method	Effectiveness*
Vasectomy	Nearly 100%
Tubal ligation	Nearly 100%
Depo-Provera	99%
Norplant	99%
Oral contraception	97%
Skin patch	97%
Vaginal ring	97%
Intrauterine device (IUD)	95%
Male condom	85%
Female condom	85%
Diaphragm with spermicide	80%
Cervical cap with spermicide	80%
Withdrawal	75%
Rhythm method	70%

*Effectiveness is the percentage of women who do not become pregnant in one year while using this method of birth control.

Condoms act by preventing sperm from being deposited in the vagina. A *male condom* is a thin sheath of latex rubber that is slipped over the penis prior to sexual intercourse. A *female condom* is a thin polyurethane bag with a flexible ring at each end. The woman inserts the bag into the vagina prior to sexual intercourse. Condoms reduce but do not eliminate the chance of infection by sexually transmitted diseases (STDs).

A **diaphragm** is a dome-shaped sheet of rubber supported by a firm but somewhat flexible ring. It is placed in the superior vagina over the cervix prior to intercourse. It prevents sperm from entering the uterus. Spermicidal substances are used along with a diaphragm.

A **cervical cap** is a thimble-shaped piece of latex rubber that fits snugly over the cervix. Spermicidal substances are used with a cervical cap. Females with abnormal Pap smears or cervical infections should not use a cervical cap.

Spermicides

A variety of **spermicides** are available, including creams, jellies, and suppositories. These chemicals kill sperm by destroying their plasma membranes.

Behavioral Methods

The **rhythm method** requires abstinence from sexual intercourse from three days before the day of ovulation to three days after ovulation, for a total of seven days. It is based on the fact that the secondary oocyte may be penetrated by a sperm for only about 24 hours after ovulation. One difficulty with this method is determining the time of ovulation, because few women have perfectly regular cycles. The failure rate is higher in women who have irregular cycles.

Withdrawal, or **coitus interruptus,** is the removal of the penis from the vagina just prior to ejaculation. Failures of this method result from preejaculatory emission of semen or failure to withdraw before ejaculation.

Anti-Implantation Devices

An *intrauterine device (IUD)* is a small plastic and copper object that is inserted into the uterus by a physician. The IUD causes inflammation of the endometrium, preventing implantation of a preembryo. An IUD may remain in place for long periods of time, but it should be checked at regular intervals by a physician because an IUD may be spontaneously expelled from the uterus. Undesirable side effects include excessive menstrual bleeding, painful cramps, and an increased risk of pelvic inflammatory disease and infertility.

Sterilization

Sterilization surgeries may be performed on both males and females. In males, a **vasectomy** is performed by cutting and blocking the vasa deferentia within the scrotum so that sperm do not form part of the semen. The sperm disintegrate and are reabsorbed. In females, a **tubal ligation** is performed through a small abdominal incision. The uterine tubes are cut and blocked to prevent the transport of a secondary oocyte toward the uterus. The ovulated secondary oocytes disintegrate and are reabsorbed. Vasectomies and tubal ligations do not affect the production of sex hormones or the sexual response.

Induced Abortion

An **abortion** is the premature expulsion of an embryo or fetus from the uterus. Spontaneous abortions are called *miscarriages*. They usually result from hormonal disorders or serious abnormalities of the developing embryo. An *induced abortion* may be used to terminate an unwanted pregnancy. Induced abortion involves dilation of the cervix and removal of the embryo or fetus by suction or surgical means. Possible side effects of induced abortions include prolonged bleeding, perforation of the uterus, and emotional trauma.

The so-called abortion pill, *mifepristone* (RU 486), is a progesterone antagonist. It causes the endometrium to break down, thereby detaching the embryo or fetus, which is passed from the uterus after administration of prostaglandins to promote uterine contractions. Its use is limited to the first five weeks of pregnancy and requires the supervision and care of a physician.

18 CHAPTER

Development, Pregnancy, and Genetics

Sandra and Colin, a married couple in their early 40s, discover that Sandra is pregnant with their first child and are very excited. However, they are aware that later in life pregnancies are higher risk. Decades of research show that there is a higher risk of chromosomal abnormalities, most notably Down Syndrome, in fetuses conceived by men and women over 40. The likelihood of physical malformations, such as clubbed foot, or organ malfunction is also greatly increased. In light of all of these potential problems, Sandra and Colin elect to have an amniocentesis test performed at 18 weeks of development. During the test, a small amount of amniotic fluid is withdrawn from around the fetus. The fetal tissues within the fluid show no evidence of chromosomal anomalies, nor do they indicate other fetal issues such as neural tube defects. Their 3D ultrasound at 24 weeks of development also shows normal development of the major organs and limbs. With confirmation that their fetus is healthy and developing normally, Sandra and Colin breathe a little easier and move into the third trimester of the pregnancy with happy anticipation.

CHAPTER OUTLINE

Module 14
Reproductive System

SELECTED KEY TERMS

Allele An alternate form of a gene.
Embryo The developmental stage extending from the beginning of the third week through the end of the eighth week of development.
Fertilization The union of sperm nucleus and ovum nucleus; conception.
Fetus The developmental stage extending from the beginning of the ninth week of development to birth.
Gene A unit of heredity; part of a DNA molecule in a chromosome.
Genotype The genetic composition of an individual.
Germ layer One of three embryonic tissues from which all subsequent tissues develop.
Heterozygous (hetero = different) A condition in which alleles of a gene pair are different.
Homozygous (homo = the same) A condition in which alleles of a gene pair are identical.
Implantation The attachment of a blastocyst to the endometrium.
Parturition The process of giving birth.
Phenotype (phen = visible) The observable traits determined by genes.
Placenta (plac = flat) Temporary organ that allows for the exchange of substances between maternal and embryonic/fetal bloods.
Preembryo The developmental stage extending from fertilization through the end of the second week of development.
Zygote (zygo = yolk) Cell formed by the union of the sperm nucleus and ovum nucleus that leads to the development of a human being.

EACH NEW LIFE BEGINS AS A SINGLE CELL formed by the fusion of a sperm and an ovum. At the instant of its formation, this first cell contains all of the inherited information necessary for growth and development, plus all of the inherited characteristics that provide an individual's unique identity. After about 38 weeks of *prenatal* (fertilization to birth) growth and development or 40 weeks from the beginning of the last menstruation (pregnancy), an infant is born.

18.1 Fertilization and Early Development

Learning Objective

1. Describe the processes of fertilization, preembryonic development, and implantation.

Recall from chapter 17 that each **primary oocyte** undergoes the first meiotic division while still in the ovarian follicle. This division forms a **secondary oocyte** and the first **polar body,** each containing 23 chromosomes. At ovulation, the secondary oocyte and first polar body, still enclosed within a sphere of granulosa cells, are released into a uterine tube. They are slowly moved toward the uterus through peristalsis and the beating cilia of epithelial cells lining the uterine tube.

Fertilization

After semen is deposited in the vagina, sperm begin their long journey into the uterus and on into the uterine tubes. Sperm inherently swim against the slight current of fluid that flows from the uterine tubes through the uterus and into the vagina, which helps to guide sperm towards the uterine tubes. Of the millions of sperm in the semen, only a few hundred sperm actually enter the uterine tubes. Prostaglandins in semen stimulate reverse peristalsis of the uterus and uterine tubes that greatly aids the migration of sperm. Sperm reach the superior portions of the uterine tubes within one hour after sexual intercourse. Usually, only one uterine tube contains a secondary oocyte because only one secondary oocyte is usually released at ovulation. Sperm entering an empty uterine tube have no chance for fertilization.

Within the uterine tube, sperm are chemically attracted to the secondary oocyte. Many sperm cluster around the oocyte and attempt to penetrate the granulosa cells (figure 18.1). The acrosomes of the sperm release enzymes that dissolve the "glue" holding granulosa cells

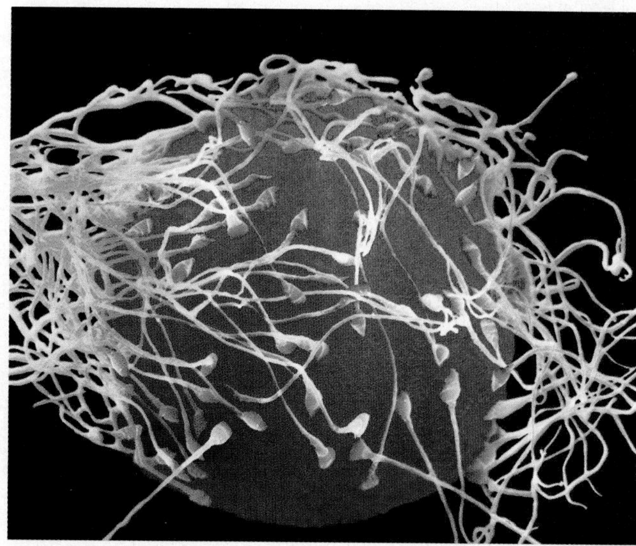

Figure 18.1 Scanning electron photomicrograph of sperm clustered on the surface of a secondary oocyte (1,200×). Only one sperm may enter the oocyte.

together so they can reach the oocyte. It takes many sperm to disperse the granulosa cells, so that one sperm can eventually wriggle between them to contact the oocyte. The acrosome then releases a different enzyme that enables the sperm to penetrate the oocyte membrane and enter the oocyte. Once this happens, changes in the oocyte plasma membrane prevent other sperm from entering (figure 18.2).

When a sperm enters the secondary oocyte, it triggers the second meiotic division, which forms the **ovum** and a second polar body. Then, the sperm nucleus and ovum nucleus unite in **fertilization** to form a **zygote,** the first cell of the infant-to-be. The zygote is the first stage of development. The zygote contains 46 chromosomes, 23 from the sperm and 23 from the ovum.

A secondary oocyte remains viable for about 24 hours after ovulation. Most sperm remain viable in the female reproductive tract for about 72 hours, although some may be viable for up to five days. Therefore, fertilization is most likely to occur when sexual intercourse occurs from three days before ovulation to one day after ovulation.

Preembryonic Development

Immediately after fertilization, the zygote begins to divide by mitotic cell division. The early cell divisions are collectively known as *cleavage.* These divisions occur so rapidly that maximum cellular growth between divisions is not possible, which results in increasingly smaller cells. During this time, the **preembryo** is carried along the uterine tube by peristalsis and the beating cilia of epithelial cells lining it. By the time the preembryo reaches the uterus, it consists of a solid ball of cells called a *morula* (mor′-u-lah) that is not much larger than the zygote.

Continued but slightly slower mitotic divisions form a larger hollow ball of cells called a **blastocyst** (blas′-to-sist). Located within the blastocyst is the *embryoblast* or *inner cell mass,* a specialized group of cells from which the **embryo** later develops. The superficial wall of the blastocyst is called the *trophoblast,* which later will form the embryonic portion of the placenta.

About the seventh day of development, the blastocyst attaches to the endometrium. Digestive enzymes, released from the trophoblast, enable the blastocyst to penetrate into the endometrium, where it is soon covered by the superficial endometrium. This entire process is called **implantation** and is completed by the fourteenth day (figure 18.3).

Clinical Insight

Identical, or *monozygotic, twins* develop from a single zygote; this means that the twins possess identical genetic characteristics. The embryoblast of the blastocyst separates completely, usually by the end of the first week of embryonic development, and results in two embryos within separate amnion sacs yet sharing a common chorion and placenta. *Fraternal,* or *dizygotic, twins* develop from two zygotes: two different secondary oocytes are fertilized by different sperm. These twins do not possess identical genetic characteristics and develop within separate amnions and chorions. Each embryo develops its own placenta, though the placentas may fuse if they are located near each other within the uterus.

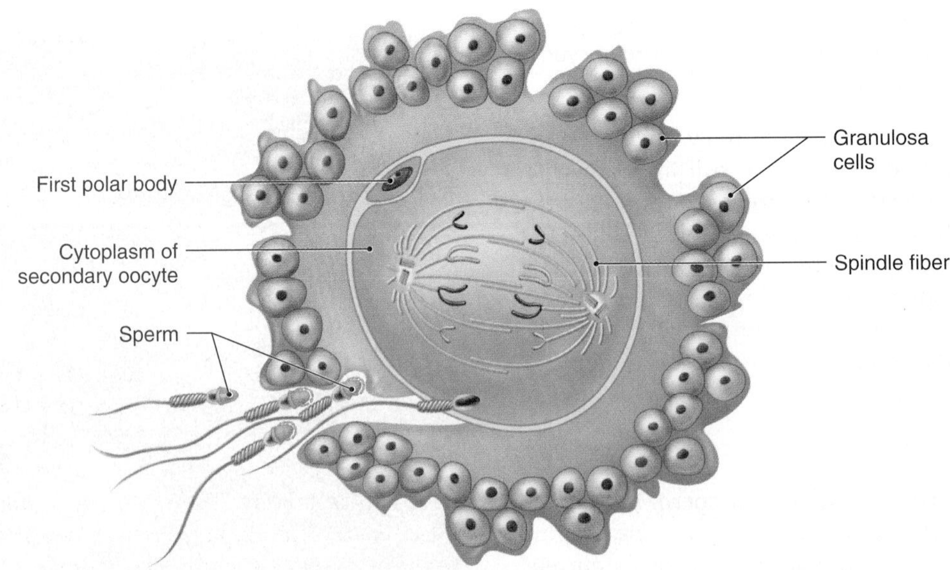

Figure 18.2 Enzymes from many sperm are required to separate the granulosa cells, enabling one sperm to penetrate the plasma membrane of the secondary oocyte. Sperm penetration triggers the second meiotic division within the secondary oocyte in order to form the ovum.

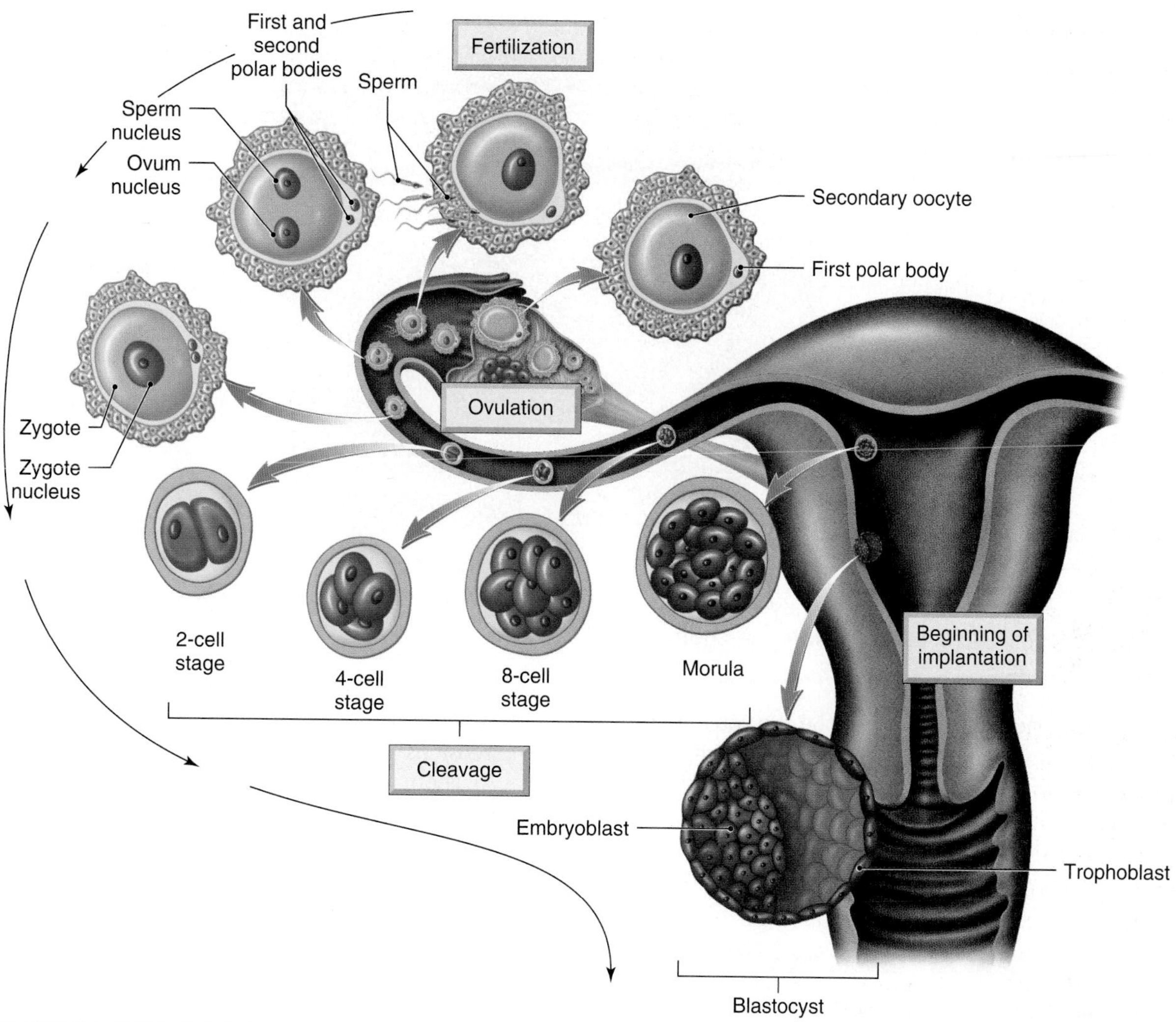

Figure 18.3 The stages of fertilization, cleavage, and implantation. Implantation begins on about the seventh day of development. APIR

 Check My Understanding

1. How do sperm reach the secondary oocyte?
2. What are the major events that occur from sperm penetration of a secondary oocyte to implantation?

18.2 Embryonic Development

Learning Objectives

2. Describe the major events of embryonic development.
3. Explain the importance of the three germ layers.
4. Identify the extraembryonic membranes and their functions.
5. Describe the structure and function of the placenta.

The embryonic stage of development begins at the start of the third week of development and is completed at the end of the eighth week. During this time, the embryo undergoes rapid development, forming the rudiments of all body organs, extraembryonic membranes, and the placenta. By the end of the eighth week, it has a distinct human appearance.

Germ Layers

After implantation, the embryoblast grows to become the *embryonic disc*, which is supported by a short stalk extending from the wall of the blastocyst. The embryonic disc consists of three embryonic tissues: ectoderm, mesoderm, and endoderm. The ectoderm forms the posterior surface of the developing embryo, while the endoderm forms the anterior surface. The mesoderm is the middle tissue layer. These embryonic tissues are called **germ layers** because all body tissues and organs are formed from them. Figure 18.4 illustrates the formation of the germ layers.

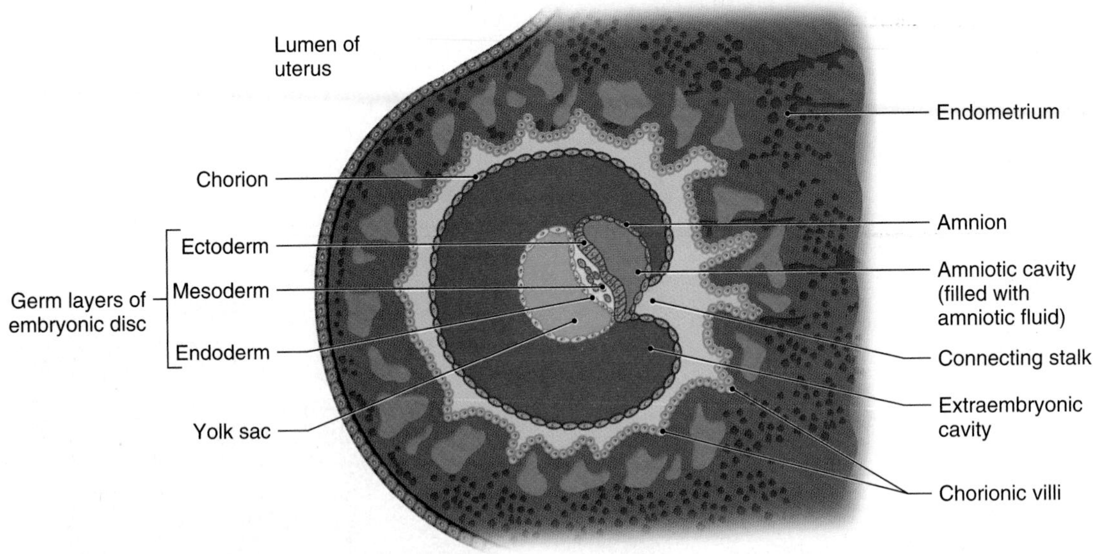

Figure 18.4 Development of the germ layers of the embryonic disc and three extraembryonic membranes—chorion, amnion, and yolk sac—in an embryo. Note the chorionic villi.

Briefly, **ectoderm** forms all of the nervous system and the epidermis of the skin. **Mesoderm** forms muscles, bones, blood, and other forms of connective tissues. **Endoderm** forms the epithelial lining of the digestive, respiratory, and urinary tracts. Table 18.1 provides a more detailed listing of the major structures formed by each primary germ layer.

Extraembryonic Membranes

While the embryonic disc is forming, slender extensions from the trophoblast grow into the surrounding endometrium, firmly anchoring the blastocyst. The trophoblast of the blastocyst is now called the **chorion** (kō′-rē-on), the most superficial extraembryonic membrane, and the slender extensions are known as **chorionic** (kō-rē-on′-ik) **villi.**

At about the same time, two other extraembryonic membranes separate from the embryonic disc. The **amnion** (am′-nē-on) is formed posterior to the embryo and the **yolk sac** is formed anterior to the embryo (figure 18.4). *Amniotic fluid* fills the *amniotic cavity*, the space between the embryonic disc and the amnion. As the embryo develops, the amnion margins move toward the anterior surface of the embryo. In a short time, the embryo is enveloped by the amnion (figure 18.5).

Amniotic fluid serves as a shock absorber for the developing embryo. It also prevents adhesions from developing

Table 18.1 Structures Formed by the Primary Germ Layers

Ectoderm	Mesoderm	Endoderm
All nervous tissue	All connective tissues including bone and cartilage	Epithelial lining of alimentary canal (except oral cavity and anal canal)
Sensory epithelium of sense organs	Skeletal, cardiac, and most smooth muscle tissue	Liver and pancreas
Pituitary gland, pineal gland, and adrenal medulla	Red bone marrow and lymphoid tissue	Epithelial lining of respiratory system (except nasal cavity and paranasal sinuses)
Epidermis of the skin including nails, hair follicles, sebaceous glands, sweat glands, and mammary glands	Blood and lymphatic vessels	Epithelial lining of urethra and bladder
Cornea, lens, and internal muscles of eye	Dermis of the skin	Thymus, thyroid and parathyroid glands
Epithelial lining of oral and nasal cavities, paranasal sinuses, and anal canal	Kidneys, ureters, gonads, and reproductive ducts	Epithelial lining of accessory reproductive glands
Salivary glands	Adrenal cortex	Epithelial lining of tonsils, auditory tubes, and tympanic cavity
Tooth enamel	Synovial and serous membranes	
Cranial and spinal meninges	Spleen	

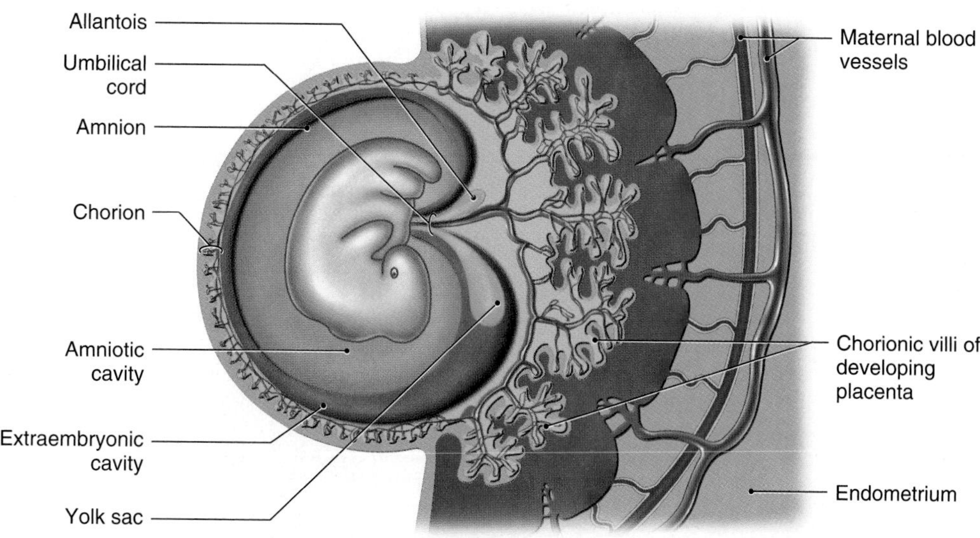

Figure 18.5 Amniotic fluid surrounds the embryo as the amnion continues to grow. The yolk sac and allantois become part of the developing umbilical cord. The chorion surrounding the embryo becomes thinner, the chorionic villi grow into the endometrium, and embryonic blood vessels extend into the chorionic villi as the placenta develops.

between various parts of the embryo. Later in development, the fetus swallows and inhales amniotic fluid and discharges dilute urine into it.

The yolk sac forms an outpocketing that becomes the **allantois** (al-lan′-to-is), the last of the extraembryonic membranes. Both the allantois and the yolk sac subsequently become part of the **umbilical** (um-bil′-i-kal) **cord,** which attaches the embryo to the placenta (figure 18.5). The yolk sac forms the early formed elements and germ cells for the embryo. It also serves as a shock absorber for the embryo, in addition to forming the primitive gut. The allantois also produces formed elements and brings umbilical blood vessels to the placenta.

Placenta

As the embryo continues to grow and the chorion enlarges, the layer of the endometrium covering the blastocyst becomes increasingly thinner. The chorionic villi in this region disintegrate, and only those in contact with the thick, spongy endometrium persist. This leads to the formation of the **placenta** (plah-sen′-tah), a disc-shaped structure formed of both embryonic and maternal tissues. The embryonic portion is formed of the chorion and chorionic villi, and the maternal portion is formed of the associated endometrium (figure 18.5).

The developing embryo is attached to the placenta by the umbilical cord. Two umbilical arteries bring embryonic blood to the placenta, and a single umbilical vein returns the blood to the embryo. There are no nerves in the umbilical cord.

The placenta provides an interface between the embryonic and maternal bloods for the exchange of water, respiratory gases, nutrients, wastes, hormones, and antibodies. The embryonic blood must receive all required substances from the mother's blood and pass metabolic wastes into the mother's blood. The placenta is usually fully functional by the end of the eighth week.

The embryonic and maternal bloods do not mix in the placenta. The maternal blood vessels open into blood-filled spaces called *lacunae* into which the embryonic blood vessels extend and branch repeatedly. This arrangement provides a large surface area for the exchange of substances between embryonic and maternal bloods. Observe this relationship in figure 18.6.

Clinical Insight

Many substances are harmful to a developing embryo and fetus because they can pass across the placenta from the maternal blood into the blood of the embryo and fetus. Such harmful substances include alcohol, cocaine, heroin, nicotine, caffeine, and many therapeutic drugs. Physicians recommend that no drugs or other chemical substances be taken during pregnancy except those that are crucial for the mother's health.

In addition, pathogens of certain maternal infections can also pass across the placenta and infect the fetus. These infections include AIDS, German measles, syphilis, and toxoplasmosis (transmitted from cat feces). All of these diseases may cause serious fetal disorders.

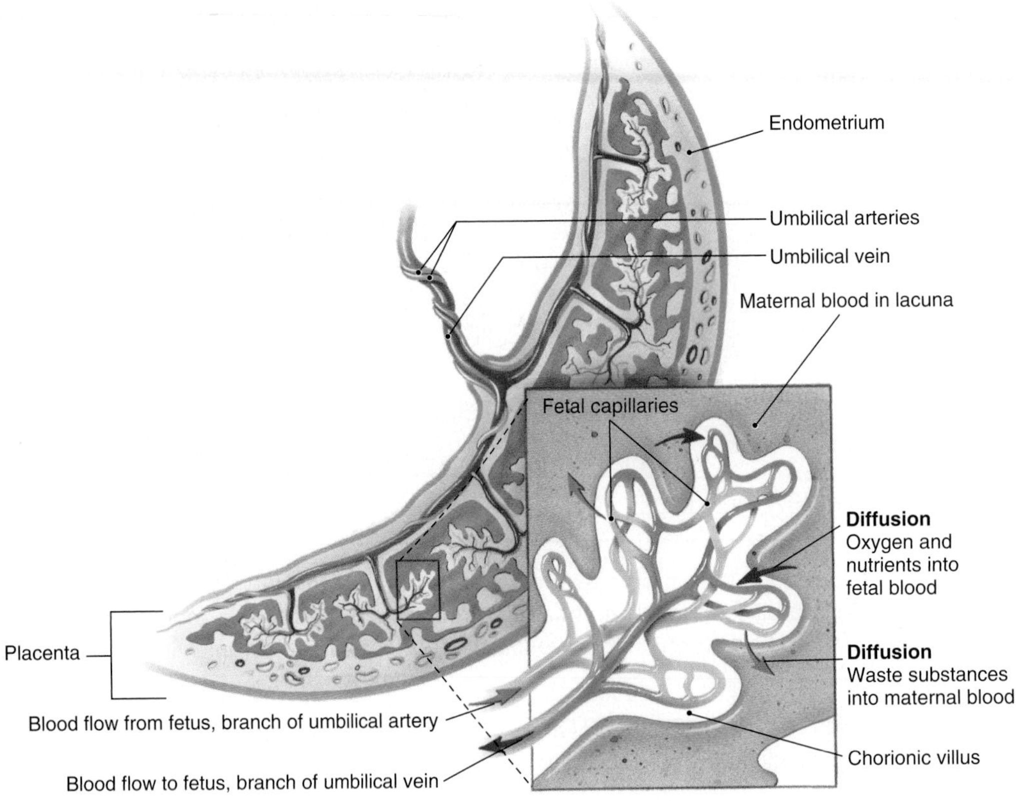

Figure 18.6 The placenta is formed of both embryonic and maternal tissues. The chorion and chorionic villi form the embryonic portion, while the endometrium of the uterine wall forms the maternal portion. Materials are exchanged between embryonic and maternal bloods by diffusion.

External Appearance

By the fourth week of development, the head and limb buds of the embryo are recognizable. By the seventh week, the rudiments of all organs are present, and the eyes and ears are visible. Figure 18.7 illustrates the visible changes in the embryo from the fourth to the seventh week.

 Check My Understanding

3. What are the three primary germ layers and what does each form?
4. What are the extraembryonic membranes, and what are their roles?
5. What composes the placenta and what is its function?

18.3 Fetal Development

Learning Objective

6. Describe the major changes that occur during fetal development.

At the beginning of the ninth week of development, the developing offspring has a distinctively human appearance and is now referred to as a **fetus** (figure 18.8). Table 18.2 highlights the changes in physical appearance that take place during fetal development.

At the start of the fetal stage, the head is as large as the body and all rudimentary organs are present. The ossification of bones begins during weeks nine through twelve and organs continue to develop. By the twelfth week, the fetus weighs about 45 grams and is about the size of a candy bar.

During the thirteenth through sixteenth weeks, the eyes and ears reach their final positions and a heartbeat may be detected with a stethoscope. During the seventeenth through twentieth weeks, fine hair (*lanugo*) covers the body and hair appears on the scalp. Sebum from sebaceous glands and dead epidermal cells form the *vernix caseosa,* which protects the skin from the digestive and urinary wastes in the surrounding amniotic fluid. Movements of the fetus may now be detectable by the mother. By the end of the twentieth week, the fetus weighs about 460 grams, or approximately one pound.

Continued growth of organ systems during the twenty-first through twenty-ninth weeks enables the

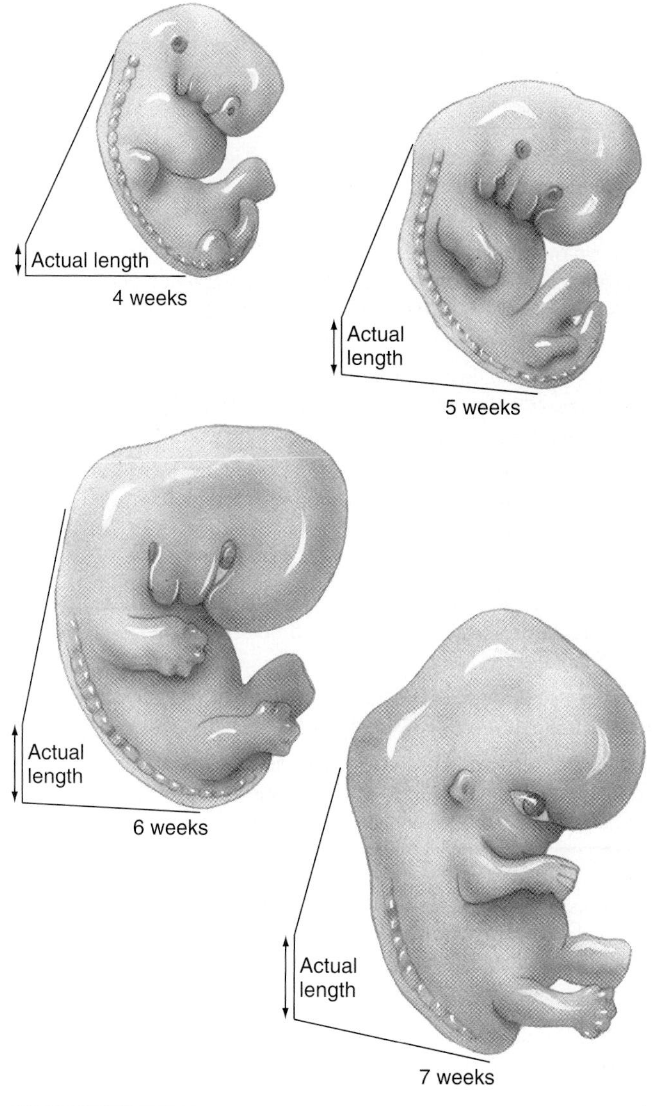

Figure 18.7 External appearance of the embryo from the fourth to the seventh week of development.

head, body, and limbs to attain infant proportions. However, the fetus is still lean, with wrinkled, rather translucent skin. By the twenty-ninth week, the fetus weighs about 1,300 grams, or approximately 3 pounds.

Deposition of adipose tissue in the subcutaneous tissue plus continued growth and development of organ systems, as well as the descent of the testes in males, occur in the eighth and ninth months. By the time the fetus is full term, the lanugo has been shed, the skin is pinkish with ample subcutaneous tissue, hair covers the scalp, and all organ systems are ready for birth. The fetus weighs about 3,400 grams or about 7.5 pounds.

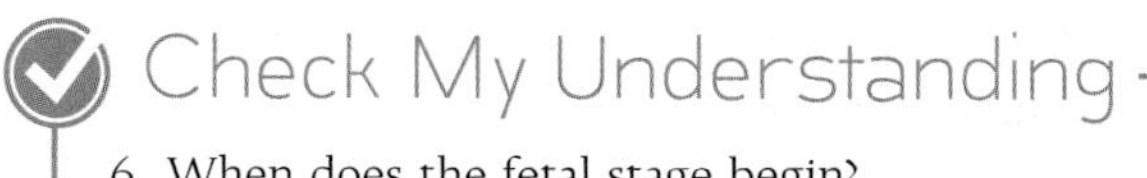

Check My Understanding

6. When does the fetal stage begin?

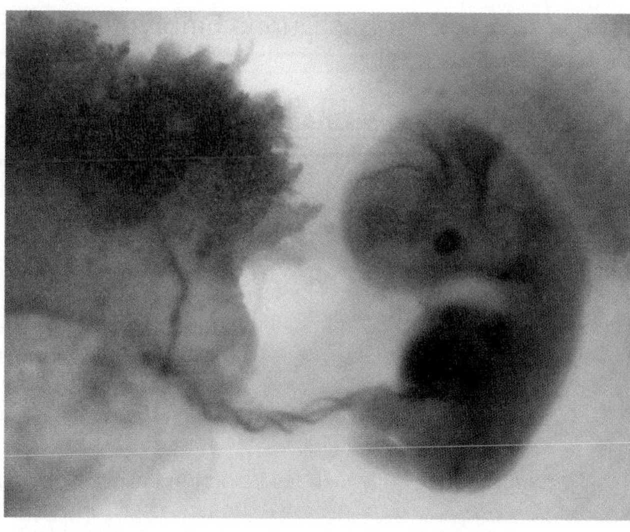

Figure 18.8 A photograph of an eight-week embryo showing the umbilical cord and the enveloping amnion.

18.4 Hormonal Control of Pregnancy

Learning Objective

7. Explain the hormonal control of pregnancy.

Without the formation of a preembryo, the corpus luteum degenerates about two weeks after ovulation as a result of the decline in luteinizing hormone from the anterior lobe of the pituitary gland. The resulting decline in blood levels of estrogens and progesterone causes the endometrium to break down and to be shed with menstruation. Pregnancy causes hormone changes that maintain the endometrium.

The trophoblast of the blastocyst (see figure 18.3) secretes **human chorionic gonadotropin (hCG)** (kō-rē-on′-ik gōn-ah-dō-trō′-pin), an LH-like hormone that maintains the corpus luteum. Recall that the trophoblast forms the chorion, which forms the embryonic portion of the placenta and is also a source of hCG. Under stimulation of hCG, the corpus luteum continues to secrete progesterone and estrogens to maintain the endometrium. Pregnancy tests are designed to detect the presence of hCG in a woman's blood or urine, usually within eight to ten days after fertilization.

Clinical Insight

If the corpus luteum stops secreting progesterone and estrogens too quickly or if the placenta is too slow in starting to secrete these hormones, a decline in their concentrations results. Such a decline may detach the placenta from the uterine wall, causing a miscarriage.

Table 18.2 Changes During Embryonic And Fetal Development

Weeks of Development	Changes
5–8 Weeks	Recognizable human shape; head as large as the body; eyes far apart; upper and lower limbs present with digits; heart possesses four chambers; tail disappears; formed elements produced by liver; organ systems present in rudimentary form
9–12 Weeks	Head is half the length of the body; nails form on digits; brain enlarges; eyes nearly fully developed but still far apart and eyelids still fused; ears formed but low set; nose bridge forms; upper limbs almost fully lengthened; heartbeat detectable; gender distinguishable by external genitalia; heartbeat detectable but not with stethoscope; ossification begins; fetus moves but not detectable by mother
13–16 Weeks	Body is larger than head; eyes and ears reach characteristic positions; facial features well developed; lips exhibit sucking movements; bones distinct; lower limbs lengthen; kidneys well formed; meconium (fetal feces) begins to form in intestines; heartbeat detectable by stethoscope
17–20 Weeks	Head more proportional to body; lanugo and vernix caseosa cover body; eyebrows, eyelashes, and head hair present; lower limbs fully lengthened; brown fat forms for future heat production; fetus assumes fetal position due to limited space; fetal movements felt by mother
21–25 Weeks	Rapid increase in weight gain; skin pink and wrinkled; surfactant formed by lungs
26–29 Weeks	Eyes open; body is lean; head and body proportional; skin still wrinkled and pinkish; subcutaneous tissue begins to form; testes begin to descend towards scrotum in males; red bone marrow begins formed element production; organ systems continue to develop
30-34 Weeks	Skin pinkish and smoother due to subcutaneous tissue deposition; testes continue to descend in males; bones still continuing to ossify; fetus assumes "head down" position
35–38 Weeks	More subcutaneous tissue is deposited; skin smoother and pinkish because melanin is not produced until skin is exposed to light; testes located within scrotum in males; lanugo is shed; vernix caseosa still present; skull bones largely ossified except at fontanelles; body larger than head

The concentration of hCG in the blood rises sharply and peaks after approximately 10 weeks of development. Blood hCG levels then decline and level off between 16 and 20 weeks, leading to the degeneration of the corpus luteum. However, by approximately twelve weeks, the placenta takes over the role of producing estrogens and progesterone, which prevents loss of the endometrium when the corpus luteum degenerates. The ovaries remain inactive during the remainder of the pregnancy because the high level of progesterone in the blood suppresses secretion of gonadotropin-releasing hormone (GnRH) by the hypothalamus, which in turn prevents the release of follicle-stimulating hormone (FSH) and luteinizing hormone (LH) from the anterior lobe of the pituitary.

The blood concentrations of estrogens and progesterone continue to increase throughout pregnancy (figure 18.9). Note that the level of estrogens increase faster than that of progesterone. As pregnancy continues, placental estrogens and progesterone stimulate development of the mammary glands in preparation for milk secretion.

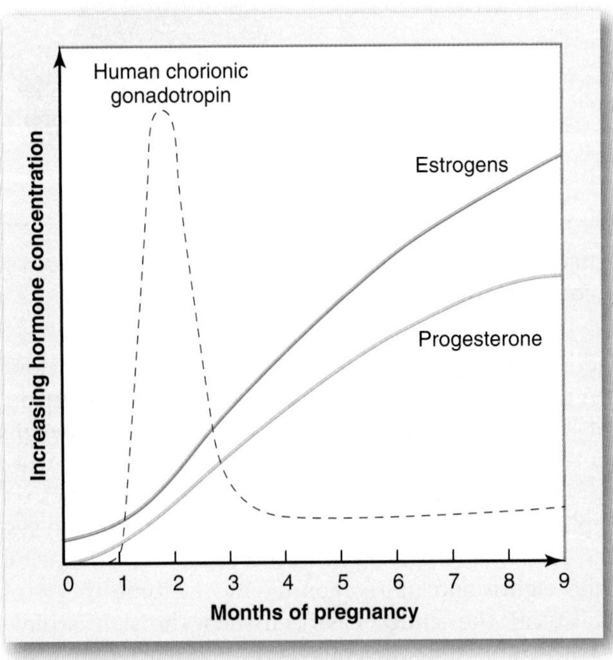

Figure 18.9 Relative concentrations of human chorionic gonadotropin (hCG), estrogens, and progesterone during pregnancy. Remember that fertilization usually occurs approximately 2 weeks after the beginning of the last menstruation.

Check My Understanding

7. What hormonal changes are triggered by the presence of an implanted embryo?

18.5 Birth

Learning Objective

8. Describe the neuroendocrine positive-feedback mechanism controlling labor.

During the latter stages of pregnancy, the blood concentration of estrogens become increasingly greater than that of progesterone, as noted earlier. Whereas progesterone inhibits uterine contractions, estrogens promote them. Therefore, there is an increasing tendency toward the onset of uterine contractions as the pregnancy approaches full term. These uterine contractions are often referred to as *Braxton Hicks contractions* or "false labor." Throughout pregnancy, the hormone **relaxin** is secreted first by the corpus luteum and then by the placenta. It helps in the development of blood vessels within the placenta, in addition to other cardiovascular changes that occur in the mother.

Birth usually occurs within two weeks of the calculated due date, which is 280 days from the beginning of the last menstruation. The birth process is called **parturition** (par-tū-rish′-un), and the events associated with parturition are collectively called **labor.** The fetus is usually in a "head down" position at this time (figure 18.10). As the fetus reaches full term, the high blood levels of estrogens override progesterone's inhibition of uterine contractions, allowing uterine contractions to occur. Pressure of the fetus on the cervix stretches the cervix, stimulating a neuroendocrine positive-feedback mechanism that promotes uterine contractions. In fact, physicians sometimes initiate labor by breaking the amnion so that increased pressure is placed on the cervix. The process of labor can be divided into three stages.

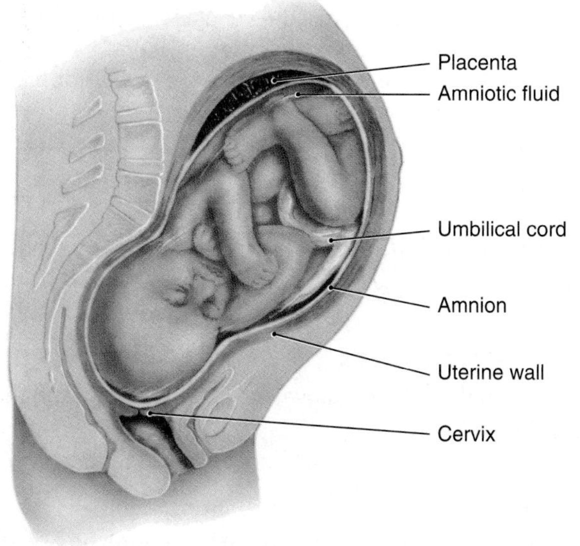

Figure 18.10 A full-term fetus positioned with the head against the cervix.

The *first stage of labor* is the dilation of the cervix. The stretching of the cervix triggers the formation of nerve impulses that are sent to the hypothalamus. When these reach a critical frequency, the hypothalamus activates the posterior lobe of the pituitary gland to release *oxytocin,* increasing its concentration in the blood. Oxytocin stimulates the characteristic rhythmic contractions that begin at the superior end of the uterus and move toward the cervix, pushing the fetus toward the *vagina,* or birth canal.

The continued contractions of the uterus force the fetus's head against the cervix, which results in greater stretching of the cervix. The increase in cervical stretching causes more nerve impulses to be sent to the hypothalamus. The hypothalamus then stimulates the posterior lobe of the pituitary to release more oxytocin into the blood. The higher blood levels of oxytocin trigger more intense and frequent uterine contractions to occur, which in turn produce greater cervical stretching. This positive-feedback mechanism will produce increasingly stronger uterine contractions until birth occurs.

Dilation of the cervix is the longest stage of labor. It may last from 6 to 12 hours, depending on the size of the fetus and whether the mother has had other children. During this time, the amnion ruptures (or "water breaks") and the cervix dilates to the size of the fetus's head (figure 18.11).

The *second stage of labor* is the delivery (expulsion) of the fetus. It usually lasts less than an hour, with contractions occurring every two to three minutes and lasting about one minute. Once the head is expelled, the rest of the body exits rather quickly.

The *third stage of labor* is the delivery of the placenta. Within 15 minutes after birth of the infant, the placenta detaches. Continued contractions expel the placenta (the afterbirth). The placenta is checked carefully to see that all of it has been removed from the uterus because any residue may cause a serious uterine infection. Detachment of the placenta produces some bleeding because endometrial blood vessels at the placental site are ruptured. However, uterine contractions compress the broken blood vessels so that serious bleeding is usually avoided. Subsequently, the uterus decreases in size rather quickly.

Clinical Insight

About 5% of births are *breech births* in which the fetus is presented buttocks first. This complicates the delivery and may require delivery by *Cesarean section.* In a cesarean section, a transverse incision is made just superior to the pubic symphysis through the walls of the abdomen and uterus, through which the infant is extracted.

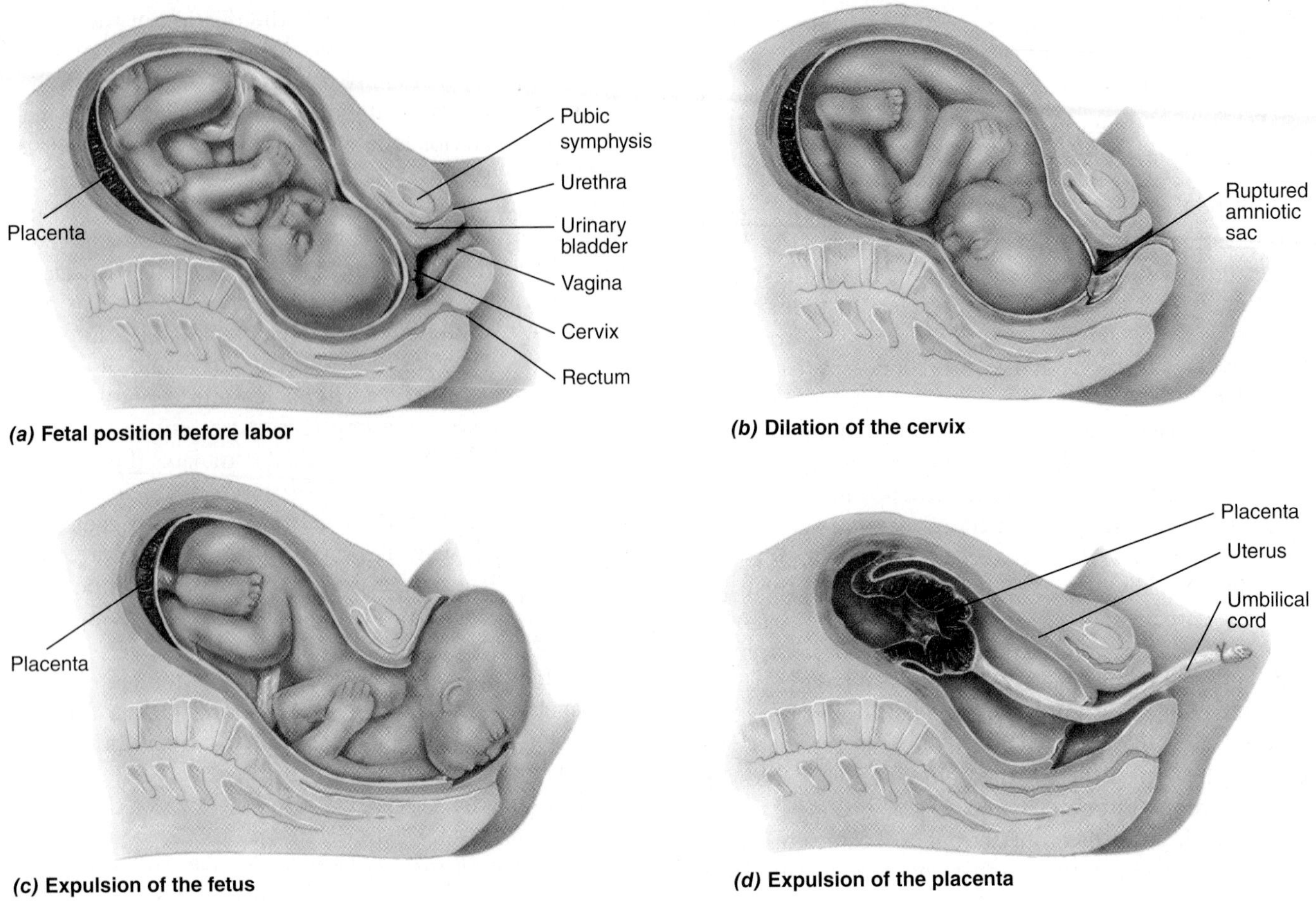

(a) Fetal position before labor

(b) Dilation of the cervix

(c) Expulsion of the fetus

(d) Expulsion of the placenta

Figure 18.11 Stages of Labor.

First Breath

Immediately after birth, the infant's nose and mouth are aspirated to remove mucus or fluid that would impair breathing. The umbilical cord is clamped and cut, separating the infant from the placenta, which has served as its prenatal respiratory organ. As carbon dioxide increases in the infant's blood, the respiratory rhythmicity center in the medulla oblongata is activated and the infant's first inspiration is stimulated.

The first breath is difficult because the lungs are collapsed. In an infant, surfactant in alveoli reduces surface tension, making the first breath and subsequent breathing easier.

Clinical Insight

Premature infants born before 24 weeks of development seldom survive. The respiratory system of such an infant is not sufficiently developed, even with the use of synthetic surfactant, to enable rapid gas exchange with blood in the lungs.

Check My Understanding

8. How does an increased level of estrogens in the blood contribute to the onset of labor?
9. How does the neuroendocrine positive-feedback mechanism control uterine contractions?

18.6 Cardiovascular Adaptations

Learning Objectives

9. Explain the fetal cardiovascular adaptations and their value to the fetus.
10. Describe the cardiovascular changes that occur in postnatal development.

Fetal circulation is quite different from adult circulation because the digestive tract, lungs, and kidneys are not functioning. Oxygen and nutrients are obtained from the maternal blood in the placenta, while carbon dioxide and

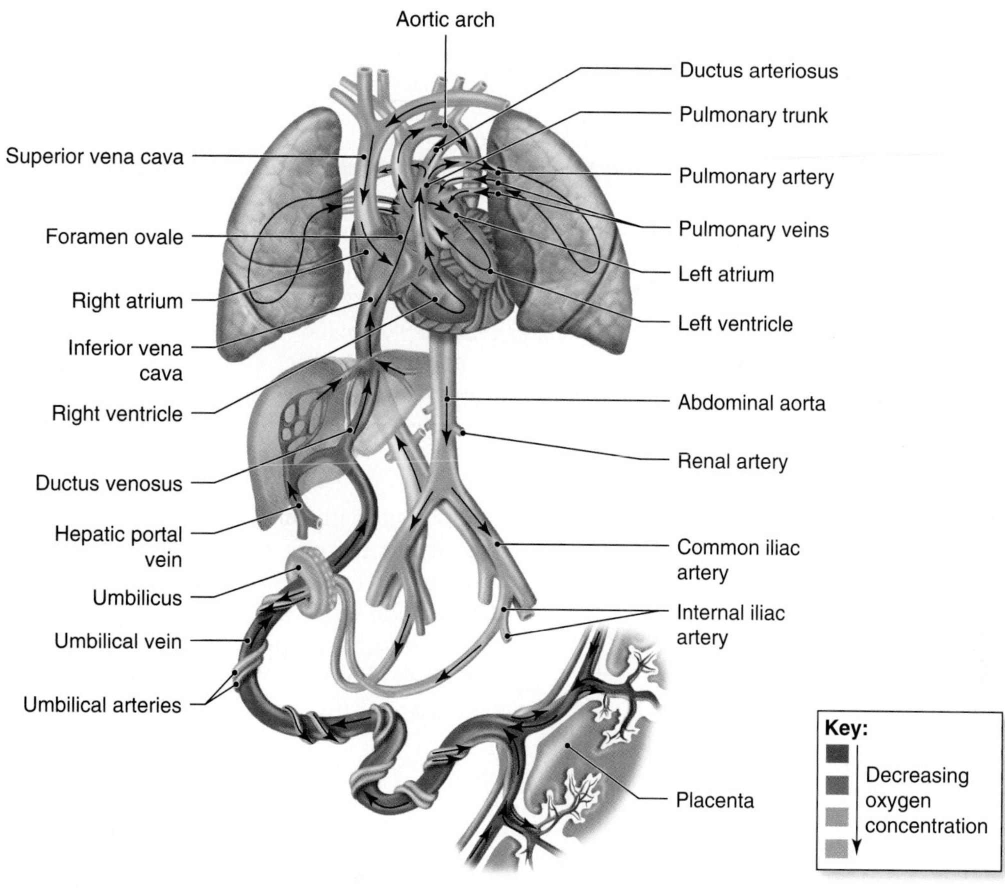

Figure 18.12 Cardiovascular Adaptations in a Fetus.

other metabolic wastes are removed via the maternal blood. The pattern of fetal circulation is an adaptation to these conditions. Birth immediately separates the infant from its supply of nutrients and oxygen and stimulates cardiovascular changes to accommodate independent living as an air-breathing human.

Fetal Cardiovascular Adaptations

Figure 18.12 illustrates the pattern of fetal circulation. Oxygenated and nutrient-rich blood is carried from the placenta to the fetus by the *umbilical vein,* which enters the fetus at the umbilicus (navel). Within the fetus, the umbilical vein passes toward the liver, where it divides into two branches. About half of the blood carried by the vessel enters the liver, while the other half bypasses the liver by flowing through the **ductus venosus** (duk′-tus ven-ō′-sus) and into the inferior vena cava. Full blood flow through the fetal liver is not necessary because the fetal intestines are nonfunctional and the mother's liver removes potentially hazardous substances before her blood enters the placenta. The oxygenated blood from the umbilical vein is mixed with deoxygenated blood in the inferior vena cava. The addition of blood from the ductus venosus increases the blood pressure within the inferior vena cava and the right atrium, which keeps the foramen ovale open.

Most of the blood entering the right atrium of the fetal heart passes directly through the **foramen ovale** (ō-vah′-lē), an opening in the interatrial septum, into the left atrium. The blood in the left atrium flows into the left ventricle and is pumped into the aorta for transport throughout the body. The blood that does enter the right ventricle is pumped through the pulmonary trunk. However, most of it bypasses the lungs by flowing through the **ductus arteriosus** (duk′-tus ar-te-rē-ō′-sus) into the aortic arch. These two lung bypasses work together to provide better oxygen and nutrient delivery to fetal tissues by providing additional blood for transport to the body. However, sufficient blood flows through the pulmonary circuit to maintain the nonfunctional lungs.

Blood is returned to the placenta by two *umbilical arteries* that branch from the internal iliac arteries. Trace the flow of blood through the fetal cardiovascular system shown in figure 18.12. Table 18.3 summarizes these fetal cardiovascular adaptations.

Table 18.3 Fetal Cardiovascular Adaptations

Structure	Function
Umbilical vein	Carries oxygenated and nutrient-rich blood from the placenta to the fetus
Ductus venosus	Carries about half of the blood in the umbilical vein into the inferior vena cava, bypassing the liver and mixing oxygenated and deoxygenated blood
Foramen ovale	Allows a large portion of the blood entering the right atrium to pass through the interatrial septum directly into the left atrium, bypassing the pulmonary circuit and providing as much oxygen and as many nutrients as possible for body cells via the systemic circuit
Ductus arteriosus	Carries most blood from the pulmonary trunk directly into the aorta, bypassing the nonfunctional lungs and providing more blood with available oxygen and nutrients for the systemic circuit
Umbilical arteries	Carry deoxygenated blood from the internal iliac arteries back to the placenta

Postnatal Cardiovascular Changes

After the infant is breathing, changes are made to convert the pattern of fetal circulation into that of an air-breathing infant. This involves closure of the foramen ovale and the constriction of all of the vessels used to get blood quickly from the umbilical vein into the aorta.

The distal portions of umbilical arteries constrict, inhibiting the flow of blood to the placenta. Subsequently, the proximal portions of the umbilical arteries persist as *superior vesical arteries* and the distal portions become *medial umbilical ligaments.* Blood continues to flow through the umbilical vein from the placenta to the newborn for about one minute, and then it constricts. It will become the *round ligament.* At the same time, the ductus venosus constricts. It will subsequently become the *ligamentum venosum* in the wall of the liver.

As the umbilical vein constricts and the pulmonary circulation becomes functional, blood pressure in the right atrium decreases, whereas blood pressure in the left atrium increases due to increased blood flow to and from the lungs. The higher blood pressure in the left atrium closes a tissue flap in the left atrium over the foramen ovale, separating the pulmonary and systemic circuits. Fusion of the tissue flap leaves a slight depression in the interatrial septum, called the *fossa ovalis,* at the site of closure. About the same time, the ductus arteriosus constricts and ultimately becomes the *ligamentum arteriosum.* These cardiovascular changes are functionally complete within 30 minutes after birth, but it takes about one year for tissue growth to make them permanent (figure 18.13).

Clinical Insight

Umbilical cord blood is rich in blood-forming stem cells, which are required for the production of formed elements. After delivery and after the umbilical cord is cut, cord blood may be collected, saved, and kept frozen and available in case it is needed by the child later in life. If the child develops leukemia later in life and needs a transfusion of blood-forming stem cells, the cord blood is available for transfusion. Because it is essentially a self-transfusion, rejection is not a problem.

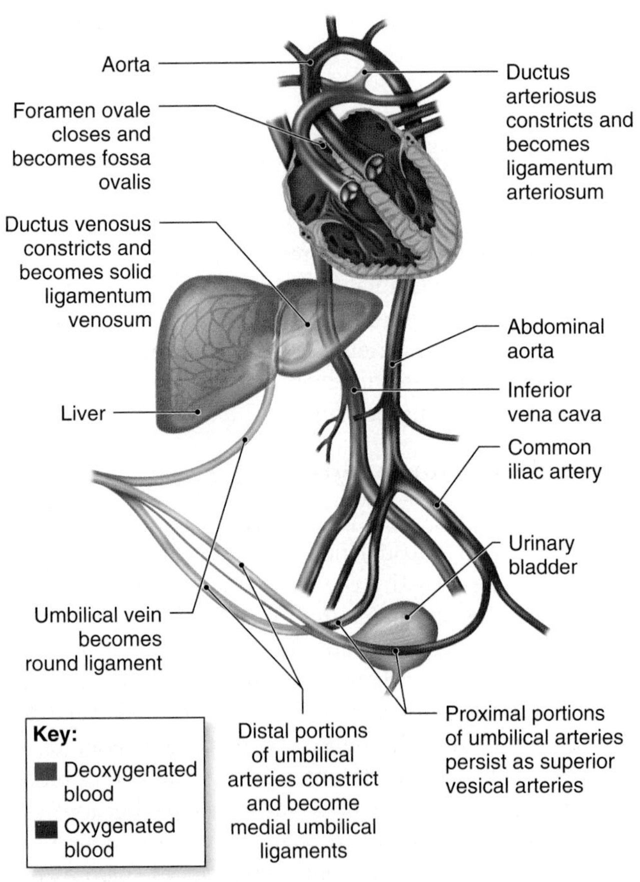

Figure 18.13 Cardiovascular Changes in a Newborn Infant.

Check My Understanding

10. What are the cardiovascular adaptations in a fetus?
11. What cardiovascular changes occur in a newborn infant?

18.7 Lactation

Learning Objective

11. Describe the control of lactation and milk ejection.

High blood levels of estrogens and progesterone during pregnancy stimulate the development of the mammary glands and enlargement of the breasts in preparation for milk secretion, or **lactation** (lak-tā′-shun). Although the mammary glands are capable of secreting milk, the high levels of estrogens and progesterone inhibit the hypothalamus so that milk secretion does not occur during pregnancy.

After birth, the blood levels of estrogens and progesterone drop dramatically, removing the inhibitory effect and enabling the hypothalamus to secrete *prolactin-releasing hormone (PRH).* PRH activates the anterior lobe of the pituitary to secrete **prolactin,** which stimulates milk production.

Prolactin stimulates the mammary glands to secrete milk, but its effects are not evident for two to three days. In the meantime, the mammary glands produce *colostrum* (kō-los′-trum), which differs from true milk by containing higher concentrations of protein and essentially no fat. The high protein content provides an added boost of essential nutrients for protein synthesis, which is needed for the continued development and growth of the infant.

The continued secretion of both prolactin and milk are maintained by mechanical stimulation of the nipples by the suckling infant. Suckling triggers the formation of nerve impulses that stimulate the hypothalamus to secrete PRH, which, in turn, causes continued prolactin secretion by the anterior lobe of the pituitary. Thus, lactation is maintained by a *positive-feedback* mechanism.

Clinical Insight

Breast-feeding seems to provide advantages for the infant, including (1) better nutrition because nutrients are easier to absorb, (2) rapid bonding due to prolonged contact with the mother, (3) antibodies that prevent digestive inflammation and provide defense against pathogens, and (4) enhanced cognitive development.

Clinical Insight

Pitocin, the brand name for oxytocin, can be administered intravenously to induce labor, reinforce ongoing labor contractions, and control postpartum bleeding and hemorrhage. The oxytocin released during breast-feeding promotes the uterine contractions needed to return the uterus to near its nonpregnant size.

Milk does not simply flow from the breasts. Instead, it is ejected after about 30 seconds of suckling by the infant. Stimulation of the nipple by suckling sends nerve impulses to the hypothalamus, which triggers the release of oxytocin by the posterior lobe of the pituitary. *Oxytocin* stimulates contraction of specialized epithelial cells surrounding the ducts and alveolar glands within the mammary glands, resulting in milk ejection or "letdown."

Milk production may continue for as long as the nipple is suckled, but the volume of milk produced gradually declines. If suckling is stopped, milk accumulates, the secretion of prolactin is inhibited, and milk production ceases in about a week.

Check My Understanding

12. How do hormones control lactation?

18.8 Disorders of Pregnancy, Prenatal Development, and Postnatal Development

Learning Objective

12. Identify the major disorders of pregnancy, prenatal development, and postnatal development.

Pregnancy Disorders

Eclampsia (ē-klamp′-sē-ah), or *toxemia of pregnancy,* is a disorder that occurs in two forms. *Preeclampsia* of late pregnancy is characterized by increased blood pressure, edema, and proteinuria (protein in the urine). The cause is unknown. If unsuccessfully treated, it may develop into *eclampsia,* a far more serious disorder that may lead to convulsions and coma. Both infant and maternal mortality are high in eclampsia. Rapid termination of the pregnancy by Cesarean section may be indicated.

An **ectopic pregnancy** is the implantation of a preembryo anywhere other than in the uterus. A common site is in a uterine tube. Treatment involves surgical removal of the embryo.

A **miscarriage** is a spontaneous abortion. Most miscarriages occur within the first twelve weeks of development as a result of gross abnormalities of the embryo or placenta. Another cause is the untimely transfer of the production of estrogens and progesterone from the corpus luteum to the placenta at approximately 12 weeks of development.

Morning sickness is characterized by nausea and vomiting upon getting up in the morning. It usually starts around the sixth week of the pregnancy and typically lasts from one to six weeks. The exact cause is unknown, though high levels of hCG and progesterone in the blood are believed to play a role. About 60% of pregnant women experience this discomfort.

Prenatal and Postnatal Disorders

Birth defects may be inherited or may be caused by a variety of *teratogens* (ter-ah′-to-jens), environmental agents that produce physical abnormalities during prenatal development. Teratogens include alcohol, illegal drugs, some therapeutic drugs, X-rays, and certain diseases such as German measles (Rubella). Generally, the earlier the embryo is exposed, the greater the defect produced. Alcohol is the most common teratogen. It produces *fetal alcohol syndrome,* which is characterized by a small head; mental retardation; facial deformities; and abnormalities of the heart, genitals, and limbs.

Physiological jaundice, a postnatal disorder, sometimes occurs in a newborn because the destruction of RBCs occurs faster than the liver can process the bilirubin. This results in excess bilirubin in the blood. Phototherapy (exposure to UV light) is a common treatment to speed up bilirubin breakdown. Jaundice may be a more serious problem in premature infants. This type of jaundice usually resolves once the newborn's liver gains full function.

Infant respiratory distress syndrome (IRDS), or *hyaline membrane disease,* is a postnatal disorder characterized by an inability to produce surfactant within the lungs of an infant. The lack of surfactant decreases the ability of the infant to successfully inflate its lungs during inspiration. IRDS is most common in premature infants, whose lungs have not yet begun or have not completed surfactant production.

Sudden infant death syndrome (SIDS), or "crib death," is a postnatal disorder characterized by the sudden death of an infant with no medical history or explanation upon autopsy. Infants are at highest risk of SIDS during sleep and, though its exact cause is unknown, risk factors such as hypoxia while sleeping, deficits in respiratory control, and nicotine exposure during development have been identified.

18.9 Genetics

Learning Objectives

13. Explain the roles of DNA, genes, and chromosomes in inheritance.
14. Describe the basic patterns of inheritance.

Genetics is the study of heredity, the passing of inherited traits from one generation to the next. The determiners of hereditary traits are located on **chromosomes,** consisting of DNA and proteins. It is the DNA that controls inheritance and directs cellular functions.

Each human body cell contains 46 chromosomes that exist as 23 unique pairs. Chromosome pairs 1 through 22 are called *autosomes* because they control most inherited traits except gender. Gender is determined by chromosome pair 23, the *sex chromosomes.* There are two types of sex chromosomes, a large X chromosome and a small Y chromosome. Males possess one X chromosome and one Y chromosome (XY). Females possess two X chromosomes (XX).

A person's chromosomes, including the sex chromosomes, may be examined by making a *karyotype.* The chromosomes in a dividing cell are photographed during metaphase (see chapter 3) and the photograph is enlarged. Then the chromosomes are cut out, matched in pairs, and arranged by size and location of the centromere. Figure 18.14 is a karyotype of a normal male. Note the X and Y chromosomes and that the chromosomes are arranged in pairs.

Sex Determination

Recall that gametes are formed by meiotic cell division, a process that places one member of each chromosome pair

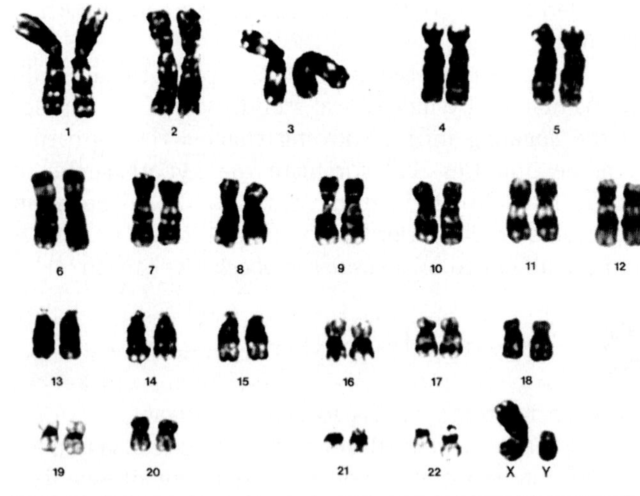

Figure 18.14 Karyotype of a human male. The only difference in a female karyotype would be the presence of a pair of X chromosomes in place of the XY pair of the male.

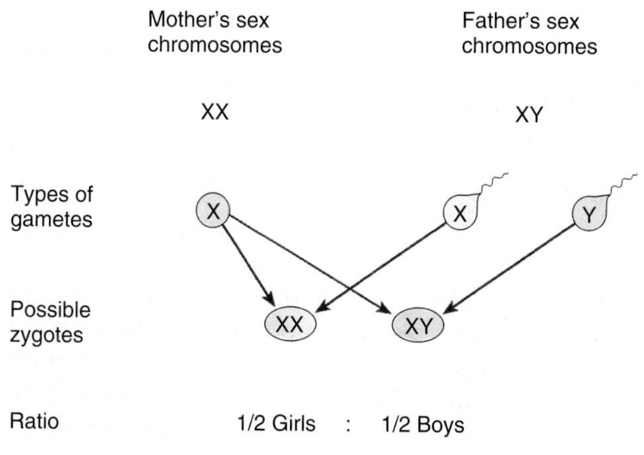

Figure 18.15 The inheritance of sex.

in each gamete. Each human gamete contains 23 chromosomes–22 autosomes and 1 sex chromosome. We will consider only the sex chromosomes here.

Because a female has two X chromosomes in her cells, all of her gametes contain an X chromosome. A male has both an X chromosome and a Y chromosome in his cells. Therefore, half of his gametes are X-bearing, and half are Y-bearing. If a secondary oocyte is fertilized by an X-bearing sperm, the child will be a girl. If a secondary oocyte is fertilized by a Y-bearing sperm, the child will be a boy. Obviously, the probability of any zygote becoming a girl (or a boy) is one-half or 50%. Figure 18.15 illustrates the determination of sex.

Genes

Recall from chapters 2 and 3 that DNA consists of a double strand of nucleotides that are joined by complementary pairing of their nitrogenous bases: adenine (A), thymine (T), cytosine (C), and guanine (G). The sequence of these bases forms the genetic code, which contains the information for producing proteins that regulate cellular functions and determines the inheritance of genetic traits.

A **gene** is a unit of inheritance. It consists of a specific sequence of DNA that codes for a unique molecule of RNA. This RNA molecule will either be directly involved in the synthesis of a polypeptide or indirectly involved in regulating the production of a polypeptide. Genes occur in a linear sequence along a chromosome and a single chromosome may contain hundreds of genes.

Because chromosomes occur in pairs, genes also occur in pairs. An inherited trait is determined by at least one pair of genes. There may be two or more alternate forms of a gene controlling the expression of a particular trait. These alternate forms are called **alleles** (ah-lēls′), and each allele affects the expression of a trait differently. So, in the simplest case, a trait is determined by one pair of alleles present in a person's cells. If the two alleles for a trait are identical, the person is **homozygous** for that trait; if they are different, the person is **heterozygous** for that trait.

Gene Expression

Each person's chromosomes contain a unique catalog of genes, the **genotype** for that person. The expression of those genes yields observable traits known as the **phenotype.** Though the phenotype is what is seen, the genotype is responsible for the inheritance and expression of those traits.

Dominant and Recessive Inheritance

Some alleles are dominant, and some are recessive. A *dominant allele* is always expressed, whereas a *recessive allele* is expressed only when both alleles are recessive.

Consider the example of skin pigmentation. Normal skin pigmentation is controlled by a dominant allele *(A)*. The absence of pigment (albinism) is controlled by a recessive allele *(a)*. Note that it requires only one dominant allele to express the dominant trait but that both recessive alleles must be present for the recessive trait to be expressed.

Genotype	Phenotype
AA	Normal
Aa	Normal
aa	Albino

Table 18.4 indicates a few human traits that are determined by dominant and recessive alleles.

Clinical Insight

From 1990 to 2003, the *Human Genome Project* was an international research project that determined the base sequence of more than 92% of the human genome and mapped the base sequences of known genes. This information has greatly accelerated genetic discoveries, such as identifying genes that cause human genetic disorders. Treatment regimens for treating genetic disorders have also advanced through development of targeted drug therapies and genetic engineering. For example, great strides have been made in preventing or modulating the effects of autoimmune disorders, such as type I diabetes mellitus, by transferring therapeutic genes into affected cells in mice. These findings have definite future human applications. Human trials involving selective destruction of the autoimmune T cells that destroy beta cells within the pancreas have also shown promising results.

Table 18.4 Examples of Traits Determined by Dominant and Recessive Alleles

Traits Determined by Dominant Alleles	Traits Determined by Recessive Alleles
Freckles	Absence of freckles
Dimples in cheeks	Absence of dimples
Dark hair	Light hair or red hair
Full lips	Thin lips
Free earlobes	Attached ear lobes
No thalassemia	Thalassemia
Feet with arches	Flat feet
Huntington disease	No Huntington disease
Astigmatism	No astigmatism
Farsightedness	No farsightedness
Panic attacks	No tendency to panic attacks
Extra fingers or toes	Normal number of digits
No cystic fibrosis	Cystic fibrosis
No hemophilia	Hemophilia*
Type A, B, or AB blood	Type O blood
Type Rh+ blood	Type Rh− blood
Normal color vision	Red–green color blindness*
No gout tendency	Gout*

*X-linked trait.

Incomplete Dominance

Incomplete dominance is a type of inheritance where the two alleles for a gene can create three different phenotypes. Each genotype–homozygous dominant, heterozygous, and homozygous recessive–has a different phenotype. An example is sickle-cell disease, a condition characterized by defective hemoglobin that cannot carry adequate oxygen. Erythrocytes with the defective hemoglobin assume a characteristic sickled or crescent shape. Sickle-cell disease occurs among people whose ancestors lived in central Africa. About 8.3% of black Americans possess the allele for sickle-cell disease.

A person who inherits both recessive alleles for sickle-cell disease (H^SH^S) produces abnormal hemoglobin, leading to the formation of sickled cells that cannot carry sufficient oxygen. Because of their shape, the sickled cells tend to plug capillaries. Symptoms include pain in joints and the abdomen and chronic kidney disease.

In the heterozygous state (HH^S), some hemoglobin molecules are normal but others are abnormal. Fortunately, few RBCs become sickled when oxygen is at normal levels and clinical symptoms are absent at such times. However, more RBCs become sickled during times of decreased blood oxygen level, a characteristic that allows detection of carriers of the sickle-cell allele. The heterozygote state affords some protective advantage against the pathogen causing malaria. The homozygous dominant genotype produces the phenotype of all normal hemoglobin.

Codominance

In some traits, both alleles are expressed and affect the phenotype. This type of inheritance is referred to as **codominance.** An example of codominance can be seen with the ABO blood group. There are three alleles involved: a dominant I^A that causes the production of the A antigen; a dominant I^B that causes the production of the B antigen; a recessive *i* that has no function. If both I^A and I^B are present, both alleles are expressed. Since the recessive *i* has no function, genotype *ii* produces neither A nor B antigens. This is called type O blood, which simply means there are no A or B antigens. The possible genotypes and phenotypes for the ABO blood group are

Genotypes	Phenotypes
I^AI^A, I^Ai	Type A blood
I^BI^B, I^Bi	Type B blood
I^AI^B	Type AB blood
ii	Type O blood

Polygenetic Inheritance

Many traits are controlled by **polygenes,** a number of different genes that may be located on the same or different chromosomes. Each gene contributes to the phenotype, though some genes may have more influence on the trait than others. To add to the complexity of polygenic inheritance, each gene involved may possess a number of different alleles. Environmental factors may also exert influence over the expression of a phenotype. For these reasons, it is difficult to predict the inheritance of polygenic traits. Examples of traits controlled by polygenes are height, skin pigmentation, and intelligence.

The ABO blood group is also governed by polygenes. The gene for the H antigen is found on chromosome 19. The H gene possesses two alleles: a dominant *H* that causes the production of H antigen and a recessive *h* that is nonfunctional. Individuals who are homozygous dominant (*HH*) or heterozygous (*Hh*) possess the H antigen. The I^A and I^B alleles, which are located on chromosome 9, produce enzymes that add to the H antigen and produce either A or B antigens. Many people mistakenly conclude that type O blood has no antigens because the *i* alleles have no function. However, most people with blood type O actually have H antigens. Individuals with genotype *hh* do not produce the H antigen and have what is called the *Bombay phenotype.* These individuals will be Type O even if their genotype contains the I^A, I^B, or both I^A and I^B alleles because, without the H antigen, A and B antigens cannot be formed.

X-Linked Traits

A few traits are determined by genes on the X chromosome. These are **X-linked,** or *sex-linked,* **traits.** Recessive X-linked traits affect males more frequently than females. Males only possess one X chromosome. If a recessive trait is carried by the X chromosome in a male, the trait will be seen. Females possess two X chromosomes. To see the recessive trait, a female must possess two recessive alleles. If the female possesses one dominant "normal" allele, the recessive trait will not be seen.

Red-green color blindness is a common X-linked recessive trait. A color-blind male inherits the allele for color blindness from his mother, who provides his X chromosome. The mother may either have normal color vision or be red-green color-blind (table 18.5). It is important to note that if the mother has normal color vision, she still possesses the allele for color blindness and is considered a carrier for the color-blindness trait.

Table 18.5 Possible Genotypes and Phenotypes for Red-Green Color Blindness, an X-Linked Trait

	Genotype	Phenotype
Females	$X^C X^C$	Normal color vision
	$X^C X^c$	Normal color vision carrier
	$X^c X^c$	Color-blind
Males	$X^C Y$	Normal color vision
	$X^c Y$	Color-blind

C = Allele for normal color vision; *c* = Allele for color blindness.

Predicting Inheritance

Parents often wonder about the chances of their child developing certain inherited traits. This can be predicted for some traits for which the inheritance pattern has been determined and if the genotypes of the parents are known. Such predictions indicate the *probability,* rather than absolute certainty, that a trait will be inherited.

Let's consider freckles. Freckles are determined by a dominant allele (*F*), and a nonfreckled phenotype is determined by a recessive allele (*f*). The possible genotypes and phenotypes are

Genotypes	Phenotypes
FF	Freckled
Ff	Freckled
ff	Nonfreckled

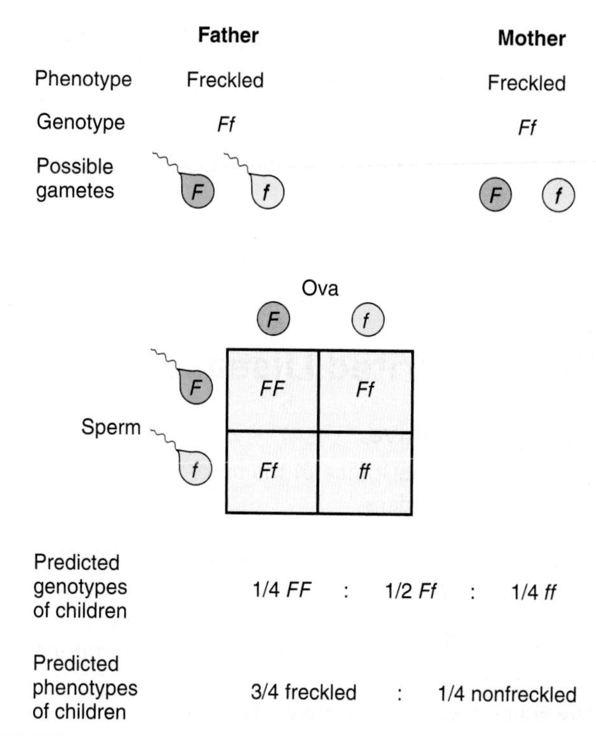

Figure 18.16 The probability of freckles being inherited by children of parents who are heterozygous for freckles.

Figure 18.16 shows how to determine the probability of the freckled or nonfreckled trait in the next generation if the genotypes of the parents are known. In this example, the parents are known to be heterozygous for freckles. What is the probability that their children will be freckled?

Because each parent is heterozygous, meiotic division during gamete formation causes half of the gametes of each parent to contain an allele for freckles (*F*), and half to carry an allele for normal pigmentation (*f*). The union of sperm and secondary oocyte occurs at random (by chance), so we must allow for all possible combinations of gametes. This is accomplished by using a *Punnett square* (a chart named after Reginald Punnett, a geneticist).

The alleles in ovum are placed along the horizontal axis, while the alleles in sperm are placed along the vertical axis. Next, the allele of each ovum is written in the squares below each ovum and the allele of each sperm is written in the squares to the right of each sperm. The Punnett square now shows all possible genotypes that may occur in the next generation.

From this information, the predicted genotype ratio may be determined. Then, knowing that the trait for freckles is dominant and that the presence of a single dominant allele (*F*) produces freckles, the predicted phenotype ratio may be determined. Note in figure 18.16 that it is possible for two heterozygous freckled parents to have a child with normal pigmentation. However, if one parent is homozygous dominant for freckles and the other is heterozygous for freckles, all children would be freckled.

The inheritance of any dominant/recessive trait may be determined in a similar manner.

Check My Understanding

13. What are the relationships among chromosomes, DNA, genes, and alleles?
14. What distinguishes the dominant/recessive pattern of inheritance?
15. Why are recessive X-linked traits expressed more often in males than in females?

18.10 Inherited Diseases

Learning Objective

15. Explain the inheritance of the more common inherited disorders.

Inherited, or genetic, diseases are caused by either chromosome abnormalities or specific alleles. The development of advanced techniques and new knowledge makes an understanding of genetic disease increasingly important.

Chromosome Abnormalities

Some genetic diseases are related to the presence of an additional chromosome or to the absence of a chromosome. These disorders result from errors that occur during meiotic cell division, causing some gametes to receive both members of a chromosome pair while other gametes receive neither member. If such gametes are involved in zygote formation, a genetic disorder occurs. The genetic damage usually is so severe that it causes a spontaneous abortion. In some cases, the effect is not lethal but disabling.

An example of a disabling genetic disorder is Down syndrome, one of the more common genetic disorders. It is caused by the presence of an extra chromosome 21, as shown in figure 18.17. Down syndrome is characterized by mental retardation, short stature, short digits, slanted eyes, and a protruding tongue. A girl with Down syndrome is shown in figure 18.18. Infants with Down syndrome are born more often to mothers and fathers over 40 years of age.

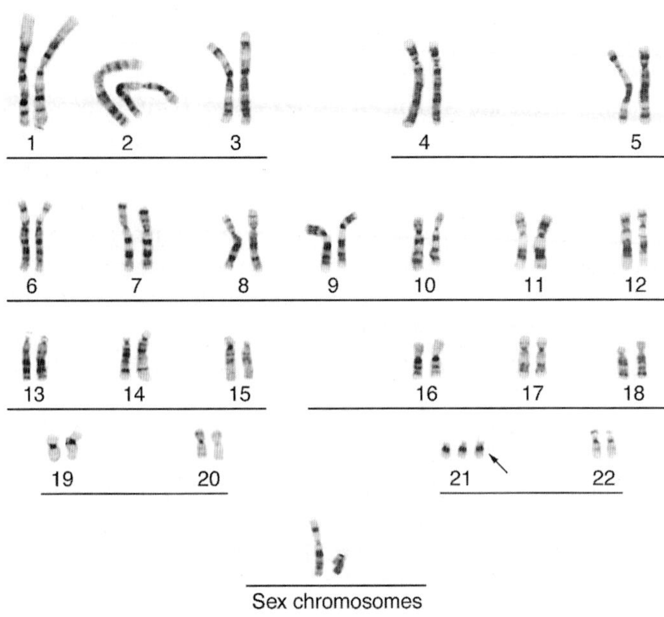

Figure 18.17 A karyotype of a male with Down syndrome caused by trisomy 21.

Figure 18.18 A girl with Down syndrome.

Single-Gene Disorders

These disorders usually affect the infant's metabolism after birth, when it must depend on its own life processes, or they may appear later in life. Consider a few examples of single-gene disorders.

Cystic fibrosis, an autosomal recessive disorder, is the most common genetic disorder among Caucasians. It is caused by a missing chloride channel on mucus-secreting cells. This causes production of thick mucus that blocks respiratory airways and leads to an early death from respiratory infections.

Phenylketonuria (PKU), an autosomal recessive disorder, is due to a missing enzyme needed to metabolize phenylalanine (an amino acid). Without treatment, mental and physical retardation result. A special diet that limits phenylalanine can prevent these effects if it is started at birth and continued to adulthood.

Tay-Sachs disease, an autosomal recessive disorder, primarily affects Jewish people of central European ancestry. An enzyme needed to metabolize a fatty substance associated with neurons is missing. The results are mental retardation, muscle weakness, seizures, and finally death, usually by two years of age.

Huntington disease, an autosomal dominant disorder, results from one or more missing enzymes needed in cellular respiration. This causes a buildup of lactic acid in neurons in the brain. Uncontrollable muscle contractions, memory loss, and personality changes begin between 30 and 50 years of age. Death occurs within 15 years after the appearance of symptoms.

Hemophilia A and *hemophilia B,* X-linked recessive disorders, result from missing clotting factors. Prolonged bleeding can be life threatening, and joints may be painfully disabled. Patients are dependent upon frequent transfusions of normal plasma or intravenous injections of the missing clotting factor.

Genetic Counseling

Prospective parents who have genetic disorders in one or both of their families may benefit from genetic counseling. By collecting genetic information, a genetic counselor can inform prospective parents of the probability of a genetic disorder's appearance in their children. Genetic information may be collected from family histories and blood tests of the prospective parents and family members. If the woman is pregnant, ultrasound may be used to detect gross fetal abnormalities, and fetal cells may be obtained for examination. Fetal cells are obtained in two ways: amniocentesis and chorionic villi sampling.

In *amniocentesis,* a hollow needle is inserted through the abdominal and uterine walls of the mother and into the amnion to draw out a sample of amniotic fluid (figure 18.19). Fetal cells in the sample are grown by tissue culture and are analyzed to see if there are chromosomal abnormalities. Also, the amniotic fluid is analyzed for the presence of specific proteins that indicate serious neural defects. Amniocentesis is usually done around the 14th week of development, when ample amniotic fluid is available for sampling without injury to the fetus.

In *chorionic villi sampling,* a narrow tube is inserted through the cervix and fetal tissue from chorionic villi is suctioned out. Chromosomes in the fetal cells are examined for abnormalities. Chorionic villi sampling may be performed at 10 weeks and chromosome examination can be done immediately.

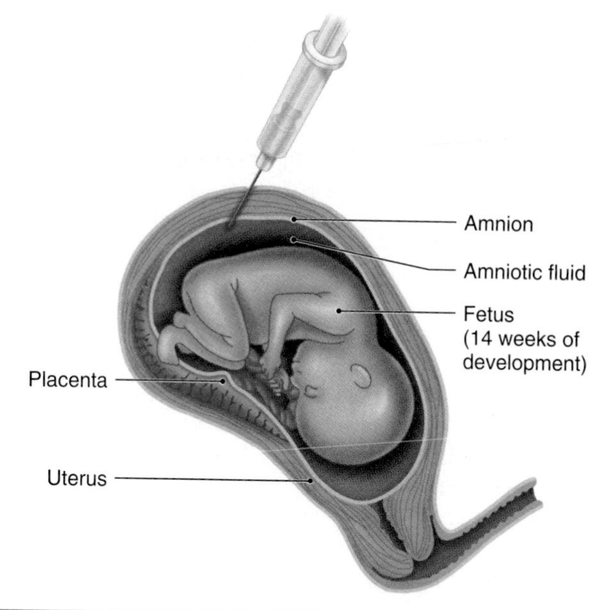

Figure 18.19 In amniocentesis, a sample of amniotic fluid is withdrawn and the suspended fetal cells are examined for genetic abnormalities.

Both amniocentesis and chorionic villi sampling have inherent risks for mother and fetus. Fetal risks seem to be greater in chorionic villi sampling.

Fetal cells can also be collected for analysis through a procedure called *fetal cell sorting.* A small amount of fetal cells enter the mother's blood supply during pregnancy. A fluorescent cell sorter can be used to identify and separate out the rare fetal cells from a maternal blood sample for analysis. This method of fetal cell collection circumvents the health risks associated with amniocentesis and chorionic villi sampling. Though these conventional methods are still more commonly used, fetal cell sorting is being used for purposes of sex determination and determination of fetal Rh status, in addition to the detection of major chromosomal abnormalities and some single-gene disorders.

Chapter Summary

18.1 Fertilization and Early Development

- Sperm usually encounter the secondary oocyte in a uterine tube. Only one sperm can enter an oocyte.
- When a secondary oocyte is penetrated by a sperm, it undergoes the second meiotic division, forming an ovum and a second polar body. Fertilization occurs with the fusion of the sperm nucleus and ovum nucleus, forming a zygote. The zygote is the first stage of development.
- The zygote undergoes cleavage divisions, forming a morula. Continued mitotic divisions form a blastocyst containing an embryoblast.
- On about the seventh day of development, the blastocyst becomes implanted in the endometrium.

18.2 Embryonic Development

- The embryonic disc separates into the three germ layers: ectoderm, mesoderm, and endoderm.

- The trophoblast becomes the chorion, the first and most superficial extraembryonic membrane. The chorionic villi will become the embryonic part of the placenta.
- The amnion develops posterior to the embryo, envelops the embryo, and becomes filled with amniotic fluid that serves as a shock absorber.
- The yolk sac develops anterior to the embryo. It subsequently branches to form the allantois. Both of these extraembryonic membranes form the early formed elements for the embryo. The allantois also brings umbilical blood vessels to the placenta.
- The placenta is formed of both embryonic and maternal tissues and is functional by the end of the eighth week. It allows the exchange of materials between embryonic and maternal bloods.
- By the seventh week, the embryo exhibits a head, a body, limbs with digits, eyes, and ears.

18.3 Fetal Development

- The embryo is called a fetus at the beginning of the ninth week, and it clearly resembles a human. Organ systems are in rudimentary form and the head is as large as the body.
- Fetal development includes development of the organ systems to functional levels, an increase in size and weight, ossification of bones, and development of distinguishable gender.

18.4 Hormonal Control of Pregnancy

- Through 10 to 12 weeks of development, the trophoblast and chorion produce hCG, which maintains the corpus luteum in the ovary. The corpus luteum produces progesterone and estrogens, which in turn maintain the endometrium.
- By approximately twelve weeks, the placenta takes over the production of progesterone and estrogens. The corpus luteum degenerates and the ovaries remain inactive for the remainder of the pregnancy.

18.5 Birth

- Relaxin from the corpus luteum and placenta promotes the growth of placental blood vessels, in addition to other cardiovascular changes in the mother.
- The high blood levels of estrogens in the late stage of pregnancy counteracts the inhibitory action of progesterone against uterine contractions.
- The onset of labor involves nerve impulses and oxytocin. Pressure of the fetus on the cervix leads to the formation of nerve impulses that are sent to the hypothalamus, which stimulates oxytocin release by the posterior lobe of the pituitary. Oxytocin stimulates uterine contractions, dilating the cervix, which triggers the formation of more nerve impulses. These interactions set up a positive-feedback mechanism that strengthens contractions until birth.
- Labor involves three stages: (a) dilation of the cervix, (b) delivery of the infant, and (c) delivery of the placenta.
- After birth, an accumulation of carbon dioxide stimulates the respiratory rhythmicity center to trigger the first breath. Surfactant in alveoli makes breathing easier.

18.6 Cardiovascular Adaptations

- The pattern of fetal circulation is adapted to a life in which the lungs and digestive system are nonfunctional and nutrients and oxygen are derived from the mother's blood via the placenta. Embryonic blood is carried to the placenta by umbilical arteries and is returned by an umbilical vein.
- Cardiovascular adaptations pass oxygenated and nutrient-rich blood as quickly as possible from the umbilical vein to the aorta to meet the needs of body cells. These adaptations include the ductus venosus, foramen ovale, and ductus arteriosus.
- When the infant starts breathing, fetal cardiovascular adaptations are eliminated to enable an efficient separation of the pulmonary and systemic circuits.

18.7 Lactation

- The high blood levels of estrogens and progesterone during pregnancy prepare the mammary glands for lactation but prevent hormone stimulation of lactation until birth.
- After birth, the low blood levels of progesterone and estrogens allow the hypothalamus to secrete PRH, which stimulates the anterior lobe of the pituitary to secrete prolactin. Prolactin stimulates the mammary glands to secrete milk.
- Colostrum is secreted first, followed by true milk after two to three days.
- Suckling stimulates the formation of nerve impulses that are sent to the hypothalamus, which (a) secretes PRH, ensuring the continued release of prolactin and milk secretion; and (b) stimulates the posterior lobe of the pituitary to secrete oxytocin, which results in milk ejection.

18.8 Disorders of Pregnancy, Prenatal Development, and Postnatal Development

- Pregnancy disorders include eclampsia, ectopic pregnancy, miscarriage, and morning sickness.
- Prenatal and postnatal disorders include birth defects, physiological jaundice, infant respiratory distress syndrome, and sudden infant death syndrome.

18.9 Genetics

- The determiners (genes) of hereditary traits are located on chromosomes. There are 46 chromosomes in human body cells: 22 pairs of autosomes and one pair of sex chromosomes.

- Sex chromosomes for females are XX; for males, XY.
- A gene is the unit of heredity. It is a portion of a DNA molecule that codes for a specific molecule of RNA. The RNA molecule is either directly or indirectly involved in the production of a polypeptide.
- Alternate forms of a gene are called alleles. A person with identical alleles for a trait is homozygous for that trait. If the alleles are different, the person is heterozygous.
- The genotype is the alleles present for a gene determining a trait. The phenotype is the observable characteristics determined by the genotype.
- Many genes have only two alleles whose expression is either by dominant/recessive inheritance or by incomplete dominance. Some traits are determined by codominance and others are determined by polygenetic inheritance.
- X-linked traits are determined by genes located on the X chromosome.
- The probability of transmitting traits to the next generation may be predicted using a Punnett square.

18.10 Inherited Diseases

- Genetic diseases may result from either chromosomal abnormalities or defective alleles.
- Prospective parents who have genetic diseases in their family histories may benefit from genetic counseling.

Self-Review

Answers are located in appendix B.

1. Usually a sperm meets a secondary oocyte in the ______.
2. Fusion of sperm nucleus and ovum nucleus is called ______.
3. A ______ is implanted in the endometrium on about the ______ day of development.
4. Secretion of ______ by the trophoblast and chorion maintains the ______, which in turn continues to secrete estrogens and ______.
5. The embryonic disc forms the three ______ that subsequently form all other tissues of the embryo and fetus.
6. The ______ contains fluid in which the embryo develops, and the chorionic villi become the embryonic part of the ______.
7. At the beginning of the ______ week of development, the developing offspring is called a ______ and it clearly has the features of a human.
8. The first and longest stage of ______ is the dilation of the cervix, which results from uterine contractions stimulated by the hormone ______.
9. In a fetus, the ______ enables blood to pass from the right atrium into the left atrium, while the ______ passes blood from the pulmonary trunk into the aorta.
10. Lactation begins after ______ from the hypothalamus stimulates secretion of ______ by the anterior lobe of the pituitary gland.
11. The units of inheritance are ______, which are small segments of ______ that make up chromosomes.
12. Humans possess ______ chromosomes in their cells, and a person possessing two X chromosomes is a ______.
13. X-linked recessive traits occur more often in ______.
14. In dominant/recessive inheritance, a recessive allele is expressed only when the person is ______ for the recessive trait.
15. It is possible to predict the ______ of a trait appearing in children if the genotypes of parents are known.

Critical Thinking

1. Is it possible for a child with type O blood to be born to parents with type O blood and type AB blood? Why or why not?
2. Describe the various roles of the placenta during pregnancy.
3. Explain why a miscarriage is likely if the placenta is too slow to take over its hormone-producing role.
4. DNA analysis can determine if a person's genome contains any of the known defective genes. At present, there are no cures or effective treatments for most of the disorders caused by these genes. Would you like to know if you possess defective genes? Explain.

ADDITIONAL RESOURCES

STUDY GUIDE 1

1. Anatomy and Physiology

Write the terms that match the phrases in the spaces at the right.

1) The study of cells ______________________

2) The study of body organization and structure ______________________

3) The study of body functions ______________________

2. Levels of Organization

a. List the levels of organization from the most complex to the simplest.

1) ______________________ 4) ______________________

2) ______________________ 5) ______________________

3) ______________________ 6) ______________________

b. Write the terms that match the phrases in the spaces at the right.

1) A coordinated group of organs. ______________________

2) Structural and functional units of the body. ______________________

3) An aggregation of similar cells. ______________________

c. Match the names of the organ systems with the phrases.

Cardiovascular	Lymphoid	Reproductive, male
Digestive	Muscular	Respiratory
Endocrine	Nervous	Skeletal
Integumentary	Reproductive, female	Urinary

1) Stomach, liver, intestines. ______________________

2) Brain, spinal cord, nerves. ______________________

3) Secretes hormones. ______________________

4) Skin, hair, nails. ______________________

5) Returns lymph to blood; provides immunity. ______________________

6) Bones, ligaments, cartilages. ______________________

7) Contraction enables movement. ______________________

8) Transports materials to and from cells. ______________________

9) Kidneys, ureters, urinary bladder. ______________________

10) Testes, penis, prostate gland. ______________________

11) Ovaries, uterine tubes, uterus, vagina. ______________________

12) Blood, heart, arteries, veins. ______________________

13) Supports the body. ______________________

14) Secretes hormones that regulate functions. ______________________

15) Regulates volume of body fluids. ______________________

16) Protects underlying tissues. ______________________

17) Rapid coordination of body functions. ______________________

18) Digests food and absorbs nutrients. ______________________

19) Gas exchange between air and blood. ______________________

20) Larynx, trachea, bronchi, and lungs. ______________________

3. Directional Terms

Provide the term that correctly completes each statement.

1) The head is ______ to the neck. ______________________

2) The hand is ______ to the wrist. ______________________

3) The skin is ______ to the muscles. ______________________

4) The mouth is ______ to the nose. ______________________

5) The elbow is ______ to the wrist. ______________________

6) The ear is on the ______ surface of the head. ______________________

7) The umbilicus is on the ______ body surface. ______________________

8) The hip is on the ______ body surface. ______________________

9) The buttocks are on the ______ body surface. ______________________

4. Body Regions

Label the body regions by placing the number of the label line in the space by the correct label.

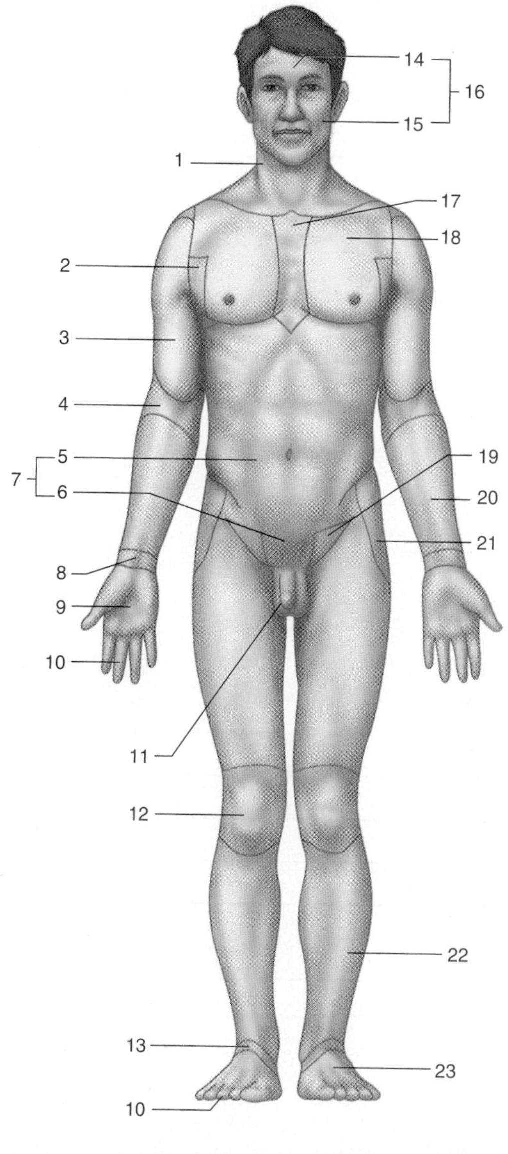

______ Abdominal
______ Abdominopelvic
______ Antebrachial
______ Antecubital
______ Axillary
______ Brachial
______ Carpal
______ Cephalic
______ Cervical
______ Coxal
______ Cranial
______ Crural
______ Digital
______ Facial
______ Genital
______ Inguinal
______ Palmar
______ Patellar
______ Pectoral
______ Pedal
______ Pelvic
______ Sternal
______ Tarsal

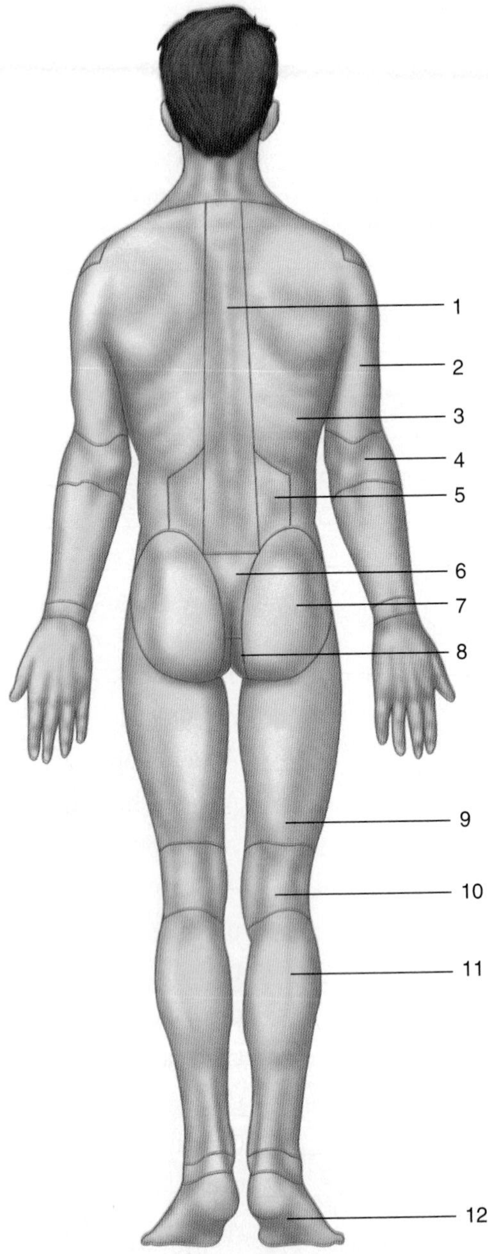

_____ Brachial
_____ Dorsum
_____ Femoral
_____ Gluteal
_____ Lumbar
_____ Olecranal
_____ Plantar
_____ Perineal
_____ Popliteal
_____ Sacral
_____ Sural
_____ Vertebral

5. Body Planes and Sections

Name the planes that match the statements.

1) Divides the body into equal left and right halves. ______________________
2) Divides the body into superior and inferior portions. ______________________
3) Divides the body into left and right portions. ______________________
4) Divides the body into anterior and posterior portions. ______________________
5) Any cut along the longitudinal axis of a structure. ______________________
6) Any cut at a 90° angle to the longitudinal axis of a structure. ______________________
7) A cut between the longitudinal axis and a 90° angle of a structure. ______________________

6. Body Cavities

a. Label the body cavities and related structures by placing the number of the label line in the space by the correct label.

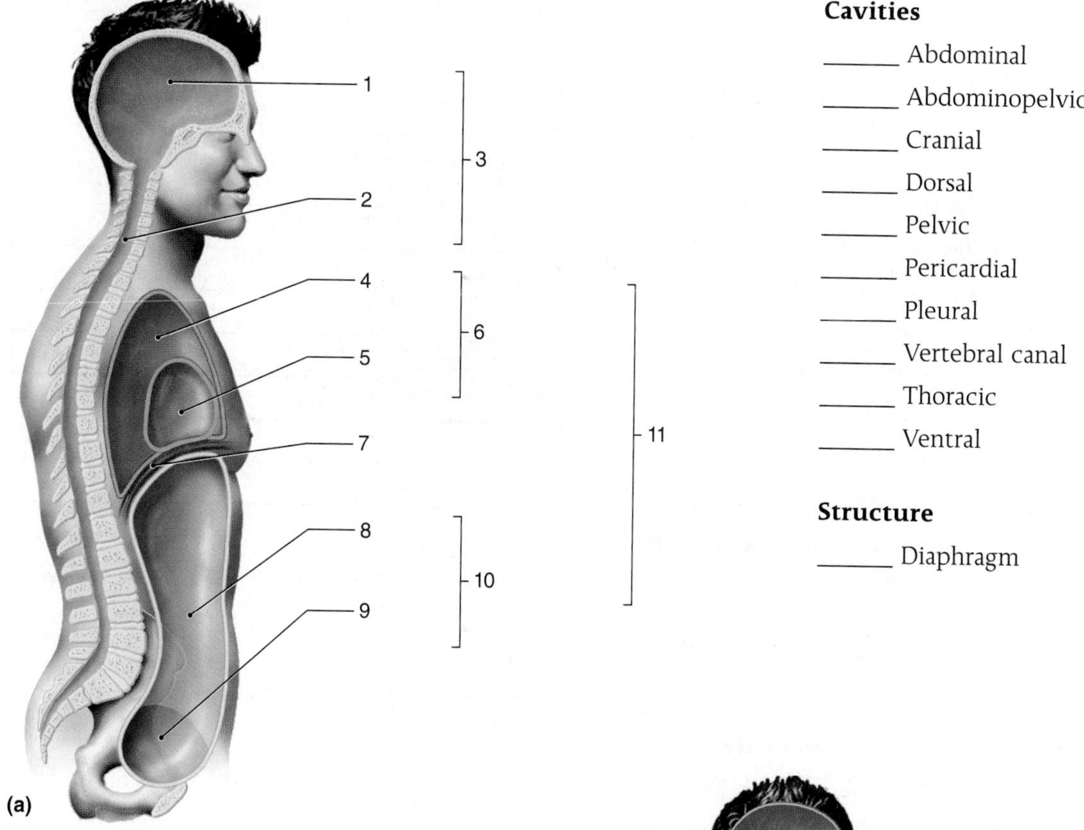

Cavities

_____ Abdominal
_____ Abdominopelvic
_____ Cranial
_____ Dorsal
_____ Pelvic
_____ Pericardial
_____ Pleural
_____ Vertebral canal
_____ Thoracic
_____ Ventral

Structure

_____ Diaphragm

Cavities

_____ Abdominal
_____ Cranial
_____ Pelvic
_____ Pericardial
_____ Pleural
_____ Vertebral

Structure

_____ Diaphragm

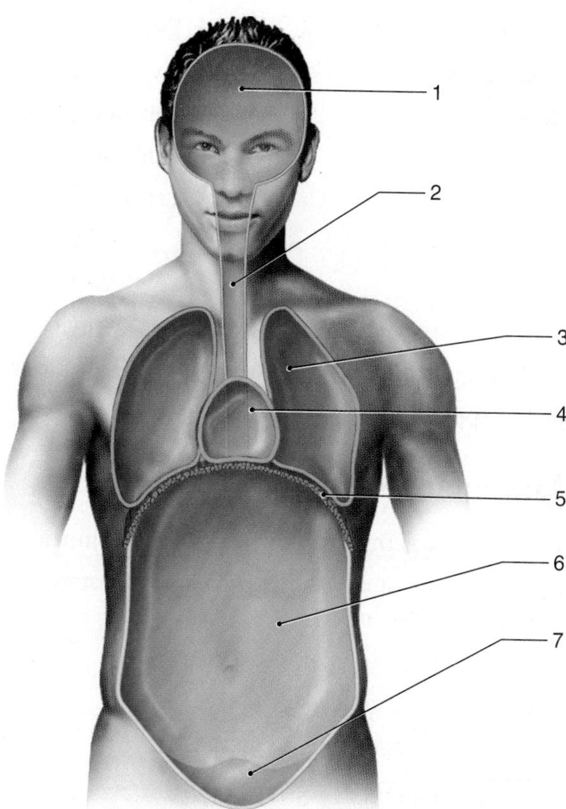

(b)

b. Place the number of the cavity in which the organ occurs in the space by the organ.

1) Abdominal	3) Pelvic	5) Thoracic, lateral parts
2) Cranial	4) Vertebral canal	6) Thoracic, mediastinum
______ Brain	______ Kidneys	______ Spinal cord
______ Gallbladder	______ Liver	______ Thymus
______ Heart	______ Lungs	______ Urinary bladder
______ Intestine, small	______ Pancreas	______ Rectum

c. Write the names of the membranes that match the statements in the spaces at the right.

1) Covers the surface of the heart. ______________________________

2) Covers the surface of the stomach. ______________________________

3) Lines the abdominal cavity. ______________________________

4) Covers the brain. ______________________________

5) Lines the thoracic cavity. ______________________________

6) Lines the vertebral canal. ______________________________

7) Covers the surface of the lungs. ______________________________

8) Forms double-membrane sac around heart. ______________________________

9) Double-layered membranes supporting abdominal organs. ______________________________

7. Abdominopelvic Subdivisions

Select the abdominopelvic quadrant and abdominopelvic region in which the following structures are located.

Quadrants	**Regions**	
1. Right upper	5. Epigastric	10. Right hypochondriac
2. Left upper	6. Hypogastric	11. Right inguinal
3. Right lower	7. Left hypochondriac	12. Right flank
4. Left lower	8. Left inguinal	13. Umbilical
	9. Left flank	

______ Gallbladder	______ Stomach
______ Spleen	______ Ascending colon
______ Rectum	______ Urinary bladder
______ Right kidney	______ Left kidney
______ Appendix	______ Pancreas

8. Maintenance of Life

Write the terms that match the statements in the spaces at the right.

1) Maintenance of a dynamic balance of substances in body fluids. ______________________________

2) Breakdown of complex substances. ______________________________

3) Synthesis of complex substances. ______________________________

4) Mechanism regulating homeostasis. ______________________________

5) Sum of the chemical reactions that occur in the body. ______________________________

6) List the five basic needs essential for human life. ______________________________

____

____

____

____

7) Components of a negative-feedback mechanism. ______________________________

____

____

9. Clinical Insights

a. A patient complains of pain in the epigastric region. What organs may be involved?

____

____

b. A patient complains of pain in the right lower quadrant. What organs may be involved?

____

____

c. How is homeostasis related to health and disease?

____

____

5) A measure of the H^+ concentration in a solution. ______________________

6) Chemicals that keep the pH of a solution relatively constant. ______________________

7) Class of compounds formed of many simple sugars joined together. ______________________

8) Type of reaction that joins two glucose molecules to form maltose. ______________________

9) Storage form of carbohydrates in the body. ______________________

10) Composed of three fatty acids and one glycerol. ______________________

11) Composed of two fatty acids and a phosphate-containing group joined to one glycerol. ______________________

12) Type of fat whose fatty acids contain no carbon-carbon double bonds. ______________________

13) Compound used to store excess energy reserves. ______________________

14) Class of lipids that includes sex hormones. ______________________

15) Class of compounds formed of 50 to thousands of amino acids. ______________________

16) Chemical bonds that determine the three-dimensional shape of proteins. ______________________

17) Bonds joining amino acids together in proteins. ______________________

18) A single-stranded nucleic acid that is involved in protein synthesis. ______________________

19) Building units of nucleic acids. ______________________

20) Steroid that tends to plug arteries when in excess. ______________________

21) Sugar in DNA molecules. ______________________

22) Primary carbohydrate fuel for cells. ______________________

23) Building units of proteins. ______________________

24) Water compartment containing about 65% of the water in the body. ______________________

25) Molecule releasing energy to power chemical reactions within cells. ______________________

26) Double-stranded nucleic acid. ______________________

27) Molecules catalyzing chemical reactions in cells. ______________________

28) Type of reaction breaking a large molecule into smaller molecules. ______________________

29) Molecule controlling protein synthesis in cells. ______________________

30) Element whose atoms form the backbone of organic molecules. ______________________

c. Match the four classes of organic compounds with the listed substances.

1) Carbohydrates 2) Lipids 3) Proteins 4) Nucleic acids

______ Amino acids	______ Nucleotides	______ Enzymes
______ Steroids	______ Monosaccharides	______ RNA
______ Glycogen	______ Triglycerides	______ DNA
______ Cholesterol	______ Starch	______ Fatty acids

d. Label the parts of the small portion of an RNA molecule shown and draw a line around one nucleotide.

_____ Nitrogen bases _____ Ribose sugars _____ Phosphate groups

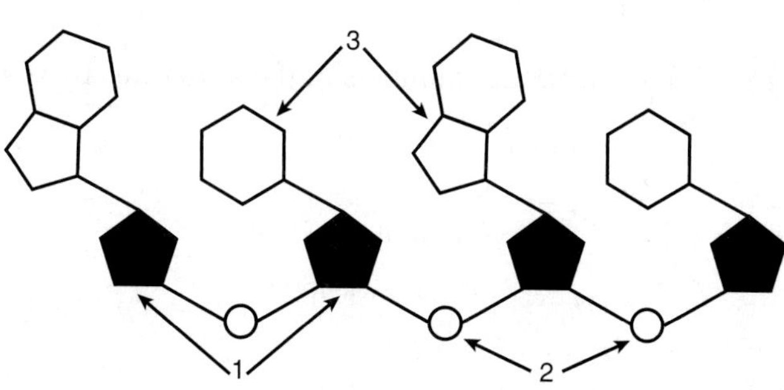

e. Show the interaction of ADP, ATP, Pi, and energy in the formation and breakdown of ATP by placing the numbers of the responses in the correct spaces provided.

1) ADP
2) ATP
3) Energy from nutrient molecules + (–Pi)
4) Energy released for cellular processes + (–Pi)

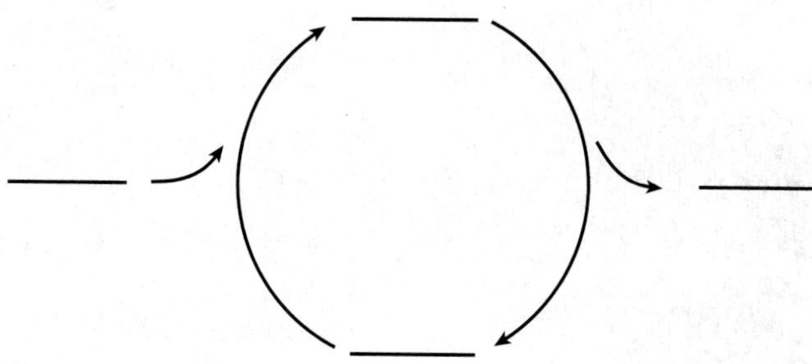

f. Explain the importance of the shape of an enzyme. ____________________

____

g. How does a change in pH change the shape of and inactivate an enzyme? ____________________

____

4. Clinical Insights 

a. Why does a diet high in saturated fats increase the risk of coronary heart disease? ____________________

____

____

____

b. A patient in a coma is brought to the emergency room. A blood test shows that he has severe hypoglycemia (abnormally low blood glucose) and acidosis. Treatment is begun immediately to increase both blood glucose and pH.

1) Why is a normal level of blood glucose important? ____________________

____

2) Why is severe acidosis a problem? ____________________

____

STUDY GUIDE 3

1. Cell Structure

a. Label the diagram of the cell by placing the numbers of the structures by the labels listed.

______ Centrioles
______ Cilia
______ Cytoplasm
______ Golgi complex
______ Lysosome
______ Microfilament
______ Microtubule
______ Microvilli
______ Mitochondrion
______ Nuclear envelope
______ Nucleolus
______ Plasma membrane
______ Rough endoplasmic reticulum
______ Ribosomes
______ Secretory vesicle
______ Smooth endoplasmic reticulum

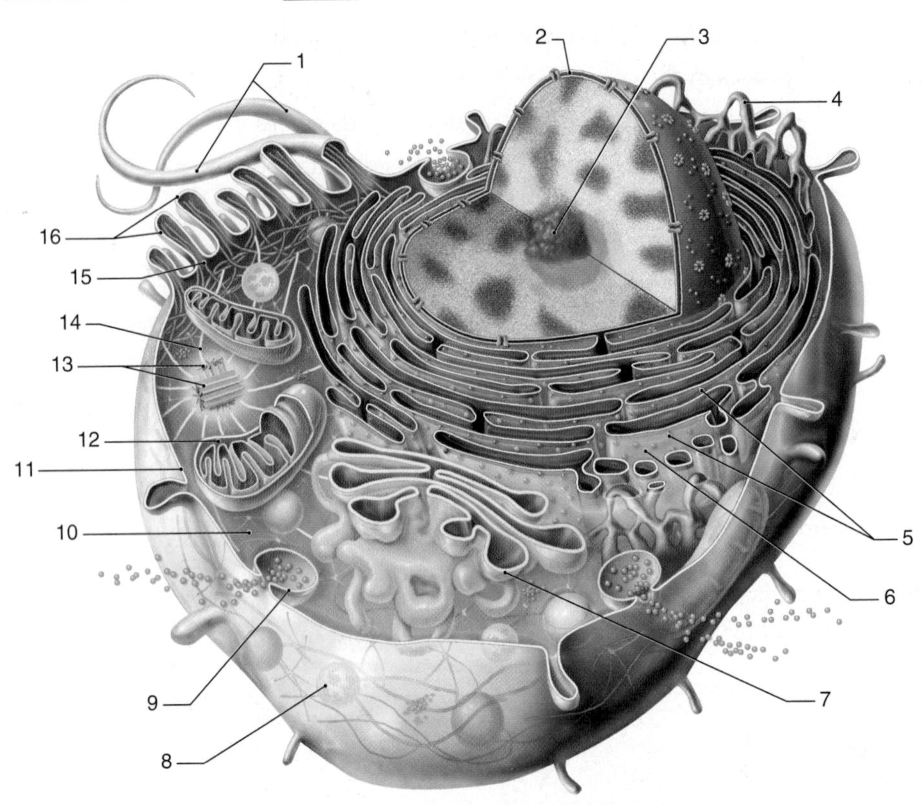

b. Write the terms that match the phrases in the spaces at the right.

1) Endoplasmic reticulum with ribosomes. ______________________________

2) Forms cytoskeleton (two answers). ______________________________
____

3) Packages materials for export from cell. ______________________________

4) Sites of protein synthesis, composed of rRNA. ______________________________

5) Composed of DNA and protein. ______________________________

6) Intranuclear site of RNA synthesis. ______________________________

7) Site of anaerobic respiration. ______________________________

8) Controls movement of materials between nucleus and cytoplasm. ______________________________

9) Endoplasmic reticulum without ribosomes. ______
10) Vesicles of digestive enzymes. ______
11) Provides motility for sperm. ______
12) Short cylinders formed of microtubules. ______
13) Semifluid material around organelles. ______
14) Short, hairlike projections that move substances across cell surfaces. ______
15) Controls movement of materials into and out of the cell. ______
16) Sites of aerobic respiration. ______
17) Contains chromosomes. ______
18) Forms channels for material transport in the cytoplasm. ______
19) Molecule determining inheritance. ______
20) Organelle controlling cell functions. ______

2. Transport Across Plasma Membranes

a. Match the terms and phrases. More than one answer may apply.

Diffusion Osmosis Phagocytosis Pinocytosis

1) Passive movement of water across the selectively permeable membrane. ______
2) A process that uses the plasma membrane to engulf large particles. ______
3) A process that uses the plasma membrane to engulf liquid droplets. ______
4) Movement of molecules from an area of higher concentration to an area of lower concentration. ______
5) Results from random motion of substances. ______

b. Identify the transport processes as either active (A) or passive (P).

______ By channel proteins ______ Osmosis
______ Diffusion ______ Pinocytosis
______ Phagocytosis ______ Exocytosis

c. Write the missing words in the spaces at the right.

1) A cell does not change in size after it is placed in a(n) ______ solution. ______
2) A cell will increase in size after it is placed in a(n) ______ solution. ______
3) A cell will decrease in size after it is placed in a(n) ______ solution. ______

4) Ions move across the plasma membrane from an area of higher concentration to an area of lower concentration by the process of ______. ______

5) Sodium and potassium ions move across the plasma membrane in a direction opposite to the concentration gradient by the process of ______. ______

6) Oxygen and carbon dioxide move across the plasma membrane by the process of ______. ______

3. Cellular Respiration

a. Write the summary equation for the cellular respiration of glucose in the blank space provided. Words may be used instead of chemical formulas. ______

______

b. Provide the term that correctly completes each statement.

1) List the products of cellular respiration. ______

______ ______

2) The primary source of energy captured in ATP. ______

c. Explain why cellular respiration is a continuous process. ______

______

______

4. Protein Synthesis

Provide the term that correctly completes each statement.

1) The genetic code consists of the sequence of bases in ______ molecules. ______

2) The genetic code is transcribed to the sequence of bases in ______ molecules. ______

3) Molecule that carries instructions for protein synthesis to ribosomes. ______

4) Molecule that carries amino acids to ribosome for addition to amino acid chain. ______

5) Small molecules that join to form a protein during translation. ______

5. Cell Division

a. Indicate the type of cell division described by the statements.

1) Provides new cells for growth and repair. ______

2) Forms sperm and ova. ______

3) Daughter cells have the same chromosome number and composition as the parent cell. ______

4) Daughter cells have half the number of chromosomes as the parent cell. ______

b. Select the phase of the cell cycle described by the statements.

Interphase Prophase Metaphase Anaphase Telophase

1) Division of the cytoplasm. ______________________________

2) Replication of chromosomes. ______________________________

3) Chromosomes appear as threadlike bodies. ______________________________

4) Chromatids move toward ends of spindle. ______________________________

5) New nuclei start to form. ______________________________

6) Occupies most of cell cycle. ______________________________

7) Chromosomes line up at equator of spindle. ______________________________

8) Cell performs its normal functions. ______________________________

c. Human body cells have 46 chromosomes. How many chromosomes are in daughter cells formed by mitotic cell division? ______________________________

d. Place the numbers (1-5) of the cell parts in the spaces by the correct label and write the names of the mitotic phases (6-9) in the spaces provide.

______ Centrioles ______ Chromosome ______ Spindle fiber

______ Centromere ______ Chromatid

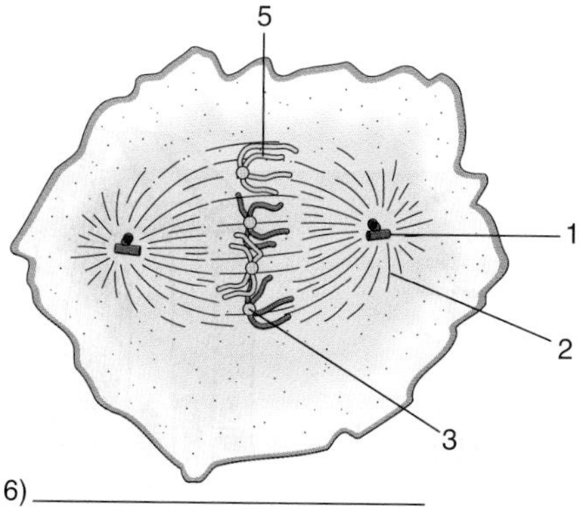

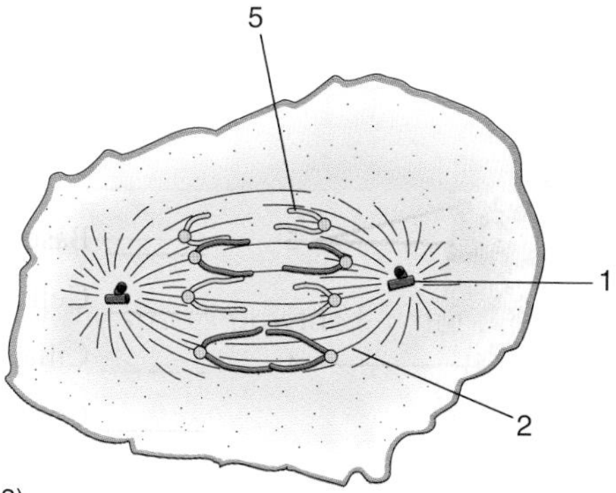

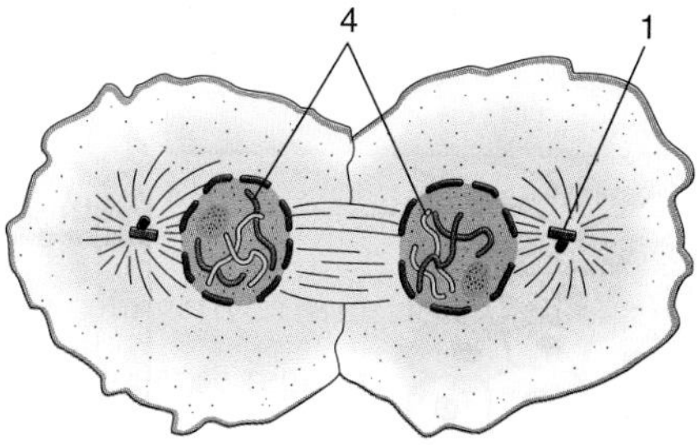

6. Clinical Insights

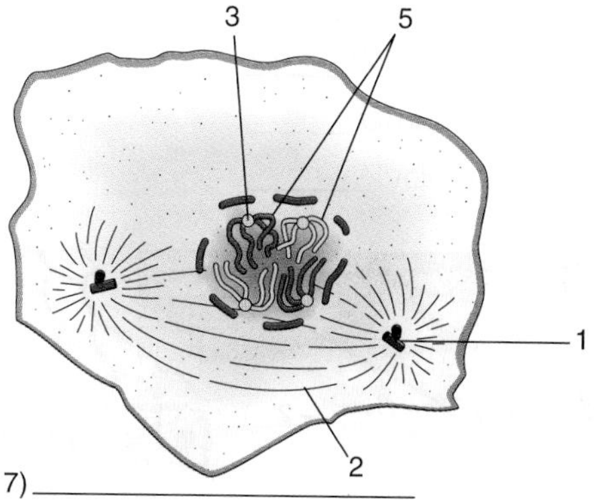

a. When you drink a glass of water, how does the water enter the blood? ______________________________

Why does this occur? ______________________________

b. Explain why a chemotherapy drug that disrupts formation of spindle fibers kills cancerous cells.

____

____

4. Special Connective Tissues

a. Place the number of each structure (1-7) in the space by the correct label and write the names of the tissues in the spaces provided (8-10).

______ Canaliculi

______ Chondrocyte in a lacuna

______ Elastic fibers

______ Central canal

______ Lamellae

______ Nonfibrous matrix

______ Osteocyte in a lacuna

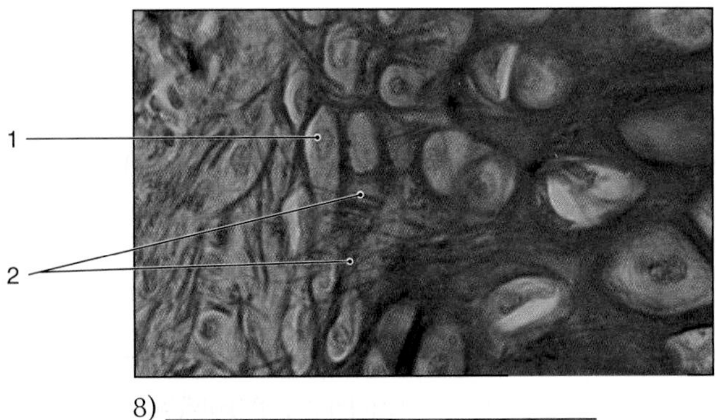

8) ______________________________

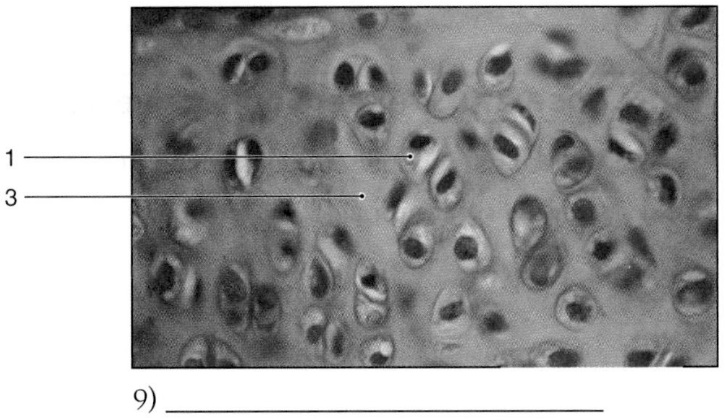

9) ______________________________

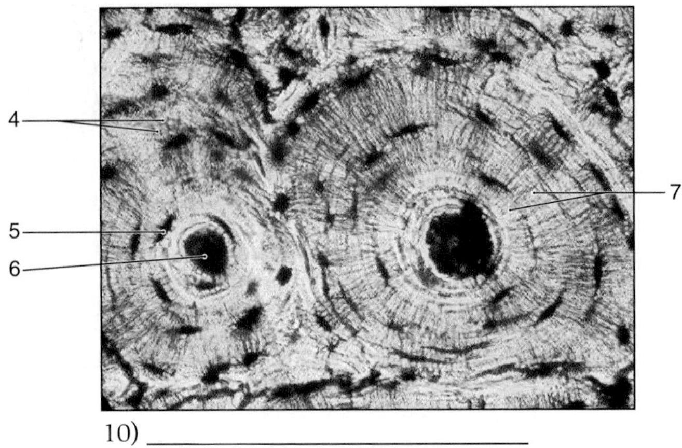

10) ______________________________

b. In each of the spaces provided, write the number of the connective tissue described by the statement.

1) Elastic cartilage　　2) Fibrocartilage　　3) Hyaline cartilage

4) Bone　　5) Blood

______ Forms intervertebral discs.

______ Forms epiglottis.

______ Forms embryonic bones.

______ Has a liquid matrix.

______ Has a smooth, glassy matrix.

______ Has fibers that impart elasticity.

______ Has a hard, rigid matrix.

______ Has dense collagen fibers to resist strong pressures.

5. Muscle Tissues

a. Place the number of each structure (1–3) in the space by the correct label and write the names of the muscle tissues in the spaces provided (4–6).

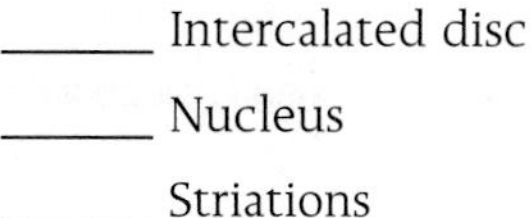
______ Intercalated disc
______ Nucleus
______ Striations

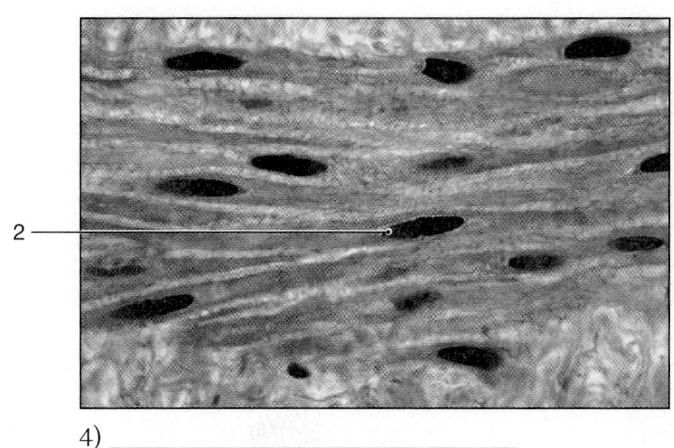

4) ____________________________

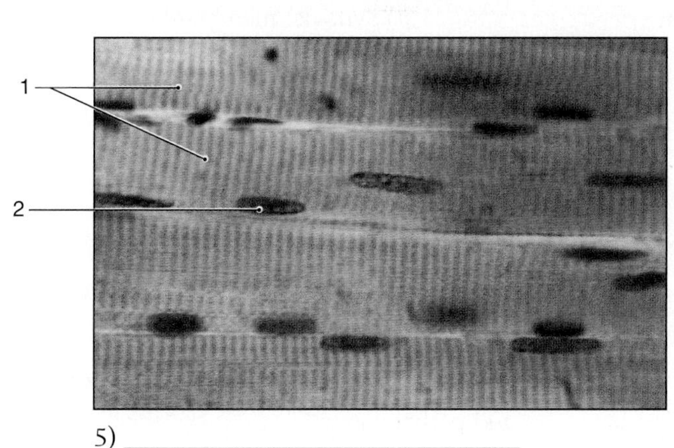

5) ____________________________

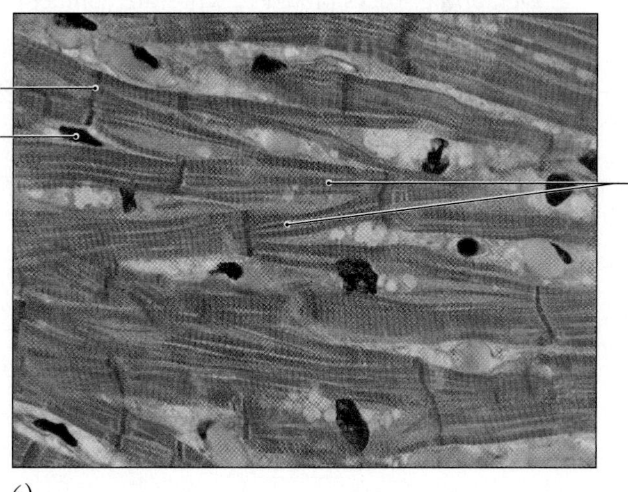

6) ____________________________

b. In each of the spaces provided, write the number of the muscle tissue(s) described by the statement.

1) Cardiac 2) Skeletal 3) Smooth

______ Voluntary
______ Involuntary (2 answers)
______ In walls of intestine
______ Slow contractions
______ Rapid contractions
______ In walls of heart

6. Nervous Tissue

a. Indicate whether each statement is true (T) or false (F).

______ Nerve cells are called neuroglia.
______ The nucleus of a neuron is located in the cell body.
______ Neurons generate and transmit nerve impulses.
______ Supporting cells in nervous tissue are fibroblasts.

b. Write the name of the cell in the space below the figure and place the number of each structure in the space by the correct label.

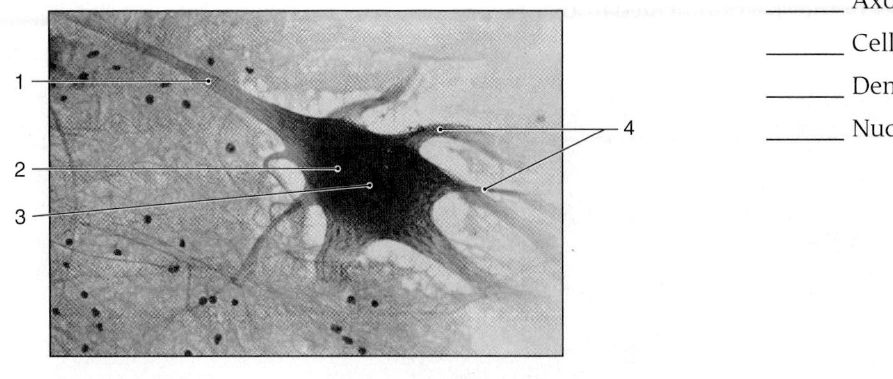

______ Axon
______ Cell body
______ Dendrites
______ Nucleus

____

7. Body Membranes

In each of the spaces provided, write the number of the membrane identified by the statement.

1) Cutaneous membrane
2) Meninges
3) Mucous membranes
4) Perichondrium
5) Periosteum
6) Serous membranes
7) Synovial membranes

______ Line the abdominal cavity and cover most abdominal organs.
______ Protective covering of the brain and spinal cord.
______ Covers the entire body.
______ Epithelial membranes. (3 answers)
______ Covers the surfaces of cartilage.
______ Line tubes and cavities that open to the external environment.
______ Line cavities of freely movable joints.
______ Covers surfaces of bones.
______ Secrete a watery fluid to reduce frictions. (2 answers)
______ Secretions help to trap foreign particles and pathogens.

8. Clinical Insights 

a. Most cancers are carcinomas. How do you explain this? ____

____

b. Judy tore a knee cartilage on a skiing vacation. Can she expect a rapid recovery? ________ Explain. ________

____

STUDY GUIDE 5

1. Functions of the Skin

List six functions of the integumentary system.

1) ______
2) ______
3) ______
4) ______
5) ______
6) ______

2. Structure of the Skin and Hypodermis

Select the structure described by each statement.

Dermis   Epidermis   Stratum basale   Stratum corneum   Subcutaneous tissue

1) Contains abundant adipose tissue. ______
2) Deep layer of the epidermis. ______
3) Composed of stratified squamous epithelium. ______
4) Contains collagen and elastic fibers. ______
5) Attaches skin to deeper tissues. ______
6) Skin layer lacking blood vessels. ______
7) Superficial layer of the epidermis. ______
8) Composed primarily of keratinocytes. ______
9) Forms new epidermal cells. ______
10) Formed of dead keratinized cells. ______
11) Provides insulation for the body. ______
12) Provides strength and elasticity of skin. ______
13) Deep layer of the skin. ______
14) Superficial cells are continuously sloughed off. ______
15) Formed of areolar and dense irregular connective tissues. ______

3. Skin Color

a. Provide the term described by each statement.

1) Three pigments that determine skin color.
   Indicate the color of each pigment. ______
   ______
   ______

2) Protects against UV radiation. ______
3) Cells producing melanin. ______
4) Increases melanin production. ______
5) Used to produce vitamin A ______

c. Label the diagram of the skull, lateral view, by placing the number of each structure in the space by the correct label.

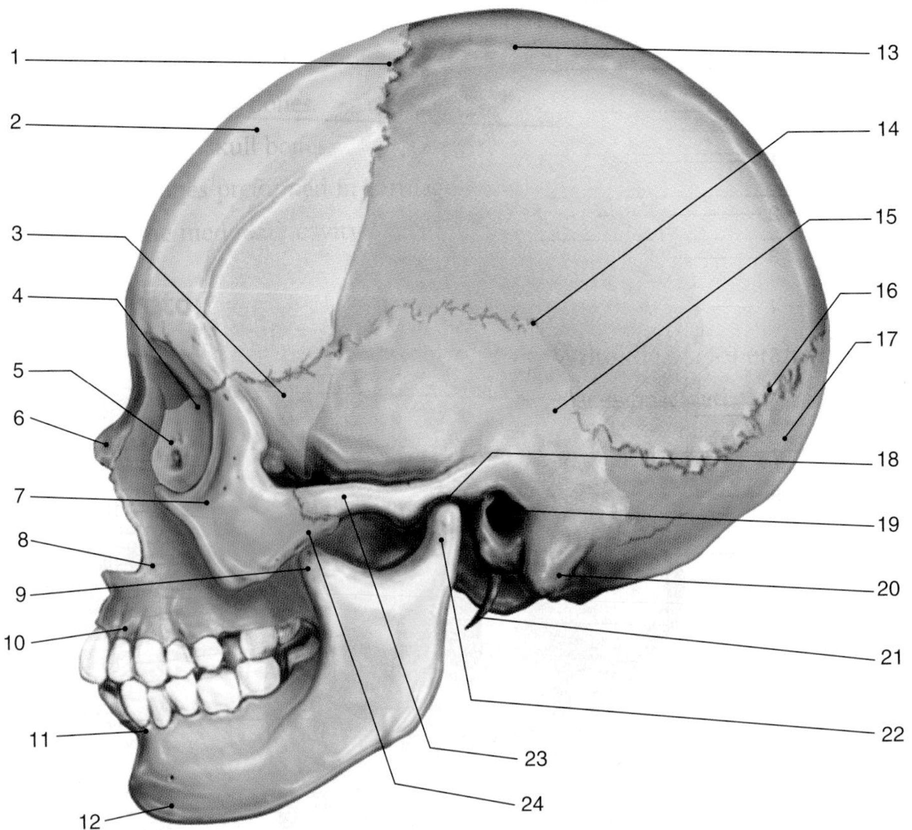

_____ Alveolar arch
_____ Alveolar process
_____ Coronal suture
_____ Coronoid process
_____ Ethmoid
_____ External acoustic meatus
_____ Frontal bone
_____ Lacrimal bone
_____ Lambdoid suture
_____ Mandible
_____ Mandibular condyle
_____ Mastoid process
_____ Maxilla
_____ Nasal bone
_____ Occipital bone
_____ Parietal bone
_____ Sphenoid
_____ Squamous suture
_____ Styloid process
_____ Temporal bone
_____ Temporal bone, zygomatic process
_____ Temporomandibular joint
_____ Zygomatic bone
_____ Zygomatic bone, temporal process

d. Write the terms that match the statements in the spaces provided.

1) Contains the foramen magnum. ____________
2) Forms anterior portion of hard palate. ____________
3) Contains external acoustic meatus. ____________
4) The seven vertebrae of the neck. ____________
5) Weight-bearing portion of a vertebra. ____________
6) Foramen through which spinal cord passes. ____________
7) Vertebrae bearing ribs. ____________
8) Number of pairs of true ribs. ____________
9) Attaches true ribs to sternum. ____________
10) First cervical vertebra. ____________
11) Fibrocartilaginous pads between vertebrae. ____________
12) Forms posterior wall of pelvis. ____________
13) Vertebrae with heaviest bodies. ____________
14) The breastbone. ____________

e. Name the group of bones that provides protection for the

1) Brain ____________ 2) Heart and lungs ____________

f. Label the vertebra by placing the number of the structure in the space by the correct label.

_____ Body
_____ Spinous process
_____ Superior articular facet
_____ Superior articular process
_____ Transverse process
_____ Vertebral arch
_____ Vertebral foramen

7. Appendicular Skeleton

a. Write the missing words in the spaces at the right.

The pectoral girdle is formed of two __1__ and two __2__. Its function is to support the upper __3__. Each __4__ articulates with the scapula at one end and the __5__ at the other. The scapulae are attached to the axial skeleton by __6__ instead of ligaments.

1) ____________
2) ____________
3) ____________
4) ____________
5) ____________
6) ____________

5. Major Skeletal Muscles

Label the muscles and associated structures in the following diagrams by writing the names of the labeled parts in the spaces provided.

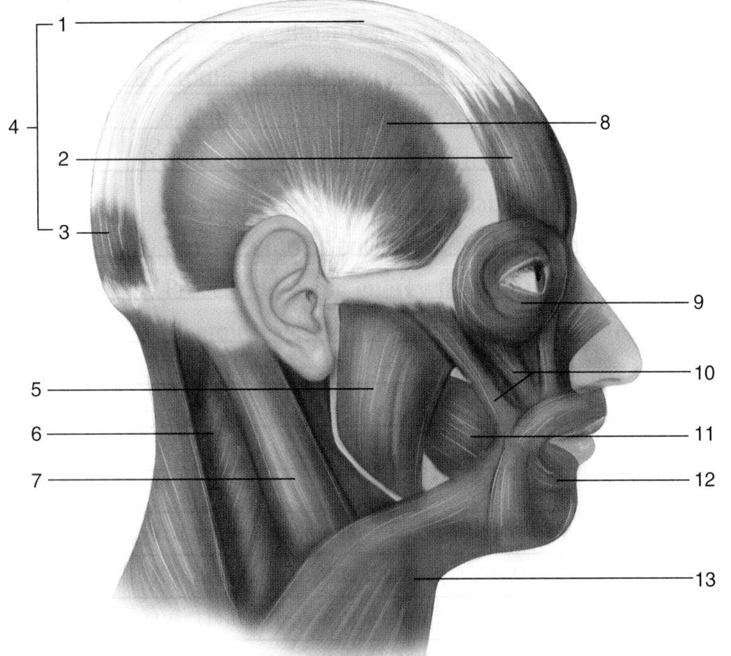

Head and Neck

1) ______________________
2) ______________________
3) ______________________
4) ______________________
5) ______________________
6) ______________________
7) ______________________
8) ______________________
9) ______________________
10) ______________________
11) ______________________
12) ______________________
13) ______________________

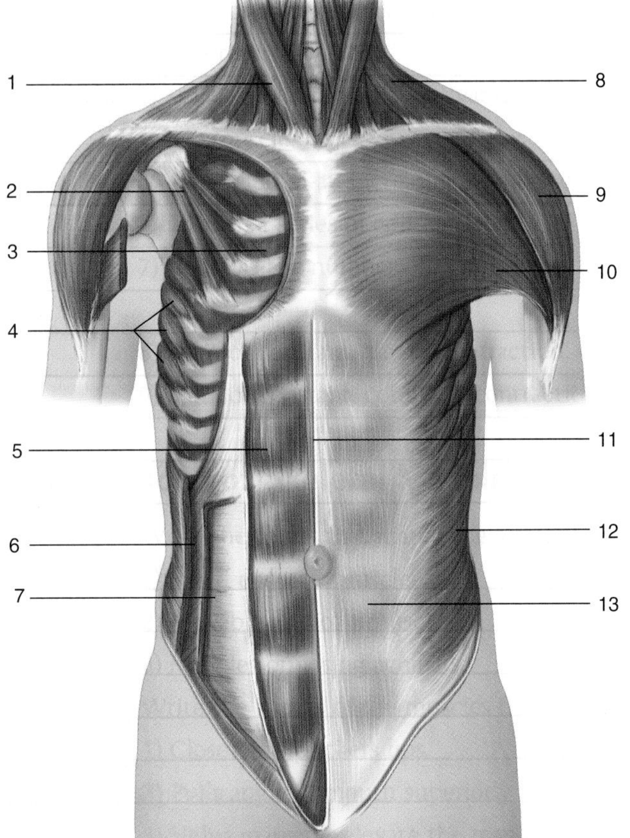

Anterior Trunk

1) ______________________
2) ______________________
3) ______________________
4) ______________________
5) ______________________
6) ______________________
7) ______________________
8) ______________________
9) ______________________
10) ______________________
11) ______________________
12) ______________________
13) ______________________

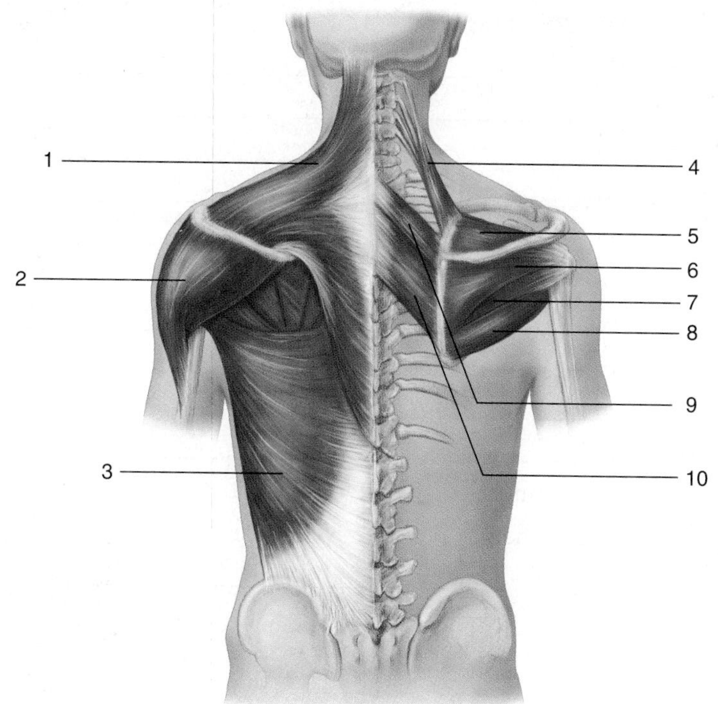

Posterior Trunk

1) ______
2) ______
3) ______
4) ______
5) ______
6) ______
7) ______
8) ______
9) ______
10) ______
11) The muscles labeled 5, 6, 7 and one muscle not shown are collectively called ______

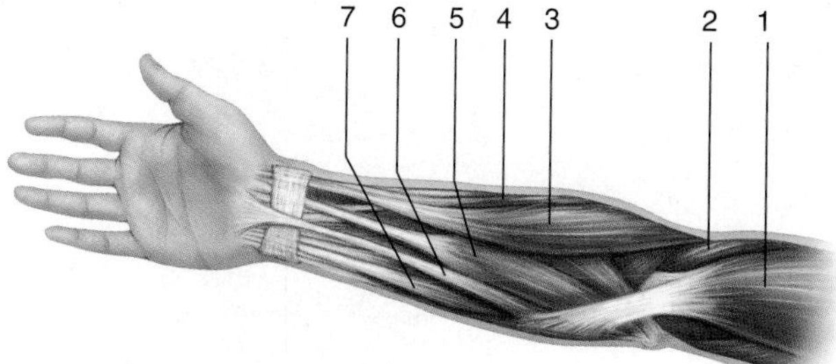

Anterior Forearm

1) ______
2) ______
3) ______
4) ______
5) ______
6) ______
7) ______

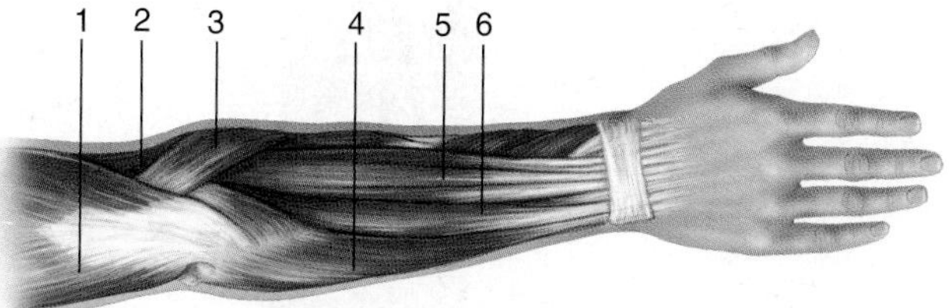

Posterior Forearm

1) ______
2) ______
3) ______
4) ______
5) ______
6) ______

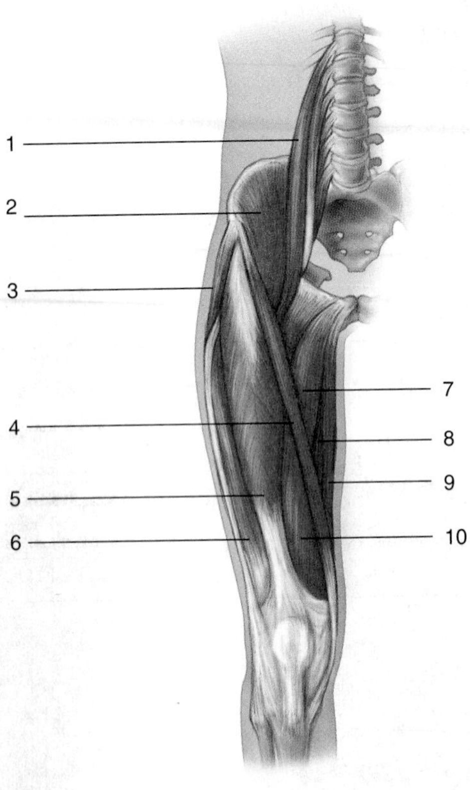

Anterior Thigh

1) ______
2) ______
3) ______
4) ______
5) ______
6) ______
7) ______
8) ______
9) ______
10) ______
11) The muscles labeled 5, 6, 10 and one other muscle not shown are collectively called the______
______

Posterior Thigh

1) ______
2) ______
3) ______
4) ______
5) ______
6) ______
7) ______
8) ______
9) The muscles labeled 3, 4, and 8 are collectively called the______
______

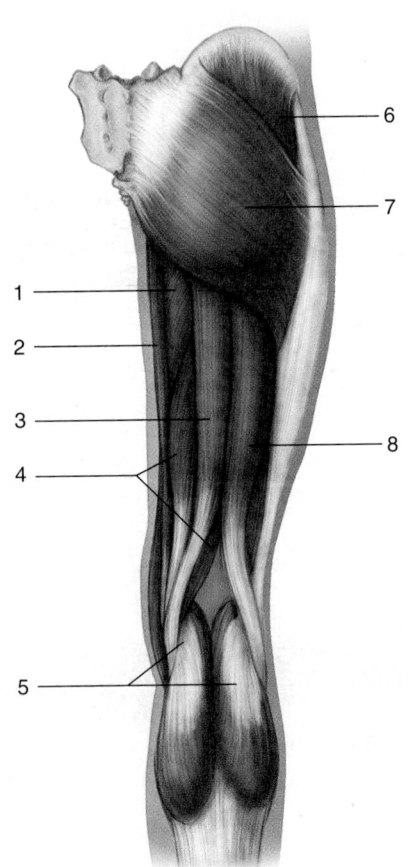

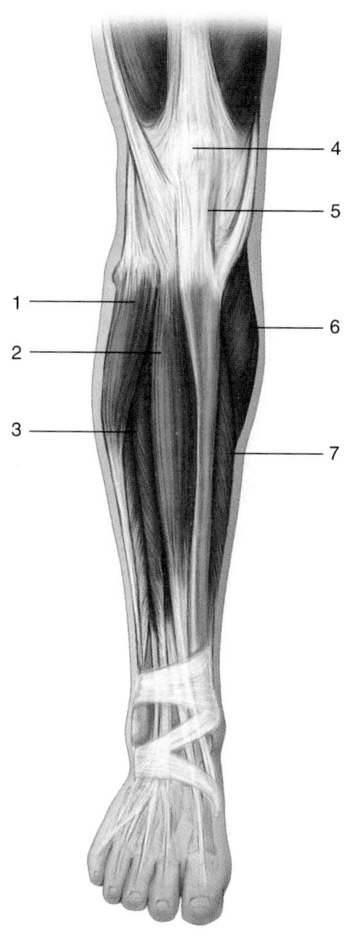

Anterior Leg

1) ____________________
2) ____________________
3) ____________________
4) ____________________
5) ____________________
6) ____________________
7) ____________________

Lateral Leg

1) ____________________
2) ____________________
3) ____________________
4) ____________________
5) ____________________
6) ____________________
7) ____________________
8) ____________________

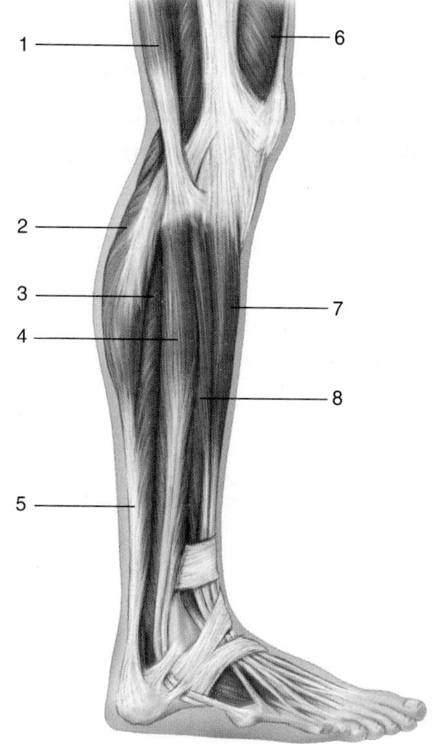

6. Disorders of the Muscular System

Write the names of the disorders in the spaces provided.

1) Inflammation of connective tissues of muscles. ______________________
2) Involuntary tetany of a muscle. ______________________
3) Antibodies attach to acetylcholine receptors, preventing normal stimulation of muscles. ______________________
4) Inflammation of muscle tissue. ______________________
5) A pulled muscle. ______________________
6) Abnormal increase of connective tissue in a muscle. ______________________
7) Viral disease that destroys somatic motor neurons and paralyzes skeletal muscles. ______________________
8) Group of diseases characterized by the progressive degeneration of muscles. ______________________
9) A bacterial disease that prevents the release of acetylcholine from terminal boutons. ______________________
10) A bacterial disease commonly called "lockjaw." ______________________
11) Sudden, involuntary weak contractions of a muscle or group of muscles. ______________________

7. Clinical Insights

a. Strenuous exercise can make muscles sore. Would heat or cold applications be best to alleviate the soreness? __________ Explain. ______________________

____

b. While playing tennis, Jim had a sudden pain on his posterior left thigh. Was this a sprain or a strain? ________ What muscles were probably involved? ______________________

____

c. Tom has been working out to build up his muscles. At the microscopic level, how does a muscle increase in size and strength? ______________________

____

____

STUDY GUIDE 8

1. Divisions of the Nervous System

a. List the anatomical subdivisions. ______________________

____

b. List the functional subdivisions. ______________________

____

1) ______________________

2) ______________________

2. Nervous Tissue

a. Write the terms that match the statements in the spaces at the right.

1) Portion of a neuron containing the nucleus. ______________________

2) Neuronal processes receiving nerve impulses from other neurons. ______________________

3) Structural type of neuron having several dendrites branching off from the cell body. ______________________

4) Functional type of neuron carrying nerve impulses toward the CNS. ______________________

5) Neuronal process carrying nerve impulses to effectors. ______________________

6) Functional type of neuron carrying nerve impulses to effectors. ______________________

7) Insulation around some axons. ______________________

8) Functional type of neuron carrying nerve impulses within the CNS. ______________________

9) Structural type of neuron having a single dendrite branching off from the cell body. ______________________

10) Structure required for regeneration of an axon. ______________________

11) Small spaces between adjacent myelin-forming cells, where the axon is exposed. ______________________

b. Write the names of the neuroglia that match the statements.

1) Line the ventricles of the brain. ______________________

2) Form myelin sheath around axons in the CNS. ______________________

3) Engulf pathogens and debris. ______________________

4) Form myelin sheath around axons in the PNS. ______________________

5) Form the blood–brain barrier. ______________________

c. Label the figure by placing the number of the structure by the correct label. Draw arrows to show direction of nerve impulse.

_____ Axon
_____ Cell body
_____ Dendrite
_____ Myelin sheath
_____ Myelin sheath gaps
_____ Nucleus
_____ Nucleus of Schwann cell
_____ Terminal boutons

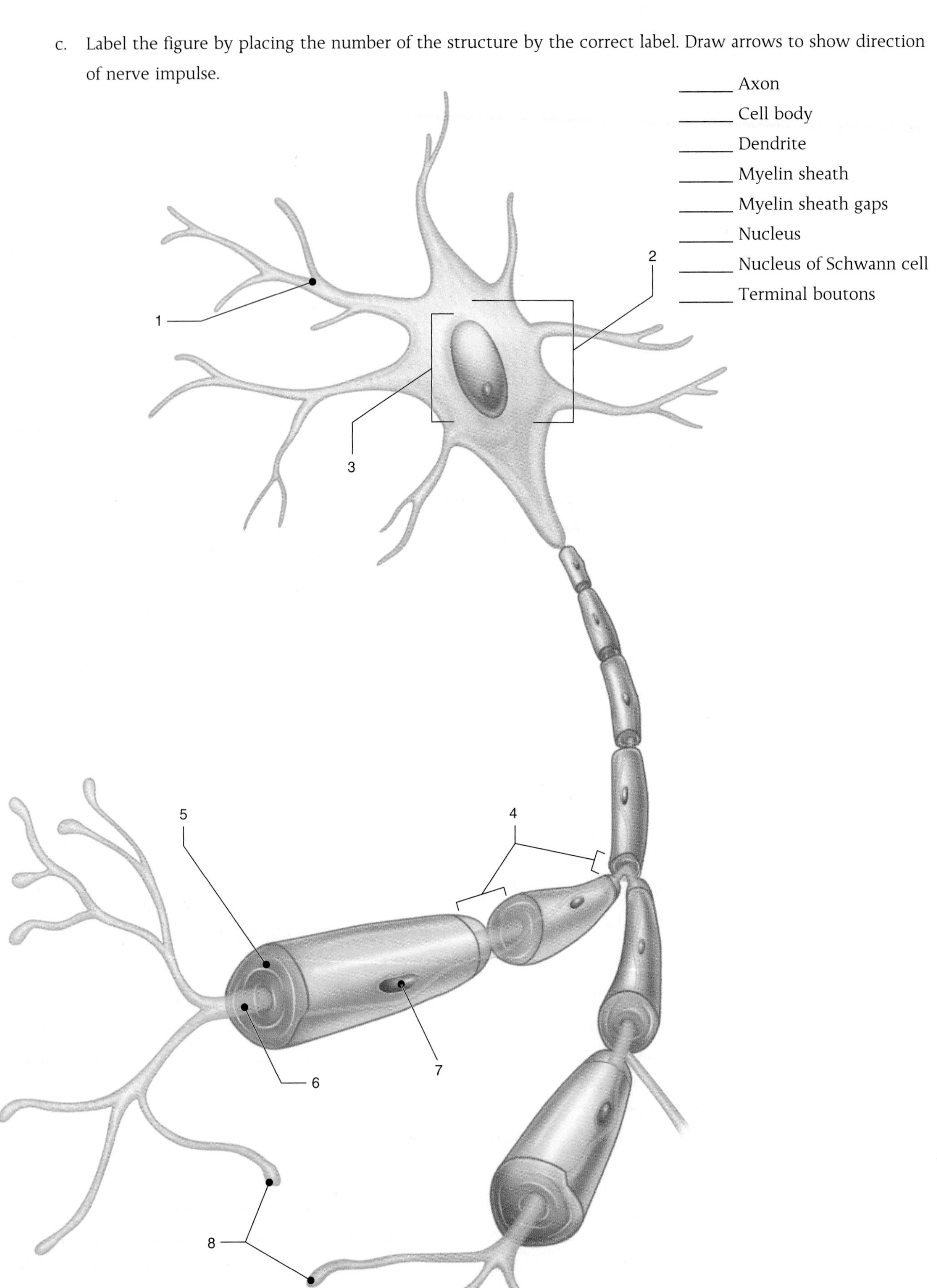

3. Neuron Physiology

a. Write the terms that complete the sentences in the spaces at the right.

In a resting neuron, __1__ (cation) concentrations are high in the ECF and low in the cytosol, whereas __2__ (cations) have the opposite distribution. These differences, in addition to the negatively charged proteins and ions in the cytosol, __3__ the plasma membrane, meaning there are a net excess of __4__ charges on the ECF side and a net excess of __5__ charges on the cytosol side. The result is a __6__ of −70mV.

A threshold stimulus makes the membrane permeable to __7__ ions that rapidly diffuse into the neuron, which depolarizes the membrane forming a nerve __8__. A depolarized membrane is repolarized when __9__ diffuse out of the neuron. The __10__ then reestablishes the resting state distribution of ions. When a nerve impulse is formed in a portion of an axon, it triggers a wave of __11__ that conducts the nerve impulse along the axon. Nerve impulse conduction is more rapid in __12__ axons.

In neuron-to-neuron synaptic transmission, a nerve impulse reaching a terminal bouton causes the release of a __13__ into the __14__. The __15__ binds with __16__ on the postsynaptic neuron, which either promotes or inhibits the formation of a __17__. The __18__ are then removed from the synaptic cleft through reabsorption, diffusion, or decomposition by __19__.

1) ________________
2) ________________
3) ________________
4) ________________
5) ________________
6) ________________
7) ________________
8) ________________
9) ________________
10) ________________
11) ________________
12) ________________
13) ________________
14) ________________
15) ________________
16) ________________
17) ________________
18) ________________
19) ________________

b. If a drug prevents an excitatory neurotransmitter from binding to receptors of the postsynaptic neuron, will synaptic transmission occur? ______ How might such a drug be useful in a clinical situation?

____

____

____

____

4. Protection for the Central Nervous System

Write the terms that match the phrases in the spaces at the right.

1) Most superficial meninx. ________________

2) Middle meninx. ________________

3) Deepest meninx. ________________

4) Portion of skull protecting brain. ________________

5) Meningeal space containing cerebrospinal fluid. ______________________

6) Space filled with adipose tissue in vertebral canal. ______________________

7) Provides bony protection for spinal cord. ______________________

5. Brain

a. Label the figure by placing the number of the structure by the correct label.

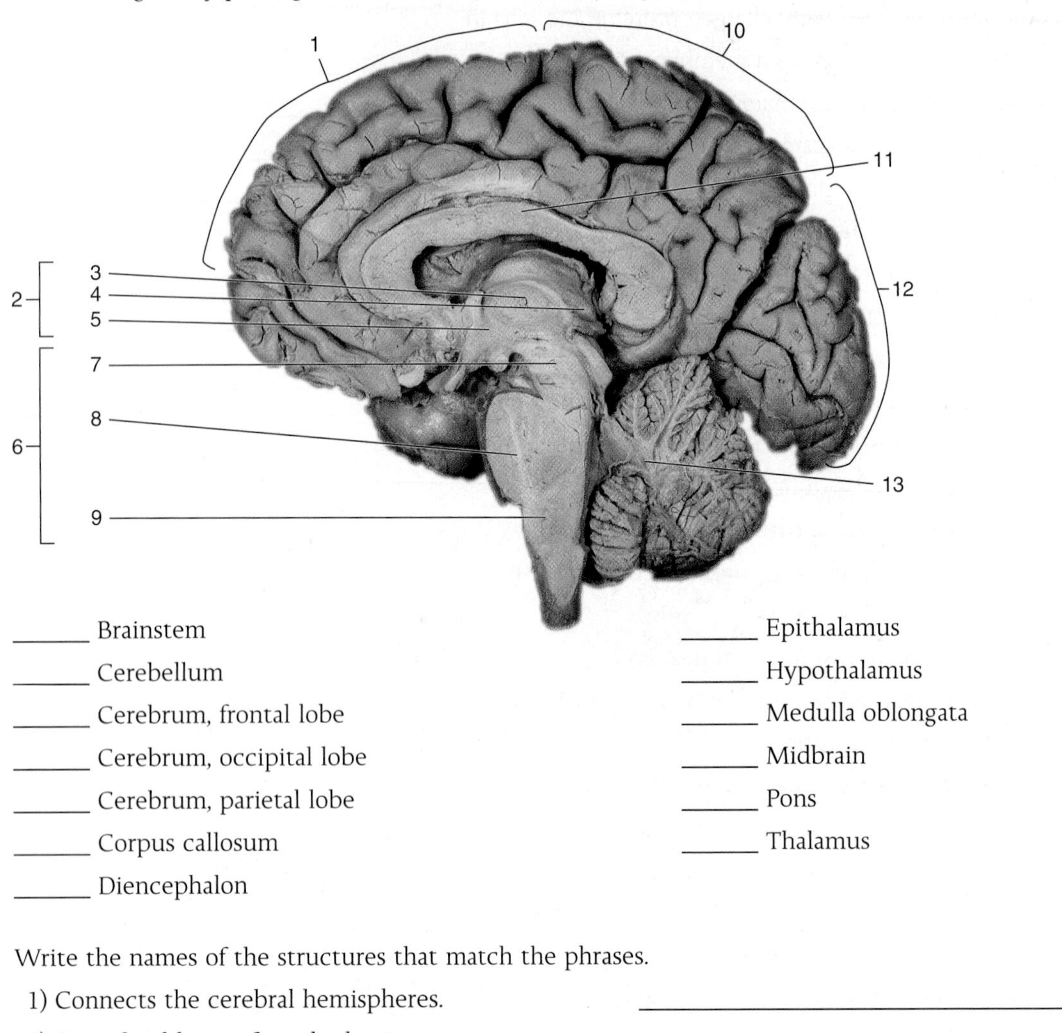

______ Brainstem	______ Epithalamus
______ Cerebellum	______ Hypothalamus
______ Cerebrum, frontal lobe	______ Medulla oblongata
______ Cerebrum, occipital lobe	______ Midbrain
______ Cerebrum, parietal lobe	______ Pons
______ Corpus callosum	______ Thalamus
______ Diencephalon	

b. Write the names of the structures that match the phrases.

1) Connects the cerebral hemispheres. ______________________

2) Superficial layer of cerebral gray matter. ______________________

3) Folds on cerebral surface. ______________________

4) Shallow grooves on cerebral surface. ______________________

5) Groove separating cerebral hemispheres. ______________________

6) Cerebral lobe anterior to central sulcus. ______________________

7) Cerebral lobe posterior to central sulcus. ______________________

8) Most posterior cerebral lobe. ______________________

9) Cerebral lobe inferior to lateral sulcus. ______________________

10) Portion of the brainstem that is continuous with the spinal cord. ______________________

11) Two lateral masses of nervous tissue connected by the interthalamic adhesion. ______________________

12) Major structure of the epithalamus. ______________________
13) Formed of pons, midbrain, and medulla oblongata. ______________________
14) Recognized by its bulblike anterior portion. ______________________
15) Superior portion of brainstem. ______________________
16) Consists of two lateral hemispheres separated by the vermis. ______________________
17) Cavities within the brain. ______________________
18) Cavity continuous with the central canal of the spinal cord. ______________________
19) Cavity between lateral masses of the thalamus and above the hypothalamus. ______________________
20) Second largest portion of the brain. ______________________

c. Match the parts of the brain with the functions described.

1) Cerebrum
2) Cerebellum
3) Hypothalamus
4) Medulla oblongata
5) Limbic system
6) Midbrain
7) Pons
8) Reticular formation
9) Thalamus

______ Controls auditory and visual reflexes.
______ Coordinates body movements, posture, and equilibrium.
______ Controls moods and emotional behavior.
______ Controls voluntary actions, will, and intellect.
______ Provides general, nonspecific awareness of sensations.
______ Provides specific interpretation of sensations.
______ Controls wakefulness and arouses cerebrum.
______ Controls heart rate and blood pressure.
______ Assists medulla oblongata in control of breathing.
______ Major center controlling homeostasis.

d. Match parts of the cerebrum with the functions described.

1) Prefrontal area
2) Motor speech area
3) Corpus callosum
4) Left cerebral hemisphere
5) Occipital lobe
6) Precentral gyrus
7) Premotor area
8) Postcentral gyrus
9) Temporal lobe
10) Right cerebral hemisphere

______ Awareness of sensations from the skin.
______ Center for visual sensations.
______ Center for sound sensations.
______ Controls ability to speak.
______ Primary control area for contractions of skeletal muscles.
______ Hemisphere controlling verbal and computational skills in most people.
______ Hemisphere controlling artistic and spatial skills in most people.
______ Conducts nerve impulses between cerebral hemispheres.

STUDY GUIDE 9

1. Sensations

a. Match the structures with the statements that follow.

1) Cerebral cortex   2) Nerve   3) Sensory receptor

_____ Carries nerve impulses.

_____ Forms sensory nerve impulses.

_____ Interprets nerve impulses as sensations.

_____ Decreases nerve impulse formation when repeatedly stimulated.

_____ Projects sensation back to region where nerve impulses seem to originate.

_____ Sensitive to a particular type of stimulus.

_____ Exhibits sensory adaptation.

b. Define sensory adaptation. ______________________________

____

2. General Senses

a. Match the responses with the terms or statements that follow.

1) Cold receptors
2) Nociceptors
3) Warm receptors
4) Tactile corpuscles
5) Lamellated corpuscles
6) Proprioceptors

_____ Pain receptors.

_____ Touch receptors.

_____ Most sensitive to temperatures over 25°C.

_____ Most sensitive to temperatures under 20°C.

_____ Stimulated by tissue damage.

_____ Detect deep pressure and stretch.

_____ Located in superficial portion of dermis.

_____ Located in muscles and tendons.

_____ Temperature receptors closest to epidermis.

_____ Touch receptor abundant in hairless skin.

_____ Involved in maintenance of erect posture.

b. What is referred pain? ______________________________

____

____

3. Taste and Smell

Write the terms described by the statements in the spaces at the right.

1) Microscopic organs containing taste receptors. ______________________________

2) Sensory receptors located in olfactory epithelium. ______________________________

3) Name the five confirmed basic taste sensations. ______________________________

4) Number of odors that can be detected by the average person. ______________________________

4. Ear Structure

a. Label the figure by placing the number of the structure in the space by the correct label.

_____ Auditory tube
_____ Auricle
_____ Cochlea
_____ External acoustic meatus
_____ Incus
_____ Malleus
_____ Oval window
_____ Round window
_____ Semicircular canals
_____ Stapes
_____ Tympanic cavity
_____ Tympanic membrane
_____ Vestibulocochlear nerve

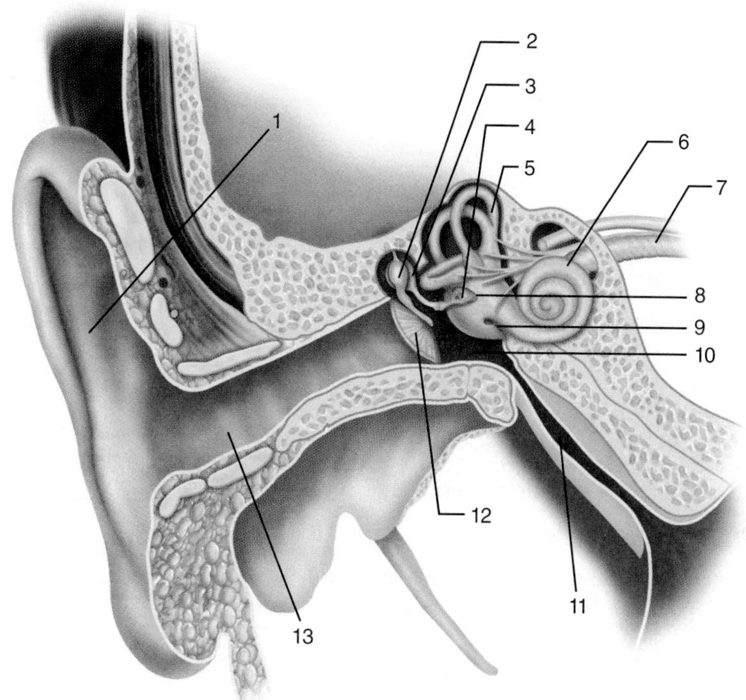

b. Write the terms that match the statements in the spaces at the right.

1) Part of the bony labyrinth that contains sensory receptors for

a) hearing. _____

b) static equilibrium (2 answers). _____ _____

c) dynamic equilibrium (3 answers). _____ _____ _____

2) Fills membranous labyrinth. _____

3) Fills the space between the bony and membranous labyrinths. _____

4) Fills tympanic cavity. _____

c. Label the figure by placing the number of the structure in the space by the correct label.

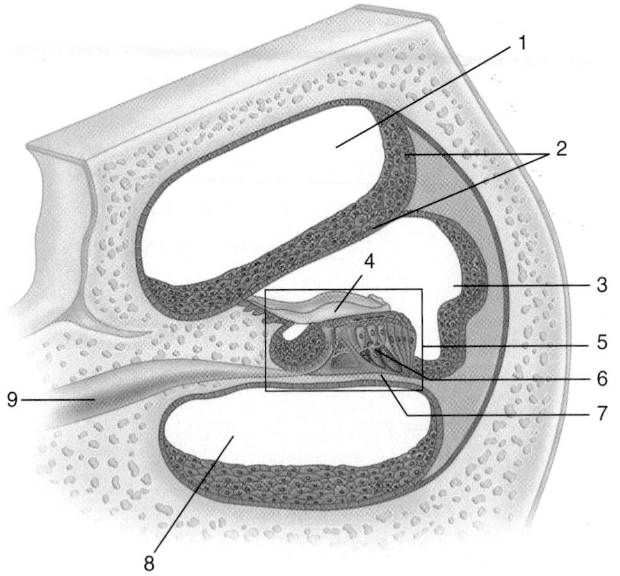

______ Basilar membrane
______ Cochlear duct
______ Cochlear branch of CN VIII
______ Cochlear hair cell
______ Scala tympani
______ Scala vestibuli
______ Spiral organ
______ Tectorial membrane
______ Vestibular membrane

5. Hearing

a. Write the terms that match the statements in the spaces at the right.

1) Receptor organ for hearing. ______________________
2) Receptor cells for hearing. ______________________
3) Carry vibrations from tympanic membrane to oval window. ______________________
4) Directs sound waves to tympanic membrane. ______________________
5) Contains fibers of increasing length. ______________________
6) Membrane-covered opening at end of scala tympani. ______________________
7) Membrane struck by sound waves. ______________________
8) Membrane that determines pitch of sound. ______________________
9) Membrane contacting stereocilia of receptor cells. ______________________
10) Allows air to enter tympanic cavity. ______________________

b. Write the terms that complete the sentences in the spaces at the right.

Sound waves enter the __1__ and strike the __2__, causing it to vibrate. This vibration is transmitted by the __3__ to the __4__ that fills the scala vestibuli and scala tympani. Oscillating movements of this fluid cause comparable vibrations of portions of the __5__ membrane and the __6__ that rests upon it. This causes the cochlear hair cells to contact the __7__, which stimulates them to form __8__ that are carried to the brain by the __9__ nerve. The hearing areas in the __10__ lobes interpret these nerve impulses as sound sensations.

1) ______________________
2) ______________________
3) ______________________
4) ______________________
5) ______________________
6) ______________________
7) ______________________
8) ______________________
9) ______________________
10) ______________________

6. Equilibrium

Write the terms that match the statements in the spaces at the right.

1) Sensory organ for static equilibrium. ______________________
2) Chambers containing sensory organs for static equilibrium. (2 answers) ______________________ ______________________
3) Force stimulating vestibular hair cells of macula. ______________________
4) Sensory organ for detecting rotational motion. ______________________
5) Locations of sensory organs for detecting rotational motion. ______________________
6) Fluid moving the ampullary cupula when head is turned. ______________________
7) Part of brain controlling equilibrium. ______________________

7. Accessory Structures of the Eye

Write the terms that match the statements in the spaces at the right.

1) Lines eyelids and covers anterior sclera. ______________________
2) Group of muscles that move the eye. ______________________
3) Secretes tears. ______________________
4) Collect tears at medial corner of eye. ______________________
5) Carries collected tears into nasal cavity. ______________________

8. Eye Structure

a. Write the terms that match the statements in the spaces at the right.

1) Fills anterior chamber. ______________________
2) Pigmented layer of the eyeball. ______________________
3) Substance holding retina against choroid. ______________________
4) Opening in center of iris. ______________________
5) Fluid mostly responsible for intraocular pressure in eye. ______________________
6) Protective external fibrous coat of eye. ______________________

b. Label the figure by placing the number of the structure in the space by the correct label.

______ Anterior chamber (filled with aqueous humor)
______ Choroid
______ Ciliary body
______ Cornea
______ Fovea centralis
______ Iris
______ Lens
______ Optic disc (CN II)
______ Optic nerve
______ Posterior chamber (filled with aqueous humor)
______ Pupil
______ Retina
______ Sclera
______ Ciliary zonule
______ Vitreous chamber (filled with vitreous body)

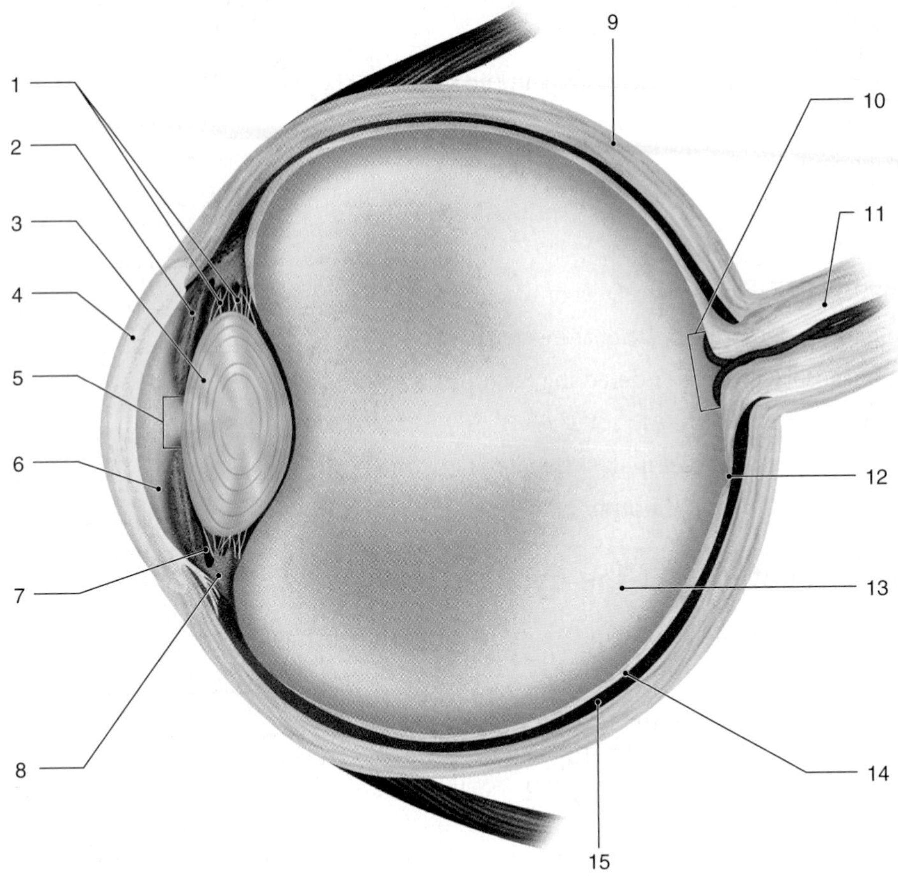

9. Vision

Write the terms that match the statements in the spaces at the right.

1) Contains photoreceptors. ______
2) Site of sharpest color vision. ______
3) Clear window through which light enters eye. ______
4) Controls amount of light entering eye. ______
5) Focuses light rays on retina. ______
6) Layer containing blood vessels. ______
7) Changes shape of the lens. ______
8) Retinal area lacking photoreceptors. ______
9) Opening within the iris that light rays pass through. ______
10) Absorbs excessive light in eye. ______
11) Receptors for black and white vision. ______
12) Receptors for color vision. ______
13) Receptors absent in fovea centralis. ______
14) Carries nerve impulses from retina to brain. ______
15) Where medial axons cross over. ______
16) Light-sensitive pigment in rods. ______
17) Colors of light absorbed by three types of cones. ______
18) Vitamin required for rhodopsin synthesis. ______

10. Disorders of the Special Senses

Write the disorders described in the spaces at the right.

1) An infection called "pink eye." ____________
2) Cloudiness of the lens. ____________
3) Acute infection of the middle ear. ____________
4) Deafness due to exposure to loud noises. ____________
5) Deafness correctible by hearing aids. ____________
6) Results from unequal curvatures of lens or cornea. ____________
7) Loss of all taste function. ____________
8) Perception of an odor that is not present. ____________
9) Decreased elasticity of the lens. ____________
10) Nausea due to repeated stimulation of equilibrium receptors. ____________
11) Group of disorders producing nausea, dizziness, and tinnitis. ____________
12) Cancer of immature retinal cells. ____________
13) Crossed eyes. ____________

11. Clinical Insights

a. An older patient calls the office and complains of pain at the base of the neck, left shoulder, and left arm. What is the probable cause of the pain? ______ Why was the pain localized in these areas? ____________

____

____

What would you advise the patient to do? ____________

b. An audiometry test verifies that a college student has a decreased sense of hearing. Discussion brings out that he has been working as an audio engineer at a local hip hop club for three years. What relationship probably exists between his job and his hearing loss? ____________

____

____

STUDY GUIDE 11

1. General Characteristics of Blood

Write the answers that match the statements in the spaces at the right.

1) pH range of the blood. ________________
2) Liquid portion of blood. ________________
3) Blood cells and platelets. ________________
4) Percentage of blood formed by liquid portion. ________________
5) Percentage of blood formed by RBCs, WBCs, and platelets. ________________
6) Average hematocrit in adult males. ________________
7) Average hematocrit in adult females. ________________
8) Basic function of blood. ________________

2. Red Blood Cells

Write the answers that match the phrases or statements in the spaces at the right.

1) Shape of RBCs. ________________
2) Primary protein in RBCs. ________________
3) Primary function of RBCs is transport of ______. ________________
4) Iron-containing portion of hemoglobin. ________________
5) Healthy range of RBCs/μl of blood in males. ________________
6) Healthy range of RBCs/μl of blood in females. ________________
7) RBC-forming tissue in children and adults. ________________
8) Hormone stimulating RBC production. ________________
9) Organs releasing the hormone that stimulates RBC production. (2 answers) ________________ ________________
10) Organ producing intrinsic factor. ________________
11) Vitamins required for RBC production. (2 answers) ________________ ________________
12) Intrinsic factor enables absorption of ______. ________________
13) Stem cells from which RBCs originate. ________________
14) Average life span of RBCs. ________________
15) Organs where old RBCs are destroyed and heme breakdown occurs. (2 answers) ________________ ________________
16) Phagocytic cells destroying RBCs. ________________
17) Portion of heme that is reused. ________________
18) Portion of heme that is excreted. ________________

3. White Blood Cells

a. Write the answers that match the statements in the spaces at the right.

1) Stem cells from which WBCs originate. ______
2) Healthy range of WBCs/μl of blood. ______
3) Basic function of WBCs is ______. ______
4) Where most functions of WBCs occur. ______
5) Group of WBCs with visible cytoplasmic granules. ______
6) Group of WBCs lacking visible cytoplasmic granules. ______
7) WBCs with lavender-staining granules. ______
8) WBCs with blue-staining granules. ______
9) Largest WBCs. ______
10) Smallest WBCs. ______
11) WBCs with red-staining granules. ______
12) Form 20% to 40% of WBCs. ______
13) Migrate into tissues to become macrophages. ______
14) First WBCs attracted from blood into damaged tissues. ______
15) WBCs that move into tissues to complete cleanup of tissue damage. ______
16) Form 40% to 60% of WBCs. ______
17) Release histamine in allergic reactions. ______
18) Neutralize heparin to reduce inflammation. ______
19) Life span of WBCs. ______
20) Destroy parasitic worms. ______
21) Produce antibodies. ______
22) Compose 2% to 8% of WBCs. ______
23) Compose 0.5% to 1.0% of WBCs. ______
24) Compose 1% to 4% of WBCs. ______
25) Two major phagocytic WBCs. ______ ______

b. Use colored pencils to draw these WBCs as they appear after Wright staining.

Neutrophil **Eosinophil** **Basophil** **Monocyte** **Lymphocyte**

4. Platelets

Write the answers that match the statements in the spaces at the right.

1) Life span of platelets. ______
2) Typical range of platelets per μl of blood. ______

3) Cells that fragment to form platelets. ______________
4) Two functions of platelets. ______________
____

5. Plasma

Write the answers that match the statements in the spaces at the right.

1) Constitutes over 90% of plasma. ______________
2) General term for dissolved substances. ______________
3) Most abundant plasma proteins. ______________
4) Plasma proteins that are antibodies. ______________
5) Plasma protein converted into fibrin. ______________
6) Plasma proteins transporting lipids. (3 answers) ______________
____
____
7) Plasma proteins helping to regulate pH and osmotic pressure of the blood. ______________
8) Organ forming most plasma proteins. ______________
9) Nitrogenous wastes of protein breakdown. (2 answers) ______________
____
10) Collective term for inorganic ions in the plasma. ______________

6. Hemostasis

Write the answers that match the statements in the spaces at the right.

1) Three processes of hemostasis. ______________
____
____
2) Constriction of damaged blood vessel. ______________
3) Formed elements that temporarily plug break in damaged blood vessel. ______________
4) Substance released by platelets and damaged blood vessel walls that starts clotting process. ______________
5) Electrolyte required for clotting to occur. ______________
6) Threadlike strands forming a blood clot. ______________
7) Cells that enter clot to form new dense irregular connective tissue and repair damage. ______________
8) Enzyme converting fibrinogen into fibrin. ______________

7. Human Blood Types

Write the answers that match the statements in the spaces at the right.

1) Location of antigens used in blood typing. ______________
2) Location of antibodies against blood-typing antigens. ______________

3) Antigen(s) in type A blood. ____________
4) Antigen(s) in type AB blood. ____________
5) Antigen(s) in type O blood. ____________
6) Antibodies in type B blood. ____________
7) Antibodies in type AB blood. ____________
8) Antibodies in type O blood. ____________
9) Antibodies in Rh– blood of person sensitized to the Rh antigen. ____________
10) Caused by maternal anti-Rh antibodies binding with Rh antigens on fetal RBCs. ____________

8. Disorders of the Blood

Write the answers that match the statements in the spaces at the right.

1) Reduced ability to form blood clots. ____________
2) Reduced capacity to carry oxygen. ____________
3) An excessive concentration of RBCs. ____________
4) Infection of lymphocytes by Epstein-Barr virus. ____________
5) Anemia due to a deficiency of iron. ____________
6) Cancer producing excess of WBCs. ____________
7) Anemia due to inability to absorb vitamin B12. ____________
8) Anemia due to excessive bleeding. ____________
9) Anemia due to sickling of RBCs. ____________
10) Anemia due to premature rupture of RBCs. ____________
11) Anemia due to loss of red bone marrow. ____________
12) Fetal blood contains erythroblasts. ____________

9. Clinical Insights

a. What ABO and Rh blood type could a person with type O+ blood safely receive? ____________
Explain. ____________

b. Mary's blood type is A–. She is at the hospital for delivery of her second child, and her first child is A+. The attending physician wants blood available in case the infant exhibits hemolytic disease of the newborn upon delivery. What blood type should he order? ____________
Explain. ____________

c. Chemotherapy destroys cancer cells because it disrupts cell division in rapidly dividing cells. What impact would chemotherapy have on the production of formed elements? ____________
Explain. ____________

STUDY GUIDE 12

1. Anatomy of the Heart

a. Write the correct labels in the spaces at the right.

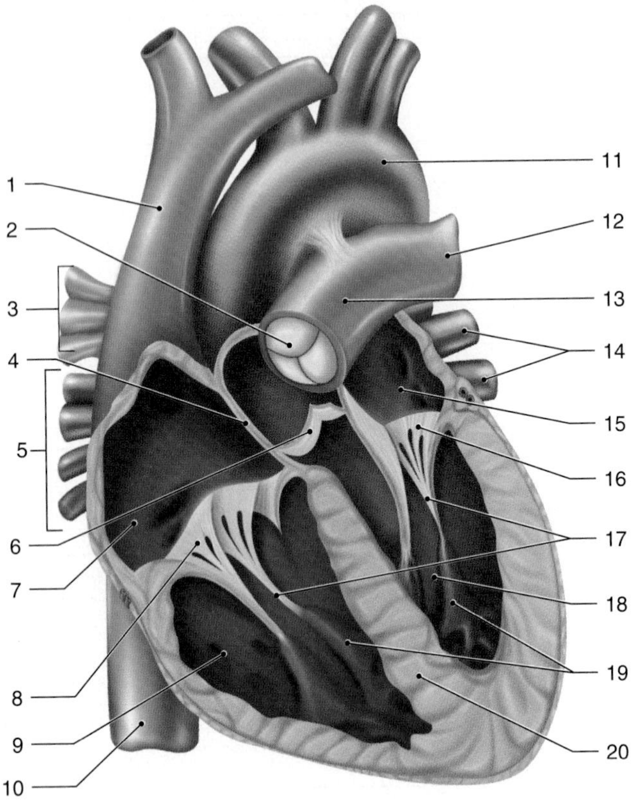

1) ______________________
2) ______________________
3) ______________________
4) ______________________
5) ______________________
6) ______________________
7) ______________________
8) ______________________
9) ______________________
10) ______________________
11) ______________________
12) ______________________
13) ______________________
14) ______________________
15) ______________________
16) ______________________
17) ______________________
18) ______________________
19) ______________________
20) ______________________

b. Write the names of the structures that match the statements in the spaces at the right.

1) Receives blood from venae cavae. ______________________
2) Receives blood from pulmonary veins. ______________________
3) Separates ventricles. ______________________
4) Prevents backflow of blood from right ventricle into right atrium. ______________________
5) Prevents backflow of blood from left ventricle into left atrium. ______________________
6) Prevents backflow of blood from aorta into left ventricle. ______________________
7) Prevents backflow of blood from pulmonary trunk into right ventricle. ______________________
8) Restrain cusps of AV valves. ______________________
9) Pumps blood into pulmonary trunk. ______________________
10) Pumps blood into aorta. ______________________

2. Cardiac Cycle

Write the terms or the names of the structures that match the statements in the spaces at the right.

1) Contraction phase of the ventricles. ______
2) Relaxation phase of the ventricles. ______
3) Valves closing to produce first heart sound. ______
4) Valves closing to produce second heart sound. ______
5) Valves open during ventricular systole. ______
6) Valves closed during ventricular systole. ______
7) Valves open during ventricular diastole. ______
8) Valves closed during ventricular diastole. ______

3. Heart Conduction System and Electrocardiogram

Write the terms or the names of the structures that match the statements in the spaces at the right.

1) Small fibers carrying impulses to ventricular myocardium. ______
2) Pacemaker of the heart. ______
3) Thick fibers extending from AV node. ______
4) Transmits impulses to atria and AV node. ______
5) Transmits impulses to AV bundle. ______
6) Wave caused by depolarization of ventricles. ______
7) Wave caused by repolarization of ventricles. ______
8) Wave caused by depolarization of atria. ______

4. Regulation of Heart Function

a. Write the answers that match the statements in the spaces at the right.

1) Term for the volume of blood pumped from each ventricle in 1 min. ______
2) Term for the volume of blood pumped from each ventricle in one contraction. ______
3) If the heart rate is 65 beats/min and the stroke volume is 70 ml/beat, what is the cardiac output? ______
4) Autonomic center controlling heart rate and force of contraction. ______
5) ANS division whose nerve impulses increase heart rate. ______
6) ANS division whose nerve impulses decrease heart rate. ______
7) Gender with faster heart rate. ______
8) Nerve carrying parasympathetic axons to the heart. ______

b. Match the effect on heart rate with the factors listed.

1) Increases 2) Decreases 3) No effect

______ Epinephrine	______ Excess blood K^+	______ Insufficient blood Ca^{2+}
______ Old age	______ Acetylcholine	______ Excitement
______ Fever	______ Anxiety	______ Thyroxine
______ Physical conditioning	______ Increase in blood pressure	______ Norepinephrine

5. Types of Blood Vessels

Write the terms or the names of the structures that match the statements in the spaces at the right.

1) Composed of endothelium and a layer of areolar connective tissue. ______
2) Vessels with thickest walls. ______
3) Vessels with valves. ______
4) Carry blood from capillaries to heart. ______
5) Carry blood from heart to capillaries. ______
6) Vessels exchanging materials with tissues. ______
7) Smallest and most numerous vessels. ______
8) Force moving fluid from blood to interstitial fluid. ______
9) Force moving fluid from interstitial fluid to blood. ______
10) Process moving dissolved substances between capillaries and interstitial fluid. ______

6. Blood Flow and Blood Pressure

a. Write the answers that match the statements in the spaces at the right.

1) Systemic vessel with fastest blood flow. ______
2) Vessels with slowest blood flow. ______
3) Systemic vessel with greatest blood pressure. ______
4) Primary force moving blood. ______
5) Two additional forces that help return venous blood to the heart. ______ ______
6) Average healthy systolic blood pressure. ______
7) Average healthy diastolic blood pressure. ______
8) Autonomic center controlling diameter of blood vessels. ______
9) The term for the difference between the systemic and diastolic blood pressures. ______
10) Effect on precapillary sphincters by a local decrease in oxygen and pH. ______
11) Effect on precapillary sphincters by sympathetic nerve impulses. ______

b. Indicate whether the following conditions cause an *increase* (+) or *decrease* (−) in blood pressure.

______ An increase in peripheral resistance.

______ A marked decrease in blood volume.

______ A decrease in cardiac output.

______ Dilation of a great many arterioles.

______ A significant increase in plasma proteins.

______ Sympathetic nerve impulses to arterioles.

______ Constriction of most arterioles.

______ An increase in heart rate.

7. Circulation Pathways

Trace the pathway of blood from a ventricle of the heart to the organ indicated and back to an atrium of the heart. Write the names of the correct heart chambers, arteries, and veins in the blanks.

1) Right superficial little finger.

Left ventricle → ____________________ → ____________________ →
____________________ → ____________________ →
____________________ → ____________________ → right little finger
→ ____________________ → ____________________ →
____________________ → ____________________ → ____________________
→ ____________________ atrium.

2) Small intestine.

____________________ ventricle → ____________________ → ____________________
→ small intestine → ____________________ →
____________________ → liver → ____________________ →
____________________ → right atrium.

8. Systemic Arteries

Label the figure by writing the names of the numbered arteries in the spaces.

1) ______________________
2) ______________________
3) ______________________
4) ______________________
5) ______________________
6) ______________________
7) ______________________
8) ______________________
9) ______________________
10) ______________________
11) ______________________
12) ______________________
13) ______________________
14) ______________________
15) ______________________
16) ______________________
17) ______________________
18) ______________________
19) ______________________
20) ______________________
21) ______________________
22) ______________________
23) ______________________
24) ______________________
25) ______________________
26) ______________________
27) ______________________
28) ______________________
29) ______________________
30) ______________________

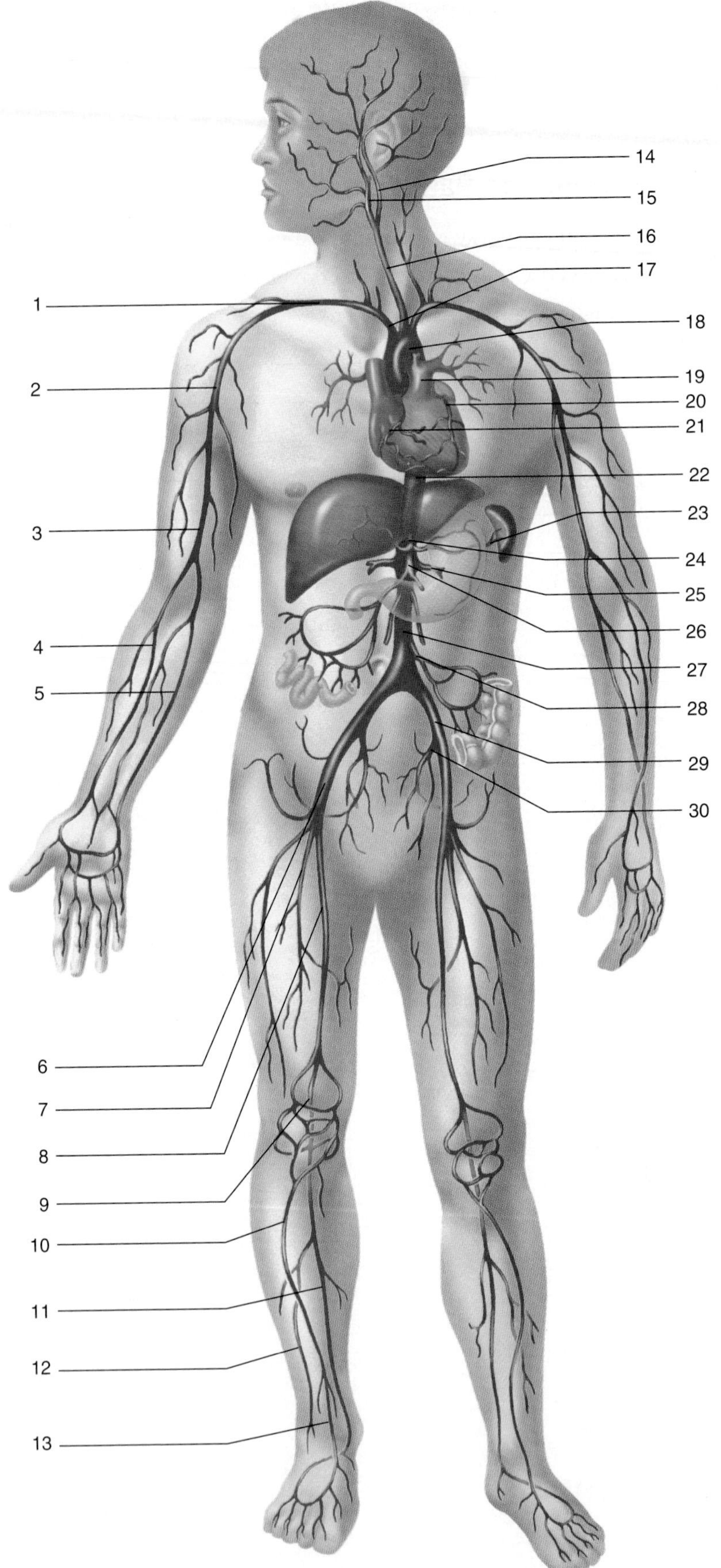

9. Systemic Veins

Label the figure by writing the names of the numbered veins in the spaces.

1) ____________
2) ____________
3) ____________
4) ____________
5) ____________
6) ____________
7) ____________
8) ____________
9) ____________
10) ____________
11) ____________
12) ____________
13) ____________
14) ____________
15) ____________
16) ____________
17) ____________
18) ____________
19) ____________
20) ____________
21) ____________
22) ____________
23) ____________
24) ____________
25) ____________
26) ____________
27) ____________
28) ____________
29) ____________
30) ____________
31) ____________

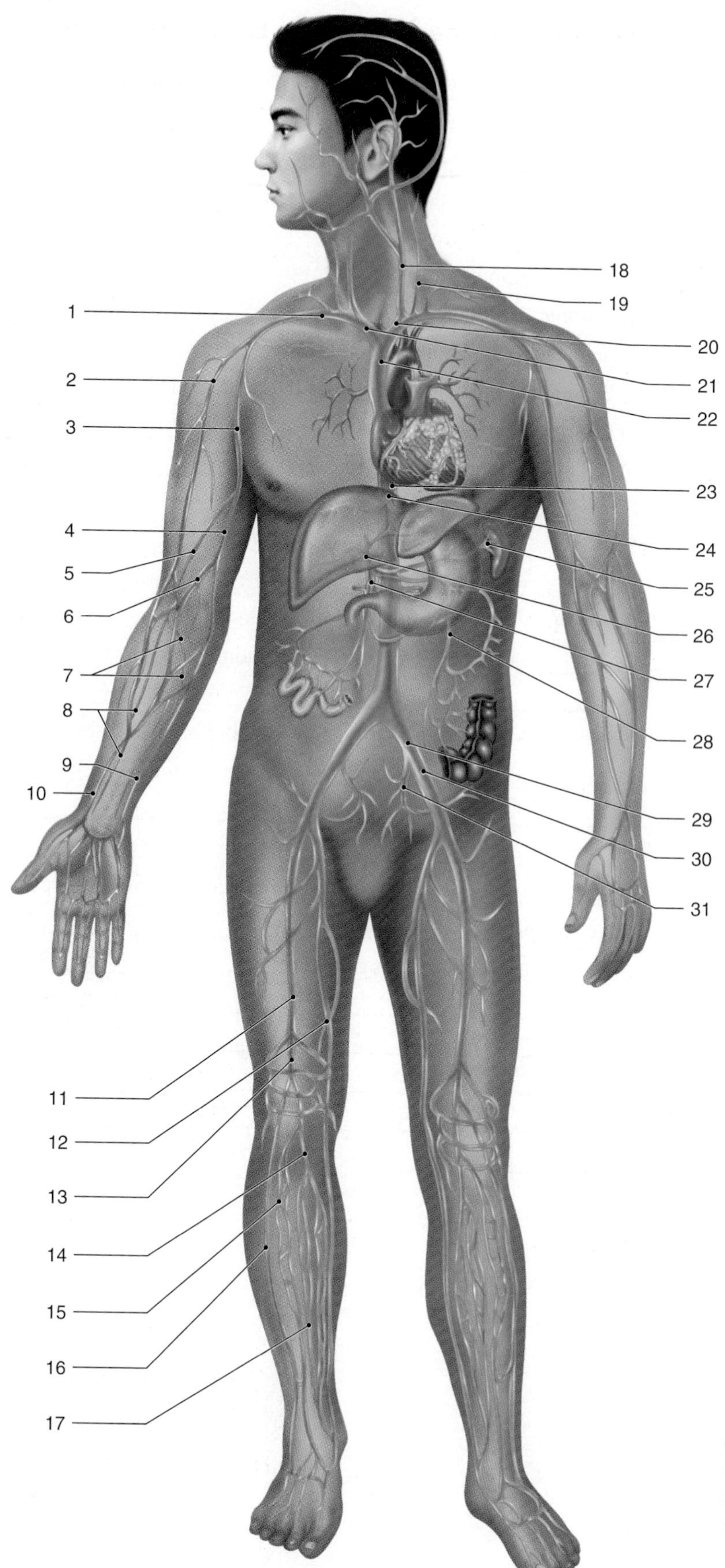

10. Disorders of the Heart and Blood Vessels

Write the disorders described by the statements in the spaces at the right.

1) Unusual heart sounds. ____________________
2) Hardening of the arteries. ____________________
3) Death of a portion of the myocardium. ____________________
4) Abnormal heart rhythm. ____________________
5) Inflammation of a vein. ____________________
6) Chronic high blood pressure. ____________________
7) Swollen veins due to defective valves. ____________________
8) Balloonlike enlargement of blood vessel. ____________________
9) Fatty deposits along the walls of arteries. ____________________
10) Edema of lungs, viscera, legs, and feet. ____________________

11. Clinical Insights

a. A 60-year-old man complains of chest pain during moderate exercise. The pain goes away after he rests for a while. What is the likely cause of the pain? ____________________

Without treatment, what complications may arise? ____________________

____

b. An accident victim has lost considerable blood. His blood pressure is only slightly below normal, and his pulse rate is elevated. How is the body compensating for the loss of blood? ____________________

____

____

c. A patient has a blood clot in the right femoral vein. If a part of the clot should break loose, where is it likely to lodge? ____________________

Would this be a serious complication? ____________ Explain. ____________________

____

____

____

____

STUDY GUIDE 13

1. Lymph and Lymphatic Vessels

a. Write the answers that match the statements in the spaces at the right.

1) Fluid within interstitial spaces. ______
2) Fluid within lymphatic vessels. ______
3) Smallest lymphatic vessels. ______
4) Lymphatic vessels draining large body regions. ______
5) Forms wall of lymphatic capillaries. ______
6) Lymphatic duct draining superior right portion of the body. ______
7) Lymphatic duct draining superior left portion of the body and all tissue inferior to the diaphragm. ______
8) Prevent backflow of lymph. ______
9) Provide forces that move lymph. (4 answers) ______ ______ ______ ______
10) Receives lymph from thoracic duct. ______
11) Empties into right subclavian vein. ______
12) Vessels collecting interstitial fluid. ______
13) Source of interstitial fluid. ______

b. Explain the value of the lymphoid system collecting interstitial fluid and returning it to the blood.

______
______
______

2. Lymphoid Organs

Write the answers that match the statements in the spaces at the right.

1) Site of origination of all lymphocytes cells in adults. ______
2) Grouped along larger lymphatic vessels. ______
3) Bilobed gland located superior to the heart. ______
4) Large lymphoid organ located near stomach. ______
5) Organs that filter lymph. ______
6) Lymphoid organ that filters blood. ______
7) Site where T cells become immunocompetent. ______
8) Vessels carrying lymph into a lymph node. ______
9) Vessels carrying lymph out of a lymph node. ______
10) Contains a reserve supply of RBCs and platelets. ______
11) Hormones that promote T cell maturation. ______

3. Lymphoid Tissues

Write the answers that match the statements in the spaces at the right.

1) Tonsil on the base of the tongue. ____________________

2) Tonsil located posterior to nasal cavity. ____________________

3) Lymphoid nodules within the mucosae of respiratory, digestive, urinary, and reproductive tracts. ____________________

4. Nonspecific Resistance

Match the type of nonspecific resistance with the statements.

1) Mechanical barriers
2) Chemicals
3) Phagocytosis
4) Inflammation
5) Fever

______ Skin
______ Mucus
______ Lysozyme
______ Mucous membranes
______ Acidic pH
______ Gastric juice
______ Interferon
______ Produces edema
______ Pus formation
______ Flow of tears
______ Release of histamine
______ High body temperature
______ Attracts neutrophils and monocytes
______ Tissue macrophage system
______ Flow of saliva
______ Neutrophils and macrophages
______ Increases local blood supply
______ Clot seals off pathogens
______ Speeds up body processes
______ Pathogens are engulfed and digested
______ Complement fixation

5. Immunity

a. Indicate whether the following statements are true (T) or false (F).

1) Immunity is resistance against specific pathogens. ____________________

2) Bacterial toxins are neutralized by cell-mediated immunity. ____________________

3) Diseased body cells are destroyed by antibody-mediated immunity. ____________________

4) Immunity requires lymphocytes to distinguish between self and nonself molecules. ____________________

5) Antigens are molecules that cause an immune response. ____________________

6) Undifferentiated lymphocytes are produced in the spleen. ____________________

7) All lymphocytes differentiate in the thymus. ____________________

8) The majority of lymphocytes in the blood are T cells. ____________________

9) Differentiation of lymphocytes occurs throughout life. ____________________

10) T cells provide cell-mediated immunity. ____________________

11) B cells provide antibody-mediated immunity. ____________________

12) Lymphocyte receptors for specific antigens are inherited. ____________________

13) Lymphocyte receptors are formed by contact with specific antigens. ____________

14) There are thousands of different types of B and T cells, and each type responds to a different specific antigen. ____________

15) Immunity depends upon lymphocytes whose receptors fit with a specific antigen. ____________

16) Immunity involves the interaction of lymphocytes, antigens, and antigen-presenting cells. ____________

17) At any one time, either cell-mediated immunity or antibody-mediated immunity is at work–never both at the same time. ____________

18) Proliferation of immunocompetent lymphocytes occurs in the thymus. ____________

b. Write the words that complete the sentences describing cell-mediated immunity in the spaces at the right.

Cell-mediated immunity begins when a/an __1__ engulfs a foreign antigen and displays part of the __2__, in combination with its own self __3__, on its plasma membrane. A T cell, which recognizes this __4__, binds to the antigen and the APC self proteins and becomes __5__. The activated T cell undergoes repeated mitotic divisions to form a __6__ of identical T cells. An activated helper T cell yields a clone of active helper T cells and __7__ T cells. An activated cytotoxic T cell yields a clone of active __8__ T cells and memory T cells.

Active helper T cells secrete __9__ that stimulate phagocytosis and activity of B and T cells. __10__ T cells are essential in starting an immune response by T cells and B cells. Active __11__ T cells bind to cells with the targeted antigen and release a lethal dose of chemicals to kill the cell. After the pathogen has been destroyed, the dormant __12__ T cells remain to launch a quicker and stronger attack if the same antigen reappears in the body.

1) ____________
2) ____________
3) ____________
4) ____________
5) ____________
6) ____________
7) ____________
8) ____________
9) ____________
10) ____________
11) ____________
12) ____________

c. Write the words that complete the sentences describing antibody-related immunity in the spaces at the right.

Antibody-mediated immunity begins when a foreign antigen binds to __1__ of a B cell. The B cell engulfs and digests the antigens and displays part of the antigens on its __2__ along with its own self antigens. When a __3__ T cell binds to this antigen and self protein complex, it secretes __4__ that activate the B cell. The activated B cell divides repeatedly producing a __5__ of identical B cells. Most of the B cells in the clone become __6__, but some become __7__ B cells.

Plasma cells rapidly produce and release __8__, which circulate in blood and other body fluids binding to the targeted antigens so they are more easily destroyed by __9__ and by other means. After the pathogen has been destroyed, __10__ B cells remain to launch an even quicker and stronger attack if the same antigen reenters the body.

1) ____________________
2) ____________________
3) ____________________
4) ____________________
5) ____________________
6) ____________________
7) ____________________
8) ____________________
9) ____________________
10) ____________________

d. Match the antibodies with the statements. More than one answer may apply.

IgA IgD IgE IgG IgM

1) Most abundant antibody in the blood. ____________________
2) Fixes complement to antigens. ____________________
3) Serves as receptors on B cells. ____________________
4) Involved in allergic reactions. ____________________
5) Transferred to child via mother's milk. ____________________
6) Transferred to fetus via placenta. ____________________
7) Binds with antigens. ____________________
8) Protects mucous membranes. ____________________
9) Neutralizes toxins. ____________________

6. Immune Responses

a. Match the immune responses with the statements.

1) Primary immune response 2) Secondary immune response

______ Occurs when an antigen is encountered for the first time.
______ Occurs in subsequent encounters with same antigen.
______ Results from activation of memory cells.
______ The more rapid and intense response.

b. Match the types of immunity with the statements.

1) Naturally acquired active
3) Naturally acquired passive
2) Artificially acquired active
4) Artificially acquired passive

______ Immunity from antibodies received in mother's milk.
______ Immunity from a vaccine of dead pathogens.
______ Immunity after having the disease and recovering.
______ Immunity from injected antibodies.

7. Rejection of Organ Transplants

Write the answers that match the statements in the spaces at the right.

1) The type of antigen on the surface of WBCs that must be matched between organ donor and recipient. ______________________
2) The type of drugs given to prevent rejection. ______________________

8. Disorders of the Lymphoid System

Write the answers that match the statements in the spaces at the right.

1) Microscopic worms plug lymphatic vessels. ______________________
2) An abnormally intense immune reaction. ______________________
3) HIV destroys helper T cells. ______________________
4) A tumor of lymphatic tissue. ______________________
5) Inflammation of the tonsils. ______________________
6) Lymphocytes attack own body tissues. ______________________
7) Allergy attack that involves entire body. ______________________
8) Transmitted via blood exchanges and sexual intercourse. ______________________

9. Clinical Insights

a. Mary is a grocery checker with no evidence of heart disease. She complains that when she comes home from work, her feet and legs are swollen and sometimes painful. In the morning, the swelling is gone. How do you explain this? ______________________

b. The AIDS virus attacks helper T cells. Explain how this, in time, causes immunodeficiency. ______________________

c. Infants typically receive a series of three DTaP injections (vaccinations against three infectious diseases) followed by a booster shot at four to six years of age. Explain the value of the booster shot. ______________________

STUDY GUIDE 14

1. Structures of the Respiratory System

a. Label the parts of the upper respiratory tract by placing the numbers of the structures by the correct labels.

_____ Auditory tube
_____ Epiglottis
_____ Esophagus
_____ Hyoid
_____ Inferior nasal concha
_____ Laryngopharynx
_____ Larynx
_____ Lingual tonsil
_____ Middle nasal concha
_____ Nasal cavity
_____ Nasopharynx
_____ Nostril
_____ Olfactory mucosa
_____ Oral cavity
_____ Oropharynx
_____ Palate, hard
_____ Palate, soft
_____ Palatine tonsil
_____ Pharyngeal tonsil
_____ Superior nasal concha
_____ Sinus, frontal
_____ Sinus, sphenoidal
_____ Tongue
_____ Trachea

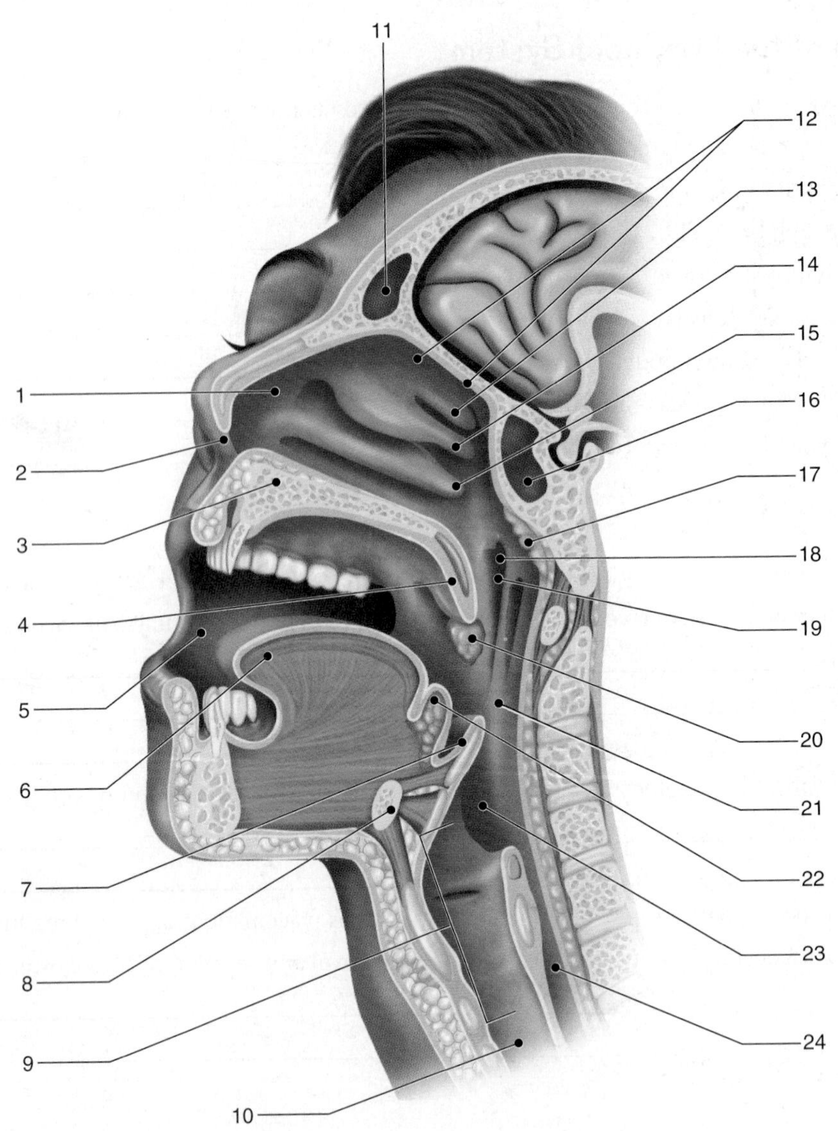

b. Label the parts of the lower respiratory tract by placing the numbers of the structures by the correct labels.

______ Bronchus, lobar (2 answers)	______ Left main bronchus	______ Right main bronchus
______ Bronchus, segmental (2 answers)	______ Left superior lobe	______ Right middle lobe
______ Diaphragm	______ Parietal pleura	______ Right superior lobe
______ Larynx	______ Pleural cavity	______ Trachea
______ Left inferior lobe	______ Right inferior lobe	______ Visceral pleura

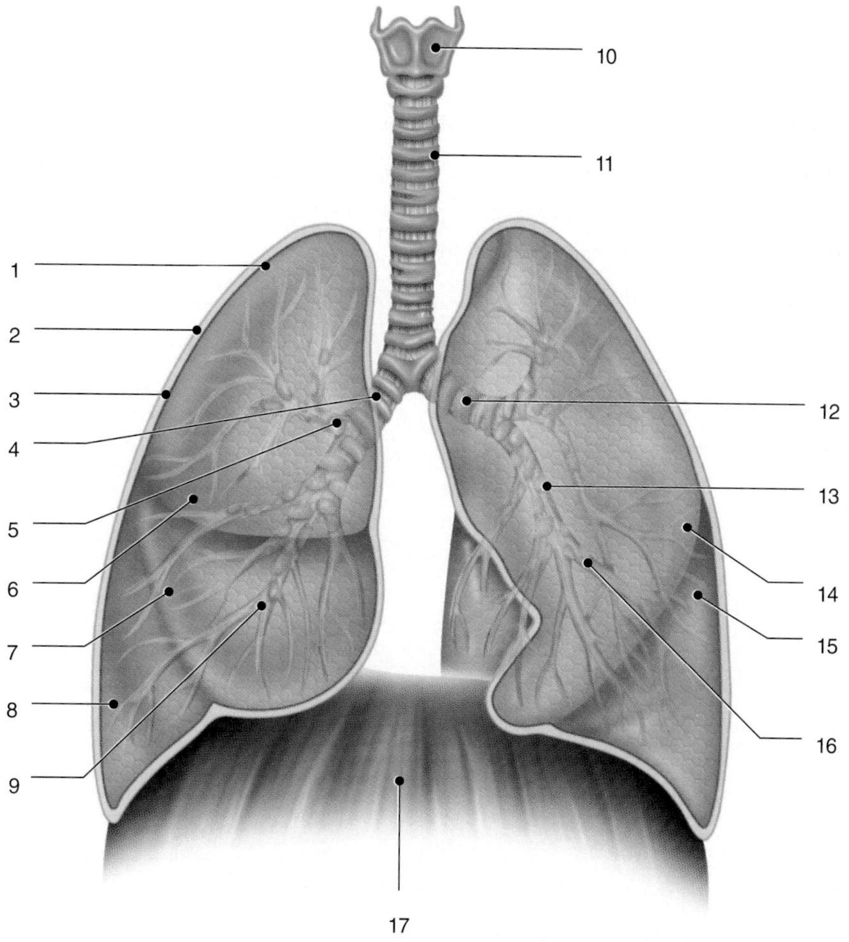

c. Write the words that complete the sentences in the spaces at the right.

Air enters the nasal cavity via the __1__, and it is __2__, __3__, and __4__ as it flows over the nasal mucosae. Mucus and entrapped particles are moved by __5__ to the __6__ and are swallowed.

1) ______________________
2) ______________________
3) ______________________
4) ______________________
5) ______________________
6) ______________________

d. Write the names of the structures that match the statements in the spaces at the right.

1) Openings allowing air to enter nose. ______________________
2) Internal chamber of nose. ______________________
3) Separates oral and nasal cavities. ______________________
4) Increase surface area of nasal cavity. ______________________

5) Connects nasal cavity with larynx. ______________________
6) Air-filled cavities in skull bones. ______________________
7) Lymphoid tissue in superior pharynx. ______________________
8) Lymphoid tissue at junction of pharynx and oral cavity. ______________________
9) Cartilaginous box containing vocal folds.
10) Opening between vocal folds. ______________________
11) Cartilage forming Adam's apple. ______________________
12) Folds over larynx opening in swallowing. ______________________
13) Air passageway extending from larynx to bronchi. ______________________
14) Bronchi that enter lungs. ______________________
15) Support walls of bronchi. ______________________
16) Lines air passageways larger than bronchioles. ______________________
17) Tiny air sacs at ends of alveolar ducts. ______________________
18) Membrane covering external surface of lung. ______________________
19) Membrane lining internal wall of thoracic cage. ______________________
20) Potential space between pleurae. ______________________

2. Breathing

a. Write the words that complete the sentences in the spaces at the right.

Air flows into the lungs during __1__ and out of the lungs during __2__. During breathing, air flows from an area of __3__ pressure to an area of __4__ pressure. During quiet inspiration, contraction of the __5__ and the __6__ causes an increase in the __7__ of the thoracic cavity and lungs, which decreases the air __8__ within the lungs. Air flows into the lungs because of the __9__ atmospheric pressure. When the muscles of inspiration relax, the __10__ of the thoracic cavity and lungs is decreased, which increases the air __11__ within the lungs. Air flows out of the lungs because of the __12__ air pressure within the lungs.

1) ______________________
2) ______________________
3) ______________________
4) ______________________
5) ______________________
6) ______________________
7) ______________________
8) ______________________
9) ______________________
10) ______________________
11) ______________________
12) ______________________

b. Indicate whether each statement is true (T) or false (F).

______ Negative pressure in the pleural cavity is necessary for expiration.

______ Surfactant prevents collapse of alveoli.

______ Forceful expiration involves contraction of abdominal muscles and external intercostals.

______ Pneumothorax causes collapse of a lung.

______ Breathing exchanges air between the atmosphere and the alveoli of the lungs.

3. Respiratory Volumes and Capacities

Match the respiratory volumes and capacities with the statements.

1) Expiratory reserve volume
2) Inspiratory reserve volume
3) Residual volume
4) Tidal volume
5) Total lung capacity
6) Vital capacity

______ Volume of air exhaled in quiet expiration.
______ Air that remains in the lungs after maximum forceful expiration.
______ Volume forcefully exhaled after quiet expiration.
______ Volume forcefully inhaled after quiet inspiration.
______ Maximum volume forcefully exhaled after maximum forceful inspiration.
______ Averages about 500 ml.
______ Averages about 5,800 ml.

4. Control of Breathing

Write the terms that match the statements in the spaces at the right.

1) Name the two respiratory control centers located in the medulla oblongata. ______________________
2) Name the respiratory control center in the pons. ______________________
3) Controls rhythmic, quiet breathing. ______________________
4) Integrates sensory inputs and sends nerve impulses to the VRG to modify breathing cycles. ______________________
5) Receives input from higher brain areas and relays nerve impulses to the DRG to modify breathing cycles. ______________________
6) Controls voluntary changes in the breathing cycle. ______________________
7) Occurs when the VRG sends nerve impulses to the muscles of breathing. ______________________

5. Factors Influencing Breathing

Indicate whether each statement is true (T) or false (F).

______ A fever increases the breathing rate.
______ The respiratory rhythmicity center detects blood levels of carbon dioxide and oxygen.
______ An increase in the blood H^+ concentration increases the breathing rate.
______ A mild increase in blood oxygen concentration decreases the breathing rate.
______ An increase in blood CO_2 concentration increases the breathing rate.
______ Higher brain centers can permanently override the action of the respiratory rhythmicity center.
______ Chemoreceptors for blood oxygen are located in the carotid and aortic bodies.
______ A very low blood oxygen concentration increases the breathing rate.

6. Gas Exchange

Write the words that complete the sentences in the spaces at the right.

The exchange of respiratory gases occurs	1) ______
by __1__. In comparison to the air in the	2) ______
alveoli, blood returning to the lungs has a lower	3) ______
concentration of __2__ and a higher concentration	4) ______
of __3__. Therefore, oxygen diffuses from the __4__	5) ______
into the __5__, and carbon dioxide diffuses from	6) ______
the __6__ into the __7__. Blood leaving the lungs	7) ______
is __8__-rich and __9__-poor.	8) ______
In comparison to concentrations in tissue	9) ______
cells, blood entering systemic capillaries has	10) ______
a lower concentration of __10__ and a higher	11) ______
concentration of __11__. Therefore, oxygen	12) ______
diffuses from the __12__ into the interstitial fluid	13) ______
before entering the __13__, and carbon dioxide	14) ______
diffuses from the __14__ into the interstitial fluid	15) ______
before entering the __15__. Blood leaving systemic	16) ______
capillaries is __16__-rich and __17__-poor.	17) ______

7. Transport of Respiratory Gases

a. Use symbols (O_2 and CO_2) and arrows to show the direction of diffusion for respiratory gases between blood in systemic capillaries and tissue (not lung) cells.

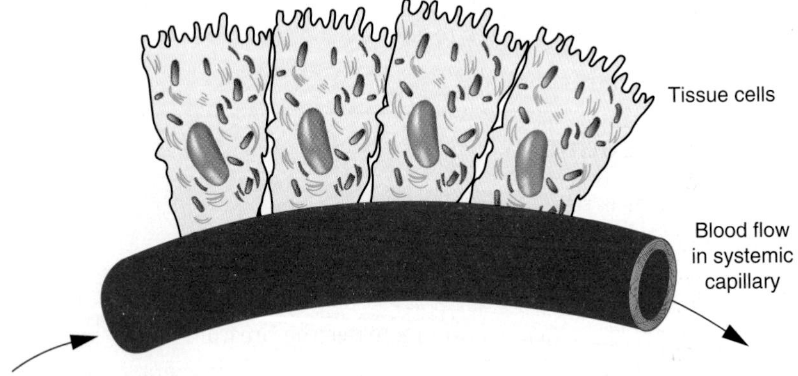

b. Write the terms that match the statements in the spaces at the right.

1) Compound used to transport most oxygen. ______
2) Cell in which most oxygen is transported. ______
3) Substance that carries most CO_2. ______
4) Combination of hemoglobin and CO_2. ______
5) Enzyme speeding up reaction of CO_2 and H_2O to form carbonic acid. ______
6) Cell in which most carbonic acid is formed. ______

c. Indicate whether each statement is true (T) or false (F).

______ Hemoglobin can carry both O_2 and CO_2 at the same time.

______ Oxygenated blood carries some carbon dioxide.

______ Deoxygenated blood carries some oxygen.

______ Hemoglobin loads or unloads oxygen, depending on the surrounding oxygen concentration.

______ Oxygen and carbon dioxide compete for the same binding site on the hemoglobin molecule.

______ Blood loads or unloads CO_2 depending upon the surrounding carbon dioxide concentration.

8. Disorders of the Respiratory System

Write the disorders that match the statements in the spaces at the right.

1) Accumulation of fluid in the lungs. ______________________

2) Inflammation of the bronchi. ______________________

3) Rupture of alveoli due to exposure to airborne irritants. ______________________

4) Acute inflammation of alveoli due to viral or bacterial infection. ______________________

5) Collapse of alveoli in infants due to an insufficient amount of surfactant. ______________________

6) Wheezing, labored breathing due to constriction of bronchioles. ______________________

7) Viral disease characterized by fever, chills, aches, and coldlike symptoms. ______________________

8) Disorder characterized by a reduction of the respiratory surface area and decrease in the expiratory reserve volume. ______________________

9) Inflammation of nasal mucosae. ______________________

10) Blockage of a small artery in lung by a transported blood clot. ______________________

9. Clinical Insights

a. One treatment for hyperventilation is having the patient breathe into a paper bag. How does this reestablish normal breathing? ______________________

____

b. A newborn infant, born a month early, is having difficulty breathing and is placed under an O_2 hood. What is the probable problem? ______________________

Should the infant receive pure oxygen or an oxygen–carbon dioxide mixture? ______________________

Explain. ______________________

____

STUDY GUIDE 15

1. Digestion: An Overview

Indicate the substances that perform these roles in chemical digestion.

1) Added to complex, nonabsorbable food molecules to split them into smaller, absorbable nutrient molecules. ______________________

2) Speed up hydrolysis of food molecules. ______________________

2. Alimentary Canal: General Characteristics

a. Label the parts of the digestive system by placing the numbers of the structures in the spaces by the correct labels.

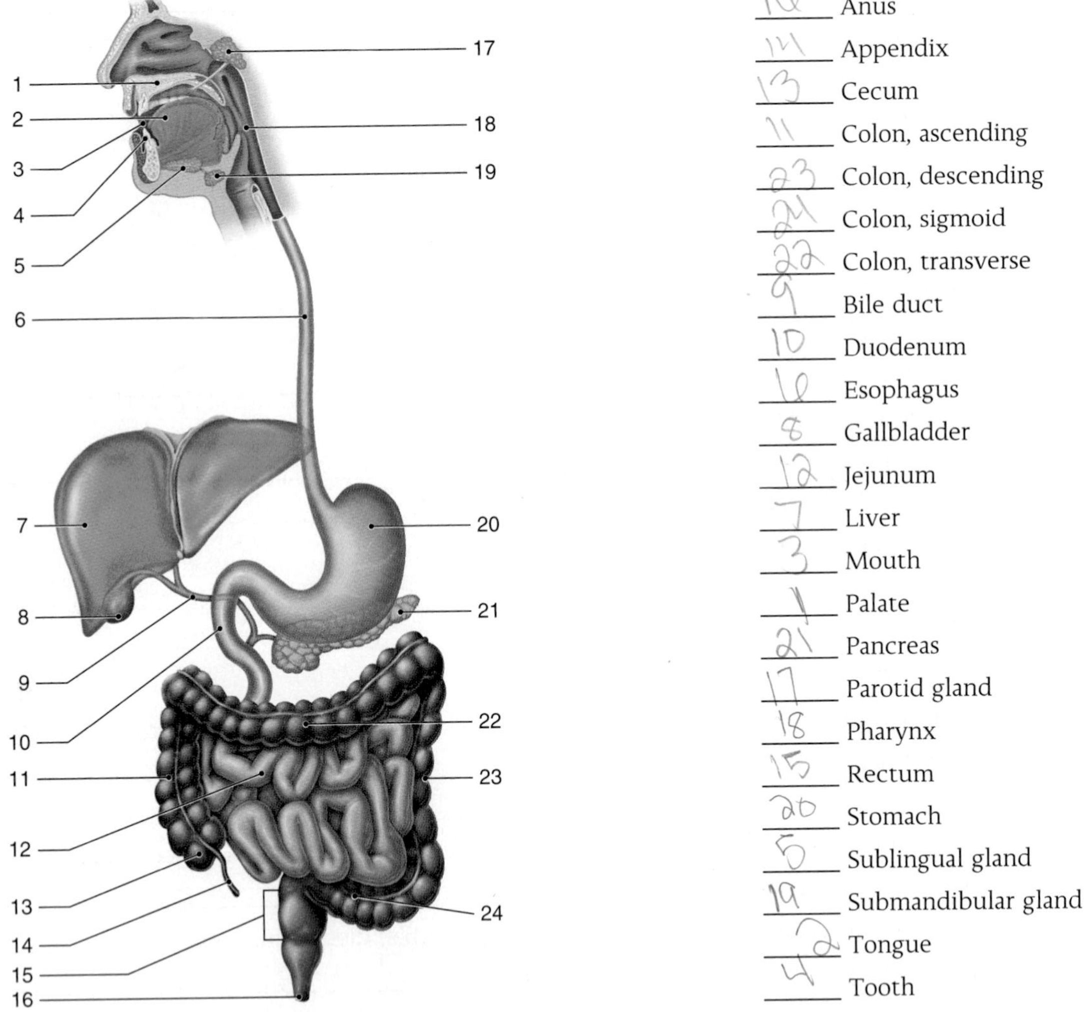

16 Anus
14 Appendix
13 Cecum
11 Colon, ascending
23 Colon, descending
24 Colon, sigmoid
22 Colon, transverse
9 Bile duct
10 Duodenum
6 Esophagus
8 Gallbladder
12 Jejunum
7 Liver
3 Mouth
1 Palate
21 Pancreas
17 Parotid gland
18 Pharynx
15 Rectum
20 Stomach
5 Sublingual gland
19 Submandibular gland
2 Tongue
4 Tooth

b. List the layers of the wall of the alimentary canal from the deepest to the most superficial.

1) ______________________ 3) ______________________

2) ______________________ 4) ______________________

c. What contractions propel food through the alimentary canal? ______________________

3. Mouth

a. Label the figure by placing the numbers of the structures in the spaces by the correct labels.

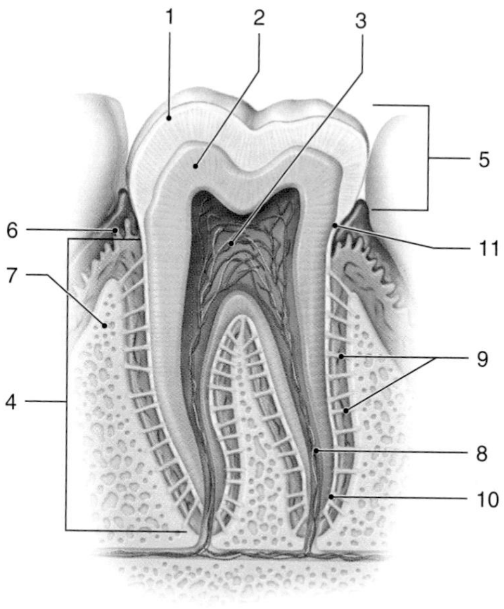

_____ Alveolus
_____ Cement
_____ Crown
_____ Dentin
_____ Enamel
_____ Gingiva
_____ Neck
_____ Periodontal ligament
_____ Pulp cavity of the crown
_____ Root
_____ Root canal

b. Write the terms that match the statements in the spaces at the right.

1) Form lateral walls of the mouth. _______________
2) Separates oral and nasal cavities. _______________
3) Manipulates food during chewing. _______________
4) Tiny projections containing taste buds. _______________
5) Number of deciduous and permanent teeth. (2 answers) _______________
____
6) Teeth used to bite off pieces of food. _______________
7) Teeth used to grasp and tear food. _______________
8) Teeth used to crush and grind food. _______________
9) Three pairs of salivary glands. _______________
____
____
10) Cleanses and lubricates mouth. _______________
11) Salivary enzyme acting on polysaccharides. _______________
12) Product of polysaccharide digestion in mouth. _______________
13) Saliva secretion is stimulated by a ______-feedback mechanism. _______________

4. Pharynx and Esophagus

Write the terms that match the statements in the spaces at the right.

1) Tube carrying food to the stomach. ____________________
2) Relaxes to let food enter stomach. ____________________
3) Carries food from mouth to esophagus. ____________________
4) Covers laryngeal opening in swallowing. ____________________

5. Stomach

Write the terms that match the statements in the spaces at the right.

1) Region of stomach joining esophagus. ____________________
2) Region of stomach joining duodenum. ____________________
3) Glands of mucosa secreting gastric juice. ____________________
4) Hormone stimulating gastric secretion. ____________________
5) Hormones inhibiting gastric secretion. ____________________
____
6) Autonomic nerve impulses stimulating gastric secretion. ____________________
7) Gastric enzyme acting on triglycerides. ____________________
8) Acid in gastric juice. ____________________
9) Gastric enzyme acting on proteins. ____________________
10) Gastric enzyme curdling milk proteins. ____________________
11) Products of gastric protein digestion. ____________________
12) Gastric substance enabling absorption of vitamin B12 by small intestine. ____________________

6. Pancreas

Write the terms that match the statements in the spaces at the right.

1) Carries pancreatic juice from pancreas to duodenum. ____________________
____
2) Two hormones stimulating secretion of pancreatic juice. ____________________
____
3) Source of these hormones. ____________________
4) Pancreatic enzyme acting on polysaccharides. ____________________
5) Product of pancreatic polysaccharide digestion. ____________________
6) Pancreatic enzyme acting on triglycerides. ____________________
7) Products of pancreatic triglyceride digestion. ____________________
____
8) Pancreatic enzyme acting on proteins. ____________________
9) Products of pancreatic protein digestion. ____________________

7. Liver

a. Label the figure by placing the numbers of the structures in the spaces by the correct labels.

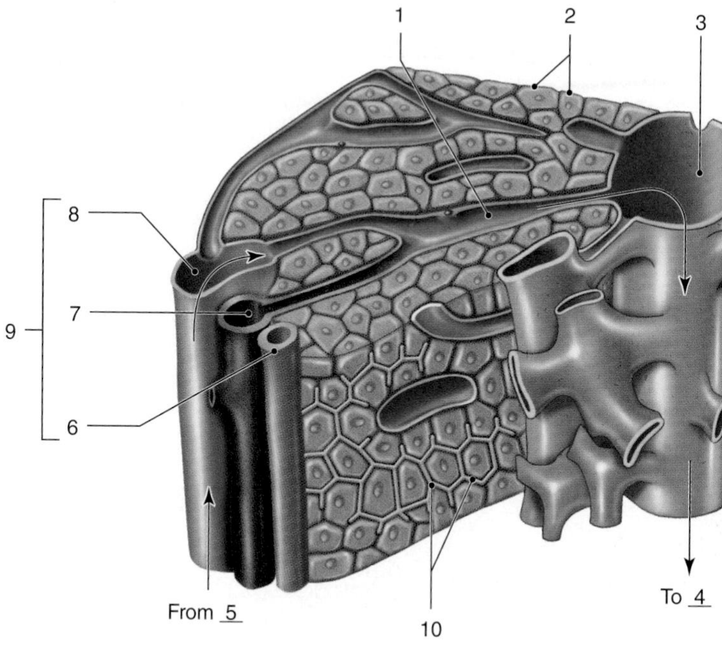

_____ Interlobular bile ductule
_____ Branch of the hepatic artery proper
_____ Branch of the hepatic portal vein
_____ Central vein
_____ Intestines
_____ Hepatocytes
_____ Hepatic sinusoid
_____ Hepatic triad
_____ Hepatic vein

b. Write the terms that match the statements in the spaces at the right.

1) Mineral stored in the liver that is needed for hemoglobin production. _____
2) Vessels carrying blood to liver:
   a) carries oxygenated blood. _____
   b) carries deoxygenated, nutrient-rich blood. _____
3) Vessel carrying blood from liver. _____
4) Structural and functional units of the liver. _____
5) Spaces in a hepatic lobule receiving blood from branches of the hepatic artery proper and hepatic portal vein. _____
6) Vessels in hepatic lobules that merge to form the hepatic vein. _____
7) Carbohydrate stored in liver. _____
8) Secretion formed by liver. _____
9) Stores excess bile. _____
10) Carries bile to duodenum. _____
11) Hormone contracting gallbladder. _____
12) Bile component emulsifying fats. _____
13) Bile component from hemoglobin breakdown. _____

8. Small Intestine

a. Label the figure by placing the numbers of the structures in the spaces by the correct labels.

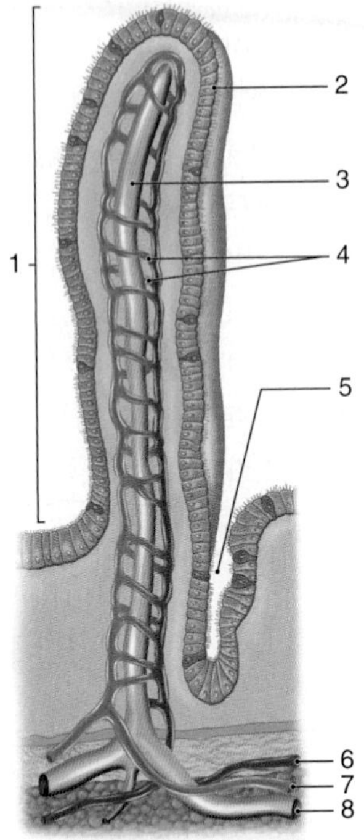

1
2
3
4
5
6
7
8

_____ Arteriole
_____ Blood capillary network
_____ Intestinal gland
_____ Lacteal
_____ Lymphatic vessel
_____ Simple columnar epithelium
_____ Venule
_____ Villus

b. Write the terms that match the statements in the spaces at the right.

1) Segment continuous with the stomach. _______________
2) Segment continuous with the cecum. _______________
3) Membranes supporting small intestine. _______________
4) Relaxes to allow chyme to enter the small intestine. _______________
5) Secretion of intestinal glands. _______________
6) Fingerlike projections of the mucosa. _______________
7) Microscopic folds of exposed epithelial cell membranes. _______________
8) Hormone released by mucosa due to presence of lipid-rich chyme. _______________
9) Hormone released by mucosa due to presence of acidic chyme. _______________
10) _____ nerve impulses stimulate an increase in the rate of intestinal secretions. _______________
11) Enzyme acting on sucrose. _______________
12) End products of sucrose digestion. _______________
____
13) Enzyme acting on lactose. _______________
14) Enzyme acting on maltose. _______________
15) End product of maltose digestion. _______________
16) End products of lactose digestion. _______________
____
17) Enzyme acting on peptides. _______________
18) End products of peptide digestion. _______________

c. Write the terms that complete the sentences in the spaces at the right.
Monosaccharides and amino acids are absorbed into the __1__ networks of the __2__. To be absorbed, monoglycerides and fatty acids are first reunited to form __3__ inside __4__ cells. Clusters of triglycerides are coated with protein, forming __5__ that enter the __6__ of the __7__. These protein-coated lipid clusters are then passed from lymph into blood at the __8__ vein.

1) ____________
2) ____________
3) ____________
4) ____________
5) ____________
6) ____________
7) ____________
8) ____________

9. Large Intestine

Write the terms that match the statements in the spaces at the right.

1) Pouchlike first part of large intestine. ____________
2) External opening of large intestine. ____________
3) Colon segment along left side of abdominopelvic cavity. ____________
4) Colon segment along right side of abdominopelvic cavity. ____________
5) S-shaped colon segment continuous with rectum. ____________
6) Wormlike extension of cecum. ____________
7) Involuntarily controlled anal sphincter. ____________
8) Voluntarily controlled anal sphincter. ____________
9) Decompose undigested materials. ____________
10) Fluid absorbed by large intestine. ____________
11) Relaxes to allow chyme to enter cecum. ____________
12) Reflex activated by filling of rectum with feces. ____________

10. Nutrients: Sources and Uses

Write the terms that match the statements in the spaces at the right.

1) Process producing ATP from energy foods. ____________
2) Plant polysaccharide providing fiber. ____________
3) Preferred energy source for body cells. ____________
4) Organs regulating blood glucose levels. ____________
____
5) Most common lipids in the diet. ____________
6) Type of fats common in animal foods. ____________
7) Type of fats common in plant foods. ____________
8) Lipid abundant in egg yolks. ____________
9) Lipid used to form steroid hormones. ____________
10) Lipid forming much of plasma membranes. ____________
11) Molecules transporting lipids in blood. ____________
____
12) Chemical converted by the liver into urea. ____________
13) Amino acids that cannot be made by liver. ____________

14) Water-soluble vitamin needed to synthesize proteins and form antibodies. ____________________

15) Vitamin essential for the formation of prothrombin. ____________________

16) Mineral required for blood clotting and bone formation. ____________________

11. Disorders of the Digestive System

Write the names of the disorders that match the statements.

1) Inflammation of the large intestine that permanently changes intestinal tissues. ____________________

2) Self-induced starvation due to an abnormal concern about weight control. ____________________

3) Decay of the teeth due to acids formed by certain oral microorganisms. ____________________

4) Dry, hard feces making defecation difficult. ____________________

5) Crystallization of cholesterol in bile within the gallbladder. ____________________

6) Replacement of destroyed hepatocytes by dense irregular connective tissue. ____________________

7) Repeated overeating and purging. ____________________

8) Inflammation of the liver. ____________________

9) Inflammation, bleeding, and degeneration of the gingivae, alveolar processes, and alveolar arch. ____________________

10) Watery feces due to excessive peristalsis. ____________________

11) Enlarged and inflamed veins in anal canal. ____________________

12) Inflammation of the appendix. ____________________

13) Inflammation of the peritoneum. ____________________

14) Inflammation of colon diverticula. ____________________

12. Clinical Insights

a. Severe diarrhea in infants or small children can be a life-threatening event. Explain why. ____________________

____

____

b. A patient is found to have a gastric ulcer. Antibiotics and a drug to reduce the secretion of gastric juice are prescribed. Explain the basis for the prescriptions. ____________________

____

____

____

What serious results may occur with an untreated ulcer? ____________________

____

c. A patient is admitted to the emergency room complaining of severe and spasmodic pain in the epigastric region, and the whites of his eyes are yellowish. He informs the physician that he has had similar but milder pains after meals for four to six weeks. What is the likely problem and the likely solution? ____________________

____

____

STUDY GUIDE 16

1. Urinary System, General

a. Label the figure by placing the numbers of the structures in the spaces by the correct labels.

______ Abdominal aorta	______ Kidney	______ Ureter
______ Inferior vena cava	______ Renal artery	______ Urethra
______ Hilum	______ Renal vein	______ Urinary bladder

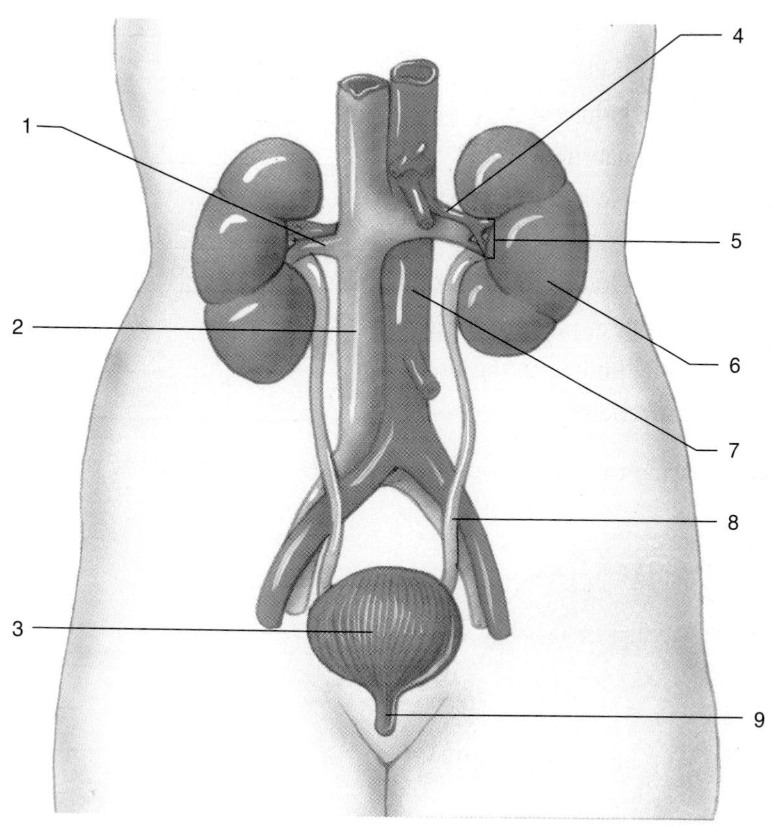

b. Write the names of the organs that match the functions in the spaces at the right.

1) Stores urine temporarily. ______________________

2) Produces urine. ______________________

3) Carries urine to the external environment. ______________________

4) Carries urine away from the kidneys. ______________________

5) Maintains composition and volume of body fluids. ______________________

2. Functions of the Urinary System

Indicate whether each statement is true (T) or false (F).

______ The urinary system helps regulate blood electrolyte levels.

______ The urinary system regulates respiration.

______ The urinary system produces hormones.

______ Urea, uric acid, and creatinine are carbohydrate wastes.

3. Anatomy of the Kidneys

a. Write the names of the structures that match the statements in the spaces at the right.

1) Superficial layer containing renal corpuscles. ______________________
2) Layer containing renal pyramids. ______________________
3) Funnel-like cavity continuous with ureter. ______________________
4) Receptacles surrounding renal papillae. ______________________
5) Thin layer of connective tissue enveloping kidney. ______________________
6) Capillaries in renal corpuscle. ______________________
7) Functional units of the kidneys. ______________________
8) U-shaped segment of renal tubule. ______________________
9) Segment of renal tubule joined to glomerular capsule. ______________________
10) Segment of renal tubule joined to a collecting duct. ______________________
11) Formed of modified cells at point of contact between the ascending limb with the afferent and efferent glomerular arterioles. ______________________

b. Label the figure by placing the numbers of the structures in the spaces by the correct labels.

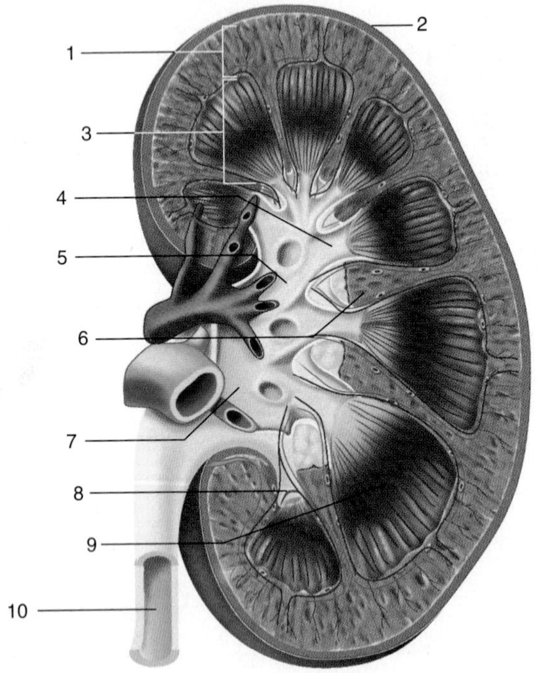

1
2
3
4
5
6
7
8
9
10

_____ Fibrous capsule
_____ Major calyx
_____ Minor calyx
_____ Renal column
_____ Renal cortex
_____ Renal medulla
_____ Renal papilla
_____ Renal pelvis
_____ Renal pyramid
_____ Ureter

c. Label the figure by placing the numbers of the structures in the spaces by the correct labels.

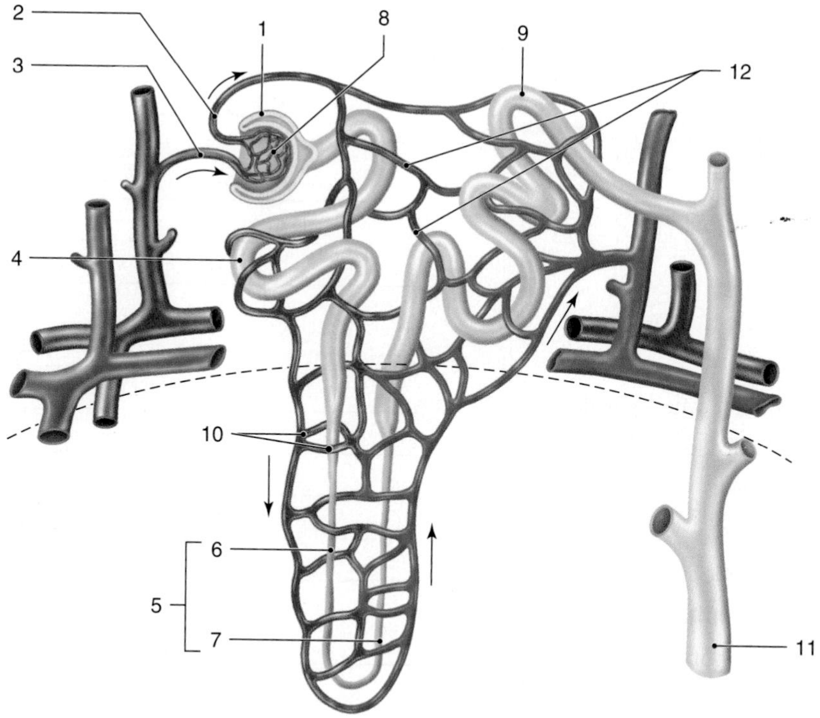

_____ Ascending limb
_____ Afferent glomerular arteriole
_____ Collecting duct
_____ Descending limb
_____ Distal convoluted tubule
_____ Efferent glomerular arteriole
_____ Glomerular capsule
_____ Glomerulus
_____ Nephron loop
_____ Peritubular capillary
_____ Proximal convoluted tubule
_____ Vasa recta

4. Urine Formation

a. Write the words that complete the sentences in the spaces at the right.

Urine formation begins with __1__, which forces water and diffusible solutes from blood plasma into the glomerular __2__. As the tubular fluid moves along the renal tubule, useful solutes are returned to blood in the __3__ by __4__ and certain solutes are moved from blood into the tubular fluid by __5__. Concentrated urine is formed by the reabsorption of __6__ from the collecting duct and DCT.

1) ____________________
2) ____________________
3) ____________________
4) ____________________
5) ____________________
6) ____________________

b. Write the words that complete the sentences in the spaces at the right.

A decrease in the glomerular filtration rate causes the __1__ complex to secrete __2__, which triggers the __3__ __4__ mechanism. The end product of these reactions is __5__, which increases systemic blood pressure by __6__ arterioles, stimulating __7__ secretion by the posterior lobe of the pituitary gland, and stimulating __8__ secretion by the adrenal cortex.

1) ____________________
2) ____________________
3) ____________________
4) ____________________
5) ____________________
6) ____________________
7) ____________________
8) ____________________

c. Write the terms that match the statements in the spaces at the right.

1) Three homeostatic processes that maintain the glomerular filtration rate. ____________________
____
____

2) Plasma component that cannot pass through glomerular pores. ____________________

3) Force producing filtration. ____________________

4) Fluid in glomerular capsule. ____________________

5) Recovery of needed materials from tubular fluid into the blood. ____________________

6) Volume of glomerular filtrate formed per day. ____________________

7) Method of transport of sodium ions. ____________________

8) Method of transport of water. ____________________

9) Method of transport of glucose. ____________________

10) Method of transport of chloride ions. ____________________

11) Process moving substances from blood in the peritubular capillary into the tubular fluid. ____________________

d. Indicate whether each statement is true (T) or false (F).

______ Urine contains waste and excessive materials removed from the blood.

______ Urine formation depends upon maintenance of the blood pressure within the peritubular capillaries.

______ Most of the tubular fluid volume is secreted.

______ Negatively charged ions and positively charged ions are electrochemically attracted to each other.

______ The active reabsorption of sodium ions increases the rate of water reabsorption by osmosis.

5. Excretion of Urine

a. Write the terms that match the statements in the spaces at the right.

1) Carry urine to urinary bladder. ____________________

2) Muscle in wall of urinary bladder. ____________________

3) Method of urine transport by ureters. ____________________

4) Type of muscle tissue composing the internal urethral sphincter. ____________________

5) Type of muscle tissue composing the external urethral sphincter. ____________________

6) Type of muscle tissue in walls of ureters. ____________________

b. Write the words that complete the sentences in the spaces at the right.

The accumulation of __1__ stretches the urinary bladder wall, which triggers the __2__ reflex. This reflex causes rhythmic involuntary contractions of the __3__ and opens the involuntarily controlled __4__ urethral sphincter. If the voluntarily controlled __5__ urethral sphincter is relaxed, __6__ occurs.

1) ____________________
2) ____________________
3) ____________________
4) ____________________
5) ____________________
6) ____________________

6. Characteristics of Urine

Indicate whether each statement is true (T) or false (F).

_______ The color of urine is due to the presence of urea.

_______ The pH of urine is usually slightly acidic.

_______ Normal urine is never alkaline.

_______ Normal urine has a specific gravity greater than 1.000.

_______ Normal urine does not contain proteins or hemoglobin.

7. Maintenance of Blood Plasma Composition

a. Write the terms that match the statements in the spaces at the right.

1) Hormone promoting water reabsorption. _______________

2) Hormone promoting reabsorption of Na^+. _______________

3) Hormone promoting secretion of K^+. _______________

4) Hormone decreasing blood level of Ca^{2+}. _______________

5) Hormones increasing blood level of Ca^{2+}. _______________

6) Hormone inhibiting sodium reabsorption. _______________

b. Indicate whether each statement is true (T) or false (F).

_______ Cellular metabolism does not affect plasma composition.

_______ Plasma composition is changed by the work of kidneys.

_______ Electrolytes are totally reabsorbed into the blood.

_______ About 99% of water in the tubular fluid is reabsorbed.

_______ Water is lost from the body only in urine.

_______ Urine volume is reduced when water intake is curtailed.

_______ Perspiring heavily may reduce the volume of urine.

_______ Electrolyte balance is largely maintained by the active reabsorption of negatively charged ions.

_______ Urea is formed by the kidneys from amine groups.

_______ Buffers are chemicals in body fluids that either combine with or release hydrogen ions.

_______ The production of carbon dioxide by metabolizing cells tends to make the blood more alkaline.

_______ Kidneys help to regulate the pH of body fluids by secreting excess hydrogen ions into the glomerular filtrate.

_______ Water and electrolyte balance in body fluids is essential for normal cell functioning.

_______ ADH is released by the posterior lobe of the pituitary gland when the water concentration of the blood is reduced.

_______ Aldosterone is secreted by the adrenal cortex when the concentration of K^+ in the blood is reduced.

_______ ADH increases the permeability of the vasa recta to water.

_______ Electrolyte concentrations in the blood affect the movement of water into cells by osmosis.

_______ Atrial natriuretic peptide promotes the reabsorption of sodium ions and the excretion of water to decrease blood volume.

8. Disorders of the Urinary System

Write the names of the disorders matching the statements in the spaces at the right.

1) Inflammation of the glomeruli. ______________________
2) Inflammation of the urinary bladder. ______________________
3) Inflammation of nephrons and renal pelvis. ______________________
4) Excessive urine production. ______________________
5) Kidney stones. ______________________
6) Inflammation of the urethra. ______________________
7) Deposits of uric acid in the renal pelvis. ______________________
8) Characterized by uremia. ______________________
9) Characterized by protein in the urine. ______________________

9. Clinical Insights

a. A 60-year-old woman comes to the clinic with severe edema of her lower limbs. A diuretic is prescribed, and she is placed on a salt-free diet. She is also advised to take a 30-minute walk each morning and afternoon and to elevate her feet higher than her head for 20-minute periods morning and afternoon. Explain how the diuretic will reduce her edema. ______________________

____

____

____

Explain how the salt-free diet will help reduce her edema. ______________________

____

____

____

____

____

____

Explain how elevating her feet and walking will help reduce her edema. ______________________

____

____

b. Explain why women develop cystitis more frequently than men. ______________________

____

1. Male Reproductive System

a. Label the figure by placing the numbers of the structures by the correct labels.

______ Bulbo-urethral gland
______ Corpus cavernosum
______ Corpus spongiosum
______ Ejaculatory duct
______ Epididymis
______ Glans penis
______ Penis
______ Prepuce
______ Prostate gland
______ Scrotum
______ Seminal vesicle
______ Testis
______ Urethra
______ Urinary bladder
______ Vas deferens

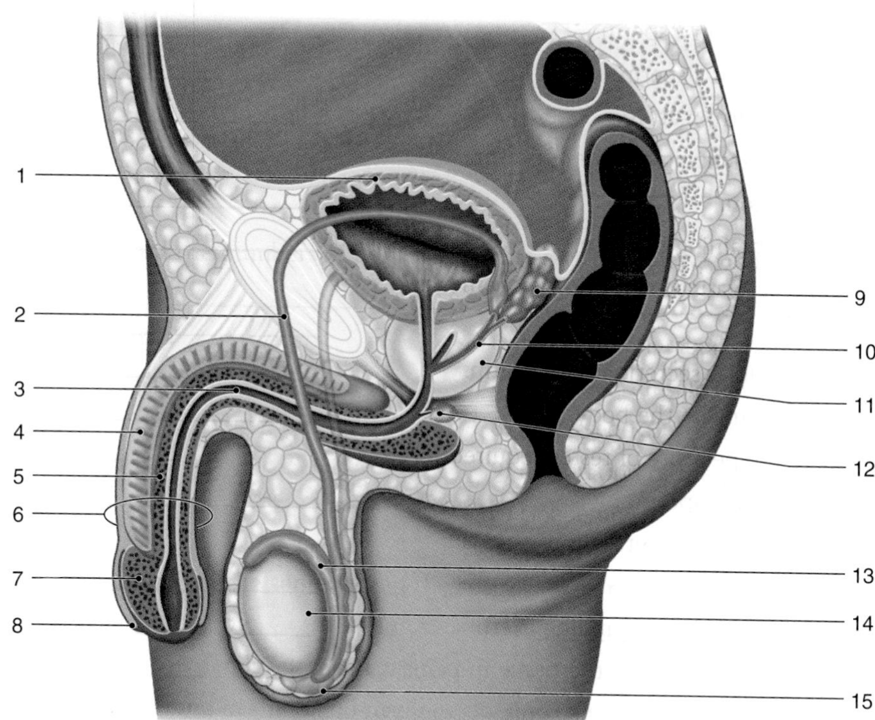

b. Trace the path of sperm from a testis to the external environment by placing the numbers of the ducts in the spaces below.

1) Ejaculatory duct
2) Epididymis
3) Urethra
4) Vas deferens

Testis → ________ → ________ → ________ → ________ → the external environment.

6. Hormonal Control of Reproduction in Females

a. Match the hormones listed with the following statements.

Estrogens GnRH Progesterone FSH LH

1) Secreted by the granulosa cells. ______
2) Stimulates maturation of female reproductive organs. ______
3) Maintains endometrium in pregnancy. ______
4) Develops female secondary sex characteristics. ______
5) Secreted by corpus luteum. ______
6) Secreted by the hypothalamus. ______
7) Stimulates development and function of corpus luteum. ______
8) First secretion starts onset of puberty. ______
9) Stimulates development of ovarian follicles. ______
10) Prepare endometrium for pregnancy. ______
______
11) High blood concentrations inhibit GnRH secretion. ______
12) Promotes thickening of endometrium. ______
13) Promotes formation of blood vessels in endometrium. ______

b. Write the words that complete the sentences in the spaces at the right.

Between puberty and __1__, a woman experiences one reproductive cycle per month consisting of an ovarian cycle and a __2__ cycle. These cycles are controlled by __3__ and average about __4__ days in length. A cycle is started by the secretion of __5__ by the hypothalamus, which activates the release of __6__ and a small amount of __7__ by the anterior lobe of the pituitary. FSH stimulates the development of a primary __8__ and the secretion of __9__ by the granulosa cells. __10__ promotes the thickening of the endometrium. The increasing production of estrogens triggers a sharp increase in __11__ secretion and a lesser increase in FSH production, leading to __12__ on day 14. Under stimulation by LH, the remnants of the mature ovarian follicle become a __13__ that secretes a high level of __14__ and estrogens, which together prepare the __15__ of the uterus to receive a preembryo. The high level of __16__ inhibits secretion of GnRH, preventing development of additional ovarian follicles. If pregnancy does not occur, the corpus luteum degenerates and the blood levels of estrogens and __17__ rapidly decline, resulting in breakdown of the endometrium leading to __18__ and secretion of GnRH starting a new reproductive cycle.

1) ______
2) ______
3) ______
4) ______
5) ______
6) ______
7) ______
8) ______
9) ______
10) ______
11) ______
12) ______
13) ______
14) ______
15) ______
16) ______
17) ______
18) ______

7. Mammary Glands

Indicate whether each statement is true (T) or false (F).

_______ Mammary glands are specialized for milk and hormone production.

_______ Breasts contain areolar connective tissue but little adipose tissue.

_______ Alveolar glands occur in lobes of mammary glands.

_______ Estrogens stimulate the production of milk.

_______ A pigmented areola surrounds a protruding nipple.

_______ Mammary glands are present but are nonfunctional in males.

_______ Breasts are formed deep to the pectoralis muscles.

8. Birth Control

Indicate whether each statement is true (T) or false (F).

_______ Contraceptives are designed to prevent union of sperm and secondary oocyte.

_______ Contraception and birth control are synonymous.

_______ Spermicides act by killing the sperm.

_______ Progesterone in the "pill" inhibits GnRH secretion.

_______ Spermicides are usually used with barrier methods.

_______ It is impossible to catch STDs when using a condom.

_______ Diaphragms and cervical caps are about equally effective.

_______ The rhythm method relies on knowing when ovulation occurs.

_______ A condom is a barrier contraceptive.

_______ Pregnancy is not possible when using a condom.

_______ The "pill" is the most effective contraceptive.

_______ There are no undesirable side effects of hormonal birth control.

_______ An IUD prevents implantation of an embryo.

_______ Induced abortion is a contraceptive procedure.

_______ A tubal ligation prevents ovulation.

_______ Diaphragms and cervical caps are barrier contraceptives.

_______ A vasectomy prevents the ejaculation of semen.

_______ Use of an IUD may cause pelvic inflammatory disease.

_______ Withdrawal is more effective than a condom.

_______ Undesirable side effects may result from induced abortion.

9. Disorders of the Reproductive Systems

Write the disorders that match the statements in the spaces provided.

Male Disorders

1) Inability to maintain an erection. _______________

2) Common reproductive cancer in males. _______________

3) Inflammation of the prostate glands. _______________

4) Inability to produce sufficient viable sperm. _______________

5) Causes constriction of urethra in about one-third of older males. _______________

Female Disorders

1) Physical pain during menstruation. ______
2) Growth of endometrial tissue outside the uterus. ______
3) Absence of menstruation without pregnancy. ______
4) Associated with highly absorbent tampons. ______
5) Inability to become pregnant. ______
6) Infection in reproductive organs and/or pelvic cavity. ______
7) Physical and emotional distress just prior to menstruation. ______
8) Caused by toxin formed by *S. aureus*. ______

Sexually Transmitted Diseases

1) Results from infection with herpes simplex virus type 2. ______
2) Fatal disease resulting from HIV infection. ______
3) Two bacterial diseases that may lead to sterility in females. ______ ______
4) Characterized by chancre in first stage. ______
5) Caused by HPV. ______
6) Curable with antibiotics. ______ ______
7) Viral diseases for which there are no cures. ______ ______

9. Clinical Insights

a. Failure of the testes to descend into the scrotum (cryptorchidism) causes sterility in males. Explain why. ______

b. Secondary amenorrhea in female athletes results from strenuous activity, which blocks the hypothalamic regulation of reproduction. How does this stop the reproductive cycles? ______

Women with amenorrhea produce little, if any, estrogens, which causes osteoporosis (bone loss). Why are such women deficient in estrogens? ______

c. Without effective treatment, a bacterial sexually transmitted disease usually leads to pelvic inflammatory disease (PID) in females. How does this happen? ______

STUDY GUIDE 18

1. Fertilization and Early Development

a. Write the terms that complete the sentences in the spaces at the right.

A __1__ oocyte containing __2__ chromosomes is released at ovulation, and it is enveloped by several layers of __3__ cells. After entering a __4__ tube, it is slowly carried toward the __5__ through peristalsis and the beating __6__ of the cells lining the tube. The oocyte remains viable for about __7__ hours after ovulation. Sperm deposited in the __8__ swim into the uterus and on into the __9__ tubes. The sperm usually encounter the secondary oocyte in the __10__ portion of a uterine tube. Sperm remain viable in the female reproductive tract for about __11__ hours.

The __12__ of the sperm releases enzymes that disperse the __13__ cells surrounding the secondary oocyte. Once a __14__ enters the secondary oocyte, chemical changes in the __15__ prevent other sperm from entering. The secondary oocyte immediately completes the __16__ meiotic division, forming an __17__ and another polar body, each containing __18__ chromosomes. Union of __19__ and __20__ completes fertilization, forming a __21__ containing __22__ chromosomes.

1) ______
2) ______
3) ______
4) ______
5) ______
6) ______
7) ______
8) ______
9) ______
10) ______
11) ______
12) ______
13) ______
14) ______
15) ______
16) ______
17) ______
18) ______
19) ______
20) ______
21) ______
22) ______

b. Write the terms that match the statements in the spaces at the right.

1) Collective term for the early mitotic divisions of the preembryo. ______
2) Solid ball of cells formed by cleavage. ______
3) Hollow ball of cells. ______
4) Mass of cells within the blastocyst. ______
5) Superficial wall of the blastocyst. ______
6) Embedding of blastocyst in endometrium. ______
7) Length of preembryonic stage. ______
8) Length of full-term pregnancy. ______
9) Length of prenatal growth and development. ______

2. Embryonic Development

a. In each of the spaces provided, write the number (1-3) of the germ layer from which the structures develop.

1) Ectoderm 2) Mesoderm 3) Endoderm

_____ Pituitary gland
_____ Dermis
_____ Nervous system
_____ Skeletal muscle
_____ Lining of digestive tract
_____ Liver and pancreas
_____ Epidermis
_____ Salivary glands and tooth enamel
_____ Kidneys and gonads
_____ Lining of respiratory tract

b. Write the terms that match the statements in the spaces at the right.

1) Becomes the chorion. _____________________
2) Connects embryo to placenta. _____________________
3) Form early embryonic formed elements. _____________________
____
4) Serves as shock absorber for fetus. _____________________
5) Membrane surrounding embryo/fetus. _____________________
6) Fingerlike projections from chorion that penetrate endometrium. _____________________
7) Site of exchange of materials between embryonic and maternal bloods. _____________________
8) Name given to the embryo at beginning of ninth week. _____________________
9) Fluid in which the embryo develops. _____________________
10) Developmental stage between second and eighth weeks. _____________________

c. Label the figure by placing the numbers of the structures in the spaces by the correct labels.

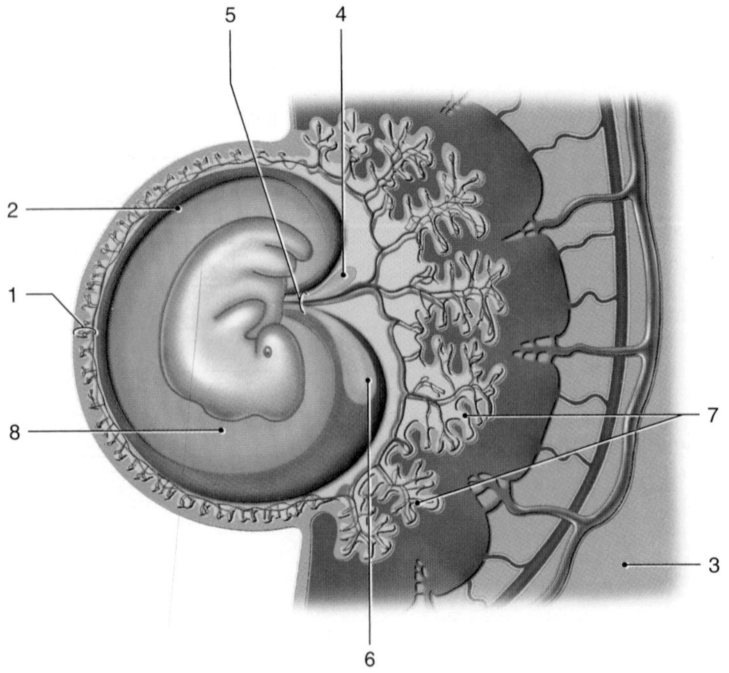

_____ Allantois
_____ Amnion
_____ Amniotic cavity
_____ Chorion
_____ Chorionic villi
_____ Endometrium
_____ Umbilical cord
_____ Yolk sac

3. Fetal Development

Write the appropriate numbers indicating the weeks of development that match the statements in the spaces at the right.

1) Vernix caseosa forms. ____________________
2) Testes begin to descend in males. ____________________
3) Fetus assumes the "head down" position. ____________________
4) Heartbeat present but not detectable with stethoscope. ____________________
5) Surfactant production in lungs begins. ____________________
6) Lanugo is shed. ____________________
7) Head is as large as body. ____________________
8) Meconium begins to form in intestines. ____________________

4. Hormonal Control of Pregnancy

Write the terms that match the statements in the spaces at the right.

1) Hormone secreted by the trophoblast. ____________________
2) Maintains the corpus luteum for the first 10 to 12 weeks. ____________________
3) Hormones maintaining the endometrium. ____________________
____
4) Hormone detected by pregnancy tests. ____________________
5) Takes over secretion of estrogens and progesterone from approximately week 12 to birth. ____________________
6) Prevents GnRH secretion by hypothalamus during pregnancy. ____________________
7) Primary source of estrogens and progesterone for the first 10 to 12 weeks. ____________________
8) Two hormones that prepare mammary glands for milk secretion after birth. ____________________
____

5. Birth

a. Label the figure by placing the numbers of the structures in the spaces by the correct labels.

1
2
3
4
5
6

______ Amnion
______ Amniotic fluid
______ Cervix
______ Placenta
______ Umbilical cord
______ Uterine wall

b. Write the terms that match the statements in the spaces at the right.

1) Hormone stimulating development of placental blood vessels. ____________________

2) Hormone that inhibits uterine contractions during pregnancy. ____________________

3) Hormone that sensitizes uterine muscles for starting contractions as birth nears. ____________________

4) Term for the events associated with the birth process. ____________________

5) Hormone starting and maintaining uterine contractions. ____________________

6) Receives nerve impulses triggered by stretching of the cervix. ____________________

7) Releases oxytocin into blood. ____________________

8) Longest stage of labor. ____________________

9) Stage of labor when the infant is born. ____________________

10) Stage when the placenta is expelled. ____________________

11) Name for the birth process. ____________________

c. Write the words that complete the sentences in the spaces at the right.

As the time of birth approaches, the high blood concentration of __1__ overrides the inhibitory effect of __2__ on uterine contractions so that such contractions are possible. The __3__ feedback mechanism controlling labor seems to be started by pressure of the fetus on the __4__, which triggers the formation of __5__ that are carried to the hypothalamus. The __6__ stimulates the posterior lobe of the pituitary to release __7__, which stimulates uterine __8__. The continued dilation of the __9__ increases the frequency of __10__ sent to the hypothalamus, which, in turn, stimulates the posterior lobe of the pituitary to release more __11__, which increases the strength and frequency of uterine __12__. This pattern of positive feedback produces increasingly stronger contractions until the fetus is __13__. Shortly after birth, uterine contractions cause the detachment and expulsion of the __14__.

When the __15__ is cut, the level of __16__ increases in the infant's blood, stimulating the __17__ center in the medulla oblongata to trigger the first breath. After the first breath, breathing becomes easier because __18__ in the alveoli keeps the __19__ open.

1) ____________________
2) ____________________
3) ____________________
4) ____________________
5) ____________________
6) ____________________
7) ____________________
8) ____________________
9) ____________________
10) ____________________
11) ____________________
12) ____________________
13) ____________________
14) ____________________
15) ____________________
16) ____________________
17) ____________________
18) ____________________
19) ____________________

6. Cardiovascular Adaptations

a. Write the terms that match the statements relating to fetal circulation in the spaces at the right.

1) Opening between left and right atria. ____________________

2) Carries deoxygenated blood from fetus to placenta. ____________________

3) Carries blood from umbilical vein to inferior vena cava, bypassing the liver. ____________________

4) Carries blood from pulmonary trunk to aortic arch. ____________________

5) Carries oxygenated blood from the placenta to the fetus. ____________________

b. Write the words that complete the sentences regarding fetal circulation in the spaces at the right.

The fetal blood receives oxygen and nutrients from __1__ blood in the placenta. Oxygenated blood is carried from the placenta by the __2__ vein that enters the fetus at the __3__. This vessel divides near the liver, allowing about half of the oxygenated blood to pass through the __4__, bypassing the liver, and mixing with deoxygenated blood in the inferior __5__. When this mixed blood enters the __6__ atrium, most of it passes through the __7__ into the __8__ atrium and then into the __9__ ventricle. Contraction of the ventricle pumps blood into the __10__ for transport to body cells. Blood entering the __11__ ventricle is pumped into the pulmonary trunk. However, most of it bypasses the lungs by flowing through the __12__ into the aortic arch. The two lung bypasses work together to __13__ blood flow to body cells. A small amount of blood is carried by __14__ arteries to the nonfunctional lungs and returned to the left __15__. Blood is returned to the placenta by two __16__ arteries.

1) ____________________
2) ____________________
3) ____________________
4) ____________________
5) ____________________
6) ____________________
7) ____________________
8) ____________________
9) ____________________
10) ____________________
11) ____________________
12) ____________________
13) ____________________
14) ____________________
15) ____________________
16) ____________________

c. Write the terms that match the statements relating to postnatal circulatory changes in the spaces at the right.

1) Remnant of the umbilical vein. ____________________

2) Remnants of the umbilical arteries. ____________________

3) Remnant of the ductus venosus. ____________________

4) Remnant of the ductus arteriosus. ____________________

7. Lactation

a. Write the terms that match the statements in the spaces at the right.

1) Two hormones preparing mammary glands for lactation. ____________________

____

2) Hormone stimulating lactation. ____________________

3) Secretes prolactin-releasing hormone. ________________

4) Secretes prolactin. ________________

5) First secretion of mammary glands. ________________

6) Two hormones whose high blood levels inhibit secretion of PRH. ________________

____

7) Hormone stimulating milk ejection. ________________

b. Write the words that complete the sentences in the spaces at the right.

After birth, the drop in the blood levels of __1__ and __2__ allows the hypothalamus to secrete __3__, which stimulates secretion of __4__ by the anterior lobe of the pituitary, promoting lactation. __5__, the first secretion of the mammary glands, is rich in __6__ and contains essentially no __7__. True __8__ secretion starts within two to three days.

Suckling stimulates formation of __9__ that are carried to the hypothalamus, causing it to secrete __10__, which continues production of prolactin in order to maintain __11__. The hypothalamus also stimulates the posterior lobe of the pituitary to release __12__, which stimulates __13__ through the contraction of specialized epithelial cells within the mammary glands.

1) ________________
2) ________________
3) ________________
4) ________________
5) ________________
6) ________________
7) ________________
8) ________________
9) ________________
10) ________________
11) ________________
12) ________________
13) ________________

8. Disorders of Pregnancy, Prenatal Development, and Postnatal Development

Write the terms that match the statements in the spaces at the right.

1) Implantation of preembryo at a site other than the uterus. ________________

2) Spontaneous abortion. ________________

3) Increased blood pressure, edema, and convulsions or coma in late pregnancy. ________________

4) Nausea and vomiting in early pregnancy. ________________

5) Sudden death with no medical history or explanation. ________________

6) Bilirubin is produced faster than the liver can process it. ________________

7) Caused by insufficient surfactant in alveoli. ________________

8) May result from fetal exposure to X-rays, alcohol, and illegal or legal drugs. ________________

9. Genetics

a. Write the terms that match the statements in the spaces at the right.

1) Number of chromosomes in human body cells. ________________

2) Number of chromosomes in human gametes. ________________

3) Sex chromosomes in a female. ______________________
4) Sex chromosomes in a male. ______________________
5) A unit of inheritance. ______________________
6) Alternate forms of a gene. ______________________
7) Condition in which both alleles for a trait are identical. ______________________
8) Condition in which the alleles for a trait are different. ______________________
9) An allele that is always expressed. ______________________
10) An allele that is expressed only when a dominant allele is absent. ______________________
11) A type of gene expression where the two alleles for a gene can create three different phenotypes. ______________________
12) A type of inheritance where both alleles are expressed and affect the phenotype. ______________________
13) A type of inheritance involving many genes that contribute to the phenotype. ______________________
14) The observable traits. ______________________
15) The alleles controlling the expression of a trait. ______________________
16) Traits whose alleles occur on the X chromosome. ______________________
17) Type of cell division that separates chromosome pairs into gametes. ______________________

b. Indicate the genotypes for the following traits.

1) Heterozygous freckled. ______________________
2) Homozygous freckled. ______________________
3) Homozygous nonfreckled. ______________________
4) Color-blind male. ______________________
5) Normal color vision, carrier female. ______________________
6) Color-blind female. ______________________
7) Homozygous type A blood. ______________________
8) Type AB blood. ______________________
9) Type O blood. ______________________
10) Heterozygous type B blood. ______________________

c. Indicate the possible genotypes of gametes that can be formed by parents with these genotypes.

1) Homozygous freckled. ______________________
2) Heterozygous freckled. ______________________
3) Homozygous nonfreckled. ______________________
4) Color-blind male. ______________________
5) Normal color vision, carrier female. ______________________
6) Color-blind female. ______________________
7) Heterozygous type A blood. ______________________
8) Type AB blood. ______________________

d. Indicate the predicted phenotype ratios for the following matings.
   1) Homozygous freckled × homozygous nonfreckled. ______________________
   2) Heterozygous freckled × homozygous nonfreckled. ______________________
   3) Type AB blood × type O blood. ______________________
   4) Heterozygous type A blood × type O blood. ______________________
   5) Color-blind mother × normal color vision father. ______________________

10. Inherited Diseases

Indicate whether each statement is true (T) or false (F).

______ Genetic disease may be caused by the presence of an extra chromosome.
______ Recessive sex-linked traits appear more frequently in females since they have two X chromosomes.
______ It is possible to examine fetal cells for chromosome abnormalities.
______ Some genetic diseases caused by specific alleles do not show up until adulthood.
______ Down syndrome is caused by trisomy 7.
______ Amniocentesis is used to obtain a sample of amniotic fluid for examination.
______ Genetic counseling may be helpful for prospective parents with genetic disease in their family histories.

11. Clinical Insights

a. When the sperm count in semen falls below 20 million/ml, male infertility results. How do you explain this? ______________________

b. Physicians advise women to avoid all drugs (legal and illegal) during pregnancy. What is the basis for this advice? ______________________

c. What problems would occur if a newborn's foramen ovale failed to close? ______________________

d. Why can monozygotic twins receive blood transfusions from each other without difficulty, whereas dizygotic twins often cannot? ______________________

e. Mary and Joe have discovered that they are both heterozygous for sickle-cell disease. They want to know what the chance is that their children will inherit sickle-cell disease. What would you advise them? ______________________

APPENDIX A

Keys to Medical Terminology

MEDICAL TERMS MAY CONSIST OF three basic parts: a prefix, a root word, and a suffix. All terms have a root word, but some terms may lack either a prefix or a suffix. A **prefix** is the first portion of a term and comes before the root word. A **suffix** comes after the root word and is the last portion of a term. Both the prefix and the suffix modify the meaning of a root word, and they may be used with many different root words. The **root word** is the main portion of the term. Root words often occur at the beginning of a term, but they also may end a term or may be sandwiched between a prefix and a suffix. When determining the meaning of a term, you start with the suffix, then move to the prefix, and finally consider the root word. Consider these examples.

1. **Laryngitis** becomes **laryng/itis** when broken into its component parts.

   laryng- = the root word meaning *larynx*
   -itis = the suffix meaning *inflammation*

   Thus, the meaning of laryngitis is *inflammation of the larynx.*

2. **Endogastric** becomes **endo/gastr/ic** when broken into its component parts.

   endo- = the prefix meaning *within*
   gastr- = the root word meaning *stomach*
   -ic = the suffix meaning *pertaining to*

   Thus, the meaning of endogastric is *pertaining to within the stomach.*

The parts of a term are linked together in a way that aids pronunciation. This often requires the use of *combining vowels.* For example, when linking *gastr-* and *-pathy* to form a term meaning disease of the stomach, the vowel *o* is inserted to form *gastropathy.*

Some terms consist of more than one root word. For example, *gastr/o/enter/o/col/itis* means inflammation of the stomach, intestine, and colon.

You can see that once you know the meaning of common prefixes, root words, and suffixes, understanding medical terminology becomes much easier.

In the sections that follow, some common prefixes, suffixes, and root words are listed along with examples to help you understand medical terminology.

Singular and Plural Endings

Most medical terms are derived from Greek and Latin words. Therefore, changing from singular to plural is done by changing the ending of the term rather than by adding an *s* or *es* or changing a *y* to *ies* as in English terms. Examples of singular and plural endings are

Singular Ending	Plural Ending	Example
-a	-ae	pleura; pleurae
-en	-ena	lumen; lumena
-is	-es	testis; testes
-ma	-mata	carcinoma; carcinomata
-um	-a	epicardium; epicardia
-ur	-ora	femur; femora
-us	-i	glomerulus; glomeruli
-x	-ces	appendix; appendices

Common Prefixes

Prefix	Meaning	Example
a-	without, not	a/sepsis: sterile; without germs
ab-	away from, from	ab/duct: carry away from
ad-	to, toward	ad/duct: carry toward
an-	without, not	an/ergia: without energy
ante-	before	ante/cibum: before meals
anti-	against	anti/histamine: against histamine
bi-	two	bi/lateral: on two sides
bio-	life	bio/logy: study of life
brachy-	short	brachy/gnathia: shortness of the lower jaw
brady-	slow	brady/cardia: slow heart rate
cent-	hundred	centi/meter: 1/100 of a meter
circum-	around	circum/oral: around the mouth
co-, com-, con-	with, together	com/press: squeeze together
de-	from, down	de/congest: reduce congestion
dia-	through	dia/rrhea: flow through
dis-	apart	dis/infect: free from infection
dys-	bad, difficult	dys/pepsia: difficult digestion
ect-	external, outer	ecto/derm: outer skin
en-	in, on	en/cranial: in the cranium
end-	within	endo/crine: secrete within
epi-	upon	epi/dermis: upon the skin
ex-	out, away from	ex/halation: to breathe out
extra-	outside of, in addition to	extra/ocular: outside the eye
hemi-	half	hemi/plegia: paralysis of one-half of the body
hyper-	above, over	hyper/trophy: excessive growth
hypo-	below, under	hypo/dermic: under the skin
infra-	below, beneath	infra/orbital: below the orbit
inter-	between	inter/cellular: between cells
intra-	within	intra/cellular: within cells
kil-	thousand	kilo/gram: 1,000 grams
macr-	large	macro/cyst: large cyst
mal-	bad, ill, poor	mal/ady: disease, disorder
mes-	middle	meso/nasal: middle of the nose
meta-	after, beyond	meta/tarsals: beyond the tarsals
micr-	small	micro/colon: abnormally small colon
milli-	one-thousandth	milli/gram: 1/1,000 of a gram
multi-	many	multi/cellular: having many cells
neo-	new	neo/plasm: new growth
ob-	against, in the way of	ob/scure: indistinct, hidden
olig-	few	oligo/spermia: few sperm
onc-	tumor	onco/genic: tumor-causing
per-	through	per/forate: to make holes
peri-	around	peri/osteum: around a bone
poly-	many	poly/morphous: many forms
post-	after	post/ocular: behind the eye
pre-	before, in front of	pre/mature: before maturation
presby-	old	presby/cardia: old heart
pro-	before, in front of	pro/chondrial: before cartilage
re-	again, back	re/flex: bend back
retr-	backward, behind	retro/nasal: back part of nose
semi-	half	semi/lunar: half moon
sub-	under	sub/cutaneous: under the skin
super-	above, superior	super/acute: strongly acute
supra-	above, superior	supra/nasal: above the nose
sym-	together, with	sym/physis: growing together
syn-	together, with	syn/dactyly: fusion of fingers
tachy-	fast	tachy/cardia: rapid heart rate

Common Suffixes

Suffix	Meaning	Example
-algia	pain	neur/algia: pain in a nerve
-centesis	puncture to aspirate fluid	amnio/centesis: puncture amnion to obtain a sample of amniotic fluid
-cide	kill	bacterio/cide: substance killing bacteria
-cis	cut	in/cision: a cut into
-cyte	cell	erythro/cyte: red blood cell
-dynia	pain	entero/dynia: intestinal pain
-ectomy	cut out	append/ectomy: procedure to cut out the appendix
-emesis	vomiting	poly/emesis: much vomiting
-emia	blood	an/emia: without blood
-gnosis	knowledge	pro/gnosis: foreknowledge
-gram	record	myo/gram: muscle record
-graphy	making a record	myo/graphy: making a record of muscle action
-iasis	abnormal condition	candid/iasis: *Candida* infection
-itis	inflammation	sinus/itis: inflammation of sinuses
-lepsy	seizures	narco/lepsy: seizures of numbness
-logy	study of	bio/logy: study of life
-lysis, -lytic	breakdown, dissolve	myo/lysis: breakdown of muscles
-megaly	enlargement	nephro/megaly: kidney enlargement
-oid	resembling	ov/oid: resembling an egg
-oma	tumor	carcin/oma: cancerous tumor
-osis	abnormal condition	nephr/osis: abnormal kidney condition
-ostomy	make an opening	ile/ostomy: opening into the small intestine
-pathy	disease	neuro/pathy: disease of nerves
-penia	deficiency, poor	leuko/penia: deficiency of white blood cells
-pepsia	digestion	dys/pepsia: poor digestion
-philia	attraction, love	acido/philic: attracted to acid
-phobia	abnormal fear	acro/phobia: fear of heights
-plasia	formation	hypo/plasia: deficient formation
-plasty	make, shape	angio/plasty: shaping a blood vessel
-plegia	paralysis	para/plegia: paralysis of lower body and both legs
-pnea	breath	brady/pnea: slow breathing
-rrhea	discharge, flow	pyo/rrhea: pus discharge
-soma, -some	body	chromo/some: colored body
-stasis	control, stop	hemo/stasis: stop bleeding
-therapy	treatment	thermo/therapy: heat therapy
-tomy	to cut	laparo/tomy: to cut into the abdomen
-trophy	development	hyper/trophy: excessive development
-uria	urine	glucos/uria: glucose in the urine

Common Root Words

Root Word	Meaning	Example
acr-	extremity, peak	acro/phobia: fear of heights
aden-	gland	aden/oma: tumor of a gland
angi-	blood vessel	angio/pathy: diseased vessel
arthr-	joint	arthr/itis: inflammation of joints
brachi-	arm	brachi/al: pertaining to the arm
carcin-	cancer	carcin/oma: cancerous tumor
card-	heart	cardio/logy: study of the heart
carp-	wrist	carp/al: pertaining to the wrist
cephal-	head	cephal/ic: pertaining to the head
cervic-	neck	cervic/al: pertaining to the neck

Common Root Words *continued*

Root Word	Meaning	Example
chole-	bile	chole/cyst/itis: inflammation of the gallbladder
chondr-	cartilage	chondro/cyte: cartilage cell
colp-	vagina	colpo/dynia: vaginal pain
cost-	rib	cost/algia: rib pain
crani-	skull	cranio/malacia: softening of the skull
cutan-	skin	sub/cutan/eous: under the skin
cyan-	blue	cyan/osis: bluish skin color
cyst-	bladder	cyst/algia: pain in bladder
cyt-	cell	cyto/logy: study of cells
dactyl-	fingers, toes	dactylo/megaly: abnormally large fingers or toes
dent-	tooth	denti/form: toothlike
derm-, dermato-	skin	hypo/derm/al: under the skin
dors-	back	dors/al: pertaining to the back
duct-	carry	ad/duct: carry toward
edema	swelling	edema/tous: swollen
encephal-	brain	encephal/itis: inflammation of the brain
enter-	intestine	enter/dynia: intestinal pain
erythr-	red	erythro/cyte: red blood cell
esthe-	sensation, feeling	an/esthe/tic: substance causing an absence of sensation
esthen-	weakness	my/esthenia: muscle weakness
febr-	fever	febr/ile: feverish
gastr-	stomach	gastro/spasm: stomach spasm
gen-	to produce	patho/gen: disease-causing agent
gingiv-	gum	gingiv/itis: inflammation of the gums
glu-, glyc-	sugar	hyper/glyc/emia: excessive blood sugar
gynec-, gyno-	female	gyneco/logy: study of female disorders
hem-, hemato-	blood	hemo/genesis: blood formation
hepat-	liver	hepat/ectomy: removal of the liver
hist-	tissue	histo/logy: study of tissues
hom-, home-	same	homo/sexual: attracted to the same sex
hydr-, hydra-	water	hydra/tion: gaining water
kerat-	horny, cornea	kerat/osis: condition of abnormal horny growths
lacrim-	tear	lacrim/al: pertaining to tears
lact-	milk	lacta/tion: producing milk
lapar-	abdomen	laparo/tomy: to cut into the abdomen
laryng-	larynx	laryngo/pathy: disease of the larynx
later-	side	uni/lateral: one-sided
leuc-, leuk-	white	leuko/cyte: white blood cell
lingu-	tongue	lingu/iform: tongue-shaped
lip-	lipids, fat	lip/oid: similar to fat
lith-	stone	oto/lith: ear stone
mamm-	breast	mammo/gram: X-ray of breast
melan-	black	melan/in: black skin pigment
men-	monthly, mensis	meno/pause: cessation of menses
metr-	uterus	myo/metr/ium: muscle layer of uterus
morph-	shape, form	morpho/logy: study of shape
my-	muscle	myo/card/itis: inflammation of the heart muscle
myel-	marrow, spinal cord	myel/algia: pain of the spinal cord or its membranes
nas-	nose	nas/al: pertaining to the nose
nephr-	kidney	nephr/itis: inflammation of a kidney
neur-	nerve	neur/ectomy: excision of a nerve
odont-	tooth	odonto/pathy: disease of the teeth
oo-	egg, ovum	oo/genesis: formation of ova
orchid-	testis	orchid/itis: inflammation of a testis
oss-, oste-	bone	osteo/malacia: softening of bones
ot-, aur-	ear	oto/lith: ear stone aur/icular: pertaining to the ear

Common Root Words ***continued***

Root Word	Meaning	Example
path-	disease	patho/logy: study of disease
pect-	chest	pecto/ral: pertaining to the chest
ped-	child	ped/iatrics: medical specialty dealing with children's disorders
pep-, peps-	digest	pep/tic: pertaining to digestion
phag-	eat	phago/cyt/osis: engulfment of particles by cells
pharyng-	throat, pharynx	pharnygo/rrhea: discharge from the pharynx
phleb-	vein	phleb/itis: inflammation of a vein
pneum-	air	pneumo/thorax: air in the chest
pneumon-	lung	pneumono/pathy: disease of a lung
proct-	rectum	procto/col/itis: inflammation of the rectum and colon
pseud-	false	pseudo/hernia: false rupture
psych-	mind	psycho/genic: originating in the mind
pulmo-, pulmon-	lung	pulmon/ary: pertaining to a lung
py-	pus	pyo/cele: cavity containing pus
pyel-	kidney pelvis	pyelo/gram: X-ray of kidney pelvis
quadr-	four	quadri/plegia: paralysis of both upper and lower limbs
rhin-	nose	rhin/itis: inflammation of the nose
salping-	uterine tube	salping/ectomy: removal of a uterine tube
scler-	hard	scler/osis: hardening
sect-	cut	sect/ion: process of cutting
sept-	presence of microbes	septic/emia: infection of blood
sten-	narrow	sten/osis: narrowed condition
strict-	draw tight	con/strict/ion: draw tightly together
therm-	heat	hypo/thermia: low body temperature
thorac-	chest, thorax	thoraco/dynia: chest pain
thromb-	clot	thromb/us: a blood clot
tox-	poison	tox/in: poisonous substance
vas-	vessel	vaso/dilation: expansion of a vessel
viscer-	internal organ	viscero/genic: originating in the internal organs
vita-	life	vita/l: essential for life

APPENDIX B

Answers to Self-Review Questions

Chapter 1

1. physiology
2. cardiovascular
3. nervous
4. distal
5. appendicular
6. popliteal
7. femoral
8. dorsal; vertebral
9. right upper; epigastric
10. mediastinum
11. peritoneum (parietal peritoneum)
12. homeostasis

Chapter 2

1. protons
2. ion
3. compound
4. covalent
5. water
6. lower
7. synthesis
8. monosaccharides
9. amino acids
10. nucleotides
11. triglycerides
12. ATP

Chapter 3

1. plasma membrane
2. DNA
3. mitochondria
4. ribosomes
5. nucleolus
6. endoplasmic reticulum
7. diffusion
8. carrier-mediated diffusion
9. cellular respiration
10. messenger RNA
11. mitosis

Chapter 4

1. one
2. little
3. pseudostratified ciliated columnar
4. simple columnar
5. stratified squamous
6. matrix
7. collagen
8. adipose
9. fibrocartilage
10. cardiac
11. smooth
12. neuroglia
13. mucous

Chapter 5

1. basale
2. keratin
3. dermis
4. subcutaneous tissue
5. papillae
6. sebaceous
7. eccrine
8. arrector pili
9. reduces
10. wrinkles and sagging skin
11. athlete's foot
12. calluses; corns

Chapter 6

1. support; protection
2. epiphyses; spongy
3. nutrient foramen
4. intramembranous
5. periosteum
6. axial
7. mandible; temporal
8. atlas; occipital
9. costal cartilages
10. pectoral girdle
11. humerus; ulna; radius
12. ilium; ischium; pubis
13. femur; tibia; patella
14. hinge; ball-and-socket
15. osteoporosis

Chapter 7

1. fibers (muscle cells)
2. tendons
3. sarcomere
4. acetylcholine
5. thin
6. insertion
7. masseter
8. rectus abdominis
9. latissimus dorsi
10. deltoid
11. triceps brachii
12. gluteus maximus
13. quadriceps femoris
14. gastrocnemius

Chapter 8

1. axon
2. interneurons
3. sodium
4. neurotransmitter
5. insula
6. frontal
7. prefrontal
8. hypothalamus
9. medulla oblongata
10. cerebellum
11. subarachnoid
12. posterior
13. anterior
14. autonomic
15. sympathetic

Chapter 9

1. hair root plexus
2. referred pain
3. taste buds
4. olfactory
5. sensory adaptation
6. hearing; saccule; semicircular canals; utricle
7. endolymph
8. spiral organ; basilar
9. auditory ossicles
10. saccule; utricle
11. cornea; lens
12. cones; color
13. ciliary body
14. vitreous body
15. optic chiasma

Chapter 10

1. endocrine; hormones
2. receptors
3. nonsteroid
4. negative-feedback

5. hypothalamus
6. TSH; adrenal cortex; FSH; LH
7. thyroid hormones; thyroid gland
8. calcitonin; parathyroid hormone
9. medulla
10. aldosterone
11. glucagon
12. estrogens; progesterone; testosterone

Chapter 11

1. 45
2. hemoglobin; red blood cells
3. hemacytoblasts
4. oxygen; erythropoietin
5. liver
6. white blood cells
7. neutrophils; monocytes
8. basophils
9. eosinophils
10. lymphocytes
11. plasma
12. platelets
13. fibrinogen; fibrin
14. antigen(s)
15. B+, B−, O−, O+

Chapter 12

1. pericardium
2. left atrium
3. tricuspid; right atrium
4. atrial; systole
5. SA; AV
6. sympathetic; parasympathetic
7. arteries; veins
8. arteries; capillaries
9. diffusion; filtration
10. right; trunk; arteries; veins; left
11. brachiocephalic trunk; internal carotid artery
12. thoracic; celiac trunk, common hepatic artery; hepatic artery proper
13. hepatic portal vein; sinusoids; hepatic vein; inferior
14. popliteal vein; external iliac vein; common iliac vein; inferior
15. axillary vein; subclavian vein; brachiocephalic vein; superior

Chapter 13

1. interstitial fluid
2. lymph; subclavian
3. lymph nodes; lymphocyte
4. spleen; red blood
5. nonspecific
6. neutrophils; macrophages (monocytes)
7. inflammation
8. thymus
9. antigen-presenting; helper T
10. helper; cytotoxic; memory
11. antibody; cell
12. helper T; B; T
13. plasma; antibodies
14. naturally acquired active
15. vaccination; secondary

Chapter 14

1. filtered; mucous membrane
2. larynx; glottis
3. cartilages; bronchioles
4. alveoli
5. pleural cavity
6. surfactant
7. atmospheric; intra-alveolar
8. tidal; 500
9. medulla oblongata; DRG; PRG
10. carbon dioxide; hydrogen ions
11. oxygen; carbon dioxide
12. oxyhemoglobin; bicarbonate ions

Chapter 15

1. enzymes; nutrient molecules
2. mucosa
3. peristalsis
4. dentin; enamel
5. salivary amylase; maltose
6. peristalsis; lower esophageal sphincter
7. pepsin; proteins
8. parasympathetic; gastrin
9. secretin; cholecystokinin; cholecystokinin
10. pancreatic amylase; pancreatic lipase; trypsin
11. small intestine; monosaccharides; amino acids
12. chylomicrons
13. large intestine
14. liver
15. vitamins

Chapter 16

1. cortex
2. medulla; collecting ducts and nephron loops
3. nephrons
4. glomerular; systemic
5. glomerular capsule; filtrate
6. tubular fluid; peritubular capillaries
7. actively; passively; osmosis
8. urine
9. renin
10. antidiuretic; collecting ducts
11. aldosterone; potassium
12. 60
13. ureters; peristalsis
14. urethra
15. internal; external

Chapter 17

1. testes; scrotum
2. seminiferous tubules; spermatogenesis
3. epididymis; ejaculatory duct; urethra
4. blood
5. LH; testosterone
6. ovaries; oogenesis
7. secondary oocyte; uterine tube
8. secondary oocyte; polar body
9. vagina
10. GnRH; LH
11. estrogens
12. corpus luteum; endometrium
13. decrease; menstruation
14. ovulation; implantation
15. estrogens; testosterone

Chapter 18

1. uterine tube
2. fertilization
3. blastocyst; seventh
4. human chorionic gonadotropin (hCG); corpus luteum; progesterone
5. germ layers
6. amnion; placenta
7. ninth; fetus
8. labor; oxytocin
9. foramen ovale; ductus arteriosus
10. prolactin-releasing hormone; prolactin
11. genes; DNA
12. 46; female
13. males
14. homozygous
15. probability

GLOSSARY

A

abdomen The anterior portion of the trunk located between the diaphragm and pelvis.

abdominal cavity The portion of the abdominopelvic cavity between the diaphragm and the pelvis.

abdominopelvic cavity The portion of the ventral body cavity inferior to the diaphragm.

abdominopelvic quadrants The four divisions of the abdominopelvic cavity formed by a median plane and a transverse plane through the umbilicus.

abdominopelvic regions The nine divisions of the abdominopelvic cavity formed by the intersection of two sagittal and two transverse planes.

abduction The movement of a body part away from the midline.

abortion The removal of an embryo or fetus from the uterus prior to birth.

absorption The uptake of substances by cells; the process by which nutrients pass from the alimentary canal into the blood.

accessory organs Organs that assist the functions of primary organs.

accommodation Adjustment of the lens to focus an image on the retina of the eye.

acetylcholine (ACh) A neurotransmitter secreted from the terminal bouton of many neurons.

acetylcholinesterase An enzyme promoting the breakdown of acetylcholine in synaptic clefts.

acid A substance that ionizes in water, releasing hydrogen ions.

acidosis Condition of arterial blood below pH 7.35.

acne Plugged hair follicles that form pimples due to infection by certain bacteria.

acquired immunodeficiency syndrome (AIDS) A progressive decrease in immune capability caused by infection of T cells and macrophages with HIV.

acromegaly A disorder caused by hypersecretion of growth hormone after bone growth is complete.

active immunity Immunity derived from activation of B cells and T cells by an invasion of a pathogen.

active site The location on an enzyme where the chemical reaction occurs.

active transport A process that requires the use of energy to move substances across plasma membranes.

Addison disease An endocrine disorder caused by a hyposecretion of hormones by the adrenal cortex.

adduction The movement of a body part toward the midline.

adenine A nitrogen base of nucleic acids that pairs with thymine in DNA and uracil in RNA.

adenosine diphosphate (ADP) A molecule used to form adenosine triphosphate.

adenosine triphosphate (ATP) Chemical energy storage molecule in the body.

adipose tissue Loose connective tissue containing large numbers of fat-storing adipocytes.

adrenal cortex The superficial portion of an adrenal gland.

adrenal gland An endocrine gland located on the superior surface of each kidney.

adrenal medulla The deep portion of an adrenal gland.

adrenocorticotropic hormone (ACTH) A hormone secreted by the anterior lobe of the pituitary gland that stimulates the adrenal cortex to secrete hormones.

adult stem cells Partially specialized cells capable of producing several types of specialized cells.

aerobic respiration The part of cellular respiration that requires oxygen and mitochondria.

afferent glomerular arteriole The arteriole carrying blood to the glomerulus of a nephron.

age-related macular degeneration (AMD) The destruction of the macula resulting in loss of vision in the center of the visual field.

ageusia A loss of taste function.

agglutination The clumping of red blood cells in an antigen-antibody reaction.

agonist A muscle whose contraction moves a body part.

agranulocyte A type of white blood cell that lacks visible cytoplasmic granules.

albumin An abundant plasma protein that helps transport substances in blood, and maintain the pH and osmotic pressure of blood.

aldosterone A mineralocorticoid hormone produced by the adrenal cortex that regulates potassium and sodium concentrations in the blood.

alimentary canal The tube through which food passes from the esophagus to the anus.

alkaline Pertaining to a base.

alkalosis Condition of arterial blood above pH 7.45.

allantois An extraembryonic membrane that forms as an outpocketing from the yolk sac.

allele An alternate form of a gene.

allergen A foreign substance capable of stimulating an allergic reaction.

allergy An abnormally intense immune reaction.

all-or-none response The type of response by muscle cells and neurons to stimulation; total response or no response.

alopecia Excessive hair loss.

alveolar ducts Tiny air passages that open into alveoli.

alveolar gas exchange The exchange of oxygen and carbon dioxide between the air in alveoli and the blood in alveolar capillaries.

alveolus An air sac in a lung.

Alzheimer disease A disorder caused by a loss of cholinergic neurons in the brain and characterized by loss of memory.

amenorrhea The absence of menstruation.

amino acid The building unit of proteins.

amnion The extraembryonic membrane that envelops the embryo and fetus.

amphiarthrosis A slightly movable joint.

ampulla The expanded portion of a semicircular canal.

amylase An enzyme that catalyzes the digestion of starch and glycogen.

anaerobic respiration The part of cellular respiration that does not require oxygen and occurs within the cytosol.

anaphase The stage of mitosis in which chromatids of replicated chromosomes separate and move to opposite poles of the cell.

anatomy The study of body organization and structure.

androgens Generic term for hormones related to testosterone.

anemia The decreased ability of the blood to carry oxygen.